SUPEROXIDE AND SUPEROXIDE DISMUTASE IN CHEMISTRY, BIOLOGY AND MEDICINE

This book is dedicated to Professor Alessandro Rossi-Fanelli.

SUPEROXIDE AND SUPEROXIDE DISMUTASE IN CHEMISTRY, BIOLOGY AND MEDICINE

Proceedings of the 4th International Conference on Superoxide and Superoxide Dismutase held in Rome, Italy, 1–6 September 1985

Edited by
Giuseppe Rotilio

Program Committee
J.V. Bannister (U.K.), B. Bielski (U.S.A.), G. Cohen (U.S.A.), E.M. Fielden (U.K.), L. Flohe (F.R.G.), I. Fridovich (U.S.A.), S. Marklund (Sweden), G. Rotilio (Italy), H.M. Steinman (U.S.A.)

1986

ELSEVIER SCIENCE PUBLISHERS
AMSTERDAM · NEW YORK · OXFORD

ISBN 0 444 80797 7

Published by:
Elsevier Science Publishers B.V. (Biomedical Division)
P.O. Box 211
1000 AE Amsterdam
The Netherlands

Sole distributors for the USA and Canada:
Elsevier Science Publishing Company Inc.
52 Vanderbilt Avenue
New York, NY 10017
USA

Library of Congress Cataloging in Publication Data

International Conference on Superoxide and Superoxide
Dismutase (4th : 1985 : Rome, Italy)
Superoxide and superoxide dismutase in chemistry,
biology, and medicine.

Includes bibliographies and index.
1. Superoxide--Congresses. 2. Active oxygen in the
body--Congresses. 3. Superoxide dismutase--Congresses.
I. Rotilio, G. II. Title. [DNLM: 1. Superoxide--
congresses. 2. Superoxide Dismutase--congresses.
W3 IN1943AM 4th 1985s / QU 140 I61 1985s]
QP535.O1I474 1985 574.19'258 86-9002
ISBN 0-444-80797-7 (U.S.)

Printed in The Netherlands

PREFACE

The Fourth International Conference on Superoxide and Superoxide Dismutase was held in Rome, September 1st–6th, 1985. It is now a consolidated tradition that these meetings cover all the aspects of oxygen-related free radicals in pure and applied research, as well as many less strictly related topics of biological oxidations. Yet, it is clearly implied that their focus should be on biochemical mechanisms involved in the various processes included in the general term ''free-radical''. This approach unequivocally traces back to the primary historical event that actually started the biochemical age of free radical research, namely the discovery of the enzymatic function of superoxide dismutase in 1969 by McCord and Fridovich. This discovery not only prompted fascinating ideas on the role of superoxide and superoxide dismutase in biological system but also started the utilization of the enzyme as a privileged object and tool for studies spanning from protein structure to gene expression, from catalytic mechanisms to molecular evolution, from cell biology to pharmacology and medicine. Although the scope of this book is obviously much wider than just O_2^- and SOD, I wanted to emphasize in the title, the debt of the scientific community to that discovery. Actually the book aims to be, as the conference in fact was, an up to date survey of advanced research in the field of oxy-radicals, which may in turn be instrumental in directing future work. In this perspective the book consists of short and concise articles and tries to avoid all redundancies without being just a list of abstracts.

I appear as the sole editor of the book not simply because the meeting was organised in Rome. All the members of the Program Committee accurately programmed the scientific sessions of the meeting, and the content of the book essentially reflects their efforts and great success in this enterprise. They all deserve my deep gratitude and the reader's appreciation for the quality of this book. Any shortcomings, on the other hand, are solely my responsibility.

Rome, January 1986 | Giuseppe Rotilio

ACKNOWLEDGEMENTS

Special gratitude is expressed to the three major sponsor Institutions of the Fourth International Conference on Superoxide and Superoxide Dismutase:

International Union of Biochemists
(through its Interest Group on "Kinetics and Mechanisms of Enzymes and Metabolic Networks" chaired By Professor J. Richard),

University of Rome "Tor Vergata",

CNR (Italian National Research Council).

Special financial support, without which the Conference could not have been held, was given by the following firms:

D.D.I. Pharmaceuticals Inc., Mountain View, California

Pietro Isnardi SpA, Imperia, Italy.

Additional financial support was provided by: University of Rome "La Sapienza"; Gruenenthal, GmbH, Aachen, Germany; Ipsen, Fondation pour la Recherche Therapeutique, Paris, France; Istituto di Ricerca Angelini, Rome, Italy; L.K.B. SpA, Rome, Italy; Merck Sharp & Dohme Italia SpA, Rome, Italy; Zambeletti SpA, Milan, Italy.

The following Institutions have generously made available staff and facilities: C.N.R. Centre for Molecular Biology, Department of Biology University of Rome "Tor Vergata", Department of Biochemistry University of Rome "La Sapienza".

To all members of the Rome research group on superoxide dismutase I am grateful for the continuous collaboration throughout the innumerable organizational steps of the Conference.

Drs. Irene Mavelli and Giovanni Musci helped finalize the editing of the book.

CONTENTS

EVOLUTION AND BIOSYNTHESIS OF SUPEROXIDE DISMUTASES

MECHANISMS OF OXY-RADICAL REACTIONS

THE ROLE OF TRANSITION METALS IN OXY-RADICAL INDUCED OXIDATION PROCESSES IN BIOLOGICAL SYSTEMS

BENON H. J. BIELSKI AND DIANE E. CABELLI
Chemistry Department, Brookhaven National Laboratory, Upton, NY 11973 (USA)

INTRODUCTION

This paper describes some recent work carried out in our laboratory and discusses it in view of the roles ascribed to transition metals in oxidative processes involving active oxygen species. Some of these oxidative processes are suspected to occur in living systems where they can lead to deleterious effects. Therefore, great efforts have gone into the study of fundamental reactions/mechanisms believed capable of either initiating or, conversely, suppressing the initiation (superoxide dismutases, catalase, antioxidants, etc.) of such events [1]. As a detailed treatment of this subject would be very involved, only two topics will be discussed here. First, the nature and role of superoxide-driven Fenton chemistry in biological systems will be considered and second, the reactivity of transition metal cations in higher oxidation states with biological compounds will be examined.

THE SUPEROXIDE-DRIVEN FENTON REACTION

In studies of the ferrous ion catalyzed decomposition of hydrogen peroxide, Haber and Weiss [2] observed that if they mixed the reactants in a ratio of $2Fe^{2+} : 1H_2O_2$ the classical Fenton reaction [3] controlled the overall mechanism, yielding two Fe^{3+} for every H_2O_2 consumed:

$$Fe^{2+} + H_2O_2 \longrightarrow Fe^{3+} + OH^- + OH \qquad (1)$$

$$Fe^{2+} + OH \longrightarrow Fe^{3+} + OH^- \qquad (2)$$

If, however, they added excess hydrogen peroxide $[H_2O_2] \gg [Fe^{2+}]$, the amount of H_2O_2 consumed was significantly larger than the stoichiometric amount of Fe^{3+} formed and the system evolved molecular oxygen. To account for this they suggested that hydrogen peroxide competes with Fe^{2+} for the OH radical, and that the HO_2 radical in turn reacts with hydrogen peroxide (reaction 4, commonly known as the Haber-Weiss reaction) to yield an OH radical [2]:

$$H_2O_2 + OH \longrightarrow HO_2 + H_2O \qquad (3)$$

$$H_2O_2 + HO_2 \longrightarrow OH + H_2O + O_2 \qquad (4)$$

If correct, this mechanism would not only account for the evolution of dioxygen but also for the lower yield of Fe^{3+}.

Barb et al. [4], reinvestigated this system in much greater detail and proposed an alternative mechanism. In studying the competition between reactions (5) and (6) as a function of pH (1.35-2.65)

$$Fe^{2+} + HO_2 \xrightarrow{H^+} Fe^{3+} + H_2O_2 \quad (5)$$

$$Fe^{3+} + O_2^- \longrightarrow Fe^{2+} + O_2 \quad (6)$$

and $[Fe^{2+}]/[Fe^{3+}]$, they concluded that the controlling step is reaction (6) and that the Haber-Weiss reaction (4) is insignificant. As they assumed that the value of the $pK_{HO_2} \sim 2.0$

$$HO_2 \rightleftharpoons O_2^- + H^+ \quad (7,-7)$$

and not 4.8 [5], they also concluded that Fe^{3+} is preferentially reduced by O_2^- and not HO_2.

Despite the discrepancy in the pK_{HO_2} value used by Barb et al., their conclusions have been corroborated by some recent experiments carried out in our laboratory. While the rate of reduction of $FeSO_4^+$ by HO_2 ($k_8 < 10^3$ M^{-1} s^{-1}) was determined in strong acid media (pH 0.0, 0.56 and 1.76) in competition with Ce(III),

$$FeSO_4^+ + HO_2 \xrightarrow{H^+} Fe^{2+} + O_2 + H_2SO_4 \quad (8)$$

$$FeSO_4^+ + O_2^- \xrightarrow{2H^+} Fe^{2+} + O_2 + H_2SO_4 \quad (9)$$

$$Ce(III) + HO_2 \longrightarrow Ce(IV) + H_2O_2 \quad (10)$$

the corresponding reduction of $Fe(OH)^{2+}$ by O_2^-($k_{11} = 1.5 \times 10^8$ M^{-1} s^{-1}, [6]) near pH 7.0 was determined by direct observation of Fe(III) disappearance

$$Fe(OH)^{2+} + O_2^- \longrightarrow Fe^{2+} + OH^- + O_2 \quad (11)$$

The latter experiments could be carried out because of the discovery of a "time window" (the time between formation of Fe(III) and the onset of its gelling and precipitation) by Stuglik and Zagorski [7] in pulse radiolytic experiments.

As is apparent from the magnitudes of k_8 and k_{11}, reduction of Fe(III) by O_2^- is the predominant pathway even at pH zero. Utilizing the relationship in Eq. (I)

$$\frac{k_8}{k_5} = \frac{k_{11}}{k_5} \frac{K_{HO_2}}{K_H + (H^+)} \quad (I)$$

and the ratios of k_8/k_5 measured by Barb et al. [4], with our direct measurement of k_{11}, $k_5 = 1.2 \times 10^6$ M^{-1} s^{-1} [8] and the accepted value for $\log K_H = -3.05$ (hydrolysis constant for Fe(III), [9]), a value of $pK_{HO_2} = 4.3$ is obtained. This is in relatively good agreement with the presently accepted value of 4.8 [5].

In view of the collective evidence against the occurrence of reaction (4) in the absence [10] and presence of metal cations, the Haber-Weiss reaction should be considered as insignificant under our experimental conditions. Therefore, the O_2^- induced process responsible for generation of OH radicals in the presence of hydrogen peroxide and Fe(II)/Fe(III), described by the above given mechanism, is actually superoxide-driven Fenton chemistry.

Numerous observations of hydroxylation and/or degradation products in in vivo experiments, suggest occurrence of Fenton-type chemistry in biological systems that are exposed to a flux of hydrogen peroxide and/or superoxide radicals in presence of metal ions/complexes (any metal ion/complex which upon reaction with H_2O_2 yields an OH radical can be considered to undergo Fenton type chemistry). As can be deduced from known rates of reactions, the rate controlling step in an O_2^- driven Fenton mechanism is most likely reaction (12) which, for ordinary ferrous salts, reacts

$$M^{n+} + H_2O_2 \longrightarrow M^{(n+1)+} + OH^- + OH \qquad (12)$$

at $k \sim 40\text{-}50\ M^{-1}\ s^{-1}$ in acid media [4,11]. In view of the various competitive reactions that can occur in such a system, in addition to reactions of catalase and superoxide dismutases that remove H_2O_2 and O_2^- very efficiently (k's of the order of 10^7 and $10^9\ M^{-1}\ s^{-1}$) in physiological conditions, reaction (12), in order to be an effective OH radical source, must be accelerated. This can often be achieved by increasing the pH which may cause hydrolysis of simple iron salts (e.g., $Fe_{aq}^{2+} \rightarrow FeOH_{aq}^{+}$) to a more reactive form or through complexation of the metal ion with appropriate ligands (e.g. Fe^{2+} when complexes with EDTA, HEDTA or NTA reacts with H_2O_2 at rates $k_1 = (1\text{-}4) \times 10^4\ M^{-1}\ s^{-1}$ [12]).

In addition to generating OH radicals, the Fenton-type system yields metal cations/complexes in higher oxidation states. The latter which can be strong and relatively long lived oxidizing agents, are also often generated directly by reactions between metal ions/complexes and HO_2/O_2^- without the involvement of H_2O_2 [13]. As catalase and superoxide dismutases, singly or jointly, do not always suppress a given oxidative process, these may be due to strong oxidizing agents other than the OH radical. Recent results on Mn(II)/Mn(III) model systems [13,14] suggest that at least in part oxidative processes can occur upon direct interaction of the transition metal in a higher oxidation state and a biological compound.

THE REACTIVITY OF Mn(III)/Fe(III) WITH BIOLOGICAL COMPOUNDS

Studies of the oxidation of organic compounds by transition metals in higher oxidation states (Fe^{3+}, Mn^{3+}, etc.) under homogeneous conditions have in the past been restricted to the low pH range because of their relative instability in neutral media. At higher pH's and in absence of strong complexing ligands these cations have been shown to undergo rapid hydrolysis and/or disproportionation reactions that result in the formation of interfering precipitates. To overcome these difficulties, use has been made in our laboratory of the "time windows" (vide infra) that exist in pulse radiolysis experiments.

Mn(III) and Fe(III) complexes were generated by pulse radiolysis via the following reactions:

$$Mn(II)/Fe(II) + OH \xrightarrow{N_2O} Mn(III)/Fe(III) \quad (13)$$

$$Mn(II) + O_2^- \xrightarrow{O_2,HCOO^-} MnO_2^+ \xrightarrow{fast} Mn^{3+} \quad (14)$$

To avoid direct interaction between OH radicals and the biological compounds (which react at near diffusion controlled rates), these studies were carried out in presence of relatively high concentrations of metal(II) ions. In experiments in which the metal ion was oxidized by O_2^-, much lower concentrations of metal(II) ions could be used because of the lower rates of reaction between HO_2/O_2^- and the given biological compound (see Table 1).

TABLE I

REACTIVITY OF O_2^-/HO_2, Mn(III) AND Fe(III) COMPLEXES WITH SELECT BIOLOGICAL COMPOUNDS

Reactant	Substrate	pH	k, $M^{-1}\ s^{-1}$
O_2^-,[5]	AH^-	7.0	5.0×10^4
Mn(III)-pyrophosphate[a]	AH^-	7.0	1.4×10^4
Mn(III)-phosphate[a]	AH_2/AH^-	4.7	1.4×10^6
Fe(III)-acetate[a]	AH^-	7.0	fast*
HO_2/O_2^-,[15]	Trolox	7.0	2.0×10^3
Mn(III)-pyrophosphate,[15]	Trolox	7.0	5.0×10^2
Mn(III)-phosphate[b]	Trolox	5.0	1.0×10^5
Fe(III)-acetate[b]	Trolox	7.0	$\ll 10^3$ **
HO_2/O_2^-,[16]	NADH	7.0	5.0×10^2***
Mn(III)-pyrophosphate,[17]	NADPH	5.0	0.8
Mn(III)-phosphate[a]	NADH	5-7	chain reaction****
Fe(III)-acetate[a]	NADPH	5-7	$\ll 10^3$

a) Unpublished data; b) Cabelli DE, Bielski BHJ, Arch. Biochem. Biophys. (submitted); *A complex is formed that then dissociates at $k \sim 10^3\ s^{-1}$; **Limiting rate; ***Value extrapolated from data given in ref. [16]; ****$k \sim 10^2\ M^{-1}\ s^{-1}$, measured in presence of O_2.

The compounds chosen for this study, ascorbic acid, trolox (6-hydroxy-2,5,7,8-tetramethylchroman-2-carboxylic acid) an α-tocopherol analogue, and NADH/NADPH, are typical biological reducing agents. Two conclusions can be drawn from the data on the reactivity of Mn(III)-complexes with the aforementioned compounds. First the reactivity of these metal species is generally enhanced over that of HO_2/O_2^-, particularly where Mn(III) is loosely liganded, i.e. Mn(III)-phosphate. Secondly as Mn(III)-complexes have much longer lifetimes than HO_2/O_2^- they can diffuse longer distances to potential target sites.

In comparison preliminary data on simple iron complexes indicate that even at pH 7, they are as expected much less reactive than similar manganese compounds. The rates given in Table I, represent only upper limits as precipitation of Fe(III) near neutrality interferes with the measurement of slow reactions. Overall this underscores the point that in the iron system, the ligand effect should be profound as the ligand serves not only to increase the reactivity of the Fe(III) itself but also serves to stabilize the cation in that oxidation state.

ACKNOWLEDGEMENTS

This research was supported by NIH Grant R01 GM23656-09 and was carried out at Brookhaven National Laboratory, which is operated under contract DE-AC02-76CH00016 to the U.S. Department of Energy.

REFERENCES

1. Oberly, LW, ed (1982) Superoxide Dismutase. CRC Press, Boca Raton, FL, USA, Vol. 1
2. Haber F, Weiss J (1934) Proc R Soc London Ser A147:332-351
3. Fenton HJH, Jackson H (1899) J Chem Soc (London) 75:1-11
4. Barb WG, Baxendale JH, George, P, Hargrave KR (1951) Trans Faraday Soc, I:462-500, II:591-616
5. Bielski BHJ, Cabelli DE, Arudi RL, Ross AB (1985) J Phy Chem Ref Data (in press)
6. Rush JD, Bielski BHJ (1985) J Phys Chem (in press)
7. Stuglik Z , Zagorski ZP (1981) Radiat Phys Chem 17:229-233
8. Jayson GG, Parsons BJ and Swallow AJ (1973) J Chem Soc Faraday Trans 1 69:236-242
9. Schneider W (1984) Comments Inorg. Chem., Comments Mod. Chem. Part A, Vol. III, 4, 205-223
10. Weinstein J, Bielski BHJ (1979) J Am Chem Soc 101:58-62 (and references therein)

11. Hardwick TJ (1957) Can. J. Chem. 35:428-436

12. Borggaard OK, Farver O, Andersen VS (1971) Acta Chem. Scand. 25:3541-3543

13. Cabelli DE, Bielski BHJ (1984) J. Phys. Chem. 88, I:3111-3115, II: 6291-6294

14. Archibald FS, Fridovich I (1982) Arch. Biochem. Biophys. 214:452-463.

15. Bielski BHJ (1985) Philosophical Trans. Roy. Soc. B (in press).

16. Nadezhdin A, Dunford HB (1979) J. Phys. Chem. 83:1957-1961.

17. Bielski BHJ, Arudi RL, Cabelli DE (1984) In: Bors W, Saran M, Tait D (eds) Oxygen Radicals in Chemistry and Biology, Walter de Gruyter & Co., Berlin-New York, pp. 1-15

Superoxide and Superoxide Dismutase in Chemistry,
Biology and Medicine; G. Rotilio, editor

CATALYSTS OF THE HABER-WEISS REACTION. STUDIES WITH ENZYMATICALLY AND RADIOLYTICALLY GENERATED SUPEROXIDE

CHRISTINE C. WINTERBOURN, HARRY C. SUTTON* and GLENN VILE

Pathology Department, Christchurch Hospital, Christchurch (New Zealand) and
*D.S.I.R. Institute of Nuclear Sciences, Lower Hutt (New Zealand).

Although $OH^{\cdot}$ radicals are frequently implicated in a wide range of pathological processes, if and how they are formed biologically is still a subject of much debate (1-3). It is now generally agreed that transition metal catalysis is vital for $OH^{\cdot}$ production (2,4), and much interest has focussed on the iron-catalysed Haber-Weiss reaction (or as it is alternatively called, the $O_2^{-\cdot}$-driven Fenton reaction). It is widely accepted that when this reaction is catalysed by iron-(EDTA), $OH^{\cdot}$ is produced, via the following mechanism:

$$Fe^{2+}(EDTA) + H_2O_2 \longrightarrow Fe^{3+}(EDTA) + OH^- + OH^{\cdot} \quad [1]$$

$$Fe^{3+}(EDTA) + O_2^{-\cdot} \longrightarrow Fe^{2+}(EDTA) + O_2 \quad [2]$$

However there is less consensus regarding other iron chelates as catalysts.

We have carried out a detailed study using $O_2^{-\cdot}$ generated (a) by γ-irradiation of oxygenated water (5), and (b) by the action of xanthine oxidase on hypoxanthine and O_2 (8). Its reactions with H_2O_2 and iron complexes with potentially physiological chelators have been investigated. To detect $OH^{\cdot}$ we have measured

1. Oxidation of formate to CO_2. This occurs by a simple and efficient mechanism in which quantitative conversion of $OH^{\cdot}$ to CO_2 can be obtained:

   $$OH^{\cdot} + HCOO^- \longrightarrow CO_2^{-\cdot} + H_2O$$

   $$CO_2^{-\cdot} + O_2 \longrightarrow CO_2 + O_2^{-\cdot}$$

   In a system where $OH^{\cdot}$ is regenerated from $O_2^{-\cdot}$ (e.g. reactions 1 and 2), there is a chain sequence in which many molecules of CO_2 per initial $O_2^{-\cdot}$ can be produced.
2. Deoxyribose oxidation to thiobarbituric acid (TBA)-reactive products. This method is practically more convenient and sensitive, but the products are less well defined (10). With both methods, appropriate scavengers were used to establish the involvement of free $OH^{\cdot}$.

$O_2^{-\cdot}$ PRODUCED BY γ-IRRADIATION

The results of irradiating formate solutions (at pH 7.4) are shown in Table I. With Fe(EDTA), oxidation of formate was observed, with side reactions of $O_2^{-\cdot}$ terminating the chain so that 2.5 moles CO_2 per mole $O_2^{-\cdot}$ were produced. With the other iron chelates, at most only a few percent of this amount of CO_2 was produced. With uncomplexed iron in phosphate or Tris buffer, no catalysis of the Haber-Weiss reaction was detected. CO_2 formation with ATP, ADP,

TABLE I

HYDROXYL RADICAL PRODUCTION FROM H_2O_2, Fe^{2+} AND $O_2^{-\cdot}$ PRODUCED BY γ-IRRADIATION

Results are taken from refs. 5 and 8. Reactions were in oxygen-bubbled 20 mM phosphate pH 7.4, except for the adenine nucleotides in 5 mM Tris.

Chelator (3-10 fold excess over iron)	Formate Oxidation*		Deoxyribose Oxidation†
	Moles CO_2 produced per mole $O_2^{-\cdot}$	Relative yield of CO_2	Relative yield of TBA-reactive products
EDTA	2.5	100	100
DTPA	0.08	3	3
Citrate	0.10	4	5
ATP	0.15	6	N.D.
ADP	0.06	2	N.D.
AMP	<0.01	<0.5	N.D.
Pyrophosphate	0.06	2	3
Fe in phosphate or Tris	<0.02	<0.8	<0.5

* 10 µM Fe^{2+}, 1.4 mM H_2O_2, 200 µM $O_2^{-\cdot}$ and 10 mM formate.
† 2 µM Fe^{2+}, 0.6 mM H_2O_2, 41 µM $O_2^{-\cdot}$ and 1 mM deoxyribose.

citrate and pyrophosphate required H_2O_2 and chelated iron, and was fully inhibited by superoxide dismutase. The Haber-Weiss reaction is thus implicated, but functioning with very low efficiency. In accompanying experiments the ability of the iron chelates to undergo the Fenton reaction at pH 7.4 was investigated. $OH^{\cdot}$ production was observed with all but Fe(AMP) and unchelated iron. This implies that a major determinant of efficiency as a Haber-Weiss catalyst is the ease of reduction of the ferric chelate by $O_2^{-\cdot}$, and that this reaction is fast enough for appreciable catalysis only with Fe^{3+}(EDTA).

Our measurements of deoxyribose oxidation (8) were entirely consistent with the formate studies. The yield of TBA-reactive products was very low except in the presence of Fe(EDTA), and with no chelator added, there was no detectable reaction (Table I).

$O_2^{-\cdot}$ FROM XANTHINE OXIDASE

The above results, although agreeing with some previous studies, are contrary to other reports that iron nucleotides and iron in phosphate or Tris buffer are good Haber-Weiss catalysts (3,7,11). These, and many other like studies, used xanthine oxidase to generate $O_2^{-\cdot}$, so we also looked at this system. Firstly, with formate oxidation, under conditions similar to Fig. 1 with 2 µM iron and 50 µM EDTA added, $O_2^{-\cdot}$ from xanthine oxidase gave a good yield of CO_2, but this was decreased to less than 3% by omitting the EDTA (8), i.e. in complete agreement with the radiation experiments, there was no production of $OH^{\cdot}$ in the absence of EDTA.

We obtained quite different results for deoxyribose oxidation, however. The

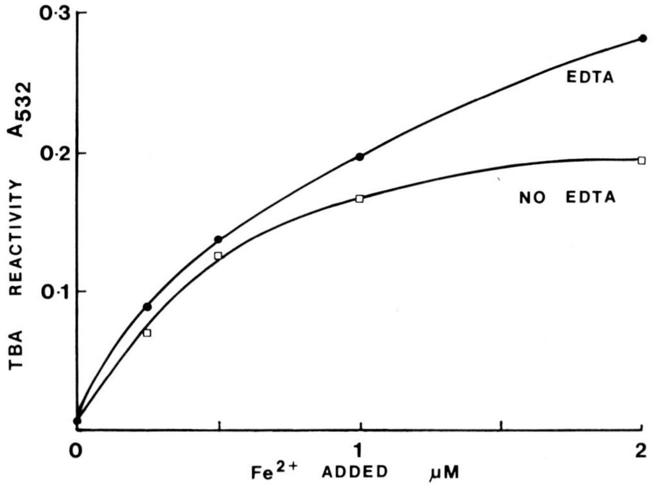

Fig. 1. Deoxyribose oxidation by hypoxanthine, xanthine oxidase and H_2O_2 in air. Conditions as for Table II.

reaction occurred to almost the same extent in the presence or absence of EDTA, either in phosphate or bicarbonate buffer (Fig. 1). Scavenger experiments performed in the presence of Fe(EDTA) gave an inhibitory pattern (Table II) compatible with the oxidant being free $OH^{\bullet}$. However, in the absence of EDTA a completely different scavenger pattern was evident: mannitol and Tris inhibited strongly but formate, benzoate and DMSO had little effect (Table II). From these results it must be concluded that xanthine oxidase-generated $O_2^{-\bullet}$, H_2O_2 and iron produce an oxidant species, but it is not free $OH^{\bullet}$.

Other chelators added to the xanthine oxidase system all progressively inhibited deoxyribose oxidation (Fig. 2). The low product yields with saturating concentrations of ATP, citrate and pyrophosphate are of the same order as these observed with radiolytic $O_2^{-\bullet}$. These effects were not due to radical scavenging, except to a minor extent at higher concentrations of ADP.

TABLE II

INHIBITION OF DEOXYRIBOSE OXIDATION BY XANTHINE OXIDASE-HYPOXANTHINE-H_2O_2

Solutions in bicarbonate or phosphate buffer pH 7.4 containing 3.5 mM deoxyribose and 0.005 U/mL xanthine oxidase were incubated for 40 min, then heated at 100°C with TBA and trichloroacetic acid as in ref. 11.

Catalyst	TBA Reactivity (A_{532})	Percent Inhibition by Formate (10 mM)	Benzoate (10 mM)	DMSO (10 mM)	Mannitol (10 mM)	Tris (2 mM)
Fe^{2+} (1μM)	0.25	8	15	6	81	71
Fe^{2+} (1μM) EDTA (100μM)	0.36	87	81	95	67	36

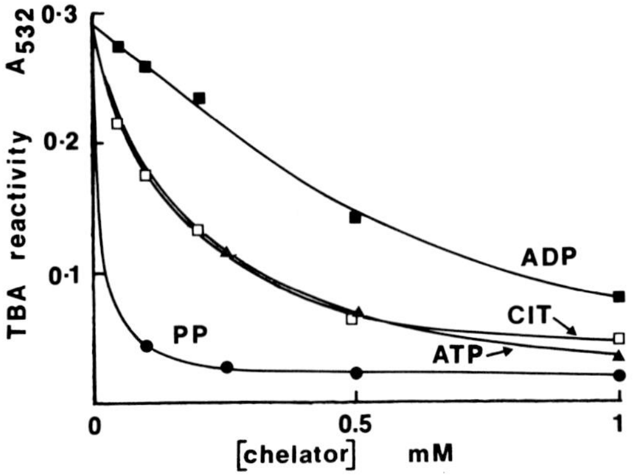

Fig. 2. Inhibition of deoxyribose oxidation by chelators. Conditions as in Table II with 7 mM deoxyribose and 1 µM Fe^{2+}.

The requirement for high chelator concentrations for maximum effect implies that they were not merely removing uncomplexed iron from solution. What then is the form of the iron? Although we have no direct evidence at this stage, taken together our results are compatible with its being bound to a high affinity site on the xanthine oxidase. It appears that this iron is able to react with H_2O_2, and $O_2^{-\cdot}$ produced by the enzyme, to produce an oxidant that is not free $OH^\cdot$. The oxidant could be considered as site-localized, reacting while associated with the enzyme, and possibly more accessible to some substrates than others.

These observations appear to rationalize a number of apparently discrepant findings in the literature. It seems likely that many authors, when studying xanthine oxidase-generated $O_2^{-\cdot}$ and iron, had insufficient chelator present to complex all the iron and were looking at the effects of our xanthine oxidase-derived oxidant, not free $OH^\cdot$. Further, the characteristics of this oxidant are reminiscent of those attributed to 'crypto-OH' (12). This species was first described with enzymatically generated paraquat and other xenobiotic radicals, which (as described below) behave very like $O_2^{-\cdot}$ in our experiments. 'Crypto-OH' may be the equivalent of our enzyme-derived oxidant.

$OH^\cdot$ PRODUCTION FROM PARAQUAT RADICALS

$O_2^{-\cdot}$ is not unique in its ability to reduce ferric compounds, and can be replaced in reaction 2 by a number of reducing agents (1,3). The paraquat radical ($PQ^{+\cdot}$) and semiquinones are such examples (9,10,13). We have observed with $PQ^{+\cdot}$ that, in contradistinction to the Haber-Weiss reaction, Fe(DTPA) as

TABLE III

OXIDATION OF DEOXYRIBOSE WITH XANTHINE OXIDASE-GENERATED PARAQUAT RADICALS

Solutions in 10 mM phosphate and 5 mM deoxyribose were incubated under N_2 for 30 min then heated with TBA and trichloroacetic acid.

Chelator (+0.5 μM Fe^{2+})	TBA Reactivity (A_{532})	Percent Inhibition by 10 mM Formate	Benzoate	Mannitol
None	1.50	25	36	72
EDTA (100 μM)	0.45	71	80	65
ATP (1 mM)	0.25	80	75	62
ADP (2 mM)	0.35	73	76	60
Pyrophosphate (1 mM)	0.40	58	70	48
Citrate (1 mM)	0.25	76	76	64

well as Fe(EDTA) can catalyse $OH^{\cdot}$ production with reasonable efficiency. This is true of radicals generated by radiation and with xanthine oxidase (9,10). However, as with O_2^{-}, only radicals generated enzymatically cause deoxyribose oxidation in the absence of EDTA, and no formate oxidation to CO_2 is observed under the same conditions. $PQ^{+\cdot}$ from the two sources also differ in reactivity with O_2.

Further studies of enzymatically-generated $PQ^{+\cdot}$ have shown that its reaction with H_2O_2 and iron is analogous to that of $O_2^{-\cdot}$. Whereas inhibition by scavengers in the presence of Fe(EDTA) or Fe(DTPA) is as expected for a reaction of free $OH^{\cdot}$, when only iron is added, mannitol but not formate or benzoate is strongly inhibitory (Table III). Thus, the oxidant formed in the absence of chelator again does not have the characteristic reactivity of free $OH^{\cdot}$. Similar results have been obtained using ferredoxin reductase and NADPH to generate $PQ^{+\cdot}$, and using adriamycin (which is reduced to the semiquinone under these conditions) instead of paraquat. This, therefore, appears to be a fairly general phenomenon.

The other chelators decreased deoxyribose oxidation, but not to the very low levels seen with $O_2^{-\cdot}$ (Table III). The same high concentrations were required for maximum effect, however. Scavengers, when added in the presence of these chelators had the same effect as in the presence of Fe(EDTA), giving the characteristic inhibitory pattern for a reaction of $OH^{\cdot}$ (Table II). We conclude, therefore, that as with $O_2^{-\cdot}$, iron chelation inhibits formation of the xanthine oxidase-derived oxidant. With $PQ^{+\cdot}$ as distinct from $O_2^{-\cdot}$, however, the chelates can then participate in Fenton-type reactions.

CONCLUSIONS

1. Although Fe(EDTA) is an efficient Haber-Weiss catalyst, none of the likely physiological iron chelates that we examined are. Thus, the relevance of this reaction as a model of $O_2^{-\cdot}$-toxicity must be viewed with reservation.
2. This restriction does not necessarily apply to the Fenton reaction driven by other reducing agents, e.g. paraquat radicals. Iron complexes with physiological compounds such as ADP, ATP or citrate, if present, could catalyse $OH^{\cdot}$ production by this mechanism. Whether this occurs, and contributes to the biological toxicity of compounds such as paraquat and quinones, depends on other factors, e.g. oxygen tension and availability of iron. The possibility that organic radicals produce an oxidant like $OH^{\cdot}$, rather than just acting as a source of $O_2^{-\cdot}$, warrants consideration.
3. Xanthine oxidase-derived $O_2^{-\cdot}$ reacts with H_2O_2 and low concentration of iron to give an oxidant species that differs in reactivity from free $OH^{\cdot}$. It appears that the iron is bound to a high affinity site on the enzyme. If the necessary iron is available to xanthine oxidase under physiological conditions, this oxidant could be formed when $O_2^{-\cdot}$ is generated. Intracellular chelators, e.g. ADP or ATP, may not be sufficiently concentrated to prevent the reaction. If they are, however, neither this oxidant nor $OH^{\cdot}$ should be produced in appreciable amounts. This reaction may be particularly relevant in the tissue damage seen in ischemia and reperfusion, in which xanthine oxidase is implicated.
4. Reactions seen with xanthine oxidase are not always applicable to other sources of $O_2^{-\cdot}$, and care should be taken in extrapolating experimental results obtained with different radical sources.
5. An oxidant with similar properties to that produced from xanthine oxidase and $O_2^{-\cdot}$ is also produced from other enzymatically generated radicals.

REFERENCES

1. Winterbourn CC (1981) Biochem J 198:125-131
2. Halliwell B, Gutteridge JMC (1984) Biochem J 219:1-14
3. Winterbourn CC (1984) In: Rotilio G, Bannister JV (eds) Life Chemistry Reports. Harwood Academic Publishers GmBH, New York, Suppl 2 pp 1-7
4. Aust SD, Morehouse LA, Thomas CE (1985) J Free Radicals Biol Med 1:3-25
5. Sutton HC (1985) J Free Radicals Biol Med (In Press)
6. Baker MS, Gebicki JM (1984) Arch Biochem Biophys 234:258-264
7. Graf E, Mahoney JR, Bryant RG, Eaton JW (1984) J Biol Chem 259:3620-3624
8. Winterbourn CC, Sutton HC (1985) (Sumitted for publication)
9. Sutton HC, Winterbourn CC (1984) Arch Biochem Biophys 235:106-115
10. Winterbourn CC, Sutton HC (1984) Arch Biochem Biophys 235:116-126
11. Flitter W, Rowley DA, Halliwell B (1983) FEBS Lett 158:310-312
12. Youngman RJ, Elstner EF (1981) FEBS Lett 129:265-268
13. Winterbourn CC, Gutteridge JMC, Halliwell B (1985) J Free Radicals Biol Med 1:43-49

Superoxide and Superoxide Dismutase in Chemistry, Biology and Medicine; G. Rotilio, editor

COPPER-ZINC AND MANGANESE SUPEROXIDE DISMUTASES INHIBIT THE SUPEROXIDE-DRIVEN FENTON REACTION AT TWO DIFFERENT STAGES

JOHN M.C. GUTTERIDGE
Division of Antibiotics and Chemistry, National Institute for Biological Standards and Control, Holly Hill, Hampstead, London NW3 6RB (UK)

INTRODUCTION

In the absence of high energy radiation, hydroxyl radicals ($OH^{\bullet}$) are probably formed in biological systems only when transition metal ions participate in a Fenton reaction (Equation 1).

$$Fe^{2+} + H_2O_2 \longrightarrow Fe^{3+} + OH^- + OH^{\bullet} \quad \text{(Eq.1)}$$

By far the most important transition metal ions in biological systems are iron and copper, whose ability to donate and accept single electrons make them ideally suited for reactions with molecular oxygen. Normally, these reactions proceed in controlled steps allowing the safe elimination of reduced oxygen intermediates such as superoxide (O_2-) and hydrogen peroxide (H_2O_2) (1-2). However, when iron and possibly copper are 'ill-placed' through excessive intake or release from damaged cells, oxygen radical damage can then result.

Iron and copper ions cannot exist free in biological solutions at physiological pH and must therefore bind to various components present. The affinity of iron-binding may be high as in the case of apotransferrin or low as occurs with albumin. Copper ions, however, readily ligate to almost all amino groups present in biological fluids. Techniques for the detection and measurement of loosely-bound iron and copper ions in biological fluids have been developed and applied to various fluids taken from human subjects (3-7). In most cases metal-binding does not prevent the metal ions from undergoing redox cycles, essential for Fenton chemistry, it merely prevents the formation of hydroxyl radicals in free solution (8).

RESULTS AND DISCUSSION

Site-specific $OH^{\bullet}$ radical damage can be produced by both copper (9) and iron (10). Such damage is, by definition, often difficult to protect against and may, in the absence of EDTA be uncharacteristic of an $OH^{\bullet}$ radical reaction (10-12). However, in the presence of EDTA a radiomimetic pattern of scavenging of $OH^{\bullet}$ in free solution is seen (Table 1).

Recently, it was observed that the superoxide-driven Fenton reaction, in which an EDTA-Fe^{3+} complex is used to damage deoxyribose in the presence of a superoxide generating system, was inhibited by superoxide dismutase (SOD) at two different stages (13). SOD inhibits the first stage (Eq. 2) by preventing O_2^- from reducing an EDTA-Fe^{3+} complex and in the second stage (Eq. 3) prevents formation of $OH^{\bullet}$ from EDTA-Fe^{3+} and hydrogen peroxide, produced during the dismutation reaction (Eq. 2), by a mechanism which must be similar to that of the first stage (13).

Stage 1. $EDTA\text{-}Fe^{3+} + 2O_2^- \xrightarrow[SOD]{2H^+} EDTA\text{-}Fe^{3+} + H_2O_2 + O_2$ (Eq.2)

Stage 2. $EDTA\text{-}Fe^{3+} + H_2O_2 \longrightarrow EDTA\text{-}Fe^{2+} + 2H^+ + O_2^-$ (Eq.3)

TABLE 1

THE EFFECT OF HYDROXYL RADICAL SCAVENGERS ON DAMAGE TO DEOXYRIBOSE BY A FERROUS SALT IN THE PRESENCE AND ABSENCE OF EDTA

	TBA-reactivity A532 nm	% Inhib.	TBA-reactivity A532 nm + EDTA(0.22mM)	% Inhib.
Blank (No Fe^{2+})	0.005		0.005	
Control + Fe^{2+}(0.22mM)	0.282		0.332	
" + Formate (11 mM)	0.250	11%	0.049	85%
" + Ethanol (11 mM)	0.249	12%	0.070	79%
" + Butan-1-ol (11 mM)	0.233	17%	0.054	83%
" + Urea (1.1 mM)	0.263	7%	0.310	4%
" + Thiourea (1.1 mM)	0.144	49%	0.068	80%
" + Glucose (11 mM)	0.071	57%	0.072	78%
" + Mannitol (11 mM)	0.030	89%	0.056	83%
" + Tris (11 mM)	0.016	94%	0.052	84%

Final reaction concentrations are shown. Deoxyribose was 1.1 mM in 44 mM phosphate buffer pH 7.4. Samples were incubated at 37°C for 30 min. TBA-reactivity was developed by adding 1% (w/v) thiobarbituric acid in 0.05M NaOH and an equal volume of 2.8% (w/v) trichloroacetic acid then heating at 100°C for 10 min.

As would be expected, CuZnSOD, MnSOD and catalase substantially inhibit deoxyribose degradation mediated by a reaction mixture containing $EDTA\text{-}Fe^{3+}$, hypoxanthine and xanthine oxidase (Table 2). The product of SOD scavenging in the first stage of this reaction will be $EDTA\text{-}Fe^{3+}$ and hydrogen peroxide (Eq. 2), itself an $OH^{\bullet}$ generating system. However, as shown in Table 3, CuZnSOD, MnSOD and catalase are again effective inhibitors of deoxyribose degradation when $EDTA\text{-}Fe^{3+}$ and hydrogen peroxide are present.

These results may have an important implication for a wider protective role for the SOD enzymes. Hydrogen peroxide is produced by enzymic two-electron transfers as well as by the dismutation reaction. Its safe removal by catalase and gluathione peroxidase is essential in order to prevent interaction with transition metal complexes. Howevever, catalase is known to be absent from certain aerobic bacteria such as *Bacillus popilliae* and *Streptococcus sanguis* which respond to changes in oxygen concentration by inducing higher levels of SOD (14,15). If biological iron complexes can behave in the way shown by EDTA then, the SODs would substantially protect against $OH^{\cdot}$ formation generated by complexed iron and hydrogen peroxide.

TABLE 2

INHIBITION BY CuZnSOD AND MnSOD OF DEOXYRIBOSE DEGRADATION STIMULATED BY A SUPEROXIDE GENERATING SYSTEM

	TBA-reactivity A532 nm after 1½ h 37 C	% Inhibition
Blank (XOD only) (subtracted)	0.079	
Control (XOD + HXn + EDTA-Fe^{3+})	0.212	
" + CuZnSOD	0.056	74%
" + MnSOD	0.064	70%
" + Catalase	0.004	98%
" + Albumin	0.197	7%

Final reaction concentrations were: deoxyribose 1.4 mM, phosphate buffer pH 7.4 29 mM, hypoxanthine (HXn) 0.67 mM, proteins 0.04 mg/ml, xanthine oxidase (XOD) 0.05 units, EDTA-Fe^{3+} (1.1:1) 0.14 mM. Albumin was included as a control for non-specific radical scavenging effects.

TABLE 3

INHIBITION BY CuZnSOD AND MnSOD OF DEOXYRIBOSE DEGRADATION STIMULATED BY HYDROGEN PEROXIDE AND EDTA-Fe^{3+}

	TBA-reactivity A532 nm after 1½ h 37 C	% Inhibition
Blank (H_2O_2 only)	0.096	
Control (H_2O_2 + EDTA-Fe^{3+})	0.604	
" + CuZnSOD	0.095	84%
" + MnSOD	0.088	85%
" + Catalase	0	100%
" + Albumin	0.661	0%

Final reaction concentrations were: deoxyribose 1.4 mM, phosphate buffer pH 7.4 29 mM, H_2O_2 0.14 mM, proteins 0.04 mg/ml, EDTA-Fe^{3+} (1.1:1) 0.14 mM.

REFERENCES

1. Fridovich I (1975) Annu Rev Biochem 44: 454-458
2. Fridovich I (1983) Annu Rev Pharmacol Toxicol 23:239-257
3. Gutteridge JMC, Rowley DA, Halliwell B (1981) Biochem J 188:263-265
4. Gutteridge JMC, Rowley DA, Halliwell B, Westermarck T (1982) Lancet (ii): 459-460
5. Rowley DA, Gutteridge JMC, Blake DR, Farr M, Halliwell B (1984) Clin Sci 66:691-695
6. Gutteridge JMC, Rowley DA, Halliwell B, Cooper DF, Heeley DM (1985) Clin Chim Acta 145:267-273
7. Gutteridge JMC (1984) Biochem J 218: 983-985
8. Gutteridge JMC, Wilkins S (1983) Biochim Biophys Acta 759:38-41
9. Samuni A, Chevion M, Czapski G (1981) J Biol Chem 256:12632-12635
10. Gutteridge JMC (1984) Biochem J 224:761-767
11. Youngman RJ, Elstner EF (1981) FEBS Lett 129:265-268
12. Gutteridge JMC (1984) Biochem Pharmacol 33:1725-1728
13. Gutteridge JMC (1985) FEBS Lett 185:19-23
14. Yousten AA, Bulla LE, McCord JM (1973) J Bacteriol 113:524-525
15. DiGuiseppi J, Fridovich I (1982) J Biol Chem 257:4046-4051

© 1986 Elsevier Science Publishers B.V. (Biomedical Division)
Superoxide and Superoxide Dismutase in Chemistry,
Biology and Medicine; G. Rotilio, editor

INITIATION OF AUTOOXIDATION BY HUMAN POLYMORPHONUCLEAR LEUKOCYTES: A ROLE FOR IRON

M. J. THOMAS, C. HEDRICK, and P. SHIRLEY

Department of Biochemistry, The Bowman Gray School of Medicine, Wake Forest University, Winston-Salem, North Carolina 27103 (U.S.A.)

INTRODUCTION

Oxy-radicals and singlet oxygen ($^1O_2$) have been suggested to participate in the bactericidal response of the human polymorphonuclear neutrophil (PMN). However, more recent work [1,2] would suggest that $^1O_2$ is not formed by the phagocytosing PMN. Formation of $^1O_2$ directly from superoxide has not been observed [3], and other mechanisms that might be active in the PMN (e.g., HOCl + H_2O_2) may not proceed in the purportedly acidic phagosome. Other studies have questioned the formation of strongly oxidizing radicals by PMNs [4,5]. The oxy-radical HO• should be formed in the presence of Fe(II) and H_2O_2 via the Fenton reaction.

To further examine the role that oxy-radicals and possibly $^1O_2$ might play in bacterial killing we have studied the oxidation by PMNs of linoleyl alcohol coated onto latex beads. 1,4-Dienes like linoleyl alcohol are easily oxidized by free radical chain autooxidation giving conjugated diene hydroperoxides. However, when 1,4-dienes react with $^1O_2$ they give approximately equal amounts of conjugated and nonconjugated hydroperoxides [6]. See the following scheme. Therefore, the product distribution can be used as a measure of the relative contributions of $^1O_2$ oxidation and free radical chain autooxidation. The hydroperoxides are reduced to facilitate analysis.

MATERIALS AND METHODS

General. Linoleyl alcohol was from NuChek Prep, Inc. Carboxylated latex beads (by IDC Spheres) of 0.732 μm diameter were purchased from Duke Scientific Co. Dulbecco's phosphate buffered saline was from Grand Island Biological Co. All other chemicals were of the highest quality available from commercial sources. PMNs were prepared as reported [7] using Dextran T500 in place of plasma gel. Coated beads were prepared by gently mixing a suspension of beads with linoleyl alcohol.

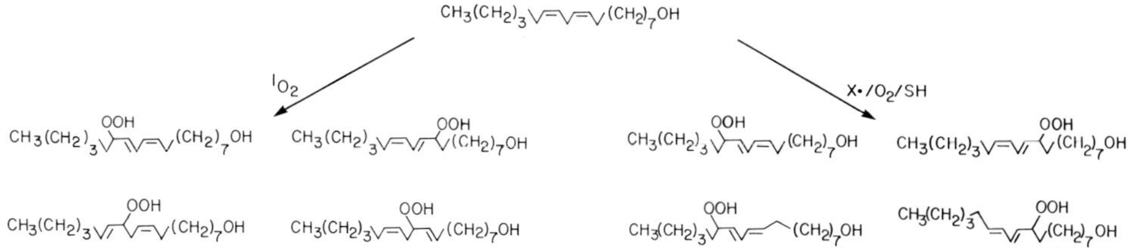

Stimulation by coated beads. Fresh PMNs (1.1×10^8 PMNs/ml) and glucose (1.1 m*M*) were divided into 3 ml aliquots preincubated for 5 min at 37°C and then treated with 0.1 ml (5×10^{10}) coated beads. After incubation the preparations were treated with 7 ml of 1 m*M* *N*-ethylmaleimide (NEM). Control cells were treated with 7 ml NEM and kept on ice. The PMNs were pelleted at 900 x *g*, washed, treated with 1 ml methanol, then extracted with 3-15 ml portions of 1:1 hexane-ether containing 0.001 weight % 2,6-di-*tert*-butyl-4-methylphenol. The residues after evaporation were reduced with methanolic $NaBH_4$. After acidic workup the organic layers were washed with water until the washings were neutral. The samples were plated onto Whatman silicar K TLC plates and developed with 1:1 hexane-ether. Areas containing oxidized products were removed and the silica gel extracted 18 h with 1:1 hexane-ether. When iron was added, beads were treated with sufficient aqueous $FeCl_3$ to give a solution 16 m*M* in Fe(III).

Mass spectrometric analysis. Gas chromatography/mass spectrometry (GC/MS) was performed on a Ribermag R10-10C quadrupole mass analyzer. Separation of the diene-diol products and linoleyl alcohol was obtained on a 7 m x 0.25 mm ID DB-1 fused silica capillary column (J & W, Inc.) that terminated in the source. Methyl stearate and 1,12-dihydroxy-9-octadecene were used as internal standards.

RESULTS

After exposing PMNs to coated beads, both the PMN extracts and a sample of beads were examined for linoleyl alcohol content. Within experimental error, the same amount of linoleyl alcohol was found in the PMN extracts as in an aliquot of coated beads. The average coating was 4.3 ± 0.3 mg/ml. The stability of hydroperoxides to activated PMNs was determined by treating PMNs with beads coated with linoleyl alcohol containing 2.5 weight % hydroperoxide prepared by $^1O_2$ oxidation of linoleyl alcohol. Diene-diols were detected in these extracts after reduction. The GC/MS detection limit was found to be approximately 1 ng/injection, equivalent to 1% oxidation of the linoleyl alcohol taken up by the PMNs. Linoleyl alcohol-coated beads were shown to stimulate all of the metabolic responses commonly associated with phagocytosing PMNs.

In the absence of added iron, no diene-diols were found in the PMN extracts or in their controls. Diene-diols were found in those samples to which exogenous iron was added (see figure). The predominant products were the conjugated diene-diols. However, significant amounts of the nonconjugated diene-diols were measured. NEM-inhibited control preparations showed only small amounts of these products.

DISCUSSION

Only in the presence of iron salts do we find oxidation of linoleyl alcohol to diene-diols. We speculate that in the presence of iron salts reduction of Fe(III)

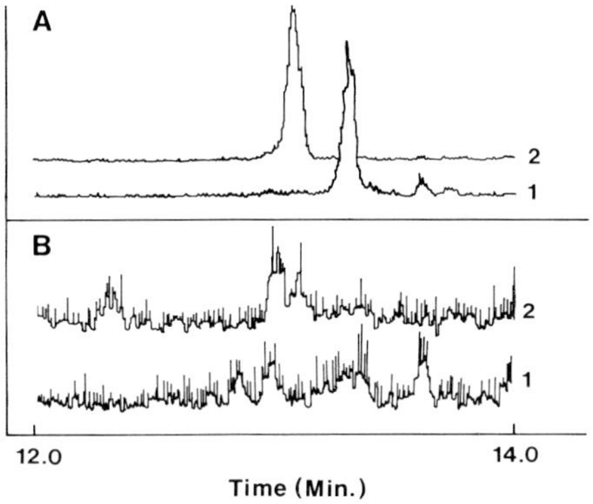

GC/MS selected ion traces of extracts from a mixture of PMNs, iron salts, and linoleyl alcohol-coated beads. Panel A, PMNs treated for 30 min at 35°C. Panel B, PMNs treated for 30 min at 4°C in the presence of 1 m*M* NEM. Trace 1 shows conjugated diene-diols while trace 2 shows nonconjugated diene-diols.

by superoxide is competitive with disproportionation and that reduction is followed by a Fenton reaction between Fe(II) and H_2O_2. These findings also imply that, if $HO_2\cdot$ is formed in the phagosome, it disproportionates faster than it can abstract H· from 1,4-dienes [8]. The formation of nonconjugated diene-diols may imply that $^1O_2$ is formed. Formation of $^1O_2$ from superoxide is impossible unless the free energy change of the reaction is greater than the 0.98 v needed to form $^1O_2$ ($^1\Delta_g$). No other hydroxylated products were observed ruling out H· abstraction from monoallyic positions as the source of the nonconjugated diene-diols.

The physiologic relevance of adding iron to PMN preparations is yet to be determined, but it may indicate a role for lactoferrin in PMN-mediated radical processes [9]. The availability of iron will certainly determine the degree of free radical participation in bacterial killing or in PMN-induced tissue injury.

ACKNOWLEDGEMENTS

We wish to acknowledge the excellent technical assistance of Mr. Michael Samuel in obtaining mass spectra. This work was supported by NIH grant GM29611.

REFERENCES

1. Foote CS, Abakerli RA, Clough RL, Shook FC (1980) Dev Biochem IIB:222-230
2. Thomas MJ, Shirley P, DeChatelet LR (1982) Abstract No. 1256, 66th Meeting of The Federation of American Societies for Experimental Biology, New Orleans, Louisiana, April 15-23
3. Foote CS, Shook FC, Abakerli RA (1980) J Am Chem Soc 102:2503-2504
4. McCay PB, Noguchi T, Fong KL, Lai EK, Poyer JL (1980) In: Pryor WA (ed) Free Radicals in Biology, Vol 4. Academic Press, New York, pp 155-186
5. Rosen H, Klebanoff SJ (1979) J Clin Invest 64:1725-1729
6. Thomas MJ, Pryor WA (1980) Lipids 15:544-548
7. DeChatelet LR, Shirley PS (1982) Inflammation 6:217-226
8. Bielski BHJ, Arudi RL, Sutherland MW (1983) J Biol Chem 258:4759-4761
9. Ambruso DR, Johnson RB (1981) J Clin Invest 67:352-360

Superoxide and Superoxide Dismutase in Chemistry, Biology and Medicine; G. Rotilio, editor

LIPID PEROXIDATION PROMOTED BY HUMAN POLYMORPHONUCLEAR LEUKOCYTES

GUNNAR CARLIN & KARL-E. ARFORS
Department of Experimental Medicine, Pharmacia AB, S-751 82 Uppsala (Sweden)

INTRODUCTION

Peroxidation of polyunsaturated fatty acids in membrane structures is considered to be an important mechanism underlying cell damage in various pathological states. The processes that cause this peroxidation are not clear and very little attention has been paid to other initiators of lipid peroxidation than xanthine oxidase and the microsomal enzyme NADPH-cytochrome P_{450} reductase. The purpose of the present study was to characterize the peroxidation of lipids caused by polymorphonuclear leukocytes (PMNL), especially in terms of iron requirement and participating radical species.

METHODS

Human PMNL from heparinized blood were prepared by conventional methods. The cells were resuspended in Dulbeccos phosphate buffer (DPB) to about 10^6 cells/ml and kept on ice. Linolenic acid micelles were prepared by sonication of 25 µl of linolenic acid in 1 ml of DPB immediately before the experiment. Folch fraction III phospholipids from bovine brain (10 mg in chloroform) were dried under argon and sonicated in 1 ml of DPB. In the peroxidation experiments 10^5 PMNL in 0.5 ml of DPB were incubated with 10 µl of the lipid suspension and test substance. It was found that linolenic acid micelles stimulated the production of oxygen radicals and thus of lipid peroxidation maximally. Addition of 2 µg of phorbol myristate acetate (PMA) had no further effect in this sytem and PMA was therefore not added routinely. However, peroxidation of phospholipid micelles required PMNL activation by PMA. Lipid peroxidation was measured as expression of thiobarbituric acid reactive species (TBARS). Superoxide dismutase (SOD) like activity was measured as the inhibition of increase in A_{550} in a reaction mixture containing 23 µM cytochrome-c, 1 mM hypoxanthine and 15 mU/ml xanthine oxidase. It has been shown earlier that the compounds tested did not interfere with the enzymatic activity of the xanthine oxidase.

RESULTS

Human PMNL promoted peroxidation of both linolenic acid and phospholipid micelles. The peroxidation required catalysis by iron ions. However, peroxidation of phospholipids required ferric iron chelated to pyrophosphate or ADP, whereas that of linolenic acid was equally well catalyzed by iron chelated to ADP, EDTA or the phosphate of the buffer. Furthermore, desferrioxamine inhibited the reac-

TABLE I

COMPARISON OF EXPRESSION OF TBARS BY LEUKOCYTE INCUBATED WITH PHOSPHOLIPIDS AND WITH LINOLENIC ACID.

Additive	Phospholipid % TBARS	Linolenic acid % TBARS
None	0.8 ± 3	35 ± 10
Pyrophosphate 800 µM + Fe^{3+} 60 µM	100 ± 9*	
+ desferrioxamine 0.8 mM	-16 ± 5	
+ SOD 60 U/ml	7 ± 1	
+ catalase 20 U/ml	100 ± 8	
ADP 800 µM + Fe^{3+} 60 µM	38 ± 6	100*
+ desferrioxamine 0.8 mM		22 ± 4
+ SOD 60 U/ml		92 ± 5
+ catalase 20 U/ml		103 ± 4
ATP 800 µM + Fe^{3+} 60 µM	5 ± 3	
EDTA 80 µM + Fe^{3+} 60 µM	7 ± 3	92 ± 5
$FeCl_3$ 60 µM	6 ± 3	100 ± 4

*to which other values in the same column are related.

tion in both systems. Catalase had no effect, while SOD inhibited peroxidation of phospholipids but not of linolenic acid (Table I).

Phospholipid peroxidation by PMNL was also catalyzed by ferritin but not by apoferritin or transferrin. The peroxidation was inhibited by desferioxamine (Table II).

TABLE II

EXPRESSION OF TBARS BY LEUKOCYTES INCUBATED WITH PHOSPHOLIPIDS IN THE PRESENCE OR IRON PROTEINS

Protein	% TBARS*
Ferritin (22 % iron)	
0.4 mg/ml	70 ± 3
0.04 mg/ml	11 ± 6
0.4 mg/ml + SOD 60 U/ml	5 ± 1
0.4 mg/ml + desferrioxamine 0.2 mM	9 ± 3
Apoferritin (0.01 % iron)	
0.4 mg/ml	0 ± 0
Transferrin (saturated, 0.15 % iron)	
1 mg/ml	7 ± 0

*100 % was the value obtained in the presence of 800 µM pyrophosphate + 60 µM Fe^{3+}. See Table I.

TABLE III

EFFECT OF VARIOUS COMPOUNDS WITH SUPEROXIDE DISMUTASE LIKE ACTIVITY ON THE EXPRESSION OF TBARS BY LEUKOCYTES INCUBATED WITH PHOSPHOLIPID AND Fe^{3+}-PYROPHOSPHATE

Compound, concentration (mM)		% TBARS	SOD-like activity*
None		100 ± 4	0
5-amino salicylic acid	1	40 ± 7	67
	0.1	49 ± 7	30
	0.01	83 ± 2	22
Paracetamol	1	11 ± 14	57
	0.1	49 ± 3	44
	0.01	–	35
Gentisic acid	1	47 ± 3	82
	0.1	72 ± 2	41
	0.01	–	21
Salicylic acid	1	91 ± 2	6

*expressed as % inhibition of cytochrome-c reduction.

Phospholipid peroxidation was inhibited by the drugs 5-aminosalicylic acid, paracetamol and gentisic acid, all of which were shown to react with xanthine oxidase derived superoxide by their ability to decrease the rate of cytochrome-c reduction (Table III).

DISCUSSION

The observed differences between peroxidation of linolenic acid and phospholipid micelles indicate two different reaction mechanisms. Peroxidation of linolenic acid was not directly superoxide-dependent, and may be similar to the well known NADPH-cytochrome P_{450} reductase dependent microsomal lipid peroxidation. In the latter case the iron is not reduced by superoxide but directly by the enzyme (1).

The peroxidation of phospholipids was inhibited by SOD and was therefore probably initiated by a superoxide-producing enzyme such as NAD(P)H-oxidase. The requirement of a diphosphate iron chelate cannot be explained at the present. A similar chelator dependence has been reported for microsomal lipid peroxidation, which is also catalyzed by pyrophosphate- or ADP-chelated iron ions while EDTA chelation is ineffective (2).

REFERENCES

1. Morehouse LA, Thomas CE, Aust SD (1984) Arch Biochem Biophys 232:366-377

2. Ernster L, Nordenbrand K, Orrenius S (1980) In: Yagi K (ed) Lipid peroxides in biology and medicine, Academic Press, NY, pp 55-80

Superoxide and Superoxide Dismutase in Chemistry, Biology and Medicine; G. Rotilio, editor

HYDROGEN PEROXIDE-ACTIVATED METHAEMOGLOBIN AND OTHER HAEMOPROTEINS AS INITIATORS OF MEMBRANAL LIPID PEROXIDATION

STELLA HAREL AND JOSEPH KANNER

Department of Food Science, A.R.O., The Volcani Center, P.O. Box 6, Bet Dagan 50250, Israel.

INTRODUCTION

In recent years the role of lipids in health and disease has received increasing attention (1). Lipid peroxidation has been implicated in a number of pathological reactions *in vivo*, such as vascular diseases, oxidative damage to DNA, cataractogenesis, metabolic activation of carcinogens and aging (2,3).

The direct oxidation of unsaturated fatty acid by molecular oxygen is spin--forbidden. Various free radicals were found to initiate lipid peroxidation (4) however, most of the authors connected hydroxyl radicals as the most important initiator of lipids radicals.

Most recently, we published that H_2O_2 activated metmyoglobin probably a porphyrin cation radical, $P^{\dot{+}}-Fe^{IV}=O$, could initiate membranal lipid peroxidation (4). This compound was also found to oxidize several other molecules such as ascorbic acid, phenol red, methional, β-carotene (5) and a series of phenols (6), uric acid (7) and to cause protein crosslinking (8).

The present study was conducted to extend our knowledge on H_2O_2-activated methemoglobin (MetHb) and other haemoproteins as initiators of lipid peroxidation.

MATERIALS AND METHODS

All the reagents were of the highest purity commercially available.

Isolation of the microsomal fraction from muscle turkey dark tissue was done by a procedure described by Apgar and Hultin (9).

Microsomes for the lipid peroxidation assay were incubated in a shaking water bath at 37°C. The reaction mixture contained 1 mg microsomal proteins per ml, the MetHb/H_2O_2 or other haemoproteins/H_2O_2 reactive compounds, 0.12 M KCl, 5 mM histidine buffer, pH 7.0. The reactions were initiated by activated MetHb with H_2O_2 or by glucose oxidase system which contained glucose oxidase, 0.04 units and glucose, 400 μM. The glucose oxidase system was adjusted for H_2O_2 generation of 1 nmole/ml/min.

Thiobarbituric acid reactive compounds (TBARS) and proteins were determined as previously described (4).

The results are means of triplicates of one microsomal preparation, in the figures each error bar I denotes the standard deviation.

RESULTS AND DISCUSSION

Initiation of membranal lipid peroxidation occurred only in the presence of both glucose oxidase system and MetHb. The glucose oxidase system demonstrated that small amounts of H_2O_2, generated continuously, could activate MetHb and lipid peroxidation efficiently, such as incubation of large amounts of H_2O_2 with MetHb (Figs. 1,2). Lipid peroxidation initiated by H_2O_2-activated MetMb or

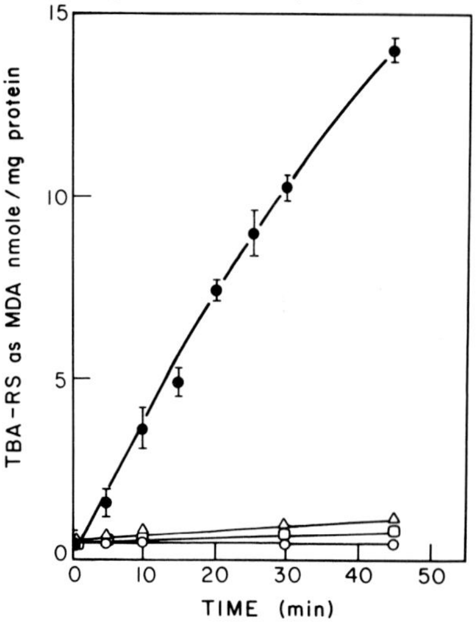

Fig. 1. The effect of glucose oxidase system (GOxS) on membranal lipid peroxidation by activated MetHb, ●, MetHb (5 μM) + GOxS; △, MetHb + catalase (40 units) + GOxS; ○, GOxS alone; □, MetHb alone.

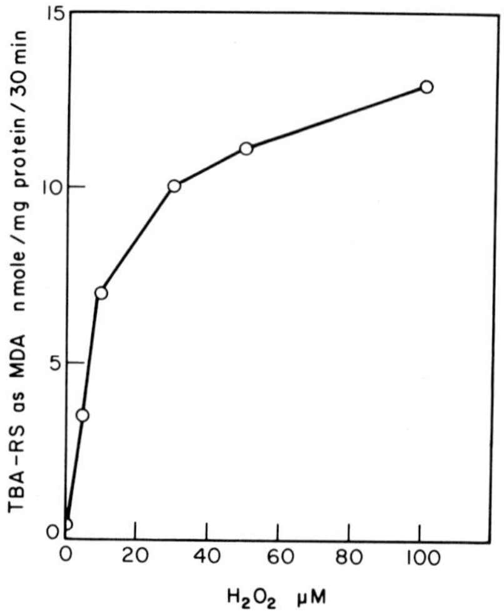

Fig. 2. Membranal lipid peroxidation by activated MetHb (5 μM) as affected by H_2O_2 concentration (reaction mixture described in Materials and Methods)

MetHb were found to be dependent on pH, showing maximum activity similar to that of the pH *in situ* of muscle tissues (6.5) and blood (7.3), respectively (Fig. 3).

Activated MetMb dependent membranal lipid peroxidation was not inhibited by several hydroxyl radical scavengers, such as ethanol, DMSO or histidine (4).

Low concentration of EDTA (100 μM) which was previously found to inhibit iron redox-cycle dependent membranal lipid peroxidation (10) failed to inhibit membranal lipid peroxidation induced by activated MetMb. Only very large amounts of EDTA was found to inhibit the reaction (Fig. 4). The antioxidative effect of EDTA at high concentration seems to stem from its characteristics to act as a weak electron donor (11,12). The relative activity of several H_2O_2-activated haemoproteins toward membranal lipid peroxidation is presented (Table 1). Cytochrome C which contain iron chelated to all six coordinations does not initiate membranal lipid peroxidation in the presence of H_2O_2.

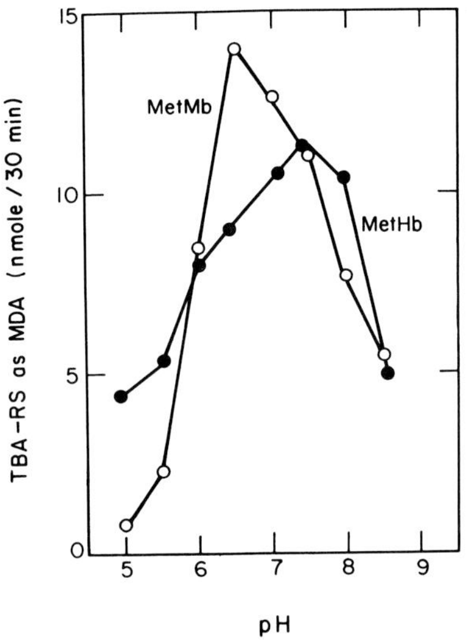

Fig. 3. Membranal lipid peroxidation by activated MetHb and MetMb as affected by pH. The reaction mixture contained MetHb (5 μM) or MetMb (30 μM) activated by H_2O_2 (30 μM) for 3 min before the addition to microsomes, 1 mg protein in 1 ml acetate or phosphate buffers (0.05 M) at 37^0.

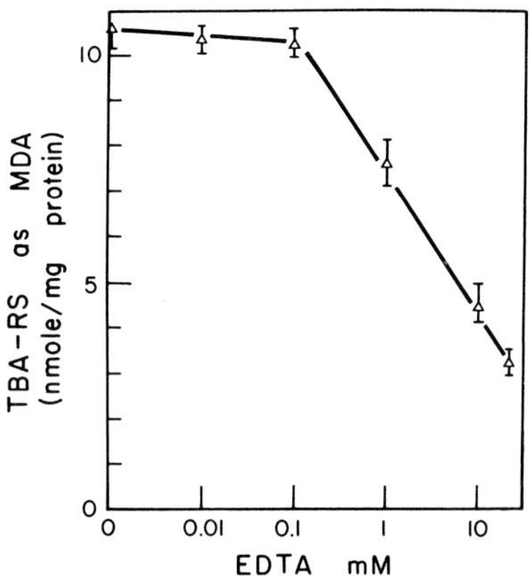

Fig. 4. Membranal lipid peroxidation by H_2O_2 (30 μM) activated MetMb (5 μM) as affected by EDTA concentration (reaction mixture described in Materials and Methods).

TABLE I

INITIATION OF MICROSOMAL LIPID PEROXIDATION BY H_2O_2 ACTIVATED HAEMOPROTEINS

Compound	TBA-RS as MDA nmole/mg protein/30 min
Metmyoglobin	15.0 ± 0.8
Methaemoglobin	8.6 ± 0.5
Peroxidase (HR)	3.8 ± 0.8
Lactoperoxidase	2.5 ± 0.6
Hemin	2.3 ± 0.4
Cytochrome C	0 -

H_2O_2 and haemoproteins at a concentration of 30 μM (MetHb, 5 μM), pH 7.0 and incubated 3 min before addition to reaction mixture.

Our results demonstrated that the interaction of haemoproteins and especially MetMb and MetHb with H_2O_2 generated a new catalyst, probably a porphyrin cation radical oxene which initiated membranal lipid peroxidation. The exact mechanism by which this catalyst initiates the generation of the lipid radical is not clear, e.g. it could be either by hydrogen abstraction or by removing an electron from the PUFA double bonds.

Our work showed that H_2O_2-activated MetMb or MetHb are powerful selective oxidants in vitro. However, more work is needed to demonstrate their activities in vivo.

REFERENCES

1. Levy RE, Rifkind BM, Dennis B, Ernst N (1979) Nutrition, Lipids and Coronary Heart Disease. Raven Press, NY, p.1
2. Yagi K (1982) Lipid Peroxides in Biology and Medicine. Academic Press, New York, pp 1-364
3. Bhuyan K, Bhuyan D (1983) In: Greenwald RA, Cohen C (eds) Oxy Radicals and their Scavenger Systems. Vol II, Cellular and Medical Aspects, pp 349-355
4. Kanner J, Harel S (1985) Arch Biochem Biophys 237: 314-321
5. Kanner J, Harel S (1985) Lipids (in press)
6. Shinga T, Imaizumi K (1975) Arch Biochem Biophys 167: 469-479
7. Ames BN, Catheart R, Schwiers E, Hochstein P (1981) Proc Natl Acad Sci USA 78: 6858-6862
8. Rice RH, Lee YM, Brown WD (1983) Arch Biochem Biophys 221: 417-427
9. Apgar ME, Hultin OH (1982) Cryobiology 19: 154-162
10. Harel S, Kanner J (1985) J Agric Food Chem (in press)
11. Friswell WR, Chung WC, MacKenzie J (1959) J Biol Chem 234: 1297-1301
12. Holstrom B (1967) Photochem and Photobiol 3: 115-119

© 1986 Elsevier Science Publishers B.V. (Biomedical Divison)
Superoxide and Superoxide Dismutase in Chemistry,
Biology and Medicine; G. Rotilio, editor

STUDIES ON THE PEROXIDATION OF VESICLES OF DIFFERENT PHOSPHOLIPID COMPOSITION

BRUNA TADOLINI, LUCIANA CABRINI, LAURA LANDI

Istituto di Chimica Biologica, Università di Bologna, Via Irnerio 48, 40126 Bologna (Italy)

INTRODUCTION

Lipid peroxidation, identified as a basic deteriorative reaction in cellular mechanism of aging and injury, results from free radicals attack to polyunsaturated fatty acid of biological membranes. It is well known that the lipid composition of membranes differs significantly depending on the subcellular organelle and on the tissue type of origin (1). It has also been established that in many biological membranes lipids are asymmetrically distributed in the transverse plane (1).

In spite of these facts, not much attention has been given to the effect that the different phospholipid composition of liposomes can exert on lipid peroxidation. Apart from an early report (2) this aspect of lipoperoxidation has been neglected.

In an attempt to ascertain whether the phospholipid composition of the liposome has any influence on peroxidation we have studied the peroxidation of vesicles of different composition.

METHODS

Liposomes were prepared by sonication as previously described (3).

Liposomal peroxidation reactions initiated by $FeCl_2$, $FeCl_3$, Fenton's reagent were sampled for malondialdehyde (MDA) (4) and lipid hydroperoxide (LOOH) (5) content.

Oxygen consumption was measured by a Clark oxygen electrode.

Free iron determination was made by the *o*-phenanthroline method (6).

RESULTS

The phospholipid vesicles tested were made of 100% phosphatidylcholine (PC), 1:1 molar ratio of PC and phosphatidylinositol (PI) and 1:1 molar ratio of PC and phosphatidic acid (PA).

Peroxidation of these vesicles, mesured as MDA formation, catalyzed by either iron or Fenton's reagent in the presence or absence of iron chelators shows different patterns (Table I).

We studied also the interaction of iron with the vesicles. The different vesicles were incubated with iron and then free ferrous iron and total free iron were determined. PC vesicles catalyze a time dependent oxidation of $FeCl_2$. The reaction depends on $FeCl_2$ and vesicle concentration. PA and PI containing vesicles interact with $FeCl_2$ in a time independent manner. When PI containing vesicles are present, the iron product of the reaction can be converted almost quantitatively to the free ferrous form by hydroxylamine.

TABLE I

PEROXIDATION OF VESICLES OF DIFFERENT PHOSPHOLIPID COMPOSITION BY EITHER IRON OR FENTON'S REAGENT IN THE PRESENCE OR ABSENCE OF IRON CHELATORS

Vesicle	Chelator	Catalyst		
		$FeCl_2 + H_2O_2$ A_{532}	$FeCl_2$ A_{532}	$FeCl_3$ A_{532}
PC	---	0.314	0.023	0.020
PC	EDTA	0.295	0.372	0.062
PC	ADP	0.351	0.028	0.038
PC	Cysteine	0.364	0.371	0.418
PC : PI	---	0.290	0.040	0.270
PC : PI	EDTA	0.377	0.339	0.139
PC : PI	ADP	0.406	0.057	0.081
PC : PI	Cysteine	0.267	0.075	0.254
PC : PA	---	0.213	0.104	0.026
PC : PA	EDTA	0.100	0.080	0.048
PC : PA	ADP	0.486	0.068	0.117
PC : PA	Cysteine	0.422	0.042	0.443

On the contrary the $FeCl_2$ and $FeCl_3$ that interacts with PA containing vesicles is undetectable by the o-phenanthroline method also in the presence of hydroxylamine and thus it is probably strongly bound.

The different interaction of the vesicles with $FeCl_2$ results in a different dependence of the peroxidation on $FeCl_2$ concentration, in the presence of hydrogen peroxide. A maximum of peroxidation is obtained at 0.1 mM $FeCl_2$ with PC and PI containing vesicles while peroxidation of PA containing vesicles linearly increases up to 0.3 mM $FeCl_2$.

The interaction of iron with the different vesicles results in oxygen consumption. PC vesicles intensely consumes oxygen when added with $FeCl_2$. Chelation of $FeCl_2$ with cysteine stimulates oxygen uptake while either EDTA or ADP reduce it. A similar inhibitory effect is obtained when $FeCl_2$ is added in the presence of H_2O_2.

A correlation exists between the oxygen consumption observed in the lipid hydroperoxides formed in the same experimental conditions.

PI containing vesicles consumes oxygen in the presence of $FeCl_2$ and the oxygen uptake is biphasic being inhibited by high $FeCl_2$ concentrations. Oxygen uptake is very low when $FeCl_2$ is incubated with PA containing vesicles. Only PI containing vesicles consume oxygen in the presence of $FeCl_3$.

DISCUSSION

The results obtained clearly indicate that the phospholipid composition of liposome has a great effect on the peroxidation of the liposome itself. These results may depend on the different nature of the hydrophylic moyeties of the vesicles. The different composition of the vesicles appears to influence the proportion of iron in the aqueous and lipid phase and also iron reactivity. It is conceivable that, as iron interaction with different chelators affects the peroxidation pathway, a similar result may be obtained by iron interaction with the different polar heads of phospholipids.

The possibility that the results observed may be ascribed either to the different composition in unsaturated fatty acids or to the lipid hydroperoxide content of the hydrophobic core seems at least in some instances, unlikely. In fact 100% PC and PA containing vesicles have a similar fatty acid composition and lipid hydroperoxide content.

ACKNOWLEDGEMENTS

This work was supported by grants from Italian C.N.R. and M.P.I.

REFERENCES

1. Nilsson OS, Dallner G (1977) Biochim Biophys Acta 464: 453-458
2. Gutteridge JMC (1977) Biochem Biophys Res Commun 74: 529-537
3. Tadolini B, Cabrini L, Landi L, Varani E, Pasquali P (1985) Biogenic Amines in press
4. Beuge JA, Aust SD (1978) Methods Enzymol 52: 302-310
5. Cavallini L, Valente M, Bindoli A (1983) Biochim Biophys Acta 752: 339-345
6. Mahler HR, Elowe DG (1954) J Biol Chem 210: 165-175

Superoxide and Superoxide Dismutase in Chemistry,
Biology and Medicine; G. Rotilio, editor

SUPEROXIDE RADICAL GENERATION IN RADIATION INDUCED AUTOXIDATION OF FATTY ACIDS

MOHAMAD I. ALSHEIKHLY and MICHAEL G. SIMIC
Center for Radiation Research, National Bureau of Standards
Gaithersburg, MD 20899, USA

INTRODUCTION

Generation of $^{\bullet}O_2H$ in nonpolar liquid phase on autoxidation of 1,4-cyclohexadiene, 1,4-CHD, was inferred by Criegee (1) from the formation of benzene. The mechanism was suggested later by Russell (2) and additional evidence provided by Howard and Ingold (3). Generation of $^{\bullet}O_2H$ from polyunsaturated fatty acids, PUFA, was not observed (3). Since in aqueous solutions at neutral pH superoxide radical, SOR, is in the $^{\bullet}O_2^-$ form (4), an additional driving force for generation of $^{\bullet}O_2^-$ would be expected because of solvation energies for $^{\bullet}O_2^-$ and H^+. Consequently, it was of interest to investigate whether SOR is generated from PUFA radicals and oxygen reaction in aqueous solutions.

Possible reactions of $^{\bullet}O_2H$ and $^{\bullet}O_2^-$ with various biochemical molecules were investigated by Gebicki and Bielski (5). They reported moderate reactivity for $^{\bullet}O_2H$, and much lesser reactivity for $^{\bullet}O_2^-$. It also was shown that $^{\bullet}O_2^-$ is practically unreactive with PUFA (5) and vitamin E (7). Hence, if $^{\bullet}O_2^-$ is generated in membranes, other protective mechanisms, such as SOD (8), would have to be involved for elimination of $^{\bullet}O_2^-$.

MATERIAL AND METHODS

Fatty acids were supplied by Sigma, formate by Baker, 1,4-cyclohexadiene and TNM by Aldrich, N_2O by Matheson.

Water was purified by a Millipore Milli-Q system. TNM was purified by washing with pure water.

Oxygen uptake was measured during radiolysis using a Clark electrode (Orion). Nitroform anion was measured spectrophotometrically in 10 cm cells.

Steady-state radiolysis was carried out on a vertical beam Co-60 source at 0.5 Gy/min. Dose-yield plots with 4 - 5 points were used for determination of G-values.

Pulse radiolysis was done with a 2 MeV, 50 ns Febetron 705 set-up.

RESULTS

Aqueous solutions of fatty acids, FA, at close to neutral pH were saturated with N_2O/O_2 mixtures in the presence of tetranitromethane, TNM, and irradiated up to dose = 15 Gy (6% conversion), at 0.5 Gy/min, using Co-60 gamma rays. The dose was determined using NBS standard procedures. Under those conditions, the radiation generated OH radical reacts with FA either by adding to a double bond or abstracting an H atom from a $-CH_2-$ group. The resulting FA radicals react with oxygen to give either peroxy radicals (addition) or superoxide radical (oxidation). It is well established that peroxy radicals do not react with TNM while superoxide radicals do (9). This was demonstrated by using pulse radiolysis of N_2O/O_2-saturated solutions of sodium formate and showing that $^{\bullet}O_2^-$ reduced TNM with a bimolecular rate constant of $1.6 \times 10^9\ M^{-1}s^{-1}$. Under pulse radiolytic conditions, $^{\bullet}O_2^-$ produced via elimination of fatty acids peroxy radicals also reduced TNM with $k = 1.6 \times 10^9\ M^{-1}s^{-1}$

$$^{\bullet}O_2^- + C(NO_2)_4 \longrightarrow C(NO_2)_3^- + NO_2 + O_2 \quad (1)$$

The nitroform, $C(NO_2)_3^-$ or NF^- absorbs at $\lambda = 350$ nm with $\varepsilon = 13{,}500\ M^{-1}cm^{-1}$.

The measurements of NF^- yields for some unsaturated fatty acid systems are shown in Table 1.

TABLE 1

NITROFORM ANION G-VALUES FOR UNSATURATED FATTY ACID AS A MEASURE OF SUPEROXIDE RADICAL YIELD

Aqueous solution of 0.1 mM UFA, 0.02 mM TNM, $N_2O + O_2$ (4:1) 1 atm. pH ~ 7, D_r = 0.5 Gy/min

Unsaturated Fatty Acid	$G(NF^-)$
Oleic Acid	1.1
Linoleic Acid	2.2
Arachidonic Acid	4.7
1,4-Cyclohexadiene*	6.5
Formate**	6.8***

* Used as symetrical model

** 10 mM

*** Measure of total free radical yield via formation of $^{\bullet}O_2^-$

DISCUSSION

The reaction of OH radical with 1,4-cyclohexadiene is expected to proceed in two modes:

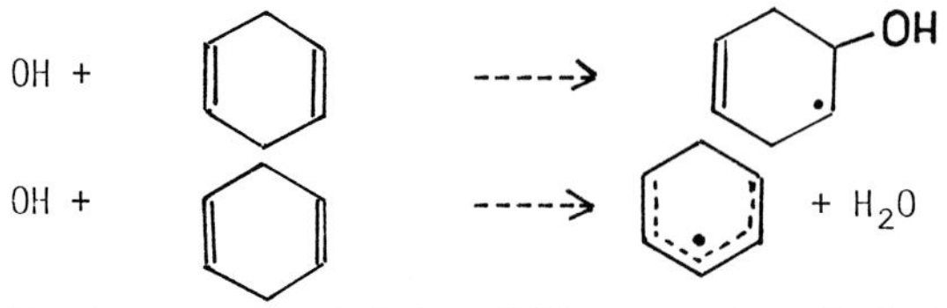

(2a)

(2b)

From the known reactivities (10), reaction (2a) is the predominant one. In the presence of oxygen, both radicals appear to give $^{\bullet}O_2^-$ because the $G(NF^-) = 6.5$ (Table 1), which is very close to the total $^{\bullet}O_2^-$ yield of $G = 6.8$.

In unsaturated fatty acids OH undergoes a reaction similar to reactions (2a) and (2b). However, in fatty acids, the attack on $-CH_2-$ groups located away from the double bond would give a radical which does not yield $^{\bullet}O_2^-$, when it reacts with oxygen. An increase of the number of double bonds in the fatty acids investigated (Table 1) also increases the fraction of reducing radicals and consequently, the NF^- (i.e., $^{\bullet}O_2^-$) yield.

It would be interesting to investigate these processes in model membrane and membrane systems.

REFERENCES

1. Criegee R (1953) In: Muller E (ed) Houben-Weyl Methoden der Organichen Chemie Vol 8, George Thieme Verlag, Stuttgart, pp 16-28
2. Russell GA (1959) J Chem Ed 36: 111
3. Howard JA, Ingold KV (1967) Can J Chem 45: 785-792
4. Bielski BHJ (1978) Photochem Photobiol 28: 645-649
5. Gebicki JM, Bielski BHJ (1981) J Am Chem Soc 103: 7020-7022
6. Bielski BHJ, Arudi RL, Sutherland MW (1983) J Biol Chem 258: 4759-4761
7. Arudi RL, Sutherland MW, Bielski BHJ (1983) In: Cohen G, Greenwald RA (eds) Oxy Radicals and their Scavanger Systems 1: pp 26-31
8. McCord JM, Fridovich (1969) J Biol Chem 244: 6049-6055
9. Henglein A, Jaspert J (1957) Z Phys Chem 12: 324-334
10. NSRDS-NBS 59 (1977) Ross AB, Farhataziz (eds), US Dept of Commerce

© 1986 Elsevier Science Publishers B.V. (Biomedical Division)
Superoxide and Superoxide Dismutase in Chemistry,
Biology and Medicine; G. Rotilio, editor

REACTION OF POLYUNSATURATED FATTY ACIDS WITH THE *tert*-BUTOXYL RADICAL IN AQUEOUS SOLUTION

MICHAEL ERBEN-RUSS, WOLF BORS, CHRISTA MICHEL, and MANFRED SARAN
Inst. Biologie, GSF Forschungszentrum, 8042 Neuherberg, FRG

INTRODUCTION

Alkoxyl radicals (1) are potential intermediates in lipid oxidation processes. Their reactions have been studied extensively; however, most of these studies have been carried out in organic solvents, because alkoxyl radicals become highly unstable with increasing solvent polarity. As it is assumed that such oxidative processes are not strictly confined to a membrane matrix, the reactivity of these species in aqueous solution is also of interest. In the following we present a pulse radiolysis study on the reaction of a model alkoxyl radical with polyunsaturated fatty acids (PUFA) - which are the major target of oxidative processes in membranes - in aqueous solution.

MATERIALS AND METHODS

All chemicals were of the highest purity available and were obtained from Merck (*tert*-butanol [^{t}BuOH], NaOH), Fluka (*tert*-butyl hydroperoxide [^{t}BuOOH], di-*tert*-butyl peroxide [(^{t}BuO)$_2$]) and Sigma (oleic, linoleic, linolenic, and arachidonic acids). Additional purification of the fatty acids by HPLC had no influence on the experimental results.

The pulse radiolysis set-up (2) consisted of a Febetron accelerator (electron pulses of 1.8 MeV, doses between 8 and 54 Gy), a Xenon lamp as analyzing light source, a spectrograph, a photomultiplier array, a transient digitizer, and a computer for experiment control and data evaluation.

RESULTS AND DISCUSSION

Solvated electrons (e^-_{aq}) in aqueous solution react very rapidly with ^{t}BuOOH to form *tert*-butoxyl (^{t}BuO$^\cdot$) radicals (1):

[1] $^{t}BuOOH + e^-_{aq} \longrightarrow {}^{t}BuO^\cdot + OH^-$ $\qquad k_1 = 1.6\times10^{10}M^{-1}s^{-1}$

These radicals decay by a fast unimolecular fragmentation reaction:

[2] $^{t}BuO^\cdot \longrightarrow {}^\cdot CH_3 + \text{acetone}$ $\qquad 10^6 s^{-1} \leq k_2 \leq 10^7 s^{-1}$

An approximate range of k_2 has been derived from ESR measurements in aqueous solution (3).

In the presence of PUFA (RH) reaction [3] may occur:

[3] $^{t}BuO^\cdot + RH \longrightarrow {}^{t}BuOH + {}^\cdot R$

Values of k_3 have been measured (4) by means of competition kinetics in benzene/(^{t}BuO)$_2$, where the unimolecular decay of ^{t}BuO$^\cdot$ is much slower than in water. As $^{t}BuO^{\cdot}$ abstracts H atoms from various sites of the PUFA chain, different alkyl radicals $^{\cdot}R$ are produced; the major one is the pentadienyl radical formed by H abstraction from doubly allylic sites:

[4] 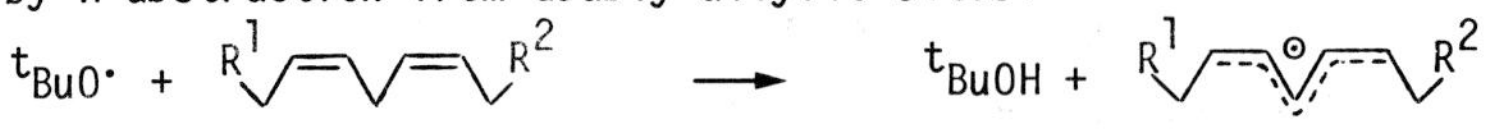

In our study, this radical was observed directly. It was identified by comparison with the pentadienyl radical produced by reaction of $^{\cdot}OH$ with PUFA (5), whose absorption spectrum (Fig.1), decay rate, and reaction rate constant with O_2 are nearly identical. Furthermore, the transient absorption was not observed when the monounsaturated oleic acid was used instead of PUFA.

Several experiments were carried out to verify that the pentadienyl radical was in fact produced by attack of $^{t}BuO^{\cdot}$ on PUFA. The strong pentadienyl absorption was not observed when $^{t}BuOOH$ was omitted, or when e^{-}_{aq} was scavenged by N_2O. $^{\cdot}CH_3$, the product of reaction [2], was specifically generated according to

[5] $CH_3Cl + e^{-}_{aq} \longrightarrow {}^{\cdot}CH_3 + Cl^{-}$ $\quad k_5 \geq 8\times10^{8} M^{-1} s^{-1}$ (6)

and did not react with PUFA. On the other hand, the pentadienyl radical could be observed when $^{t}BuO^{\cdot}$ was produced from $(^{t}BuO)_2$ instead of $^{t}BuOOH$.

Rate constants were derived from the build-up kinetics of the transients and the dependence of the pentadienyl radical yield on PUFA concentration (Fig.2). A computer program was used to simulate the measured curves of absorbance vs. time using a kinetic system comprising reactions [2], [4], [6], and [7]:

[6] $2\ {}^{t}BuO^{\cdot} \longrightarrow (^{t}BuO)_2$

[7] $^{t}BuO^{\cdot} + RH \longrightarrow {}^{\cdot}R'$

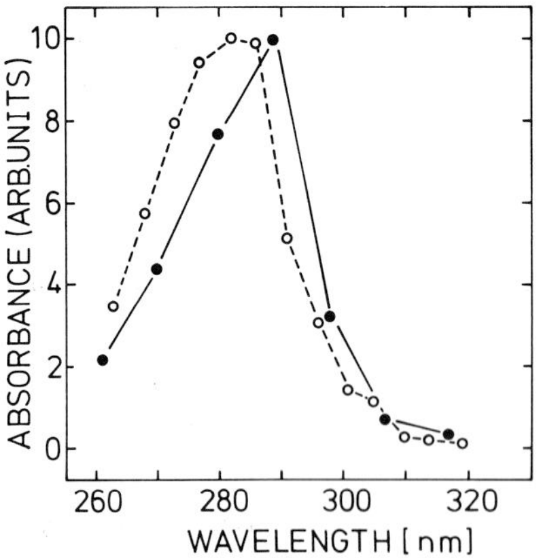

Fig. 1. Absorption spectra of the pentadienyl radical produced by reaction of PUFA with
- t-BuO· (●, solid line)
- ·OH (o, dashed line)

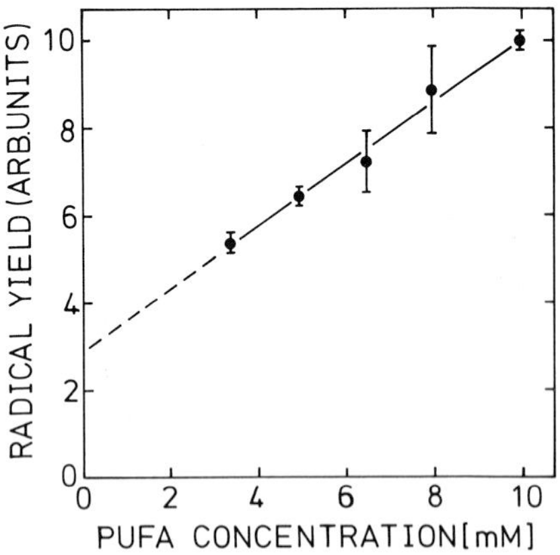

Fig. 2. Dependence of the pentadienyl radical yield on PUFA concentration. Ordinate scale: 1 arb.unit ≈ 2% of the initial radical concentration.

where ·R' symbolizes all PUFA radicals other than the pentadienyl (reaction [3] is the sum of reactions [4] and [7]). The contribution of reaction [6] (with an assumed $k_6 \geq 1x10^9 M^{-1}s^{-1}$) was found to be small but not quite negligible; k_7 - which was not accessible by our measurements - was varied within reasonable limits (cf. Table I). The reactions of $^tBuO\cdot$ with ^tBuOH or ^tBuOOH do not play any role, as was verified experimentally.

The build-up of the transient absorption was too fast to allow accurate kinetic analysis with the time resolution of our system. Therefore we cannot give exact values but rather lower and upper limits of k_2 and, consequently, also of k_4; these are listed in Table I. For comparison we have included the results of Small et al. (4); note that the rate constants in aqueous solution are at least some 2- to 3-fold higher than in non-polar solvent.

TABLE I
REACTIONS OF $^tBuO\cdot$ WITH PUFA
Rate constants $k \times 10^{-7}$ $[M^{-1}s^{-1}]$

fatty acid	$k_4^{(a)}$	$k_3^{(b)}$	$k_4^{(b)}$
linoleic	0.8 - 2.2	0.88	0.54
linolenic	3.2 - 9.0	1.30	1.09
arachidonic	5.1 - 13	2.05	1.63

(a) lower limits based on k_2= $1.0x10^6 s^{-1}$, k_6= $1x10^9\ M^{-1}s^{-1}$, k_7= 0
upper limits based on k_2= $2.5x10^6 s^{-1}$, k_6= $1x10^{10} M^{-1}s^{-1}$, k_7= $5x10^7 M^{-1}s^{-1}$
(b) Data from ref. 4

In conclusion, we have demonstrated that the reaction of the tert-butoxyl radical with PUFA in aqueous solution competes well with its unimolecular decay. This result implies that secondary chain reactions involving alkoxyl radicals and PUFA may be effective even in aqueous environment.

REFERENCES

1. Bors W, Tait D, Michel C, Saran M, Erben-Russ, M (1984) Israel J Chem 24:17-24
2. Saran M, Vetter G, Erben-Russ M, Bors W (1984) In: Bors W, Saran M, Tait D (eds) Oxygen Radicals in Chemistry and Biology. De Gruyter, Berlin, pp 91-94
3. Gilbert BC, Marshall PDR, Norman ROC, Pineda N, Williams PS (1981) JCS Perkin II 1392-1400
4. Small RD, Scaiano JC, Patterson LK (1979) Photochem Photobiol 29:49-51
5. Patterson LK, Hasegawa K (1978) Ber Bunsenges Phys Chem 82:951-956
6. Balkas TI, Fendler JH, Schuler RH (1970) J Phys Chem 74:4497-4505

Superoxide and Superoxide Dismutase in Chemistry, Biology and Medicine; G. Rotilio, editor

GENERATION AND OBSERVATION OF FATTY ACID PEROXYL RADICALS BY PULSE RADIOLYSIS

WOLF BORS, MICHAEL ERBEN-RUSS and MANFRED SARAN
Inst. Biologie, GSF Forschungszentrum, 8042 Neuherberg, FRG

INTRODUCTION

Peroxyl radicals (ROO$^{\cdot}$) are major intermediates of enzymatic and auto-oxidation reactions of polyunsaturated fatty acids. Their formation proceeds via pentadienyl radicals which add oxygen to yield peroxyl radicals containing two conjugated double bonds adjacent to the dioxygen group (1). This structural feature might lead to kinetic properties different from those of normal aliphatic ROO$^{\cdot}$ species. Since it has also been proposed that phenolic antioxidants scavenge these radicals preferentially at the lipid-water interface (2), we determined the kinetic parameters of such reactions in aqueous solution, using pulse radiolysis combined with kinetic spectroscopy. Data are presented for two naturally occurring flavone antioxidants, which were the most suitable target molecules out of six investigated phenols.

MATERIALS AND METHODS

Linoleic acid (LH, Sigma) or enzymatically prepared 13-hydroperoxylinoleic acid (13-LOOH) were precursors of the fatty acid peroxyl radicals. Phenolic radical scavengers were the flavones kaempferol (Fluka) and quercetin (Serva).

The pulse radiolysis set-up consists of a Febetron 705 accelerator, a Xenon light source, a spectrograph/photomultiplier arrangement and a Wang 2200 VP computer (3). Digitized kinetic traces were obtained at 15 wavelengths simultaneously and evaluated by computer simulation.

RESULTS AND DISCUSSION

Pulse-radiolytic generation of peroxyl radicals is generally achieved in solutions saturated with a mixture of N_2O and O_2:

$$[1]\quad e^-_{aq} + N_2O + H_2O \longrightarrow {}^{\cdot}OH + OH^- + N_2$$
$$[2]\quad {}^{\cdot}OH + LH \longrightarrow {}^{\cdot}L + H_2O$$
$$[3]\quad {}^{\cdot}L + O_2 \longrightarrow LOO^{\cdot}$$

Owing to the unspecific attack of $^{\cdot}OH$ radicals one always obtains a mixture of various isomeric LOO$^{\cdot}$ species.

To generate 13-LOO$^{\cdot}$ radicals selectively, we developed a method of oxidizing 13-hydroperoxylinoleic acid with azide radicals:

$$[4]\quad {}^{\cdot}OH + N_3^- \longrightarrow {}^{\cdot}N_3 + OH^-$$
$$[5]\quad {}^{\cdot}N_3 + LOOH \longrightarrow LOO^{\cdot} + N_3^- + H^+$$

A direct observation of these fatty acid peroxyl radicals is impossible due to the overlapping absorption of the conjugated diene group. Indirect evidence for their presence was obtained from the formation of strongly absorbing aroxyl radicals derived from phenolic antioxidants - which are also produced by attack of $^{\bullet}OH$ and $^{\bullet}N_3$:

[6] $^{\bullet}OH/O^- + PhOH \longrightarrow PhO^{\bullet} + H_2O/OH^-$

[7] $^{\bullet}N_3 + PhOH \longrightarrow PhO^{\bullet} + N_3^- + H^+$

The experiments had to be performed in alkaline solutions (pH 11.5) to enhance the solubility of both the phenols and the fatty acid, respectively its hydroperoxide. Since $^{\bullet}OH$ is partially dissociated at this pH ($^{\bullet}OH \rightleftharpoons O^- + H^+$), the pertinent rate constants had to be determined under these conditions (Table I).

TABLE I
RATE CONSTANTS WITH HYDROXYL AND AZIDE RADICALS
(pH 10.5 - 11.5; average SEM-values ± 10%)

Substrate	Rate Constants x 10^{-9} $[dm^3mol^{-1}s^{-1}]$ k_6	k_7
quercetin	4.30	6.60
kaempferol	3.65	6.95 (a)
linoleic acid	5.60 (b)	not determined
13-hydroperoxylinoleic acid	12.75 (b)	1.3 (c)

(a) value determined at pH 10.5, all others at pH 11.5;
(b) values from competition with KCNS ($k_{\cdot OH/O^-} = 7.25 \times 10^9$ $dm^3mol^{-1}s^{-1}$);
(c) average value from competition experiments with quercetin and kaempferol.

Evaluation of the kinetic parameters of the peroxyl radicals with the phenolic scavengers was impeded by the fact that the observed aroxyl radicals are formed and decay by various competing reactions:

[8] $LOO^{\bullet} + PhOH \longrightarrow PhO^{\bullet} + LOOH$

[9] $LOO^{\bullet} + PhO^{\bullet} \longrightarrow$ adduct formation

[10] $PhO^{\bullet} + PhO^{\bullet} \longrightarrow$ dismutation

[11] $LOO^{\bullet} + LOO^{\bullet} \longrightarrow$ non-radical products

Using a kinetic model comprising these reactions, the kinetic parameters given in Table II were calculated. Reaction [11] with a rate constant of $\leq 5 \times 10^8$ $dm^3mol^{-1}s^{-1}$ had a negligible effect on the overall kinetics.

TABLE II

RATE CONSTANTS OF AROXYL RADICAL FORMATION AND DECAY IN THE PRESENCE OF LINOLEIC ACID PEROXYL RADICALS (pH 11.5).

Substrate	System	Rate Constants x 10^{-7} [$dm^3 mol^{-1} s^{-1}$] k_8	k_9	k_{10}
quercetin	$N_3/LOOH/N_2O$	5.3	21	0.23
kaempferol	"	3.0	33	9.3
quercetin	$LH/N_2O + O_2$	4.0	28	0.13
kaempferol	"	2.7	91	14.5

The data in Table II reveal only minor differences between the two flavones except for the higher stability of the quercetin aroxyl radical (k_{10}) and evidently a faster addition of peroxyl radicals other than 13-LOO• at kaempferol (k_9). There is only a marginal effect of oxygen on the decay of the aroxyl radicals, as has been found for the α-tocopheryl radical (4).

The relative high value of k_8 demonstrates the radical-scavenging efficiency of the two flavone antioxidants in aqueous alkaline solution. In comparison, a value of $2x10^5$ $dm^3mol^{-1}s^{-1}$ was determined for α-tocopherol in organic solvents (5). Finally, the almost diffusion-controlled rate constant of the peroxyl-aroxyl radical recombination (k_9) demonstrates the general importance of this adduct formation for the reaction of phenols with peroxyl radicals (5,6).

REFERENCES

1. Porter NA, Wujek DG (1984) J Am Chem Soc 106:2626-29
2. Barclay LRC, Locke SJ, MacNeil JM, VanKessel J (1984) J Am Chem Soc 106: 2479-81
3. Saran M, Vetter G, Erben-Russ M, Bors W (1984) In: Bors W, Saran M, Tait D (eds) Oxygen Radicals in Chemistry and Biology. de Gruyter, Berlin, pp 91-94
4. Doba T, Burton GW, Ingold KU, Matsuo M (1984) JCS, Chem Comm 461-462
5. Winterle J, Dulin D, Mill T (1984) J Org Chem 49:491-495
6. Boozer CE, Hammond GS, Hamilton CE, Sen JN (1955) J Am Chem Soc 77:3233-37

Superoxide and Superoxide Dismutase in Chemistry,
Biology and Medicine; G. Rotilio, editor

THE SUPEROXIDE MEDIATED OXIDATION OF WEAKLY ACIDIC HYDROCARBONS IN APROTIC MEDIA

ARYEH A. FRIMER and TOVA FARKASH-SOLOMON
Department of Chemistry, Bar-Ilan University, Ramat-Gan 52100, Israel

INTRODUCTION

The recent interest in the organic chemistry of superoxide anion radical ($O_2^{\cdot-}$)[1,2] has led to a great deal of discussion regarding the mechanism of the $O_2^{\cdot-}$ mediated oxidation of weakly acidic hydrocarbons. Diphenylmethane (pK_a = 34) for example is oxidized by $O_2^{\cdot-}$ to benzophenone. Two general mechanisms are possible. On the one hand, $O_2^{\cdot-}$ may be acting as a radical, abstracting a labile hydrogen atom in the initiation step of a free radical autoxidation. Indeed, radical mechanisms were commonly assumed to be the predominant mode of $O_2^{\cdot-}$ action by the early researchers in the field. However, recent thermodynamic calculations by Liebman and Valentine[3] have revealed that $O_2^{\cdot-}$ is expected to be an extremely poor hydrogen atom abstractor.

Sawyer[4], on the other hand, has demonstrated that $O_2^{\cdot-}$ is quite effective as a base, particularly in aprotic media, and can promote deprotonation of weakly acidic oxygen acids such as alcohols (pK_a of n-butanol is 33 in DMF[5]). Although it is well known that it is far easier to deprotonate a heteroacid than a carbon acid irrespective of pK_a, nevertheless the effectiveness of $O_2^{\cdot-}$ in proton removal suggests the possibility of an alternate mode of oxidation of carbon acids. Namely, $O_2^{\cdot-}$ would function as a base in a base catalyzed autoxidation which proceeds via the corresponding carbanion.

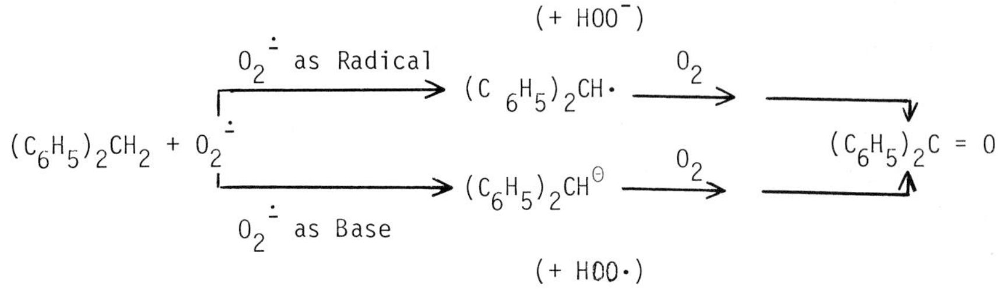

RESULTS

To study this problem, we synthesized a series of ring-substituted and methylene-deuterated diphenylmethanes and reacted them competitively with $O_2^{\cdot-}$ (generated from KO_2/18-crown-6 polyether) in benzene. Relative rate constants of reaction (k_{rel}) were determined by following the disappearance of the competing substrates by gas chromatography in conjunction with the equation (1), where

$$k_{rel} = k_A/k_B = \log (A_o/A_t) \div \log (B_o/B_t) \quad (1)$$

A_o, A_t, B_o and B_t are the GLC peak areas of diphenylmethanes A and B (normalized by the peak area of the inert internal standard) prior to the addition of $O_2^{\dot{-}}$ (t=0) and at time t. The relative rate constants (k_{rel}) correlated best (r = 0.993) with σ^- giving a ρ value of 3.96 ± 0.16. For the corresponding oxygenation mediated by t-butoxide, the ρ obtained was 1.77 ± 0.41 (r = 0.959). The primary deuterium isotope effect (k_H/k_D) on the superoxide reaction of diphenylmethane and its 4,4'-dichloro analog was 2.36 and 2.14 respectively. The rate of reaction was found to be linearly proportional to the crown ether concentration and no reaction occurred in its absence.

DISCUSSION

From the above data the following conclusion can be drawn:

1) The reaction is homogeneous - No reaction occurs in the absence of crown ether.

2) The reaction is first order in substrate - This is clear from the fact that equation 1, derived for competition reactions first order in substrate, yield k_{rel} values which are essentially invariant throughout the duration of the reaction.

3) The reaction is first order in solubilized superoxide - The solubilized $O_2^{\dot{-}}$ may either be paired with the potassium cation-crown ether complex [$C(K^+)$], or exist as a free ion (Equations 2 and 3, respectively).

$$C + KO_2 \longrightarrow C(K^+)O_2^{\dot{-}} \quad (2) \qquad C + KO_2 \longrightarrow C(K^+) + O_2^{\dot{-}} \quad (3)$$

In the former, case, the concentration of the solubilized superoxide will be linearly proportional to the initial concentration of crown ether. Equation 3, on the other hand, requires the concentration of $O_2^{\dot{-}}$ to be proportional to the square root of the initial crown ether concentration. Of these two possibilities we favor the former for two reasons. First, ion pairing is the expected state of aggregation in a non-polar aprotic medium such as benzene. More importantly the linear proportionality between the rate and the crown ether concentration is readily accommodated by equation 2 if the rate of reaction is first order in superoxide. For equation 3 to be operative, the reaction would have to be second order in superoxide in order to compensate for the aforementioned square root relationship between the superoxide and crown ether concentration. However, as will be shown later, the physical data (substituent and isotope effects) are inconsistent with a mechanism in which two molecules of $O_2^{\dot{-}}$ are involved up to the rate determining step.

4) <u>The rate determining step involves carbanion formation at the benzylic carbon</u> - This is clear from the excellent correlation between log k_{rel} and σ^- with a ρ of nearly 4. Nevertheless, several mechanisms are consistent with the observed substituent effect on the reaction.

MECHANISM 1	MECHANISM 2	MECHANISM 3
$RH + O_2^{\cdot -} \overset{Fast}{\rightleftharpoons} R\cdot + HOO^-$	$RH + O_2^{\cdot -} \overset{Fast}{\rightleftharpoons} R^- + HOO\cdot$	$RH + O_2^{\cdot -} \xrightarrow{Slow} R^- + HOO\cdot$
$R\cdot + O_2^{\cdot -} \xrightarrow{Slow} R^- + O_2$	$R^- + O_2 \xrightarrow{Slow}$ Products	$R^- + O_2 \xrightarrow{Fast}$ Products

However, the magnitude of the isotope effect ($\sim$ 2.25) rules out the rapid pre-equilibria required by mechanisms 1 and 2 which should be much less sensitive to deuteration. Thus the kinetic data, namely the correlation with σ^- , the magnitudes of ρ and the primary deuterium isotope effect, are only consistent with the reaction sequence described in Mechanism 3.

5) <u>The transition state for the proton abstraction occurs late</u> - A plot of log k_{rel} vs ΔpK_a yields a Brönsted α value of 0.69 ($r = 0.993$). The magnitude α suggests that the proton abstraction involves a late transition state. The aforementioned deuterium isotope effect of $\sim$ 2.25 could be consistent with either an early or late transition state. The α value, however, demonstrates that the latter is indeed the case.

CONCLUSION[6]

The above data indicates that the superoxide mediated oxidation of diarylmethanes and presumably other weakly acidic carbon acids involves <u>proton</u> transfer from substrate to $O_2^{\cdot -}$ in the rate determining step with the resulting benzylic carbanion undergoing subsequent oxygenation. In closing we note that the relevance of the aprotic media work to biological systems has been confirmed by reacting $O_2^{\cdot -}$ with organic reagents intercalated into the bilayer of liposomes.[7]

REFERENCES

1. Frimer AA (1982) In: Oberley LW (ed) Superoxide Dismutase, Chemical Rubber Co., Boca Raton, Florida, Vol. 2, pp 83-125.
2. Frimer AA (1983) In: Patai S (ed) The Chemistry of Functional Groups-Peroxides, Wiley, Chichester, pp 429-461.
3. Liebman JF, Valentine JS (1983) Israel J. Chem 23: 439-441.
4. Sawyer DT, Gibian MJ (1979) Tetrahedron 35: 1471-1481
5. Nanni EJ, Stallings MD, Sawyer DT (1980) J Am Chem Soc 102: 4481-4485
6. Compare with the conclusions of Lee-Ruff E, Timms N (1980) Can J Chem 58: 2138-2141.
7. Frimer AA, Hameiri-Buch J (1985) unpublished results.

Superoxide and Superoxide Dismutase in Chemistry, Biology and Medicine; G. Rotilio, editor

INTERACTION OF PEROXYL RADICAL ADDUCTS OF DNA BASES WITH REDUCTANTS

P. O'NEILL AND S. DAVIES

MRC Radiobiology Unit, Chilton, Oxon, UK.

INTRODUCTION

The hydroxyl radical ($^{\bullet}OH$) has been implicated in several reactions and processes which may lead to chemical changes to essential biomolecules resulting in the subsequent loss of biological information. Indeed the toxicity of oxygen and various cytotoxic agents is often discussed in terms of conversion of superoxide to the $^{\bullet}OH$ radical by processes (Fenton Chemistry) which may involve metal ions (1). One of the cells natural defences against oxygen toxicity is considered to be superoxide dismutase which efficiently removes $O_2^{\bullet -}$. Antioxidants are also implicated in mechanisms (2) whereby they act by interactions with cellular peroxyl-radicals which may be involved in chain reactions.

The present study was undertaken in order to investigate the electrophilic character (3,4) of the radicals produced on interaction of O_2 with the $^{\bullet}OH$-adducts of nucleosides.

EXPERIMENTAL

The 4.3 MeV linear accelerator and the optical detection and analysis systems have previously been described (5). All solutions (2 mmol dm^{-3} phosphate, pH7) were saturated with O_2 (B.O.C., Medical grade) in order to investigate the properties of the resulting peroxyl radicals of the $^{\bullet}OH$-adducts of deoxynucleosides. $O_2^{\bullet -}$, produced on scavenging of e^-aq, is expected to be inactive (3,4) on the timescale of the experiments.

RESULTS AND DISCUSSION

Interaction of peroxyl radicals of pyrimidines with reductants

It has previously been shown (3,4,6) that the resulting $^{\bullet}OH$-adducts of the pyrimidines interact with O_2 to yield the corresponding peroxyl radicals ($RO_2^{\bullet}$). The bimolecular rate constants for interaction of the peroxyl radicals of thymidine (2.2 mmol dm^{-3}) and deoxycytidine (2.0 mmol dm^{-3}) with a series of reductants (100-400 µmol dm^{-3}) were determined from the dependence of the first order rate constants for formation of the one-electron oxidised reductant on the concentration of reductant and are presented in Table 1. The N-(1) substituted pyrimidines were used in order to stabilise the peroxyl radical adducts to elimination of $O_2^{\bullet -}$ (3,4). From these observations it is inferred that the peroxyl radical adducts of the deoxynucleosides are electrophilic in character. The yields of peroxyl radicals (see Table 1) with oxidising properties (based

TABLE I

RATE CONSTANTS AND YIELDS FOR INTERACTION OF REDUCTANTS WITH PEROXYL RADICAL ADDUCTS OF DEOXYNUCLEOSIDES

Reductants	T	dC	dG
		$k \times 10^{-7}/dm^3\ mol^{-1}\ s^{-1}$	
N,N,N',N'-Tetramethyl p-phenylenediamine	8.3	6.8	150 (52)*
ABTS	1.3 (55)*	1.2 (45)*	
Ascorbate	2.5	1.8	
Hydroquinone	0.65	1.1	
p-Methoxyphenol**	< 0.1	< 0.1	

* Yields expressed as a percentage of the yield of $^{\cdot}OH$-radicals.

** The phenoxyl radical was not observed to be formed.

upon the variation of yield with dose extrapolated to zero dose) may be accounted for in terms of those $^{\cdot}OH$-base adducts with reducing properties (3,4) interacting with O_2 to yield peroxyl radicals with oxidising properties and that ∿ 20% of the $^{\cdot}OH$ radicals interact at the sugar moiety to yield radicals which do not interact with the reductants ($k < 10^7\ dm^3\ mol^{-1}\ s^{-1}$, based upon the reactivity of aliphatic peroxyl radicals (7,8)).

<u>Interaction of peroxyl radicals of purines with reductants</u>

The sites of addition, redox properties and yields of the various OH-adducts of deoxyguanosine (dG) and deoxyadenosine (dA) have been documented (5,8-10). On pulse irradiation of 10^{-3} mol dm^{-3} dG solution containing TMPD (60-100 μmol dm^{-3}) and saturated with O_2, the yield and rate of formation of $TMPD^{\cdot +}$ (see Table 1) are the same as those determined (9) in the absence of O_2. From these observations it is proposed that O_2 does not compete efficiently with the reductants for the OH-adducts of dG with oxidising properties ($k < 10^7\ dm^3\ mol^{-1}\ s^{-1}$) and that if O_2 interacts with the reducing OH-adducts of dG, the resulting peroxyl radical does not have oxidising properties.

With dA (1.9 mmol dm^{-3}), the yield of $ABTS^{\cdot +}$ produced on oxidation of ABTS (49-200 μmol dm^{-3}) by one of the OH-adducts of dA in the presence of O_2 represents 30% of the yield of $^{\cdot}OH$ radicals and is consistent with the known yield of the C(2)/C(4) $^{\cdot}OH$ adducts of dA (8,10) which have also been shown to interact with O_2 (6,10). Furthermore it is inferred that those OH-adducts of dA with reducing properties do not yield peroxyl radicals with electrophilic properties consistent with the observations with dG, (k < 10^6 dm^3 mol^{-1} s^{-1}).

SUMMARY

It has been demonstrated that only certain types of OH radical adducts of DNA bases interact with O_2 to produce peroxyl radicals with oxidising properties. Indeed those OH-adducts of DNA bases with oxidising properties (8-10) may not interact with O_2 (k < 10^7 dm^3 mol^{-1} s^{-1}) especially with the purine bases. Furthermore the purine OH-adducts with reducing properties may not interact efficiently with O_2. With dA and the pyrimidines, peroxyl radicals produced on interaction of O_2 with the $^{\bullet}$OH-adducts are electrophilic in character based upon their interaction with anti-oxidants. From the variation of the rate constants (Table 1), it is inferred that the process may involve electron transfer. This latter interaction may be important in protection against subsequent interactions of the peroxyl radicals such as H-atom abstraction as has been recently inferred (11). Such interactions may have implications for oxygen toxicity and to the cytotoxic effects of DNA binding agents. The question arises as to the importance of such electrophilic peroxyl radicals being involved in the initiation of cellular damage and its subsequent expression?

ACKNOWLEDGEMENTS

We thank Dr. E.M. Fielden for helpful discussion.

REFERENCES

1. Halliwell B, Gutteridge JMC (1984) Biochem J 219:1-14.
2. Doba T, Burton GW, Ingold KU, Matsuo M (1984) JCS Chem Comm 461-462.
3. Fujita S, Steenken S (1981) J Amer Chem Soc 103:2540-2545.
4. Hazra DK, Steenken S (1983) J Amer Chem Soc 105:4380-4386.
5. O'Neill P, Chapman PW (1985) Int J Radiat Biol 47:71-80.
6. Willson RL (1970) Int J Radiat Biol 17:349-358.
7. Packer JE, Willson RL, Bahnemann D, Asmus K-D (1980) JCS Perkin Trans II 296-299.
8. O'Neill P, Fielden EM (1985) In: 'Proceedings of NATO advanced study institute on Radiation Carcinogenesis and DNA alterations', submitted.
9. O'Neill P (1983) Radiat Res 96:198-210.
10. O'Neill P, Chapman PW, Papworth DG (1985) Life Chem Reports 3:62-69.
11. Gee CA, Kittridge KJ, Willson RL (1985) Brit J Radiol 58:251-256.

Abbreviations: T = thymidine, dC = deoxycytidine, dG = deoxyguanosine, dA = deoxyadenosine.

Superoxide and Superoxide Dismutase in Chemistry,
Biology and Medicine; G. Rotilio, editor

RADIOLYSIS STUDY OF THE REACTION OF DESFERRIOXAMINE WITH O_2^- FREE RADICALS

CATHERINE RIBIERE [(1)], DOMINIQUE SABOURAULT [(1)], JAMAL SINACEUR [(1)] and ROGER NORDMANN [(1)]
CHANTAL HOUEE-LEVIN [(2)] and CHRISTIANE FERRADINI [(2)]
[(1)] Laboratoire de Biochimie (Paris-Ouest) et INSERM U 72, [(2)] Laboratoire de Chimie Physique
Université Paris V, 45, rue des Saints-Pères 75270 Paris Cedex 06 (France)

INTRODUCTION

Beside the clinical use of desferrioxamine (DFO) in the prevention and the treatment of iron-overload, this iron-chelating agent is frequently used to inhibit the iron-catalyzed production of hydroxyl radicals (OH) by superoxide generating systems. We have previously shown that DFO could inhibit the reduction of cytochrome c or NBT by O_2^- free radicals produced by the xanthine-xanthine oxidase system (1) suggesting a reaction between DFO and O_2^-. To confirm the latter hypothesis, we studied the reaction of DFO with O_2^- generated by γ and pulse radiolysis.

MATERIAL AND METHODS

Desferrioxamine methane sulfonate was a gift from Ciba-France. DFO titrations were made by adding ferric chloride solutions dissolved in sulfuric solutions 0.1 mol.l^{-1}. The ferrioxamine complex thus formed was dosed by absorption spectrophotometry at 430 nm (ϵ = 2300 mol^{-1}. l. cm^{-1}).

γ -irradiations were made in a ^{60}Co irradiator. Pulse-radiolysis experiments were made using a Febetron 708 double beam accelerator (800 keV-pulse duration 4 ns at mid-height, optical path 1 cm). The O_2^- initial concentration varied between 10^{-5} mol. l^{-1} and 10^{-4} mol. l^{-1} (ca. 16 to 160 Gy).

RESULTS

<u>γ-radiolysis.</u> $[DFO]_o$ varied between 10^{-4} and 4.3 x 10^{-4}mol. l^{-1} (pH 7.2, O_2 atmosphere). DFO disappearance is independent of $[DFO]_o$ and of dose rate (results not shown). This suggests that all O_2^- are scavenged by DFO. The product of this irradiation could also be measured by the oxidation of I^- giving I_3^- in the conditions of (2). The initial yield was independent of $[DFO]_o$ and of the dose rate. However no H_2O_2 was detected by reaction according to (3).

<u>Pulse-radiolysis.</u> $[DFO]_o$ was equal to 10^{-4} or 2 x 10^{-4} mol. l^{-1} (pH 7.2, O_2 atmosphere). Figure 1 shows the absorption spectrum of the DFO free radical produced by the reaction of O_2^- with the drug.

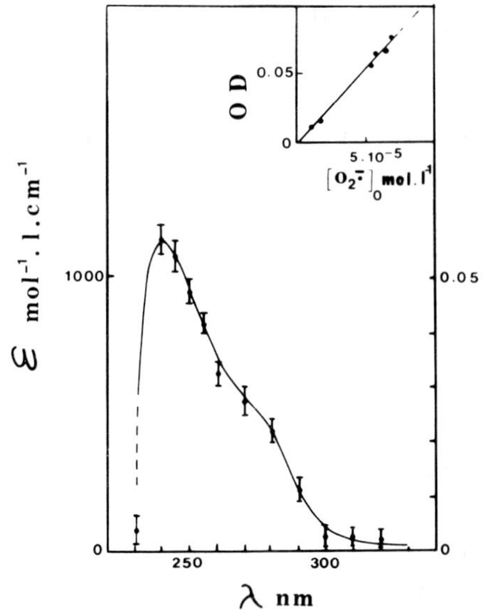

Fig. 1

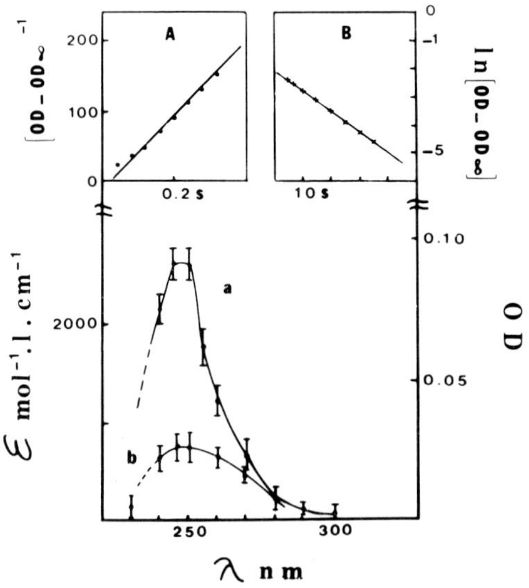

Fig. 2

Fig. 1. Absorption spectrum of desferrioxamine free radical produced by reaction with O_2^-, 25 ms after the pulse.
Dose $\simeq$ 80 Gy; $[DFO]_o = 2 \times 10^{-4}$ mol. l^{-1}; phosphate buffer 6×10^{-2} mol.l^{-1}, pH 7.2; O_2 atmosphere; $[HCOO^-] = 0.16$ mol. l^{-1}.
Inset : Variation of OD_{245} measured 25 ms after the pulse with $[O_2^-]_o$.
Fig. 2. Absorption spectra after desferrioxamine decay, (a) between 600 and 900 ms after the pulse; (b) 24 s after the pulse.
Dose $\simeq$100 Gy; $[DFO]_o = 2 \times 10^{-4}$ mol. l^{-1}; phosphate buffer 6×10^{-2} mol. l^{-1}, pH 7.2; O_2 atmosphere; $[HCOO^-] = 0.16$ mol.l^{-1}.
Inset A : Second order plotting for free radical decay, measured at 250 nm, in the same conditions as above.
Inset B : First order plotting for final product rearrangement, measured at 250 nm, in the same conditions as above.

The absorbances of the whole spectrum were the same for $[DFO]_o = 10^{-4}$ mol. l^{-1} or 2×10^{-4} mol. l^{-1}, they were proportional to $[O_2^-]_o$ (from 10^{-5} mol. l^{-1} to 10^{-4} mol. l^{-1}) as it is shown on inset of figure 1. The rate constant of this reaction was determined in the following conditions : $[DFO]_o = 10^{-4}$ mol. l^{-1}, $[O_2^-]_o = 1.1 \times 10^{-5}$ mol. l^{-1} and $[DFO]_o = 2 \times 10^{-4}$ mol. l^{-1}, $[O_2^-]_o = 6 \times 10^{-5}$ mol. l^{-1}. In these two conditions the kinetics obeyed to pseudo first-order law with an apparent constant proportional to $[DFO]_o$:

$$k_{O_2^- + DFO} = (1.35 \pm 0.05) \times 10^{6} \text{ mol. } l^{-1}. s^{-1}.$$

The free radical decay is made up of two reactions. The first one is a second-order disproportionation since NMR analysis of the final product showed that no dimer was formed (figure 2, spectrum a) : $2\ k = (9.70 \pm 0.05) \times 10^4\ mol^{-1}.l.\ s^{-1}$. This disproportionation is followed by a first-order reaction leading to spectrum b of figure 2, with $k = 0.069 \pm 0.005\ s^{-1}$.

DISCUSSION

Our results show that DFO reacts directly with O_2^- with a high rate constant at pH 7.2 ($k = 1.35 \times 10^6\ mol^{-1}.\ l.\ s^{-1}$). This confirmed our first assumption deduced from the inhibition of cytochrome c reduction by the xanthine-xanthine oxidase system (1). The rate constant measured by pulse-radiolysis is higher than the one calculated using competition with the rate of cytochrome c reduction (4). However the use of this last biochemical method as a way of determining a kinetic scheme for reactions with O_2^- has been criticized (5). This discrepancy may be explained by an additional reaction of a product formed by DFO and O_2^- with cytochrome c.

As for the action of DFO in systems producing free radicals it should be taken into account that DFO reacts rapidly with O_2^-. This new property may explain the constant inhibitory effect of DFO in vitro on $OH^\bullet$ production and its ability to prevent in vivo the toxic effects of ethanol via oxygen radicals on some cellular components (6, 7).

ACKNOWLEDGEMENTS

This work was supported by grants from Institut National de la Santé et de la Recherche Médicale and Université René Descartes (Faculté de Médecine Paris-Ouest). We thank J. Després for her technical assistance in the set up of this paper.

REFERENCES

1. Sinaceur J, Ribière C, Nordmann J, Nordmann R (1984) Biochem Pharmacol 33 : 1693-1694
2. Allen AO, Hochanadel CJ, Ghormley JA, Davies TW (1952) J Phys Chem 56 : 575-583
3. Pick E, Keisari Y (1980) J Immun Meth 38 : 161-170
4. Halliwell B (1985) Biochem Pharmacol 34 : 229-233
5. Diguiseppi J, Fridovich I (1980) Arch Biochem Biophys 203 : 145-150
6. Antebi H, Ribière C, Sinaceur J, Abu-Murad C, Nordmann R (1984) In : Bors W, Saran M, Tait D (eds) Oxygen Radicals in Chemistry and Biology. Walter de Gruyter, Berlin, New-York, pp 757-760.
7. Ribière C, Sinaceur J, Sabourault D, Nordmann R (1985) Alcohol 2 : 31-33

Superoxide and Superoxide Dismutase in Chemistry,
Biology and Medicine; G. Rotilio, editor

KINETICS AND MECHANISM OF THE INTERACTION OF $O_2^{\bar{\cdot}}$ WITH ASCORBIC ACID AND ALPHA-TOCOPHEROL

I.B.AFANAS'EV, V.V.GRABOVETSKII, N.S.KUPRIANOVA, and V.I.GUNAR
All-Union Vitamin Research Institute, Nauchny proezd 14A, 117246, Moscow (USSR)

Mechanism of the interaction of superoxide ion and dioxygen with vitamin E and ascorbic acid is of considerable biological significance. Therefore we studied the interaction of electrochemically generated $O_2^{\bar{\cdot}}$ with alpha-tocopherol (TH), alpha-tocopheryl acetate (T-Ac), and ascorbic acid (AH_2) as well as the mechanism of the oxidation of their anions by dioxygen.

In aprotic medium (acetonitrile) a main product in the reaction of $O_2^{\bar{\cdot}}$ with TH was the tocopheryl anion (T^-) with an absorption maximum at 323 nm. Under the conditions studied, T^- was unstable and rapidly transformed into the neutral free radical T· (with a wide maximum at 400-420 nm).

$$O_2^{\bar{\cdot}} + TH \longrightarrow HOO\cdot + T^- \quad (1) \qquad T^- + O_2 \longrightarrow T\cdot + O_2^{\bar{\cdot}} \quad (2)$$

As it might be expected, the oxidation of T^- was accelerated when the solution was saturated with oxygen. The rate of reaction (1) was determined by following spectrophotometrically the disappearance of TH; $k_1 = 0.59 \pm 0.23\ M^{-1}s^{-1}$.

The reaction of $O_2^{\bar{\cdot}}$ with T-Ac was studied during the initial period only. It was suggested that in this case, as in the reactions with other esters, $O_2^{\bar{\cdot}}$ splits off the ester group via reaction (3) [1]. The rate of reaction (3) was determined by

$$O_2^{\bar{\cdot}} + T\text{-}Ac \rightleftharpoons CH_3C(\dot{O})(T)OO^- \longrightarrow CH_3C(O)OO\cdot + T^- \quad (3)$$

following the disappearance of $O_2^{\bar{\cdot}}$ at 249 nm and T-Ac at 275 nm; $k_3 = 0.060 \pm 0.015\ M^{-1}s^{-1}$. In accord with the mechanism proposed, this value depends on [T-Ac].

Mixing the acetonitrile solutions of $O_2^{\bar{\cdot}}$ and AH_2 led to instantaneous disappearance of reagents and the appearance of a maximum at 272 nm corresponding to the ascorbate anion, AH^-. AH^- was the main product of this reaction not only in acetonitrile but also in aqueous solution and mixed water-acetonitrile solvents. Therefore the interaction of $O_2^{\bar{\cdot}}$ with AH_2 also proceeds via a deprotonation mechanism (reaction (4)). Under the conditions studied, AH^-

$$O_2^{\bar{\cdot}} + AH_2 \longrightarrow HOO\cdot + AH^- \quad (4)$$

was unstable due to reaction (5). It was found that the k_5 value

$$AH^- + O_2 \longrightarrow AH\cdot + O_2^{\cdot-} \qquad (5)$$

was considerably smaller in mixed solvents than in pure water or acetonitrile (Fig.1).

For elucidation of the mechanism of AH^- oxidation, we studied the effect of Fe^{3+} ions on oxidation rates. It was found that in aqueous solution the oxidation rate sharply increased in the presence of Fe^{3+} ions and was close to zero when EDTA was added (Fig.2). Therefore it can be suggested that the oxidation of AH^- in aqueous solution proceeds only with the participation of Fe^{3+} ions. In contrast to that the rate of AH^- decay in mixed solvents and acetonitrile solution was not affected neither by Fe^{3+} ions nor by EDTA.

Thus, although alpha-tocopherol and ascorbic acid are good radical scavengers, $O_2^{\cdot-}$ is not able to abstract from them a hydrogen atom; in both cases the reaction proceeds via a deprotonation mechanism to form corresponding anions. Earlier, Sawyer et al.[2] has proposed that $O_2^{\cdot-}$ reacts with ascorbic acid by the concerted transfer of a hydrogen atom and a proton (reaction (6)). But

$$O_2^{\cdot-} + AH_2 \longrightarrow A^{\cdot-} + H_2O_2 \qquad (6)$$

the predominant formation of AH^- under all experimental conditions seems to rule such a mechanism out.

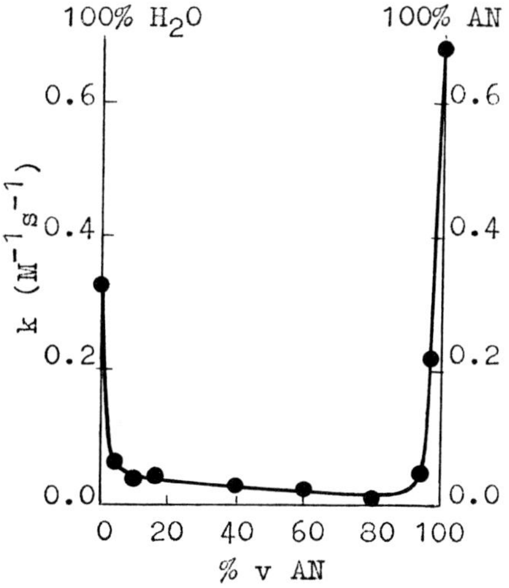

Fig. 1. Dependence of the rate constant for ascorbate oxidation on the solvent composition.

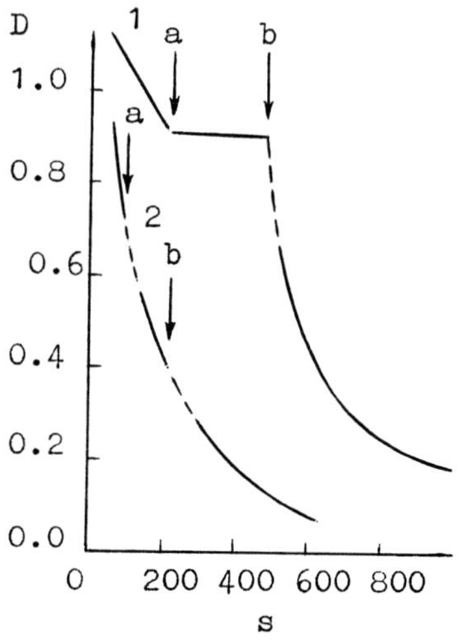

Fig. 2. Kinetics of the AH^- oxidation in water (1) and acetonitrile (2). a) EDTA was added. b) $FeCl_3$ was added.

In accord with data for aqueous solution [3], our results indicate the low reactivity of alpha-tocopherol in the reaction with $O_2^{\cdot}$ in acetonitrile.

From the experiments with different solvents, it follows that the catalytic oxidation of ascorbate (reaction (7)) proceeds only

$$AH^- + O_2 \overset{Fe^{3+}}{\rightleftharpoons} \text{complex} \longrightarrow A^{\cdot} + H^+ + O_2^{\cdot} \qquad (7)$$

in pure water, as neither EDTA nor Fe^{3+} ions did not affect the reactions rates in acetonitrile and mixed solvents. It might be explained by the fact that the solvation of AH^- in water supresses direct electron transfer to dioxygen, and therefore the reaction can proceed only in the presence of Fe^{3+} ions capable to form a complex with the ascorbate anion. On the other hand in acetonitrile a one-electron transfer process between AH^- and O_2 (reaction (5)) is realized. Both mechanisms are supressed in mixed solvents owing to the hydration of AH^- and the formation of unreactive complexes of Fe^{3+} with acetonitrile.

From our results follows that in living cells ascorbic acid can manifest its pro-oxidant properties in aqueous phase only in the presence of Fe^{3+} ions, whereas in lipid membranes a direct one-electron transfer between AH^- and O_2 to form $O_2^{\cdot -}$ is possible.

REFERENCES

1. Afanas'ev IB, Kuprianova NS (1983) Int J Chem Kinet 15:1063-1068
2. Sawyer DT, Chiericato G, Tsuchiya T (1982) J Am Chem Soc 104:6273-6278
3. Arudi RL, Sutherland MW, Bielski BH (1983) In: Cohen G, Greenwald RA (eds) Oxy Radicals and Their Scavenger Systems. Elsevier, Amsterdam, 1 pp 26-31

Superoxide and Superoxide Dismutase in Chemistry,
Biology and Medicine; G. Rotilio, editor

COLLAGEN BREAKDOWN BY SUPEROXIDE ANION IN GAMMA AND PULSE RADIOLYSIS.

JEAN-CLAUDE MONBOISSE[+], MONIQUE GARDES-ALBERT[•], ALAIN RANDOUX[+], CHRISTIANE FERRADINI[•] and JACQUES-PAUL BOREL.

+ Lab. Biochimie, ERA CNRS 959, Univ. Reims, 51 rue Cognacq Jay, F51095 REIMS, France - • Lab. Chimie-Physique, Univ. R. Descartes, 45 rue des Saints-Pères, F75270 PARIS, France.

INTRODUCTION

Extracellular matrix is the main site where inflammatory phenomena develop. Collagen fibrils are destroyed during the degradative stage of inflammation. In addition to well known proteolysis phenomena, mechanisms involving oxy free radicals have to be considered. Greenwald and Moy (1) demonstrated that acid soluble collagen is sensitive to oxy free radicals generated by the xanthine oxidase-hypoxanthine system. In a previous study of the effects of superoxide anion $O_2^{\cdot-}$ on collagen fibrils reconstituted in vitro, we demonstrated that the helical parts of fibrils were attacked and small peptides liberated into the incubation medium (2,3).

The aim of this work is the quantitative study of the reaction of $O_2^{\cdot-}$ on collagen fibrils by gamma and pulse radiolysis.

MATERIAL AND METHODS

Gamma radiolysis and evaluation of the degradation of collagen

Gamma radiolysis was carried out with a ^{60}Cobalt source with a total activity of 230 Ci and an irradiation intensity of 1.6×10^{18} $eV.ml^{-1}.h^{-1}$ ($25.6\ Gy.h^{-1}$).

Delipidated acid soluble calf skin collagen was reconstituted into fibrils at the final concentration 1 mg per ml by dialysis versus a 0.067 M phosphate buffer pH 7.4 containing 0.1 M sodium formate (3). Experiments were performed under either air, oxygen or N_2O at atmospheric pressure, for periods ranging from 15 to 60 min.

The degradation of collagen fibrils was followed and measured by determination of small 4-hydroxyproline-containing-peptides (4-hyp-peptides) liberated in the incubation medium. At the end of irradiations, the suspension of collagen fibrils was centrifuged at 30,000 g for 30 min, the supernatant precipitated by adding ethanol to 80 % (v/v). The supernatant obtained by centrifugation at 30,000 g for 30 min was hydrolyzed in 6 M HCl at 105°C for 18 h. Proline and 4-hyp were evaluated by a quantitative fluorometric technique (4).

Control by incubation with 2×10^{-4} M H_2O_2 were performed.

Gamma radiolysis of bovine serum albumin

In a control experiment, bovine serum albumin dissolved in the same buffer was submitted to the same treatment. The supernatant after alcool precipitation was hydrolyzed and its content in amino acids checked by the ninhydrin reaction (5).

Pulse radiolysis

Experiments of pulse radiolysis were achieved with an electron accelerator Febetron 708, with an electron energy 800 keV, pulse duration 4×10^{-9} s at half height, total dose about 20 krad (200 Gy). The decrease of $O_2^{\cdot-}$ in the presence of collagen fibrils was followed by spectrophotometry at $\lambda = 245$ nm and compared to its decrease without collagen at pH 7.4.

RESULTS AND DISCUSSION

Gamma radiolysis

Gamma radiolysis determined a liberation of 4-hyp-peptides that was linear with the doses (fig. 1). For a given radiation dose, the amount of 4-hyp-peptides increased with the initial collagen concentration, suggesting the existence of a competition between $O_2^{\cdot-}$ dismutation and its reaction on collagen fibrils.

Under air, the rate of collagen breakdown was identical with that observed under oxygen. Under N_2O, only formate radical $COO^{\cdot-}$ were produced : there was no modification of the collagen (data not shown).

When collagen fibrils were incubated with H_2O_2, no 4-hyp-peptides were found in the supernatant. The fragmentation of collagen into small peptides did not depend on H_2O_2.

Pulse radiolysis

On the fig. 2, graph A represents the experimental decay of $O_2^{\cdot-}$ in the presence of collagen 3.2×10^{-6} M and graph B is the control without collagen calculated for the same initial dose at pH 7.4, with a determined dismutation kinetic constant of $4.2 \times 10^5\ mol^{-1}.l.s^{-1}$.

In the presence of collagen, the rate of $O_2^{\cdot-}$ decay is increased and can be explained by the following kinetic scheme =

(1) $O_2^{\cdot-} + HO_2^{\cdot} \xrightarrow{H^+} H_2O_2 + O_2$ $\qquad 2\,k_1 \times 10^{pK_A-pH} = 4.2 \times 10^5\ M^{-1}.s^{-1}$

(2) $O_2^{\cdot-} + \text{collagen} \longrightarrow \text{products}$ $\qquad k_2$

We checked by simulating $O_2^{\cdot-}$ decay at 245 nm, using an iteration program and the kinetic scheme (1) (2), several values of k_2. The best fit to experimental results (n = 6 determinations) was :

$$k_2 = (2.43 \pm 0.07) \times 10^6\ M^{-1}.s^{-1}.$$

These results demonstrated that $O_2^{\cdot-}$ degrades collagen at the contrary to other proteins, for instance bovine serum albumin (data not shown).

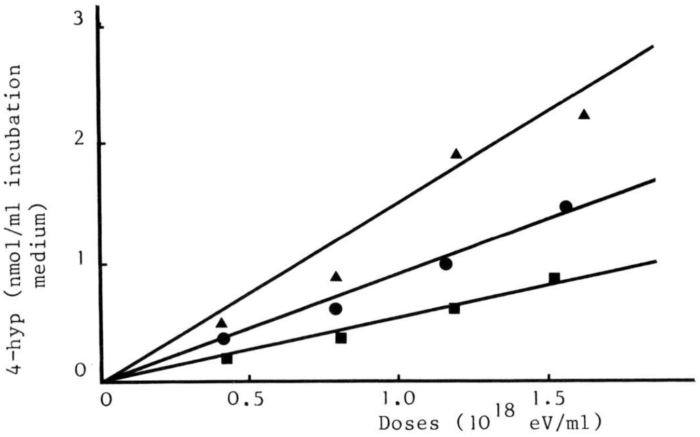

Fig. 1. Effect of various irradiation doses on collagen breakdown measured by evaluation of 4-hyp-peptides.
Collagen concentration :■—■: 1.1×10^{-6}M ; ●—●: 2.2×10^{-6}M ;▲—▲: 3.2×10^{-6}M
0.067 M phosphate buffer containing 0.1 M sodium formate, pH 7.4, air.
Dose rate : 1.6×10^{18} eV/ml/h (25 Gy/ml).

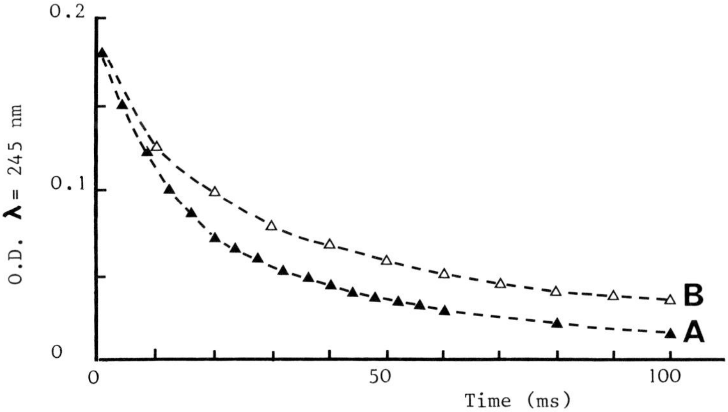

Fig. 2. $O_2^{\bar{.}}$ decay in the presence (A) or in the absence (B) of collagen fibrils. Collagen concentration : 3.2×10^{-6}M. 0.067 M phosphate buffer containing 0.1 M sodium formate, pH 7.4, air. $[O_2^{\bar{.}}]_0 = 9.4 \times 10^{-5}$M, $\varepsilon_{245}\ O_2^{\bar{.}} = 1910\ M^{-1}.cm^{-1}$, l = 1 cm.

REFERENCES

1 - Greenwald RA, Moy WW (1979) Arth Rheum 22 : 251-259

2 - Monboisse JC, Braquet P, Randoux A, Borel JP (1983) Biochem Pharmacol 32 : 53-58

3 - Monboisse JC, Poulin G, Braquet P, Randoux A, Ferradini C, Borel JP (1984) Int J Tissue React VI : 385-390

4 - Bisker A, Pailler V, Randoux A, Borel JP (1982) Anal Biochem 122 : 52-57

5 - Rosen H (1957) Arch Biochem Biophys 67 : 10

© 1986 Elsevier Science Publishers B.V. (Biomedical Division)
Superoxide and Superoxide Dismutase in Chemistry,
Biology and Medicine; G. Rotilio, editor

COPPER DEPENDENT CONTROL OF THE ENZYMIC AND PHAGOCYTE INDUCED DEGRADATION OF HYALURONIC ACID, SYNOVIAL FLUID AND CYTOCHROME C

HANS-JÜRGEN HARTMANN, ALFRED GÄRTNER, ULRICH WESER

Anorganische Biochemie, Physiologisch-Chemisches Institut der Universität Tübingen, Hoppe-Seyler-Str. 1, D-7400 Tübingen, FRG

INTRODUCTION

Very little is known on the metabolism of copper on a molecular level. As free Cu(II)-ions can undergo many uncontrolled reactions, copper has to be kept thoroughly chelated, for example , in polynuclear tetrahedral Cu(I)-thiolate clusters of Cu-thionein (1). Different reaction mechanisms should be considered in the metabolism of thionein-Cu (2,3). Firstly, a direct Cu(I)-transfer was shown. Secondly, an enzymically controlled oxidative breakdown of the Cu(I)-thiolate chromophore is proposed leading to biochemically essential Cu(II).

In inflammmatory processes there is a marked rise in serum copper concentration (4). The same phenomenon is observed in synovial fluid of patients suffering from rheumatoid arthritis (5). The elevated copper level which is suggested to possibly originate from hepatic Cu-thionein is accompanied by an increased number of polymorphonuclear leukocytes (PMN). These cell release excited oxygen species and hypochlorite which are reputed to be harmful to healthy tissue and body fluids (6,7). Both enzymically and PMN cell produced excited oxygen species are able to depolymerize hyaluronic acid, a constituent of synovial fluid (8). The course of the depolymerization was studied in the presence of $CuSO_4$, Cu-thionein and ceruloplasmin. Furthermore, bleaching of cytochrome c by phagocyte-generated hypochlorite served as a convenient assay to measure the activity of the PMN's (7).

RESULTS

Oxidation of Cu-thionein by enzymically generated H_2O_2

Several H_2O_2-producing enzymes were able to oxidize the Cu(I)-thiolate chromophore of both yeast Cu-thionein and bovine liver Cu,Zn-thionein leading to the formation of Cu(II) (2). The oxidation by xanthine oxidase in the presence of xanthine and hypoxanthine as substrates was studied in more detail. The Cu(I)-thiolate degradation and the concomitant generation of Cu(II) was measured magnetically (EPR) and spectropolarimetrically by monitoring the characteristic chiroptical properties of yeast Cu-thionein between 400 and 320 nm. During the oxidative decay the Cotton extrema were diminished. It should be pointed out that the oxidation process occurs only in the presence of a strong copper chelator i.e. EDTA.

Inhibition of xanthine oxidase by low M_r copper-compounds and Cu-thionein

The enzymic activity of xanthine oxidase (0.1 µM) in the presence of both EDTA and low M_r copper compounds of different origin as well as Cu-thionein was quantified by measuring the urate formation. The hypoxanthine/xanthine oxidase reaction remained unaffected. By contrast, omission of EDTA was inhibitory. Complete inhibition of the xanthine oxidase activity was noticed in the presence of 5 µM Cu. This result is consistent with earlier observations (9) (Fig. 1).

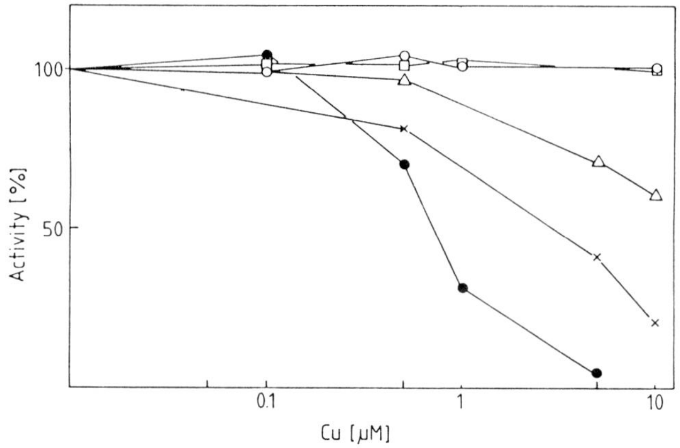

Fig. 1. Xanthine oxidase activity in the presence of $CuSO_4$ and Cu-thionein. $CuSO_4$ + EDTA (○); $CuSO_4$ (●); yeast Cu-thionein (△); yeast Cu-thionein + EDTA (□); hepatic Cu,Zn-thionein (X). 0.1 µM xanthine oxidase, 100 µM hypoxanthine.

Unlike the vertebrate Cu-thionein the more stable Cu(I)-thiolate chromophore of yeast copper-thionein exerts a diminished inhibitory effect. The inhibition of xanthine oxidase is reversible. After removal of copper using strong chelators, i.e. EDTA the enzymic activity is nearly fully recovered.

Depolymerization of hyaluronic acid and synovial fluid

The xanthine oxidase-dependent oxidation of xanthine or hypoxanthine decomposes hyaluronic acid. It was of interest whether or not Cu-thionein would prevent this enzymically induced degradation. Successful inhibition of the depolymerization occurred after preincubation of the Cu-thionein with hypoxanthine/xanthine oxidase to ensure oxidation of the Cu(I)-thiolate chromophores. $CuSO_4$ alone was immediately active. In the presence of

EDTA all Cu(II) was removed and the degradation proceeded uninhibited.

When fresh bovine joint synovial fluid from healthy subjects was used no significant decrease in viscosity could be seen. It was suggested that the biogenic Cu(II) (2-4 µM) was sufficient to inhibit the depolymerization process. Blocking of possible free coordinating sites of the biogenic Cu(II) employing EDTA resulted in the usually observed degradation of the synovial fluid.

Copper and PMN's

Translation of the above enzymic reaction onto a cellular level was performed. The interaction of different copper compounds with activated porcine blood PMN's was investigated. These cells produce excited oxygen species which are able to depolymerize hyaluronic acid. Virtually no degradation was noticed in the presence of 30 µM $CuSO_4$. The reactivity of Cu-thionein was distinct although less pronounced compared to $CuSO_4$. This phenomenon was attributed to the incomplete formation of Cu(II) from Cu(I)-thionein. Masking of Cu(II) by EDTA restored the viability of the PMN cells.

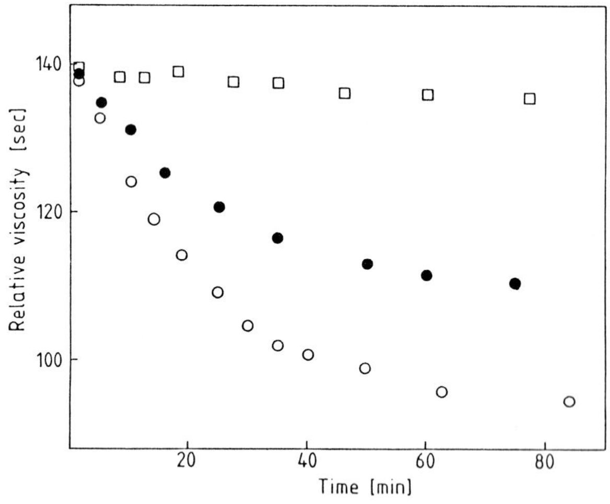

Fig. 2. Leukocyte dependent degradation of hyaluronic acid in the presence of $CuSO_4$ (□), and yeast Cu-thionein (●); control (○). 1.3 ml hyaluronic acid (2.27 mg/ml), 0.7 ml PMN suspension ($77 \cdot 10^6$ cells).

Surprisingly, ceruloplasmin (25 µM Cu) inhibited the PMN-dependent degradation of hyaluronic acid almost completely. Unlike Cu-thionein this copper protein is suggested to act stoichiometrically with superoxide.

Activated PMN's are known to bleach the heme chromophore of e.g. cytochrome c by the formation of ClO^- (6). No bleaching was seen using porcine blood leukocytes. By way of contrast, neutrophils from rat peritoneal cavity were much more active and were able to destroy the heme structure of the metalloprotein. Both $CuSO_4$ and Cu-thionein diminished this destructive process up to 40-80% in the presence of 1-100 µM Cu.

DISCUSSION

The generation and involvement of excited oxygen species in inflammmatory diseases are recognized as established facts. An attempt was made to combine both these findings and the phenomenon of the rise of extracellular copper levels. The reactivity of copper with the flavoenzyme xanthine oxidase was examined. Cu(II) was shown to be released from Cu(I)-thionein. On the other hand, bivalent Cu itself inhibits the enzyme. Activated PMN's displayed a similar reactivity on a cellular level. In the course of an inflammatory event these cells will continuously generate oxygen free radicals and hypochlorite. The control of the overall rate has been shown to be effective in the presence of copper. The reactivity of ceruloplasmin on the PMN cells is of special interest. It could possibly be seen as a regulator of the activity of leukocytes. We do not know the exact mode of action of copper. Whether or not the coordination side and/or the superoxide dismutase mimetic activity can be assigned to the observed phenomenon remains open.

ACKNOWLEDGEMENT

We are indebted to the Breuninger Stiftung, Stuttgart.

REFERENCES

1. Rupp H, Weser U (1978) Biochim Biophys Acta 533:209-226
2. Hartmann HJ, Gärtner A, Weser U (1984) Z Physiol Chem 365:1355-1359
3. Hartmann HJ, Gärtner A, Weser U (1985) Clin Chim Acta (in the press)
4. Deuschle U, Weser U (1985) Progr Clin Biochem Med 2 (in the press)
5. Scudder PR, McMurray W, White AG, Dormandy TL (1978) Ann Rheum Dis 37:71-72
6. Babior BM (1978) Engl J Med 298:721-725
7. Albrich JM, McCarthy CA, Hurst JK (1981) Proc Natl Acad Sci 78:210-214
8. McCord JM (1974) Science 185:529-531
9. Bergel F, Bray RC (1959) Biochem J 73:182-192

Superoxide and Superoxide Dismutase in Chemistry,
Biology and Medicine; G. Rotilio, editor

HYALURONIC ACID DEGRADING REACTIONS UNDER HYPERTHERMIC CONDITIONS AND EFFECTS OF THIOLS

RITA M FINK, EDMUND LENGFELDER

Institute of Radiation Biology, Ludwig-Maximilians-University,
Schillerstraße 42, D-8000 Munich 2, G.F.R.

INTRODUCTION

Hyaluronic acid (HA), an alternate copolymer of N-acetyl-D-glucosamine and D-glucoronate, is an important structural component of connective tissue (1). Previous studies on the degradation of HA *in vitro* have shown that activated oxygen species, especially the highly reactive hydroxyl radical, seem to be the predominant degrading component (2-5).

Oxygen radicals may also play an essential role in the mechanism of cytotoxicity and the thermosensitization by thiols. Some of the thiols used as radical scavengers or radioprotecting agents in chemical and biological systems have been shown to be cytotoxic especially at low concentrations (6,7).

Elevated temperatures of 40 - 44°C (hyperthermia) alone or in combination with radiation and chemotherapy, respectively, have proved to be therapeutically useful as a modality in the treatment of malignant disease (8).

The aim of this study was to investigate the effect of thiols on hyaluronic acid degradation by chelated iron and copper under hyperthermic conditions.

MATERIALS AND METHODS

Hyaluronic acid degradation was determined viscometrically using a Schott Micro Ubbelohde viscometer in a temperature controlled water bath ($\pm$ 0.05°C). The results are expressed as the percentage decrease in initial HA viscosity due to a particular substance or reaction (9). Each value is the mean of at least three determinations. All experiments were performed in 66 mM phosphate buffer at pH 7.4. Hyaluronic acid (from human umbilical cords) was from Serva, Heidelberg, thiols from Sigma, Munich.

RESULTS

Influence of transition metal/EDTA complexes on HA degradation under hyperthermic conditions

In presence of either 0.1 mM $FeCl_3$/EDTA or $FeSO_4$/EDTA we found a yield of HA degradation in which the latter was more efficient. However, in contrast to ferrous or ferric EDTA complexes, 0.1 mM $CuCl_2$/EDTA was not able to degrade hyaluronic acid. $FeCl_3$ without EDTA could not induce HA degradation, although $FeSO_4$ could.

Higher temperatures increased the HA degradation rate caused by $FeCl_3$/EDTA and $FeSO_4$/EDTA (data shown for $FeCl_3$/EDTA in the figure). Increasing the temperature from 37°C to 43.5°C resulted in an additional HA degradation of about 10 % in case of $FeCl_3$/EDTA and about 5 % in case of $FeSO_4$/EDTA.

Effects of thiols on HA degradation by chelated transition metals and influence of hyperthermia

The addition of thiols to the reaction mixture containing HA and $FeSO_4$/EDTA or $FeCl_3$/EDTA, respectively, led to a marked increase in HA degradation. Compared to the effects of $FeSO_4$/EDTA and $FeCl_3$/EDTA which showed similar efficiency, the system with $CuCl_2$/EDTA and thiol caused only a moderate enhancement of the HA degrading process. The concentration of the thiols was 0.4 mM and the concentration of the transition metal/EDTA complexes was 0.1 mM in the reaction mixtures.

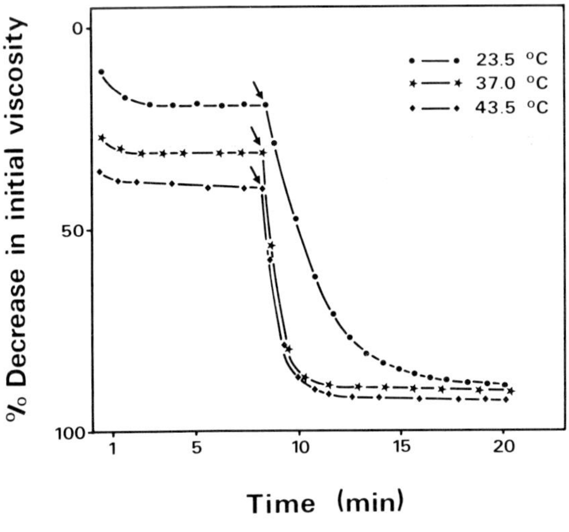

Figure: HA degradation by $FeCl_3$/EDTA at various temperatures and the effect of cysteamine (MEA). The oxygen gassed reaction mixture contained in 3 ml: 66 mM phosphate buffer pH 7.4, 2 mg HA, 0.1 mM $FeCl_3$/EDTA and 0.4 mM MEA, injected at the time indicated by arrows. The temperatures investigated were 23.5°C (●), 37°C (★) and 43.5°C (◆).

The extent of the stimulation of degradation also depended on the type of thiol as shown in the table. During the interaction with each of the three transition metal complexes, cysteamine (MEA) as well as cysteine (CYS) increased the degradation rate to the highest extent compared to the other thiols investigated dithiothreitol (DTT), glutathione (GSH) and N-acethylcysteine (NAC). The ability of these thiols to generate HA degrading species was markedly reduced in the sequence DTT, NAC to GSH.

Although the thiols were applied in a low concentration (0.4 mM), in which they have been shown to exhibit high cytotoxicity (6), thiols alone were not able to degrade hyaluronic acid.

The addition of thiols to non chelated ferrous or ferric iron could neither induce HA degrading reactions in case of $FeCl_3$ nor yield in a further stimulation in case of $FeSO_4$.

TABLE

EFFECTS OF THIOLS ON HA DEGRADATION BY CHELATED TRANSITION METALS

0.4 mM thiols were added to the oxygen gassed reaction mixtures containing 66 mM phosphate buffer pH 7.4, 2mg HA and 0.1 mM chelated transition metal in 3 ml volume at 37°C. Symbols for thiols are cysteamine (MEA), cysteine-HCl (CYS-HCl), cysteine free base (CYS-fb), dithiothreitol (DTT), glutathione (GSH) and N-acethylcysteine (NAC).

Experimental conditions	% decrease in initial HA viscosity					
	MEA	CYS-HCl	CYS-fb	DTT	NAC	GSH
$FeCl_3$/EDTA *	31	47	41	45	46	44
+ thiol **	87	65	63	55	51	49
$FeSO_4$/EDTA *	52	53	52	54	48	51
+ thiol **	83	68	65	64	51	55
$CuCl_2$/EDTA *	3	4	4	4	4	4
+ thiol **	10	22	31	11	9	8

* % decrease in initial HA viscosity by transition metal complex without thiol 8 min after start of reaction

** % decrease in initial HA viscosity 2 min after thiol addition

The enhancement of HA degradation by the addition of thiols in presence of Fe^{3+}/EDTA or Fe^{2+}/EDTA could only be increased to a small extent by hyperthermic conditions. As shown in the figure, increasing the temperature from 23,5°C to 37°C led to a marked stimulation of HA degradation rate, whereas hyperthermia at 43.5°C could not induce a significant enhancement of the HA degradation mediated by Fe^{3+}/EDTA and MEA under the experimental conditions.

DISCUSSION

Hyaluronic acid degradation is an appropriate parameter to investigate the production and reactivity of ·OH like species (3,4). Our experiments demonstrate that thiols, which are thought to be essential components of the non-enzymatic protective system in the defense of radical damages, could promote the generation of highly oxidizing species in the presence of chelated ferric, ferrous and copper-II compounds. The potential of the thiols investigated to stimulate the production of ·OH like species, which are able to degrade hyaluronic acid, corresponds with their initial rate of oxygen consumption (7).

Increasing the temperature led to an additional increase in the production of HA degrading species by the interaction of thiols with EDTA complexed transition metals.

The most efficient thiols were MEA and CYS, while GSH, one of the most important cellular thiols, exhibited only a minor capacity to promote the generation of highly reactive radicals. These results support the hypothesis that thiol cytotoxicity is connected with an interaction of trace metals and peroxide to generate hydroxyl radicals, as suggested in previous studies (7).

REFERENCES

1. Comper WD, Laurent TC (1978) Physiol Rev 58:255-315
2. McCord JM (1974) Science 185:529-531
3. Betts WH, Cleland LG, Gee DJ, Whitehouse MW (1983) In: Cohen G, Greenwald RA (eds) Oxy Radicals and their Scavenger Systems. Elsevier, New York, Vol 1, pp 95-100
4. Kreisl C, Lengfelder E (1984) In: Michelson AM, Bannister JV (eds) Oxidative Damage and Related Enzymes. Series: Life Chemistry Reports. Harwood Academic Publishers, London, pp 81-86
5. Issels RD, Fink RM, Lengfelder E (1985) paper submittet to Radiat Res
6. Issels RD, Biaglow JE, Epstein L, Gerweck LE (1984) Cancer Res 44:3911-3915
7. Biaglow JE, Issels RD, Gerweck LE, Varnes ME, Jacobson B, Mitchell JB, Russo A (1984) Radiat Res 100: 298-312
8. Hahn GM (1982) In: Hahn GM (ed) Hyperthermia and Cancer. Plenum Publishing Corp, New York, pp 232-242
9. Del Maestro R (1979) Ph D Thesis, University of Uppsala, Uppsala, Sweden

Superoxide and Superoxide Dismutase in Chemistry, Biology and Medicine; G. Rotilio, editor

SUPEROXIDE DISMUTASE ACTIVITY OF SOME COPPER PHENANTHROLINE COMPLEXES AND THE MECHANISM OF THE OXIDATION OF THE VARIOUS CUPROUS COMPLEXES BY H_2O_2

SARA GOLDSTEIN AND GIDON CZAPSKI

Department of Physical Chemistry, The Hebrew University of Jerusalem, Jerusalem 91904, Israel

INTRODUCTION

A large number of copper compounds have been tested for the rate at which they react with O_2^-, some of them were found able to catalyze its dismutation with an efficiency similar to that of Cu-Zn SOD (1-3).

In view of the role of Cu(II) in cleaving DNA in the presence of 1,10-phenanthroline(OP), a reducing agent and O_2 or H_2O_2 (4-5), we have decided to study the kinetics and mechanism of the reactions of some copper complexes of OP and substituted OP with O_2^- and H_2O_2.

MATERIAL AND METHODS

A Varian 7715 accelerator was used for all kinetic measurements. The pulse duration ranged from 0.1-1.5 µsec with 200 mA current of 5 MeV electrons. Kinetic data were obtained by monitoring the formation and the decay of Cu(I) peak in the maximum of its absorption in the visible region.

The Cu(I) complexes were generated in oxygenated solutions containing 0.02 M formate ions and 1 mM buffer phosphate at pH 7. Under these conditions all the radicals formed by irradiation are converted into O_2^-. We were not able to follow the decay of O_2^- directly at 250 nm, since OP absorbs very strongly in this region.

All solutions were prepared in distilled water which was further purified by a Millipore reagent grade water system. All chemicals employed were of analytical grade and were used as received. Stock solutions were prepared by dissolving $CuSO_4$ with 2.2 equivalents of the ligand (L).

RESULTS

CuL_2^{2+} as Catalysts for Superoxide Dismutation. Using excess concentrations of CuL_2^{2+}, we determined the rate constant of reaction (1) (Table):

$$CuL_2^{2+} + O_2^- \xrightarrow{k_1} CuL_2^+ + O_2 \qquad (1)$$

Under the conditions where $[CuL_2^{2+}]_o < [O_2^-]_o$ we determined the rate constant of reaction (2) (Table):

$$CuL_2^+ + O_2^- + 2H^+ \xrightarrow{k_2} CuL_2^{2+} + H_2O_2 \qquad (2)$$

Under this condition the yield of CuL_2^+ was less than the initial concentration of CuL_2^{2+} and therefore we assume that both CuL_2^{2+} and CuL_2^+ compete for O_2^-. Assuming steady state for CuL_2^+ and knowing the maximum extinction coefficient of CuL_2^+, we determined k_2 from equation (3):

$$OD_{max} = \frac{k_1 \ell \varepsilon_{max}}{k_1 + k_2} [CuL_2^{2+}]_o \qquad (3)$$

Knowing the values of k_1 and k_2 we have calculated the 'turnover' rate constant, k_{cat}, of the various copper complexes (Table).

$$-\frac{d[O_2^-]}{dt} = \frac{2k_1k_2}{k_1+k_2} [CuL_2^{2+}]_o[O_2^-] = k_{cat}[CuL_2^{2+}]_o [O_2^-]$$

TABLE

THE VALUES OF k_1, k_2 and k_{cat} OF THE VARIOUS COPPER COMPLEXES

L	k_1, $M^{-1}sec^{-1}$	k_2, $M^{-1}sec^{-1}$	k_{cat}, $M^{-1}sec^{-1}$
5-methyl-OP	2.3×10^9	2.3×10^8	4.18×10^8
OP	1.93×10^9	2.95×10^8	5.1×10^8
5-chloro-OP	1.55×10^9	2.12×10^8	1.86×10^8
5-nitro-OP	1.55×10^9	8.25×10^8	1.08×10^3
2,9-dimethyl-OP	3.0×10^8	2.35×10^8	2.63×10^8
2,9-dimethyl-4,7-diphenyl-OP	9.6×10^7	2.1×10^7	3.45×10^7

The Oxidation of CuL_2^+ by H_2O_2: When $[H_2O_2]$ was present in excess relative to $[(OP)_2Cu^+]$ and $[O_2]$, the decay of $(OP)_2Cu^+$ was found to be first order with respect to $[(OP)_2Cu^+]$ and $[H_2O_2]$ and independent of $[HCO_2^-]$ at 0.1 μsec pulse duration. During the oxidation of $(OP)_2Cu^+$ by H_2O_2 a chain reaction might take place

initiation: $(OP)_2Cu^+ + H_2O_2 \xrightarrow{k_3} X$ (3)

propagation: $X + HCO_2^- \xrightarrow{k_4} CO_2^- + \text{products}$ (4)

$CO_2^- + O_2 \xrightarrow{k_5} O_2^- + CO_2$ (5)

$O_2^- + (OP)_2Cu^{2+} \xrightarrow{k_6} (OP)_2Cu^+ + O_2$ (6)

termination: $(OP)_2Cu^+ + X \xrightarrow{k_7} 2(OP)_2Cu^{2+} + \text{products}$ (7)

where X is $OH^\cdot$ or $(OP)_2Cu^{3+}$ or $(OP)_2Cu\ H_2O_2^+$ (6-7), assuming steady state for X, CO_2^- and O_2^- we obtain:

$$-\frac{d[(OP)_2Cu^+]}{dt} = \frac{2k_3k_7[H_2O_2][(OP)_2Cu^+]^2}{k_4[HCO_2^-] + k_7[(OP)_2Cu^+]} \qquad (8)$$

In the case where X is OH$^\cdot$, $k_4 = 3x10^9\ M^{-1}sec^{-1}$ (8) and under our experimental conditions $k_4[HCO_2^-] >> k_7[(OP)_2Cu^+]$ and therefore equation (8) yields second order dependence on $[(OP)_2Cu^+]$ and dependence on $[HCO_2^-]^{-1}$ in contrast to our observations. Therefore we assume that X might be either $(OP)_2Cu^{3+}$ or $(OP)_2CuH_2O_2^+$ and that these species react very slowly with HCO_2^-. We obtain equation (9) which agrees with our results:

$$-\frac{d[(OP)_2Cu^+]}{dt} = 2k_3[(OP)_2Cu^+][H_2O_2] = k_{obsd}[(OP)_2Cu^+][H_2O_2] \qquad (9)$$

We have determined k_{obsd} to be 1620 $M^{-1}sec^{-1}$, 930 $M^{-1}sec^{-1}$ and 440 $M^{-1}sec^{-1}$ for 5-methyl-OP, OP and 5-nitro-OP, respectively.

From these results we conclude that during the oxidation of $(OP)_2Cu^+$ by H_2O_2 initially no OH$^\cdot$ is formed, but either $(OP)_2Cu^{3+}$ or $(OP)_2CuH_2O_2^+$ as active intermediates. We cannot distinguish between these two species as their properties are unknown, and their kinetic behaviour is identical.

ACKNOWLEDGEMENT

This research was done under contract #1409 of the Council of Tobacco Research.

REFERENCES

1. Klug D, Rabani J (1976) J. Phys. Chem. 80:588-591
2. Younes M, Lengfelder E, Zienau S, Weser U (1975) BBRC 81:576-580
3. Brigelius R, Spottl R, Bors W, Lengfelder E, Saran M, Weser U (1974) FEBS Lett. 47:72-75
4. Marshall LE, Graham DR, Reich KA, Sigman DS (1981) Biochem. 20:244-250
5. Que BG, Doweny KM, So AG (1980)Biochem. 19:5987-5991
6. Ponganis KV, DeAraujo MA, Hodges HL (1980) Inorg. Chem. 19:2704-2709
7. Johnson GRA, Nazhat NB, Saadalla-Nazhat RA (1985) J. Chem. Soc. Chem. Commun. 407-408
8. Farhataziz, Ross AB (1977) Natl. Stand. Ref. Data Ser. Natl. Bur. Stand. No. 59.

© 1986 Elsevier Science Publishers B.V. (Biomedical Division)
Superoxide and Superoxide Dismutase in Chemistry,
Biology and Medicine; G. Rotilio, editor

THE SOD ACTIVITY VERSUS TOXICITY OF TERNARY COMPLEXES OF SOME COPPER COMPLEXES WITH DNA

GIDON CZAPSKI AND SARA GOLDSTEIN

Department of Physical Chemistry, The Hebrew University of Jerusalem, Jerusalem 91904, Israel

INTRODUCTION

It has been demonstrated that degradation of double-stranded DNA occurs in the presence of Cu(II) ions, 1,10-phenanthroline (OP), a reducing agent and O_2 (1-2). The copper complex of 5-NO_2-OP was found to be more effective than OP in cleaving DNA, while that of 2,2-bipyridyl (bipy) did not degrade DNA under the same conditions (1-3). The mechanism of the cleavage reaction was assumed to be through a site specific mechanism where the bound Cu(I) complex to DNA is oxidized by H_2O_2 to form the OH$^\cdot$ radical at the binding site. This mechanism explains the toxicity of the OH$^\cdot$ formed as it is formed at the binding site and can hardly be scavenged.

DISCUSSION

The reaction scheme for bound and free copper complexes (CuL_2^{2+}, CuL_2^{+}, DNA≡CuL_2^{2+}, DNA≡CuL_2^{+}) includes the following steps:

$$CuL_2^{2+} \text{ or } DNA{\equiv}CuL_2^{2+} + O_2^- \longrightarrow CuL_2^{+} \text{ or } DNA{\equiv}CuL_2^{+} + O_2 \quad (1)$$

$$CuL_2^{+} \text{ or } DNA{\equiv}CuL_2^{+} + H_2O_2 \longrightarrow CuL_2^{2+} \text{ or } DNA{\equiv}CuL_2^{2+} + OH^- + OH^\cdot \quad (2)$$

$$DNA \text{ or } DNA{\equiv}CuL_2^{2+} + OH^\cdot \longrightarrow \text{damage} \quad (3)$$

In addition the reoxidation of Cu(I) by O_2^- can occur:

$$CuL_2^{+} \text{ or } DNA{\equiv}CuL_2^{+} + O_2^- + 2H^+ \longrightarrow CuL_2^{2+} \text{ or } DNA{\equiv}CuL_2^{2+} + H_2O_2 \quad (4)$$

If we analyze the sequence of these reactions we find that reactions (1) and (2) describe the catalysis of the Haber-Weiss reaction which yields the deleterious OH$^\cdot$, while reactions (1) and (4) describe the catalysis of the dismutation of O_2^- and hence prevent the formation of OH$^\cdot$. Thus a copper compound can serve either as a catalyst of the Haber-Weiss reaction or as a catalyst of the dismutation of O_2^-. It turns out that whether the copper compound catalyzes the dismutation of O_2^- or catalyzes the Haber-Weiss reaction depends on the competition between reactions (2) and (4). It has been shown that $k_4/k_2 > 10^5$ for the various free copper complexes (4) and since in most biological systems $[H_2O_2]/[O_2^-] < 10^3$ (5), the dismutation of O_2^- predominates and we do not expect any appreciable production of OH$^\cdot$ through the Haber-Weiss reaction and no toxicity is expected (6).

The question is how is O_2^- toxic in the presence of these copper complexes? We have shown that CuL_2^{2+} and CuL_2^{+} bind to DNA to form ternary complexes (7):

$$DNA + CuL_2^{2+} \rightleftharpoons DNA{\equiv}CuL_2^{2+} \quad (5)$$

$$DNA + CuL_2^{+} \rightleftharpoons DNA{\equiv}CuL_2^{+} \quad (6)$$

It turns out that these ternary complexes react very slowly with O_2^-, if at all, while the rates of the oxidation of free and bound CuL_2^+ by H_2O_2 are very similar (4,7). $DNA{\equiv}CuL_2^+$ is formed through the reduction of trace amounts of CuL_2^{2+} present in equilibrium (5) by O_2^-. In the system where CuL_2^{2+} is bound to DNA, the rate of reaction (2) exceeds by much that of reaction (4) and therefore CuL_2^{2+} in the presence of DNA catalyzes the Haber-Weiss reaction.

The difference observed in the cleavage efficiencies for the various chelators depends on the equilibria constants K_5 and K_6 together with the equilibrium constant of the ligand dissociation from bound CuL_2^+ (K_7), since $DNA{\equiv}CuL^+$ is not oxidized by O_2^-, O_2 and H_2O_2 (7).

$$DNA{\equiv}CuL_2^{+} \rightleftharpoons DNA{\equiv}CuL^{+} + L \quad (7)$$

It turns out that no ligand dissociation takes place from bound CuL_2^+ in the case of 5-NO_2-OP while K_7 is very high in the case of bipy in comparison to that of OP. For bipy the values of K_5 and K_6 are relatively small compared to those of 5-NO_2-OP and OP. Therefore for bipy the concentration of free CuL_2^+ (according to the experimental conditions in ref. 1 and 2) exceeds that of bound CuL_2^+ and most of the oxidizing radicals are formed in the bulk. In the case of OP and 5-NO_2-OP most of CuL_2^+ is bound to DNA and thus the majority of the $OH^\cdot$ is formed at the binding site in a site specific mechanism.

In summary the cleavage efficiency depends on whether the majority of reaction (2) occurs in the bulk or at the binding site. This depends on the values of the equilibria constants K_5, K_6 and K_7, together with the concentration of free ligand and DNA.

ACKNOWLEDGEMENT

This work was done under contract #1409 of the Council of Tobacco Research.

REFERENCES

1. Marshall LE, Graham DR, Reich KA, Sigman DS (1981) Biochem. 20:244-250
2. Que BG, Doweny KM, So AG (1980) Biochem. 19:5987-5991
3. Reich KA, Marshall LE, Graham DR, Sigman DS (1981), J. Am. Chem. Soc. 103: 3582-3584
4. Goldstein S, Czapski G (1985) Inorg. Chem. 24:1087-1092
5. Boveris A (1977) Adv. Exp. Med. J. Biol. 78:67-82
6. Goldstein S, Czapski G (1985) J. Free Radicals in Biol. & Med. 1 (in print)
7. Goldstein S, Czapski (1985) submitted for publication.

Superoxide and Superoxide Dismutase in Chemistry, Biology and Medicine; G. Rotilio, editor

SOD-LIKE ACTIVITY OF COPPER-HISTIDINE IN PLASMA AND BUFFER

BENGT GERDIN, GUNNAR CARLIN & RICHARD DJURSÄTER

Department of Experimental Medicine, Pharmacia AB, Uppsala and Department of Surgery, University Hospital, Uppsala (Sweden)

INTRODUCTION

Several low molecular weight copper chelates such as Cu-salicylate (1) and Cu--penicillamine (2) have been shown to dismutate superoxide with reaction rates similar to CuZnSOD. Some have been found to be antiinflammatory in various models (e.g. 3). In a series of studies, Williams et al. (e.g. 4,5) investigated several aspects of the use of low molecular weight copper chelates. They found that copper bound effectively to albumin in plasma and that histidine was one of few chelators that could compete for the copper, while Cu-acetylsalicylate and Cu-indomethacin, for example, would lose their copper to albumin. In the present study the superoxide dismutase-like activity of Cu-histidine complexes was investigated in buffer and plasma.

MATERIAL AND METHODS

SOD-like activity was assayed as reduction of cytochrome-c. The reaction mixture contained 25 µM cytochrome-c, 1 mM hypoxanthine and 50 U/ml catalase in a total volume of 0.58 ml. The reaction was initiated by addition of 5 µl of xanthine oxidase (1.7 U/ml). The change in absorbance at 550 mM, at 37°C, was recorded for 2 min, during which period the change was linear, and the slope was calculated. The slope of the control experiment, using buffer as test substance, was set at 100 % activity. The test sample was 25 µl of buffer or plasma preincubated for 30 min at room temperature with the specified concentrations of Cu^{2+} and histidine. Note that all concentrations given in the results refer to the 25 µl test sample and not to the final concentration in the cuvette.

Inhibition of xanthine oxidase was assayed as oxygen consumption, using a Clark electrode.

RESULTS

Addition of 10 mM histidine and various concentrations of Cu^{2+} resulted in Cu^{2+}--dose-related SOD activity (Fig. 1). Maximum activity was attained at 0.1 mM Cu^{2+}. In the absence of histidine, Cu^{2+} was inactive in plasma. In buffer, free Cu^{2+} inhibited xanthine oxidase, which interfered with the assay.

Addition of 0.05 mM Cu^{2+} and various concentrations of histidine to plasma resulted in histidine-dose-related SOD-like activity which reached its maximum at 10 mM histidine (Fig. 2). In the presence of increasing concentrations of histidine,

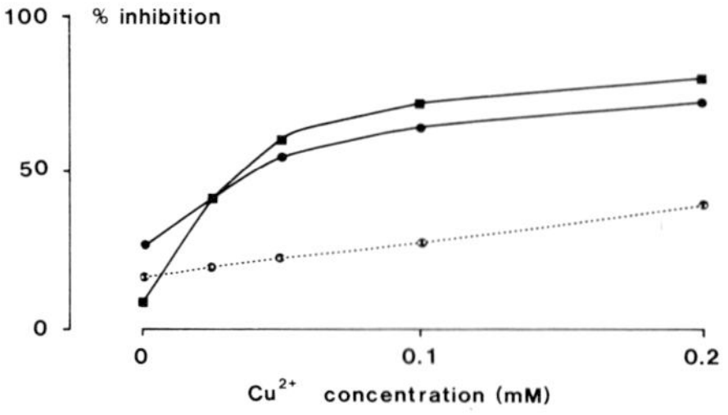

Fig. 1. SOD-like activity of 10 mM histidine and various concentrations of Cu^{2+} in buffer (■—■) and plasma (●—●), and of Cu^{2+} without histidine in plasma (●---●).

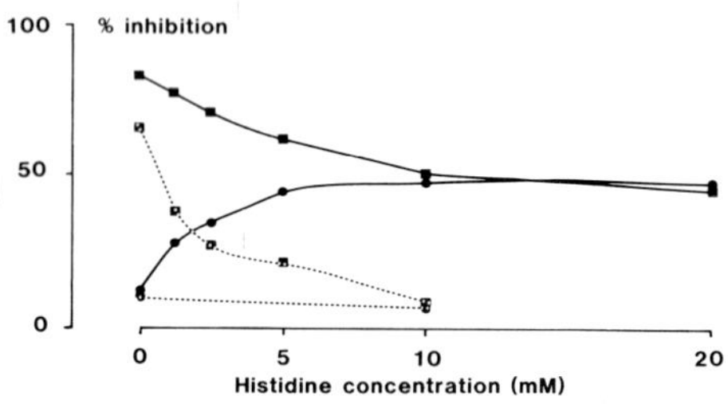

Fig. 2. SOD-like activity of 0.05 mM Cu^{2+} and various concentrations of histidine in buffer (■—■) and plasma (●—●). Xanthine oxidase inhibition in buffer (■---■) and plasma (●---●).

and with formation of increasing amounts of Cu^{2+}-histidine complexes, the xanthine oxidase inhibition of free Cu^{2+} decreased. In plasma no xanthine oxidase inhibition due to free Cu^{2+} was seen, probably on account of formation of Cu^{2+} complexes with plasma components.

At a constant Cu^{2+}/histidine ratio of 1/100 in buffer, maximal SOD-like activity was attained at 0.05 mM Cu^{2+} and 5 mM histidine (Fig. 3). In plasma, five times as high a concentration of histidine was required to achieve maximal SOD-like activity.

High concentrations of CuZnSOD resulted in complete inhibition of cytochrome-c reduction (Fig. 4), a finding which may be compared with the 75 % inhibition obtainable by an optimal Cu^{2+}/histidine ratio.

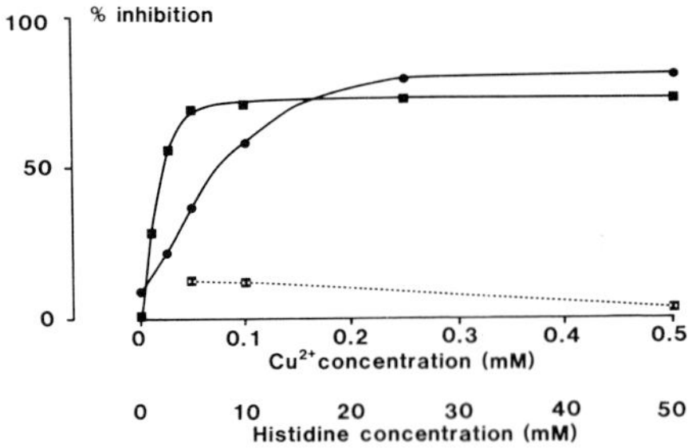

Fig. 3. SOD-like activity of various concentrations of Cu^{2+}-histidine at a constant ratio of 1:100 in buffer (■—■) and plasma (●—●). Xanthine oxidase inhibition in buffer (■---■).

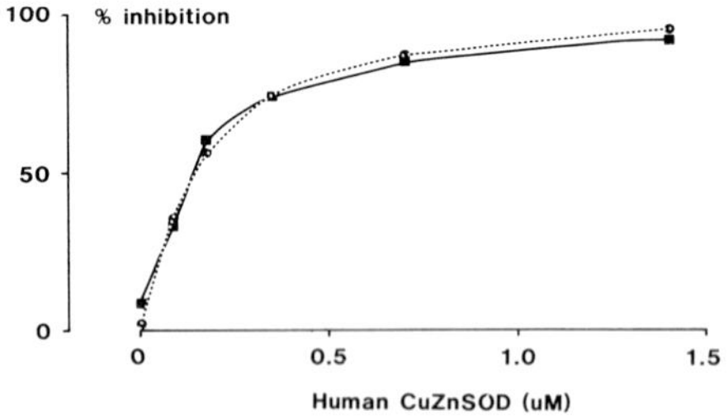

Fig. 4. SOD activity of human CuZn--SOD at various concentrations in buffer (■—■) and plasma (●—●).

DISCUSSION

The present study has indicated that in the presence of excess histidine, Cu^{2+}--histidine complexes are formed in buffer and plasma that have SOD-like activity. In plasma maximum activity was reached at 0.25 mM Cu^{2+} and 25 mM histidine. It was shown, further, that Cu^{2+}, but not Cu^{2+}-histidine or Cu^{2+}-plasma complexes, inhibited xanthine oxidase. At low concentrations of Cu^{2+} and histidine in plasma, Cu^{2+} seems to be partitioned between inactive plasma and active histidine complexes. CuZnSOD was about 170 times more potent as a superoxide scavenger than Cu^{2+}-histidine in plasma, as calculated per molecule of Cu, since 0.15 µM SOD (= 0.3 µM Cu) and 0.05 mM Cu^{2+}-histidine both inhibited the cytochrome-c reduction by 50 %.

The results indicate that the SOD activity of Cu^{2+}-histidine complexes could form the basis for attempts to develop an alternative pharmacologically active non-protein superoxide dismutase.

REFERENCES

1. Weser U, Richter C, Wendel A, Younes M (1978) Bioorg Chem 8:201-213
2. Lengfelder E, Fuchs C, Younes M, Weser U (1979) Biochem Biophys Acta 567: 492-502
3. Sorenson JRJ (1976) J Med Chem 19: 135-148
4. Jackson GE, May PM, Williams DR (1978) J inorg nucl Chem 40:1189-1194
5. Fiabine AM, Williams DR (1978) J inorg nucl Chem 40:1195-1200

Superoxide and Superoxide Dismutase in Chemistry, Biology and Medicine; G. Rotilio, editor

ON THE SUPEROXIDE DISMUTASE ACTIVITY OF CYTOCHROME OXIDASE FROM *PSEUDOMONAS AERUGINOSA*

ANUSH V. KARAPETIAN, MARGARITA G. KAMALIAN, ROBERT M. NALBANDYAN
Institute of Biochemistry, Academy of Sciences, Armenian SSR, Yerevan, 375044 USSR.

It is known that many terminal copper-containing oxidases, blue or non-blue, soluble on membraneous, possess also certain levels of the superoxide dismutase (SOD) activity (1-3). The significance of the SOD activity of copper oxidases is unclear yet, although one may speculate that it is connected with the possible formation of superoxide radicals during the reduction of oxygen. However some copper complexes of low-molecular weight also have the SOD activity (4,5). It was of interest, therefore, to test whether other terminal oxidases, besides copper-containing, have the SOD activity. It should be mentioned that the SOD activity of the main mitochondrial terminal oxidase, cytochrome *c*-oxidase, which contains both copper and haems aa_3, is due to copper rather than to the haems (3).

There are, however, many terminal oxidases which contain only haem prosthetic groups. Thus, cytochrome cd_1 or cytochrome *o* are terminal in a number of microorganisms. In particular, the terminal oxidase of *Pseudomonas aeruginosa* and other denitrifying bacteria are well known to contain two different haem groups (haems *c* and d_1) and they are lacking of copper (6,7). We have studied, therefore, the SOD activity of cytochrome oxidase (nitrite reductase, cytochrome cd_1) from Ps. aeruginosa. The results obtained are presented in this communication.

MATERIALS AND METHODS

Cytochrome oxidase and cytochrome c-551 from Pseudomonas aeruginosa (strain IFO 3080) were prepared essentially according to Parr S.R. et al. (8). Both preparations were electrophoretically homogeneous. The oxidase in its completely oxidized form had the spectral purity index, A_{411}/A_{280}, of 1.17 and the turnover number of the enzyme determined at 22°C in 0.01 M phosphate buffer, pH 6.0, with cytochrome c-551 as a substrate, was of 120 $M{\cdot}min^{-1}$. The spectral purity index, A_{280}/A_{551}, of reduced cytochrome c-551 was of 1.14.

Haem d_1 was extracted from cytochrome oxidase by the treatment of the enzyme by acidified acetone (9). The extracted haem d_1 and the remaining moiety obtained after its extraction (cytochrome *c*), were further purified essentially according to (10). However the additional step of gel-filtration of haem d_1 through Sephadex LH-20 was introduced.

Routinely, SOD activities of the native cytochrome oxidase, haem $\underline{d}_1$ and the preparation containing only haem $\underline{c}$ were assayed according to Nishikimi et al. (11). In some experiments the alkaline hydrogen peroxide solution in which superoxide radicals are generated was also used as a test system of the SOD activity following concentrations of superoxide radicals by their EPR spectral intensities (12). During studies of effects of inorganic anions on this activity, samples of the enzyme were preincubated with anions at 22°C for 30 min. For calculations of concentrations of cytochrome oxidase, haem $\underline{d}_1$ and cytochrome $\underline{c}$ the generally accepted values of molar extinctions of main absorbtion bands were used (7, 10, 13).

RESULTS AND DISCUSSION

It was found that cytochrome oxidase inhibited the reaction of the formazan formation which arises during interaction of superoxide radicals with tetrazolium (11). This SOD activity was further verified by direct measurements of the superoxide radical concentration (12), which showed that it decreased sharply under the effect of the enzyme.

The dependence of the inhibition of the formazan formation on the concentration of cytochrome oxidase from Ps. aeruginosa shows that 50% inhibition is observed with 4.10^{-8} M enzyme. The similar inhibition with cuprozinc SOD was observed at concentration of 5.10^{-9}M (11). Thus, the SOD activity of the oxidase was about one order of magnitude lower than that of cuprozinc SOD. At the same time cytochrome oxidase as an enzyme with the SOD activity was at least as active (if not more active) as iron-or manganese -containing SODs from different sources (14, 15).

The SOD activity of cytochrome oxidase determined by the formazan method (11) was found to be cyanide sensitive. 5.10^{-6}M cyanide completely inhibits the activity. Among other anions tested, SCN^- and F^- were also inhibitors, however they were significantly less effective than cyanide; N_3^- was not inhibitory. It was of interest that iron SODs from some sources were found to be cyanide sensitive.

In further experiments we assayed separately the SOD activities of haem $\underline{d}_1$ extracted from cytochrome oxidase and the remaining haem $\underline{c}$-containing moiety of the enzyme. It was concluded from this study that the SOD activity of cytochrome oxidase from Ps. aeruginosa was completely due to haem $\underline{d}_1$ rather than haem $\underline{c}$ of the enzyme. The SOD activity of haem $\underline{d}_1$ was found to be also cyanide sensitive.

All iron-containing SOD described till now are known to belong to non-haem iron proteins (17). Thus, the evidence that cytochrome oxidase from Ps. aeruginosa and haem $\underline{d}_1$ possess marked SOD activity is, as far as we know, the first demonstration of SOD activity of haemoproteins.

Many results (17-19) indicate that haems _c_ and $\underline{d}_1$ in the cytochrome oxidase (nitrite reductase) reaction have different roles. In particular, it should be noted that haem $\underline{d}_1$ seems to be the oxygen and nitrite binding site of the enzyme. Removal of haem $\underline{d}_1$ from the enzyme results in the inhibition of both cytochrome oxidase and nitrite reductase activities. It is reasonable to suggest that oxygen intermediates should be formed at the haem $\underline{d}_1$ moiety rather than at the haem _c_ of the enzyme.

Recent work (20) shows that, similar to blue or non-blue copper terminal oxidases, the copper-containing nitrite reductase from _Rhodopseudomonas spheroides_ also has a SOD activity which is as high as of one third that of cuprozinc SOD. Thus, SOD activity has been demonstrated for several types of terminal oxidases including blue or non-blue copper oxidases, copper-haem $\underline{aa}_3$ oxidases and haems $\underline{cd}_1$ oxidases lacking of copper.

REFERENCES

1. Goldstein JM, Kaplan HB, Edelson HS and Weissman G (1979) J Biol Chem 254, 4040-4045.
2. Hamilton G, Libby RD, Hartzell ChR (1973), Biochem Biophys Res Commun 55, 333-340.
3. MarkossianKA, Pogossian GG, Paitian NA and Nalbandyan RM (1978), Biochem Biophys Res Commun 81, 1136-1143.
4. Joester KE, Yung G, Weber U, Weser U (1972) FEBS Letters 25, 25-28.
5. Lengfelder E, Fuchs Ch, Jounes M, Weser U (1979) Biochim Biophys Acta 567, 492-502.
6. Kuronen T, Ellfolk N (1972) Biochim Biophys Acta, 275, 308-318.
7. Horio T, Higashi T, Sasagawa M, Kusai K, Nakai M and Okunuki K (1960), Biochem J 77, 194-201.
8. Parr SR, Barber D, Greenwood C, Phillips BW and Meeling J (1976), Biochem J, 157, 423-430.
9. Yamanaka T, Okunuki K (1963) Biochim Biophys Acta, 67, 407-416.
10. Hill KE, Wharton DC (1978) J Biol Chem, 253, 489-495.
11. Nishikimi M, Rao NA, Jagi K (1972) Biochem Biophys Res Commun, 46, 849-854.
12. Symonyan MA, Nalbandyan RM (1979) Biochim Biophys Acta, 583, 279-286.
13. Horio T, Higashi T, Yamanaka T, Matsubara H and Okunuki K (1961) J Biol Chem 236, 944-951.
14. Eldred GE and Hoffert JR (1981) Anal. Biochem, 110, 137-143.
15. Gregory EM, Dapper ChH (1983) Arch Biochem Biophys 220, 293-300.
16. Fridovich I (1982) in Superoxide Dismutase, v. 1 (Oberley LW ed.) pp. 69-77, CRC Press.
17. Shimada H, Orii J (1976) J Biochem, 80, 135-140.

18. Walsh TA, Johnson MK, Thomson AJ, Barber D, Greenwood C (1981) J Inorg Biochem 14, 1-14.
19. Yamanaka T, Okunuki K (1963) Biochem Z 338, 62-72.
20. Michalski WP, Nicholas DJD (1985) Biochim Biophys Acta, 828, 130-137.

Superoxide and Superoxide Dismutase in Chemistry, Biology and Medicine; G. Rotilio, editor

SUPEROXIDE DISMUTASE ACTIVITY OF MANGANESE CHELATES

J. EPP, S. FAIRCHILD, G. ERICKSON AND W.H. KOPPENOL

Department of Chemistry, University of Maryland Baltimore County, Catonsville, Maryland 21228 (U.S.A.)

INTRODUCTION

The absence of any of the known superoxide dismutases in *Lactobacillus plantarum* led to the discovery that this organism uses manganese in relatively high concentrations to dispose of the superoxide anion (1). The mechanisms of the reactions of superoxide with complexes of manganese and sulphate, formate, phosphate and pyrophosphate have been studied in detail (2). Various Mn-porphyrins also appear to catalyze the dismutation reaction (3). With EDTA and NTA as ligands, Mn^{2+} was found to be oxidized by superoxide *via* an inner sphere mechanism (4). Mn^{2+}_{aq} is not expected to be a catalyst; the standard reduction potential of the couple Mn^{3+}/Mn^{2+} is 1.51 V, which prevents oxidation of Mn^{2+} by $O_2^{-}\cdot$.

We report here on the superoxide dismutase activity of complexes of manganese with multidentate ligands.

MATERIALS AND METHODS

Horse cytochrome *c*, Type VI, was obtained from Sigma. The concentration of cytochrome *c* was determined from the difference in absorption between the ferric and ferrous form at 550 nm. All chemicals were of analyzed reagent grade, purchased from J.T. Baker, Sigma and Kodak. Irradiations were carried out in a ^{137}Cs γ-source (Gammator B, Kewaunee Scientific Co.), courtesy of Dr. S. Rosenberg. The dose rate was 3.2 Gray (0.32 krad) min^{-1}, as determined by Fricke dosimetry in the 25 ml erlenmeyers used for the experiments. In the presence of sodium formate and oxygen, all radical species formed upon irradiation of water are converted quantitatively to superoxide. This anion reduces ferricytochrome *c* with rate constants of 1.4 and $1.2 \times 10^6\ M^{-1}s^{-1}$ at pH 6 and 7, respectively, at low ionic strength (5).

In the presence of a metal chelate with catalytic activity less than the expected amount of ferricytochrome *c* will be reduced. The decrease in ferricytochrome *c* concentration per superoxide formed will be:

$$\frac{dc^{3+}}{dO_2^{-}\cdot} = \frac{k_1[c^{3+}]}{k_1[c^{3+}] + k_2[ML]} \qquad [1]$$

in which k_1 is the bimolecular rate constant for the reaction of superoxide and

*For abbreviations see Table I

ferricytochrome *c* (c^{3+}), and k_2 the same for the reaction of superoxide and metal chelate (ML). $dc^{3+}/dO_2^-\cdot$ equals c_c^{2+}/c_o^{2+} in which c_c^{2+} represents the concentration of ferrocytochrome *c* after irradiation in the presence of a catalyst, and c_o^{2+} the same in the absence of a catalyst. Defining P as c_c^{2+}/c_o^{2+}, then

$$1 - P = \frac{k_2[ML]}{k_1[c^{3+}] + k_2[ML]} \qquad [2]$$

The quantity 1-P is the fraction of ferrocytochrome *c* missing due to the presence of a catalyst. Plots of 1-P *vs.* [ML] should be curved, approaching 1.0 asymptotically. Catalytic rate constants below 10^6 $M^{-1}s^{-1}$ cannot be detected.

The solutions contained 2 mM Na,K-phosphate, pH 6.0, 2 mM sodium formate, 13-15 μM ferricytochrome *c*, 0.2-3 μM Mn^{2+}, ligand (see Table I) in 1000 fold excess, and were irradiated for 2 minutes (6.2 Gray).

RESULTS

Of the metal chelates examined (Table I), only MnNTA, MnHEDTA and MnEDDA appear to have superoxide dismutase activity, as shown in Fig. 1. MnNTA and MnEDDA behave as predicted while the results for MnHEDTA follow a straight line. The activity decreases with pH, as shown in Fig. 2 for MnNTA. Manganous ions, probably as phosphate and formate complexes, were found to decrease the amount of ferricytochrome *c* reduced, but at such high concentration (10-100 μM Mn^{2+}) the effect is not necessarily catalytic (see Ref. 2). The ligands alone were unable to inhibit the reduction of ferricytochrome *c*, up to concentrations of 3 mM.

TABLE I

MANGANESE CHELATES

Mn^{2+} chelated by	Abbreviation	k_2 at pH 6
Nitrilotriacetic acid	NTA	2.0×10^7
Ethylenediaminediacetic acid	EDDA	1.9×10^7
(N-hydroxyethyl)ethylenediaminetriacetic acid	HEDTA	0.7×10^7 (?)
Ethylenediaminetetraacetic acid	EDTA	$< 10^6$
Diethylenetriaminepentaacetic acid	DTPA	$< 10^6$
Ethylenediaminedi(O-hydroxyphenyl)acetic acid	EHPG	$< 10^6$

DISCUSSION

Those complexes which possess activity have a coordination site available or, in the case of MnHEDTA, a sixth ligand which is loosely bound. This is in agreement with the notion that the first step of the catalysis, namely oxidation of

Mn^{2+}, takes place *via* an inner sphere mechanism (4). Chelates in which Mn^{2+} is totally encapsulated do not possess activity.

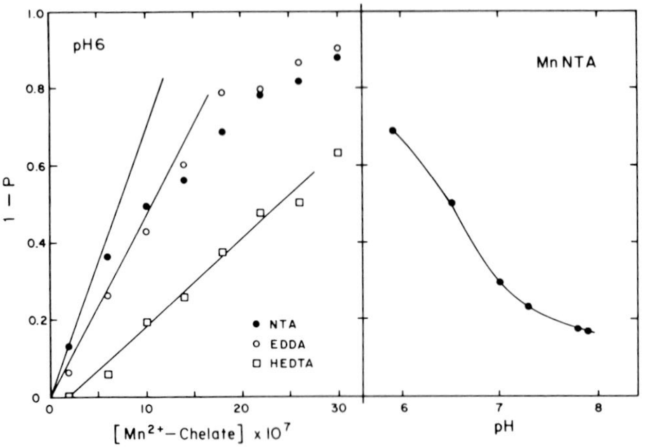

Fig. 1. Superoxide dismutase activities of MnNTA, MnEDDA and MnHEDTA at pH 6. Conditions are described under Materials and Methods.

Fig. 2. Dependence of superoxide dismutase activity of MnNTA on pH; uncorrected for the small decrease in ferricytochrome *c* reactivity.

The free metal ion concentration is extremely low; for MnEDDA, which has an association constant of 10^7 M^{-1} (6), we calculate $[Mn^{2+}]$ to be 10^{-10} M over a concentration range of $2x10^{-7}$ to $3x10^{-6}$ Mn^{2+}, when EDDA is present in a 1000-fold excess. Given the activity of Mn^{2+} itself and that the inhibition of ferrocytochrome *c* formation was larger than what would have resulted from the formation of a one-to-one Mn^{2+}-$O_2^{-}\cdot$ complex, we conclude that MnNTA and MnEDDA do catalyze the dismutation reaction, while MnHEDTA might form long-lived complexes with superoxide.

ACKNOWLEDGEMENTS

This research was supported by the Petroleum Research Fund, administered by the American Chemical Society.

REFERENCES

1. Archibald FS, Fridovich I (1982) Arch Biochem Biophys 214:452-463
2. Cabelli DE, Bielski BHJ (1984) J Phys Chem 88:6291-6294
3. Faraggi M (1984) In: Bors W, Saran M, Tait T (eds) Oxygen Radicals in Chemistry and Biology. De Gruyter, Berlin, pp 419-424
4. Lati J, Meyerstein D (1978) JCS Dalton 1105-1118
5. Koppenol WH, Buuren KJH van, Butler J, Braams R (1976) Biochim Biophys Acta 499:157-168
6. Martell AE, Smith RM (1982) Critical Stability Constants Vol. 5 Plenum Press, p. 45

Superoxide and Superoxide Dismutase in Chemistry,
Biology and Medicine; G. Rotilio, editor

RATE CONSTANTS FOR THE REACTION OF SUPEROXIDE WITH GLUCOSE OXIDASE AND ITS FLAVOSEMIQUINONE

ROBERT F. ANDERSON

Cancer Research Campaign Gray Laboratory, Mount Vernon Hospital, Northwood, Middlesex HA6 2RN (England)

VINCENT MASSEY, LAWRENCE M. SCHOPFER

Department of Biological Chemistry, University of Michigan, Ann Arbor, Michigan 48109 (USA)

INTRODUCTION

The primary free radical species produced upon the irradiation of cells are quickly scavenged by cellular constituents. Subsequent thermodynamically controlled electron transfer reactions occur which would be expected to transiently increase the concentration of superoxide ($O_2^{\bullet-}$) in aerated cells and also increase the concentration of flavosemiquinones ($FH^{\bullet}$), due to the high abundance and redox potentials of flavoproteins. Electron transfer from one-electron reduced NAD^+ (1) and AMP (2) to flavins has been demonstrated using pulse radiolysis. Further it has been found that $FH^{\bullet}$ and $O_2^{\bullet-}$ react to form a transient hydroperoxide species in the case of free flavins (3,4) and for the enzyme glucose oxidase (5). In this paper we present rate data for the reaction of $O_2^{\bullet-}$ with both glucose oxidase (GOx-F) and its flavosemiquinone (GOx-$FH^{\bullet}$).

RESULTS AND DISCUSSION

1) <u>Reaction of $O_2^{\bullet-}$ with GOx-$FH^{\bullet}$</u> Both $O_2^{\bullet-}$ and GOx-$FH^{\bullet}$ species were produced simultaneously as described previously (3) by pulse radiolysis (10-90 Gy in 0.5 μs) of aerated solutions of GOx-F (40 μmol dm^{-3}) containing formate ions (0.1 mol dm^{-3}, pH 6.0). By comparing the known absorption spectrum of GOx-$FH^{\bullet}$ (5,6) with that observed under the above conditions it was found that *ca.* 10% of the radical yield formed GOx-$FH^{\bullet}$. The species decayed rapidly with first-order kinetics to the flavin 4a-hydroperoxide (GOx-FOOH) presumably by reacting with the excess $O_2^{\bullet-}$ formed, reaction (1). The GOx-FOOH species in turn decayed with a half-life of 2 ms (5) to yield GOx-F (and H_2O_2).

$$GOx\text{-}FH^{\bullet} + O_2^{\bullet -} \longrightarrow GOx\text{-}FOOH \qquad (1)$$

The rate of GOx-FH$^{\bullet}$ decay increased linearly with radiation dose (figure 1) and by using $[O_2^{\bullet -}] = 0.55\ \mu mol\ dm^{-3}\ Gy^{-1}$ we calculate k(1) = 8.5×10^8 $dm^3\ mol^{-1}\ s^{-1}$. This rate constant is slightly higher than that measured for free flavins (3) but a higher ionic strength was employed in the present study.

2) Reaction of $O_2^{\bullet -}$ with GOx-F On pulsing GOx-F (40 $\mu mol\ dm^{-3}$) in solutions containing O_2 (130 $\mu mol\ dm^{-3}$) and formate ions (0.1 $mol\ dm^{-3}$, pH 6.0) it was found that following the fast decay of GOx-FH$^{\bullet}$ through reaction (1) a slow formation of GOx-FH$^{\bullet}$ (figure 2, insert) was observed (spectrum not shown). It was noted earlier (5) that this slow reaction was initiated by $O_2^{\bullet -}$ and is consistent with the reaction

$$O_2^{\bullet -} + GOx\text{-}F \longrightarrow O_2 + GOx\text{-}FH^{\bullet} \qquad (2)$$

The measured rate (figure 2) is only slightly dependent on dose (probably due to the perturbation of first order kinetics by the dismutation of $O_2^{\bullet -}$) with k(2) = ca. $10^6\ dm^3\ mol^{-1}\ s^{-1}$. This electron transfer reaction is understood thermodynamically as the one-electron reduction potentials (E^1) of glucose oxidase at pH 5.3 and 9.3 are -63 mV and -200 mV (6) compared to -210 mV and -400 mV for free flavins (7). Thus, unlike for free flavins, electron transfer from $O_2^{\bullet -}$ ($E^1\ O_2/O_2^{\bullet -}$ = -155 mV at pH 7 for 1 $mol\ dm^{-3}$ O_2 (8)) to glucose oxidase is favourable for pH $\leqslant 7$.

Rate constant k(2) is several orders of magnitude smaller than the rate constant for $O_2^{\bullet -}$ with superoxide dismutases. This implies that high potential flavoproteins are unlikely to act as effective $O_2^{\bullet -}$ dismutases per se although other flavoproteins of different structures and higher E^1 need to be studied. However, k(1) is high implying that reaction (1) may well be a significant pathway for the removal of $O_2^{\bullet -}$ in irradiated cells.

ACKNOWLEDGMENT
This work is financially supported by The Cancer Research Campaign.

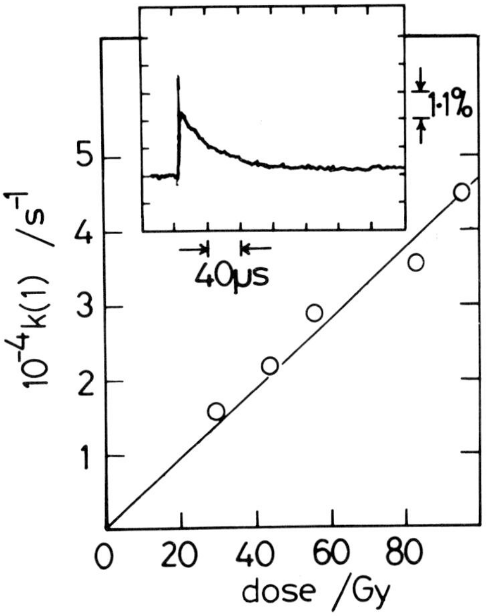

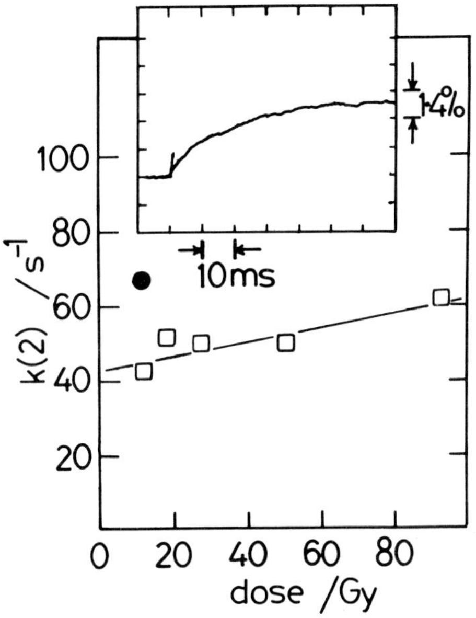

Fig.1. First-order rate constants for the disappearance of transient absorbing at 560 nm formed on the pulse radiolysis of solutions (pH 6) containing glucose oxidase (40 μmol dm^{-3}), sodium formate (0.1 mol dm^{-3}) and oxygen (0.26 mmol dm^{-3}). Insert: Oscilloscope trace following 47 Gy dose.

Fig.2. First-order rate constants for the appearance of the transient as in Fig. 1. but with different concentrations of oxygen (0.13 mmol dm^{-3}) and glucose oxidase (□, 40 μmol dm^{-3}; ●, 65 μmol dm^{-3}) Insert: Oscilloscope trace following 18 Gy dose (□).

REFERENCES

1. Anderson RF (1976) Ber. Bunsenges. physik. Chem. 80:969-972

2. Land EJ, Swallow AJ (1969) Biochemistry 8:2117-2125

3. Anderson RF (1981) In: Rodgers MAJ, Powers EL (eds) Oxygen and Oxy-radicals in Chemistry and Biology. Academic, New York, pp 597-600

4. Anderson RF (1982) In: Massey V, Williams CH (eds) Flavins and Flavoproteins. Elsevier, Amsterdam, pp 278-283

5. Anderson RF, Massey V, Schopfer LM (1984) In: Bray RC, Engel PC, Mayhew SG (eds) Flavins and Flavoproteins. Walter de Gruyter, Berlin, pp 573-576

6. Stankovich MT, Schopfer LM, Massey V (1978) J. Biol. Chem. 253:4971-4979

7. Anderson RF (1983) Biochim. Biophys. Acta 722:158-162

8. Ilan YA, Meisel D, Czapski G (1974) Isr. J. Chem. 12:891-895

© 1986 Elsevier Science Publishers B.V. (Biomedical Division)
Superoxide and Superoxide Dismutase in Chemistry, Biology and Medicine; G. Rotilio, editor

EFFECT OF Cu^{+2} AND SOD ON REDUCTION OF FERRICYANIDE BY SUPEROXIDE RADICALS

M.A. SYMONYAN, R.M. NALBANDYAN
Institute of Biochemistry, Academy of Sciences Armenian SSR, Yerevan, 375044, USSR

Superoxide radicals are known to possess reductive properties. In fact, many methods of the SOD-activity determination are based on reductive properties of superoxide radicals. Thus, they are able to reduce some organic and inorganic compounds, among them ferricyanide (1-3). The reaction may be expressed by a equation:

$$Fe(CN)_6^{-3} + O_2^- \text{ --- } Fe(CN)_6^{-4} + O_2 \qquad 1)$$

Evidently, the reaction should be inhibited by SOD (4). Furthermore, according to the generally accepted concept the reduction of metals (copper, iron or manganese) in the active site of SODs by superoxide radicals is the first step of the dismutation mechanism. In this connection the question arises whether any metals are able to catalyze the reaction in which superoxide radicals participate as reductants. In this study we considered effects of a number of metal ions on the reaction 1).

We have used alkaline hydrogen peroxide solutions as a source of superoxide radicals (5). Generally, 1% of purified peroxide (6) was adjusted to pH 11.0 and heated at 80-90°C for 1-2 min. This results in the steady-state concentration of the radicals of 10^{-5} M. At room temperature this level was practically unchanged during at least 60 min. Aliquotes of this solution were added to freshly prepared ferricyanide solutions, and the drop of their absorbance at 420 nm followed. In parallel experiments superoxide radical concentrations of the aliquotes were determined by EPR.

The titration data showed that one mole of ferricyanide was reduced by one mole of superoxide radicals. The addition of cuprozinc SOD (10^{-8}M) to the system superoxide-ferricyanide results in inhibition of the ferricyanide reduction. When 10^{-7} - 10^{-4}M Cu^{+2} were added to alkaline hydrogen peroxide solutions the concentration of superoxide radicals was significantly decreased. Also the EPR-signal of Cu^{+2} disappears under effect of O_2^-. Thus, Cu^{+2} is reduced by superoxide radicals:

$$Cu^{+2} + O_2^- \text{ ——— } Cu^{+1} + O_2$$

However, when 10^{-10} - 10^{-8}M Cu^{+2} were added to the system superoxide -ferricyanide the bleaching of the 420 nm band of ferricyanide was sharply accelerated (Fig. 1). At low concentrations of Cu^{+2} the rate depends linearly on concentrations of Cu^{+2} (Fig. 2). The reduction of ferricyanide by superoxide radicals was eliminated in the presence of EDTA or EGTA.

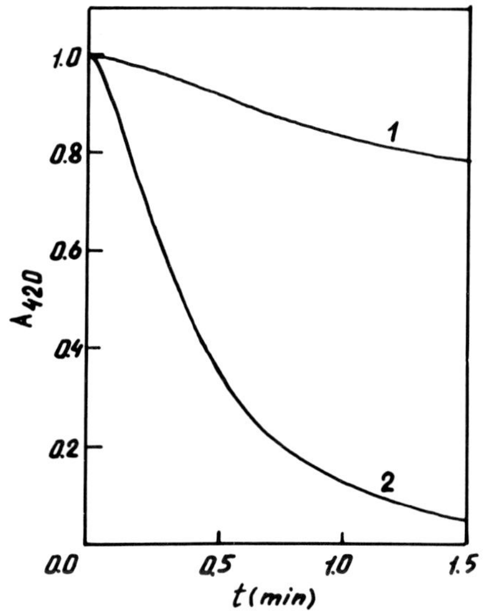

Fig. 1. The effect of Cu^{+2} on the reduction of ferricyanide by superoxide radicals ($3x10^{-7}$M). 1- Without Cu^{+2}, 2- In the presence of $2.5x10^{-9}$M Cu^{+2}.

Further experiments showed that other metal ions in concentrations of 10^{-8}-10^{-6}M were unable to substitute copper in the catalytic process. Thus, the effect of Cu^{+2} was specific among other metals.

Using catalytic properties of Cu^{+2} in the reduction of ferricyanide by superoxide radicals we have developed a simple kinetic method of determination of copper.The method allows to analyze the copper content in aqueous solutions at the level 10^{-10} - 10^{-9}M.

In our mind, the catalytic cycle of copper may be represented by equations 2 and 3

$$Cu^{+2} + O_2^- \text{ —— } Cu^{+1} + O_2 \qquad 2)$$

$$Cu^{+1} + Fe^{+3} \text{ —— } Cu^{+2} + Fe^{+2} \qquad 3)$$

It is expected from these data that trace amounts of hydrated copper in SOD solutions should accelerate the reduction of metals in the enzyme by superoxide radicals.

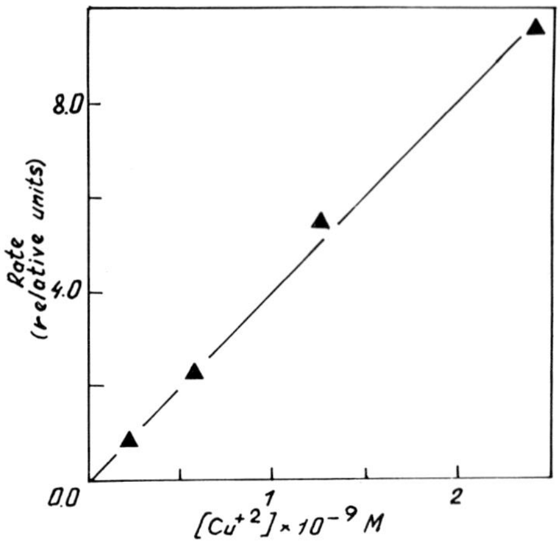

Fig. 2. Dependence on concentration of Cu^{+2} of the rate of ferricyanide reduction by superoxide radicals.

REFERENCES

1. McCord JM, Fridovich I (1969) J Biol Chem, 244, 6049-6055.
2. Nishikimi M, Rao NA, Jagi K (1972) Biochem Biophys Res Commun, 46, 839-854.
3. Frimer AA (1982) In "Superoxide dismutase" v. 2. Ed. Oberley LW, CRC Press, Florida, USA.
4. Symonyan MA (1982) Biochem Biophys Res Commun, 111, 1751-1756.
5. Symonyan MA, Nalbandyan RM (1979) Biochim Biophys Acta, 583, 279-286.
6. Schumb WC, Satterfield CN, Wentworth RL (1958). In "Hydrogen peroxide". Amer Chem Soc Monograph series. Reinold NEW YORK.

NAD-DIMER REACTION WITH CYTOCHROME C.

L.AVIGLIANO, V.CARELLI, A.CASINI, A.FINAZZI-AGRO'*, F.LIBERATORE AND A.ROSSI.

Institutes of Biochemistry and Pharmaceutical Chemistry, University of Rome "La Sapienza" and *Department of Experimental Medicine and Biochemical Sciences, University of Rome "Tor Vergata", Rome, Italy.

INTRODUCTION

Recently, we presented conclusive evidence concerning the structure and the number of the dimeric species from anaerobic electrochemical one-electron reduction of NAD^+. A detailed analysis has in fact led to the conclusion that the product largely contains three diasteroisomeric dimers of the 4,4'-tetrahydrobipyridil type, $(NAD)_2$ (1).

The back oxidation of $(NAD)_2$ to NAD^+, accomplished by chemical (2), electrochemical (3), photochemical (4) and enzymatic (5) methods, indicates the occurrence of an "in vitro" redox cycle $NAD^+/(NAD)_2$ analogous to that of the $NAD^+/NADH$ couple.

The present paper deals with the aerobic reaction between $(NAD)_2$ and cytochrome c Fe^{+++} both in the presence and absence of superoxide dismutase (SOD) and with its possible exploitement for the determination of SOD.

MATERIAL AND METHODS

$(NAD)_2$ was prepared and purified as previously described (1). Cytochrome c (Type VI) was obtained from Sigma, St. Louis, USA. Pure bovine copper-zinc superoxide dismutase was a gift of Dr. D.Cocco, University of Rome. Purified Mn-superoxide dismutase from B. Stearothermophilus was a gift of Dr. J.V.Bannister, University of Oxford. The reduction of cytochrome c was followed by monitoring the absorbance at 550 nm. Anaerobic experiments were performed in a Thunberg-type cuvette after four cycles of vacuum-oxygen-free argon.

RESULTS

$(NAD)_2$ reacts with ferric cytochrome c both in the presence and absence of oxygen, yielding NAD^+ and ferrous cytochrome c. The redox reaction proceeds with clear bimolecular kinetics. The rate shows a dependence on the ionic strength of the medium as expected for a reaction between two oppositely charged species. At 25°C the bimolecular rate constant is 34.5 M^{-1} s^{-1} in 0.1 M potassium phoshate buffer pH 7.2. The rate is almost the same both in the presence and absence of oxygen and the stoichiometry is about 2 mols of cytochrome c reduced per mol of $(NAD)_2$ oxidized. The addition of superoxide dismutase reduces the rate of reaction between $(NAD)_2$ and cytochrome c if the experiment is carried out in the presence of oxygen, while the stoichiometry is changed to a ratio of about 1:1. No effect of superoxide dismutase can be seen in the absence of oxygen.

On the above findings, the following reaction pathways should be operating:

Reaction in absence of SOD

$$(NAD)_2 + \text{cyt. c } Fe^{+++} \longrightarrow NAD^+ + NAD^{\cdot} + \text{cyt. c } Fe^{++}$$

$$NAD^{\cdot} + O_2 \longrightarrow NAD^+ + O_2^{\overline{\cdot}}$$

$$O_2^{\overline{\cdot}} + \text{cyt. c } Fe^{+++} \longrightarrow O_2 + \text{cyt. c } Fe^{++}$$

Reaction in presence of SOD

$$(NAD)_2 + \text{cyt. c } Fe^{+++} \longrightarrow NAD^+ + NAD^{\cdot} + \text{cyt. c } Fe^{++}$$

$$NAD^{\cdot} + O_2 \longrightarrow NAD^+ + O_2^{\overline{\cdot}}$$

$$2\ O_2^{\overline{\cdot}} + 2\ H^+ \longrightarrow H_2O_2 + O_2$$

The effect of superoxide dismutase on the aerobic $(NAD)_2$-cytochrome c reaction is linearly dependent on the enzyme concentration in the range 1-7 μg (Fig. 1). A similar pattern was also observed with bacterial Mn-superoxide dismutase. In order to differentiate this enzyme from copper-zinc superoxide dismutase, the reaction can be carried out in the presence of 0.5 mM KCN, which completely inhibits copper-zinc superoxide dismutase. Therefore a method for the determination of SOD can be based on its inhibitory effect on the reduction of cytochrome c. The present method offers some advantages over similar indirect assays of superoxide dismutase; namely the use of only two reagents and the possibility of working at neutral pH.

This study confirms the previous suggestion (4,5) that $(NAD)_2$ can be utilized as a clean source of superoxide radicals at physiological pH values, since the NAD radicals, generated by its cleavage, readily reduce oxygen at diffusion controlled rate (6).

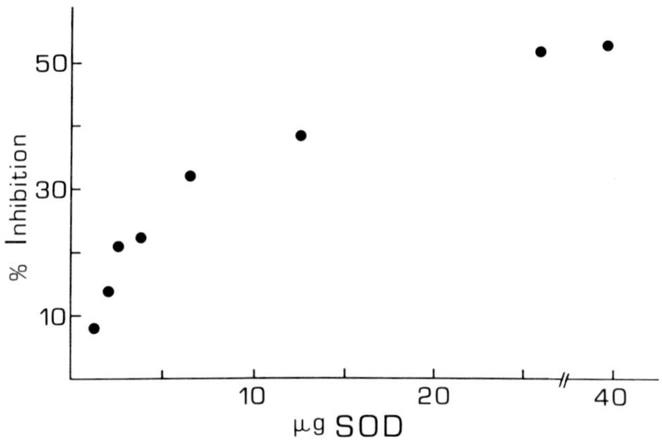

Fig. 1. Inhibition of the reaction between $(NAD)_2$ and cytochrome c in air by superoxide dismutase. The incubation mixture contained 120 µM cytochrome c, 80 µM $(NAD)_2$ and 0.1 mM EDTA in K-phosphate buffer pH 7.2 at 25°C in air.

REFERENCES

1. Carelli V, Liberatore F, Casini A, Mondelli R, Arnone A, Carelli I, Rotilio G, Mavelli I (1980) Bioorg Chem 9:342-350

2. Burnett RW, Underwood AL (1968) Biochemistry 7:3328-3333

3. Carelli I, Rosati R, Casini A (1981) Electrochim Acta 26:1695-1697

4. Avigliano L, Carelli V, Casini A, Finazzi-Agrõ A, Liberatore F (1983) Biochim Biophys Acta 723:372-375

5. Avigliano L, Carelli V, Casini A, Finazzi-Agrõ A, Liberatore F (1985) Biochem J 226:391-395

6. Bielski BHJ, Chan PC (1975) J Biol Chem 250:318-321

Superoxide and Superoxide Dismutase in Chemistry, Biology and Medicine; G. Rotilio, editor

HYDROXIDE-INDUCED ACTIVATION OF DIOXYGEN BY 1,2-DISUBSTITUTED HYDRAZINES, PYRIDINE AND ANTHRAQUINONE: MODEL REACTIONS FOR THE BIOGENESIS OF SUPEROXIDE ION AND HYDROGEN PEROXIDE.

DONALD T. SAWYER, LEE SPENCER, THOMAS S. CALDERWOOD AND JULIAN L. ROBERTS, JR.
Department of Chemistry, Texas A&M University, College Station, Texas, 77843, USA.

Activation of dioxygen is accomplished in biological systems by its reduction to superoxide anion ($O_2^{\overline{\cdot}}$) and hydrogen peroxide (H_2O_2).[1] One example is the xanthine/xanthine oxidase cycle,[2,3] whereby the xanthine oxidase protein catalyzes the oxidation to urate, and reduction of dioxygen to approximately 85% H_2O_2 and 15% $O_2^{\overline{\cdot}}$. Although xanthine oxidase has become the system of choice for enzymic production of superoxide ion, little is known of the precise chemical mechanism for the oxygen activation step.

Recent studies in our laboratories on the mechanism of oxygen activation have shown that under alkaline conditions the combination of oxygen with 1,2-disubstituted hydrazines,[4] dihydrophenazine or dihydrolumiflavin[5] results in autoxidation of the substrate with production of both H_2O_2 and $O_2^{\overline{\cdot}}$. The addition of OH^- to dioxygen and anthraquinone in acetonitrile,[6] and to dioxygen in pyridine[7] results in the formation of significant quantities of H_2O_2 and $O_2^{\overline{\cdot}}$. Finally, substantial yields of $O_2^{\overline{\cdot}}$ and hydroxy radical (•OH) are produced from the addition of OH^- to a solution of H_2O_2 in pyridine.[8]

RESULTS AND DISCUSSION

Hydroxide Ion-Induced Autoxidation of Diphenylhydrazine.[4] The addition of 0.5 mM tetraethylammonium hydroxide (TEAOH) to an oxygen-saturated (8 mM) solution of 5 mM PhNHNHPh in acetonitrile induces its rapid autoxidation to yield one mole of azobenzene and one mole of H_2O_2 per mole of PhNHNHPh.

Stoichiometric addition of TEAOH to PhNHNHPh causes the oxidation potential for the amine to shift from +0.6V vs SCE to -1.1V vs SCE (characteristic of the amine anion). This value is sufficiently negative to bring about the reduction of O_2 to $O_2^{\overline{\cdot}}$ (-0.7V vs SCE in Me_2SO)[9] via eq. 1-2.

$$OH^- + PhNHNHPh \longrightarrow PhN^-NHPh + H_2O \quad (1)$$

$$PhN^-NHPh + O_2 \longrightarrow PhN{=}NPh + HO_2^- \quad (2)$$

The resulting products then undergo a partial conproportionation reaction with PhNHNHPh.

$$PhNHNHPh + PhN{=}NPh + 2HO_2^- \longrightarrow 2H_2O_2 + 2PhNN^{\bar{\cdot}}Ph \qquad (3)$$

The resultant anion radical reacts rapidly with O_2 to give PhN=NPh and $O_2^{\bar{\cdot}}$, which induces the further oxidation of PhNHNHPh.

$$PhNN^{\bar{\cdot}}Ph + O_2 \xrightarrow{\text{fast}} PhN{=}NPh + O_2^{\bar{\cdot}} \qquad (4)$$

$$PhNHNHPh + O_2^{\bar{\cdot}} \xrightarrow{k,\ 100M^{-1}s^{-1}} PhNN^{\bar{\cdot}}Ph + H_2O_2 \qquad (5)$$

Compounds such as PhNHNHPh are simple models for reduced flavins. The mechanism of eq. 1-5 may be relevant to the fractional yields of H_2O_2 and $O_2^{\bar{\cdot}}$ that result from the flavin-mediated activation of O_2 in xanthine oxidase. The mode of action of the enzyme may involve oxidation of xanthine to uric acid by the Mo^{VI} cofactor with the concomitant generation of OH^-. In turn the base-catalyzed autoxidation of the reduced flavin cofactor (as outlined by eq. 1-5) would result in the formation of H_2O_2 and $O_2^{\bar{\cdot}}$.

The One-Electron Reduction of Pyridine by OH^-. The addition of OH^- to pyridine (py)[7] in the presence of oxygen results in formation of a brown diagmagnetic polymer and low yields of $O_2^{\bar{\cdot}}$. The similarity of the polymeric product with that obtained by direct addition of OH·[10] is consistent with a mechanism in which the initial step is a base-induced autoxidation (eq.6).

$$OH^- + O_2 + py \longrightarrow O_2^{\bar{\cdot}} + 1/n[py(OH\cdot)]_n \qquad (6)$$

The low yield of $O_2^{\bar{\cdot}}$ is due to its hydrolysis

$$O_2^{\bar{\cdot}} + H_2O \longrightarrow HO_2^{\cdot} + OH^- \qquad (7)$$

and the subsequent disproportionation of $HO_2^{\cdot}$

$$2HO_2^{\cdot} \longrightarrow H_2O_2 + O_2 \qquad (8)$$

Thus, the overall process is the reduction of O_2 to H_2O_2 by OH^-.

One-Electron Reduction of Anthraquinone by Hydroxide Ion. Electrochemical studies[6] have shown that when two equivalents of anthraquinone (AQ) are combined with 0.5 mM OH^-, the product solution contains equal amounts of $AQ^{\bar{\cdot}}$ and AQ. An anodic peak at +2.1V vs. SCE is characteristic of H_2O_2 with a peak current corresponding to 0.2 mM H_2O_2. A mechanism that is consistent with these data is outlined by eq. 9 and 10.

$$AQ + OH^- \rightleftharpoons AQ(OH)^- \qquad (9)$$

$$AQ(OH)^- + AQ \longrightarrow AQ^{\bar{\cdot}} + A\dot{Q}(OH) \rightarrow AQ + 1/2\ H_2O_2 \qquad (10)$$

In the first step AQ reacts with OH^- in a rapid reversible formation of an addition complex, $[AQ(OH)]^-$. Semiquinone ($AQ^{\bar{\cdot}}$) formation occurs in a slower second step via an electron-transfer reaction with a second molecule of AQ. The resultant $[A\dot{Q}(OH)]$ radical then dimerizes and decomposes to produce H_2O_2 and regenerate anthraquinone. In the presence of oxygen, $AQ^{\bar{\cdot}}$ undergoes autoxidation to produce $O_2^{\bar{\cdot}}$.

$$AQ^{\bar{\cdot}} + O_2 \rightleftharpoons O_2^{\bar{\cdot}} + AQ \qquad (11)$$

The generation of H_2O_2 from the combination of anthraquinone and hydroxide ion may have important applications to the wood-pulp industry.[11] The reduction of quinones by OH^- has been postulated to be a crucial step in photosynthetic O_2 evolution.[12]

Base-Induced Generation of Superoxide Ion and Hydroxyl Radical From Hydrogen Peroxide. Substantial yields of $O_2^{\bar{\cdot}}$ are obtained when OH^- is added to hydrogen peroxide in pyridine. The proposed reaction mechanism involves an initial proton abstraction followed by electron transfer (eq. 12 and 13).

$$H_2O_2 + OH^- \longrightarrow HO_2^- + H_2O \qquad (12)$$

$$H_2O_2 + HO_2^- \longrightarrow O_2^{\bar{\cdot}} + H_2O + {}^{\cdot}OH \qquad (13)$$

The base-induced decomposition of H_2O_2 provides a convenient method to produce $O_2^{\bar{\cdot}}$ in pyridine. It also provides a means to generate ·OH, which, in solvents like CH_3CN and H_2O, has sufficient lifetime to react with substrates.

All of the preceding processes involve base-induced activation of dioxygen. The mechanism of the activation step involves either (a) a proton abstraction followed by the reduction of O_2 by the substrate anion or (b) direct reduction of the substrate by hydroxide ion with subsequent reduction of O_2. Although this step is thermodynamically "uphill" ($E^0_{\cdot OH/OH^-}$, +1.2V vs SCE in MeCN), the reaction is driven by stabilization of ·OH via covalent bonding to the substrate. This process is akin to stabilization of ·OH in systems that contain metal ions by formation of the so-called, "crypto-hydroxyl" radicals.[13]

The demonstrated reactivity of OH^- with quinones and reduced flavins in matrices with membrane-like properties may have implications for a variety of naturally occurring electron transfer processes, of which photosynthesis and the xanthine oxidase cycle are but two examples.

ACKNOWLEDGMENT

This work was supported by the National Science Foundation under grant CHE-8212299 and by the National Institutes of Health under grant GM-36289.

REFERENCES

1. Nozaki M, Yamamoto S, Ishimura Y, Coon MJ (1982) In: Oxygenases and Oxygen Metabolism. Academic Press, New York
2. Fridovich I, Handler P (1960) J Biol Chem 236:1836-1840
3. McCord JM, Fridovich I (1969) J Biol Chem 244:6049-6055
4. Calderwood TS, Johlman CL, Roberts JL Jr, Wilkins CL, Sawyer DT (1984) J Am Chem Soc 106:4683-4687
5. Sawyer DT, Calderwood TS, Johlman CL, Wilkins CL (1985) J Org Chem 50:1409-1412
6. Roberts JL Jr, Sugimoto H, Barrette WC, Sawyer DT (1985) J Am Chem Soc 107:4556-4557
7. Srivatsa GS, Sawyer DT (1985) Inorg Chem 24:1732-1734
8. Roberts JL Jr, Morrison MM, Sawyer DT (1978) J Am Chem Soc 100:329-330
9. Sawyer DT, Valentine JS (1981) Acc Chem Res 14:393-400
10. Dorfman LF, Adams GE (1973) Natl Stand Ref Data Ser (US Natl Bur Stand) NSRDS-NBS 46:20 (SD catalog No C13.48:46)
11. Haggan J (1984) Chem Eng News 62:20-23
12. Webber AN, Spencer L, Sawyer DT, Heath RL FEBS Lett (in press)
13. Elstner EF, Osswald W, Konze JR (1980) FEBS Lett 121:219-221

Superoxide and Superoxide Dismutase in Chemistry, Biology and Medicine; G. Rotilio, editor

EFFECTS OF SOD AND CATALASE ON THE AUTOXIDATION OF THE TRYPTOPHAN METABOLITE 3-HYDROXYANTHRANILIC ACID

JAMES A. DYKENS, STEPHEN GENE SULLIVAN AND ARNOLD STERN

Department of Pharmacology, New York University Medical Center, 550 First Ave, New York, NY, USA 10016

INTRODUCTION

A number of naturally occurring metabolites of tryptophan are known to be neurotoxic (1) and carcinogenic (2-5), including 3-hydroxyanthranilic acid (3-HAT), which normally arises at a branch point in the kynurenine pathway of tryptophan catabolism (6) and has been linked to bladder and breast carcinomas (2-5). 3-HAT undergoes base-catalyzed autoxidation (7) yielding the carcinogenic (3,8) pigment cinnabarinic acid, superoxide anion, and hydrogen peroxide (9-13). It has been proposed (13) that cinnabarinate formation is a two step process: an initial autoxidation of 3-HAT (accompanied by $O_2^{\cdot-}$ production) to an anthranilyl radical which oxidizes to the quinoneimine, followed by a condensation between the quinoneimine and 3-HAT to generate cinnabarinate and H_2O_2. However, H_2O_2 reacts with cinnabarinate, bleaching and destroying the pigment (13-15). Accordingly, we have examined the effects of the enzymes SOD and catalase on 3-HAT autoxidation.

MATERIAL AND METHODS

Since 3-hydroxyanthranilic acid (Sigma Chemical Co) is insoluble at pH values near neutrality, a 187.5 mM stock solution was prepared daily in 0.1 M HCl. Catalase (88,823 units/mg, CalBiochem-Behring) and superoxide dismutase (SOD; Sigma Chemical Co) were prepared in 100 mM Na Phosphate buffer (pH = 7.4). Autoxidation of 3-HAT was initiated by increasing the pH *via* addition of small (40 µl) aliquots of the stock solution to 3 ml of 100 mM sodium phosphate buffer, giving a final 3-HAT concentration of 2.5 mM. The pH of the buffer was precisely adjusted to 7.4 using 0.10 N HCl or NaOH. 3-HAT autoxidation was monitored using a Cary 14 spectrophotometer by following the appearance of cinnabarinic acid (mM ε_{455} = 23) (16). Hypoxic buffer (pO_2 < 1 torr) was prepared by sparging with N_2 gas prior to transfer to an air-tight cuvette. All reaction mixtures were continually stirred.

RESULTS AND DISCUSSION

Formation of cinnabarinate during 3-HAT autoxidation requires molecular oxygen: no cinnabarinate, or any other compound with absorbance in the visible region, is detected in the absence of oxygen for as long as the reaction mixture is hypoxic (Fig. 1). However, cinnabarinate formation begins immediately upon exposure of the reaction buffer to air and then proceeds at rates typical of controls lacking enzymic additions.

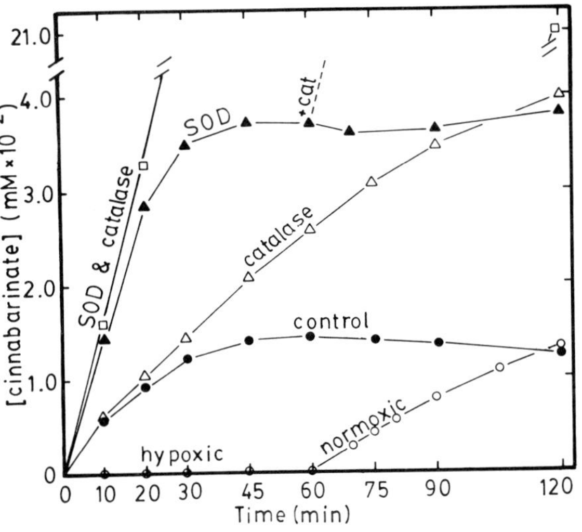

Figure 1. Rates of cinnabarinate formation during autoxidation of 2.5 mM 3-HAT at pH = 7.4 with indicated additions.

The rate of cinnabarinate formation in the control slows after 10-20 min and reaches a plateau by 45 min, suggesting a steady state between cinnabarinate formation and bleaching by H_2O_2. Although excess catalase (1000 U/ml) does not accelerate autoxidation, cinnabarinate formation proceeds linearly for at least 1 h and never obtains a plateau, indicating that catalase prevents cinnabarinate bleaching by H_2O_2. When excess SOD (660 U/ml) is present, the rate of cinnabarinate formation is much faster than both controls and reactions containing catalase. SOD may act by increasing the steady state concentration of the anthranilyl radical, perhaps by preventing the reverse reaction of anthranilyl radical plus $O_2^{\bar{\cdot}}$ yielding 3-HAT and O_2. With excess SOD the reaction slows after 20 min and the rate of cinnabarinate formation reaches a plateau at an absorbance typically 2-3 times higher than controls. Addition of catalase during this plateau results in rates of cinnabarinate formation identical to those observed when SOD alone or SOD and catalase are present at the start of the reaction, reinforcing the suggestion that the plateau reflects a steady state between cinnabarinate formation and bleaching by H_2O_2. When both SOD and catalase are present cinnabarinate formation proceeds in a linear fashion for at least 2 h at which time the cinnabarinate concentration is 0.21 mM, approximately 20 percent of the theoretical maximal yield.

ACKNOWLEDGEMENTS

This research was supported by N.I.H. grant No. ES 03425.

REFRENCES

1. Schwartz R, Whetsell, W, Mangano, M (1983) Science 219:317-318.
2. Allen, MJ, Boyland E, Dukes CE, Horning ES, Watson JG (1957) Br J Cancer 11:212-228.
3. Bryan GT, Brown RR, Price JM (1964) Cancer Res 24:596
4. Ehrhart H, Ried W, Benoit W, Dorfler H (1968) Z Krebsforsch 71:1-6
5. Fujii K, Watanabe M (1980) J Cancer Res Clin Oncol 96:163-168
6. Bender D (1975) Amino Acid Metabolism. Wiley-Interscience, London
7. Dykens JA, Sullivan SG, Stern A (1985) in preparation
8. Boyland E (1960) Acta Unio Int Cancrum 16:273
9. Ehrensvard G, Liljekvist J, Heath RG (1960) Acta Chem Scand 14:2018-2088
10. Subba Rao PV, Vaidyanathan CS (1966) Biochem J 99:317-322
11. Savage N, Prinz W (1977) Biochem J 161:551-554
12. Tomoda A, Shirasawa E, Nagao S, Minami M, Yoneyama Y (1984) Biochem J 222:755-760
13. Ogawa H, Nagamura Y, Ishiguro I (1983) Hoppe-Seyler Z Physiol Chem 364:1507-1518
14. Ogawa H, Nagamura Y, Ishiguro I (1983) Hoppe-Seyler Z Physiol Chem 364:1059-1066
15. Ryall RL, Ryall RG, Howells AJ (1976) Insect Biochem 6:135-142
16. Ishiguro I, Nagamura Y, Hara A (1971) Yakugaka Zasshi 91:760-765

Superoxide and Superoxide Dismutase in Chemistry,
Biology and Medicine; G. Rotilio, editor

OXY-PEROXIDASE-CATALYSED BIOACTIVATION OF THE ANTITUMOR DRUG 9-HYDROXY-ELLIPTICINE IN VITRO

EMMANUELLE VOISIN, TAM HA, BERNARD MEUNIER and CHRISTIAN AUCLAIR.

Laboratoire de Biochimie-Enzymologie, INSERM U 140, CNRS LA 147, Institut Gustave Roussy, 94800 Villejuif, France.
Laboratoire de Pharmacologie et de Toxicologie Fondamentales du CNRS, 31000 Toulouse, France.

INTRODUCTION

9-Hydroxyellipticine is the head of a series of cytotoxic and antitumor agents which exhibits the ability to intercalate between DNA base pairs (1) and may undergo oxidative activation generating reactive oxidizing species including oxy-radicals and the electrophilic quinoneimine 9-oxo-ellipticine (2,3). The oxidative activation may occur through both oxidase and peroxidase reactions respectively catalysed by superoxide dimutase (SOD) and peroxidase (Px) according to the following overall reactions :

$$9\text{-OH-E} + O_2 \xrightarrow{\text{SOD}} 9\text{-OXO-E} + H_2O_2 \quad (1)$$

$$9\text{-OH-E} + H_2O_2 \xrightarrow{\text{Px}} 9\text{-OXO-E} + 2H_2O \quad (2)$$

In the presence of thiol group-containing nucleophiles such as reduced glutathione or cystein the peroxidase-catalysed oxidation of 9-OH-E and derivatives, results in the formation of the corresponding covalent adducts through a Michael addition reaction at the C10 position on the ellipticine ring (4) :

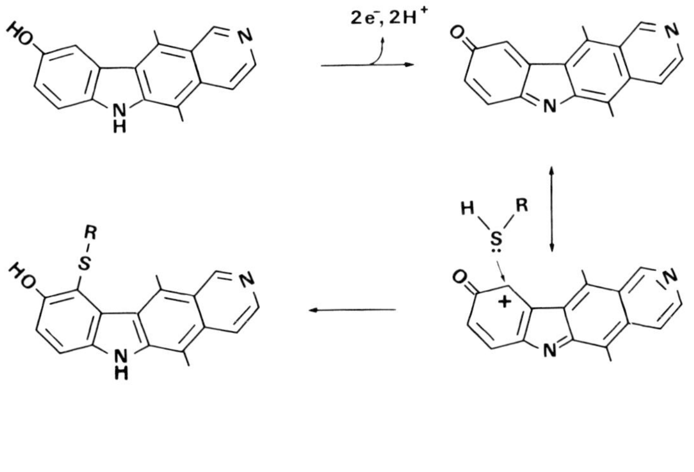

The glutathione and cystein conjugates are devoid of toxicity on mice and can be considered as the detoxifying products of the electrophilic quinone. The glutathione conjugate of the clinically used drug 9-hydroxy-ellipticinium (9-OH-E+) was detected in biological fluids of patients under chemotherapy (5) as well as in vitro, in cultured cells and in human erythrocytes (unpublished). The present work has been undertaken to determine to what extent SOD could be involved as a catalyst in the formation of the thiols-9-OH-Es conjugates either through the oxidase reaction (eq.1) alone or through an oxiperoxidase mechanism involving sequentially reactions 1 and 2. The occurence of such a catalytic effect could emphasize the role of SOD in the biotransformation of oxidizable drugs and xenobiotics.

MATERIALS AND METHODS

9-OH-E was provided by the SANOFI Co. (France). SOD, horse radish peroxidase (HRP), and cystein were from Sigma Chemical Co. (USA).

The cystein-9-OH-Es conjugates were identified by mass spectrometry (DCI system) and HPLC (Waters chromatograph, C18 μ-Bondapak column, reverse phase).

RESULTS

Figure 1A shows a typical chromatogram obtained using reverse phase HPLC system of a mixture containing 9-OH-E and the cystein-9-OH-E conjugate (cys-9-OH-E). Figure 1B shows the kinetics of the formation of cys-9-OH-E in various experimental conditions.

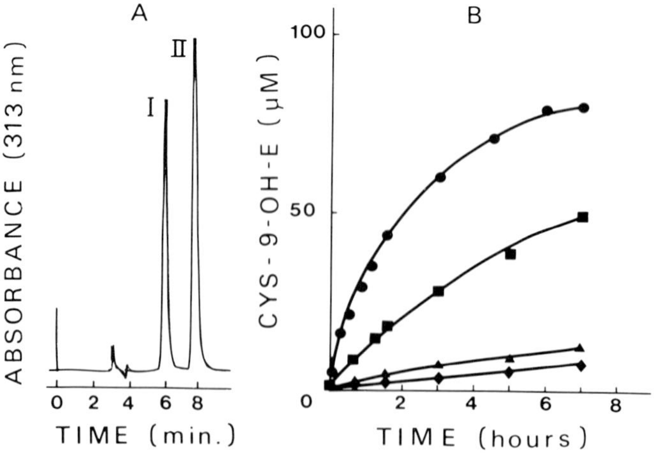

Fig. 1. A) HPLC profile of a mixture containing 9-OH-E and cys-9-OH-E. B) Kinetics of the formation of cys-9-OH-E. Experiments were performed at 37°C in 0.01 M phosphate buffer (pH 7) containing 80 μM 9-OH-E, 1 mM cystein, 10 nM HRP and 20 μg/ml SOD. ●—● : complete mixture, ■—■ : -SOD, ▲—▲ : -HRP, ◆—◆ : -HRP and -SOD.

In the presence of 9-OH-E and cystein, a very limited amount of conjugate is generated (spontaneous autoxidation). The addition of SOD alone results in a significant but little increase in the cys-9-OH-E production rate. In contrast, the addition of peroxidase alone results in a great increase, and the addition of both enzymes in a stronger increase in the generation rate of the conjugate, markedly more than the additive effect of each enzyme (synergetic effect). Figure 2A shows the effect of increasing concentrations of SOD on the amount of cys-9-OH-E produced, in the absence and in the presence of an excess HRP.

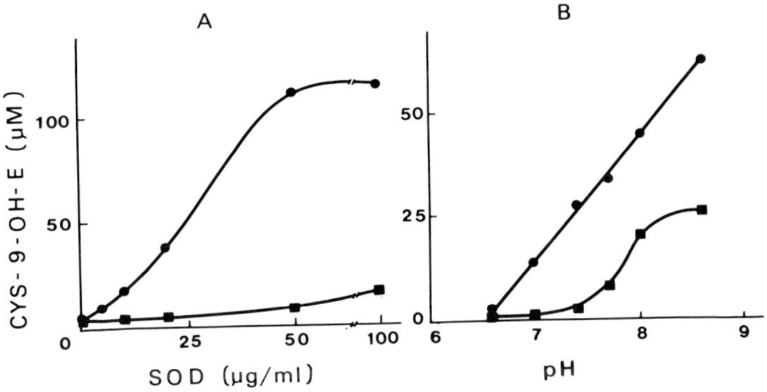

Fig. 2. A) Effect of increasing concentrations of SOD on the production of cys-9-OH-E. B) Effect of pH on the production of cys-9-OH-E. Experiments in A were performed at 37°C in 0.01 M phosphate buffer (pH 7.4) containing 80 μM 9-OH-E, 1 mM cystein, and SOD in the absence (■——■) and in the presence (●——●) of 10 nM HRP. In B, the reaction mixture was as in A, in the presence of 10 nM HRP except that pH varied. (●——●) : +SOD 20 μg/ml, (■——■) : -SOD.

In the presence of HRP, the amount of conjugate generated increases linearly as a function of SOD concentration until 50 μg/ml and reaches a plateau indicating that the limiting step is the rate of the oxidase reaction. In the absence of HRP, SOD exhibits no significant effect until relatively high concentrations. However, as indicated in figure 2B, raising the pH from 6.60 to 8.60 results in an exponential increase in the amount of cys-9-OH-E generated, followed by a plateau, in the presence of HRP alone, and in a linear increase in the presence of SOD-HRP.

DISCUSSION

The results indicated in the present paper clearly show that in the presence of HRP, SOD efficiently catalyses the oxidation of 9-OH-E leading to the generation of the cys-9-OH-E adduct. The sequence of the reactions which may account for this effect is as follows :

$$9\text{-}O\text{-}E + O_2 \longrightarrow 9\text{-}O^{\circ}E + O_2^{\circ-} \qquad (3)$$

$$9\text{-}O^{\circ}E + 9\text{-}O^{\circ}\text{-}E \longrightarrow 9\text{-}OXO\text{-}E + 9\text{-}O\text{-}E \qquad (4)$$

$$O_2^{\circ-} + O_2^{\circ-} + 2H^{+} \xrightarrow{SOD} H_2O_2 + O_2 \qquad (5)$$

$$9\text{-}OH\text{-}E + H_2O_2 \xrightarrow{HRP} 9\text{-}OXO\text{-}E + 2H_2O \qquad (6)$$

$$9\text{-}OXO\text{-}E + cys \longrightarrow cys\text{-}9\text{-}OH\text{-}E \qquad (7)$$

The increase in the dismutation rate of $O_2^{\circ-}$ (eq.5) upon the addition of SOD results in the decrease of $O_2^{\circ-}$ concentration at the steady state and consequently displaces the equilibrium in eq.3 towards the right. Such a mechanism has previously been found to be involved in the inhibition of NBT reduction by SOD (6) as well as in the reoxidation of the semiquinone radical to the quinone during the reduction of the anthracyclinones (7). However, in these two latter cases, SOD increases the reoxidation of the reducing free radical intermediate to the initial unreactive compound whereas in contrast, in the presence of 9-OH-Es, SOD increases the production of the electrophilic phenoxy radical which may either dismutate to the quinoneimine or react with any available electron donors. Accordingly, in the presence of cystein, eq.4 is unlikely and R-O° can be rereduced to 9-OH-E explaining why, in the absence of HRP, a very limited amount of adduct is generated.

It can be concluded that in physiological conditions, 9-OH-Es can be oxidized by SOD-HRP to electrophilic quinones through an oxiperoxidase process leading to covalent binding in the presence of suitable nucleophiles. It can be reasonably hypothezised that various phenols and aromatic amines can be subjected to such a biotransformation in vivo.

REFERENCES

1. Le Pecq,J.B., Dat-Xuong,N., Gosse,C. & Paoletti,C. (1974) Proc. Natl. Acad. Sci. USA 71, 5078-5082.
2. Auclair,C. & Paoletti,C. (1981) J. Med. Chem. 24,289-295.
3. Auclair,C., Hyland,K. & Paoletti,C. (1983) J. Med. Chem. 26, 1438-1444.
4. Meunier,G., Meunier,B., Auclair,C., Bernadou,J. & Paoletti,C. (1983) Tetrahedron Lett. 24, 365-368.
5. Monsarrat,B., Maftouh,M., Meunier,G., Dugue,B., Bernadou,J., Armand,J.P., Picard-Fraire,C., Meunier,B. & Paoletti,C. (1983) Biochem. Pharmacol. 32, 3887-3890.
6. Auclair,C., Torres,M. & Hakim,J. (1978) FEBS Letts. 89, 26-28.
7. Winterbourn,C.C., (1981) Arch. Biochem. Biophys. 799, 159-167.

Superoxide and Superoxide Dismutase in Chemistry,
Biology and Medicine; G. Rotilio, editor

IDENTIFICATION OF PRODUCTS OF HYDROXYL RADICAL ATTACK ON DOPAMINE

ADAM SLIVKA AND GERALD COHEN
Graduate Program in Neurobiology and Department of Neurology
Mount Sinai School of Medicine of the City University of New York
New York, New York 10029, U.S.A.

INTRODUCTION

These experiments were undertaken in order to design a procedure to detect the formation of hydroxyl radicals (.OH) in vivo within dopamine (DA; 3,4-dihydroxyphenylethylamine) neurons. A major impetus was the need to assess working hypotheses (1,2) about mechanisms associated with DA neuron senescence in Parkinson's Disease and DA neuron destruction by neurotoxic drugs, such as 1-methyl-4-phenyl-1,2,3,6-tetrahydropyridine (MPTP) or amphetamine.

The detection of .OH in vivo and its specific localization to a single cell type in the central nervous system is a formidable task. In the brain, DA is present in certain neurons that control motor function and mentation. A detection system is needed that will not reflect the presence of .OH in other cell types. We approached this problem by selecting DA as a potential target of .OH attack. DA is present in the terminals of DA neurons at an extraordinarily high

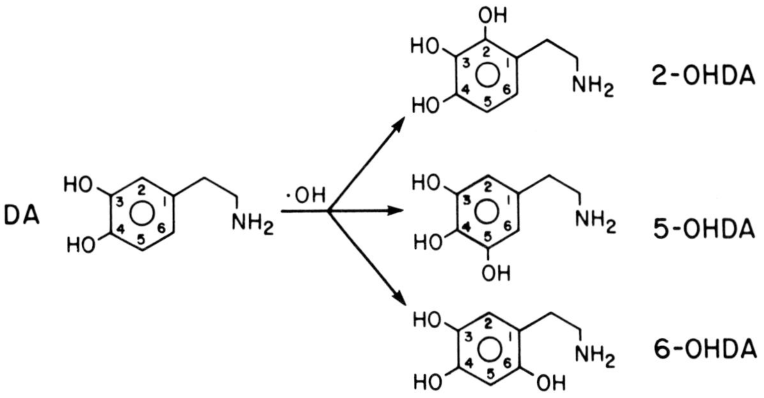

Fig. 1. Ring-monohydroxylated products of dopamine. The products are 2-hydroxydopamine (2-OHDA), 5-hydroxydopamine (5-OHDA) and 6-hydroxydopamine (6-OHDA).

concentration: The average concentration is in the range of 47 mM (3). Thus, the DA is in a good competitive position (vis-a-vis other molecules present within the cell) to react with .OH. We sought to identify the stable end products of the interaction of .OH with DA, so that their detection might serve as an indicator of the presence of intracellular .OH. Similar approaches may be feasible with other cell types (viz., other neuronal types or secretory cells) that store high concentrations of physiologically active agents that may serve as convenient targets for .OH.

Richter & Waddell (4) described transient intermediates that are formed when .OH adds to the phenyl ring of DA. Their experiments were carried out in the absence of oxygen; under these conditions, the intermediates dehydrate to form semiquinones. In the presence of oxygen and/or ferric ions, the hydroxycyclohexadienyl radical intermediates are oxidized to stable hydroxylated compounds.

We used two model systems to initiate the attack on DA at neutral pH: Fe(II)-DTPA/H202 (a Fenton system) and Fe(II)-EDTA/ascorbate (a recycling system). The final products were identified by high-performance liquid chromatography (HPLC) with electrochemical detection (5). 2-OHDA, 5-OHDA, and 6-OHDA were formed by ring-hydroxylation (Figure 1); norepinephrine was formed by beta-hydroxylation of the side chain. We hope to utilize the detection of unnatural ring-hydroxylated forms of DA to detect the generation of .OH within DA neurons. This current report is a progress report toward that goal.

METHODS:

Details of sample preparation and HPLC analyses have been published (5).

The Fenton system [Fe(II)-DTPA/H202] consisted of 200 uM ferrous ammonium sulfate, 240 uM DTPA, and 44 uM H202, in 50 mM phosphate buffer, pH 7.2, containing 1 mM DA. The H202 was added last to initiate the Fenton reaction. The iron recycling system [Fe(II)-EDTA/ascorbate] used EDTA in place of DTPA, and ascorbate (1 mM) in place of H202; the autoxidative and recycling reactions were initiated by the addition of the ferrous salt. Generally, samples were quenched by dilution in 9 volumes of 0.4 M perchloric acid prior to analyses; however, some samples from the Fe(II)-EDTA/ascorbate system were additionally analyzed by direct injection into the HPLC.

RESULTS AND DISCUSSION:

In both systems, the yields of 2-OHDA, 5-OHDA and 6-OHDA were in the approximate ratio 3:2:1. Results are shown in Table 1.

The appearance of a plateau for the products in the Fe(II)-DTPA/H202 system reflects the rapid Fenton reaction. The reaction stops when the H202, added in limiting amounts, is consumed. On the other hand, the products in the Fe(II)-EDTA/ascorbate system continued to accumulate for 30 minutes; the latter results reflect the continuous production of H202 and the recycling of the oxidized iron chelate to the reduced state.

The analysis of 6-OHDA presents a methodological problem. This substance is extremely unstable; it is readily oxidized by molecular oxygen to quinoidal products. However, the presence of 1 mM ascorbate will maintain the 6-OHDA in the reduced state to facilitate HPLC analyses (5). The ascorbate is required even under acid conditions (viz., 0.4 M perchloric acid). On the other hand, the ease of oxidation of 6-OHDA provides a valuable factor for electrochemical assays. By setting the detector at +0.2 volts, the 6-OHDA is easily oxidized and measured, while 5-OHDA and 2-OHDA give only minimal responses at this oxidizing voltage.

TABLE 1.
FORMATION OF HYDROXYLATED DOPAMINE DERIVATIVES DURING THE .OH-MEDIATED ATTACK ON DOPAMINE IN THE Fe(II)-DTPA/H202 (FENTON) AND Fe(II)-EDTA/ASCORBATE (RECYCLING) SYSTEMS.

Time (min)	2-OHDA (uM)	5-OHDA (uM)	6-OHDA (uM)
	Fenton System		
2	1.2	0.8	0.5
10	1.7	1.0	0.5
20	1.8	1.0	0.5
30	1.7	1.0	0.5
	Recycling System		
2	1.8	1.1	0.7
10	3.8	2.4	1.3
20	4.7	2.9	1.5
30	5.4	3.3	1.7

In separate experiments, the .OH scavengers dimethylsulfoxide, mannitol, or ethanol, were incorporated into the reaction mixtures. The scavengers gave a concentration-dependent suppression of the yields of hydroxylated DA products. For example, the yield of 2-OHDA was suppressed by 11% with 1 mM mannitol, 56% with 10 mM mannitol, and 92% with 100 mM mannitol.

Experiments have been performed in which ionizing radiation (10-20 Krads with a Ce-137 source) replaced the chemical systems as the source of .OH. Hydroxylated DA products were observed. Preliminary experiments with slices of mouse striatum (rich in DA nerve terminals) exposed to ionizing radiation have yielded encouraging results. The latter experiments are currently in progress.

REFERENCES

1. Cohen, G. (1983) J Neural Transmission Suppl 19:89-103
2. Cohen, G, Mytilineou C (1985) Life Sci 35:237-242
3. Anden NE, Fuxe K, Hamberger B, Hokfelt T (1966) Acta Physiol Scand 67:306-312
4. Richter HW, Waddell WH (1983) J Amer Chem Soc 105:5434-5440
5. Slivka A, Cohen G (1985) J Biol Chem (in press)

© 1986 Elsevier Science Publishers B.V. (Biomedical Division)
Superoxide and Superoxide Dismutase in Chemistry,
Biology and Medicine; G. Rotilio, editor

FORMATION OF EXCITED SPECIES DURING GLUTATHIONE OXIDATION

HERIBERT WEFERS and HELMUT SIES
Institut für Physiologische Chemie I, Universität Düsseldorf,
Moorenstr.5, D-4000 Düsseldorf, FRG

INTRODUCTION

Protection against chemically or radiation-induced free radicals by reduced glutathione (GSH), according to reaction (1) has been described (1).

$$R^{\bullet} + GSH \longrightarrow GS^{\bullet} + RH \qquad (1),$$

The superoxide anion radical $O_2^{\bar{\bullet}}$ or its protonated form $HO_2^{\bullet}$ is capable to react with GSH (2). The formation of thiyl radicals has been confirmed by ESR for the horseradish peroxidase (HRP)-mediated oxidation of cysteine in the presence of oxygen or H_2O_2 (3), with HRP Compound I and Compound II abstracting hydrogen atoms from the thiol. The fate of the GS• radical usually is discussed in terms of reactions (2) or (3) and (4).

$$GS^{\bullet} + GS^{\bullet} \longrightarrow GSSG \qquad (2)$$

$$GS^{\bullet} + GSH \longrightarrow GSSG^{\bar{\bullet}} + H^+ \qquad (3)$$

$$GSSG^{\bar{\bullet}} \longrightarrow GSSG + e^- \qquad (4)$$

$$GS^{\bullet} + O_2 \longrightarrow GSOO^{\bullet} \qquad (5)$$

In the presence of oxygen the addition of oxygen to the glutathionyl radical (reaction 5) should be taken into consideration as described for cysteine and glutathione (4-6). The nature of the oxygen addition product is not clear, however, a peroxyradical has been suggested as intermediate in cysteine radiolysis in the presence of oxygen (5).

The results presented here indicate the formation of excited species during GSH oxidation by (a) superoxide, generated during the xanthine oxidase reaction, and (b) horseradish peroxidase (HRP) in the presence of oxygen. The available evidence indicates that the excited species is singlet molecular oxygen, $^1O_2$. Moreover, the formation of an oxidation product different from GSSG, glutathione sulfonate (GSO_3^-) was shown. The results indicate the participation of reaction (5) in radical-mediated GSH oxidation in the presence of oxygen and, furthermore, point to an intermediate with a hydroperoxy group attached to a sulfur atom during the oxidation of GSH and other biologically important thiols.

METHODS

Chemiluminescence measurements were performed as described previously (7), detection of GSSG and GSO_3^- as described in (2) and (8).

RESULTS

<u>Superoxide-dependent chemiluminescence in the presence of GSH</u>

Xanthine oxidase with xanthine or acetaldehyde as substrates yields low-level

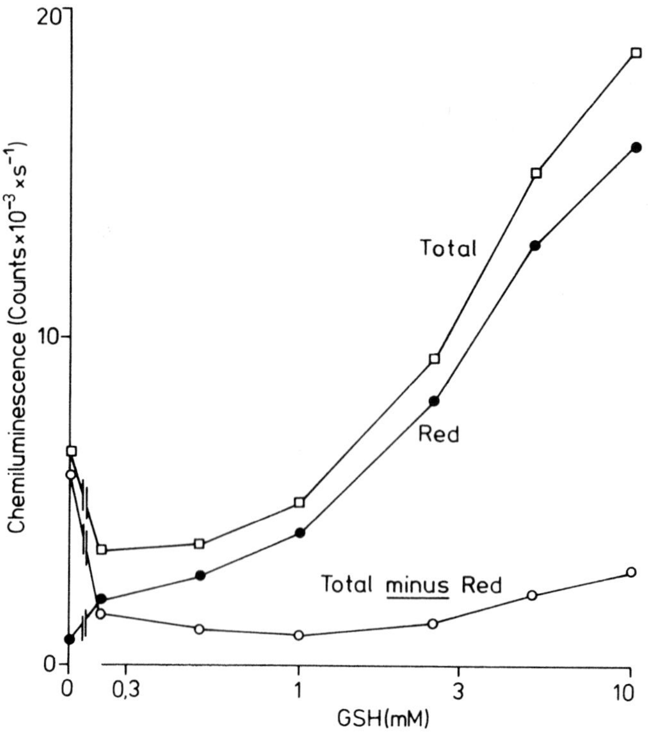

Figure 1. Glutathione-dependent chemiluminescence of the xanthine oxidase/acetaldehyde system (□): photoemission of the complete visible spectrum, (●): red photoemission (beyond 620 nm), (o): photoemission below 620 nm. XOD: 30 mU/ml, acetaldehyde: 5 mM.

chemiluminescence, as described for example by Arneson (9). However, light emission is mainly in the non-red spectral region at wavelengths where excited carbonyls emit light. Singlet oxygen , emitting light at the wavelengths 634 and 703 nm (bimolecular emission), may only contribute to an extremely low extent. This is in agreement with recent work in which chemical identification of $^1O_2$ in the xanthine oxidase reaction was not successful (10). However, as shown in figure 1, addition of GSH to a xanthine oxidase/acetaldehyde system led to pronounced emission of red light (beyond 620 nm). GSH-dependent red chemiluminescence was also detected during quinol autoxidation, where no chemiluminescence had been observed in the absence of GSH (2). This is in agreement with $^1O_2$ generation under these conditions.

Horseradish peroxidase-mediated chemiluminescence in the presence of GSH and further thiols

Addition of HRP to a GSH (5 mM) containing, oxygen saturated buffer (pH 7.4)

gives rise to red low-level chemiluminescence (8), as shown in figure 2. Other thiols also yielded chemiluminescence, e.g. cysteamine, dithioerythritol and cysteine methylester, but cysteine, N-acetyl cysteine and penicillamine gave no signal; likewise, no chemiluminescence was observed using disulfides such as GSSG, or cystamine (8).

Features of GSH-dependent chemiluminescence

Chemiluminescence of the xanthine oxidase or HRP system in the presence of GSH revealed very similar properties: both were strictly dependent on oxygen and on the GSH concentration, they were strongly inhibited by superoxide dismutase and glutathione peroxidase. OH$^\bullet$ quenchers did not diminish photoemission. Free radical quenchers such as cyanidanol ((+)-catechin) (100 µM) or antioxidants such as ascorbate (0.5 mM) abolished light emission (Fig.2). In contrast to glutathione peroxidase, catalase did not influence the chemiluminescence of the xanthine oxidase system, nor phase II of the HRP-mediated chemiluminescence. On contrary, the first phase was abolished. Glutathione S-transferase 1-2, showing peroxidase activity towards organic hydroperoxides, but not towards H_2O_2, did not abolish phase I but did phase II.

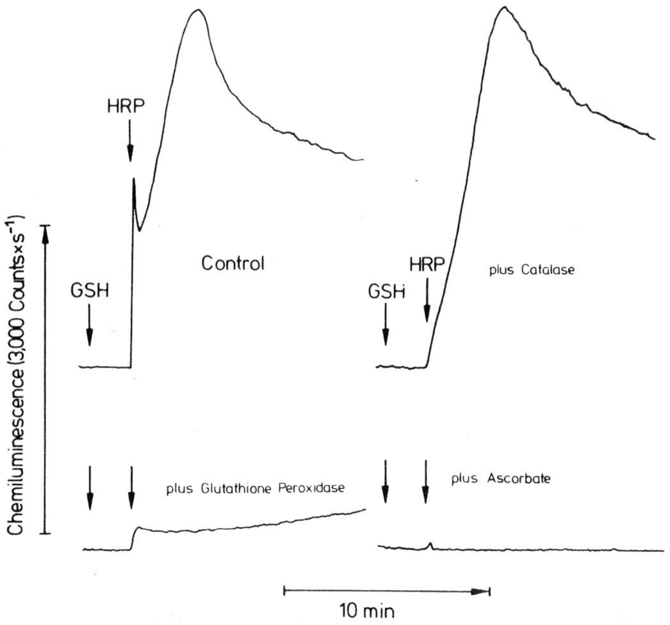

Figure 2. Horseradish peroxidase (HRP)-mediated chemiluminescence (beyond 620 nm) in the presence of glutathione. GSH (5 mM) and HRP (100 µg/ml) were added at the indicated timepoints. Catalase: 40 µg/ml, glutathione peroxidase:40 µg/ml, ascorbate: 0.5 mM.

The spectral distribution of the emitted light is indicative of singlet molecular oxygen, as are the effects of the singlet oxygen quenchers sodium azide or the effect of DABCO (shown for quinol autoxidation (2)), or the enhancement of chemiluminescence in the HRP reaction by a factor of 4.4 when H_2O was replaced by D_2O (8). Likewise, there was an increase by a factor of 1.9 in the xanthine oxidase system.

Formation of GSSG and glutathione sulfonate GSO_3^-

As shown in figure 3, parallel to the oxidation of xanthine (as measured by the formation of urate) the disulfide and the sulfonate were detected as oxidation products. The formation of both was inhibited by superoxide dismutase. GSH (5mM) led to an extra oxygen uptake, also abolished in the presence of SOD. The additional oxygen uptake was accounted for by the formation of GSSG and GSO_3^-. The sulfonate was also detected in the HRP system under conditions where phase II was observed (8).

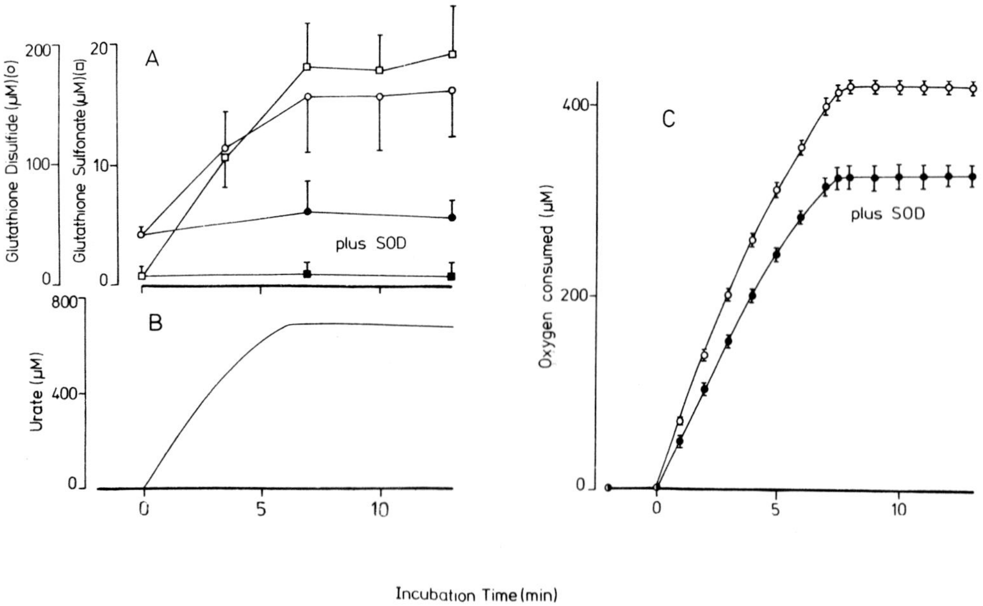

Figure 3. Formation of glutathione disulfide and glutathione sulfonate in the xanthine oxidase/xanthine system. (A) Time course of GSSG and glutathione sulfonate formation without (o,□) and with (•,■) superoxide dismutase. (B) Urate formed from xanthine. (C) Oxygen consumed without (o) and with (•) superoxide dismutase. Incubations were carried out in oxygen-saturated phosphate buffer, pH 7.0, 37°C, containing xanthine (700 μM), GSH (5 mM) and catalase (40 μg/ml). The reactions were started at zero time by addition of xanthine oxidase (30 mU/ml).

DISCUSSION

The formation of excited O_2 species, such as singlet molecular oxygen, in the systems employed suggests the involvement of oxygenated compounds derived from GSH or the GS• radical. Whereas in the case of the xanthine oxidase system GS• may be obtained by reaction (1), where R• is the superoxide anion radical, in the HRP system GS• is generated via HRP Compound I and II, as shown for cysteine (3). H_2O_2 is formed by autoxidation of the thiol. It is suggested that a peroxy species, such as GSOO•, yields the nonradical products GSSG and O_2, where the latter is (in part) $^1O_2$. A peroxy radical GSOO• will abstract hydrogen atoms (for example from GSH), yielding a hydroperoxide. Rearrangement reactions may occur, furthermore homolytic or heterolytic scission may yield products which give rise to further propagation reactions. Glutathione peroxidase, reducing hydroperoxides in a two-electron step without generating free radicals, thus inhibits chain reactions leading to the formation of excited species.

As indicated by our results, the protective role of GSH in the presence of oxygen requires further mechanisms to handle the GS• and/or the oxygen addition products such as GSOO•. These species are intermediates in reactions leading to excited species, such as singlet molecular oxygen.

ACKNOWLEDGEMENTS

This work was supported by Deutsche Forschungsgemeinschaft, Schwerpunkt "Mechanismen toxischer Wirkungen von Fremdstoffen" and by the National Foundation for Cancer Research.

REFERENCES

1. Kosower NS and Kosower EM (1978) Int Rev of Cytol 54:109-160
2. Wefers H and Sies H (1983) Eur J Biochem 137:29-36
3. Harman LS, Mottley C and Mason RP (1984) J Biol Chem 259:5606-5611
4. Schäfer K, Bonifacic M, Bahnemann D and Asmus K-D (1978) J Phys Chem 82:2777-2780.
5. Al-Thannon AA, Barton JP, Packer JE, Simis RJ, Trumbore CN and Winchester RU (1974) Int J Radiat Phys Chem 6:233-248
6. Quintiliani M, Badiello R, Tamba M, Esfandi A and Gorni G (1977) Int J Radiat Biol 32:195-202
7. Cadenas E and Sies H (1984) Methods Enzymol 105:221-231
8. Wefers H, Riechmann E and Sies H (1985) J Free Radicals Biol Med, submitted
9. Arneson RM (1970) Arch Biochem Biophys 136:352-360
10. Nagano T and Fridovich J (1985) Photochem Photobiol 41:33-37

Superoxide and Superoxide Dismutase in Chemistry, Biology and Medicine; G. Rotilio, editor

PRODUCTION OF EXCITED STATES IN THE LIPOXYGENASE REACTION AND IN MICROSOMAL PEROXIDATION.

EDY I. RIVAS, DIEGO F. MIRA, GIUSEPPE CILENTO* and ALBERTO BOVERIS
Instituto de Química y Fisicoquímica Biológicas, Facultad de Farmacia y Bioquímica, University of Buenos Aires, Junín 956, 1113 Buenos Aires, Argentina
*Instituto de Química, University of São Paulo, C.P. 20780, São Paulo, Brazil

INTRODUCTION

Lipid peroxidation produces excited triplet carbonyls and singlet oxygen, as it has been demonstrated in the lipoxygenase reaction (1,2) and in microsomal membranes (3). Since micelle-solubilized chlorophyll-a (Chl) and 9,10-dibromoanthracene (DBA) (4) and Chl bound to microsomes (3) are very efficient detectors of triplet species, they were assayed in the lipoxygenase reaction,where linoleate provided the micellar medium, and in microsomes supplemented with tert-butyl hydroperoxide. Singlet energy acceptors as 9,10-diphenylanthracene (DPA) and anthracene sulfonate (AS) were also used. The aim of this work is to determine the nature of the excited states thus generated.

MATERIALS AND METHODS

Microsomes from rat liver were prepared by ultracentrifugation in 0.25 M sucrose/ 10 mM triethanolamine HCl buffer, pH 7.4. Chemiluminescence was measured in a Hamamatsu TVC-767 photoncounter and in a Packard Tri-Carb Model 3320 liquid scintillation counter in the out-of coincidence mode.

RESULTS AND DISCUSSION

Effect of Chl and anthracene derivatives on the linoleate/lipoxygenase system

Chl, DBA and DPA, solubilized in linoleate, show absorption and fluorescence spectra similar to those obtained when solubilized in ethanol. These compounds markedly increased the emission intensity of this reaction without affecting the rate of oxygen consumption. The effect of Chl, DBA and DPA concentration upon the emission intensity shows saturation (Fig. 1A). Double reciprocal-plots were linear and gave high $k_{ET}\tau^{\circ}$ values (the product of the energy transfer constant by the lifetime of the excited donor in the absence of the acceptor) which are shown in Table I. This table also presents the efficiency of populating the fluorescent state of these acceptors per molecule of O_2 consumed (S_1/O_2) which represents the product $\phi_{chemiexcitation} \times \phi_{energy\ transfer}$.
Chl fluorescences as a consequence of energy transfer from the excited species produced in this reaction, although this compound is destroyed during the reaction (5). For a triplet donor,Chl and anthracene derivatives may be excited to

an upper triplet state, followed by intersystem crossing to the fluorescent state (4). This process is favoured by the spectral characteristics of Chl molecule, or by the presence of heavy atoms in the case of DBA and DBAS (6). Energy transfer and internal convertion to the T_1 state could also occur. Another possibility is the operation of a long range (Forster type) energy transfer mechanism, favoured by the spectral overlap between the emission spectrum of the enzymatic system,and these compounds absorption spectra. Results with DPA and AS strongly suggest the participation of singlet states. The excitation energy might populate the S_1 state by a long range Forster type energy transfer process. The difference between the $k_{ET}\tau^{\circ}$values for both compounds (ca 10^3) may be due to the fact that DPA enters the linoleate micelle (7). The proximity of the sensitizer to the short lived excited singlet species may explain the high efficiency of DPA in sensitizing this emission.

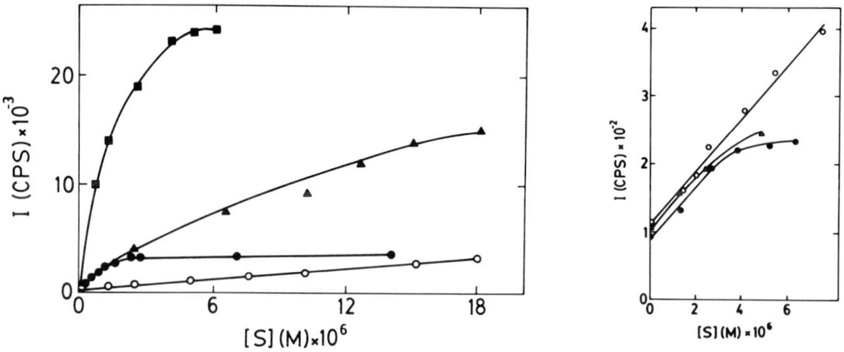

Fig. 1. Effect of Chl (■), DBA (●), DBAS (o), DPA (▲) and AS (△) concentration upon the emission intensity of: A) 0.5 mM linoleate/1 μM lipoxygenase, in potassium phosphate buffer 0.1 M, pH 6.7, at 33°C. B) microsomes (1 mg protein/ml) supplemented with 1 mM tert-butyl hydroperoxide in potassium phosphate buffer 0.1 M, pH 7.4, at 28°C.

Microsomal membranes induced by tert-butyl hydroperoxide

DBA, DBAS and AS efficiently increased the chemiluminescence intensity of this system (Fig. 1B). The $k_{ET}\tau^{\circ}$ values are given in Table I. The effect of DBA concentration upon the sensitized emission intensity shows saturation, probably due to DBA incorporation in microsomal membranes. DPA had no effect, probably due to its insolubility and to the shorter lifetime of singlet states. The energy transfer mechanisms could be the same as discussed in the case of the lipoxygenase reaction.

Mechanism(s) of excited state formation

Excited states should be formed in essentially two ways: (i) the disproportion-

TABLE I

YIELD OF S_1 SENSITIZERS MOLECULES PER OXYGEN CONSUMED AND $k_{ET}\tau°$ VALUES

Sensitizer	Linoleate/lipoxygenase $k_{ET}\tau°$(M)	S_1/O_2	Microsomes/tert butyl hydroperoxide $k_{ET}\tau°$(M)	S_1/O_2
DBA	2×10^6	7×10^{-10}	1×10^6	2×10^{-10}
DBAS	3×10^5	-	2×10^5	8×10^{-10}
DPA	7×10^5	8×10^{-10}	-	-
AS	4×10^2	-	2×10^6	-
Chl	8×10^5	3×10^{-8}	6×10^6(3)	-

ation of peroxyradicals (8) formed through the reaction of free radicals produced in side reactions,and oxygen. This process leads to the formation of conjugated triplet carbonyls and singlet oxygen. These excited carbonyls are energetically able to transfer energy to Chl (E_{S_1}= 41.5 kcal/mol) but not to DBA or to DBAS (E_{S_1}=70.2 kcal/mol) specially considering that the T_2 state of the acceptor must be populated; (ii) the formation of a dioxetane intermediate from the addition of singlet oxygen to one of the fatty acid double bound. The dioxetane cleaves giving rise to an excited aldehyde which can excite DBA, DBAS and also Chl. The sensitization by DPA and AS implies that also singlets are formed (9). These results show that lipid peroxidation produces a variety of excited species, inclusive of different multiplicity.

REFERENCES

1. Nakano M, Sugioka K, (1977) Arch.Biochem.Biophys. 181, 371-383
2. Boveris A, Cadenas E, Chance B (1980) Photobiochem.Photobiophys.1,175-182
3. Cadenas E, Sies H, Campa A, Cilento G (1984)Photochem.Photobiol.40, 661-666
4. Brunetti I L, Cilento G, Nassi L (1983) Photochem.Photobiol. 38, 511-519
5. Chauvet J P, Journeaux R, Viovy R (1974) J.Chem.Phys. 71, 879-888
6. Schmidt R, Kelm H, Brauer D (1977) Ber.Bunsenges Physik Chem. 81, 402-407
7. Ebbesen T W, Delgado O, Valla A, Giraud M, Saito Y, Tachibana H, Wada A (1982) Photochem.Photobiol. 35, 665-669
8. Russell G A (1957) J.Am.Chem.Soc. 79, 3871-3877
9. Horn K A, Schuster G B (1978) J.Am.Chem.Soc. 100, 6649-6656

ACKNOWLEDGEMENTS

This work was supported by the: CONICET, SUBCYT, FAPESP, CNPq, FINEP and the Wolkswagenerk Stiftung grants.

© 1986 Elsevier Science Publishers B.V. (Biomedical Division)
Superoxide and Superoxide Dismutase in Chemistry,
Biology and Medicine; G. Rotilio, editor

POTENTIAL METHODS FOR MEASURING OXYGEN METABOLITES IN BIOLOGICAL SYSTEMS USING DIMETHYLTHIOUREA AND DIMETHYL SULFOXIDE

WILLIAM E. CURTIS, NATALIE B. PARKER, STUART L. LINAS AND JOHN E. REPINE

Departments of Medicine and Pediatrics and the Webb-Waring Lung Institute, University of Colorado Health Sciences Center, Box C322, 4200 East Ninth Avenue, Denver, Colorado, 80262 (USA)

INTRODUCTION

Oxygen metabolites are being implicated in many diseases but their contribution often remains unclear because no good methods exist for measuring these highly reactive molecules in biological systems. We addressed this difficult problem by hypothesizing that measurements of the consumption of the highly permeable, non-toxic O_2 metabolite scavengers, dimethylthiourea (DMTU) or dimethyl sulfoxide (DMSO), could be used to help quantitate O_2 metabolites in biological systems and to help determine the mechanism of action of treatments which purportedly decrease tissue injury by decreasing O_2 metabolite concentrations (1-3).

MATERIALS AND METHODS

Aliquots from samples were injected into a gas chromatograph equipped with an integrator and a flame ionization detector. A 1/4 inch by 2 mm I.D. glass column was packed with 20% carbowax 20 M, on Supelcoport, 80/100 mesh. For DMTU or DMSO The injector temperature was varied between 260°C-350°C. The oven temperature was set initially at 120° to 155° for 2 minutes and then increased by about 30°C per minute to 155°C or 170°C for 4 minutes. Helium, hydrogen and air flow rates were set at 30, 30 and 300 ml/minute, respectively. In separate experiments (data not shown), reaction of hydrogen peroxide (H_2O_2) with DMTU was found to be complete before analysis.

DMTU or DMSO consumption in samples was calculated by the following formula:

$$\text{DMTU or DMSO consumption} = 1 - \frac{\text{DMTU or DMSO concentrations in samples after administration of } H_2O_2 \text{ or other agents}}{\text{DMTU or DMSO concentrations in samples before administration of } H_2O_2 \text{ or other agents}} \times 100$$

RESULTS

In the absence of any additions, DMTU and DMSO concentrations remained the same (no consumption). Addition of H_2O_2 caused consumption of DMTU but not DMSO while addition of H_2O_2 + Fe^{++}/EDTA caused consumption of both DMTU and DMSO (Table I).

TABLE I

COMPARISON OF DMTU OR DMSO CONSUMPTION FOLLOWING REACTION WITH H_2O_2 OR H_2O_2 AND FE^{++}/EDTA <u>IN VITRO</u>

Test conditions (Substances added to mixtures containing 14 mM DMTU or DMSO)	DMTU consumption (%)	DMSO consumption (%)
None	0*	0
H_2O_2 (30mM)	100 (3)†	13±3.7 (3)§
H_2O_2 (30mM) + Fe^{++}/EDTA (11mM/1mM)	88±3.0 (3)†	91±2.3 (3)†

* Mean ± standard error (number of determinations).

† Value significantly different ($p<0.05$) from value obtained following no additions (none).

§ Value not significantly different ($p>0.05$) from value obtained following no additions (none).

The suggested relative specificity of DMTU consumption for H_2O_2 and DMSO consumption for $\cdot OH$ was further supported when simultaneous addition of catalase (an H_2O_2 scavenger) but not mannitol, sodium benzoate or DMSO ($\cdot OH$ scavengers), decreased DMTU consumption while addition of mannitol, sodium benzoate or DMTU but not catalase inhibited DMSO consumption following addition of H_2O_2 and Fe^{++}/EDTA. In addition, elastase, oleic acid, leukotrienes, platelet activating factor or trypsin did not cause DMTU consumption in vitro.

DMTU consumption also occurs in isolated perfused rat lungs, isolated perfused rat kidneys or lung endothelial cell culture systems according to the added amounts of H_2O_2 (or glucose oxidase) and corresponding degrees of injury (4-6). Addition of catalase blocks both DMTU consumption and injury in these systems. DMTU consumption also did not occur when non-O_2 metabolite mediated injury occurred following addition of trypsin or elastase.

We have also found that reaction of H_2O_2 with DMTU yields a product which is characterizable by gas chromatography or high pressure liquid chromatography. The product peak is discrete, dependent on the added amount of H_2O_2, prevented by simultaneous addition of catalase, and not found when related compounds, such as thiourea, urea or dimethylthiourea, are mixed with H_2O_2.

DISCUSSION

The ability to measure O_2 metabolites in biological systems is the next step needed to address the importance of these toxins in health and disease. The present line of investigation is one approach to this timely issue. The most commonly used system for measuring O_2 metabolites is to assess protection following addition of superoxide dismutase (SOD) and/or catalase. However, this approach is limited, at least in part, by difficulties in getting these large molecules to the extravascular and/or intracellular sites of O_2 metabolite production and knowing that when SOD and catalase inhibit injury that they truly do so by scavenging O_2 metabolites and not by altering some other, non-O_2

metabolite dependent mechanism. Because DMTU (and DMSO) are small, highly permeable, non-toxic molecules, measuring their disappearance offers some advantages for assessing O_2 metabolites in biological systems. For example, increased amounts of DMTU (or DMSO) consumption could help confirm that increased amounts of H_2O_2 (or $\cdot OH$) are present. Correlations between the severities of injury and the amounts of DMTU consumption could also help quantitate and link toxicities to H_2O_2. In addition, DMTU (or DMSO) consumption could help determine if agents which prevent injury do so by inactivating H_2O_2 (or $\cdot OH$) or by altering some other non-H_2O_2 dependent mechanisms. For example, if an agent prevented injury but did not decrease DMTU consumption, a suspicion would be raised that the agent prevented injury by altering some non-H_2O_2 dependent mechanism. In contrast, if an agent both prevented injury and decreased DMTU consumption, the possibility that the agent was scavenging H_2O_2 and that injury was dependent on H_2O_2 would be strengthened.

A potential drawback in using DMTU (or DMSO) consumption is their relative lack of sensitivity in detecting the small concentrations of H_2O_2 (or $\cdot OH$) which might occur in physiologic systems. This limitation appears circumventable, however, because very small concentrations of DMTU (or DMSO) are measureable by using high pressure liquid chromatography (7). In addition, reaction of DMTU or DMSO with O_2 metabolites may form specific, stable, measurable products which will provide even more sensitive and/or specific markers of O_2 metabolites in biological systems. Regardless, the present approach of using DMTU (or DMSO) consumption (or refined versions of this principle) will improve determinations of the presence and significance of O_2 metabolties in biological systems.

REFERENCES

1. Repine JE, Fox RB, Berger EM (1981) J Biol Chem 256:7094-7096

2. Harada RN, Vatter AE, Repine JE (1983) Am Rev Respir Dis 128:761-762

3. Repine JE, Eaton JW, Anders MW, Hoidal JR, Fox RB (1979) J Clin Invest 64:1642-1651

4. Toth KM, Parker NB, Clifford DP, Berger EM, White JG, Repine JE (1984) Clin Res 32:88A

5. Linas S, Parker NB, Repine JE (1985) Clin Res 33:491A

6. Jackson JH, White CW, Schroeder WR, Tate RM, Clifford DP, Parker NB, Ryan JW, Repine JE (1984) Clin Res 32:528A

7. Kobayashi H, Matano O, Goto S (1981) J. Chromatogrpahy 207:281-185

Superoxide and Superoxide Dismutase in Chemistry, Biology and Medicine; G. Rotilio, editor

FLUORESCENCE PROPERTIES OF NATIVE, 4-HYDROXYNONENAL MODIFIED AND AUTOXIDIZED LOW DENSITY LIPOPROTEIN

ERNST KOLLER*, GÜNTHER JÜRGENS**, OSWALD QUEHENBERGER*, HERMANN ESTERBAUER*

*) Institute of Biochemistry, University of Graz, Schubertstraße 1, A-8010 Graz
**) Institute of Medical Biochemistry, University of Graz, Harrachgasse 21/III, A-8010 Graz

INTRODUCTION

Fluorescence measurements have been widely used as a parameter for in vitro and in vivo lipid peroxidation (1). It is assumed that malondialdehyde (MDA) generated during lipid peroxidation reacts with amino groups of phosphatidylethanolamines or proteins to fluorescent Schiff bases. Alteration of human low density lipoprotein (LDL) by reactive lipid peroxidation products such as MDA or 4-hydroxynonenal (HNE) might be involved in atherogenesis (2-4). In this present study the fluorescence properties of freshly prepared and autoxidized LDL are compared with those of HNE or MDA modified LDL. The novel technique of fluorescence topography allows to detect subtile fluorescence differences and similarities.

MATERIALS AND METHODS

Human LDL, HNE and MDA were prepared as described previously (4). Freshly prepared LDL was dialyzed against a 0.01 M phosphate buffer pH 7.4 containing 0.16 M NaCl, 1 mg/ml EDTA and 0.1 mg/ml chloramphenicol and incubated with HNE or MDA in the same buffer at 37°C for 4 hours to give the final concentrations as indicated in the figure legends. LDL autoxidation was achieved by a procedure similarly to that described by Schuh et al (5). EDTA was removed by dialysis of LDL against a buffer without EDTA. Autoxidation of LDL (1.5 mg/ml) was then performed by continued dialysis against buffer saturated with oxygen at room temperature for 22 hours. After autoxidation, the free HNE present in the dialysis buffer was determined by the HPLC method described previously (4). MDA was estimated by the thiobarbituric acid method. The apolipoprotein-B (apo-B) was isolated by extracting 3 ml of LDL solution (1.5 mg/ml) trice with 2 ml of $CHCl_3/CH_3OH$ (2:1, v/v). After washing with water the precipitated apo-B was dissolved in a 3% aqueous sodium dodecylsulfate solution. Fluorescence spectra were scanned on an Jobin Yvon spectrofluorimeter JY3 D and the fluorescence topograms were obtained on an Aminco SPF-500 spectrofluorimeter by the procedure described by Wolfbeis (6).

RESULTS AND DISCUSSION

Freshly prepared LDL shows the characteristic fluorescence of the tryptophane residues (EX 290/ EM 340 nm) and in addition a weak fluorescence in the range of 400-550 nm at an excitation wavelength of 360 nm. Modification of LDL (1.5 mg/ml)

with HNE (0.5 mM) enhances this intrinsic fluorescence significantly (Fig. 1).

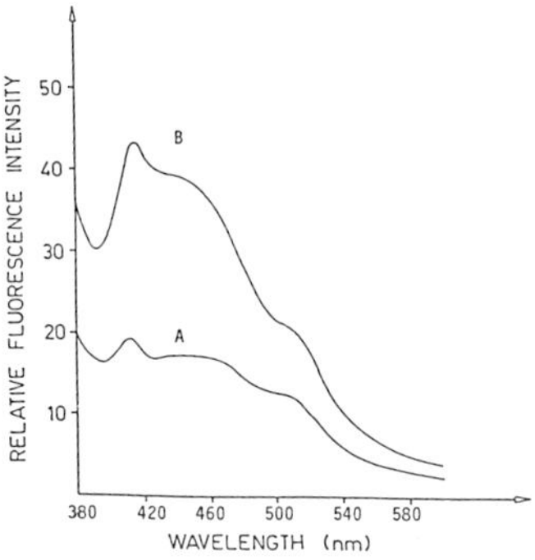

Fig. 1. Emission spectra (EX 360 nm) of (A) native and (B) HNE (0.5 mM) modified LDL (1.5 mg/ml) in 0.01 M phosphate buffer pH 7.4.

The maximum at 410 nm is due to the Raman scatter light of the solvent, whereas the broad shape of the remaining part of the spectrum is indicative for various fluorophors. More insight is given by fluorescence topography and plots of fluorescence intensity as a function of excitation and emission (= topogram). As shown in Fig. 2 HNE treatment alters the fluorescence mainly in the emission region of about 440 nm and the shape of the difference topogram indicates the presence of at least two fluorophors, i.e. EX 350/ EM 430 nm and EX 370/ EM 440 nm.

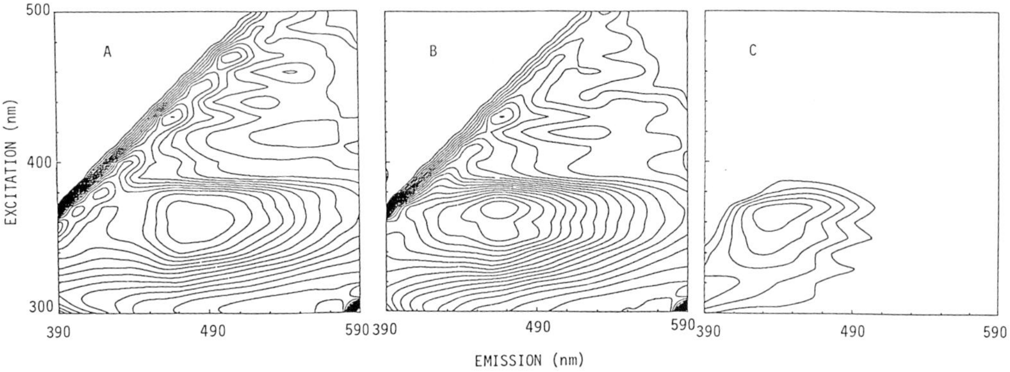

Fig. 2. Solvent corrected fluorescence topograms of (A) native, (B) HNE (1.0 mM) modified LDL (1.5 mg/ml) and (C) difference topogram of HNE modified and native LDL in 0.01 M phosphate buffer pH 7.4. Dilution of LDL samples 1:5; contour lines from 5 to 95% in 5% increments of the highest peak in (B).

Investigation of the protein and lipid part (data not shown) suggests that the major fluorescent component is the apo-B. The further studies were designed to evaluate whether fluorescence of autoxidized LDL is due to modification by HNE or

MDA. The spectra of Fig. 3 and the excitation and emission maxima (Table 1) suggest that HNE or a similar compound is responsible for the fluorescence of autoxidized LDL. This is further strengthened by the finding that free HNE is in fact formed during the course of LDL autoxidation.

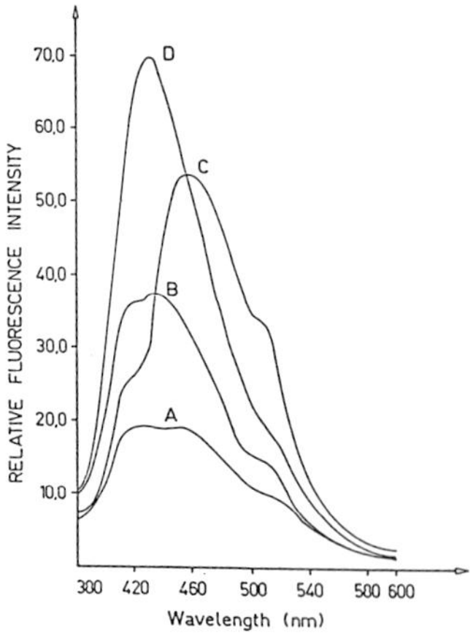

TABLE 1
EXCITATION AND EMISSION MAXIMA OF THE APO-B OF VARIOUS LDL SAMPLES

LDL sample	EX (nm)	EM (nm)
Native	350	465
HNE modified	355	425
MDA modified	400	470
Autoxidized	358	430

Fig. 3. Emission spectra (EX 360 nm) of the apo-B in 3% SDS solution of (A) native, (B) HNE (1.0 mM), (C) MDA (1.0 mM) modified and (D) autoxidized LDL (1.5 mg/ml).

In conclusion our results clearly show that (a) HNE can produce fluorescent protein and lipid chromophors, (b) the fluorescence of autoxidized LDL is not due to MDA but more likely to HNE modification. (c) Three-dimensional fluorescence spectroscopy is a powerful novel method to analyze complex fluorescent material formed by lipid peroxidation. The fluorescence topography makes possible to discriminate in biological samples between different fluorophors formed by secondary reaction of aldehydic lipid peroxidation products such as HNE or MDA.

ACKNOWLEDGEMENTS

The work was performed persuant to a contract with the National Foundation for Cancer Research, Bethesda USA.

REFERENCES

1. Tappel AL (1980) In: Pryor WA (ed) Free Radicals in Biology, Vol IV. Academic Press, New York, pp 2-47
2. Steinbrecher UP, Parthasarathy S, Leake DS, Witzum JL, Steinberg D (1984) Proc Natl Acad Sci USA 81: 3883-3887
3. Morel DW, Di Corleto PE, Chisolm GM (1984) Arteriosclerosis 4: 357-364
4. Jürgens G, Lang J, Esterbauer H (1985) Biochim Biophys Acta, in press
5. Schuh J, Fairclough GF, Haschemeyer RH (1978) Proc Natl Acad Sci USA 75: 3173-3177
6. Wolfbeis OS, Leiner M (1985) Anal Chim Acta 167: 203-215

Superoxide and Superoxide Dismutase in Chemistry,
Biology and Medicine; G. Rotilio, editor

THE DISTINCTION BETWEEN EXO- AND ENDOCELLULAR SPIN TRAPPING OF OXYGEN RADICALS.

A. SAMUNI* , A.J. CARMICHAEL, A. RUSSO, J.B. MITCHELL and P. RIESZ
(*) Molecular Biology, Hebrew University School of Medicine, Jerusalem, Israel and Radiation Oncology, National Institute of Health, Bethesda, MD. 20205, USA.

INTRODUCTION

Spin trapping and ESR detection of oxygen-derived radicals in various biological systems have been employed mainly in solutions (1). With this technique OH and O_2^- radicals undetectable by ESR under physiological conditions react with spin traps yielding long-lived radicals observable by ESR. Fewer attempts though were successful in employing this technique to whole cells (2-6). Initially the trapping of radicals within cells might appear inapplicable in view of the high concentrations of cellular constituents. In addition radical detection by spin trapping might be hampered because of: low cellular concentrations of the radicals and/or the spin traps; their compartmentation into different cellular spaces; site-specific formation of the radicals; and any decay of the spin traps and/or the spin adducts. As radicals are produced both inside and outside the cell a distinction between them is important in determining their role and location in cellular systems. We examined the feasibility of applying this technique to cells. The radicals were produced radiolytically and a discrimination between exo- and endocellular species was attempted using line broadening agents and high molecular weight scavengers which are practically excluded from the cells.

MATERIALS AND METHODS

Chemicals: The spin traps PBN, 4-POBN, PyBN and DMPO were purchased from Aldrich Chemical Co. Polyethylene glycol (PEG) from Sigma, oxalic acid, potassium oxalate, potassium dichromate from Fisher Scientific Co., Fair Lawn, N.J..

Radiation and ESR: For irradiations a ^{60}Co gamma source employing a dose rate of 51 $Gy \cdot min^{-1}$ was used. ESR spectra were monitored with a Varian E-9 X-band spectrometer (9.5 GHz) with a field modulation frequency of 100 KHz, using microwave power of 10mW and modulation amplitude of 0.4G.

Cells: Chinese hamster cells V79 line and human erythrocytes (RBC) were used. The V79 cells were harvested shortly before the experiment from long-phase monolayers, washed 4 times with and suspended in phosphate buffered saline (PBS).

RESULTS AND DISCUSSION

Effect of line broadening agents: In principle, a distinction between the ESR signals arising from endo- and exocellular species can be achieved with selective exocellular line broadening agents such as chromium oxalate (7). Since spin adducts of carbon-centered radicals are generally more stable than the adducts of oxygen-derived radicals we used the 4-POBN adduct of methyl radical, obtained upon irradiating DMSO. The six-lines signal (a_N= 15.9G ,a_H= 2.65G) characteristic of 4-POBN-CH_3 was observed but disappeared upon the addition of chromium. Later it reappeared when cells were subsequently added. This indicated the ease with which the spin adducts partition between the exo- and the endocellular environments. When RBC were irradiated in the presence of 0.1M DMPO the signal of DMPO-OH adduct was observed (Fig.1b). It was much smaller than that observed in the absence of cells (Fig.1a) and disappeared completely upon adding chromium (Fig.1c), indicating that it resulted from exocellular species. In fact chromium discriminates between exo- and endocellular free radicals according to their location rather than site of formation.

Effect of high molecular weight scavengers: Polyethylene glycols (PEG) rapidly react with OH and H radicals, primarily through H abstraction. These polymers, $H[OCH_2CH_2]_nOH$, PEG-3500 (n=68-84) are incapable of penetrating into the cell, and therefore, only scavenge radicals outside the cell. When PEG was irradiated in the presence of the spin trap PBN, the PBN-PEG signal consisted of triplet of doublets (a_N=15.85G and a_H=2.9G) was observed. Then the addition of chromium eliminated the signal even in the presence of cells, indicating that the PBN-PEG adducts were confined to the exocellular environment. Thus in the presence of DMPO and excess of PEG, DMPO is expected to trap OH radicals inside the cell but to react with the PEG secondary radicals outside the cell.

Following irradiation of 0.1M DMPO in the presence of 0.1M PEG a 9-lines signal of the DMPO-H adduct, resulting from H and e_{aq}, is observed (Fig. 1e) whereas the 1:2:2:1 quartet of the DMPO-OH is replaced by the 6-lines spectrum of DMPO-PEG spin adduct (a_N=15.75G, a_H=21.6G). The ESR spectra of DMPO-PEG (Fig.1e) and DMPO-OH (Fig. 1a) are easily distinguishable and offer a measure of the relative contributions of endo- and exocellular spin adducts to the overall ESR signal. When the RBC suspension was irradiated in the presence of both 0.1M DMPO and 0.1M PEG-3500 the ESR signal of DMPO-PEG was observed, but the DMPO-OH could not be detected (Fig. 1d).

Effect of cells on the lifetime of exocellular spin-probes: Apparently the cells facilitated the decay of the spin adducts as was found for other systems (8). We examined the effect of cells on the DMPO-OH spin adducts by adding V79 or RBC to the radiolytically formed adducts. The relatively stable DMPO-OH decayed immediately upon the addition of the cells. The decay rate increased with cell concentration, could not be blocked by cyanide, azide or by heating the cells to to 60C for 30min. The decay of the signal could not be reversed by ferricyanide.

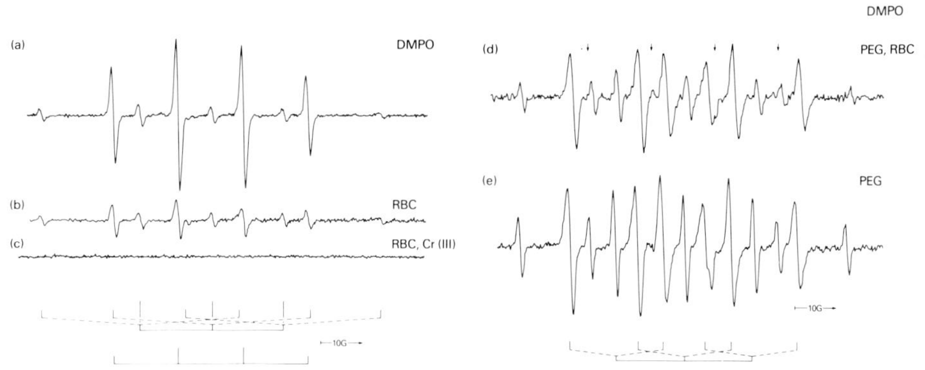

Fig. 1. ESR spectra obtained for 0.1M DMPO in PBS following gamma irradiation in the presence of: (a) control, (b) $4.1 \cdot 10^9$ RBC/ml, (c) like b but with 30mM chromium oxalate added later, (d) 0.1M PEG-3500 and $5.1 \cdot 10^9$ RBC/ml, (e) 0.1M PEG-3500. The arrows designate the locations of the 1:2:2:1 quartet due to DMPO-OH.

In conclusion: Evidently, unlike with carbon-centered radicals, the main difficulty in spin trapping oxygen-centered radicals in cells is due to the rapid decay of the resulting adducts rather than their ineffective trapping. This implies that failure to detect the spin adduct does not necessarily exclude the participation of OH or O_2^- radicals. In addition, a careful and critical review of spin trapping of OH and superoxide radicals by spin trapping inside cells would be desired.

REFERENCES

1. Janzen EG (1980) in: Pryor WA (ed) Free Radicals in Biology. Academic press, N.Y. Vol IV pp 115-154
2. Green MR and Hill HAO (1978) FEBS Lett. 100:23-26
3. Thornalley PJ, Stern A. and Bannister JV (1983) Biochem. Pharmacol. 32:3571-3575
4. Hume DA, Gordon S. Thornalley PJ and Bannister JV (1983) Biochim. Biophys. Acta 763:245-250
5. Rosen JM and Freeman BA Proc. Natl. Acad. Sci. USA (in press)
6. Kleinhans FW, Barefoot ST and Hawley DA (1984) Biophys. J. 45: 242a
7. Berg SP and Nesbitt DM (1979) Biochim. Biophys. Acta 548:608-615
8. Goldberg JS, Rauckman EJ and Rosen GM (1977) Biochem. Biophys. Res. Commun. 79:198-202

Superoxide and Superoxide Dismutase in Chemistry,
Biology and Medicine; G. Rotilio, editor

SPIN-TRAPPING STUDY ON THE FORMATION OF OXY-RADICALS DURING ENZYMIC NADPH-DEPENDENT LIPID PEROXIDATION IN RAT LIVER MICROSOMES

V.E.KAGAN, V.M.SAVOV, E.A.SERBINOVA, A.N.OSIPOV*, O.A.AZIZOVA*
Institute of Physiology, Bulgarian Academy of Sciences, Sofia 1113, Bulgaria
*N.I.Pirogov Medical Institute, Moscow, USSR

INTRODUCTION

It is suggested that various radical intermediates: products of one-electron reduction of oxygen (1), singlet oxygen (2), ferryl- and perferryl-ions (3) play an essential role in the mechanisms of initiation of the enzymic NADPH-dependent lipid peroxidation (LPO) in rat liver microsomes. The discrepancy of the available results might be explained by the indirect methods for studying the radicals. The present investigation was undertaken to study the radical intermediates of LPO in rat liver microsomal membranes with different cytochrome P-450 content and different NADPH-cytochrome P-450 reductase activity using the spin-trapping technique.

MATERIALS AND METHODS

The microsomal fraction was isolated by a routine method through differential centrifugation (105 000 g) of the liver homogenate (in 1.15% KCl). The cytochrome P-450-enriched fraction was obtained through incubation of the native microsomes in 0.05M phosphate buffer pH 7.4 with Triton X-100 (0.05% per mg protein) for 20 min at 25°C and subsequent recentrifugation at 105 000 g for 90 min at 4°C. The NADPH-cytochrome P-450 reductase-enriched fraction was obtained according to (4). The reductase activity was measured after (4). The accumulation of LPO products was determined by the malonyl dialdehyde (MDA) content (5). C-phenyl-N-tert-butyl nitrone (PBN), (kindly granted from Dr. V.Zubarev, Moscow State University) was used as a spin-trap. The ESR spectra were recorded on a Varian E-4 spectrometer (USA).

RESULTS AND DISCUSSION

ESR signals of spin-adducts (SA) of neither of superoxide nor of hydroxyl radicals were recorded during 20-min incubation of PBN (20 mM) in the LPO-induction system. This was probably due to the low rate of generation of radicals and to the instability of the SA formed (6). Thus we used an intermediate acceptor of hydroxyl radicals, ethanol (0.01M) which effectively interacted with hydroxyl radicals, forming stable SA of PBN with hydroxy-ethyl radicals (6). In the presence of ethanol we recorded SA of hydroxy-ethyl radicals with parameters of ESR spectra: a_N=16.26 G, a^H=3.2 G, H_{pp}=1.0 G (Fig.1)

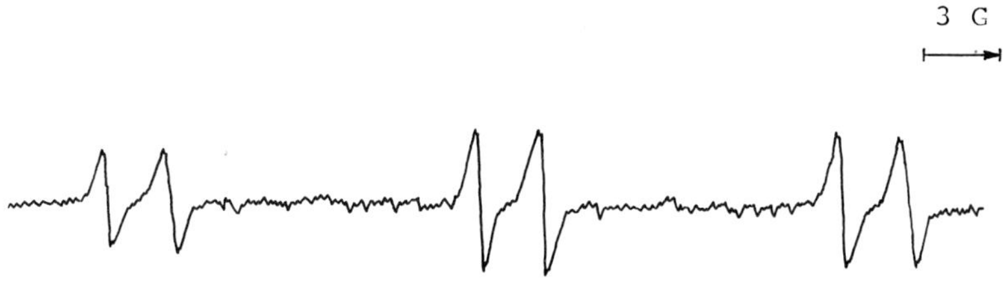

Fig.1. ESR spectrum of SA of hydroxy-ethyl radicals recorded in a system containing microsomes (2.5mg of protein in ml), NADPH (0.5mM), Fe^{2+} (50 μM), PBN (20mM) in 0.05M phosphate buffer, pH 7.4

Thermal denaturation of microsomes sharply decreased the rate of SA formation. In the absence of microsomes in the incubation medium the rate of SA formation was almost zero. When superoxide dismutase (SOD, 10 U/ml) was added to the incubation medium the magnitude of ESR signal slightly decreased and catalase (30 U/ml) strongly inhibited the formation of SA of hydroxy-ethyl radicals (Table 1). When both SOD and catalase were present in the incubation medium the signal recorded did not differ the background signal.

When PBN was added to the incubation medium the NADPH-dependent LPO was inhibited (Table 1). It might be suggested that the inhibitory effect of PBN was due to its interaction with radicals initiating NADPH-dependent LPO. A similar inhibitory effect of PBN on Fe^{2+} + ascorbate-dependent LPO was established earlier (6). The action of catalase and SOD on the NADPH-dependent MDA accumulation resembled the effect of these enzymes on the formation of SA of hydroxy-ethyl radicals (Table 1).

Data about the formation of SA and MDA in microsomal fractions with different content of cytochrome P-450 and different activity of NADPH-cytochrome P-450 reductase show: in fraction with increased reductase activity (180% of the control value) and devoid of cytochrome P-450 the rate of SA formation is 2.5 times higher than that in controls; and vice versa, in fraction with decreased reductase activity (15% of the control value) the formation of SA is strongly inhibited (to the 26% of the control value).

It might be concluded that the formation of radicals, participating in the initiation of NADPH-dependent LPO in microsomes, is effected by NADPH-cytochrome P-450 reductase. This conclusion is substantiated by data showing that the rate of accumulation of MDA in microsomal fractions with different reductase activities depends linearly on reductase activity (data not shown).

TABLE 1

EFFECTS OF SOD AND CATALASE ON THE FORMATION OF SA OF HYDROXYETHYL RADICALS AND MDA ACCUMULATION IN MICROSOMES

Additions to the LPO-induction system*	Magnitude of SA (%)	Rate of NADPH-dependent LPO (nmol MDA/mg protein/min)
No	0	2.22 ± 0.16
PBN (20 mM)	100	0.36 ± 0.08
SOD (10 U/ml)	-	1.72 ± 0.13
SOD (10 U/ml) + PBN (20mM)	85	-
Catalase (30 U/ml)	-	0.94 ± 0.10
Catalase (30 U/ml) + PBN (20mM)	47	-
SOD (10U/ml) + catalase (30U/ml)	-	0.34 ± 0.09
SOD (10U/ml) + catalase (30U/ml) + PBN 20		-

*Incubation medium contained microsomes (2.5 mg protein/ml), NADPH (0.5 mM), Fe^{2+} (50 μM), in 0.05M phosphate buffer, pH 7.4

REFERENCES

1. Blum J, Fridovich I (1983) Arch Biochem Biophys 222:1-35
2. Piette J, Chunn Y (1977) FEBS Lett 74:2-231
3. Tien M, Svingen B, Aust S (1981) Fed Proceeding 40:2-179
4. Eshourbagu N, Guselian P (1980) J Biol Chem 225:4-1279
5. Kohn H, Liversedge M (1944) J Pharmacol 82:2-193
6. Osipov A, Savov V (1981) Biofisika 26:2-193 (Russ)

© 1986 Elsevier Science Publishers B.V. (Biomedical Division)
Superoxide and Superoxide Dismutase in Chemistry,
Biology and Medicine; G. Rotilio, editor

INVESTIGATIONS ON THE IDENTITY OF THE AUTOXIDIZABLE COMPONENT OF THE MITOCHONDRIAL RESPIRATORY CHAIN

HANS NOHL[1] and WERNER JORDAN[2]

[1] Institut für Pharmakologie und Toxikologie der Veterinärmedizinischen Universität Wien, A-1030 Wien, Linke Bahngasse 11
[2] Institut für Pharmakologie, Toxikologie und Pharmazie der Tiermedizinischen Fakultät der Universität München, D-8000 München 22, Veterinärstraße 13

INTRODUCTION

Ubiquinone (Q) and cytochrome b-566 have both been postulated to cause mitochondrial $O_2^{\cdot -}$ formation by autoxidation of their reduced forms (1-3). Experimental elucidation of the autoxidizable component is impeded by the fact that all metabolic conditions so far reported to be required for an $O_2^{\cdot -}$ generation have in common a high reduction state of both candidates. Thus, it is difficult to ascertain whether a reduced Q-species or cytochome b or both together account for superoxide generation in respiring mitochondria.

The present investigation provides experimental approaches to this question by correlating separately reduced Q and cytochromes b to a one electron reduction step of molecular oxygen.

RESULTS AND DISCUSSION

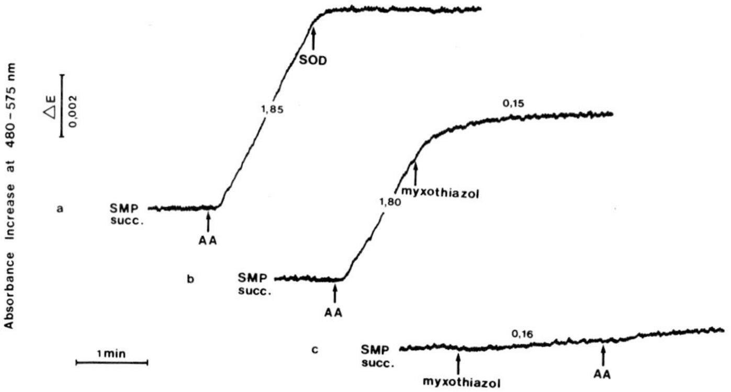

Fig. 1

Generation of superoxide radicals was followed by SOD-sensitive adrenochrome formation from epinephrine (1mmol/l). Experimental conditions: 0.9 mg of SMP from rat heart mitochondria were suspended in 0.15 mol/l KCl, 0.5 mmol/l phosphate buffer (pH 7.4) saturated with air oxygen at room temperature. Final concentrations: succinate (succ): 6 mmol/l; AA: 1 µg/ml; Myx: 0.5 µg/ml. Numbers associated with the traces represent nmol of $O_2^{\cdot -}$ /min /mg.

"Classical conditions" causing $O_2^{\cdot -}$ release from succinate respiring mitochondria in the presence of antimycin A (AA) (Fig. 1a) is also known to keep Q and b-type cytochromes in a high state of reduction. This is due to an inhibition of recycling electrons of the Q-cycle at the oxidant side of cytochrome b-562. Addition of myxothiazol (myx), a novel inhibitor which prevents electron transfer from ubiquinol to the Rieske iron sulfur center (4,5) is also capable of slowing down $O_2^{\cdot -}$ formation after a short lag phase (Fig. 1b). Inhibition of an univalent electron shuttle to oxygen in the presence of AA and myx is paralleled by a nearly complete inhibition of cytochrome b reduction (experiments not shown), thus suggesting a role for b-type cytochromes in the formation of $O_2^{\cdot -}$. Myx was not able to replace AA with respect to the initiation of $O_2^{\cdot -}$

formation. Addition of myx after succinate caused a small increase in the reduction of b type cytochromes being established more slowly and less extensively as compared to when AA instead of myx was added (exp. not shown). According to Jagow et al. (5) the analysis of this effect suggested that the extra-reduction induced by myx was due to a shift of the low to a higher potential form of cytochrome b-566. While the species causing an additional reduction of cytochrome b in the presence of myx, is unlikely to operate as a reductant of O_2, the negative redox potential of the low potential cytochrome b-566 ($E_0' = -30$ mV, ref. (6)) provides thermodynamic conditions for an univalent electron transfer to O_2. However, Myx does not only influence reduction of b-type cytochromes but was also suggested to prevent ubisemiquinone (SQ$^•$) formation in center o due to an inhibition of one electron transfer mechanism from ubiquinol to FeS-Rieske (5,7). Although a direct experimental evidence of the effect of Myx on SQ$^•$-o and its reactivity with oxygen remains to be demonstrated, the SQ$^•$-o pool was suggested to cause $O_2^{•-}$ formation in mitochondria by autoxidation (7,8). Our investigations of the influence of Myx and Myx plus AA on

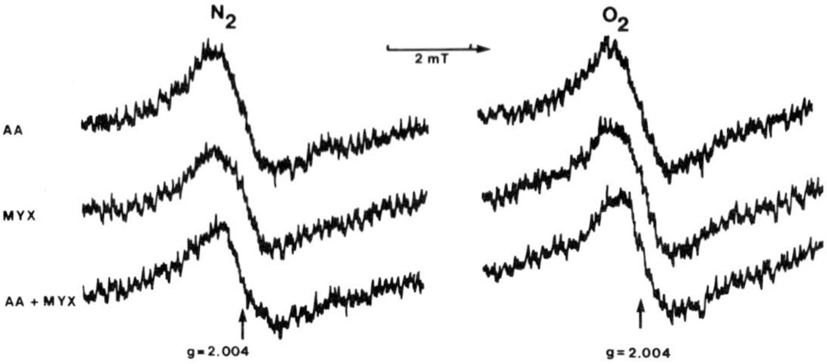

Fig. 2

Effect of AA, Myx and O_2 on ESR spectra arising from SQ$^•$-s in rat heart mitochondria at room temperature. 15 mg of mitochondrial protein was suspended in 340 µl of the reaction medium of Fig. 1 which was equilibrated with oxygen-free nitrogen or pure oxygen. AA: 1.75 µg, Myx: 0.35 µg; Microwave power: 12.5 mW; modulation: 0.05 mT; linewidth: 0.13 mT.

the existence of SQ$^•$ in succinate respiring mitochondria is demonstrated in Fig. 2. SQ$^•$ related ESR spectra were obtained despite the presence of the above inhibitors.The species detected exhibited characteristics of a SQ$^•$ associated with complex II (SQ$^•$-s). Since steady state of SQ$^•$-s was poised with a low succinate/fumarate ratio (1/100) causing very limited electron flow from succinate to Q, equilibrium should be sensitive to a possible reaction of SQ$^•$ with O_2. However, as compared by ESR spectra under N_2 and O_2, oxygen was not critical and thus, autoxidation of the SQ$^•$ pool can be excluded both in the presence and absence of the inhibitors used. The validity of these findings for SQ$^•$ pools in mitochondria besides SQ$^•$-s is supported by the idea that all reductant and oxidant centered SQ$^•$ pools of the Q-cycle are in equilibrium (9,10).

Since Q is bound in the mitochondrial membrane in such a way that the hydrophilic quinone residue is in a hydrophobic environment (11) the redox behaviour of Q might be expected to be similar to its behaviour in aprotic solvents. Fig. 3a demonstrates SQ$^•$ related ESR spectra obtained in a reaction system consisting of acetonitrile and a $O_2^{•-}$ generating system (KO_2 in crown ether). The ubisemiquinone radical remained stable for several hours, despite the presence of oxygen, while protonation of the aprotic reaction system immediatly caused the nearly complete disapperance of SQ$^•$ related signals (Fig. 3b) associated with the simultaneous existence of $O_2^{•-}$ (Fig. 2c).

Assuming that thermodynamic conditions for a reaction of redox cycling ubiquinone with oxygen is similar to that observed in aprotic solvents, autoxida-

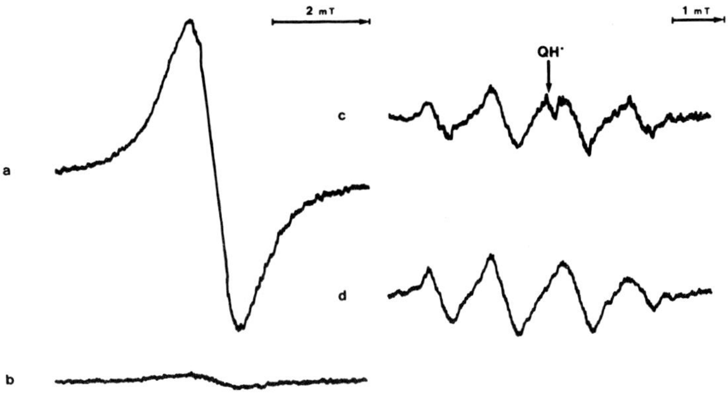

Fig. 3

ESR spectra of SQ˙ which were generated in a reaction system equilibrated with oxygen in the absence (a) and presence (b) of protons. The assignment of spectrum (c) to a DMPO-superoxide adduct (indicative of the existence of free $O_2^{\cdot-}$ radicals) was ensured by comparing this spectrum with spectrum (d) obtained in the same acetonitrile/water mixture as used in (c) by $O_2^{\cdot-}$ adduct formation with the spin trap DMPO from decomposing KO_2. The arrow in (c) indicates a superimposed single line spectrum arising from remaining SQ˙. Ubiquinone-0: 6 mmol/l; 5 µl of water was mixed to 300 µl of the reaction medium of (b); DMPO: 150 mmol/l. Microwave power: 12 mW; modulation: 0.1 mT (a,b), 0,04 mT (c,d).

tion of mitochondrial SQ˙ is not likely to occur due to the absence of protons in the phopholipid bilayer. Thus, $O_2^{\cdot-}$ formation in mitochondria appears not to result from autoxidizing SQ˙, as long as protons for charge compensation are hindered from penetrating hydrophobic domains where Q is operating.

In conclusion, our investigations exclude SQ˙-s as a site for an electron shuttle from the respiratory chain to O_2. The characterization of SQ˙-o as an alternative or as the major $O_2^{\cdot-}$ source in mitochondria remains speculative as long as experimental evidence is lacking which demonstrates that this SQ˙ species operates in a protic domaine which permits autoxidation, and that steady state concentrations of this $O_2^{\cdot-}$ source are not in equilibrium with the SQ˙-s pool of the Q-cycle. The involvement of low potential cytochrome b-566 in mitochondrial $O_2^{\cdot-}$ formation has to be considered.

REFERENCES

1. Cadenas E, Boveris A, Ragan C, Stoppani AOM (1977) Arch Biochem Biophys 180: 248-257
2. Loschen G, Azzi A, Flohe L (1973) FEBS Lett 33: 84-88
3. Nohl H, Hegner D (1978) Eur J Biochem 82: 563-567
4. Jagow vG, Engel WD (1981) FEBS Lett 136: 19-24
5. Jagow vG, Ljungdahl PO, Graf P, Ohnishi T, Trumpower BL (1984) J Biol Chem 259: 6318-6326
6. Wikström MKF (1973) Biochim Biophys A 301: 151-193
7. Ksenzenko M, Konstantinov AA, Khomutov GB, Tikhonov AN, Ruuge EK (1983) FEBS Lett 155: 19-24
8. Turrens JF, Alexandre A, Lehninger AL (1985) Arch Biochem Biophys 237: 408-414
9. Berry EA, Trumpower BL (1985) In: Lenaz G (ed) Coenzyme Q. J Wiley & Sons NY pp 365-389
10. Nohl H, Jordan W, Youngman RJ (1985) Adv Free Radical Biol Med (submitted for publication)
11. Chance B (1972) In: Azzone GF, Carafoli E, Lehninger AL, Quagliariello E, Siliprandi N (eds) Biochemistry and Biophysics of Mitochondrial Membranes. Academic Press NY pp 85-99

Superoxide and Superoxide Dismutase in Chemistry, Biology and Medicine; G. Rotilio, editor

CHARACTERIZATION OF CYTOCHROME $\underline{b}_{558}$, THE PROTEIN INVOLVED IN THE PRODUCTION OF SUPEROXIDE BY NEUTROPHILIC GRANULOCYTES

RENÉ LUTTER,[1] MARGRIET L.J. VAN SCHAIK,[2] ROB VAN ZWIETEN,[1] RON WEVER,[3] SIMON P.J. ALBRACHT,[3] LOUIS J. MEERHOF,[1] MIC N. HAMERS[1] and DIRK ROOS[1]
[1]Central Laboratory of the Netherlands Red Cross Blood Transfusion Service and Laboratory of Experimental & Clinical Immunology of the University of Amsterdam; [2]Pediatric Clinic, Academic Medical Center, Amsterdam; [3]Laboratory of Biochemistry, B.C.P. Jansen Institute, University of Amsterdam, The Netherlands

INTRODUCTION

During phagocytosis, neutrophilic granulocytes generate large amounts of reduced oxygen species, e.g. superoxide. It has been suggested that the enzyme system responsible for the reduction of oxygen consists of several distinct redox components, viz. a flavoprotein (1), a $\underline{b}$ cytochrome (2) and ubiquinone-50 (3). The actual existence of such a short electron transport chain and the role of each of its components in the reduction of oxygen to superoxide is still far from clear, however.

At present, the best evidence for a role in the "oxidase system" concerns the $\underline{b}$ cytochrome (4,5), tentatively named cytochrome $\underline{b}_{558}$. Therefore, this protein has been purified from human neutrophils (6,7). The biochemical characteristics of the purified cytochrome $\underline{b}$ preparations from the various groups differ in a number of respects. Our final preparation consists of one -hydrophilic- polypeptide with a minimal molecular mass of 127 kDa and one protoheme IX per polypeptide.

The binding of carbon monoxide (CO) to the heme is weak; therefore, there is no firm evidence for a direct interaction between the heme and oxygen (7,8). On the other hand, a purified cytochrome $\underline{b}_{558}$ preparation (7), previously reduced with dithionite, is rapidly reoxidized upon gassing with oxygen. To examine the presence of other intrinsic redox groups in cytochrome $\underline{b}_{558}$ we have used the technique of electron paramagnetic resonance (EPR). The results are discussed in view of the mechanism of superoxide formation by neutrophils.

RESULTS AND DISCUSSION

The EPR spectrum of cytochrome $\underline{b}_{558}$

EPR spectroscopy of a purified cytochrome $\underline{b}_{558}$ preparation revealed a typical low-spin heme signal with a g-value of 3.31 (not shown) and a complex signal around g = 2.0 (Fig. 1). Part of the complex signal (marked by a dotted line in

Fig. 1) is resistant to reduction with dithionite and has two peaks with apparent g-values of 2.109 and 2.032. All other peaks disappeared upon reduction. The relative absorbances in each set of peaks (i.e. the dithionite-resistant and the -sensitive signal) behaved similarly at variable microwave power and temperature, indicating that each signal is derived from only one paramagnetic component.

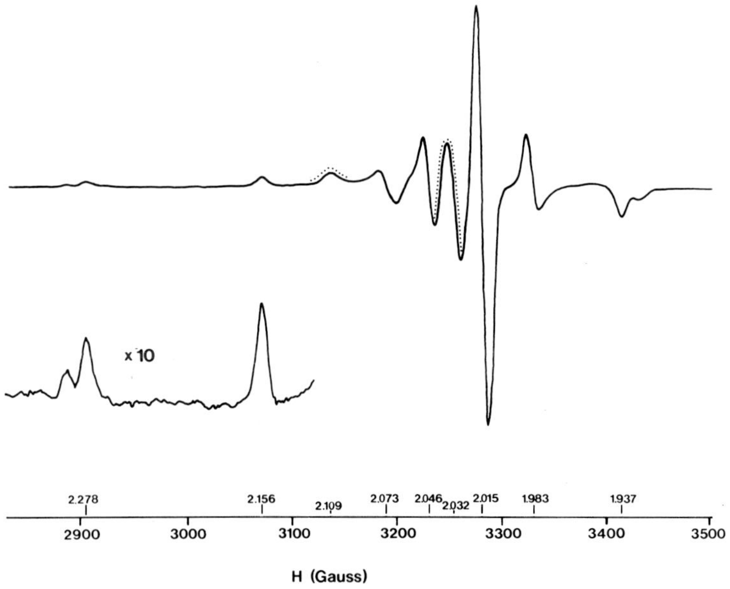

Fig. 1. EPR spectrum of purified cytochrome $\underline{b}_{558}$ (30 μM; heme to protein ratio, 0.94:1) in 100 mM potassium phosphate pH 7.2. Conditions of EPR spectroscopy; frequency: 9.258 GHz, microwave power: 2 mW; modulation amplitude: 10 G; scanning rate: 125 G. min^{-1}; time constant: 1.0 s; temperature: 18° K.

Characterization of the EPR signals

The EPR characteristics of the dithionite-insensitive signal are indicative for a Fe-S cluster (see Fig. 2). Indeed, the presence of acid-labile sulfur was confirmed with the methylene blue method (9). The determination of Fe is in progress.

The dithionite-sensitive signal is attributed to copper. The signal could be generated by computer simulation, assuming a hyperfine splitting $|A_{xy}|$ of 43 Gauss. The doublet signal, shown in the 10x magnified spectrum (bottom left, Fig. 1), is due to the natural occurring isotopes ^{63}Cu and ^{65}Cu. The presence of copper was confirmed by atomic absorption. The stoichiometry of the redox components is shown in Table 1.

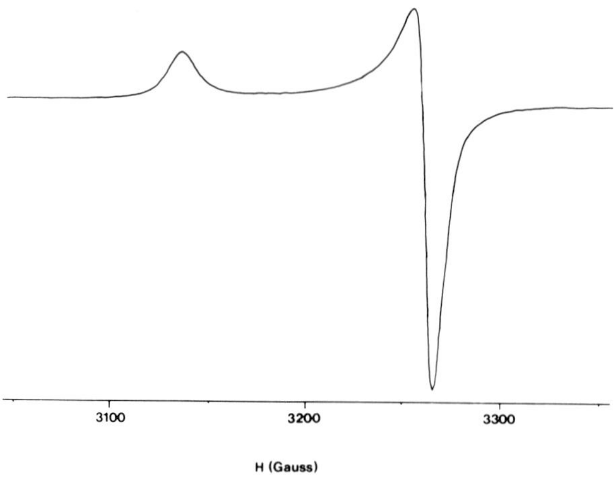

Fig. 2. The EPR spectrum of dithionite-reduced cytochrome $\underline{b}_{558}$ showing the signal of the iron-sulfur cluster. Conditions of EPR spectroscopy are similar to those in Fig. 1.

TABLE I. STOICHIOMETRY OF HEME, COPPER AND IRON-SULFUR IN CYTOCHROME $\underline{b}_{558}$

Heme	Copper	Iron-sulfur cluster
12.8 μM[1]	n.d.[3]	8.3 μM[5]
14.0 μM[2]	32 μM[4]	30 μM[6] (acid-labile sulfur)

[1]From the redox spectrum (7). [2]From the alkaline pyridine hemochrome spectrum (7). [3]In this particular preparation we did not determine the amount of copper by integration. In other preparations the amount of copper determined by integration and by atomic absorption are similar. [4]By atomic absorption. [5]By integration of the EPR signal; possibly not all iron-sulfur is detected by EPR due to a partial reduced state of the iron-sulfur cluster. [6]In the acid-labile sulfur assay (9); in case of 2Fe-2S this is equal to 15 μM iron-sulfur cluster.

Is copper the site of reduction of oxygen?

After abolishing the copper signal with dithionite, subsequent gassing with oxygen results in the reappearance of 60% of the copper signal. If oxygen binds directly to the copper, binding of nitrogen oxide (NO) to copper might be expected, which is anticipated to result in a change of the EPR spectrum. However, anaerobic incubation with about 1 mM NO did not have an effect on the copper signal (not shown).

In conclusion, cytochrome $\underline{b}_{558}$ contains a heme, a copper and an iron-sulfur cluster (2Fe-2S) in an apparent stoichiometry of 1:2:1, respectively. There are as yet no data supporting a redox function for the iron-sulfur cluster. Both the heme and copper react in an electron pathway with oxygen but neither of these components is likely to be the oxygen acceptor. Preliminary data indicate that the copper in cytochrome $\underline{b}_{558}$ is closely associated with a distinct prosthetic group. Perhaps this additional group interferes with the binding of NO to the copper, or perhaps this group functions in the binding of oxygen.

In our opinion, cytochrome $\underline{b}_{558}$ should not be regarded as a simple electron carrier protein but as a complex redox protein, comparable to xanthine oxidase. Cytochrome $\underline{b}_{558}$ contains sufficient redox centers to mediate the reduction of oxygen to superoxide. Obviously, this does not rule out the involvement of other redox components. Nevertheless, the possibility exists that cytochrome $\underline{b}$ is the oxidase itself.

REFERENCES

1. Babior BM, Kipnes RS (1977) Blood 50:517-524
2. Segal AW, Jones OTG (1978) Nature (Lond) 276:515-517
3. Sloan EP (1981) J Exp Med 153:1316-1328
4. Segal AW, et al. (1983) N Engl J Med 308:245-251
5. Hamers MN, et al. (1984) Nature (Lond) 307:553-555
6. Harper AM, et al. (1984) Biochem J. 219:519-527
7. Lutter R, et al. (1985) J Biol Chem 260:2237-2244
8. Cross AR, et al. (1982) Biochem J 204:479-485
9. Suhara K, et al. (1975) Anal Biochem 68:632-636

ENZYME STRUCTURE AND MECHANISMS

Superoxide and Superoxide Dismutase in Chemistry,
Biology and Medicine; G. Rotilio, editor

ANATOMY OF AN ENZYME: COMPUTER GRAPHICS VIEWS OF CU,ZN SUPEROXIDE DISMUTASE

ELIZABETH D. GETZOFF, ARTHUR J. OLSON, AND JOHN A. TAINER

Molecular Biology Dept., Research Institute of Scripps Clinic, La Jolla, CA 92037 USA

INTRODUCTION

A high resolution x-ray crystal structure gives detailed information about the positions of all the atoms in space. However, it is extremely difficult to understand and analyze a protein structure from its atomic coordinates. Instead models are needed to express the numerical results, and the type of model required depends upon the nature of the question being asked. In practice, interpretation and understanding of three dimensional models can be greatly aided by animation. Here we present a sketchbook of computer graphic models of bovine erythrocyte Cu,Zn superoxide dismutase (SOD) illustrating some of the principle aspects of the enzyme's anatomy as related to its stability and function. This sketchbook represents excerpts from an animated 16mm movie on the structure of SOD (1). Using the computer graphics language GRAMPS (2) and the molecular modeling program GRANNY (3), both molecular and arbitrary objects could be displayed simultaneously to better examine structural features. All of these computer graphic models are based upon the 3-dimensional structure of bovine erythrocyte SOD determined by X-ray crystallography in the laboratory of Profs. Dave and Jane Richardson of Duke University (4-6, and ref. therein).

RESULTS AND DISCUSSION

The enzyme exists and functions as a dimer. Each subunit of the dimer has a molecular weight of about 16,000 and comprises 151 amino acid residues and 2 metal ions (about 1100 non-hydrogen atoms or 2200 atoms). Only 14 residues are completely buried to solvent. Fig. 1 shows the solvent accessible surface developed by Connolly (8), which identifies the areas of the enzyme with which a water molecule or superoxide radical could come in contact.

Each subunit has a molecular volume of 20,600 $Å^3$. The dimer has about 12,390 $Å^2$ of exposed molecular surface with some 1038 $Å^2$ of surface on each subunit being buried by the dimer contact. The two metal ions are located in the bottom of a channel formed by the topography of the molecular surface. The Zn ion is not accessible to solvent, while the Cu ion has approximately 5 $Å^2$ of exposed surface area. The overall rough dimensions of the dimer, illustrated in Fig. 2, are 33 Å by 36 Å by 67 Å. Put in perspective, these dimensions represent about 1/6th the volume of an immunoglobulin molecule, 1/10,000th of the diameter of a bacteria and 1/100,000th diameter of a red blood cell. The concentration of SOD in red blood cells is estimated to be about 10^{-5} molar or about 1,000 copies per cell.

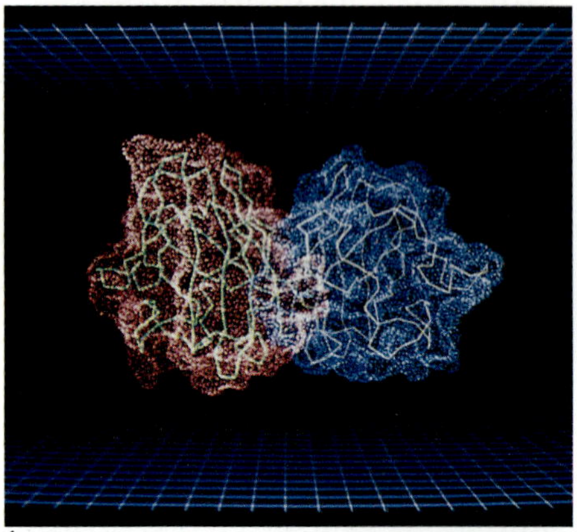

1

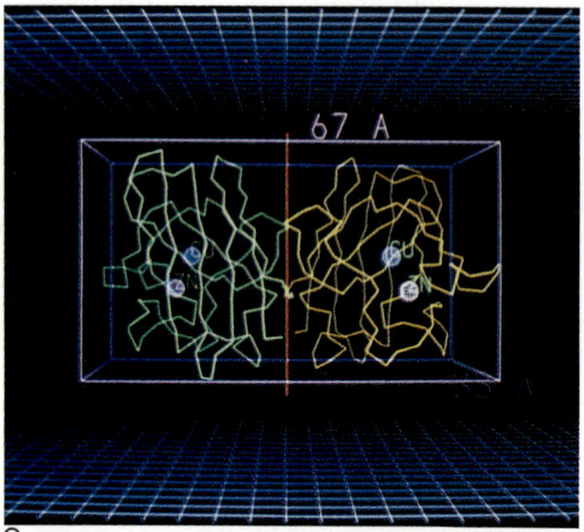

2

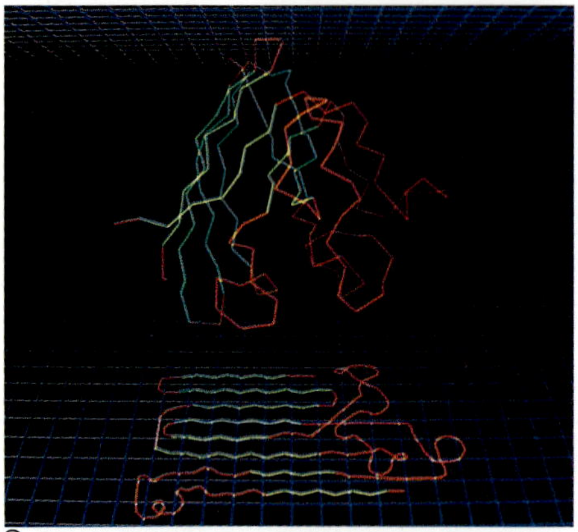

3

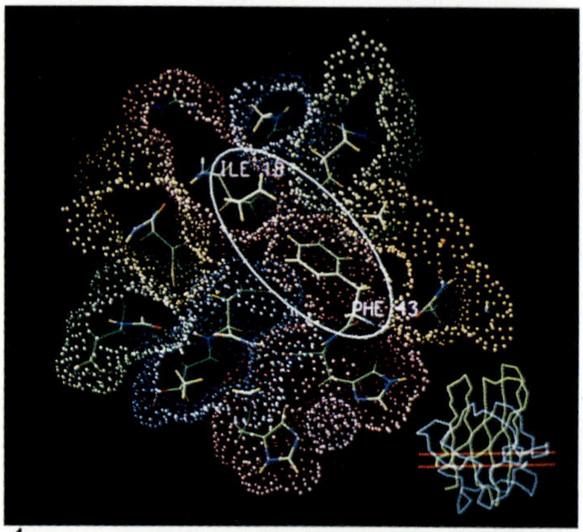

4

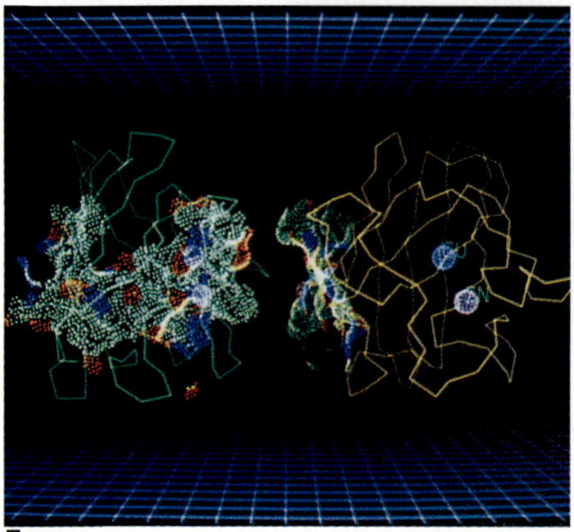

5

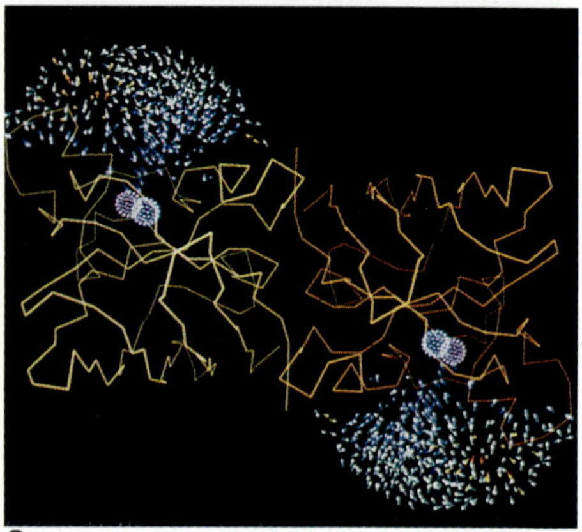

6

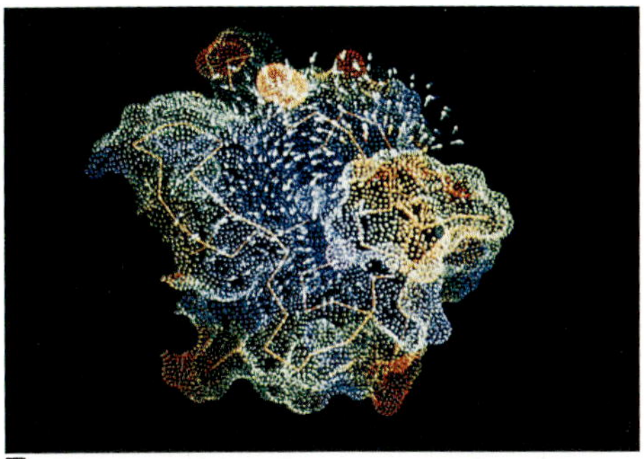

7

8

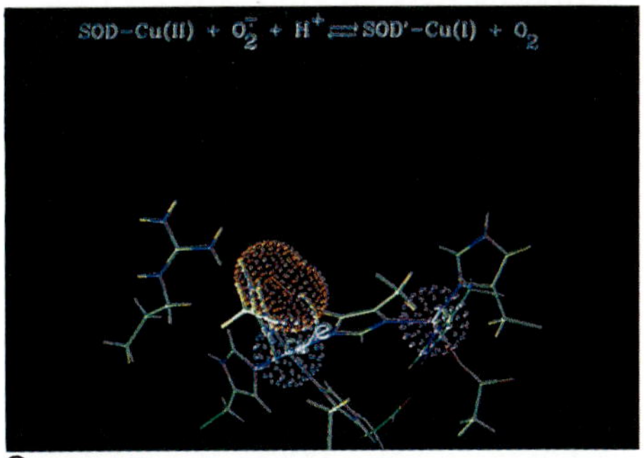

9

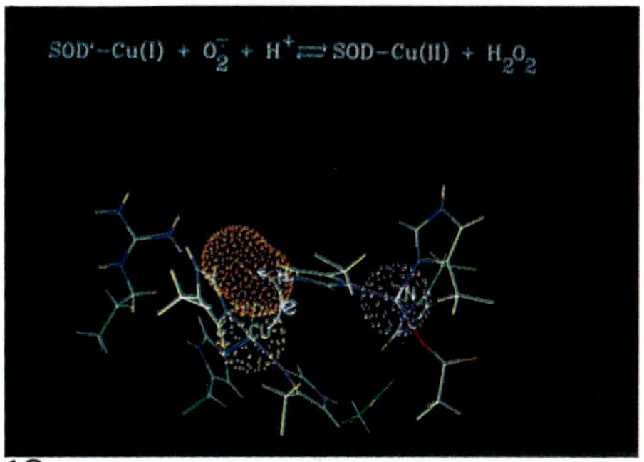

10

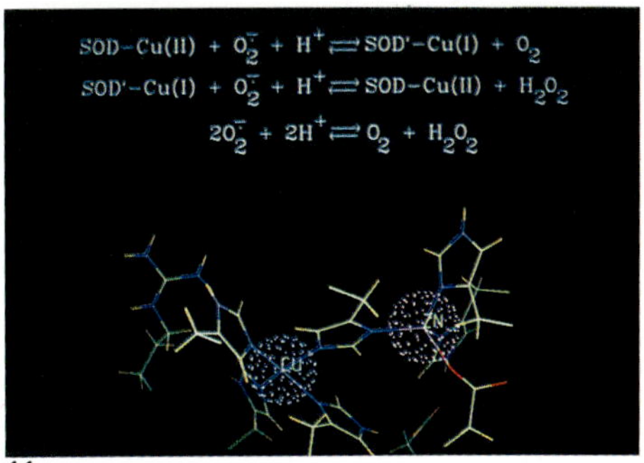

11

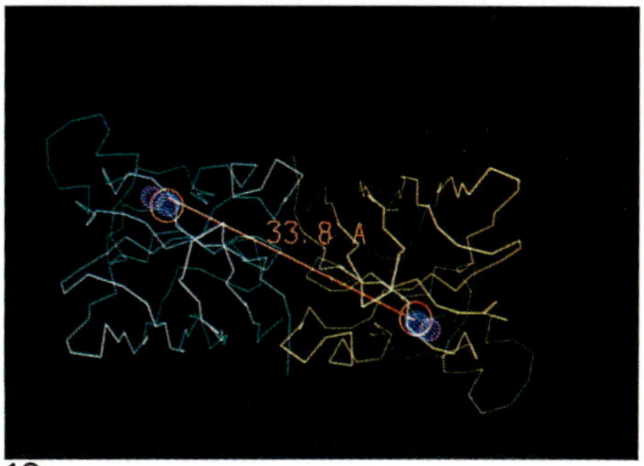

12

Fig. 1. Solvent accessible molecular surface of the dimer. The transparency of the surface (dots) allows the alpha carbon backbone to be seen.

Fig. 2. Molecular dimensions of the dimer. Spheres represent metal ions: copper (blue) and zinc (purple). The red line indicates the 2-fold axis of symmetry relating the 2 subunits of the dimer.

Fig. 3. Topology of beta-strands and loops. Beta-strands are shown in green and loops in red on the 3-dimensional model (above) and a schematic diagram (below).

Fig. 4. Close packing within the beta-barrel. An ellipse highlights the cross barrel contact between opposing beta strands surfaced in red. Bonds are colored by atom type: oxigen red, carbon green, nitrogen blue, hydrogen yellow. The gnomon at the lower right indicates the position of the cross-section.

Fig. 5. Hydrophobic/hydrophilic patterns at the dimer interface. Surface contributed by hydrophilic atoms (oxygen red and nitrogen blue) is scattered throughout the hydrophobic region contributed by carbon atoms (green).

Fig. 6. Arrangement of the electrostatic fields around the dimer.

Fig. 7. Electrostatic potential mapped onto the molucular surface of the SOD subunit. Blue surface is most positive, green is relatively neutral, and red is most negative. Red ellipsoid represents the incoming superoxide.

Fig. 8. Electrostatic field in the active site.

Fig. 9. Oxidation of superoxide to molecular oxygen.

Fig. 10. Reduction of superoxide to hydrogen peroxide.

Fig. 11. Two step reaction equation.

Fig. 12. Arrangement of active sites in the SOD dimer.

The polypeptide fold of the SOD subunit consists primarily of two large loops and 8 extended antiparallel β-strands (see Fig. 3). The overall topology is characterized as a Greek-key β barrel. This folding motif is also observed in the protein structures of immunoglobulin domains, plant lectins and in the capsid subunits of small spherical viruses. Greek-key β barrel structures tend to be structurally quite stable. Bovine SOD is among the most thermally stable of the currently categorized globular proteins. The structural basis for this relatively great stability can be seen by close examination of intra-subunit and inter-subunit stereochemical interactions.

The β-barrel structure has four very regular extended β strands on one side and four β strands in more twisted or irregular conformations. The two sides of the β barrel might also be interpreted as a β sandwich or a flattened barrel. The active site of the enzyme is located on the irregular or twisted side of the β barrel between two large loops. Calculating separate surfaces for each β strand reveals the packing between strands in cross sections perpendicular to the long axis of the β barrel (Fig. 4). Most of the contacts are made between adjacent strands around the barrel. Two particular contacts are noteworthy because they tie together the two opposite sides of the β barrel with orthogonal cross bracing due to the twist of the barrel and the particular strands joined. These cross barrel interactions stabilize the β barrel fold considerably. In addition, these two interactions involve residues adjacent to Cu ligands and thus anchor the active site through the β barrel to the more regular side of the structure. One such contact between Ile 18 and Phe 43 adjacent to Cu ligand His 44 is illustrated in Fig. 4. The second cross barrel contact, between Ile 33 and Val 117, ties Cu ligand His 118 to the opposite side of the β barrel.

The stability conferred upon the enzyme by the dimer contact can be studied by examination of the surface of each of the subunits that is buried or hidden from solvent in the dimer interface. The stability of the dimer contact is, in fact, comparable in strength to the stability of the individual folded subunit. This is reasonable since similar surface areas are buried between the two halves of the β barrel by the protein fold and between the two subunits by the dimer contact. Roughly 1/10th of the external molecular surface of a subunit is buried in the dimer contact. Cross sections through the dimer interface showed the complementarity of fit between the two buried surfaces, the interdigitation of side chains from one subunit to the other, and the twist of the dimer contact. By pulling the two subunits of the dimer apart, we can characterize patterns of hydrophobicity/hydrophilicity at each subunit interface (Fig. 5). For the most part the interface is continuous and not interrupted by gaps. Hydrophilic regions contributed by oxygen and nitrogen atoms are scattered throughout the interface. An edge-on view of the surface at the dimer interface clearly shows the 60 degree twist from front to back (Fig. 5), which increases the amount of buried surface in the contact.

Electrostatic potentials, calculated from partial charges assigned to each of the atoms in the structure, can also be mapped onto the molecular surface (5,7). At the dimer interface, the center of the pattern is electrostatically neutral and only the edges show significant electrostatic potential. The surfaces are electrostatically complementary to one another and essentially seal the edges of the dimer contact at the protein-water interface.

The independent catalytic sites of the two subunits are located on opposite sides of the dimer with the two Cu ions 33.8 Å apart (Fig. 6,12). Color-coding the solvent accessible surface for the entire subunit according to electrostatic potential (Fig. 7) shows a unique feature of the molecule: the large positive area surrounding the catalytic site. On the opposite side of the subunit, away from the active site, the distribution of electrostatic charge shows no definite pattern, i.e. there are no concentrated large patches of either positive or negative electrostatic potential. The characteristic positive area covering the active site channel represents approximately 1/10th of the surface area of the subunit and is composed almost exclusively of sequence-conserved amino acid residues.

Electrostatic field vectors calculated and displayed above the catalytic site indicate the direction in which a negative charge would flow in the field (Fig. 8). Major contributions to the electrostatic field are made by invariant residues, and both computational and experimental evidence point to the importance of these residues in the activity of the enzyme. Vectors were calculated from the derivative of the electrostatic potential at points lying on concentric spheres radiating out from the oxygen coordination site on the Cu. The direction of flow of the field is away from the sides and down into the center of the channel where the reaction takes place. In cross section the channel funnels downward stepwise from a large shallow depression about 24 Å across, to a deeper well about 10 Å wide and 5 Å deep, then narrows to less than 4 Å forming a complementary pocket for the superoxide anion directly above the Cu ion. At its greatest depth the channel is roughly 12 Å from top to bottom.

The atomic structure has implications for the mechanism proposed from biochemical studies (Fig. 9-11). In Fig. 9. all of the liganding residues and the essential Arg 141 are shown skeletally along with the Cu and Zn ions. One unique feature of this enzyme active site is the bridge formed between the Zn and Cu ions by the His 61 side chain. This bridge can be seen to be made and broken as the reaction proceeds (Fig. 10). The proton transfer from His 61 to the superoxide explains the pH independence of the reaction, i.e. this crucial proton does not come from solvent. Finally, reconstructing the enzymatic dimer and showing the animated electrostatic field vectors emphasizes the electrostatic fields around the two active sites in the dimer (Fig. 6) cover a large solid angle surrounding SOD, making it extremely efficient at attracting superoxide radicals.

ACKNOWLEDGEMENTS

This computer graphic analysis draws heavily from ref. 5 and also includes a number of ideas from other collaborators, particularly David Richardson and Jane Richardson and others from their laboratory. This computer graphics view of SOD was supported in part by a Grant from DDI Pharmaceuticals, Inc.

REFERENCES

1. Olson, A. J., Getzoff,. E. D., and Tainer, J. A. (1984). *Terms of Entrapment: Structure and Function of Superoxide Dismutase.* 8 minute Computer Graphic Film, copyright Research Institute of Scripps Clinic.

2. O'Donnell, T. J., and Olson, A. J., (1981). *Computer Graphics* 15:133.

3. Connolly, M. L., and Olson, A. J., (1985). *Computers & Chem.* 9:1-6.

4. Tainer, J. A., Getzoff, E. D., Beem, K. M., Richardson, J. S. and Richardson, D. C., (1982). *J. Mol. Biol.* 160, 181.

5. Getzoff, E.D. (1982). Ph.D. Dissertation, Duke University, Durham, N.C.

6. Tainer, J. A., Getzoff, E. D., Richardson, J. S. and Richardson, D. C. (1983). *Nature* 306, 284.

7. Getzoff, E. D., Tainer, J. A., Weiner, P. K., Kollman, P. A., Richardson, J. S. and Richardson, D. C. (1983). *Nature* 306, 287.

8. Connolly, M. L., (1983). *Science* 221:709-713.

Superoxide and Superoxide Dismutase in Chemistry,
Biology and Medicine; G. Rotilio, editor

A NEW MECHANISM OF ACTION OF CU, ZN-SUPEROXIDE DISMUTASE

ROMAN OSMAN

Department of Physiology and Biophysics, Mount Sinai School of Medicine, New York, N.Y. 10029 USA

INTRODUCTION

We have recently proposed an alternative mechanism of action for the enzyme Cu, Zn-superoxide dismutase (SOD) (1-2). The mechanism is based on quantum chemical calculations of the interaction of superoxide radical anion with a model of the active site of SOD. We have shown that, unlike in the conventional mechanism (3-7), the role of Arg-141 is not merely to position the superoxide anion in the active site (8) but rather to change its redox chemistry in such a way that the superoxide forms a stable complex with the copper without reducing it. The stable intermediate can oxidize another superoxide in an outer-sphere mechanism in which the copper ion becomes reduced. In this reduced form a proton transfer is induced to superoxide which, followed by an electron transfer from the copper, produces a hydroperoxide anion and restores the oxidized form of the enzyme. An important consequence of the new mechanism is that there is no need to invoke the dissociation of the Cu-NE2 (His 61) bond and the protonation of the resulting imidazolate anion, as proposed in the conventional mechanism (3-8). This is consistent with the conclusion made by Tainer et al. (8), based on the crystal structure of the enzyme that large rearrangements of the ligands during catalysis are unlikely.

We report here the results of a computational simulation of the dissociation of the Cu-NE2 (His 61) bond under various conditions. In order to account partially for the effect of the protein on such a process its influence is included through the inhomogenous electrostatic field it generates in the vicinity of the active site.

COMPUTATIONAL DETAILS

The model of the active site of SOD was constructed from the crystallographic coordinates of the enzyme and is identical to that in ref. 1. In the calculations in which the superoxide was replaced by a water molecule, the oxygen of the water was positioned 2.2A° from the copper ion and its lone pair was directed toward the NH_4^+ that served as a model of Arg-141. All calculations were performed with the methods and basis sets described in ref. 1.

The inclusion of the electric field of the protein was achieved by calculating the modified nuclear attraction integrals in the presence of point charges positioned in the place of the atoms of the entire protein. Hydrogens were added to the atoms of the protein with the program HYDRO written by M. Levitt and partial charges were assigned to these atoms (9). All atoms that were at a distance smaller than 1.5A° from the model of the active site were deleted. This resulted in a total of 2122 point charges that represent the atoms of the protein.

RESULTS AND DISCUSSION

The computational simulation was designed to test whether the complex between superoxide and the enzyme, proposed in the new mechanism (1), is stable with respect to the dissociation of the Cu-NE2 (His 61) bond (8,10). In this simulation it was also possible to investigate the effect of reducing the copper on the stability of this bond. We have therefore investigated the changes in the energy of three molecules upon stretching the Cu-NE2 (His 61) bond. The first was the model of the active site alone (Fig. 1A); in the second a water molecule was placed near the copper to simulate the environment in the native enzyme (Fig. 1B); the third molecule was the stable intermediate between the model of the active site and superoxide anion (Fig. 1C). Inspection of Fig. 1A and 1B indicates that stretching the Cu-NE2 (His 61) in both oxidized forms, i.e., without and with the water molecule, is energetically unfavorable. In the absence of the water molecule stretching the bond by 0.5 A° requires 23.7 Kcal/mole whereas in its presence only 17.8 Kcal/mole are required for the same stretch. The reduction of these molecules has a profound effect on the dissocation of the Cu-NE2 (His 61) bond; it makes this process energetically favorable. However, while in the absence of the water molecule the bond can only stretch by 0.2 A° before the energy starts rising, in the presence of the water molecule stretching the bond by as much as 0.5 A° stabilizes the compound by 4.9 Kcal/mole. These results are in agreement with the observed pH dependence of the reduction potential of the enzyme (11,12) which led to the suggestion that the Cu-NE2 (His 61) bond dissociates as a consequence of the reduction of the enzyme. Similar calculations on the complex of the model of the active site with a superoxide ion indicate that the stretching of the Cu-NE2 (His 61) bond in this complex is energetically unfavorable. Results shown in Fig. 1C demonstrate that stretching this bond by 0.5 A° requires 12.2 Kcal/mole. Furthermore, inspection of the charge and spin distribution in this complex indicates that the stretching does not induce the reduction of the copper by superoxide. Thus, these calculations lend further support to the new mechanism by indicating that the complex with superoxide is stable with respect to the breaking of the Cu-NE2(His 61) bond.

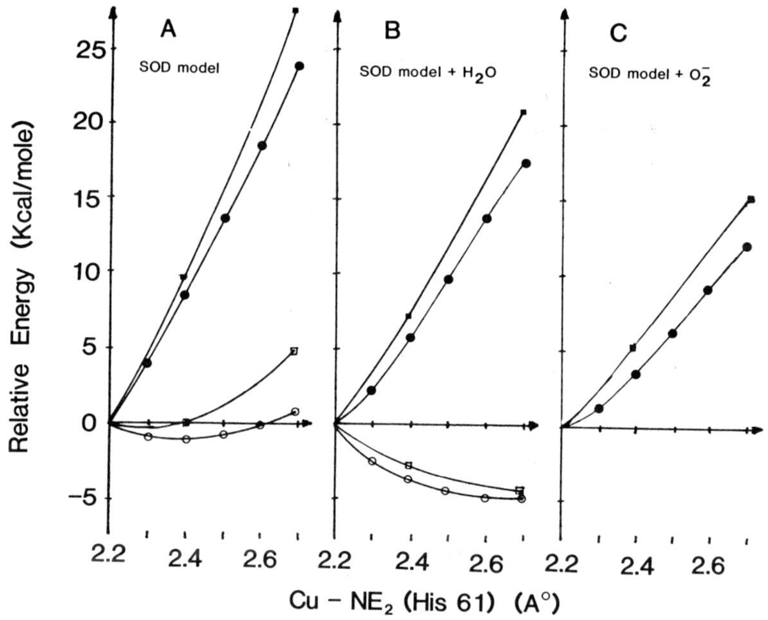

Fig.1. The effect of stretching the Cu-NE2 (His 61) bond on the energies of the oxidized (filled symbols) and reduced (unfilled symbols) models in the absence (circles) and the presence (squares) of the protein.

Inspection of Fig. 1 shows also the effect of the protein on the stretching of the Cu-NE2 (His 61) bond. In general the effect of the protein is to increase the energy required to stretch the bond. In the SOD model with water (Fig. 1B) the presence of the protein does not change the qualitative conclusions, whereas in the absence of the water (Fig. 1A) in the reduced form of the enzyme even a small stretch is energetically unfavorable. This again emphasizes the important role that the water molecule plays in the structural changes that the enyzme undergoes after reduction (11,12). The effect of the protein on the stretching of the Cu-NE2 (His 61) bond in the complex with O_2^- is to destabilize the stretched forms; the effect at longer stretches is larger that at shorter changes. These results are in agreement with the qualitative conclusion that the breaking of the Cu-His 61 bond cannot be induced by the interaction of superoxide with the enzyme.

In conclusion we demonstrated in this work that quantum mechanical simulations predict correctly the effect of reduction on the dissociation of the Cu-NE2 (His 61) bond. They also explain the importance of the water molecule in the active site and its contribution to the dissociation of this bond. Most importantly the simulation of the same process in the complex of the enzyme with superoxide does not indicate that such a dissociation is induced by superoxide and is in full support of our previously proposed new mechanism of action of SOD.

REFERENCE

1. Osman R, Basch H (1984) J Am Chem Soc 106:5710-5714.
2. Osman R, Basch H (1983) In: Cohen G, Grunewald RA (eds) Oxy Radicals and Their Scavenging Systems. Vol I: Molecular Aspects. Elsevier, Amersterdam, pp 187-192.
3. Klug D, Rabani J, Fridovich I (1972) J Biol Chem 247:4839-4842.
4. Fielden EM, Robert PB, Bray RC, Lowe DJ, Mautner, GN, Rotilio G, Calabrese L (1974) Biochem J 139:49-60.
5. Hodgson EK, Fridovich I (1975) Biochemistry 14:5297-5299.
6. Fee JA (1981) In: Sigel H (ed) Metal Ions in Biological Systems. Marcel Dekker, New York. Vol 13, pp. 259-298.
7. Strothkamp KG, Lippard SJ (1982) Acc Chem Res 15:318-326.
8. Tainer JA, Getzoff ED, Richardson JS, Richardson DC (1983) Nature (London) 306:284-287.
9. Hayes DM, Kollman PA (1976) J Am Chem Soc 98:3335-3345; ibid. 98:7811-7816.
10. McAdam ME, Fielden EM, Lavelle F, Calabrese L, Cocco D, Rotilio G (1977) Biochem J 167:271-274.
11. Fee JA, Dicorleto PD (1973) Biochemistry 12:4893-4899.
12. Lawrence GD, Sawyer DT (1979) Biochemistry 18:3045-3050.

© 1986 Elsevier Science Publishers B.V. (Biomedical Division)
Superoxide and Superoxide Dismutase in Chemistry,
Biology and Medicine; G. Rotilio, editor

pH AND IONIC STRENGTH EFFECTS ON SOD ACTIVITY. A COMPARATIVE STUDY OF OX, PIG, SHEEP AND SHARK Cu,Zn SUPEROXIDE DISMUTASES.

D. BARRA°, F. BOSSA°, L. CALABRESE°, C. CAPO°°, A. GALTIERI°°°, A. LANIA°°°, I. MAVELLI°, G. NATOLI°°, G. ROTILIO°°, M.E. SCHININA'°

°)Dept. Biochemistry and C.N.R. Center for Molecular Biology, University of Rome La Sapienza, Rome, Italy;
°°) Dept. Biology, University of Rome Tor Vergata, Rome, Italy;
°°°) Dept. Organic Chem. and Biochem., University of Messina, Messina, Italy.

INTRODUCTION

The activity of Cu,Zn SODs is dependent on ionic strength (1) and alkaline pH (2) in a typical way that points to functional role of charged amino acid residues, in particular lysines (3). Covalent modification of lysines has been a successful approach to answer this question (4). In this report we show that comparative analysis of SODs carrying a different protein net charge may give further insight into the problem of electrostatic interactions in the mechanism of superoxide dismutases.

MATERIALS AND METHODS

Ox, pig and shark SODs were purified as previously described (5-7). Sheep SOD was isolated by the standard procedure (5) with appropriate modification of the last chromatographic step. Its isoelectric point was found to be 7.6 by electrofocusing. Sequence studies on pig and sheep SODs were performed on metal-free, carboxymethylated samples; the proteins were digested with trypsin, chymotrypsin or S. aureus protease and the peptide mixtures purified by high-performance liquid chromatography. Sequence data were obtained mainly by manual dansyl-Edman degradation; details of these procedures are given in ref. 8.

SOD activity was assayed polarographically (2). Ionic strength was changed by addition of perchlorate (1).

RESULTS AND DISCUSSION

The experimental design used in this work exploits the fact that all Cu,Zn SODs show a significant activity decrease around pH 10. Among other factors, deprotonation of activity-important lysine residues has been suggested to play a role in this pH-dependence, on the basis of results obtained with lysine-modified bovine and porcine enzymes (9). Porcine SOD, however, is significantly different from bovine SOD as far as the pH range of the alkaline inactivation is concerned. The activity decrease is already detectable at pH 8, while it appears only above pH 10 in the bovine enzyme (9). Nevertheless the amino acid sequences of the two enzymes are identical in the fragment where the activity

-linked charged side-chains are located (8). On the other hand electric potential field calculations (W.H. Koppenol, personal communication) show that the absence of Glu 75 in porcine SOD makes two areas, approximately 12 Å away from the copper sites, which have a positive potential, unlike in bovine SOD. According to a generally accepted theory, the negative charge elsewhere on the protein surface reinforces the attraction by the positive channel around the copper. This is not entirely true in the case of pig SOD, since at neutral pH and low ionic strength the ratio between the catalytic constants of bovine and porcine SODs is bovine/porcine = 0.72 (6). At pH 9.5 bovine/porcine = 1.20 (Fig. 1), at pH 10.0 is 1.3 (6) and at pH 10.2, is approximately near 1 (Fig.1). The change of this ratio indicates that the local charge is optimal to activity at pH 7.6 and pH 10.0 for the porcine and bovine enzyme respectively, while it tends to be the same for either enzyme above pH 10. This is confirmed by ionic strength effects (Fig. 1) which are pronounced for both enzymes at pH 9.5 while are nearly absent at pH 10.2, suggesting neutralization of surface positive charges at this pH.

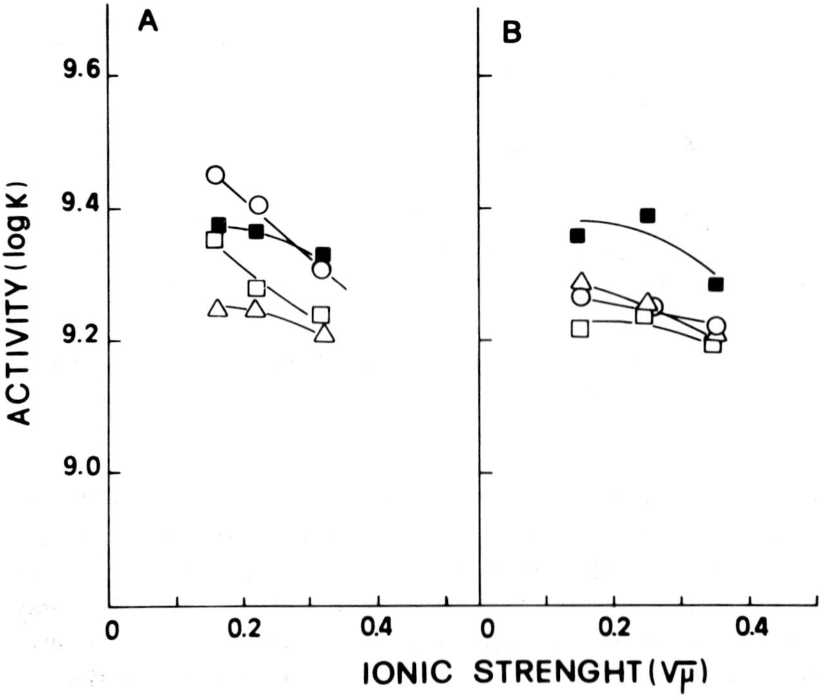

Fig. 1. Plot of catalytic constants versus ionic strength at pH 9.6 (panel A) and pH 10.2 (panel B) for ox (O-O), pig (□-□), sheep (Δ-Δ) and shark (■-■) SODs.

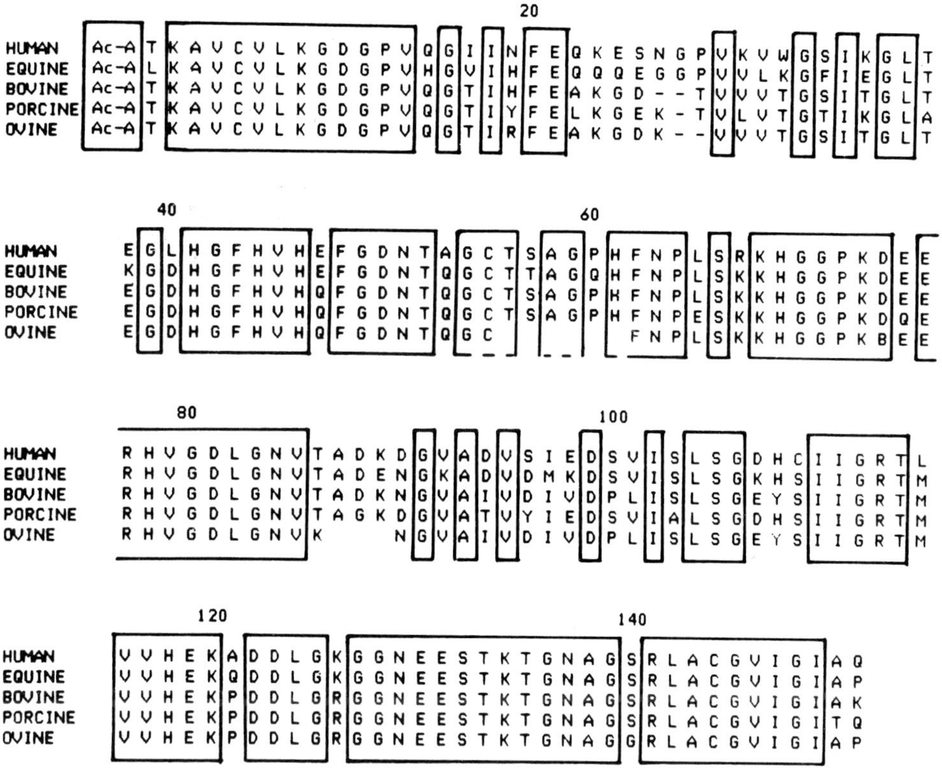

Fig. 2. Comparison of primary structures of mammalian Cu,Zn SODs (10-13). Identical residues are boxed. Gaps in the sequence of the sheep enzyme relate to undetermined sequences.

The above results support the concept that conservation of charged residues in the fragment 110-150 of the sequence (8), which contains all the proposed activity-important amino acid side chains (3), is not the only factor that regulates pH and ionic strength effects in Cu,Zn SODs. The observed differences between the two enzymes are better related to the different pI's (approx. pH 5 and pH 6 for the bovine and porcine SODs, respectively) which apparently reflect the changes of electric field distribution over the protein surface. Therefore this kind of study was extended to shark and sheep SODs, more acidic (pI = 4.6) and basic (pI = 7.6) than ox and pig SODs, respectively.

Fig. 2 reports a partial sequence of sheep SOD, which however is enough to indicate conservation of the activity-linked charged residues. Fig. 1 shows that pH and ionic strength effects confirm the concepts outlined above. The acidic shark enzyme seems to be better suited to withstand high ionic strength and increase of pH around pH 10. The sheep enzyme displays at low ionic strength and lower pH the same activity as the ox and pig enzymes do at high ionic strength and higher pH. These data suggest that a proper gradient between the charge of the active site channel and that of the rest of the molecular surface seems to improve the catalytic efficiency. Electric field calculations are not available yet for these enzymes and will let the charge distribution be compared to that present in the porcine and bovine SOD molecules. It is also clear that less localized charge effects, rather than the conservation of single residues in the active site channel, are important in determining the pH range where the enzyme activity is sensitive to changes of proton and salt concentrations.

REFERENCES

1. Rigo A, Viglino P, Rotilio G and Tomat R (1975), FEBS Letters 50, 86-88.
2. Rigo A, Viglino P and Rotilio G (1975) Anal Biochem 68, 1-8.
3. Getzoff ED, Tainer JA, Weiner PK, Kollman PA, Richardson JS and Richardson DC (1983), Nature 306, 287-290.
4. Ludd A and Fridovich I (1982), J Biol Chem 257, 11443-11447.
5. McCord JM and Fridovich I (1969) J Biol Chem 244, 6049-6055.
6. Marmocchi F, Argese E, Rigo A, Mavelli I, Rossi L and Rotilio G (1983) Mol Cell Biochem 51, 161-164.
7. Galtieri A, Natoli G, Lania A and Calabrese L (1985) Comp Biochem Physiol in press.
8. Schininã ME, Barra D, Simmaco M, Bossa F and Rotilio G (1985) FEBS Letters 186, 267-270.
9. Argese E, Rigo A, Viglino P, Orsega E, Marmocchi F, Cocco D and Rotilio G (1984) Biochim Biophys Acta 787, 205-207.
10. Barra D, Martini F, Bannister J.V., Schininã ME, Rotilio G, Bannister WH, and Bossa F (1980), FEBS Letters 120, 53-56.
11. Sherman L, Dafni N, Lieman-Hurwitz J and Groner Y (1983) Proc Natl Acad Sci USA 80, 5465-5469.
12. Lerch K and Ammer D (1981) J Biol Chem 256, 11545-11551.
13. Steinman HM, Naik VR, Abernethy JL and Hill RL (1974) J Biol Chem 249, 7326-7338.

© 1986 Elsevier Science Publishers B.V. (Biomedical Division)
Superoxide and Superoxide Dismutase in Chemistry,
Biology and Medicine; G. Rotilio, editor

NMR STUDIES OF THE ANION BINDING SITES OF OXIDIZED AND REDUCED BOVINE COPPER-ZINC SUPEROXIDE DISMUTASE

JOAN SELVERSTONE VALENTINE AND DUARTE MOTA DE FREITAS

Department of Chemistry and Biochemistry and Molecular Biology Institute, University of California, Los Angeles, California 90024 and Department of Chemistry, Loyola University of Chicago, Chicago, Illinois 60626 (USA)

INTRODUCTION

Bovine erythrocyte copper-zinc superoxide dismutase (Cu,Zn-SOD) in its oxidized form has been shown by x-ray crystallography (1) to be a dimer of two equivalent subunits, each subunit containing one copper(II) ion and one zinc(II) ion held in close proximity by a bridging imidazolate ligand from a histidyl residue of the protein. This protein has been found to be a very effective catalyst of superoxide disproportionation ($2O_2^- + 2H^+ \rightarrow O_2 + H_2O_2$) and it has therefore been proposed that the function of the protein is to act as a protective agent against deleterious reactions of superoxide generated in the cell in vivo (2).

Since superoxide, O_2^-, is an anion and since it is believed to react with both the oxidized and the reduced forms of the protein, it is of interest to know how other anions interact with both forms of the protein and how they influence the superoxide dismutase (SOD) activity. We present here a description of our most recent results concerning the interaction of phosphate anion with the oxidized protein and of chloride anion with the reduced protein as studied by NMR techniques.

MATERIALS AND METHODS

Bovine Cu,Zn-SOD was chemically modified at Arg-141 with phenylglyoxal and purified by the method of Malinowski and Fridovich (3). Succinylation of lysine residues of bovine Cu,Zn-SOD was carried out according to Marmocchi et al. (4). Native and modified SOD's were reduced by addition of sufficient sodium dithionite to bleach the color of the protein.

^{31}P NMR measurements were made at 81.02 MHz on a Bruker WP200 NMR spectrometer at the ambient probe temperature (32 °C) or at higher temperatures with a variable temperature unit. ^{35}Cl NMR measurements were made at 8.73 MHz on a JEOL FX90 NMR spectrometer equipped with a multinuclear probe. These measurements were carried out at both ambient probe temperature (32 °C) or at higher temperatures with a variable temperature unit. ^{1}H NMR spectra were obtained at 500 MHz on a Bruker WM500 NMR spectrometer at the Southern California Regional NMR Facility.

RESULTS

The binding of phosphate anion to bovine Cu,Zn-SOD was followed by the changes in the linewidth at half-height of the phosphate resonance as measured by ^{31}P NMR. Addition of 0.4 mM native bovine Cu,Zn-SOD to a 4 mM phosphate solution at pH 6.3 caused the linewidth to increase from 5.0 to 107 Hz. Much less line broadening was observed with sample solutions containing phosphate in the presence of arginine- (27 Hz) or lysine-modified (28 Hz) proteins. We attribute all of the line broadening observed to the effect of the paramagnetic Cu(II) ion in the protein since no line broadening was observed in sample solutions containing phosphate and fully reduced, i.e., Cu(I)-containing proteins. A decrease in the phosphate resonance linewidth with increasing temperature was observed for phosphate in the presence of the native and modified proteins indicating fast exchange between protein-bound phosphate and free phosphate in solution. The interaction of other biological phosphates with bovine Cu,Zn-SOD was also studied by ^{31}P NMR. Broadening of the phosphate resonances of the sodium salt of ATP was observed upon addition of native bovine Cu,Zn-SOD. The γ- and β-phosphate resonances of ATP were broadened considerably more than the α-phosphate resonance indicating that ATP interacts with the protein predominantly via the γ-phosphate. No appreciable change in the spectrum of ATP was observed upon addition of arginine-modified SOD.

Binding of Cl^- to reduced native Cu,Zn-SOD and derivatives was monitored by the linewidth at half-height of the Cl^- resonance as measured by ^{35}Cl spectroscopy. The Cl^- linewidth in solutions containing 0.1 M NaCl and reduced native Cu,Zn-SOD (0.21 mM) in 0.25 M Hepes in the pH range 5.0 - 7.0 was approximately 80 Hz. Similar experiments carried out in the equivalent concentrations of NaCl and reduced arginine- or lysine-modified Cu,Zn-SOD in the same pH range gave linewidths of approximately 55 and 35 Hz respectively while the apoprotein gave a linewidth of approximately 30 Hz. Studies of the temperature dependencies of these resonances demonstrated that the linewidths decreased with increasing temperature indicating that protein-bound chloride was in fast exchange with free chloride in solution. Similar experiments in 0.05 M phosphate buffer but at the same ionic strength gave Cl^- linewidths almost as sharp as that of free chloride in solution (approximately 20 Hz). We interpret the sharpness of the chloride resonance in the presence of phosphate as being due to competition between chloride and phosphate for the same binding sites.

The influence of Cl^- on the C-2 proton resonances of histidyl residues of reduced native Cu,Zn-SOD as well as the reduced lysine- and arginine-modified derivatives was studied by ^{1}H NMR. We found that the C-2 proton resonances of histidyl imidazoles of reduced native SOD and reduced lysine-modified SOD were significantly shifted upon addition of chloride whereas that anion had no effect upon the spectrum of the reduced arginine-modified protein, even at chloride concentrations as high as 1 M.

DISCUSSION

Addition of native bovine Cu,Zn-SOD to solutions of phosphate ion caused a much larger increase in the ^{31}P linewidth than did addition of either the arginine- or lysine-modified derivatives at the same concentrations. We attribute this difference to a lowered affinity for phosphate in the modified derivatives. It is important to note, however, that phenylglyoxal modification of bovine Cu,Zn-SOD alters only one residue per subunit, i.e. Arg-141, and that the overall charge on the protein does not change. Since the modification is limited to that one residue, our experiments imply that Arg-141 is a major site for interaction of phosphate with the native protein (5). Succinylation of lysine residues to form the lysine-modified derivative, by contrast, causes a large increase in the overall negative charge of the protein and consequently a significant decrease in the affinity of this derivative for all anions has been noted (6).

In the case of the reduced protein, Fee and Ward (7) had previously observed that the nuclear relaxivities of solutions of Cl^- were higher in the presence of reduced Cu,Zn-SOD than in the presence of the apoprotein and that addition of CN^- lowered the relaxivity of reduced protein to that of the apoprotein. The binding of halides to reduced Cu,Zn-SOD was also studied by high-field ^{1}H NMR (8). Five resonances were assigned to C-2 protons of histidyl imidazole ligands at the metal-binding site and three of these were observed to shift upon addition of halide ions. Although both of these observations are in agreement with coordination of halide ions to Cu(I), they are also consistent with binding of halide ions to a protein side chain close to the metal-binding region.

In order to differentiate between binding to Cu(I) and binding to a protein side chain, we repeated these experiments using the arginine- and lysine-modified proteins. We found that the reduced arginine- and lysine-modified proteins caused considerably less broadening of the Cl^- resonance when measured under the same conditions as were used for the native protein. However, this result indicated only that arginine and lysine residues were interacting with chloride under our experimental conditions. It did not rule out the possibility that chloride was also binding to Cu(I). However, our observation that the C-2 proton resonances of histidyl imidazoles of reduced native and reduced lysine-modified SOD were shifted upon addition of Cl^- whereas this anion had no effect on the spectrum of the arginine-modified SOD, even in 1 M chloride, suggests strongly that the major site for chloride ion in the reduced protein is Arg-141 and not the Cu(I) ion as had been previously concluded (7).

It is our conclusion that Arg-141 plays a significant role in the interaction of anions with both the oxidized and the reduced forms of Cu,Zn-SOD. It appears likely, therefore, that Arg-141 plays a significant role in the mechanism of interaction of superoxide with both forms of the protein.

ACKNOWLEDGMENT

This work was supported by grant GM-28222 from the National Institutes of Health.

REFERENCES

1. Tainer JA, Getzoff ED, Beem KM, Richardson DC, Richardson J (1982) J Mol Biol 160:181

2. Fridovich I (1975) Annu Rev Biochem 44:147

3. Malinowski DP, Fridovich I (1979) Biochem 18:5909

4. Marmocchi F, Mavelli I, Rigo A, Stevanato, R, Bossa F, Rotilio G (1982) Biochem 21:2853

5. Mota de Freitas D, Valentine JS (1984) Biochem 23:2079

6. Cocco D, Mavelli I, Rossi L, Rotilio G (1983) Biochem Biophys Res Commun 111:860

7. Fee JA, Ward RL (1976) Biochem Biophys Res Commun 71:427

8. Cass AEG, Hill, HAO, Hasemann U, Johansen JT (1978) Carlsberg Res Commun 43:439

Superoxide and Superoxide Dismutase in Chemistry, Biology and Medicine; G. Rotilio, editor

STRUCTURAL STUDIES ON CYANIDE AND AZIDE DERIVATIVES OF BOVINE SUPEROXIDE DISMUTASE USING EXAFS SPECTROSCOPY AND THE IMPORTANCE OF MULTIPLE SCATTERING INTERACTIONS IN THE EXAFS FITS.

by

Ninian J. Blackburn[1], S. Samar Hasnain[2], P. F. Knowles[3] and Richard W. Strange[1].

[1]Department of Chemistry, U.M.I.S.T., Manchester, M60 1QD,
[2]S.E.R.C. Daresbury Laboratory, Warrington, Cheshire, WA4 4AD,
[3]Astbury Department of Biophysics, University of Leeds, LS2 9JT.

INTRODUCTION

Anionic derivatives of the Cu/Zn superoxide dismutase have attracted considerable interest with regard to their mode of binding at the active site and the resultant inhibition of enzyme activity. The high resolution crystal structure [1] provides a detailed description of the structure of the active site of the native protein in which the tetrahedrally distorted copper atom is bridged to the zinc atom via his 61, and coordinated to three additional histidines, 44, 46, and 118, but no crystal structure is as yet available for an inhibited derivative nor for the reduced (Cu(I)) form of the enzyme. However, derivatives of SOD have been extensively investigated by spectroscopic means. EPR, [2] NMR [3] and ENDOR [2] studies have provided strong evidence that anions bind at equatorial positions and increase the symmetry of the site towards an axially symmetric structure. We have recently published a number of EXAFS studies [4-6] on native and reduced forms of the enzyme which have demonstrated the potential of X-ray Absorption among the armoury of spectroscopic methods. Here we report details of EXAFS analysis of cyanide and azide derivatives.

MATERIALS AND METHODS

Superoxide dismutase was isolated by the method of McCord and Fridovich [7]. Cyanide and azide derivatives were prepared by addition of an excess of chelex treated potassium cyanide and sodium azide, equilibrated to pH 7.5 in 1 mM potassium phosphate buffer. Protein EXAFS was recorded at ambient temperature as the fluoresence excitation spectrum at the S.E.R.C. Daresbury Synchrotron Radiation Facility operating at an energy of 1.8 GeV, 150 mA for the cyano derivative, 2.0 GeV, 250 mA for the azido derivative. Data analysis was accomplished as described previously [5].

RESULTS AND DISCUSSION

Exafs analysis of the outermost shells of C and N atoms of the imidazole rings of Cu-imidazole complexes including $[Cu(imid)_4](NO_3)_2$ [6], native SOD and its reduced form, have previously been found to give rise to anomalously short distances [5]. For this reason, analysis of the cyano and azido derivatives

was first carried out on Fourier filtered data back-transformed over the range 0.3-3.5 A. Fig 1(A-C) shows fits for three different models of the copper coordination. Each model assumes three coordinated imidazole groups with contributions to the EXAFS from the $3N_{\alpha}$ and $6C_{\beta}$ atoms. A: 3 imidazoles at 1.95 A with no other coordinated groups; B: as A but with an additional 1C from coordinated CN^- at 1.99 A; C: as A plus 1C at 2.00 A and one other low Z scatterer at the longer distance of 2.22 A. Visual comparison as well as consideration of a least squares fitting parameter indicates that model B is a much poorer fit than A or C, with C having a slightly lower fit index than A (0.299 vs 0.385 for k^3 weighted fits plotted from 45-440 eV).
There is overwhelming evidence in the literature for coordination of CN^- to copper derived from EPR, ENDOR, NMR, and spin echo spectroscopies [2]. Therefore model A can be discarded in favour of model C. The data then indicate that in addition to coordinated CN^- a low Z atom is coordinated at 2.22 A. Since NMR relaxation measurements show conclusively that solvent water is displaced upon cyanide binding, we suggest that the 2.22 A group is imidazole. This would imply that cyanide binding reorients the coordination sphere such that one imidazole ligand is displaced from an essentially equatorial to an axial position, with cyanide occupying the vacant equatorial site.

Fig 2(A and B) shows Fourier filtered EXAFS for the azido derivative. EXAFS analysis requires 4 first shell low Z scatterers at 1.98 A which are N atoms in the fit. The interesting feature of the azido data is the low intensity of the peak at 3.0 A in the transforms. The fit shown in Fig 2A includes 4C atoms (2 imidazole groups) around this distance while the fit shown in Fig 2B has 6C (3 imidazoles). The latter is visually and quantitatively worse than the former, and would suggest that only two imidazoles remain coordinated in the equatorial plane of the copper. There is evidence in the literature for the formation of 2:1 azide complexes derived from spectrophotometry and EPR [8,9]. The present data would seem to support the formation of a 2:1 azide derivative.

TABLE 1

Parameters Used in the EXAFS Fits of Figs 1 and 2.

CYANO			AZIDO		
Atom	R/(A)	$2\sigma^2/(A^2)$	Atom	R/(A)	$2\sigma^2/(A^2)$
3 N	1.95	0.012	4 N	1.98	0.010
1 C	1.99	0.014			
1 N	2.22	0.020	1 N	2.29	0.020
3 C	2.89	0.016	2 C	2.94	0.019
3 C	3.27	0.014	2 C	3.02	0.019
6N(C)	3.82	0.020	6N(C)	3.80	0.016

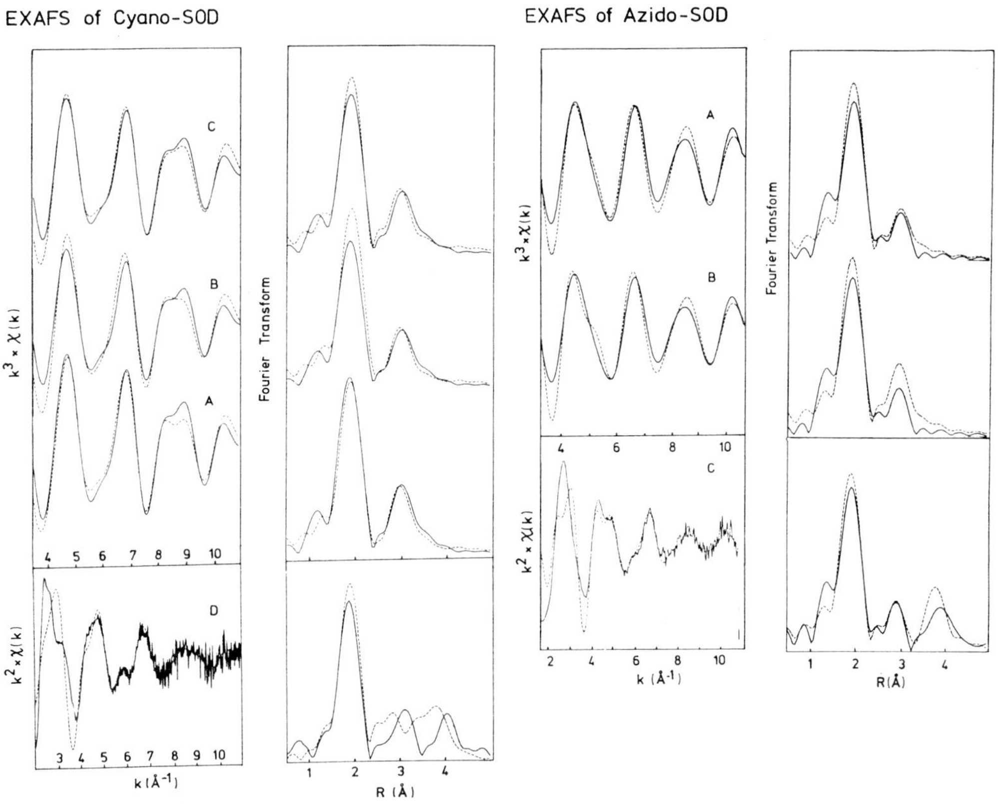

Figure 1 and 2. EXAFS and Fourier Transforms for Cyano and Azido SOD. For details see the text and Table 1.

Fig 1D and 2C show the raw unfiltered data for cyano and azido derivatives respectively. The parameters used in these final fits are the best fit parameters from the Fourier filtered data together with additional outer shell C and N atoms, and these are tabulated in Table 1. As mentioned above, it is difficult to fit the outermost shells and the distances required to give acceptable fits to the EXAFS are anomalously short for a planar imidazole ring coordinated to copper at the distance indicated by the first shell Cu-N interaction. The Fourier transforms also show a poor and foreshortened fit in the outer shell region. Below k=4 the fit is lost altogether.

It has been suggested that these anomalous effects are due to multiple scattering interactions among the atoms of the imidazole rings. Accordingly we have used a recently developed multiple scattering theory [10] to calculate all double and triple

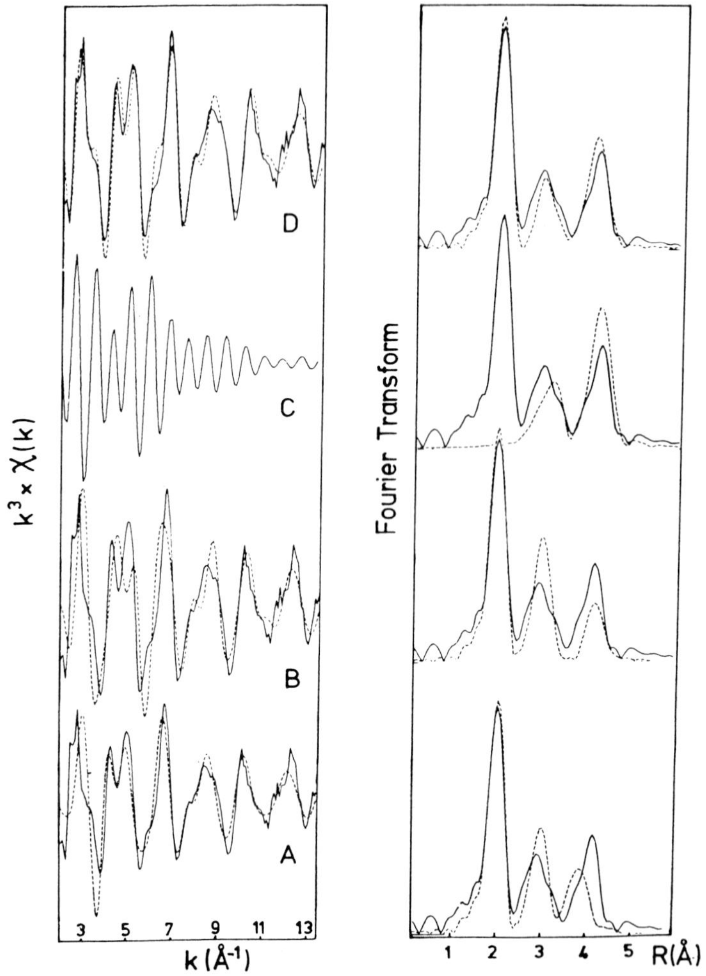

Figure 3. Comparison of Single and Multiple Scattering fits for the model compound tetraimidazolecopper(II) nitrate. For details see the text and Table 2.

scattering interactions in the model compound $[Cu(imid)_4](NO_3)_2$. Fig 3(A-D) and Table 2 show the results; Fig3A shows the best single scattering fit but this results in short outershell distances (Table 2); the poor fit shown in Fig 3B is the result of performing a single scattering calculation of the EXAFS with all atoms placed at their crystallographic distances; Fig 3C shows the calculated contribution to the EXAFS from all double and triple scattering events; Fig 3D shows the result of adding in the multiple scattering wave (3C) to 3B, i.e. with all atoms now at their crystallographic positions. The resultant fit is excellent throughout the complete range of k allowing the low k EXAFS to be fit perfectly. Thus the structure in this latter near edge region (often called the XANES region) can be accounted for by low order multiple scattering interactions.

TABLE 2

Comparison of Single and Multiple Scattering Fits. Crystallograhic values are taken from reference 11.

Atom Shell	R (Cryst)	R (Single S.)	R (Mult. S.)
4 N	2.00	1.99	1.98
4 C	2.99	2.94	2.93
4 C	3.05	3.04	3.02
4 N	4.14	3.88	4.07
4 C	4.18	3.77	4.18

These results indicate that it is possible to interpret both the EXAFS and the XANES region, and obtain the rich amount of structural information present in the latter, using a relatively simple multiple scattering theory, full details of which will be published shortly. The near edge region of the cyano and azido derivatives Fig 1D and 2C show prominent differences which probably result in part from strong multiple scattering interactions within the linear Cu-CN and Cu-NNN units, and interpretation of which will almost certainly give additional information on their mode of binding. Furthermore, extension of the fitting range to very low energies where the Debye Waller term is less effective will allow more precise determination of coordination numbers, and will be an important advance in resolving ambiguities in analysis of lower signal to noise protein data. Further studies aimed at extending the multiple scattering calculations to the protein systems are underway.

Acknowledgements. We would like to thank S.E.R.C. for financial assistance and the Daresbury Laboratory for provision of facilities for EXAFS measurement and interpretation.

REFERENCES

1. Tainer J. A., Getzoff E. D., Beem K. M., Richardson J. S.,and Richardson D. C., (1982), J. Mol. Biol., **160**, 181-217.
2. Fee J. A., (1981), in Metal Ions in Biological Systems, **13**, 259-298, and references therein.
3. Boden N., Holmes M. C., and Knowles P. F., (1979), Biochem. J., **177**, 303=309.

4. Blackburn N. J., Hasnain S. S., and Knowles P. F., (1984), in EXAFS and Near Edge Structure, eds. Hodgson, Hedman, and Penner-Hahn, Springer-Verlag, New York, pp 124-129.
5. Blackburn N. J., Hasnain S. S., Binsted N., Diakun G. P., Garner C. D., and Knowles P. F., (1984), Biochem. J., **219**, 985-990.
6. Blackburn N. J., Hasnain S. S., Diakun G. P., Knowles P. F., Binsted N., and Garner C. D., (1983), Biochem. J., **213**, 765-768.
7. McCord J. M., and Fridovich I., (1969), J. Biol. Chem., **244**, 6049-6055.
8. Fee J. A., and Gaber B. P., (1972), J. Biol. Chem., **247**, 60-65.
9. Strothkamp K. G., and Lippard S. J., (1984), Biochemistry, **20**, 7488.
10. Gurman S. J., and Binsted N., (1985), SRS Bulletin, (Daresbury Report No 85/337A), 20.
11. McFadden D. L., McPhail A. T., Garner C. D., and Mabbs F. E., (1976), J. Chem. Soc. Dalton Trans., 47-52.

AN EXAFS STUDY AT THE COPPER K EDGE OF THE DERIVATIVES OF BOVINE SUPEROXIDE DISMUTASE MODIFIED IN THE ZINC SITE.

ALESSANDRO DESIDERI, SILVIA MORANTE, DINA COCCO, GIUSEPPE ROTILIO

Dipartimento di Biologia e Fisica, II Università di Roma, Gruppo di Chimica Biologica e Strutturistica Chimica, Università di Roma "La Sapienza", Roma, Italia.

INTRODUCTION

X-ray diffraction studies (1) on crystals of native bovine Cu,Zn superoxide dismutase have revealed a copper binding site in which four imidazole groups (belonging to His 44, His 46, His 61 and His 118) and one water molecule coordinate to the Cu(II). One of the histidines (His 61) contributes an imidazolate bridge between the copper and the zinc. The zinc ion is surrounded by two more histidines and one aspartate residue. The zinc atom can be selectively removed (2) or substituted for (3) by Co(II) to give rise to two different derivatives, the zinc-depleted and the Cu,Co enzymes, which show the same activity as the native enzyme in the pH 5-6 range.

Very little is known about the structure of the copper site in the Cu,Co derivative as the presence of the cobalt renders both metals EPR undetectable over the full temperature range 1.5-300 K (4). This magnetic effect, however, does not proof the existence of the bridging imidazole group in the Cu,Co derivative. Some re-arrangement is known to occur in the zinc-depleted enzyme as its EPR spectrum becomes more axial with respect to that of the native enzyme (5), but the structural details of this symmetry change are still obscure. It would be interesting to compare the structural differences of these three samples that are functionally indistinguishable, and the EXAFS technique seems to be particularly suitable to this aim. EXAFS data, in fact, are not affected by the presence of magnetic interactions between metal ions and can in principle give information on the distance of the ligands surrounding the absorbing atoms, to a precision of 0.02 Å (6).

MATERIAL AND METHODS

Cu,Zn superoxide dismutase from bovine red blood cells, the Cu,Co and the zinc-depleted enzymes were prepared as previously described (2,3,7).

EXAFS spectra, at the copper K edge, were performed at the Frascati synchrotron radiation source (PULS), operating at an energy of 1.5 GeV with an average current of 50 mA. The spectra were recorded in the fluorescence mode, using a nickel filter on the photomultiplier. Samples were enzyme solutions with an approximately 5 mM concentration of Cu(II). Data analysis was made on EXAFS spectra that were an average of 10 different scans (time integration for each point approximately 2 sec). The quality of the samples was controlled before and after the irradiation by measuring the enzyme activity and the EPR and optical spectra of the samples. All parameters were not modified by the treatment.

RESULTS AND DISCUSSION

The Fourier transforms of the EXAFS spectra of the copper

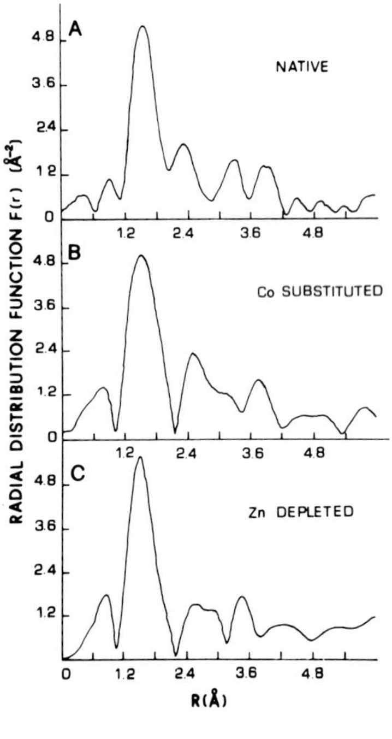

Fig. 1. Fourier transform of the EXAFS spectrum of native (A), Cu,Co (B) and zinc-depleted (C) enzymes.

site in the three different derivatives of superoxide dismutase are reported in Fig. 1. Each peak is due to the presence of the shells sorrounding the absorbing atom at different distances. The actual distance between the absorbing atom and the shell of the scattering atoms is given by the sum of the R value of the maximum of each peak plus the phase shift. The value of the phase shift can be obtained from the EXAFS spectrum of a model compound of known interatomic distances, which, in this case, can reasonably be identified with the native enzyme with the distances given by the X-ray crystallographic study (1). The differences in distance for the three first shells between the native enzyme and the Cu,Co and zinc-depleted derivatives respectively, are reported in Table I.

TABLE I
DIFFERENCE FOR THE DISTANCE OF THE FIRST THREE SHELLS BETWEEN THE NATIVE (R_n) AND THE METAL SUBSTITUTED (R_x) ENZYMES.

Derivative	$\Delta R(R_n - R_x)$*		
	1st shell	2nd shell	3rd shell
Cu,Co	+ 0.02	-0.02	-0.02
Zinc-depleted	+ 0.01	-0.06	-0.09

* The error on ΔR is $\pm$ 0.02 Å.

The results indicate that, beside the experimental error, the copper site in the Cu,Co enzyme retains the same structure as in the native one. This result can be taken as an evidence that the imidazolate bridge is present in the Cu,Co derivative. On the other hand, the copper site in the zinc-depleted enzyme displays some significant structural differences. In fact, whilst the distance of the 1st shell is unchanged, that of the 2nd and 3rd shell increases with respect to the native enzyme. Such a pattern suggests that the absence of the zinc brings about a different orientation of the imidazole rings surrounding the copper, in line with the altered symmetry of the EPR spectrum (5). Previous evidence for easier copper mobilization in this derivative (2), may reflect this different coordination.

REFERENCES

1. Tainer JA, Getzoff ED, Beem KM, Richardson JS, Richardson DC (1982) J Mol Biol 160: 181-217.
2. Valentine JS, Pantoliano MW, McDonnel PJ, Burger AR, Lippard SJ (1979) Proc Natl Acad Sci USA 76: 4245-4249.
3. Calabrese L, Rotilio G, Mondovì B (1972) Biochim Biophys Acta 263: 827-829.
4. Rotilio G, Calabrese L, Mondovì B, Blumberg WE (1974) J Biol Chem 249: 3157-3160.
5. Fee JA, Briggs RG (1975) Biochim Biophys Acta 400: 439-450.
6. Teo BK (1981) In: Teo BK, Joy DC (eds) EXAFS Spectroscopy: Technique and Applications. Plenum, New York, pp. 13-58.
7. McCord JM and Fridovich I (1969) J Biol Chem 244: 6049-6055.

Superoxide and Superoxide Dismutase in Chemistry, Biology and Medicine; G. Rotilio, editor

CHARACTERIZATION OF HUMAN Cu, Zn SUPEROXIDE DISMUTASE ISOLATED BY ION-EXCHANGE AND METAL CHELATE AFFINITY CHROMATOGRAPHY

R.J. WESELAKE*, A. PETKAU** AND A.D. FRIESEN*

*Winnipeg Rh Institute Inc., University of Manitoba, Winnipeg, Canada R3T 2N2 and **Medical Biophysics Branch, Atomic Energy of Canada Limited Research Company, Pinawa, Manitoba, Canada, ROE 1LO

INTRODUCTION

Superoxide dismutases (SOD) are metallo-enzymes which catalyze the dismutation of the superoxide radical (O_2^-) to O_2 and H_2O_2[1]. Human erythrocyte Cu, Zn SOD or erythrocuprein is a dimeric molecule composed of subunits with ~M_r 16,000 held together by non-covalent forces[2]. Charge micro-heterogeneity of human Cu, Zn SOD has been shown using electrophoresis[3,4] and isoelectric focusing[4]. Isoelectric points ranging from pI 4.6 to 5.2 have been reported for the enzyme[3,4,5]. The advent of metal chelate affinity chromatography[6] has resulted in new possibilities for isolation of therapeutically useful proteins such as SOD[7]. We have developed a step-wise chromatographic process for isolating erythrocyte human Cu, Zn SOD using DEAE-Sepharose chromatography in combination with copper chelate affinity chromatography[8]. In this report, the charge micro-heterogeneity of the enzyme, isolated by this process, is investigated by isoelectric focusing and chromatofocusing.

MATERIALS AND METHODS

Enzyme Assays

SOD, isolated by copper chelate chromatography[8], was assayed by two methods, based on the ability of the enzyme to inhibit cytochrome c reduction by O_2^-[9,10]. SOD also was isolated according to a classical procedure[9]. Protein was assayed according to Lowry *et al*[11].

Isoelectric Focusing and Crossed Immunoelectrofocusing

Electrofocusing was conducted in 0.5mm polyacrylamide gels with Ampholine pH 4-6 (LKB-Produkter AB, Bromma, Sweden) at +10°C. Gels were stained for SOD activity[12] and protein[13]. Precipitin peaks in crossed immunoelectrofocusing[14] were visualized by activity staining[12]

Chromatofocusing

This was conducted at room temperature with a Pharmacia (Uppsala, Sweden) Mono-P column using the manufacturer's FPLC system.

RESULTS AND DISCUSSION

Specific activities of SOD preparations ranged from 3,400-3,800 units/mg and both enzyme assays[9,10] gave similar results. Enzyme activity was abolished with 3mM KCN, consistent with the identity of Cu, Zn SOD[1]. A single component was identified by SDS gel electrophoresis and analytical gel filtration chromatography[8]. In contrast, electrofocusing revealed charge micro-heterogeneity in the human Cu, Zn SOD preparations. Figure 1A shows at least 4 bands of activity ranging in pI from about 5.1 to 4.9, designated components I, II, III and IV, respectively. Micro-heterogeneity of the human enzyme has been demonstrated previously by starch gel electrophoresis[3,4]. This suggests that carrier ampholytes do not interact with human Cu, Zn SOD to induce micro-heterogeneity.

Protein staining of enzyme bands (Figure 1, B, C, D) indicated banding patterns that were similar to that visualized by the activity stain (Figure 1A). SOD purified from 13-day-old or fresh erythrocytes were both similar (Figures B, C). This similarity also was apparent for enzyme purified according to a classical procedure[9] (Figure 1D). In this case, the erythrocytes were stored for over one month prior to enzyme extraction.

Crossed immunoelectrofocusing (Figure 1E) indicated that components I and II comprised about 85% of the total area under the precipitin peaks. Component II was, however, more dominant than component I. Presumably these

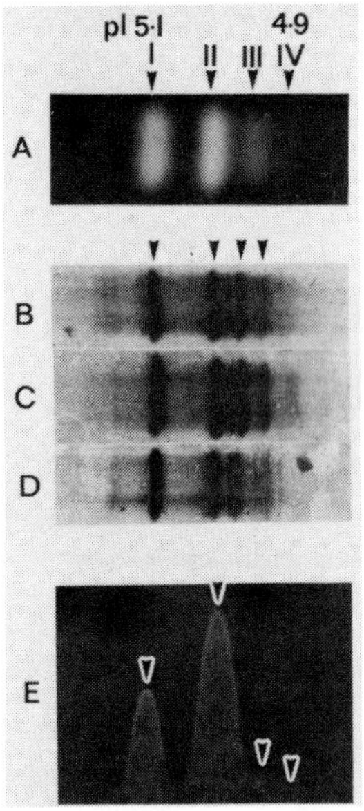

Fig. 1. Electrofocusing of SOD. A. Activity stain (0.1μg). B. Protein stain of SOD (3μg) from 13-day-old erythrocytes. C. SOD (3μg) from fresh erythrocytes. D. SOD (3μg) prepared classically[9]. E. Crossed immunoelectrofocusing of SOD (0.5μg).

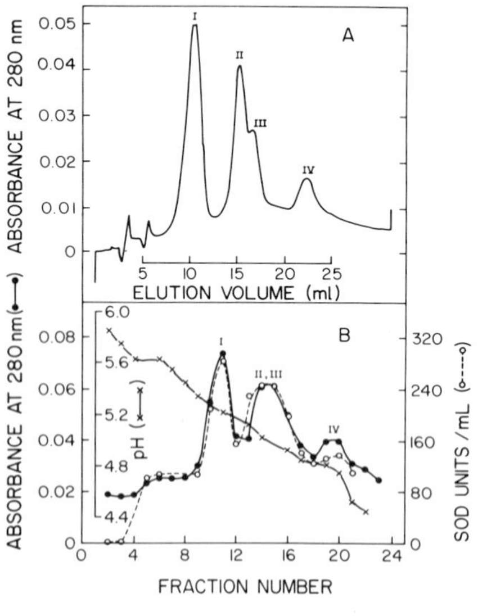

Fig. 2. Chromatofocusing of SOD (0.9mg). A. Integrator tracing of elution profile. B. Analyses of fractions (1ml). Starting buffer: 25mM piperazine-HCl, pH 6.3. Elution Buffer: 1:10 Polybuffer 74, pH 4.5. Flow rate = 50 ml/hr.

two major forms of SOD correspond to the two charge isomers of human Cu, Zn SOD isolated by Bannister _et al_[15].

Chromatofocusing also was used to demonstrate charge micro-heterogeneity of SOD. The integrator tracing in Figure 2A indicated 4 protein peaks with apparent pIs ranging from about 5.2 to 4.8 (Figure 2B). Presumptive SOD peaks II and III, however, were not resolved following analysis of chromatofocusing fractions (Figure 2B). In contrast to crossed immunoelectrofocusing (Figure 1E), the chromatofocusing profile (Figure 2) suggests that peak I, rather than peak II, is the most dominant component of human Cu, Zn SOD. This discrepancy might be related to differential antigenic reactivity of the SOD components, or to non-uniform absorption characteristics of the components.

In summary, human Cu, Zn SOD of high specific activity can be isolated by ion-exchange and metal chelate affinity chromatography. Analysis of enzyme preparations for charge micro-heterogeneity suggested that there are at least 4 components of Cu, Zn SOD, and that either fresh or stored erythrocytes are good sources of enzyme.

ACKNOWLEDGEMENTS

The authors thank Sharon Chesney and Brenda Peltz for excellent technical assistance; and the Canadian Red Cross for supplies of packed erythrocytes.

REFERENCES

1. Fridovich I (1974) Advan Enzymol 41:35
2. Hartz JW, Deutsch HF (1972) J Biol Chem 247:7043
3. Stansell MJ, Deutsch HF (1965) J Biol Chem 240:4306
4. Hartz JW, Deutsch HF (1969) J Biol Chem 244:4565
5. Kimmel JR, Markowitz H, Douglas MB (1959) J Biol Chem 234:46
6. Porath J, Carlsson J, Olsson I, Belfrage G (1975) Nature 258:598
7. Huber W, Menander-Huber H (1980) Clin Rheum Dis 6:465
8. Weselake RJ, Chesney SL, Petkau A, Friesen AD unpublished results
9. McCord JM, Fridovich I (1969) J Biol Chem 244:6049
10. Hyland K, Voisin E, Banoun H, Auclair C (1983) Anal Biochem 135:280
11. Lowry OH, Rosebrough NJ, Farr AL, Randall RJ (1951) J Biol Chem 193:265
12. Beachamp C and Fridovich I (1971) Anal Biochem 44:276
13. Merril CR, Goldman D, Sedman SA and Ebert MH (1981) Science 211:1437
14. Soderholm J, Smyth CJ, Wadstrom T (1975) Scand J Immunol 4, Suppl 2:107
15. Bannister WH, Anastasi A, Bannister JV (1977) In: Michelson AM, McCord JM, Fridovich I (eds) Superoxide and Superoxide Dismutase, Academic Press, New York, pp 107-128

Superoxide and Superoxide Dismutase in Chemistry, Biology and Medicine; G. Rotilio, editor

CHARGE ISOMERISM AND AGE DEPENDENCY OF BOVINE ERYTHROCYTE Cu_2Zn_2SUPEROXIDE DISMUTASE

ALFRED GÄRTNER, MONIKA SCHROTH-POLLMANN, ULRICH WESER

Anorganische Biochemie, Physiologisch-Chemisches Institut der Universität Tübingen, Hoppe-Seyler-Str. 1, D-7400 Tübingen, FRG

INTRODUCTION

Purified erythrocyte Cu_2Zn_2superoxide dismutase has a two banded pattern during both gel electrophoresis and isoelectric focusing. This phenomenon was attributed to the heterogeneity of two distinct SOD species. Although there are many studies on those charge isomers, molecular data on the observed heterogeneity are lacking (1). Due to the rapid age dependent inactivation of erythrocyte Cu_2Zn_2superoxide dismutase the occurrence of the charge isomerism was ascribed to an ageing of one native SOD species in vivo. The high flux of reactive oxygen species in the erythrocyte has led to the conclusion that hydrogen peroxide may be the inactivating compound. This phenomenon was examined in more detail.

RESULTS AND DISCUSSION

The aqueous isolation of erythrocyte Cu_2Zn_2superoxide dismutase was substantially improved and successfully employed. The overall yield of purified homogeneous enzyme was raised from 25% to 50% of originally present SOD. In the course of electrophoresis a three band pattern is observed. Only two charge isomers I and II are resolved during both chromatofocusing or electrofocusing. The observed gel electrophoretical three band pattern as well as the doublet in isoelectric focusing were initially seen and when the finally purified material was examined. Thus, the loss of a possible third component during the isolation can be excluded. The isoelectric points of charge isomers I and II were 5.2 and 4.9, respectively, and support previous work on the chloroform/ethanol treated enzyme (2). The intensity of charge isomer II is significantly lower using aqueously isolated Cu_2Zn_2superoxide dismutase (Table I).

Possible age dependence of the isoelectric focusing of enzymes obtained from old and young cows were not seen. There is strong evidence that charge isomer II cannot be considered a converted charge isomer I. Neither X-ray photoelectron spectroscopy nor amino acid analyses reveiled the presence of oxidized sulfur in the respective protein. Regardless of age the circular dichroism of charge isomer I at 261 nm was twice as high compared to that of charge isomer II. Upon heating the aqueously isolated protein to 77°C a transient increase of

TABLE I

PHYSICOCHEMICAL DATA OF Cu_2Zn_2SUPEROXIDE DISMUTASE CHARGE ISOMERS

Variable	Aqueous isolation		Cu_2Zn_2SOD organic solvent pretreatment	Charge isomer I + H_2O_2
	Charge isomer I	II		
Electronic absorption $\epsilon_{259}(mol^{-1}cm^{-1})$	9840 ± 50	9820 ± 50	9840 ± 50	7500 ±50
Circular dichroism				
θ_{208} $(\frac{deg \cdot cm^2}{dmol})$	-6000	-6000	-6000	-6000
θ_{261}	20000	10000	20000	-
θ_{295}	-	-	-	10000
EPR				
$g_{\perp}$	2.062	2.061	2.062	2.052
$g_{\parallel}$	2.263	2.262	2.263	2.262
$A_{\parallel}$ (cm^{-1})	0.014	0.013	0.014	0.013
XPS-spectroscopy	no oxidized sulfur			partial oxidation of sulfur
Amino acid analysis	no loss of histidine			loss of histidine
Electrophoresis	two-band	one-band	three-band	four-band
Isoelectric focusing	one-band	one-band	two-band	three-band
Superoxide dismutase activity (U/mg) NBT assay	1700	1600	1700	600
Copper content (µg/mg protein)	3.6	3.4	3.7	1.9

the specific enzymic activity by 150% was measured. This increase was not observed when freeze dried material was used.

Hydrogen peroxide was used as a model system for the generation of charge isomer II from charge isomer I. After treatment of isomer I with H_2O_2 at pH 7.1 the band intensity of charge isomer II increased dramatically both during isoelectric focusing and/or gel electrophoresis. However, the spectroscopical data differed markedly.

CONCLUSION

It is concluded, that this isomer is not generated by hydrogen peroxide inactivation of charge isomer I in vivo. Aqueously isolated Cu_2Zn_2superoxide dismutase from red blood cells of 1-2 years old cows was identical to the enzyme obtained from 12-15 year old species. Hence, ageing of the erythrocyte was not attributed to an ageing of superoxide dismutase.

ACKNOWLEDGEMENT

This study was aided by DFG grant We 401/17-1.

REFERENCES

1. Gärtner A, Schroth-Pollmann M, Weser U (1985) Inorg Chim Acta 107:117-125
2. Civalleri L, Pini C, Rigo A, Federico R, Calabrese L, Rotilio G (1982) Mol Cell Biochem 47:3-9

© 1986 Elsevier Science Publishers B.V. (Biomedical Division)
Superoxide and Superoxide Dismutase in Chemistry,
Biology and Medicine; G. Rotilio, editor

REVERSIBLE ACYLATION OF Cu, Zn SUPEROXIDE DISMUTASE BY CITRACONIC ANHYDRIDE.

MARMOCCHI F.°, MAVELLI I.°°, ROSSI L.°°°, BOSSA F.°°, and ROTILIO G.°°°
° Inst. Biol. Chemistry Univ. Camerino, Camerino - Italy, °°Inst. Biol. Chemistry Univ. Rome and CNR Center for Molecular Biology, Rome - Italy, °°° Dept. Biology 2nd Univ. Rome "Tor Vergata", Rome - Italy.

INTRODUCTION

Modification of lysine ε-amino groups by the acylating reagent citraconic anhydride leads to drastic changes of the protein charge which can influence both the function and the quaternary structure of the protein. Because of its reversible nature, citraconic anhydride can be used as a not-denaturing dissociating agent in dissociation-reassociation studies (1). In this study the reaction of bovine erythrocyte Cu,Zn superoxide dismutase with citraconic anhydride is described.

MATERIALS AND METHODS

Bovine erythrocyte Cu, Zn superoxide dismutase (BESOD) was purified as previously described (2). Bovine erythrocyte carbonic anhydrase was obtained as a by-product of the BESOD preparation method. Horse heart cytochrome C type III, horse muscle myoglobin type I and bovine serum albumin were purchased from Sigma Chemical Co.

Citraconylation was performed as previously described (3) with minor modifications. Citraconic anhydride (2-methylmaleic anhydride, Fluka) was added in small portions to BESOD (15 mg), in 0.1 M sodium phosphate + NaOH (pH = 9) over 60 min at 25°C under continuous stirring. During the course of the reaction the pH was maintained at pH = 9.0 $\pm$ 0.4 by addition of 5 M NaOH. The excess of reagent was removed by extensive dialysis.

Decitraconylation was performed by lowering the pH of the protein solution to pH = 4.0 upon addition of 1 M HCl followed by incubation at 40°C for 12 hours, or exhaustive dialysis against 0.3 mM $CuSO_4$ and $ZnSO_4$ in HCl at pH = 4 and room temperature for 48 h. Finally the pH of the protein solution was gradually raised to pH = 9 by either addition of sodium phosphate + NaOH or dialysis.

Evaluation of the extent of lysine side chain modification,

polyacrylamide gel electrophoresis (PAGE) and 0.1 % SDS-PAGE were carried out as described (4). Isoelectric focusing in the pH range 3.5-9 was performed in a LKB 2117 Multiphor System. SOD activity was assayed polarografically (5). EPR spectra were recorded at X band and liquid N_2 temperature by a Varian E-9 Spectrometer.

RESULTS AND DISCUSSION

Fig. 1 (Panel A) shows PAGE gels of citraconylated (CT-BESOD) and de-citraconylated (de-CT-BESOD) BESOD in comparison with the native protein. CT-BESOD showed a single fast anodic band indicating extensive citraconylation of lysines as confirmed by amino acid analysis. After decitraconylation, in spite of full recovery of the original lysine content, the protein only partially recovered the original electrophoretic behaviour as indicated by the presence of a multiple band pattern, with increased anodic mobility. Furthermore, the isoelectric focusing analysis of de-citraconylated protein showed a multiple pattern of bands (Fig. 1,

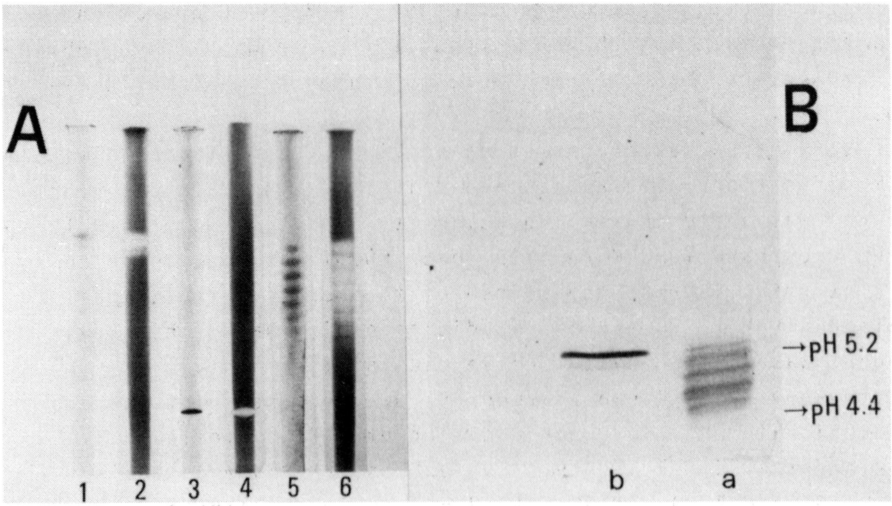

Fig. 1. PAGE (Panel A) of native BESOD (1-2) CT-BESOD (3-4) and de-CT-BESOD (5-6). Gels 1, 3, 5 were stained for protein, gels 2, 4, 6, for enzyme activity. Isoelectric focusing (panel B) of de-CT-BESOD (a) and native BESOD (b).

Panel B), located between pH = 5.2 (p.I. of the native protein) and pH = 4.4, thus confirming that modification of surface protein charge still occurs in the fully de-CT-BESOD. On the other hand no loss of copper was detected and the EPR spectrum was substantially unaltered throughout the treatment. Therefore, the band multiplicity appears to be related to charge modification not residing on lysines, possibly amide groups. If this is the case with the band multiplicity observed after de-citraconylation, this system might be a model for the electromorphs of natural SOD's (6).

Fig. 2 shows the results obtained by 0.1% SDS-PAGE; the relative enzyme activity of the various samples is also reported. It appears that, similarly to the succinylated BESOD (4), CT-BESOD shows a catalytic activity approximately 10% that of the native enzyme and a mobility in 0.1% SDS-PAGE comparable with that of myoglobin ($M_r \simeq 16,000$). The latter data suggests that the inte-

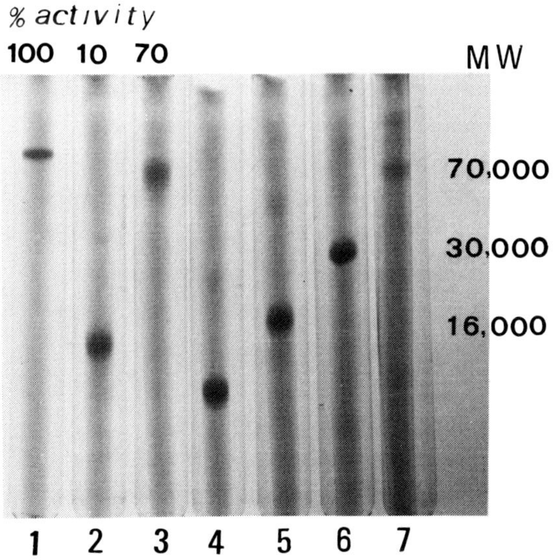

Fig. 2. 0.1% SDS-PAGE and enzyme activity (%) of native BESOD, 1; CT-BESOD, 2; de-CT-BESOD, 3. Gels 4-7 correspond to CT-cytochrome C, CT-myoglobin, CT-carbonic anhydrase and CT-bovine serum albumin respectively. The untreated proteins showed similar mobility.

raction between protein subunits is weaker than in the unreacted protein, the treatment with 0.1% SDS being not such as to bring about dissociation of native dimers (4). On the other hand de-CT-BESOD behaves as the native protein, moving as an apparent tetramer, and its enzyme activity is recovered to about 70% the value typical of the native enzyme.

These observations confirm that acylated SOD's have a less tight interaction between the two subunits (4) although the geometry of the copper site is not modified by the treatment.

In the case of CT-BESOD, the reaction is reversible and this makes the derivative suitable for the production of SOD hybrids.

ACKNOWLEDGEMENTS

This work was supported by the C.N.R. Special Project "Chimica Fine" contract n° 83.00385.95.

REFERENCES

1. Bindels J.G., Misdom L.W. and Hoenders H.J. (1985) Biochim. Biophys. Acta 828: 255-260.
2. McCord J.M., Fridovich I. (1969) J. Biol. Chem. 244, 6049-6055.
3. Atassi M.Z., Habeeb F.S. (1972) Methods in Enzymology XXV: 546-552.
4. Marmocchi F., Mavelli I., Rigo A., Stevanato R., Bossa F., Rotilio G. (1982) Biochemistry 21: 2853-2856.
5. Rigo A., Viglino P., Rotilio G. (1975) Anal. Biochem. 68: 1-8.
6. Civalleri L., Pini C., Rigo A., Federico R., Calabrese L., Rotilio G. (1982) Mol. Cell. Biochem. 47: 3-9.

FORMATION OF A METAL-DEFICIENT SUPEROXIDE DISMUTASE WITH ONE COPPER-COBALT PAIR PER DIMER

S. L. JEWETT, S. HUSETH, W. CALLAHAN
Department of Chemistry, California State University, Northridge, CA 91330

INTRODUCTION

Cobalt has been substituted for zinc in reconstituting the apoenzyme of superoxide dismutase and has proved to be a valuable spectroscopic probe for obtaining information about the enzyme active site (1-4). In this study evidence is reported indicating the formation of a metal-deficient enzyme having one copper-cobalt pair per dimer after reconstituting the apoenzyme by the simultaneous addition of equimolar copper and cobalt at pH 5.0.

MATERIALS AND METHODS

Superoxide dismutase (SOD) was purchased from Diagnostics Data, Inc. and reagents for polyacrylamide gel electrophoresis were purchased from Bio-Rad. Cobalt (II) chloride and copper (II) acetate were obtained from Mallinkrodt.

Apoenzyme was prepared at 4 °C by dialysis against the following buffers at pH 3.8: 0.01 M acetic acid and 0.01 M EDTA, 0.01 M acetic acid and 0.01 $MgCl_2$, 0.01 M acetic acid. Final dialysis was against 0.05 M acetic acid at pH 5.0 (μ = 0.1, NaCl). Preparations contained 0.5% copper and 5.0% zinc. Protein was assayed by the Lowry method using bovine serum albumin as standard (5). Electrophoresis was carried out as described by Jewett et al.(6). Reconstitution at pH 5.0 was accomplished by adding various ratios of equimolar copper and cobalt per active site. A DWZ Aminco Spectrophotometer was used for visible spectra and a Spectromagnetic Industries Model 12-700 ESR Spectrometer was used for esr measurements.

RESULTS

Equimolar copper and cobalt added simultaneously to the apoenzyme of superoxide dismutase at pH 5.0 at ratios of 0.1 to 0.4 per active site give a metal-deficient dismutase as a distinct band traveling between the apoenzyme and the native enzyme in alkaline polyacrylamide gel electrophoresis. The distribution of apoenzyme, native-like enzyme, and a metal-deficient enzyme is shown in Figure 1 as a function of added metal ions. The visible spectra are shown in Figure 2, indicating a progressive shift in absorption from 580 nm to 600 nm. The appearance of the metal-deficient band in electrophoresis coincides with this shift in wavelength maximum. The wavelength maximum at 580 nm is characteristic of a cobalt-only enzyme and the maximum at 600 nm is characteristic of an enzyme having copper-cobalt pairs in the active sites (3,4). The conclusion from these data is that

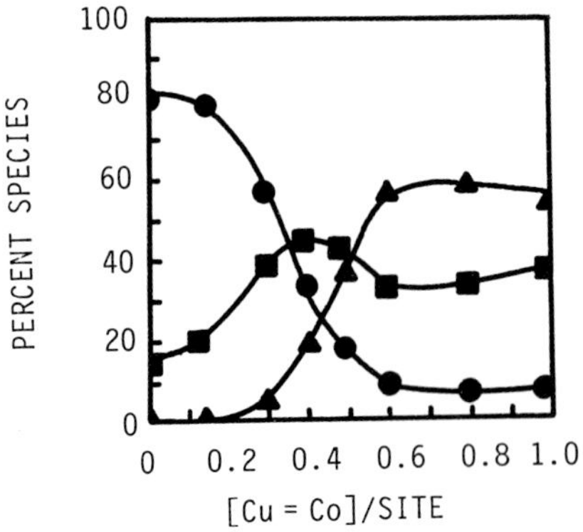

Fig. 1. Electrophoretic distribution of native enzyme (▲); apoenzyme (●); and metal-deficient enzyme (■) in the reconstitution of apoenzyme with equimolar copper and cobalt at pH 5.0

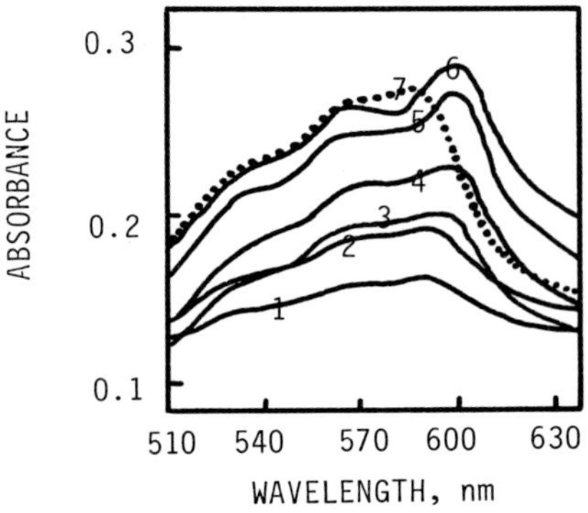

Fig. 2. Absorption spectra of apoenzyme reconstituted at various ratios of equimolar copper and cobalt per active site at pH 5.0. (1) 0.15; (2) 0.30; (3) 0.40; (4) 0.50; (5) 0.60; (6) 0.80; (7) cobalt-only enzyme

the metal-deficient species has one copper-cobalt pair per dimer.

The data from the electrophoretic pattern and the visible spectra were treated according to the equation $A_{600} = \varepsilon_1 c_1 + \varepsilon_1 c_2$. The absorbance at 600 nm is assumed to be a linear combination of the amount of metal-deficient enzyme (c_1) and its extinction coefficient (ε_1) and the amount of native-like enzyme (c_2) and its extinction coefficient (ε_2). C_1 and c_2 were determined from the total protein and the relative amounts of each species determined from electrophoresis. Figure 3

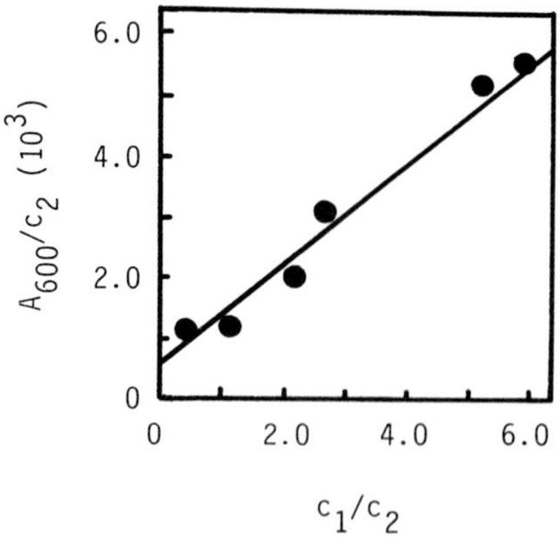

Fig. 3. Analysis of visible absorbance data at 600 nm according to the equation $A_{600}/c_2 = \varepsilon_1 c_1/c_2 + \varepsilon_2$. C_1 and c_2 are the concentrations of metal-deficient enzyme and native-like enzyme, respectively.

indicates a plot of this equation to obtain $\varepsilon_1 = 750 \pm 90\ M^{-1}\ cm^{-1}$ from the slope and $\varepsilon_2 = 650 \pm 50\ M^{-1}\ cm^{-1}$ from the y-intercept. The similarity between these extinction coefficients suggests that subunit interactions determine the chemical environment in the active sites.

Esr data indicate a progressive change in $A_{//}$ from 153 G at low ratios of added metal ions to 138 G at the higher ratios. The electrophoreses of samples at low ratios show little metal-deficient enzyme even though the esr indicates a protein containing copper. In addition at the lower metal ion ratios, the visible spectra indicate the presence of cobalt-only enzyme. The most likely interpretation of the esr and visible data at the low ratios of metal ions is that copper and cobalt add to the apoenzyme but not in the same site. The $g_{\perp}$ peak height measurements do not show a linear increase with added copper consistent with what is expected for copper-cobalt pairs being incorporated into the apoenzyme and resulting in decreased $g_{\perp}$ measurements from antiferromagnetic coupling of the copper and cobalt (7). The esr data, and the visible and electrophoretic data indicate that the metal-deficient enzyme is not formed at lower ratios of added metal ions; however appreciable quantities of the species are formed before native enzyme forms indicating that is it a precursor for the formation of native enzyme.

ACKNOWLEDGEMENT

Faculty Research Council, and the Student Projects Committee, California State University, Northridge.

REFERENCES

1. Fee JA (1973) J Biol Chem 248:4229-4234
2. Calabrese L, Cocco L, Morpurgo L, Mondovi B, Rotilio G (1975) FEBS Lett 59:29-31
3. Fee JA (1977) In: Michelson AM, McCord JM, Fridovich I (eds) Superoxide and Superoxide Dismutases. Academic Press, London, pp 173-192.
4. Rotilio G, Calabrese L (1977) In: Michelson AM, McCord JM, Fridovich I (eds) Superoxide and Superoxide Dismutases. Academic Press, London, pp 193-198.
5. Fee JA (1973) Biochim Biophys Acta 295:87-95
6. Jewett SL, Latrenta GS, Beck CM (1982) Arch Biochem Biophys 215:116-128
7. Moss TH, Fee JA (1975) Biochem Biophys Res Commun 66:799-808

Superoxide and Superoxide Dismutase in Chemistry,
Biology and Medicine; G. Rotilio, editor

NOVEL STABILIZATION OF A METALLOPROTEIN-TERNARY COMPLEX: REACTION OF COPPER-ZINC SUPEROXIDE DISMUTASE WITH DIETHYLDITHIOCARBAMATE IN POLYACRYLAMIDE GELS AND ITS APPLICATION TO THE STUDY OF THE REACTION OF THE DISMUTASE WITH HYDROGEN PEROXIDE

SANDRA L. JEWETT
Department of Chemistry, California State University, Northridge, CA 91330

INTRODUCTION

Polyacrylamide gels of bovine liver copper-zinc superoxide dismutase electrophoresed at pH 8.5 in TRIS-glycine develop a stable yellow color at the dismutase bands upon incubation with diethyldithiocarbamate dissolved in the electrophoresis buffer. The area under scans of the peaks at 448 nm is unchanged after storage of the gels at 4 °C for at least one month and probably longer. This indicates that copper is not lost from the protein in the gel. The implication of these observations is that the ternary dithiocarbamate complex is stabilized in the polyacrylamide matrix. The 448 nm absorption corresponds to the absorbance maximum of the copper-dithiocarbamate complex in solution. The area under the peaks in scans of the gels is a linear function of the amount of dismutase in the peak. This method is applied to the study of the reaction of the dismutase with hydrogen peroxide, a reaction which causes a loss of copper and an increase in heterogeneity (1). The application of this procedure to other metalloproteins will depend upon the stability of the copper-dithiocarbamate complex and the resistance of the copper to removal during the incubation of the gels with the dithiocarbamate.

MATERIALS AND METHODS

Bovine liver superoxide dismutase (SOD) was purchased from Diagnostics Data, Inc. A major band comprised 90% of the total protein. Chemicals for gel electrophoresis were purchased from Bio-Rad. Diethyldithiocarbamate (DDC) was purchased from Sigma. Protein concentrations were determined by the method of Lowry using bovine serum albumin as standard (2). For unmodified preparations of the enzyme, an extinction coefficient at 258 nm of $10.3 \times 10^{3}\ M^{-1}\ cm^{-1}$ (3) was used.

Electrophoresis was carried out as described by Jewett et al. (4) in 0.05 M TRIS - 0.38 M glycine at pH 8.5. To determine the relative amounts of protein in bands, gels containing 2 µg to 10 µg of SOD were stained with Coomassie Brilliant Blue G-250 (5). In order to stain for copper, gels containing 0.1 mg to 0.9 mg of SOD were incubated for 24 hours at room temperature or at 4 °C with 0.01 M DDC dissolved in the electrophoresis buffer, then stored at 4 °C. To determine the amount of protein in a particular band stained for copper with DDC, the total protein loaded onto the gel was multiplied by the relative amount of protein in the band determined from areas under the bands in gels stained with Coomassie Blue.

Solutions of the dismutase (1.5×10^{-4} M) were reacted for 24 hours at 25 °C with hydrogen peroxide at pH 10.0 in 0.048 M sodium carbonate buffer with 0.1 mM EDTA. Concentrations of hydrogen peroxide in aqueous stock solutions ($\varepsilon_{240\ nm}$ = 43.6 M^{-1} cm^{-1} (6)) were determined immediately before reactions with the dismutase.

RESULTS

The formation of a stable DDC complex of SOD in polyacrylamide gels is characterized by an absorbance maximum at 448 nm comparable to the maximum of the Cu^{+2}-$(DDC)_2$ complex in solution as well as the maximum of the SOD-DDC complex in solution (7). As seen in Figures 1 and 2 the area under peaks in scans of the gels at 448 nm is a linear function of the amount of dismutase in the peak.

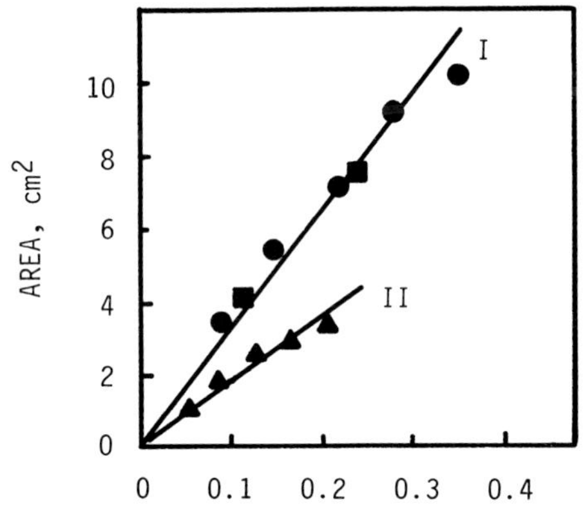

Fig. 1. Area of peaks in DDC-stained gels of SOD treated with 1.0 H_2O_2 per active site. Major band I (●); minor band II (▲); untreated native enzyme (■).

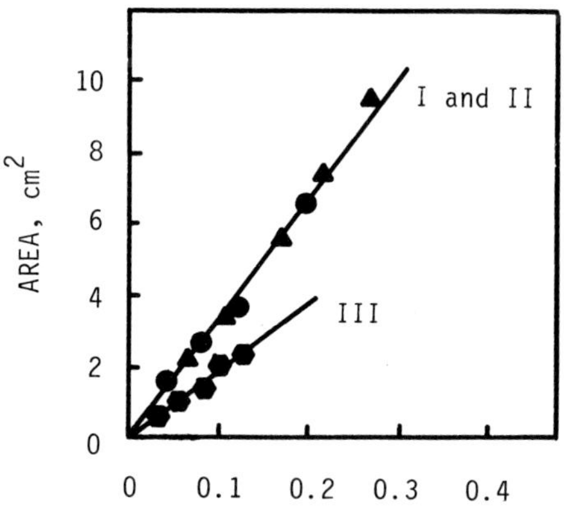

Fig. 2. Area of peaks in DDC-stained gels of SOD treated with 4.0 H_2O_2 per active site. Band I (●); band II (▲); and band III (⬢).

This staining method was used to probe the copper content of the dismutase in the reaction with hydrogen peroxide. This reaction leads to the loss of copper and an increase in heterogeneity of the enzyme (1). At 1.0 peroxide per active site (Figure 1) the relative copper content of band I (mobility comparable to native enzyme, 34 cm^2/mg) and band II (faster moving component, 18 cm^2/mg) suggest that band II has half the copper of band I. At a higher ratio of added peroxide (Figure 2) bands I and II appear to have the same copper content. A third band III (faster than II, 19 cm^2/mg) appears at this higher ratio of added peroxide and has half the copper of bands I and II. For an intermediate ratio of 2.0 peroxide per active site, the slope of the line for band II (25 cm^2/mg) indicates an intermediate copper content as if it were a mixture of forms containing one and two copper ions per enzyme. The reaction with hydrogen peroxide, though leading to the loss of copper, also leads to other modifications which alter the mobility of the enzyme. Both changes contribute to the heterogeneity and give rise to species that coelectrophorese in two different systems (5). The nature of this reaction is under further investigation.

The long term stability of the SOD-DDC complex in gels was unexpected since it is known that DDC removes copper from the enzyme (7,8). However, this complex is stable in gels for at least a month, making it particularly useful in the analysis of copper in the dismutase in gels. This method should be capable of determining the copper content in other proteins. One limitation of this method involves the fact that milligram (0.1 to 1.0 mg) quantities of protein must be loaded onto gels in order to obtain a band that can be scanned at 448 nm.

ACKNOWLEDGEMENTS

Faculty Research Council and Students Projects Committee, California State University, Northridge.

REFERENCES

1. Jewett SL (1983) Inorg Chim Acta 79:144-145
2. Fee JA (1973) Biochim Biophys Acta 295:87-95
3. Rotilio G, Calabrese L, Bossa F, Barra D, Agrò AF, Mondovi B (1972) Biochemistry 11:2182-2187
4. Jewett SL, Latrenta GS, Beck CM (1982) Arch Biochem Biophys 215:116-128
5. Blakesley RW, Boezi JA (1977) Anal Biochem 82:580-582
6. Misra HP, Fridovich I (1977) Arch Biochem Biophys 183:511-515
7. Misra HP (1979) J Biol Chem 254:11623-11628
8. Cocco D, Calabrese L, Rigo A, Marmocchi F, Rotilio G (1981) Biochem J 199:675-680

STUDIES INVOLVING THE REDUCTION OF SOD-Cu^{2+} BY CO_2^- RADICALS

DIANE E. CABELLI AND BENON H. J. BIELSKI
Chemistry Department, Brookhaven National Laboratory, Upton, NY 11973 (USA)

INTRODUCTION

Pulse radiolysis and other radiolytic techniques have proven very potent tools in the elucidation of the mechanism of O_2^- dismutation by superoxide dismutases [1]. Some recent work in this laboratory involved a systematic study of the reduction of SOD-Cu^{2+} (bovine liver) by CO_2^- radicals and the kinetics of the interaction between the resultant SOD-Cu^+ and peroxide, a process that is known to inactivate the enzyme and thought to generate a bound OH radical [2,3].

DETERMINATION OF THE G VALUE

Water, when radiolyzed, yields the following species [4]

$$H_2O \rightsquigarrow OH(2.75),\ e_{aq}^-(2.65),\ H(0.65),\ H_2O_2(0.70),\ H_2(0.45) \qquad (I)$$

where the values in brackets are G values, that is the number of molecules produced per 100 eV of energy dissipated in the system. In N_2O saturated systems containing formate the primary radicals all react to yield, ultimately, CO_2^-, where $G(CO_2^-)$ ~6.2 [5].

The change in OD at 680 nm upon ^{60}Co γ irradiation of an N_2O-saturated solution containing 0.01 M formate and 50 μM SOD-Cu^{2+} at pH 6.8 is plotted versus irradiation time; see Figure 1.

As SOD-Cu^{2+} contains two coppers [1] per SOD unit, assumptions that both Cu^{2+} centers are reducible by CO_2^- and that each contributes half of the observed absorbance lead to the mechanism given by reactions (1) and (2) and a

$$\text{SOD-}Cu^{2+},Cu^{2+} + CO_2^- \longrightarrow \text{SOD-}Cu^{2+},Cu^{+} + CO_2 \qquad (1)$$

$$\text{SOD-}Cu^{2+},Cu^{+} + CO_2^- \longrightarrow \text{SOD-}Cu^{+},Cu^{+} + CO_2 \qquad (2)$$

$G(-(\text{SOD-}Cu^{2+})) = 5.8$. This corroborated the observation in the pulse radiolysis experiments where the magnitude of the absorbance at 250 nm indicated that $G(CO_2^-)$ ~ 6.0, as expected, while the corresponding loss of absorbance at 680 nm indicated that $G(-(\text{SOD-}Cu^{2+}))$ ~ 3.0 per copper.

RATE STUDIES

The rate of reaction (1) was measured by pulse radiolysis under pseudo first-order conditions (0.1-0.01 M formate, pH 6.8, N_2O saturated, 25.1 °C); $k_3 = (7.9 \pm 1.8) \times 10^8\ M^{-1}\ s^{-1}$ (see Table I).

This rate constant was calculated assuming the stoichiometry given in reaction 3.

$$\text{SOD-}Cu^{2+},Cu^{2+} + 2CO_2^- \longrightarrow \text{SOD-}Cu^{+},Cu^{+} + 2CO_2 \qquad (3)$$

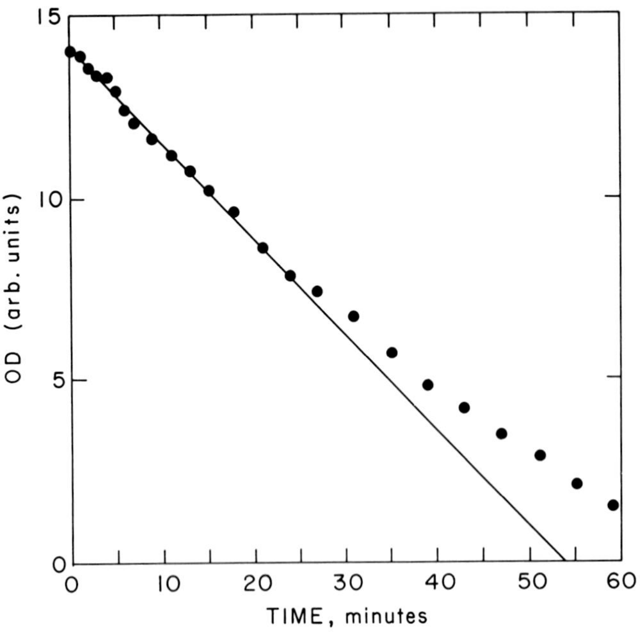

Fig. 1. Change in OD vs irradiation time; measured at 680 nm.

TABLE I
RATE OF REACTION OF SOD-Cu^{2+} WITH CO_2^-

[SOD-Cu^{2+}], μM	k_3, M^{-1} sec^{-1}	[SOD-Cu^{2+}], μM	k_3, M^{-1} sec^{-1}
100	(0.90±0.10) x 10^9	18	(0.63±0.05) x 10^9
50	(1.06±0.15) x 10^9	12.5	(0.71+0.11) x 10^9
25	(0.61±0.16) x 10^9		

In contrast to SOD-Cu^{2+} reduction by H_2O_2, radiolytic reduction by CO_2^- can be carried virtually to completion, with very little loss of activity [6].

REACTION BETWEEN Cu(II)SOD AND H_2O_2

Peroxide is known to reduce and inactivate SOD-Cu^{2+}; the latter process was postulated to occur by a Fenton-type reaction that results apparently in a bound OH radical [2]

$$\text{SOD-Cu}^{2+} + H_2O_2 \longrightarrow \text{SOD-Cu}^{+} + O_2^- + 2H^+ \quad (4)$$

$$\text{SOD-Cu}^{+} + H_2O_2 \longrightarrow \text{SOD-Cu}^{2+}\text{-OH} + OH^- \quad (5)$$

It was suggested that the interaction of the bound OH with the histidines near the copper center(s) results in this inactivation [3]. While previous studies measured the rate of inactivation of the enzyme [2,3,7], we were interested in measuring the rate of reaction (5) itself. Formate is known to protect the enzyme against deactivation by peroxide [2], suggesting that it can get into the active site and scavenge the OH radical before OH can react with the histidine groups. Therefore, if the enzyme is reduced in the presence of formate and N_2O, addition of peroxide and nitroblue tetrazolium (NBT^{2+}) to this system results in reaction (5), followed by:

$$OH \text{ (or SOD-Cu}^{2+}\text{-OH)} + HCO_2^- \longrightarrow OH^- + CO_2^- + H^+ \qquad (6)$$

$$2CO_2^- + NBT^{2+} \longrightarrow 2CO_2 + MF^+ \qquad (7)$$

where $k_6 = 3 \times 10^9$ M^{-1} sec^{-1} [8] and $k_7 = 6.4 \times 10^9$ M^{-1} sec^{-1} [9], both clearly very fast compared to reaction (5). The rate of formation of monoformazan (MF^+), easily monitored at 530 nm ($\varepsilon = 25{,}400$ M^{-1} cm^{-1} [9]), will yield 2 k_5. Our preliminary results indicate that the reaction is first order in peroxide ($[H_2O_2]$ = 3-18 mM, [SOD] = 20 μM, pH 7.0), $k_5 \sim (2\pm1)$ M^{-1} sec^{-1}, in reasonable agreement with previously determined rates of inactivation of the enzyme in the presence of H_2O_2 ($k_{inact} = 6.7$ M^{-1} sec^{-1} [2] and $k_{inact} = 3.1$ M^{-1} sec^{-1} [3] at pH 10 and pH 9 respectively). As $k_4 \gg k_5$, reactions (4) and (5) imply that the enzyme can serve as an OH radical generator; indications of this were found in that $[MF^+]$ produced was greater than that predicted from the stoichiometry of reactions (5-7) alone.

ACKNOWLEDGEMENTS

This research was supported by NIH Grant R01GM23658-09 and was carried out at Brookhaven National Laboratory, which is operated under contract DE-AC02-76CH00016 to the U.S. Department of Energy.

REFERENCES

1. Steinman HM (1982) In: L. W. Oberly (ed) Superoxide Dismutase. CRC Press, Boca Raton, FL, USA, Vol. 1, pp. 11-68
2. Hodgson EK, Fridovich I (1975) Biochemistry 14:5294-5303
3. Bray RC, Cockle SA, Fielden EM, Roberts PB, Rotilio G, Calabrese L (1974) Biochem J B9:43-48
4. Schwarz HA (1981) J Chem Ed 58:101-105
5. Bielski BHJ (1978) Photochem Photobiol 28:645-649
6. Michelson AM, Puget K (1981) In: W. Bors, M. Saran, D. Tait (eds) Oxygen Radicals in Chemistry and Biology. Walter de Gruyter, Berlin, pp. 831-842
7. Chelack WS, Petkau A (1981) Biochim Biophys Acta 660:83-90
8. Farhataziz, Ross AB (1977) NSRDS-NBS 59
9. Bielski BHJ, Shiue GG, Bajuk S (1980) J Phys Chem 84:830-833

Superoxide and Superoxide Dismutase in Chemistry, Biology and Medicine; G. Rotilio, editor

REDUCTION AND OXIDATION OF BOVINE SUPEROXIDE DISMUTASE BY H_2O_2

ADELIO RIGO[*], PAOLO VIGLINO[+], MARINA SCARPA[*] and GIUSEPPE ROTILIO[°]
[*]Institute of Biological Chemistry, University of Padova, Padova, Italy
[+]Department of Spectroscopy, University of Venice, Venice, Italy
[°]Institute of Biological Chemistry, University of Rome, Rome, Italy

INTRODUCTION

The reaction of H_2O_2 with oxidized Cu, Zn superoxide dismutase (SOD) has been studied in the past years (1-4) and it has been demonstrated to proceed by the reduction of the copper ion followed by inactivation of the enzyme. However, on the basis of the redox potentials, H_2O_2 appears to be more efficient as an oxidant than as a reductant (5). In an attempt to get a better knowledge of the redox chemistry of the enzyme-bound copper ions, we have re-investigated the reaction between SOD and H_2O_2. Some preliminary observations are reported here.

MATERIAL AND METHODS

All chemicals were of reagent grade. Native SOD was purified from bovine erythrocytes according to McCord and Fridovich (6). H_2O_2 was standardized by measuring the height of its polarographic wave. The concentration of the oxidized SOD (ECu^{++}) was monitored by NMR measurements of the paramagnetic contribution to the water protons relaxation rate (T_{1p}^{-1}) as previously reported (7), while the enzyme activity was measured by polarography (8). $NaBH_4$ was utilized to prepare the reduced SOD (ECu^{+}) (9). All experiments were performed at 24°C in solutions equilibrated with air.

RESULTS

Plots A and B of Fig. 1 are the time courses of the ECu^{++} concentration in an experiment with 4 mM H_2O_2 added both to native and reduced SOD at pH 7. From this Figure it appears that ECu^{++} is partially reduced, while ECu^{+} is partially oxidized by H_2O_2. The analysis of the plots A and B shows that the oxidation and the reduction of SOD by H_2O_2 are not simple kinetic processes and the final level of ECu^{++}, $(ECu^{++})_e$, appears to be practically independent

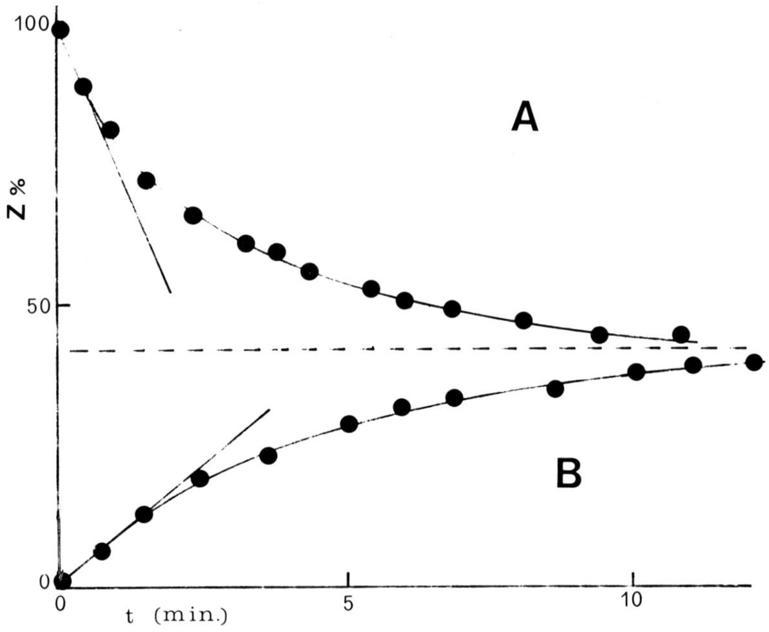

Fig. 1

Reduction and oxidation of 0.15 mM SOD by 4.0 mM H_2O_2 in 0.1 M phosphate buffer, pH 7.0.

of the initial oxidation state of the enzyme. Furthermore the increase of H_2O_2 concentration up to a factor of 4 had little influence on $(ECu^{++})_e$. Similar results were also obtained in experiments carried out in the pH range 5.5-7.0. No experiment was performed at pH $>$ 7 since, at these pH values, the oxidation of ECu^+ by O_2 becomes competitive with the oxidation and reduction of SOD by H_2O_2. The $(ECu^{++})_e$ values, measured in experiments performed at pH 5.5, 6.3 and 7.0, are reported in Table 1 together with some kinetic parameters. The enzyme inactivation was negligible during times of experimental interest. In fact the SOD activity was always higher than 80% at pH 7.0 and was 100% at pH $\leq$ 6.3.

TABLE 1

KINETIC PARAMETERS OF THE SYSTEM SOD-H_2C_2.

SOD and H_2O_2 concentrations were 0.15 and 4.0 mM respectively.

pH	$(ECu^{++})_{initial}$ %	$(ECu^{++})_e$ %	k_1* $M^{-1}s^{-1}$	k_3* $M^{-1}s^{-1}$	k_1/k_3*	k_1/k_3**
5.5	100	46	0.066	-	1.60	1.54
5.5	12	40	-	0.041		
6.3	100	39	0.32	-	2.13	2.15
6.3	9	37	-	0.15		
7.0	100	41	0.56	-	1.80	2.00
7.0	9	37	-	0.31		

* From initial rates.

** From eqn. 7 (the average values of $[ECu^{++}]_e$ were used).

DISCUSSION

The experimental data show that, in the presence of H_2O_2, besides the reduction of SOD, a reaction of oxidation occurs and, as a consequence, the $[ECu^{++}]_e$ corresponds to a steady-state due to an equilibrium between opposite reactions. Although, under our experimental conditions, $[ECu^{++}]_e$ was relatively higher than that previously reported (2), this behaviour is in accordance with the findings of incomplete reduction of the native SOD by H_2O_2. However the presence of unreacted SOD was attributed to impurities or to inhomogeneity of the enzyme. Since it appears likely that both oxidation and reduction of SOD by H_2O_2 are one-electron reactions, a plausible mechanism is:

$$1) \quad ECu^{++} + H_2O_2 \underset{k_{-1}}{\overset{k_1}{\rightleftharpoons}} ECu^{+} + O_2^{-} + H^{+}$$

$$2) \quad ECu^{++} + O_2^{-} \xrightarrow{k_2} ECu^{+} + O_2$$

$$3) \quad ECu^{+} + H_2O_2 + H^{+} \underset{k_{-3}}{\overset{k_3}{\rightleftharpoons}} ECu^{++} + OH^{\circ} + H_2O$$

On thermodynamic basis, considering the E°' values of one-electron reactions of H_2O_2 (5), equilibria 1 and 3 are left-displaced. Since, under our experimental conditions, the reactions 1 and 3 appear to occur from left to right, the driving force should be the disappearence of the O_2^-, via reaction 2, and of the highly reactive OH° radical through reactions with some components of the system. In other words, the reduction and oxidation of SOD are kinetically controlled by the one-electron reactions of H_2O_2, see eqns. 1 and 3, and are thermodinamically controlled by the E°' of the couples (O_2/H_2O_2) and (H_2O_2/H_2O) since the final products are O_2 and H_2O respectively. Furthermore, being $k_{-1} = k_2$ (10), it appears that the O_2^- generated in reaction 1 reacts mainly with ECu^{++} at $Z > 0.50$, where $Z = [ECu^{++}]/([ECu^+] + [ECu^{++}])$, increasing the reduction rate of the enzyme, while at $Z > 0.50$ the O_2^- reacts mainly with ECu^+ increasing the oxidation rate. These observations explain qualitatively the $[ECu^{++}]$ time courses of Fig. 1. According to eqns. 1-3 we can write:

$$4)\ d[ECu^{++}]/dt = -k_1[ECu^{++}][H_2O_2] - k_2[ECu^{++}][O_2^-] + k_{-1}[ECu^+][O_2^-] + k_3[ECu^+][H_2O_2]$$

$$5)\ d[O_2^-]/dt = k_1[ECu^{++}][H_2O_2] - k_{-1}[ECu^+][O_2^-] - k_2[ECu^{++}][O_2^-]$$

Let $d[O_2^-]/dt = 0$, from eqns. 4 and 5 we obtain:

$$6)\ [H_2O_2]^{-1}\, dZ/dt = -2k_1Z^2 - k_3Z + k_3$$

At the equilibrium we can write:

$$7)\ [ECu^{++}]_e = Z_e = (-1 + \sqrt{1 + 8R})\ /\ 4R$$

where $R = k_1/k_3$, from which the k_1/k_3 ratio can be calculated. On the other hand the limits of eqn. 6 for $Z \rightarrow 1$ and $Z \rightarrow 0$ are the initial rates of reduction $(v_{red})_\circ$ and oxidation $(v_{ox})_\circ$ of SOD by H_2O_2, that is $(v_{red})_\circ = 2k_1[H_2O_2]$ and $(v_{ox})_\circ = k_3[H_2O_2]$. The experimental values of k_1 and k_3 calculated from the initial rates, and their ratio R, together with the R values obtained from eqn. 7 are reported in Table 1. Although the k_1 and k_3 values so calculated are subjected to large errors, their ratio appears in agreement with the R values calculated from the Z_e values. Furthermore, it should be remarked that, no significative loss of enzyme activity was observed in the oxidation process of SOD, when, according to reaction 3, OH° radicals are generated. Therefore it appears that the probability of the reaction between OH° radicals and active site histidines, which is responsible for SOD inactivation (2), is relatively low. If this hypothesis is correct, some oxidation and reduction cycles of the copper ion in the enzyme should occur to get a substantial enzyme inactivation. These considerations and the kinetic scheme reported above, which correspond to the overall reaction:

8) $2\ H_2O_2 \longrightarrow H_2O + O_2$

are in line with a catalatic action of SOD. Finally it should be noted that the similar values of k_1 and k_3 seem to support the hypothesis that the oxidation and the reduction of the copper in the enzyme are mainly controlled by factors that are independent of the oxidation state of the copper ion.

REFERENCES

1. Rotilio G, Morpurgo L, Calabrese L, Mondovì B (1973) Biochim Biophys Acta 302: 229-235

2. Bray R, Cockle S, Fielden M, Roberts P, Rotilio G, Calabrese L (1974) Biochem J 139: 43-48.

3. Hodgson E, Fridovich I (1975) Biochemistry 14: 5294-5299.

4. Fuchs H, Borders C (1983) Biochem Biophys Res Commun 116: 1107-1113.

5. Fee J, Valentine J (1977) In: Michelson A, McCord J, Fridovich I (eds) Superoxide and superoxide dismutases. Academic Press, London, pp 19-60.

6. McCord J and Fridovich I (1969) J Biol Chem 244: 6049-6055.

7. Terenzi M, Rigo A, Franconi C, Mondovì B, Calabrese L, Rotilio G (1974) Biochim Biophys Acta 351: 230-236.

8. Rigo A, Viglino P, Rotilio G (1975) Anal Biochem 68: 1-8.

9. Viglino P, Scarpa M, Cocco D, Rigo A (1985) Biochem J 229: 87-90.

10. Fielden M, Roberts P, Bray R, Lowe D, Mautner G, Rotilio G and Calabrese L (1974) Biochem J 139: 49-60.

Superoxide and Superoxide Dismutase in Chemistry,
Biology and Medicine; G. Rotilio, editor

LOW-TEMPERATURE ESR STUDIES ON REACTIVITY OF SOD WITH PRIMARY PRODUCTS OF WATER RADIOLYSIS IN AQUEOUS SOLUTIONS AND GELS

ANDRZEJ PLONKA, DIANA METODIEWA
Institute of Applied Radiation Chemistry, 93-590 Lodz, Poland

ALOJZY ZGIRSKI, WANDA LEYKO
Institute of Biochemistry and Biophysics, 90-273 Lodz, Poland

INTRODUCTION

All species of protein-bound copper have the same general function consisting in electron transfer to oxygen. This function, however, is fulfilled through different specialized reactions which so far were followed mainly in dilute aqueous solutions.

We do not think that dilute aqueous solutions are good models of biological systems in which there is mainly the bound water. Because of that we decided to follow the specific reactions of SOD in cross-linked dextran gels which are used extensively in one of our laboratories for ESR studies at low temperature and marked change of SOD reactivity towards electrons was observed.

MATERIALS AND METHODS

Freeze-dried powder of SOD isolated from bovine blood (2900 units/mg) was supplied by Sigma Chemical Co. Its purification, activity measurements, preparation of samples for γ-irradiation and ESR measurements were described previously (1,2). To embed the SOD solution (10 mg/ml, 0.004 potassium phosphate buffer, pH = 7.2) in the cross-linked dextran, Sephadex G-200 (Pharmacia, Uppsala), the respective components were mixed in amounts corresponding to the sorption capacity of Sephadex.

RESULTS

The spectrum of native SOD in polycrystalline ice at 77 K ($g_\perp$ = 2.0682, $g_\parallel$ = 2.298, $A_\parallel$ = 134 G) was discussed recently (1). The effect of embedding of SOD into cross-linked dextran, shown in Fig. 1B ($g_\perp$ = 2.0663, $g_\parallel$ = 2.2498, $A_\parallel$ = 117 G), consist in the change from rhombic to axial symmetry. Irradiation of embedded SOD at 77 K in the dark results in formation of trapped OH radicals and carbon-based radicals derived from cross-linked dextran (Fig. 2A). Photobleaching with wide band visible light (up to 20 min.) has no effect on this spectrum. Subsequent heat treatment for 5 min. at succesively higher temperatures reveals the oxygen-based radicals we are interested in, cf. Fig. 2C ($g_{\perp}$ = 2.006, $g_{\parallel}$ = 2.267). Similar radicals were observed under these experimental conditions for SOD in polycrystalline ice (1).

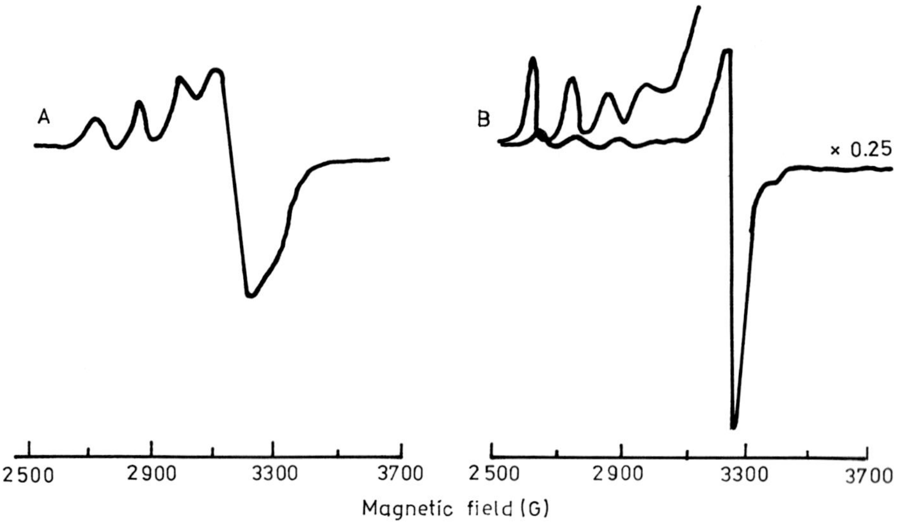

Fig. 1. ESR spectra recorded at 77 K for SOD in frozen 0.001 M potassium phosphate buffer (A) and in frozen cross-linked dextran gel (B).

The changes of the low-field part of the ESR spectrum of SOD accompanying those in the high-field part are shown in Fig. 3. It is worth to note that thermal annealing from 170 to 270 K does not affect quantitatively the ESR signal (Fig. 3I).

The effect of embedding of SOD into cross-linked dextran consists in the change of reactivity towards electrons generated radiolytically at 77 K. In frozen pure cross-linked dextran gel γ-irradiated at 77 K in dark the trapped electrons are observed. They are completely scavenged by the embedded SOD (Fig. 2A). The scavenged electron is attached to the active site of SOD and modifies the environment of copper atoms (cf. Fig. 3). Upon thermal annealing to 170 K these changes are removed, cf. Fig. 3A and 3I, and formation of superoxide radicals is observed, cf. Fig. 2C.

One possible explanation of these phenomena is that in the case of SOD embedded in the cross-linked dextran gels electrons produced at 77 K are scavenged by the modified active sites from which they are transferred to oxygen molecules at higher temperature.

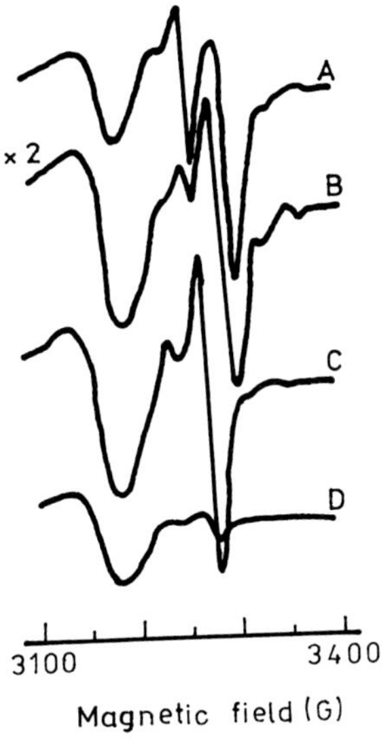

Fig. 2. ESR spectra recorded at 77 K for SOD embedded into cross-linked dextran gel: γ-irradiated at 77 K (A), and thermally annealed for 5 min. at 120 (B), 170 (C) and 210 K (D).

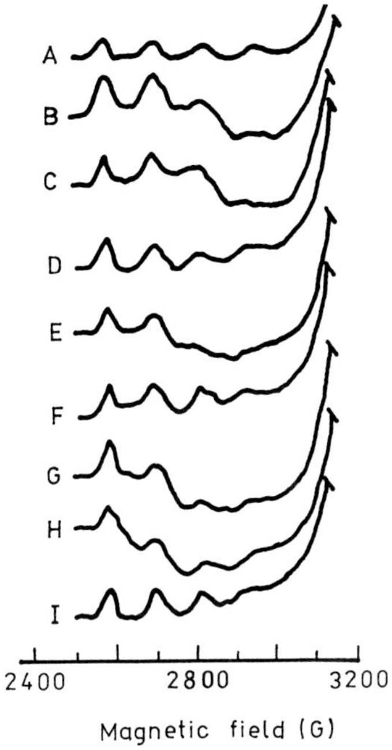

Fig. 3. Low-field parts of ESR spectra recorded at 77 K for SOD embedded into cross-linked dextran gel before (A) and after (B) γ-irradiation at 77 K in dark; photobleached with wide band visible light (C), thermally annealed for 5 min. at 100 (D), 110 (E), 120 (F), 130 (G), 150 (H) and 170 K (I).

Thus formed superoxide radicals remain trapped in the form of the complex $[E\text{-}Cu^{2+}...HO_2]$, like in polycrystalline ice (1). Further thermal annealing up to 270 K leads to the disappearance of that complex without changes in the oxidation state of copper ions, which was assumed to be the evidence of dismutation (2).

ACKNOWLEDGEMENT

This work was performed under contract R.III.13.2.

REFERENCES

1. Plonka A, Metodiewa D, Jeziorowska A (1978) Biochem Biophys Res Comm 81: 1344-1351
2. Plonka A, Metodiewa D, Zgirski A, Hilewicz M, Leyko W (1980) Biochem Biophys Res Comm 95: 978-985

© 1986 Elsevier Science Publishers B.V. (Biomedical Division)
Superoxide and Superoxide Dismutase in Chemistry, Biology and Medicine; G. Rotilio, editor

PULSE RADIOLYSIS STUDY OF SUPEROXIDE DISMUTASE IN HUMAN BLOOD POLYMORPHONUCLEARS

C. PASQUIER*, C. MARQUETTY*, C. HOUEE**, C. FERRADINI**, J. HAKIM*
*Laboratoire d'Hématologie, Faculté Bichat, Université Paris 7, 16 rue Huchard, 75018 Paris (France).
** Université Paris 5, 45 rue des Saints Pères, 75006 Paris (France).

INTRODUCTION

Superoxide dismutase (SOD) catalyses the dismutation of $O_2^{\bar{\cdot}}$, superoxide anion, in all cells, according to the reaction :

$$O_2^{\bar{\cdot}} + O_2^{\bar{\cdot}} + 2H^+ \longrightarrow H_2O_2 + O_2$$

In all eukariotic cells, two types of SOD are present. One the CuZn SOD is cyanide sensitive, and the other, the Mn SOD, is cyanide insensitive ; they are located in the cytosol and the mitochondria respectively.

In blood polymorphonuclear cells (PMN) the measurements of the cytosolic Cu-Zn-SOD and of the Mn SOD located in the 27,000 g sedimentable fraction are often complex. In the cytochrome c assay (1,2) for superoxide dismutase, the oxidation of ferrocytochrome c by myeloperoxidase, and the reduction of ferricytochrome c by diaphorase mimick the inhibition of reduction of ferricytochrome c (3) and a diminution of superoxide dismutase activity (4) respectively. Therefore, it seems necessary to use a direct technique such as pulse radiolysis (5) to produce $O_2^{\bar{\cdot}}$ which can be directly detected by rapid spectrophotometry at 245 nm. It is possible then to measure kinetic constants of SOD present in complex medium such as PMN homogenate or subcellular fractions.

MATERIAL AND METHODS

A Febetron 708 was used as an electron pulse source. The duration of the pulse was 4 nanoseconds. In aerated solutions containing 0.1 M formate, as scavenger of $OH^{\cdot}$ and $H^{\cdot}$, 0.2 mM diethylenediamine pentaacetic acid (DETAPAC) and 2 mM carbonate pH 10, the yield of $O_2^{\bar{\cdot}}$ produced after the electron pulse was 6 molecules/100 ev, giving an initial concentration of 10^{-4}M ($E = 2000\ M^{-1}s^{-1}$).

Human CuZn superoxide dismutase from Sigma was used as reference. The polymorphonuclear neutrophils were isolated from 100 ml of blood by employing sequential 0.2 % dextran sedimentation for 30 minutes at 25°C, ficoll-paque gradient centrifugation, and hypotonic lysis of erythrocytes (30 s. with cold water).Final cell suspensions in Krebs-Ringer phosphate buffer at 25×10^6 PN/ml were broken with a Lox-press at 300 psi pressure.

A subcellular fraction was obtained by 27,000 g centrifugation. Total homogenate, 27,000 supernatant (cytosol) and the subcellular fraction were diluted in the buffer pH 10, at two concentrations. The reaction kinetics of SOD with $O_2^{\bar{\cdot}}$ was observed by following the decay of the radical at 245 nm.

RESULTS

In the absence of SOD, $O_2^{\bar{\cdot}}$ decays following a pH and concentration dependent mechanism (2nd order). The rate constant measured, under our experiment set up, was $2.22 \pm 0.33\ 10^3\ M^{-1}S^{-1}$. Addition of SOD was followed by a change of the rate equation to a pseudo first ordre kinetics.

$$\frac{dO_2^{\bar{\cdot}}}{dt} = - \text{Kcat (SOD)} (O_2^{\bar{\cdot}}) = - \text{Kapp} (O_2^{\bar{\cdot}})$$

Kcat is the slope obtained by plotting Kapp experimentaly measured versus SOD concentration ($5.10^{10}M$ to $6.10^{-8}M$). Here, under our experimental conditions, Kcat $= 1.58 \pm 0{,}16\ \ 10^8M^{-1}S^{-1}$. This value is low compared to bovine erythrocyte SOD constant which is $2.3\ 10^9M^{-1}S^{-1}$ (6). We have used this value as a reference to measure SOD concentration in PMN fractions.

Polymorphonuclear SOD measurement. Kapp was measured in the two dilutions of each sample in order to assure that the Kapp was measured in an adequate concentration of the enzyme. A fitting was done with calculated and experimental curves. A control blank was measured with boiled PMN : a second order kinetics, similar to that with the buffer, was obtained. Each fraction was assayed with 1 mM KCN so as to determine Mn SOD.

TABLE 1

Kapp, TOTAL SOD AND Mn SOD (WITH 1 mM KCN) IN PMN HOMOGENATE, CYTOSOL AND SUB-CELLULAR FRACTION.

In total homogenate, CuZn SOD is the difference between total SOD and Mn SOD ; its value here was $18.3\ 10^{-10}$ moles/10^6PN which is close to that of the cytosol (13.1 and 19.4 10^{-10}moles/10^6 PN). In the subcellular fraction, the Mn SOD which is about 2.10^{-10} moles/10^6 PN is lower than that found in total homogenate. We loose 50 per cent of Mn SOD when fractionating cells.

	Number of cells	Kapp	moles/10^6PN
Homogenate			
Total SOD	$0.6\ 10^6$	0.21	$21.8\ 10^{-10}$
	$2\ 10^6$	0.77	$24.0\ 10^{-10}$
+ KCN 1 mM Mn SOD	$2\ 10^6$	0.15	$4.6\ 10^{-10}$
Cytosol			
CuZn SOD	$0.86\ 10^6$	0.18	$13.1\ 10^{-10}$
	$1.70\ 10^6$	0.53	$19.4\ 10^{-10}$
+ KCN 1 mM	$1.70\ 10^6$	second order	0
Subcellular fraction			
Mn SOD	$1.24\ 10^6$	0.03	$1.52\ 10^{-10}$
	$3.10\ 10^6$	0.11	$2.26\ 10^{-10}$
+ KCN 1 mM	$3.10\ 10^6$	0.09	$1.80\ 10^{-10}$

CONCLUSION

We have shown that CuZn SOD and Mn SOD in PMN can be measured by pulse radiolysis with a sensitivity of 10^{-10}M. No other reagents are present in the assay medium that could eventually interfere with $O_2^{\cdot-}$ as in the cytochrome c assay, where diaphorases and myeloperoxidase present in the complex medium can lead to erroneous results.

Pulse radiolysis, indeed, is a sensitive and reliable technique which allows to localize and determine Mn SOD in subcellular fractions of PMN.

REFERENCES

1. McCord JM, Fridovich I (1969) J Biol Chem 244 : 6049-6055.
2. Salin ML, McCord JM (1974) J Clin Invest 54 : 1005-1009.
3. Rest RF, Spitznagel JK (1977) Biochem J 166 : 145-153.
4. Green TR, Wu DE (1985) FEBS 2116, 179, 1 : 82-86.
5. Rotilio G, Bray RC, Fielden EM (1972) Biochem Biophys Acta 268 : 605-609.
6. Brunori M, Rotilio G (1984) In : Colowick SP, Kaplan NO (eds) Oxygen radicals in biological systems. Methods in Enzymology, Lester Packer, vol 105 p 30.

© 1986 Elsevier Science Publishers B.V. (Biomedical Division)
Superoxide and Superoxide Dismutase in Chemistry,
Biology and Medicine; G. Rotilio, editor

THE ACTIVE CENTERS OF *T. thermophilus* Mn SUPEROXIDE DISMUTASE AND *E. coli* Fe SUPEROXIDE DISMUTASE: CURRENT STATUS OF THE CRYSTALLOGRAPHY

WILLIAM C. STALLINGS, KATHERINE A. PATTRIDGE AND MARTHA L. LUDWIG

Biophysics Research Division, University of Michigan, 2200 Bonisteel Blvd., Ann Arbor, MI 48109 (USA)

INTRODUCTION

Studies of the crystal structures of Fe and Mn superoxide dismutases are in progress at The University of Michigan (1-3). The polypeptide chains of these proteins are remarkably homologous two-domain structures; the fold of the Mn enzyme from *T. thermophilus* is illustrated in Figure 1. Details of the molecular geometry of the Mn(III) center have been established at 2.4 Å resolution and a molecular model of the protein has been assembled. Using this model to interpret the 3.1 Å structure of *E. coli* dismutase, we have further demonstrated that the structural homology of the Fe and Mn proteins extends to the level of their active center environments (4). Here we describe the current status of our crystallographic studies of the active centers of these enzymes and present initial results from our study of Mn(II) superoxide dismutase.

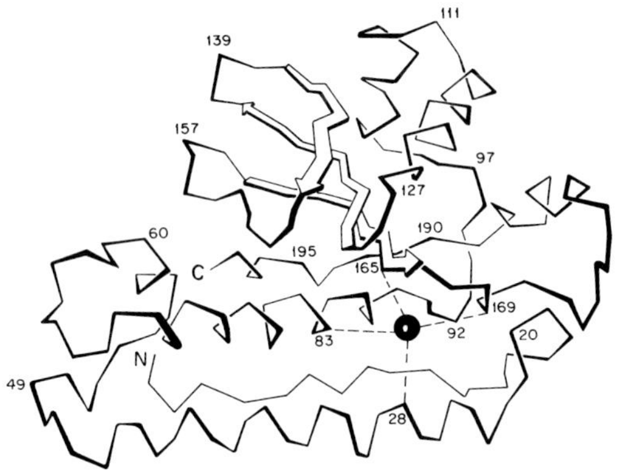

Fig. 1. C_α representation of *T. thermophilus* Mn superoxide dismutase; the protein is folded into two domains. Principal features of the secondary structure in the N-terminal domain are two crossing helices, each of which provides one metal ligand (His 28 and His 83). The remaining protein ligands are from the C-terminal domain which is of mixed α + β structure; the third protein ligand (Asp 165) is near the C-terminus of the β sheets and is separated from the fourth ligand (His 169) by only three intervening residues. The domain connector is a curved segment of polypeptide (residues 92-98) which partially encloses residues 187-192 near the C-terminus. The secondary structures are almost identically reproduced in the backbone folds of *E. coli* and *P. ovalis* Fe dismutases (1,5) except that a short segment of helix (residues 55-62) in the connection between the long crossing helices of the N-terminal domain is deleted in the dimeric Fe enzymes. This helix is involved in the stabilization of the *T. thermophilus* Mn dismutase tetramer.

DESCRIPTION OF THE ACTIVE CENTER OF *T. thermophilus* Mn DISMUTASE

The primary structure of *T. thermophilus* Mn dismutase has not yet been chemically determined. In building our 2.4 Å model of the protein, residues were initially assigned from a consensus sequence (see ref. 3) constructed from known primary structures of Mn dismutases (6-10). The consensus sequence was then modified with insertions, deletions and substitutions as indicated by the appearance of the electron density map; a point to be stressed is that all of the ligands as well as residues which surround the coordination complex are from regions which appear to be highly conserved in the known primary structures. The quality of the map, combined with the homology of the sequences and the characteristic shapes of the active site residues, leaves little ambiguity in the assignment and identification of the ligands and other active site residues.

As illustrated in Figure 2, Mn(III) is pentacoordinate in the resting, oxidized enzyme with the ligands arranged at the vertices of a distorted trigonal bipyramid; His 28 and His 83 are ligands from the N-terminal domain with Asp 165 and His 169 from the C-terminal domain. The histidine ligands all coordinate the metal ion via their N3 imidazole nitrogens and only one of the carboxyl oxygens of the coordinating Asp residue appears to ligand the metal; a water molecule serves as the fifth ligand. His 28 and the water are at axial positions. The conformations of two of the ligands are of special interest:

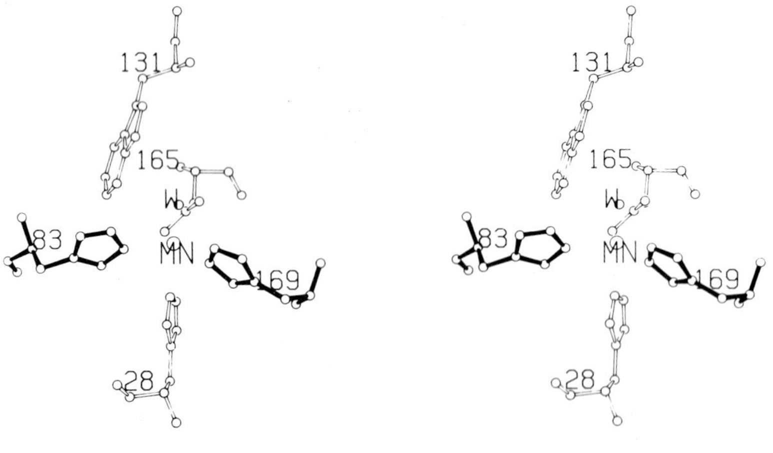

Fig. 2. The Mn(III) ligand cluster in *T. thermophilus* Mn superoxide dismutase. The distorted trigonal bipyramidal coordination by His 28, His 83, Asp 165, His 169 and water is described in the text. The bonds of His 83 and His 169 are darkened; it is postulated that additional ligands, such as substrate anions, may bind to the protein *trans* to Asp 165 at the face of the metal ion left exposed by the orientations of the imidazole rings of these residues. The close approach of the indole ring of Trp 131 to Mn is also illustrated.

the imidazole rings of His 83 and His 169 are approximately normal to the trigonal plane and their orientations (Cf. Figure 2) leave a face of the five-coordinate metal ion exposed and available for the coordination of an additional ligand. We have suggested that it is at this exposed face, and not by replacement of the water ligand, that substrate anions may bind (3,4).

Data for the structure determination were collected on crystals maintained at pH 7. If it can be assumed that the liganding histidines are uncharged and that Asp 165 titrates normally, then the Mn(III)-ligand cluster bears a net charge of +2 at this pH. In view of this, what is surprising about the active center environment is its apolar nature. The ligand cluster is surrounded by an array of aromatic and imidazole ring systems (Cf. Figure 3) which includes His 32, His 33, Tyr 36, Phe 86, Trp 87, Trp 131, Trp 167 and Tyr 182; the orientation of the aromatic rings of 36 and 131 is partially controlled by the side chain amide group of Gln 150 which links these residues by hydrogen bonding. The only charged residue which penetrates the nearby surroundings of the coordinated metal ion is contributed from another subunit across an interface of the tetramer: the carboxyl oxygens of Glu 168 from the neighboring subunit appear to hydrogen bond to backbone nitrogen atoms, N168 and N169. Tyr 172 of the adjacent subunit also appears to form a strong hydrogen bond to His 32. The catalytic significance of these interactions, which could form the basis of kinetically detectable site-site cooperativity, remains to be established.

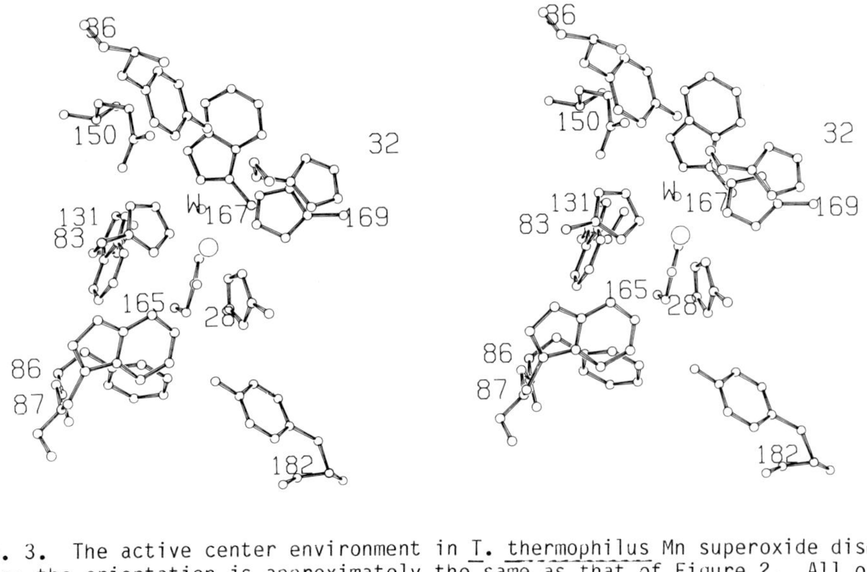

Fig. 3. The active center environment in _T. thermophilus_ Mn superoxide dismutase; the orientation is approximately the same as that of Figure 2. All of the residues in this figure are conserved in the known primary structures of Mn superoxide dismutases. His 32 and Tyr 36 appear to be strategically positioned near the ligand cluster at its exposed face and may therefore directly participate in catalysis.

DESCRIPTION OF THE ACTIVE CENTER OF *E. coli* Fe SUPEROXIDE DISMUTASE AT 3.1 Å RESOLUTION

The crystal structures of Fe superoxide dismutase from *E. coli* (1) and from *P. ovalis* (5) have revealed that Fe(III) is also coordinated by four protein ligands which are contributed from regions of structure equivalent to those which contribute Mn(III) ligands in Mn dismutase. However, only the first, third and fourth protein ligands of Fe dismutases can be identified from partial sequences. The first ligand is assigned as His 26 from comparisons of the appearance of the electron density with N-terminal sequences (11,12). In Japan, Yamakura, Isobe and their associates have determined the sequences of a number of peptide fragments from the *P. ovalis* Fe protein; included among the fragments is a 37 residue polypeptide whose sequence can be easily aligned by inspection with the four known primary structures of Mn dismutases in the region which contains the third and fourth protein ligands. The alignment (4) has allowed us to assign the third and fourth ligands in Fe dismutases as Asp and His residues, respectively, which are homologous to Asp 165 and His 169 in the *T. thermophilus* Mn enzyme. Only the identity of the second Fe dismutase ligand remains to be established from sequences.

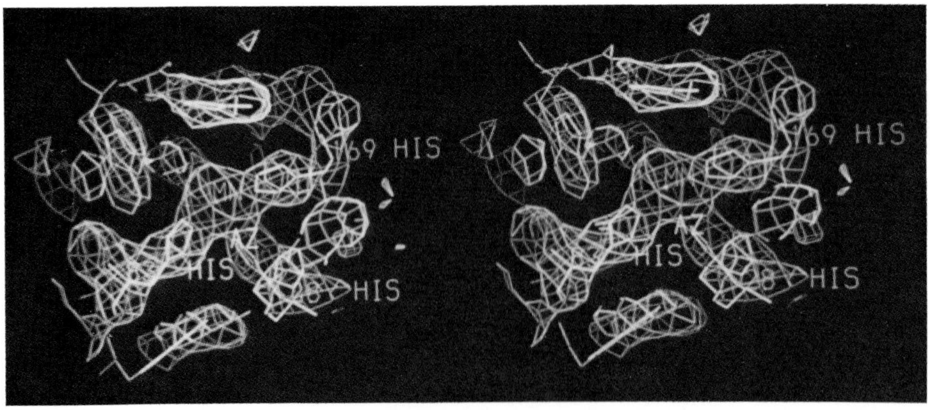

Fig. 4. The model of the Mn(III) ligand cluster oriented in the 3.1 Å *E. coli* Fe dismutase electron density map. See text. The fit of Trp 167 (above the metal), Trp 131 (above the metal to the left) and Trp 87 (below the metal to the left) can be seen in this display. As in Mn dismutase, a water ligand can be accommodated in the bulge in the density above the position of the metal ion. Display of the model in the electron density is also used to interpret the binding of azide to Fe dismutase; azide binds at the exposed face of the metal ion (see also Figures 2 and 3) apparently by increasing the coordination number from 5 to 6.

In lieu of assembling a model of Fe dismutase at 3.1 Å resolution, we have used the 2.4 Å model of Mn dismutase to interpret electron density in the metal center of the Fe protein (2,4). Our method was to determine the rigid-body transformation matrix which rotates the model of *T. thermophilus* Mn dismutase into the electron density of the *E. coli* Fe protein; the transformation matrix was derived by a least-squares procedure which compares the electron densities of the two proteins. The fit of the model to the Fourier map in the vicinity of the active center is illustrated in Figure 4. We do not believe that at 3.1 Å resolution it is possible to identify the second Fe dismutase ligand definitively, but the shape of its electron density at this resolution is at least consistent with its assignment as a histidine residue. As in the Mn(III)-coordination complex of the *T. thermophilus* protein (3), a water ligand can be lodged in the distinct bulge in the electron density surrounding the metal ion. Thus three and perhaps all four of the protein ligands in Fe dismutase are homologous to those of Mn dismutase. The coordination geometry also appears to be preserved with His 26 and the water bulge at apical positions.

Also indicated in Figure 4 is the position of the center of mass of a bound azide anion detected in difference Fourier maps calculated from diffraction data measured on crystals soaked with the inhibitor (1). The anion appears to bind to the Fe(III) ion at its exposed face and without displacement of the water ligand. The binding site can be described as a cavity formed from the van der Waals surfaces of residues which include His 30 and Tyr 34. These residues are homologous to His 32 and Tyr 36 in the *T. thermophilus* Mn protein and have been definitively assigned from partial sequences of the *E. coli* Fe protein (13). The rigid-body rotation of the Mn(III) active center environment, described above, moves the homologous His and Tyr residues into equivalent electron density in the Fe protein; in fact all of the imidazole and aromatic ring systems which surround the Mn(III)-ligand complex are maneuvered into correspondingly shaped Fe dismutase electron density by this transformation. Hence, in both Fe and Mn dismutases, the ligand clusters are enclosed by a structurally equivalent, apolar environment. This is an interesting, if frustrating, conclusion because differences in the environments would be expected in order to account for the specificity in metal ion binding exhibited by most of the Fe and Mn dismutases studied (14). We therefore anticipate possible structural differences which we cannot detect in this stage of our analysis such as the identity of the second Fe dismutase ligand or a Gln → Glu substitution in Fe dismutase at the residue homologous to Gln 150 in Mn dismutase.

COMMENTS ON REDUCED SUPEROXIDE DISMUTASES

Data to 2.4 Å resolution have been collected on crystals of _T. thermophilus_ Mn superoxide dismutase reduced by dithionite. The intensities were measured at the University of California, San Diego area detector facility and included 71410 observations of 26684 independent reflections which were scaled with an R_{sym} of 0.038. The residual comparing the diffraction data of the oxidized and reduced proteins ($R = \Sigma|I_{ox}-I_{red}|/\Sigma I_{red}$) is 0.102. A difference Fourier calculated with coefficients, $F_{red}-F_{ox}$, and isomorphous replacement phases (3) was essentially featureless in the vicinity of the active center and its surrounding environment; such a Fourier map should reveal conformational changes if they accompany reduction of the crystalline protein. The map echoed our earlier results on ascorbate-reduced _E. coli_ Fe superoxide dismutase at much lower (3.5 Å) resolution.

Bull and Fee have presented evidence that reduction of _E. coli_ Fe dismutase is linked to the uptake of one proton (15), and in their overall mechanism for the dismutation reaction they suggest a step in which the proton titrates a metal ligand, prompting its dissociation from the coordination sphere. An analogous step has been suggested for the Cu/Zn dismutase mechanism (16). In our crystallographic studies of the reduced enzymes, we find no evidence for the movement of a metal ligand. Since it is not impossible that lattice forces may lock the reduced and oxidized proteins into similar structures and prevent the ligand dissociation which may accompany reduction in solution, we are prompted by these findings to attempt de novo crystallization and structure solution of the reduced enzyme. Nevertheless we suggest that mechanisms which do not require ligand dissociation also be considered.

ACKNOWLEDGEMENTS

This research was supported by NIH grants GM 16429 to M.L. Ludwig and GM 21519 to J.A. Fee. The Detector Facility at UCSD is supported by RR 01644. We thank D. Rapley for preparation of the manuscript and S. McGrath for assistance with the figures.

REFERENCES

1. Stallings WC, Powers TB, Pattridge KA, Fee JA, and Ludwig ML (1983) Proc Nat Acad Sci USA 80:3884-3888
2. Stallings WC, Pattridge KA, Strong RK, and Ludwig ML (1984) J Biol Chem 259:10695-10699
3. Stallings WC, Pattridge KA, Strong RK, and Ludwig ML (1985) J Biol Chem in press
4. Stallings WC, Pattridge KA, Strong RK, Ludwig ML, Yamakura F, Isobe T, and Steinman HM (1985) in "The Patterson Symposium" (J.P. Glusker, ed.) in press

5. Ringe D, Petsko GA, Yamakura F, Suzuki K, and Ohmori D (1983) Proc Nat Acad Sci USA 80:3879-3883

6. Sato S, and Harris JI (1977) Eur J Biochem 73:373-381

7. McAdam ME, Lavelle F, Fox RA, and Fielden EM (1977) Biochem J 165:81-87

8. Brock CJ, Harris JI, and Sato S (1976) J Mol Biol 107:175-178

9. Walker JE, Auffret AD, Brock CJ, and Steinman HM (1980) in "Chemical and Biochemical Aspects of Superoxide and Superoxide Dismutase" (Bannister JV, and Hill HAO, eds.) pp. 212-222, Elsevier/North Holland, Amsterdam

10. Barra D, Schinina ME, Simmaco M, Bannister JV, Bannister WH, Rotilio G, and Bossa F (1984) J Biol Chem 259:12595-12601

11. Steinman HM, and Hill RL (1973) Proc Nat Acad Sci USA 70:3725-3729

12. Harris JI, Auffret AD, Northrup FD, and Walker JE (1980) Eur J Biochem 106:297-303

13. Steinman HM, Tarr G, and Fee JA (1985) Unpublished results

14. Kirby T, Blum J, Kahane I, and Fridovich I (1980) Arch Biochem Biophys 201:551-555

15. Bull C, and Fee JA (1985) J Amer Chem Soc 107:3295-3304

16. McAdam ME, Fielden EM, Lavelle F, Calabrese L, Cocco D and Rotilio G (1977) Biochem J 167:271-274

Added note: Partial sequencing of iron superoxide dismutase from *Photobacterium leiognathi* (Barra, D., Schinina, M.E., Bossa, F. and Bannister, J.V. (1985) FEBS Letters 179:329-331) demonstrates that all of the residues which are metal ligands in manganese dismutases are conserved in an iron superoxide dismutase.

© 1986 Elsevier Science Publishers B.V. (Biomedical Division)
Superoxide and Superoxide Dismutase in Chemistry,
Biology and Medicine; G. Rotilio, editor

PRIMARY STRUCTURE STUDIES ON THE IRON SUPEROXIDE DISMUTASE FROM P. LEIOGNATHI

D. BARRA[1], M.E. SCHININA'[1], W.H. BANNISTER[2], J.V. BANNISTER[3] AND F. BOSSA[1]

[1] Institute of Biochemistry and CNR Center for Molecular Biology, University of Rome, Italy; [2] Nuffield Dept. of Clinical Biochemistry, Radcliffe Infirmary and [3]Inorganic Chemistry Laboratory, University of Oxford, England.

INTRODUCTION

Superoxide dismutase (EC 1.15.1.1) is widely distributed in the animal and plant kingdom. Three different metalloforms of the enzyme are known (1). These are the copper/zinc, manganese and iron forms. The distribution of the three types of superoxide dismutase is considered to be a characteristic of the evolutionary stage of the organism and also of the organelle with which it is associated. In general terms this can be stated to be that the copper/zinc enzyme is essentially a eukaryotic enzyme and the iron enzyme is essentially a prokaryotic enzyme. The manganese enzyme is also present in prokaryotes. It is, however, also found in the mitochondrial matrix of eukaryotes. The iron superoxide dismutase is considered to be the most primitive form of the enzyme. This is because it is the form present in anaerobic bacteria. In view of the importance of determining the primary structure of superoxide dismutases from an evolutionary standpoint, we report structural studies on the iron superoxide dismutase from Photobacterium leiognathi; the primary structure has only been determined for a number of copper/zinc and manganese superoxide dismutase (2). The determination of the structure of an iron enzyme is also important because the crystal structure has been determined for the iron enzyme from E. coli (3) and Pseudomonas ovalis (4) and the iron and manganese enzymes have been shown to be structural homologs from crystallographic data (5) and may have identical ligands to the metal (6).

METHODS

Iron superoxide dismutase was isolated from P. leiognathi by the method of Yamakura (7). Apoprotein was prepared by incubation with 70% formic acid for 2 h at 40° followed by dialysis against 50% acetic acid and lyophilization. After digestion, peptides were separated either directly by HPLC of first by fractionation on an appropriate Sephadex column followed by final purification on HPLC, depending upon the expected complexity of the peptide mixture. Analytical techniques for sequence determination were carried out as described in (8).

RESULTS AND DISCUSSION

The partial amino acid sequence deduced from analysis of peptides derived from

```
                                                          *            33
B. stearoth.    P F E L P A L P Y P Y D A L E P H I D K E T M N I H H T K H H N T
E. coli         S Y T L P S L P Y A Y D A L E P H F D K Q T M E I H H T K H H Q T
S. cerevisiae   K V T L P D L K W D F G A L E P Y I S G Q I N E L H Y T K H H Q T
Human liver     K H S L P D L P Y D Y G A L E P H I N A Q I M Q L H H S K H H A A
P. leiognathi   A F E L P A L P F A M N A L E P H I S Q E T L E Y H Y G K H H N T

                                                                          73
Y V T N L N A A L E G H P D L Q N K S L E E L L S N L E A L P E S I R T A V R N
Y V N N A N A A L E S L P E F A N L P V E E L I T K L D Q L P A D K K T V L R N
Y V N G F N T A V D Q F Q E L S D L L A K E P S P A N A R K M I A I Q Q N I K F
Y V N N L N V T Q E K Y Q E A L A K ———————— G D V T A Q I A L Q P A L K F
Y V V K L N G L V E G T - E L A E K S L E E I I - K . . . . . . . . . . . . . .

              *                                                          113
N G G G H A N H S L F W T I L S P —— N G G G E P - T G E L A D A I N K K F G S
N A G G H A H N S L F W K G L K K ——— G T T L Q —— G D L K A A I E R D F G S
H G G G F T N H C L F W E N L A P E S Q G G G E P P T G A L A K A I D E Q F G S
N G G G H I N H S I F W T N L S P —— N G G G Q P - K G E L L E A I K R D F G S
. . . . . . N H T F Y W N C L A P —— N A G G Z P - T G E V A A A I Z K A F G S

                                                                         153
F T A F K D E F S K A A A G R F G S G W A W L V V N ——— N G E L E I T S T P N
V D N F K A E F E K A A A S R F G S G W A W L V —— L K - G D K L A V V S T A N
L D E L I K L T N T K L A G V Q G S G W A F I V K N L S N G G K L D V V Q T Y N
F D K F K Q K L T A A S V G V Q G S G W L G F —— N - K Q R G H L Q I A A C P N
F A E F K A K F T B - S A I N F G S S W T W L V K B A - B G S - L A I V B T S B

                                          *       *                      193
Q D S P I M E G K T ————————— P I L G L D V W E H A Y Y L K Y Q N R R P E Y I
Q D S P L M G E A I S G A S G F P I L G L D V W E H A Y Y L K F Q N R R P D Y I
Q D T V T G P L V ——————————— P L V A I D A W E H A Y Y L Q Y Q N K K A D Y F
Q D - P L Q G T T —————— G L I P L L G I D V W E H A Y Y L Q Y K N V R P D Y L
A G C P I T Z G V T ————————— P L L T V D L W E H A Y Y I D Y R N L R P S Y M

A A F W N V V N W D E V A K R Y S E A K A K
K E F W N V V N W D E A A A R F - A A K - K
K A I W N V V N W K E A S R R F D
K A I W N V I N W E N V T E R Y M A C K - K
D G F W A L V N W D F V S - K N L A —— A
```

Fig. 1. The partial amino acid structure of P. leiognathi iron enzyme is aligned with the sequences of other manganese superoxide dismutases. —— indicate gaps introduced to obtain maximal homology. .... indicate undetermined portions of sequence of the iron enzyme. Boxes: conserved residues ; * metal ligands.

tryptic, chymotryptic, <u>S. aureus</u> protease and peptic digests of <u>P. leiognathi</u> iron superoxide dismutase is reported in Fig. 1. The enzyme has an exposed N-terminal residue and the peptides sequenced represent over 90% of the total sequence. Since the manganese and iron containing superoxide dismutases are considered to be homologous proteins, the partial primary structure of the iron enzyme aligned with the published sequences of <u>B. stearothermophilus</u>, <u>E. coli</u>, yeast and human liver manganese forms are together presented in Fig. 1. The alignment of the peptides is based on homology considerations which often appear obvious. In contrast to the bacterial enzymes, the iron superoxide dismutase contains cysteine residues which are also present in the eukaryotic manganese enzyme. Of the six tryptophan residues so far identified, five align with tryptophans in all the manganese superoxide dismutases sequenced; the remaining residue is conserved in the prokaryotic and substituted in the eukaryotic manganese protein. Also conserved in the iron and in the manganese enzymes is the position of most the glycine, lysine and proline residues which suggest similar three-dimensional packing in these molecules, in agreement with crystallographic data that the two isoenzymes are structural homologs (5). Another interesting feature is the alignment of the ligands to the metals in these proteins. The crystal structure of the manganese superoxide dismutase from <u>T. thermophilus</u> at 2.4 A indicates that the ligands to the metal are His, His, Asp and His (5) which, in the absence of a primary structure of this enzyme, correspond to positions 26, 81, 175 and 179 in the aligned sequences of manganese enzymes. His 81 in <u>E. coli</u> is in position 80, but this appears to be an error in the sequence. The alignment with the iron superoxide dismutase indicates that the ligands to the manganese are also present in identical positions in this isoenzyme.

REFERENCES

1) Parker M.W., Schininà M.E., Bossa F. and Bannister J.V. (1984) Inorg. Chim. Acta 91 : 307-317

2) Parker M.W., Bossa F., Barra D., Bannister W.H. and Bannister J.V. (1986) this volume

3) Ringe D., Petsko G.A., Yamakura F., Suzuki K. and Ohmori D. (1983) Proc.Natl. Acad. Sci. USA 80 : 3879-3883

4) Stallings W.C., Powers T.B., Pattridge K.A., Fee J.A. and Ludwig M.L. (1983) Proc. Natl. Acad. Sci. USA 80 : 3884-3888

5) Stallings W.C., Pattridge K.A., Strong R.K. and Ludwig M.L. (1984) J. Biol. Chem. 259 : 10695-10699

6) Barra D., Schininà M.E., Bossa F. and Bannister J.V. (1985) FEBS Lett. 179 : 329-331

7) Yamakura F. (1976) Biochem. Biophys. Acta 422 : 280-294

8) Barra D., Schininà M.E., Simmaco M., Bannister J.V., Bannister W.H., Rotilio G. and Bossa F. (1984) J. Biol. Chem. 259 : 12595-12601

ON THE MECHANISM OF THE IRON-CONTAINING SUPEROXIDE DISMUTASE FROM *E. Coli* *

JAMES A. FEE and TATSURO YOSHIDA

Los Alamos National Laboratory, Los Alamos, New Mexico 87545, U.S.A.

CHRISTOPHER BULL

W. R. Grace and Co., Columbia, Maryland 21044, U.S.A.

PETER O'NEILL and E. MARTIN FIELDEN

M.R.C. Radiobiology Unit, Chilton, Didcot, Great Britain

INTRODUCTION

Considerable progress has recently been made in elucidating the structure- function relations of the iron-containing superoxide dismutase (FeSD) from *Escherichia coli* : A three-dimensional structure to 3 A resolution was reported (1,2), the *sod* B gene has been cloned and over-expression of the protein observed (3,4), and a comprehensive mechanism has been proposed(5). In this communication we briefly review the previous mechanistic work, describe the effect of D_2O on the kinetic parameters, and present a preliminary analysis of the effect of added salts and ionic strength on the activity of *E. coli* FeSD.

EXPERIMENTAL

Kinetic Methods. In this work stopped-flow and pulse radiolysis techniques are employed for studying superoxide dismutase mechanisms. These methods have been described in detail elsewhere: The stopped-flow method (5,6) allows one to generate very concentrated aqueous solutions of O_2^- ($\leq$ 7 mM) but is limited to relatively long observation times (~10 ms). The pulse radiolysis technique is limited to relatively low initial O_2^- concentrations (~250 µM) but allows observation of the O_2^- on the microsecond time scale; the methods have complementary features.

Data reduction. The general expression for the decay of O_2^- in the presence of superoxide dismutases is

$$\frac{-d[O_2^-]}{dt} = k_1[O_2^-] + 2k_2[O_2^-]^2 + \frac{2[enz]TN[O_2^-]}{K_m + [O_2^-]} \qquad (1)$$

where the first two terms on the right accurately describe O_2^- decay in the absence of

*This work was supported by U.S.P.H.S. Grant GM35189.

catalyst while the third term accounts for the contribution of the enzyme to superoxide dismutation. We have utilized two practical methods for obtaining the kinetic parameters of the superoxide dismutase: In pulse radiolysis studies $[O_2^-]$ is kept sufficiently low that, during the experiment,

$$k_1[O_2^-] + 2k_2[O_2^-]^2 \ll \frac{2TN(enz)\,[O_2^-]}{K_m + [O_2^-]} \tag{2}$$

and

$$\left[\frac{-d[O_2^-]}{dt}\right]_{obs} = \frac{2TN[enz][O_2^-]}{K_m + [O_2^-]} \tag{3}.$$

In the stopped-flow work, however, this was not possible and an alternative approach was taken. First, all kinetic traces ($[O_2^-]$, t) were converted to velocity traces ($d[O_2^-]/dt$, t). Second, k_1 and k_2 were accurately determined, in the absence of enzyme, for each experimental condition. This allowed the corrected velocity data to be related exclusively to the kinetic parameters of the superoxide dismutase, i.e., TN and K_m.

$$\left[-\frac{d[O_2^-]}{dt}\right]_{obs} - k_1[O_2^-] - 2k_2[O_2^-]^2$$

$$= \left[-\frac{d[O_2^-]}{dt}\right]_{corr} = \frac{2[enz]TN[O_2^-]}{K_m + [O_2^-]} \tag{4}$$

Note that this relationship is of standard Michaelis-Menten form.

BRIEF REVIEW OF THE MECHANISM OF FeSD

Early work with the Cu/ZnSD showed that K_m was large compared to ~0.25 mM. FeSD however, could be saturated at moderate $[O_2^-]$ (7), and Bull and Fee (5) studied the variation of K_m and TN over a wide range of temperature, pH, and ionic environment. A minimal kinetic scheme was proposed(Scheme I).

$$E_{ox} \underset{k_{-1}}{\overset{k_1O_2^-}{\leftrightarrow}} [E_{ox}\text{-}O_2^-] \overset{k_2}{\rightarrow} E_{red} + O_2 \qquad (5)$$

$$E_{red} \underset{k_{-3}}{\overset{k_3O_2^-}{\leftrightarrow}} [E_{red}\text{-}O_2^-] \overset{k_4}{\rightarrow} E_{ox} + H_2O_2 \qquad (6)$$

Scheme 1

in which steps 2 and 4 are considered irreversible and protons have been ignored. Steady-state analysis of this scheme led to the following expressions for

$$TN = (1/k_2 + 1/k_4)^{-1} \qquad (7),$$

$$K_m = \frac{(1/k_2 + 1/k_4)^{-1}}{\left[\dfrac{1}{\mathbf{k_r}} + \dfrac{1}{\mathbf{k_o}}\right]^{-1}} \qquad (8),$$

and

$$TN/K_m = \left[\frac{1}{\mathbf{k_r}} + \frac{1}{\mathbf{k_o}}\right]^{-1} \qquad (9)$$

where

$$\mathbf{k_r} = k_1[k_2/(k_{-1} + k_2)] \qquad (10)$$

$$\mathbf{k_o} = k_3[k_4/(k_{-3} + k_4)] \qquad (11).$$

There are two limiting cases:

Case I. $K_m >> [0_2^-]$ leads to

$$-d[O_2^-]/dt = 2\,(1/\mathbf{k_r} + 1/\mathbf{k_o})^{-1}\,[enz]\,[O_2^-]$$

$$= 2\,TN/K_m\,[enz]\,[O_2^-] \qquad (12)$$

corresponding to a second-order process, and pseudo first order decay of O_2^- will be observed. The rate determining process is the binding of O_2^- in the vicinity of the active site.

Case II. $K_m << [O_2^-]$

$$-d\,[O_2^-]/dt = 2[enz](1/k_2 + 1/k_4)^{-1}$$

$$= 2[enz]\,TN \qquad (13)$$

corresponding to a first-order process and zero-order disappearance of O_2^- will be observed. The rate determining processes are the first-order events occuring at the active site.

Examination of the pH dependence of the K_m and TN of FeSD revealed an interesting behavior of the enzyme(5) which is summarized as follows:

Turnover number (TN). This is accurately independent of pH in the range 7.8 -11 and has a value near 26,000 s^{-1} at 25 °C. It is also independent of anionic inhibitors such as azide, fluoride, and perchlorate. TN is, however, increased by general acids such as NH_4^+, glycine and others. These observations led to the suggestion that the source of protons in the overall dismutation reaction is H_2O.

Kinetic affinity (K_m). The dependence of K_m on pH and ionic composition proved to be very complicated. Typically, K_m is independent of pH below 8.5 and has a value near 100 μM. Its variation with pH is accurately described by the expression

$$K_m^{obs} = K_m^{o}\,(1 + 10^{pH - pKa}) \qquad (14)$$

where K_m^o corresponds to K_m^{obs} below pH=8 and pK_a ~9. Careful analysis revealed that ionizing groups on both oxidized and reduced proteins had pK_a ~9.

Inhibition by Anions. K_m is also dependent on the presence of ionic inhibitors such as N_3^- and F^- which act competitively and are known to bind to the Fe^{3+} form of the enzyme. For these anions, it was consistently found that $K_d = 3.0\ (\pm\ 10\%)\cdot K_I$. This was interpreted to mean that N_3^- and F^- inhibit by binding only to the oxidized protein. By contrast, ions such as ClO_4^- and SCN^- have $K_d \sim K_I$ suggesting that they bind with similar affinity to both oxidized and reduced forms of the protein.

Detailed mechanism. Based on these and other observations a mechanism for FeSD was proposed(5).

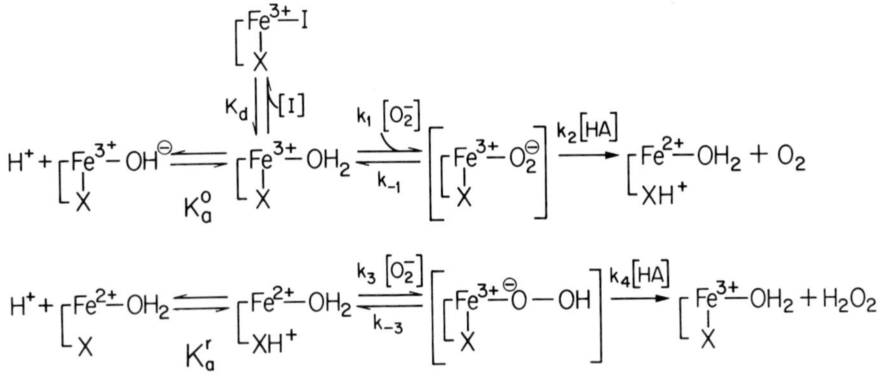

Scheme II

This mechanism is consistent with present knowledge.

In this brief communication, we extend support for the idea that H_2O is the source of protons in the dismutation reaction and describe new data on the effect of increasing ionic strength on the activity of FeSD.

RESULTS AND DISCUSSION

If H_2O is acting during catalysis as a general acid in donating a proton to developing basic centers on the enzyme, then those processes should be slowed when H_2O is replaced by D_2O. Values of TN, K_m and TN/K_m obtained under identical conditions in H_2O and D_2O are presented in Table I. The ratio, TN_H/TN_D ~ 3, is fully consistent with general acid donation of a proton during catalysis. Specific acid

TABLE I

Effect of Deuterium Oxide on the Kinetic parameters of FeSD

	Turnover Number (s^{-1})[a]	K_m (μM)[a]	TN/K_m ($M^{-1}s^{-1}$)
H_2O	19,000	94	2.0×10^8
D_2O	7,600	48	1.6×10^{8}[b]

[a]Experimental conditions: $[O_2^-]_0$ ~1.5 mM, 0.1 M TAPS; pH = 8.5 (pD = 8.9, pH_{obs} = 8.5); [FeSD] = 0.7 μM (~75 % active); 0.1 mM DETAPAC; T =25 °C.

[b]This number is slightly low because $[D_3O^+]$ was somewhat less than $[H_3O^+]$ in the two experiments which are being compared; it is likely that the pK_a affecting K_m is lowered by D_2O. TN, however, is independent of pH.

catalysis would yield a value of ~1.14 and TN would vary with $[H_3O^+]$, which it does not. This and previous results are consistent with the conclusion that the ultimate source of protons in the dismutation process is H_2O and H_3O^+ *per se* cannot participate in the reaction.

We have performed pulse radiolysis experiments to examine the effects of increasing ionic strength, μ, on the activity of FeSD. At low O_2^- concentrations ($[O_2^-] << K_m$), the observed rate of decay, TN/K_m, corresponds to the rate at which O_2^- binds to the protein's surface. At very low ionic strengths, μ = 5 mM to ~15 mM, $\log(TN/K_m)$ appears to increase while above ~15 mM, it decreases (Fig. 1). We have interpreted these data as resulting from two overlapping effects: first, the classical influence of ionic strength on a reaction between like-charged molecules as developed by DeBye and Huckel

$$k_{obs} = k_o Q \qquad (15)$$

where $Q = 10^{2Az_1z_2\sqrt{\mu}}$, Z_1 and Z_2 are the formal charges of the reactant species and the DeBye-Huckel constant, A, has a value of 0.509 at 25 °C. Second, the physical binding of the ClO_4^- anion to a site near the Fe in such manner to interfere with the approach of O_2^-. Since inhibition appears to result from a relatively weak interaction between the negatively charged inhibitor and protein species, one would also expect K_d to depend on μ, i.e.,

$$K_d = K_d^o Q \qquad (16).$$

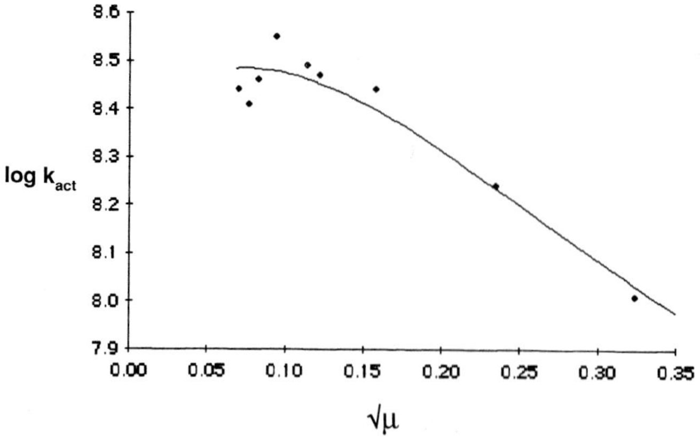

Figure 1. The influence of ionic strength on the rate constant for dismutation of O_2^- (9.8 μmol dm^{-3}) by FeSD at pH 8.5 using $NaClO_4$ (0 - 100 mol dm^{-3}) to adjust μ. All solutions contained 100 μmol dm^{-3} EDTA, 0.1 mol dm^{-3} ethanol and 0.55 mmol dm^{-3} pyrophosphate.

When expressions (15) and (16) are combined, the predicted effect of μ on activity is given by

$$k_{act} = k_o Q (Q/(Q + [ClO_4^-]/ K_d^o)) \qquad (17).$$

When the data were subjected to a non-linear least squares fitting routine in which k_o, K_d^o and Z_1Z_2 were allowed to float, the solid line in Fig. 1 was obtained. The resulting values were log k_o = 8.48, K_d^o = 34 mM, and Z_1Z_2 = 0.02. The data were equally well fitted with Z_1Z_2 fixed anywhere in the range of $\pm$ 1. Previously measured values of K_d for ClO_4^- to oxidized FeSD are 36 (5) and ~ 20 mM(8).

Since FeSD is a negatively charged molecule having pI = 4.5(9), one would expect the rate of reaction with O_2^- to increase with increasing ionic strength. The present data suggest that ionic strength effects are completely suppressed even under these extremely dilute ionic conditions. The observed decrease in activity appears entirely due to association of ClO_4^- with the previously proposed anion binding pocket(8) near the iron atom. Similar results were reported previously by Benovic et al(9) who measured steady-state activity at much higher salt concentrations($\mu \leq 0.6$ M).

We conclude with a general statement regarding charge distribution on the enzyme's surface and "electrostatic facilitation" of the superoxide dismutation reaction. Within a cell, the O_2^- anion will be at a concentration many orders of magnitude lower than other anions such as Cl^-, $P_i^{2-,1-}$, PP_i, and organic acids. Placing the active site within a positively charged region of the surface will have the effect of attracting all negatively charged ions into this region. Given the adverse concentration ratio, this will tend to discriminate against a specific attraction of O_2^- to the active site.

REFERENCES

1. Stallings, W. C., Powers, T. B., Pattridge, K. A., Fee, J. A., and Ludwig, M. L. (1983) Proc. Nat'l. Acad. Sci. U. S. A. **80**, 3884 - 3888.
2. Ringe, D., Petsko, G. A., Yamakura, F., Suzuki, K., and Ohmori, D. (1983) Proc. Nat'l Acad. Sci. U. S. A. (1983) **80**, 3879 - 3883.
3. Nettleton, C. J. Bull, Baldwin, T. O., and Fee, J. A. (1984) Proc. Nat'l. Acad. Sci. U. S. A. **81**, 4970 - 4973.
4. Sakamoto, H. and Touati, D. (1984) J. Bacteriol. **159**, 418 - 420.
5. Bull, C. and Fee, J. A. (1985) J. Am. Chem. Soc. **107**, 3295 - 3304.
6. Bull, C. McClune, G. J., and Fee, J. A. (1983) J. Am. Chem. Soc. **105**, 5290 - 5300.
7. Fee, J. A. McClune, O'Neill, P. and Fielden, E. M. (1981) Biochem. Biophys. Res. Commun. **100**, 377 - 384.
8. Fee, J. A., McClune, G. J., Lees, A. C., Zidovetski, R. and Pecht, I. (1981) Israel J. Chem. **21**, 54 - 58.
9. Slykhouse, T. O. and Fee, J. A. (1976) J. Biol. Chem. **251**, 5472 - 5477.
10. Benovic, J. Tillman, Cudd, A. and Fridovich, I. (1983) Arch. Biochem. Biophys. **221**, 329 - 332.

ACKNOWLEDGEMENT

We thank Miss S. Davies and Miss C. J. Nettleton for expert technical assistance.

Superoxide and Superoxide Dismutase in Chemistry, Biology and Medicine; G. Rotilio, editor

INACTIVATION OF IRON-SUPEROXIDE DISMUTASE BY HYDROGEN PEROXIDE

FUMIYUKI YAMAKURA, DAIJIRO OHMORI, AND KOJI SUZUKI
Department of Chemistry, Juntendo University, Narashino, Chiba 275, (JAPAN)

INTRODUCTION

Among three reported types of superoxide dismutase (SOD), Cu, Zn-SOD and Fe-SOD are inactivated by H_2O_2, a product of the dismutation reaction, whereas Mn-SOD is not inactivated by H_2O_2 [1]. The inactivation of the Cu, Zn-SOD has been attributed to the reduction of the enzyme-bound Cu^{2+} to Cu^+ by H_2O_2, followed by a Fenton's type reaction of the Cu^+ with additional H_2O_2 to form $Cu^{2+}OH\cdot$. This could oxidatively attack an adjacent histidine[2,3,4]. On the other hand, Kanematsu and Asada[5] showed that Fe-SOD from *Chromatium* was not protected against inactivation with H_2O_2 by the compounds which were reported to protect Cu, Zn-SOD against the inactivation with H_2O_2. Recently we reported that tryptophan, histidine, and cysteine residues were modified during inactivation with H_2O_2[6]. From this evidence, we speculate that the mechanism of inactivation of Fe-SOD by H_2O_2 should be somewhat different from that of the Cu, Zn-SOD. The purpose of this study is to clarify the mechanism of inactivation of Fe-SOD with H_2O_2 by using kinetic analysis and EPR measurement.

MATERIALS AND METHODS

Hydrogen peroxide was from Mitsubishi Gas Co, Tokyo. Superoxide dismutase activity was determined by the xanthine oxidase/cytochrome c method[7]. Fe-SOD was purified from *Pseudomonas ovalis* as described previously[7]. Iron content of the purified SOD was 1.5 g atom/mol of enzyme. Inactivation of the enzyme was carried out at 25°C under conditions given in the figure legends and as reported previously[6]. EPR spectra were recorded at 77 K on a JES-FE3XG spectrometer.

RESULTS AND DISCUSSION

In our previous experiments on the inactivation of Fe-SOD with H_2O_2, we showed that the rate of inactivation was a pseudo-first-order process depending upon the concentration of H_2O_2[6]. Fig. 1 shows the effect of pH on the pseudo-first-order rate constant for the inactivation of the enzyme by 0.42 mM H_2O_2. The rate constant increased with increasing pH with an inflection point around pH 8.5, suggesting that some ionization group(s), which affects the enzyme inactivation rate, has a pKa in this range. Fig.2 shows a double reciprocal plot of the rates of inactivation against various concentration of H_2O_2 (1.26-11.3 mM) at pH 7.8, 8.3 and 9.0. The straight line in Fig. 2 shows that the inactivation proceeds according to Michaelis-Menten enzyme kinetics. The value of the intercept of y axis gives the maximum rate constant for inactivation (κmax) and the value of the intercept of x axis gives the total H_2O_2 concentration at which the enzyme is half-saturated (K-inact). The κmax increased progressively with increasing pH, from 3.3 min^{-1} at pH 7.8 to 21 min^{-1} at pH 9.0, whereas K-inact remained constant around 30 mM at all pH values tested. These data indicate that the ionization group(s)(pKa ~8.5) does not participate in the binding of H_2O_2 to the enzyme but affects the maximum inactivation rate of the enzyme.

The inactivation process also gives rise to a marked transformation of the EPR spectra of the enzyme. Fig. 3 illustrats EPR spectra of the enzyme obtained after exposure to H_2O_2 for various times. The native enzyme showed the anisotropic signals of g_z=4.7, g_x=4.2, g_y=3.9, which arrise from a middle Kramers doublet of high spin ferric iron with a rhombic distortion, and a minor signal at g=4.3 which is assumed to be rhombic symmetrical ferric iron (Fig. 3,A,[8,9]). The anisotropic signals were markedly decreased after 10 s incubation of the enzyme with H_2O_2 and almost disappeared by the further incubation. The minor signal at g=4.3 seems to be unchanged during inactivation of the enzyme. A new signal of free radical type at g=2.006 appeared after 10 s incubation of the enzyme. The amount of the free radical signal was increased at 165 s incubation of the enzyme but decreased at 600 s incubation (Fig.3,C,D). After incubation of the enzyme with H_2O_2 for 10, 165, and 600 s , the remaining activities were 82, 31, and

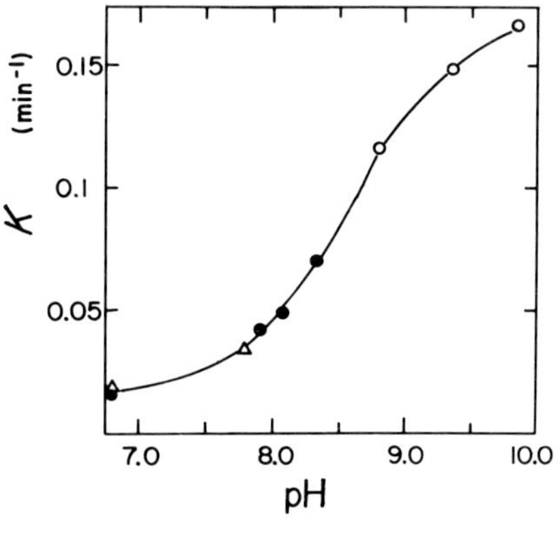

Fig. 1 Effect of pH on the pseudo-first-order rate constant (κ) for inactivation of the Fe-SOD by H_2O_2.
Fe-SOD (13 µM) was incubated with 0.42 mM H_2O_2 in 25 mM potassium phosphate buffer (△), sodium pyrophosphate buffer (●), or sodium carbonate buffer (○), at the various pH indicated in the figure.

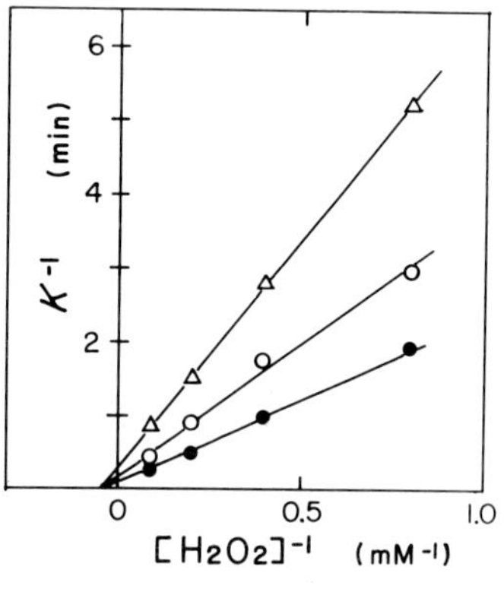

Fig. 2 A plot of κ^{-1}vs $[H_2O_2]^{-1}$ at pH 7.8, 8.3, and 9.0. The Fe-SOD (13 µM) in 25 mM sodium pyrophosphate buffer at pH 7.8 (△), pH 8.3 (○), and pH 9.0 (●) were incubated with various concentration of H_2O_2 .

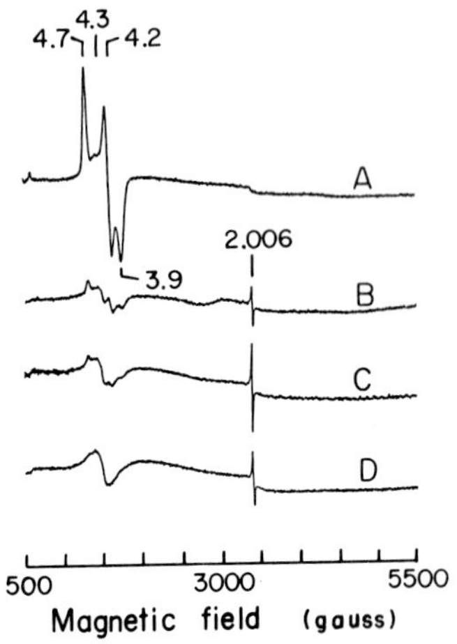

Fig. 3 EPR spectra of Fe-SOD.
The enzyme concentration was 130 μM in 25 mM sodium pyrophosphate buffer pH 7.8. Spectrum (A) was obtained by native Fe-SOD. Spectra (B), (C), and (D) were obtained after incubation of the native enzyme with 5 mM H_2O_2 for 10 s, 165 s, and 600 s, respectively. The spectra were recorded under the following condition: microwave frequency 9.209 GHz, modulation amplitude 10 G, microwave power 8 mW, and gain 500.

7.6 % of the original activity, respectively. The iron content in the 600 s incubated sample remained over 40 % of that of the native enzyme. From this evidence we propose that 1) inactivation presumably occurs after ferric iron has been reduced to ferrous iron. 2) A free radical species appears during inactivation of the enzyme. 3) Since there was no ferric iron signal other than the minor g=4.3 signal, the iron valency of the inactivated enzyme might be mostly in Fe^{2+} form. This third proposal leads us to consider that the inactivation of Fe-SOD with H_2O_2 might have occurred in a reaction other than the Fenton's type reaction in which the Cu, Zn-SOD inactivation occurred[4]. Further studies are required to elucidate the detailed mechanism of the inactivation of Fe-SOD with H_2O_2.

REFERENCES

1. Steinman H (1982) In: Oberley LW (ed) Superoxide dismutase vol 1. CRC press, Florida, pp 11-69
2. Bray RC, Cockle SA, Fielden EM, Roberts PB, Rotilio G, Calabrese L. (1974) Biochem J 139:43-48
3. Blech DM, Borders Jr CL (1983) Arch Biochem Biophys 224:579-586
4. Hodgson EK, Fridovich I (1975) Biochemistry 24:5294-5298
5. Kanematsu S, Asada K (1978) Arch Biochem Biophys 185:473-482
6. Yamakura F (1984) Biocem Biophys Res Commun 122:635-641
7. Yamakura F (1976) Biocim Biophys Acta 422:280-294
8. Lavelle F, McAdam ME, Fielden EM, Roberts PB, Puget K, Michelson AM (1977) Biocem J 161:3-11
9. Asada K, Yoshikawa K, Takahashi K, Maeda Y, Enmanji K (1975) J Biol Chem 250:2801-2807

PURIFICATION AND CHARACTERIZATION OF AN IRON-CONTAINING SUPEROXIDE DISMUTASE FROM A EUKARYOTE, *Ginkgo biloba*

MARY V. DUKE and MARVIN L. SALIN
Department of Biochemistry, Mississippi State University, Mississippi State, Mississippi 39762 (USA)

INTRODUCTION

Iron-containing superoxide dismutases were once thought to be restricted to prokaryotes and eukaryotic algae (1). However, we showed the presence of this enzyme in three of 43 representative plant families (2). Additional reports of the iron enzyme in eukaryotes have recently appeared (3,4). It would seem then that the iron-containing superoxide dismutases enjoy a wider distribution than previously held. We report the purification and characterization of the iron enzyme from *Ginkgo biloba*, the most primitive of the higher land plants possessing the iron-containing superoxide dismutase.

MATERIALS AND METHODS

Biological material and reagents. Leaves of *Ginkgo biloba* L. were obtained from trees growing locally at Mississippi State University. Supplies for chromatofocusing were products of Pharmacia or Sigma, as were the protein standards. Ampholytes were obtained from Separation Sciences, Inc. Proteases were supplied by Miles Laboratories or Sigma.

Characterization. The M_r for the purified protein was estimated by gel filtration on Sephadex G-150. Subunit size was determined by SDS-PAGE (5). Determination of pI was by non-denaturing PAGE as described in the Hoefer Scientific Instruments catalogue. Amino acid analysis was performed on a Beckman 120C analyzer as previously described (6). Metal content was determined by atomic absorption on a Perkin-Elmer 2380 spectrometer after the enzyme was dialyzed extensively against 5mM potassium phosphate, pH 7.8, containing 0.5mM EDTA.

Proteolytic digestion. Peptide fragments were prepared and analyzed by the gel slice method of Cleveland *et al*. (7). Iron SOD's were electrophoresed on SDS-PAGE (10% gel). Gel slices containing iron SOD were placed in the wells of a second SDS gel (12% gel) with 0.1 μg papain or 0.1 μg *S. aureus* V-8 protease. The proteins were electrophoresed two-thirds into the stacker gel. The current was then turned off 0.5 hr to allow proteolytic digestion. After electrophoresis was completed, gels were stained with silver (8).

Chloroplast fractionation. Chloroplasts were prepared by a modified method of Mills and Joy (9); 2% PVP was included in the grinding medium. Thylakoid and stromal fractions were prepared as described previously (10).

RESULTS

A summary of the purification is seen in Table 1. Initial activity was low, and upon purification, the enzyme activity deteriorated rapidly despite varied techniques to stabilize it. Storage of the purified enzyme for 4 days at $4^{o}C$ resulted in a 50% loss of total activity. Activity of the protein was not inhibited by 1mM KCN, whereas, the enzyme's half-life of activity was 8.8 min. in 2mM H_2O_2 at $25^{o}C$. The protein was characterized, and the following values were obtained. The enzyme consisted of two isozymes (pI=5.15 and 5.25) with an M_r of 47,000 and a subunit size of 23,000. The protein had a non-descript visible spectrum and exhibited a peak at 278 nm with a molar extinction coefficient of $4.4 \times 10^4 M^{-1} cm^{-1}$. Metal analysis showed 1.4 g atoms of iron per molecule. The amino acid composition showed a greater number of glycine residues and a slightly smaller number of aromatic amino acids on average than found in *Nuphar*, *Brassica*, and *E. coli* (data not shown). A least squares difference analysis of amino acids for the iron SOD from *G. biloba*, in comparison to other eukaryotic and prokaryotic iron SOD's, showed that the *Ginkgo* enzyme was more closely related to other plant iron SOD's and to that of *E. coli* than to that of *C. fasciculata* or *E. gracilis*. The *Ginkgo* iron enzyme also showed little homology to various eukaryotic and prokaryotic mangano SOD's based upon least squares analysis (data not shown).

TABLE 1. Purification of iron-containing SOD from *Ginkgo biloba*

Step	Protein (Total mg)	Activity[b] (units)	Specific activity (units/mg)	Yield (%)	Purification (fold)
Crude supernatant[a]	16300	6700	0.41	100	1
30-80% $(NH_4)_2SO_4$	678	4590	6.77	69	17
Batch DEAE cellulose	185	4420	23.9	66	58
Gradient DEAE cellulose	57	4250	74.6	63	182
Chromatofocusing	18.9	2200	116	33	283
Hydroxyapatite	1.61	1400	870	21	2122
Sephadex G-150	0.93	516	555	8	1354

[a]Approximately 2 kg fresh weight leaves.

[b]Total iron SOD activity represented as (CN^- insensitive) - (H_2O_2 insensitive).

Figures 1 and 2 show the peptide fragments formed as a result of proteolytic digestion of E. coli, G. biloba, N. luteum, and B. campestris iron SOD's with papain and S. aureus V-8 protease. Digestion with papain (Fig. 1) showed that the E. coli iron SOD was cleaved to one 14 kd fragment, while several similar-sized peptides were visible in the eukaryotic plant lanes in addition to the 14 kd fragment. Cleavage at the aspartic and glutamic acid residues with V-8 protease (Fig. 2) again showed similarities in fragment size for the proteins with two additional larger sized fragments appearing in the E. coli iron SOD lane.

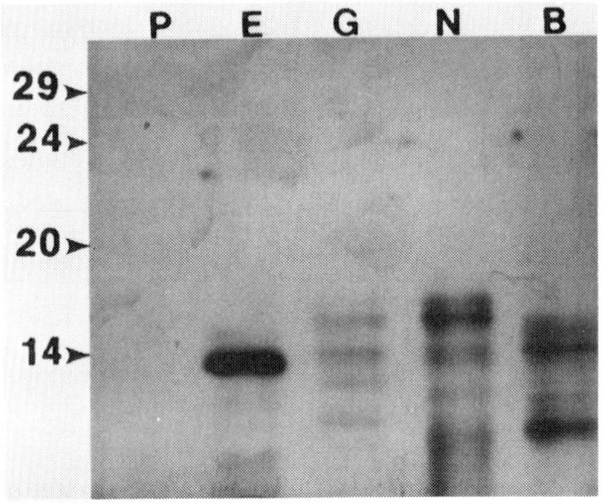

Fig. 1. SDS-PAGE of iron SOD's digested with papain and stained with silver. Lane P, 0.1 μg papain; Lanes E, E. coli iron SOD; G, G. biloba iron SOD; N, N. luteum iron SOD; B, B. campestris iron SOD; SOD's at a concentration of 3.5 μg per lane.

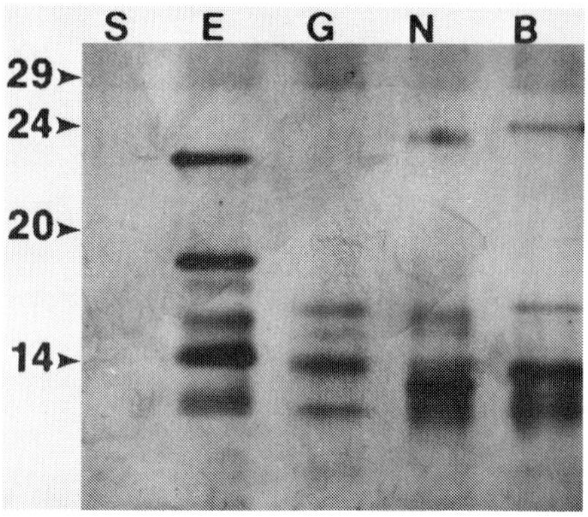

Fig. 2. SDS-PAGE of iron SOD's digested with S. aureus V-8 protease. Lane S, 0.1 μg V-8 protease; Lanes E, G, N, B, the same as in Fig. 1.

Chloroplasts were separated into their stromal (soluble) and thylakoid (membranous) components. After PAGE, the activity stained gels showed the presence of 2 iron SOD bands and one mangano SOD band in the stroma. Similar measurements of activity showed no SOD bands apparent in the thylakoid fraction. No evidence for a cuprozinc-containing SOD was found in the chloroplasts.

DISCUSSION

Iron SOD isolated from *G. biloba* appeared to be similar in size, charge, metal content, and in general amino acid composition to other eukaryotic and prokaryotic iron SOD's. Preliminary analysis of proteolytically-generated peptide fragments showed the eukaryotic iron SOD's to be similar to each other and different from *E. coli*. Elucidation of amino acid sequences of the various iron SOD's and comparative analysis of the variant and invariant sequences should yield insight into the phylogenetic origin of the eukaryotic iron SOD's.

ACKNOWLEDGEMENTS

Supported by funds from the Mississippi Agricultural and Forestry Experiment Station and the Office of Graduate Studies, Mississippi State University.

REFERENCES

1. Asada K, Kanematsu S (1978) In: Matsubara H, Yamanaka T (eds) Evolution of Protein Molecules. Japan Scientific Societies Press, Tokyo, pp 361-372
2. Bridges SM, Salin ML (1981) Plant Physiol 68:275-278
3. Sevilla F, del Rio LA, Hellin E (1984) J Plant Physiol 116:381-387
4. Kwiatowski J, Safianowska A, Kaniuga Z (1985) Eur J Biochem 146:459-466
5. Laemmli UK (1970) Nature 227:680-685
6. Salin ML, Bridges SM (1980) Arch Biochem Biophys 201:369-374
7. Cleveland DW, Fischer SG, Kirschner MW, Laemmli UK (1977) J Biol Chem 252:1102-1106
8. Merril CR, Goldman D, Sedman SA, Ebert MH (1981) Science 211:1437-1438
9. Mills WR, Joy KW (1980) Planta 148:75-83
10. Salin ML, Bridges SM (1980) Z Pflanzenphysiol 99:37-45

SUMMARY

A cyanide-insensitive SOD was purified from leaves of the eukaryote, *Ginkgo biloba*. The enzyme consisted of two equally-sized subunits and had a molecular mass of 47,000. Two isozymes were detected with isoelectric points of 5.25 and 5.15. Metal analysis showed 1.4 gram atoms of iron per molecule. The *Ginkgo* enzyme appeared to be localized in the stroma of chloroplasts. An analysis of peptide fragments obtained from proteolytic digestion showed qualitatively similar-sized fragments for iron-containing superoxide dismutases purified from other eukaryotes, *Nuphar* and *Brassica*. Differences were seen with the prokaryotic, *E. coli*, iron enzyme.

Superoxide and Superoxide Dismutase in Chemistry,
Biology and Medicine; G. Rotilio, editor

TRIS(PICOLINATO)MANGANESE(II): A CHEMICAL MODEL FOR THE MECHANISM AND FUNCTION OF MITOCHONDRIAL SUPEROXIDE DISMUTASE.

DONALD T. SAWYER, LEE SPENCER, KENNETH YAMAGUCHI AND M. STEVEN MCDOWELL.
Department of Chemistry, Texas A&M University, College Station, Texas, 77843, USA.

Superoxide dismutase (SOD) catalyzes the disproportionation of the superoxide anion ($O_2^{\bar{\cdot}}$) to hydrogen peroxide and oxygen,[1]

$$O_2^{\bar{\cdot}} + O_2^{\bar{\cdot}} + 2H^+ \xrightarrow{SOD} H_2O_2 + O_2 \qquad (1)$$

and is believed to be the natural defence against $O_2^{\bar{\cdot}}$ -initiated cell damage. Efforts to characterize the mechanism for such damage have prompted numerous investigations of $O_2^{\bar{\cdot}}$ reactivity over the past decade.[2,3] However, to date these studies have not provided chemical evidence for an authenticated cytotoxic reaction for $O_2^{\bar{\cdot}}$.

What has emerged as the most dominant characteristic of $O_2^{\bar{\cdot}}$ reactivity is its ability to act as a Brønsted base.[4] The acid-base equilibrium is driven to the right by disproportionation reactions which enhance the effective basicity[5] of $O_2^{\bar{\cdot}}$ (eq. 2-4).

$$O_2^{\bar{\cdot}} + H^+ \rightleftharpoons HO_2^{\cdot} \qquad pK_a,\ 4.88 \qquad (2)$$

$$HO_2^{\cdot} + O_2^{\bar{\cdot}} \longrightarrow HO_2^- + O_2 \qquad k,\ 1 \times 10^8 M^{-1}s^{-1} \qquad (3)$$

$$HO_2^{\cdot} + HO_2^{\cdot} \longrightarrow H_2O_2 + O_2 \qquad k,\ 8.6 \times 10^5 M^{-1}s^{-1} \qquad (4)$$

In dipolar aprotic solvents such as dimethylformamide (DMF),[6] dimethylsulfoxide[7] (Me_2SO) and acetonitrile (MeCN), $O_2^{\bar{\cdot}}$ is stable. However, proton sources again induce the rapid disproportionation of $O_2^{\bar{\cdot}}$ via formation of $HO_2^{\cdot}$.

$$O_2^{\bar{\cdot}} + H^+ \longrightarrow HO_2^{\cdot} \longrightarrow 1/2\ H_2O_2 + 1/2\ O_2 \qquad (5)$$

In DMF $HO_2^{\cdot}$ has a pK_a of 12, which means that this will be the dominant form of superoxide at physiological pH. This is important because in aprotic media the protein/lipid matrix of the mitochondrial membranes is the site of action of SOD, for which solvents such as DMF, with their aprotic polar hydrophilic character, provide excellent models. The potential for $HO_2^{\cdot}$ to inflict cell damage on such membranes has been demonstrated[8] by its ability to abstract hydrogen atoms from allylic groups in substrates such as 1,4-cyclohexadiene(1,4-CHD) and thereby initiate their autoxidation and peroxidation. Because linoleic acid and arachadonic acid esters contain allylic groups and are important components of lipids, there is reason to believe that in-situ generation of $HO_2^{\cdot}$ (via $O_2^{\cdot-} + H^+$) can initiate lipid peroxidation and autoxidation. The critical function of the SOD enzymes may then be to prevent formation of $HO_2^{\cdot}$ by the direct disproportionation of $O_2^{\cdot-}$.

Mitochondrial SOD is a manganese protein[9] for which a number of mechanisms for $O_2^{\cdot-}$ dismutation have been proposed.[10,11] The rate constant for the enzyme-catalyzed disproportionation is $3 \times 10^9 M^{-1}s^{-1}$; while that for the uncatalyzed proton-induced disproportionation of $O_2^{\cdot-}$ at pH 7 in water[13] is $1 \times 10^6 M^{-1}s^{-1}$. Thus, the apparent function of SOD is to lower the concentration of $O_2^{\cdot-}$ by at least three orders of magnitude. The tris(picolinato)manganese(II) complex $[Mn^{II}(PA)_2(PAH)(H_2O)]$ which catalyzes the disproportionation of $O_2^{\cdot-}$ in Me_2SO, represents a model catalyst for the manganese-SOD proteins.

The reactivity of $Mn^{II}(PA)_2(PAH)(H_2O)$ with $O_2^{\cdot-}$ in mixtures that contain both 1,4-CHD and H^+ has been determined, and the results have been used to develop a reasonable mechanism for SOD protection from $O_2^{\cdot-}$-initiated lipid peroxidation. Hence, we suggest that (a) $HO_2^{\cdot}$ is the active form of superoxide in-vivo and (b) the role of SOD is to remove $O_2^{\cdot-}$ before it becomes protonated and reactive with allylic substrates.

RESULTS

<u>Kinetics and Mechanism for Proton-Induced Disproportionation of $O_2^{\cdot-}$ in MeCN, Me_2SO or DMF.</u> Combination of $O_2^{\cdot-}$ with strong acids ($HClO_4$ or NH_4ClO_4) in Me_2SO or MeCN results in a disproportionation reaction that is second order with respect to $[O_2^{\cdot-}]$. The data are consistent with a rapid protonation to form $HO_2^{\cdot}$ followed by a rate-determining disproportionation.

$$H^+ + O_2^{\cdot-} \longrightarrow HO_2^{\cdot} \qquad (6)$$

$$HO_2^{\cdot} + HO_2^{\cdot} \xrightarrow{k} H_2O_2 + O_2 \qquad (7)$$

The magnitude of the second-order rate constant, k, is $10^4 M^{-1}s^{-1}$ in Me_2SO and $10^7 M^{-1}s^{-1}$ in MeCN.[6,7]

<u>The Reactivity of $HO_2^{\cdot}$ with 1,4-CHD.</u> $HO_2^{\cdot}$ abstracts hydrogen atoms from allylic hydrocarbons at rates that are competitive to $HO_2^{\cdot}$ disproportionation. Table 1 summarizes the results for the reactivity of $O_2^{\cdot-}$ with 1,4-CHD in Me_2SO.

TABLE I

OXIDATION OF 1,4-CYCLOHEXADIENE (1,4-CHD) BY $HO_2^{\cdot}$ IN Me_2SO.[a]

Proton Source (HA)	[HA], mM	$[O_2^{\cdot-}]$, mM	[1,4-CHD], mM	Reaction Efficiency,%[b]	Product Distribution, %[c]	
					benzene	1,3-CHD
		5.4	10.0	19	32	67
		10.0	20.0	18	53	47
H_2O	10	9.2	18.4	47	100	0
H_2O	100	4.0	8.1	100	30	70
H_2O	150	9.2	18.4	21	45	55
H_2O	150	20.5	40.9	63	16	84
NH_4ClO_4	5.8	5.8	10.0	20	59	41
NH_4ClO_4	2.5	2.5	10.0	17	34	66
NH_4ClO_4	3.8	7.6	10.0	7	96	4
PAH[d]	10.7	5.4	10.0	14	37	63
PAH	3.8	7.6	10.0	6	95	5
PAH	1.4	2.8	10.0	19	25	75
PAH	0.3	2.8	10.0	14	33	67
PAH	0.2	1.0	10.0	33	58	42
$HClO_4$	6.4	3.2	6.4	31	54	46
$HClO_4$	3.2	3.2	6.4	92	69	31
$HClO_4$	7.6	3.2	6.4	100	79	21
$HClO_4$	8.4	8.4	17.0	40	75	25

[a]An Me_2SO solution that contained the acid was added to a second Me_2SO solution that contained $O_2^{\cdot-}$ and 1,4-CHD. The indicated concentrations represent the initial values after mixing.

[b]100% efficiency represents the reaction of one 1,4-CHD molecule per $O_2^{\cdot-}$ added.

[c]Product yield determined by gas chromatographic analysis.

[d]PAH, picolinic acid.

A reasonable reaction scheme involves initial formation of $HO_2^{\cdot}$, with its subsequent disproportionation or attack of 1,4-CHD, (see Scheme 1). The effect of concentration on the competitive process is important;at low concentrations of $HO_2^{\cdot}$ the first-order radical process is favored over the second-order process.

Kinetics and Mechanism of The Reaction of $O_2^{\cdot-}$ With $Mn^{II}(PA)_2(PAH)(H_2O)$. Addition of $O_2^{\cdot-}$ to a solution of $Mn^{II}(PA)_2(PAH)(H_2O)$ in Me_2SO results in an initial acid base interaction that is followed by electron transfer to yield a coordinated hydroperoxy-Mn(III) adduct[15];

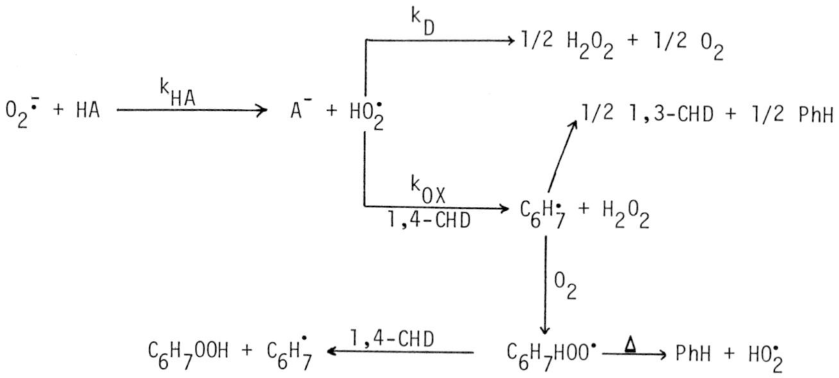

Scheme 1

$$Mn^{II}(PA)_2(PAH)(H_2O) + O_2^{\overline{\cdot}} \longrightarrow Mn^{III}(PA)_2(PAH)(O_2H) + H_2O + A^- \quad (8)$$

This then oxidizes a second $O_2^{\overline{\cdot}}$ to yield H_2O_2 and O_2 and the Mn(II) catalyst.

$$Mn^{III}(PA)_2(PAH)(O_2H) + O_2^{\overline{\cdot}} \longrightarrow Mn^{II}(PA)_2(PAH) + H_2O_2 + O_2 + A^- \quad (9)$$

Superoxide ion reacts with $Mn^{III}(PA)_2(PAH)(OH)$ by a second-order process (k, $1.4 \times 10^6 M^{-1}s^{-1}$), but this reaction pathway does not result in regeneration of an active manganese species.

Inhibition of $HO_2^{\cdot}$/RH Reaction by $Mn^{II}(PA)_2(PAH)(H_2O)$. Table II illustrates the effect on the product distribution when the Mn(II) complex is added to solutions that contain a proton source, 1,4-CHD and $O_2^{\overline{\cdot}}$. In the absence of Mn(II) the conversion of 1,4-CHD to benzene and 1,3-CHD is quantitative. However, when $Mn^{II}(PA)_2(PAH)(H_2O)$ is added to $O_2^{\overline{\cdot}}$/1,4-CHD solutions that contain H_2O or $HClO_4$ as proton sources, the yield of benzene and 1,3-CHD decreases.

The Mn(II)-picolinate complex reacts with $O_2^{\overline{\cdot}}$ before it can take up a proton from HA to form $HO_2^{\cdot}$ (and either disproportionates or abstracts a hydrogen atom from 1,4-CHD). The $Mn^{III}(PA)_2(PAH)(O_2H)$ adduct reacts with a second $O_2^{\overline{\cdot}}$ to produce O_2, which in turn reacts with the $C_6H_7^{\cdot}$ radical to produce benzene as the sole product. The $Mn^{III}(PA)_2(PAH)(OH)$ complex under the same conditions does not inhibit the production of benzene or 1,3-CHD.

DISCUSSION

The Mn^{II} $(PA)_2(PAH)(H_2O)$ complex reacts with $O_2^{\overline{\cdot}}$ at a rate that is two orders of magnitude faster than the disproportionation of $HO_2^{\cdot}$ (k, $1 \times 10^4 M^{-1}s^{-1}$). The decrease in the oxidation of 1,4-CHD by $HO_2^{\cdot}$ in the presence of $Mn^{II}(PA)_2(PAH)(H_2O)$

confirms that it competes with proton sources for $O_2^{\cdot-}$ and suppresses $HO_2^{\cdot}$ formation. A self-consistent mechanism for the function of the SOD enzymes in the prevention of lipid peroxidation is outlined in Scheme 2.

TABLE II

OXIDATION IN Me_2SO OF 1,4-CYCLOHEXADIENE (1,4-CHD) BY $HO_2^{\cdot}$ IN THE PRESENCE AND ABSENCE OF $Mn^{II}(PA)_2(PAH)(H_2O)$[a]

Proton Source (HA)	[HA], mM	$[Mn^{II}]$, mM	$[O_2^{\cdot-}]$, mM	[1,4-CHD], mM	Reaction[b,c] Efficiency, %	Yield (%)	
						PhH	1,3-CHD
H_2O	100	0	4.0	8.1	100	30	70
H_2O	100	0.4	4.0	8.1	19	100	0
H_2O	100	2.0	4.0	8.1	4	100	0
H_2O	100	4.0	4.0	8.1	9	100	0
H_2O	100	8.1	4.0	8.1	11	100	0
$HClO_4$	1.6	0	3.2	6.4	100	79	21
$HClO_4$	1.7	1.7	3.4	6.8	10	100	0
$HClO_4$	1.7	3.4	3.4	6.8	9	100	0
$HClO_4$	1.7	6.8	3.4	6.8	10	100	0

[a]An Me_2SO solution that contained the acid [and the Mn(II) complex] was added to a second Me_2SO solution that contained $O_2^{\cdot-}$ and 1,4-CHD. The indicated concentrations represent the initial values after mixing.

[b]100% represents one 1,4-CHD oxidized per $O_2^{\cdot-}$.

[c]Varying amounts of a proton source were dissolved in 5 mL of Me_2SO and mixed with 5 mL of Me_2SO stock solutions that contained $(Me_4N)O_2$ and the 1,4-CHD substrate in the absence oxygen. The $Mn^{II}(PA)_2(PAH)(H_2O)$ complex was added to the solutions that contained the proton source. After cooling in an ice-water bath, 10 mL of ice-cold water were added. The mixture was extracted with ether and $MgSO_4$ was added. A 0.5 µL portion was analyzed by capillary-column gas chromatography, and the product distribution and reaction efficiences were calculated on the basis of the integrated peak areas, and by adding a known amount of toluene or benzene as an internal standard.

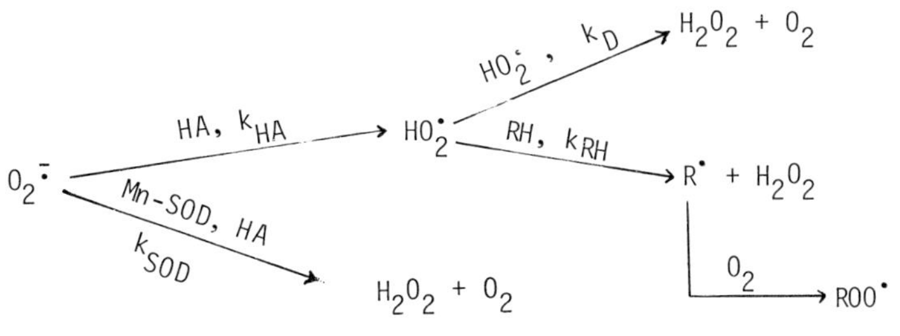

Scheme 2

In summary a likely biological function for the SOD enzymes is to remove $O_2^{\bar{\cdot}}$ and thereby preclude formation of $HO_2^{\cdot}$ and thus preventing lipid peroxidation and autoxidation. The results of Table II confirm that an SOD model complex can block formation of $HO_2^{\cdot}$, which requires that the primary step for the SOD-$O_2^{\bar{\cdot}}$ reaction be kinetically favored and dominant relative to the protonation reaction [$k_{SOD}(SOD) >> k_{HA}$] in Scheme 2.

ACKNOWLEDGMENT

This work was supported by the National Institutes of Health under grant GM-36289. Most of the experiments were performed at the University of California, Riverside, California, 92521.

REFERENCES

1. McCord JM, Fridovich I (1969) J Biol Chem 244:6049-6055
2. Fee JA (1982) Trends Biochem Sci 7:84-86 and references therein
3. Fee JA (1980) In:Spiro TG (ed) Metal Ion Activation of Dioxygen. Wiley, New York pp 209-237
4. Sawyer DT, Valentine JS (1981) Acc Chem Res 14:393-400
5. Bielski BH (1978) Photochem Photobiol 28:645-649
6. Chin DH, Chiercato G, Nanni EJ Jr, Sawyer DT (1982) J Am Chem Soc 104:1296-1299
7. McDowell MS, Sawyer DT Unpublished Results
8. Howard JA, Ingold KU (1967) Can J Chem 45:785-792
9. Fridovich I (1975) Annu Rev Biochem 44:147-159
10. McAdam ME, Fox RA, Lavelle F, Fielden EM (1977) Biochem J 165:71-74
11. Bull C, Fee JA (1985) J Am Chem Soc 107:3295-3304
12. Pick M, Rabani J, Yost F, Fridovich I (1974) J Am Chem Soc 96:7329-7333
13. Bielski BHJ, Richter HW (1977) J Am Chem Soc 99:3019-3023
14. Yamaguchi K, Sawyer DT (1985) Inorg Chem 24:971-976
15. Yamaguchi K, Sawyer DT (1985) J Am Chem Soc (in press)

© 1986 Elsevier Science Publishers B.V. (Biomedical Divison)
Superoxide and Superoxide Dismutase in Chemistry,
Biology and Medicine; G. Rotilio, editor

EFFECT OF DETERGENTS ON THE KINETICS OF GLUTATHIONE PEROXIDASE AND PHOSPHOLIPID HYDROPEROXIDE GLUTATHIONE PEROXIDASE

FULVIO URSINI, MATILDE MAIORINO, ANTONELLA ROVERI and CARLO GREGOLIN
Institute of Biological Chemistry of the University of Padova,
Via F. Marzolo 3, 35131 Padova (Italy).

INTRODUCTION

The "Peroxidation Inhibiting Protein" (PIP), identified and purified in this laboratory from different rat and pig organs (1), has been recently characterized as the new selenoenzyme "Phospholipid Hydroperoxide Glutathione Peroxidase" (PHGPX) (2). The results of the kinetic analysis were compatible with a tert-uni ping pong mechanism (2) identical to that reported for glutathione peroxidase (GPX) (3,4). In both enzymes, following the oxidation of the selenol mojety by hydroperoxides, the thiol dependent reduction takes place, without the formation of any central complex. However, while some substrates are reduced both by GPX and PHGPX, the phospholipid hydroperoxides are not substrate for GPX, independently of the physical state (bilayer or mixed micelle). The stimulation induced by Triton X-100 on the activity of PHGPX toward phospholipid hydroperoxides suggests an "interfacial character" for this enzyme. In this paper we focused the effect of ionic and non ionic detergents and fatty acids on the kinetics of both enzymes, using as peroxidic substrate either linoleic acid hydroperoxide (LAHPO) or H_2O_2 to recognize if the observed effects would be due either to a perturbation of the physical state of the lipidic substrate or to a direct interaction of the detergent with the enzyme.

MATERIALS AND METHODS

Phospholipid hydroperoxide glutathione peroxidase was purified from pig heart as previously described (2). Ox red blood cell glutathione peroxidase was from Boehringer-Mannheim. The substrate and product concentration data and the corresponding reaction rates were obtained from single progression curves (2). The digitalized absorbance readouts at 340 nm of the reaction mixture containing PHGPX or GPX, a peroxidic substrate, glutathione, glutathione reductase and NADPH were recorded by a mini computer. The BASIC routine allowed the rapid processing of the data to fit the integrated rate equation (2, 4). The enzyme concentration was measured on the basis of the Se content (2).

RESULTS AND DISCUSSION

The effect of the detergents on the two steps of the ping pong reaction could be evaluated using the linear plot of the integrated rate equation (2, 4). The oxidation rate of the active site of both enzymes, in the presence of hydroperoxides indeed, was evaluated on the basis of the slope of the linear plot from which the Dalziel coefficient Ø1 could be calculated. This coefficient corresponds to the reciprocal of K_1 (the rate constant for the oxidation of the enzyme). The reduction rate of the enzymes, on the other hand, was evaluated on the basis of $Ø_2$ calculated from the intercept on the vertical axis, that corresponds to $K_2 \times K_3/(K_2 + K_3)$ where K_2 and K_3 are the rate constants for the two consecutive reductive reactions that cannot to be separated by the steady state analysis. As indicated in Tab. 1, Triton inhibits the interaction of LAHPO with GPX. It is worth noting that Ø2 was not modified suggesting that the interaction of the oxidized enzyme with GSH was not hampered. Moreover the inhibitory effect was absent when the substrate was H_2O_2. These results are compatible with a very low or absent interaction of GPX with a substrate in Triton mixed micellar form. The deoxycholate (DOC) on the other hand did not inhibit GPX, while PHGPX was inhibited when the substrate was both LAHPO (Table 1) and H_2O_2 (not shown). This evidence and the modification of both the kinetic coefficients as well, indicates that DOC dramatically interferes with the catalytic site of the enzyme. A similar inhibition pattern was observed in the presence of micromolar concentrations of oleate. This effect is related to presence of the double bond in the fatty acid and it is independent of the steric configuration. Stearic acid indeed was not inhibitory while elaidic acid had the same effect as oleic

TABLE 1

EFFECTS OF DETERGENTS ON THE KINETICS OF PHGPX AND GPX

	PHGPX Ø1 (mM min)	PHGPX Ø2 (mM min)	GPX Ø1 (mM min)	GPX Ø2 (mM min)
LAHPO	$4.6\ 10^{-7}$	$1.4\ 10^{-4}$	$.5\ 10^{-7}$	$3.2\ 10^{-5}$
+ 0.75 mM TRITON	$4.7\ 10^{-7}$	$1.5\ 10^{-4}$	$19.3\ 10^{-7}$	$3.6\ 10^{-5}$
+ 1.2 mM DOC	$22.0\ 10^{-7}$	$5.3\ 10^{-4}$	$5.1\ 10^{-7}$	$3.2\ 10^{-5}$
+ 12 uM OLEATE	$85.0\ 10^{-7}$	$10.0\ 10^{-4}$	$5.0\ 10^{-7}$	$3.2\ 10^{-5}$
+ 0.1 mM STEARATE	$4.5\ 10^{-7}$	$1.3\ 10^{-4}$	$5.2\ 10^{-7}$	$3.1\ 10^{-5}$

The Dalziel coefficients (01 and 02) were calculated using the lienar plot of the integrated rate equation described in ref 2 and 4.

acid (not shown). Oleate inhibited PHGPX also when H_2O_2 was used as substrate and the inhibition could be prevented by the addition of Triton (Fig. 1). These results indicate that the active site of PHGPX contains a binding site for lipophylic molecules with preferential affinity for double bonds.

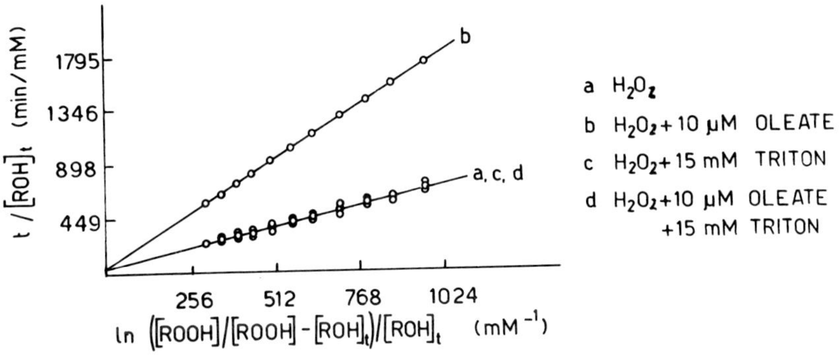

Fig. 1. Linear plot of the integrated rate equation of PHGPX (2nM) activity on H_2O_2 (4 uM). Effect of oleate and Triton.

REFERENCES

1. Ursini F, Maiorino M, Valente M, Ferri L and Gregolin C (1982) Biochim Biophys Acta 710: 197
2. Ursini F, Maiorino M and Gregolin C (1985) Biochim Biophys Acta 839: 62
3. Flohê L, Loeschen G, Gunzler W and Eichele E (1972) Hoppe Seyler Z. Physiol Chem 353: 987
4. Chaudiere J, and Tappel AL (1983) Arch Biochem Biophys 266: 448

Superoxide and Superoxide Dismutase in Chemistry, Biology and Medicine; G. Rotilio, editor

THE REACTIONS OF SUPEROXIDE RADICAL WITH CATALASE

KAZUO KOBAYASHI, NORIHIDE SHIMIZU, and KOICHIRO HAYASHI
The Institute of Scientific and Industrial Research, Osaka University, Mihogaoka 8-1, Ibaraki 567, Japan

INTRODUCTION

The reactions between O_2^- and biomolecule have received much attention from the view point of oxygen toxicity. Kono and Fridovich proposed that catalase was inhibited by O_2^- using an enzymatic source of O_2^- and that this inhibition was due to reactions of O_2^- with ferric catalase and Compound I to yield Compound III and Compound II, respectively (1). The presence of these reactions has been suggested already for peroxidase (2, 3). These reactions in catalase, however, have not been observed directly.

In order to study the kinetic behavior of the reactions of O_2^-, pulse radiolysis has been widely used. In this report, we have investigated reaction of O_2^- with ferric catalase by the use of pulse radiolysis. Here attempts were made to confirm the presence of reaction 1.

$$Fe^{3+} + O_2^- \longrightarrow Fe^{2+}-O_2 \qquad (1)$$

We also attempted to determine the rate constants of Reaction 1, and the results are compared with those obtained in peroxidase and other hemoproteins. Here physiological significance of Reaction 1 in catalase has been discussed.

RESULTS

After pulse radiolysis of bovine liver catalase with oxygen saturation, fast and slow absorption changes were observed. This suggests that two species of reactions were involved in this system. The difference spectrum obtained at 8 ms after the pulse, which corresponds to fast absorption change, is similar to that of ferric form minus Compound I of this enzyme. This result suggests that the fast absorptin change is due to the reaction of ferric enzyme with H_2O_2, as shown in Reactions 2 and 3, where H_2O_2 is formed by the dispropotionation of O_2^-.

$$2\ O_2^- \xrightarrow{2H^+} H_2O_2 \qquad (2)$$

$$H_2O_2 + Fe^{3+} \longrightarrow \text{Compound I} \qquad (3)$$

On the other hand, the difference spectrum, which is responsible for the slow absorption change, is different from those of ferric enzyme minus Compound I or Compound II. In the presence of superoxide dismutase, the formation of this compound was found to be inhibited, whereas that of Compound I was little affected. This result suggests that slow absorption change can be attributed to the reaction of O_2^- with ferric enzyme.

In order to confirm the reaction product between O_2^- with ferric enzyme, we have prepared the ferrous and oxyferrous enzymes. Both forms have not be realized spectrophotometrically. Fig. 1 shows absorption spectra of ferric catalase and its photolyzed products in the presence of 1 μM proflavin and 10 mM EDTA in a deaerated solution. Upon the addition of CO to the photolyzed product, the spectrum changed to a form with absorption maxima at 573, 542 and 418 nm. This suggests that the photolyzed product is ferrous enzyme. When O_2 was introduced into the ferrous enzyme, the absorption spectrum returned to its original ferric state. Then the reaction between ferrous enzyme and O_2 was carried out at -20°C in 33% v/v ethylene glycol in order to stabilize the intermediate.

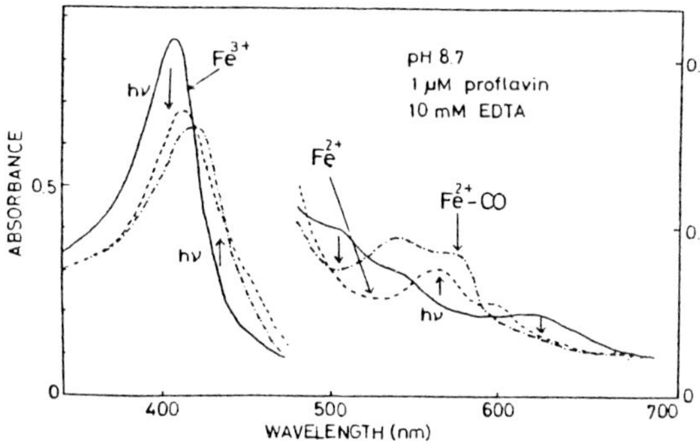

Fig. 1. Absorption spectra of ferric, ferrous and ferrous CO catalase.

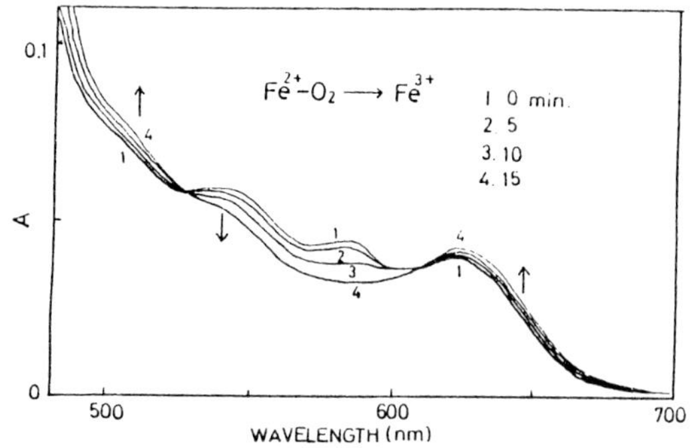

Fig. 2. Absorption spectra of oxyferrous catalase and its spectral change to ferric enzyme.

The reaction product which has absorption maxima at 584 and 544 nm was obtained, as shown in Fig. 2. Judging from its spectral characteristic, this compound is considered to be oxyferrous enzyme. The oxyferrous enzyme was rather stable at such low temperature but it was converted into ferric enzyme gradually, when the sample cuvette was left at room temperature (Fig. 2). The oxyferrous enzyme formed by the reaction of O_2 with ferrous enzyme auto-decomposed into the ferric form with a half time of 8 s at room temperature.

Fig. 3 compares the difference spectra of oxyferrous and ferrous forms minus ferric enzyme. It is noted that kinetic difference spectrum in the reaction of O_2^- with ferric enzyme is similar to oxyferrous enzyme minus ferric enzyme. Therefore it is concluded that ferric catalase reacts with O_2^- to form oxyferrous enzyme.

For determination of the rate constant of O_2^- with ferric enzyme, 1~3 μM O_2^- is generated in a solution of 80 ~ 360 μM enzyme (heme base). In these conditions, the absorption change after pulse was follow pseudo-first order kinetics. The second order rate constant of the reaction is estimated to be 2 x $10^5 M^{-1}s^{-1}$ at pH 7.4. This value increased with a decrease of pH in the range from pH 6 to 4. The pH dependence of the apparent rate constant for Reaction 1 is explained by the difference reactivity between HO_2 and O_2^- toward ferric enzyme. Similar result was obtained in the reaction of O_2^- with ferric peroxidase.

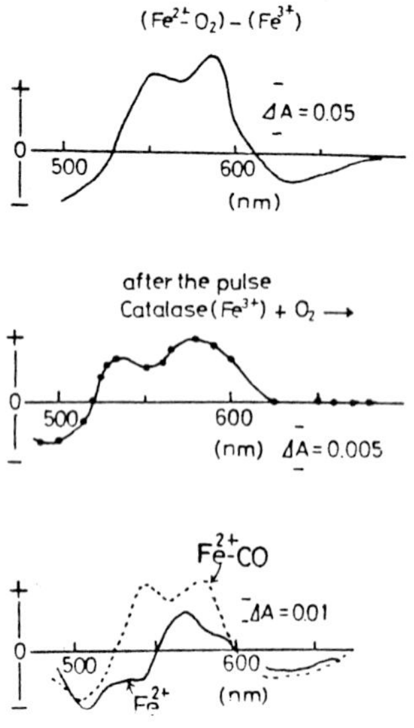

Fig. 3. Comparison of the difference spectra of the products minus ferric catalase.

Participation of O_2^- in the catalytic process of peroxidase-oxidase reaction, and with indoleamine oxygenase has been demonstrated. The second order rate constants for the binding of O_2^- to both enzymes are 10^6-10^7 $M^{-1}s^{-1}$, which is 1 order of magnitude larger than that of catalase at pH 7.0. Futhermore it should be noted that the concentration of superoxide dismutase in erythrocyte is more than 10 times that required to completely inhibit the formation of oxyferrous catalase.

REFERENCES

1. Kono Y and Fridovich I (1982) J Biol Chem 257:5751-5754
2. Bielski BHJ and Gebicki JM (1974) Biochim Biophys Acta 364: 233-235
3. Sawada Y and Yamazaki I (1973) Biochim Biophys Acta 327: 257-265

Superoxide and Superoxide Dismutase in Chemistry,
Biology and Medicine; G. Rotilio, editor

MN-CATALASE OF LACTOBACILLUS PLANTARUM: INHIBITION BY CARBONYL REAGENTS

YASUHISA KONO
Department of Bacteriology, Tottori University School of Medicine, Yonago, Tottori 683 (JAPAN)

INTRODUCTION

The aerotolerant bacterium, Lactobacillus plantarum, exhibits several peculiarities in its defenses against O_2 toxicity. It lacks superoxide dismutase (SOD), but intracellular Mn(II) at about 25 mM serves as a functional replacement for SOD (1-3). Moreover, L. plantarum, which is unable to synthesize heme, produces a nonheme-catalase. This azide-insensitive catalase has been recently isolated and found to contain manganese (4). The Mn-catalase is stable to freezing and thawing, tolerates moderate heating and is insensitive toward cyanide and azide. However, Mn-catalase was found to be irreversiblly inactivated by NH_2OH, NH_2NH_2, phenylhydrazine and isoniazide in the presence of copper ions. In this communication, exploration of mechanisms of inactivation by these inhibitors and copper ions is reported.

RESULTS

O_2-dependent inactivation by Cu(II) plus NH_2OH or NH_2NH_2. Neither NH_2OH nor NH_2NH_2, when present alone, caused any decrease of the activity of Mn-catalase. However, when the enzyme was incubated with Cu(II) plus NH_2OH or NH_2NH_2, the enzymatic activity was rapidly lost in a time-dependent manner; an effect, which was not reversed by dialysis. Chelating agents, such as EDTA and histidine, completely prevented the inactivation. This inactivation was not observed under anaerobic conditions, indicating a role for the autooxidation of NH_2OH or NH_2NH_2 in the inactivation process. In order to explore the participation of O_2^- and H_2O_2, produced during autooxidation of NH_2OH catalyzed by Cu(II) (5), in this inactivation, the effect of SOD and H_2O_2 was tested. Addition of H_2O_2 significantly enhanced the rate of inactivation under both aerobic and anaerobic conditions. Omission of Cu(II) did not affect the inactivation rate in the presence of H_2O_2, nor did the inclusion of EDTA. Addition of SOD resulted in a slight increase in the rate of inactivation. Therefore, O_2^- does not participate in the inactivation reaction, and an increased rate of inactivation is conferred by an increase in the steady state concentration of H_2O_2. The dependence of NH_2OH inactivation upon H_2O_2 was verified by exploiting Conway microdiffusion vessels.

The catalytic cycle of the enzyme is thought to involve Mn(III) and Mn(V). The H_2O_2-dependent inactivation by NH_2OH could involve univalent reduction of the Mn(V) form, generating an inactive Mn(IV) state. If this is the case, the inactive

Mn(IV) form could be reversed to the active form by oxidation or reduction. Oxidants, such as ferricyanide and persulfate were uneffective, whereas the activity was restored by dithionite or by O_2^- to an extent of 35-40%. Dithionite or O_2^- could reduce the inactive Mn(IV) to the active Mn(III) form of the enzyme. Exposure to a flux of H_2O_2 generated by glucose oxidase and glucose caused a bleaching of its visible spectrum, showing the mixture of Mn(III) and Mn(V) forms of the enzyme. When the enzyme was inactivated by exposure to NH_2OH plus H_2O_2, there was further bleaching.

O_2-independent inactivation by Cu(II) plus phenylhydrazine or isoniazid. The Mn-catalase was irreversiblly inactivated when exposed to Cu(II) plus phenylhydrazine or isoniazid. Addition of EDTA completely suppressed the inactivation. Under aerobic conditions, 0.1 mM Cu(II) caused 60% inactivation, while 0.5 mM Cu(II) caused 80% inactivation. In contrast, under anaerobic conditions, enzymatic activity was completely lost even with 0.1 mM Cu(II). Phenylhydrazine and isoniazid might be expected to reduce enzyme-Mn(V), which is formed by the reaction of the enzyme-Mn(III) with H_2O_2, and generate the inactive Mn(IV) form, as in believed to be the case in the inactivation by NH_2OH plus H_2O_2. H_2O_2 was formed by the autooxidation of phenylhydrazine catalyzed by Cu(II). This mechanism is, however, not supported by the following observations; no inactivation by phenylhydrazine plus H_2O_2 and limited inactivation by phenylhydrazine plus Cu(II) in the presence of O_2. Furthermore, exposure of inactivated enzyme to dithionite or to a photochemical source of O_2^- failed to cause recover of its activity, indicating that the inactivation was not due to the genration of the inactive Mn(IV) form of the enzyme. Potassium ferricyanide plus phenylhydrazine, which form phenyldiazene, did not cause any decrease of activity of Mn-catalase, showing no participation of phenyldiazene in the inactivation.

The pseudo-first order rate constant for irreversible inactivation (k_{obs}) approached a limit with increasing concentration of phenylhydrazine with 0.1 mM Cu(II). Graphing the reciprocal of k_{obs} against the reciprocal of phenylhydrazine concentration gave a linear plot. The rates of the inactivation were also determined at different concentrations of Cu(II) in the presence of 1 mM phenylhydrazine. The plot of k_{obs} *versus* [$CuCl_2$] was linear, indicating that the inactivation is a bimolecular process. From the slope of the plot, a value of 9.6 x 10^{-1} $mM^{-1}min^{-1}$ was calculated for the apparent second order rate constant for this inactivation.

The enzyme was treated with phenylhydrazine with and without $CuCl_2$. After excess phenylhydrazine and $CuCl_2$ were removed by exhaustive dialysis, difference absorbance spectra of the treated enzyme against the untreated native enzyme were recorded. Difference spectra revealed a change in the ultraviolet region. The spectrum of enzyme treated with phenylhydrazine plus Cu(II) showed an absorbance maximum at 243 nm and a minimum at 203 nm, while the spectrum of enzyme treated with phenylhydrazine alone, which retained full catalytic activity, showed an absorbance maximum at 275 nm and a shoulder at 225 nm. Thus, when Mn-catalase activity is inactivated by phenylhydrazine plus Cu(II), the enzyme is converted to a form different from the enzyme-phenylhydrazine complex.

The results described above are consistent with the following scheme.

$$\text{enzyme + phenylhydrazine} \overset{K_D}{\rightleftharpoons} \text{enzyme·phenyhydrazine} \xrightarrow[\text{Cu(II)} \curvearrowright \text{Cu(I)}]{k} \text{inactive enzyme}$$

The dissociation constant, K_D, and over-all rate constant, k, were calculated to be 0.72 mM and $1.75 \times 10^{-1} mM^{-1} min^{-1}$, respectively, from double reciprocal plot according to the following equation: $1/k_{obs} = 1/k + K_D/k[\text{phenylhydrazine}]$.

Kinetic data and spectral studies suggest that phenylhydrazine could bind to Mn-catalase, and Cu(II) oxidizes the enzyme·phenylhydrazine Michaelis complex to yield inactive enzyme.

Physiological role of Mn-catalase. The inhibitors of Mn-catalase described above were used to suppress the Mn-catalase in *L. plantarum* cells to explore its physiological importance. Cells were incubated with NH_2OH either aerobically or anaerobically. At any concentration of NH_2OH the loss of catalase activity in cells was higher under aerobic conditions than under anaerobic conditions. Cells whose Mn-catalase was not suppressed by NH_2OH were not killed by H_2O_2, whereas those treated with 0.1 or 1 mM NH_2OH exhibited a progressive intolerance towards H_2O_2. Thus, Mn-catalase provides a defence against the H_2O_2, which was generated by *L. plantarum* during the stationary phase of aerobic growth (4).

ACKNOWLEDGEMENT

The author wishes to thank Dr. Irwin Fridovich for helpful discussions.

REFERENCES

1. Archbald FS, Fridovich I (1981) J Bacteriol 145: 442-451
2. Archbald FS, Fridovich I (1981) J Bacteriol 146: 928-936
3. Archbald FS, Fridovich I (1982) Arch Biochem Biophys 214: 452-463
4. Kono Y, Fridovich I (1983) J Biol Chem 258: 6015-6019
5. Kono Y (1978) Arch Biochem Biophys 186: 189-195

EVOLUTION AND BIOSYNTHESIS OF SUPEROXIDE DISMUTASES

© 1986 Elsevier Science Publishers B.V. (Biomedical Division)
Superoxide and Superoxide Dismutase in Chemistry,
Biology and Medicine; G. Rotilio, editor

STRUCTURAL AND EVOLUTIONARY RELATIONSHIPS BETWEEN THE EUKARYOTIC SUPEROXIDE DISMUTASES

M.W. PARKER[1], F. BOSSA[2], D. BARRA[2], W.H. BANNISTER[3] and J.V. BANNISTER[4]
[1] Laboratory of Molecular Biophysics, University of Oxford, Oxford, England; [2] Institute of Biological Chemistry and CNR Centre for Molecular Biology, University of Rome, Rome, Italy; [3] Nuffield Department of Clinical Biochemistry, University of Oxford, Radcliffe Infirmary, Oxford, England and [4] Inorganic Chemistry Laboratory, University of Oxford, Oxford, England.

GENERAL INTRODUCTION

The enzyme superoxide dismutase (SOD) is unique in that it exists as three isoenzyme forms whose similarity is that they catalyse the same dismutation reaction: $2O_2^- + 2H^+ \rightarrow H_2O_2 + O_2$. The main important difference between the three isozymes is that they contain different metals as their functional prosthetic group (1). Eukaryotic species have two forms of the enzyme which in general contain either Cu and Zn or Mn. Two bacterial species, Photobacterium leiognathi and Caulobacter crescentus also contain the Cu/Zn enzyme (2,3). The eukaryotic SODs were found to be compartmentalised inside the cell. Intact mitochondria contain the MnSOD in the matrix space and the Cu/Zn enzyme previously thought to be exclusively localised in the cytosol was found in the intermembrane space. In rat liver the Cu/ZnSOD was also found to be lysosomal with the amount varying with the nutritional state of the animal (4). Also an examination of the subcellular distribution of the MnSOD in human liver revealed that the enzyme is not exclusively localised in the mitochondria (5). Both the Cu/Zn and the MnSOD are synthesised by the nuclear genome (6) and it is possible that the MnSOD is present in the cytosol as an active proenzyme before it is transported into the mitochondrion. Autor (7) observed that the MnSOD is synthesised as a precursor protein with an extra peptide of about 2000 daltons/subunit. However, no evidence has as yet been presented as to whether the precursor protein is an active enzyme. It is also unclear as to why mitochondria selected Mn rather than Fe as the prosthetic group metal. Prokaryotic species contain either an FeSOD or the MnSOD or both. Mitochondria are known to concentrate very efficiently both Fe and Mn (8) and if these organelles evolved from free living species then either form of the enzyme could be present. However, the MnSOD was found to be very resistant to hydroxyl radical damage in contrast to the FeSOD (9). The production of hydroxyl radicals by mitochondria via a metal catalysed Haber-Weiss reaction is a possibility since the organelle produces both O_2^- and H_2O_2. The presence of MnSOD in mitochondria is therefore an adaptation against this possibility.

INTERRELATIONSHIPS BETWEEN THE Cu/Zn SUPEROXIDE DISMUTASES

The complete primary structure has been determined for eight SODs (10-17) (Fig.1). Gaps and insertions are included to optimise sequence homologies. Only human, horse, bovine and pig SOD have an acetylated N-terminal residue. The blocked N-terminus appears to be restricted to mammals. Lower species have an exposed N-terminus. A free NH_2 terminal residue has also recently been demonstrated in the bullfrog indicating that reptiles probably also have an exposed N-terminus (18). The alignments presented show that the metal binding

```
                                                                                  *     *
HUMAN    A T K A V C V L K G D G P V Q G I I - N F E Q K E S N G P V K V W G S I K G - L T E G L H G F H V H
HORSE    A L K A V C V L K G D G P V H G V I - H F E Q Q Q E G G P V V L K G F I E G - L T K G D H G F H V H
BOVINE   A T K A V C V L K G D G P V Q G T I - H F E A K G D T - - V V V T G S I T G - L T E G D H G F H V H
PIG      A T K A V C V L K G D G P V Q G T I - V F E L K G E K - T V L V T G T I K G - L A E G D H G F H V H
FLY      V V K A V C V I N G D A K G T V F F - E Q E - - S S G T P V K V S G E V C G - L A K G L H G F H V H
YEAST    - V Q A V A V L K G D A G V S G V V - K F E Q A S E S E P T T V S Y E I A G N S P N A E R G F H I H
SWORD    V L K A V C V L R G A G E T T G T V - V F E G E G N A N A V G K G I I L K G - L T P G E H G F H V H
BACT.    - - Q D L T V K M T D L Q T G K P V G T I E L S Q N K V G V V F T P E L A D - L T P G M H G F H I H

                           *                                   *           *                 *     *
HUMAN    E F G D N T A G C T S - - - - - - - - - - - - A G P H F N P - L S R K H G G P K D E E R H V G D L G
HORSE    E F G D N T Q G C T T - - - - - - - - - - - - A G A H F N P - K S K K H G G P K E E D R H V G D L G
BOVINE   Q F G D N T Q G C T S - - - - - - - - - - - - A G P H F N P - L S K K H G G P K D E E R H V G D L G
PIG      Q F G D N T Q G C T S - - - - - - - - - - - - A G P H F N P - E S K K H G G P K D Q E R H V G D L G
FLY      E F G D N T N G C M S S - - - - - - - - - - - - G P H F N P - Y G K E H G A P V D E N R H L G D L G
YEAST    E F G D A T N G C V S - - - - - - - - - - - - A G P H F N P - F K K T H G A P T D E V R H V G D M G
SWORD    G F G D N T N G C I S - - - - - - - - - - - - A G P H F N P A - S K K H A G P K D E D R H V G D L G
BACT.    Q N G - - S - - C A S S E K D G K V V L G G A A G G H V D P E H T N K H G F P W T D D N H K G D L P

                                                                    *
HUMAN    N V T A D K D G V A D V S I E D S V I S L S G D H C I I G R T L V V H E K A D D L G K G G N E E S T
HORSE    N V T A D E N G K A D V D M K D S V I S L S G K H S I I G R T M V V H E K Q D D L G K G G N E E S T
BOVINE   N V T A D K N G V A I V D I V D P L I S L S G E Y S I I G R T M V V H E K P D D L G R G G N E E S T
PIG      N V T A G K D G V A T V Y I E D S V I A L S G D H S I I G R T M V V H E K P D D L G R G G N Q E S T
FLY      N I E A T G D C P T K V N I T D S K I T L F G A D S I I G R T V V V H A D A D D L G Q G G H E L S K
YEAST    N V K T D E N G V A K G S F K D S L I K L I G P T S V V G R S V V I H A G Q D D L G K G D T E E S L
SWORD    N V T A D A N G V A K I D I T D - K I S L T G P Y S I I G R T M V I H E K A D D L G R G G N E E S L
BACT.    A L F V S A N G L A T N P V L A P - - R L T - L K E L K G H A I M I H A G G D N - - H - - S D M P K

                       *     *
HUMAN    K T G N A G S R L A C G V I G I A Q
HORSE    K T G N A G S R L A C G V I G I A P
BOVINE   K T G N A G S R L A C G V I G I A K
PIG      K T G N A G S R L A C G V I G I T Q
FLY      S T G N A G A R I G C G V I G I A K
YEAST    K T G N A G P R P A C G V I G L T N
SWORD    K T G N A G S R L A C G V I G T - E
BACT.    A L G G G G A R V A C G V I - - - Q
```

Fig. 1. Amino Acid Sequences of Cu/Zn Superoxide Dismutases.

sites are conserved in all the enzymes. These are His48, 50, 77 and 135 as ligands for the Cu and His77, 86, 95 and Asp88 as ligands for the Zn. Also conserved is Arg158 which is considered essential for facilitating the flow of O_2^- to the active site Cu (19) and the intrachain disulphide bridge formed by Cys59 and 161. Only 18 other residues are conserved in all the eight sequences and of these 8 are Gly residues which are connected with the three dimensional tight packing of the molecule. A high degree of homology is observed between the eukaryotic Cu/ZnSODs. The lowest homology is 50% between yeast and fruitfly and the highest is 84% between pig and bovine enzymes. The bacterial enzymes exhibit a very low degree of homology with the other enzymes (Table 1). The Cu/ZnSOD in P. leiognathi was proposed to have arisen in this species as a result of gene transfer from the ponyfish Leiognathus on which the bacterium is symbiotic (20). The homology with the swordfish enzyme is however only 30%. When mutations are taken into account long stretches of closely related sequences are obtained and the weighted similarity between the two enzymes rises to 44% (21). This corresponds to 78 identical or closely related residues out of 136 residues excluding deletions and insertions.

The predicted secondary structure for all the eight Cu/ZnSODs indicates considerable identity with the secondary structure determined for the bovine enzyme

Table 1. Sequence Homologies of Cu/Zn Superoxide Dismutase

	Human	Horse	Bovine	Pig	Fly	Yeast	Swordfish	Bacterial
Human								
Horse	78							
Bovine	82	78						
Pig	82	75	84					
Fly	58	55	55	55				
Yeast	55	56	53	52	50			
Swordfish	66	66	71	67	54	53		
Bacterial	25	26	28	28	24	26	30	

Values given as percentage of sequence homologies

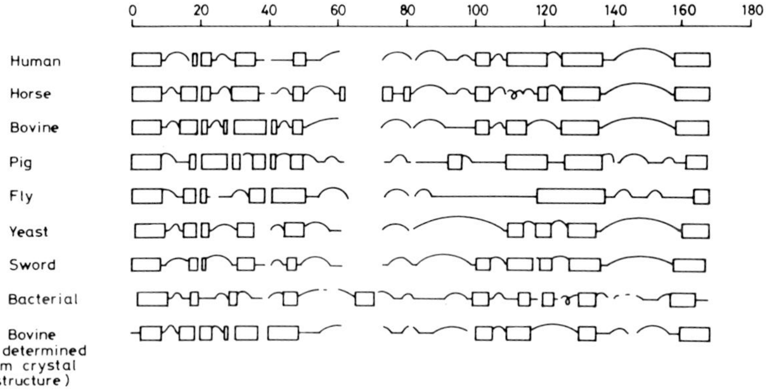

Fig. 2. Predicted Secondary Structure for Cu/Zn Superoxide Dismutases.

from the crystal structure (Fig.2). The bacterial enzyme despite having a low homology also has a similar secondary structure. The twelve residue insertion, residues 62-73, was also predicted to form a β-strand. However when an attempt was made to fit the bacterial sequence into the known bovine tertiary structure the strand could not be built in the model so as to interact appropriately with nearby strands. The insertion was built into the structure as a loop (21). The bacterial Cu/ZnSOD sequence could be fitted into the bovine structure without any serious disruptions to the fold. This evidence strongly indicates that gene transfer from the eukaryotic ponyfish to the prokaryotic P. leiognathi has taken place. The evidence for gene transfer is reinforced when predicted DNA sequences for the swordfish and bacterial SODs are compared using the known DNA sequence for the human enzyme (22) as the base sequence for the comparison. The identities obtained when deletions and insertions are excluded are given in Table 2.

Table 2 Identities Obtained from Comparison of Predicted DNA Sequences

	Human/Swordfish	Human/Bacteria	Swordfish/Bacteria
Nos. of residues	150	137	136
Nos. of identities	102 (68%)	42 (30%)	49 (36%)
Nos. of base change	26	44	47
Total nos. of residues (identities ≠ one base change)	128 (85.3%)	86 (62.8%)	96 (70.6%)

When the swordfish and bacterial enzymes are compared the number of identities are found to be 49 or 36%. However, when one base mutations are considered the number of identities rise to 96 or 70.6%.

INTERRELATIONSHIPS BETWEEN THE Mn SUPEROXIDE DISMUTASES

Compared with the Cu/Zn SODs very little work has been carried out on the MnSODs. The amino acid sequence has been determined for only four species (23-26) (Fig.3). The crystal structure of the enzyme from Thermus thermophilus

```
                                                *
B.STEAR.  P F E L P A L P V P V D A L E P H I D K E T M N I H H T K H H N T V V T N L N A A L E G H P D L Q N
E.COLI    S V T L P S L P V A Y D A L E P H F D K Q T M E I H H T K H H Q T V V N N A N A A L E S L P E F A N
YEAST     K V T L P D L K W D F G A L E P V I S G Q I N E L H V T K H H Q T V V N G F N T A V D Q F Q E L S D
HUMAN     K H S L P D L P V D V G A L E P H I N A Q I M Q L H H S K H H A A V V N N L N V T Q E K V Q E A L A
                                                                          *
B.STEAR.  K S L E E L L S N L E A L P E S I R T A V R N N G G G H A N H S L F W T I L S P - - N G G G E P - T
E.COLI    L P V E E L I T K L D Q L P A D K K T V L R N N A G G H A H N S L F W K G L K K - - - G T T L Q - -
YEAST     L L A K E P S P A N A R K M I A I Q Q N I K F H G G G F T N H C L F W E N L A P E S Q G G G E P P T
HUMAN     K - - - - - - - - G D V T A Q I A L Q P A L K F N G G G H I N H S I F W T N L S P - - N G G G Q P - Q

B.STEAR.  G E L A D A I N K K F G S F T A F K D E F S K A A A G R F G S G W A W L V V N - - - N G E L E I T S
E.COLI    G D L K A A I E R D F G S V D N F K A E F E K A A A S R F G S G W A W L V L K - - - G D K L A V V S
YEAST     G A L A K A I D E Q F G S L D E L I K L T N T K L A G V Q G S G W A F I V K N L S N G G K L D V V Q
HUMAN     G E L L E A I K R D F G S F D K F K Q K L T A A S V G V Q G S G W L G F - - N - K Q R G H L Q I A A
                                                    *       *
B.STEAR.  T P N Q D S P I M E G K T - - - - - - P I L G L D V W E H A V V L K Y Q N R R P E Y L A A F W N V V
E.COLI    T A N Q D S P L M G E A I S G A S G F P I L G L D V W E H A V V L K F Q N R R P D V I K E F W N V V
YEAST     T Y N Q D T V T G P L V - - - - - - - P L V A I D A W E H A V V L Q V Q N K K A D Y F K A I W N V V
HUMAN     C P N Q D - P L Q G T T - - - - G L I P L L G I D V W E H A V V L Q V K N V R P D Y L K A I W N V I

B.STEAR.  N W D E V A K R V S E A K A K
E.COLI    N W D E A A A R F A A A K - K
YEAST     N W K E A S R R F D - - - - -
HUMAN     N W E N V T E R V M A C K - K
```

Fig. 3. Primary Structure of Mn Superoxide Dismutases

at 2.4Å has been determined (27) and the ligands to the metal are His, His, Asp and His which when combined with the sequence of the MnSOD from B. stearothermophilus correspond to positions 26, 81, 163 and 167. In the aligned sequences presented in Fig.3 these correspond to His26, His81, Asp175 and His179. Histidine 81 in E. coli appears in position 80 but this appears to be an error in the sequence. The ligands to the Mn are also present in the Fe SOD (28) and there are strong indications that the Mn and Fe proteins are structural homologs (27). The predicted secondary structures of the MnSODs also show a great resemblance to the FeSOD structure (1). The four MnSODs exhibit varying degrees of sequence homologies (Table 3) which are not as high as those between the vertebrate Cu/ZnSODs.

Table 3 Sequence Homologies of Mn Superoxide Dismutase

	B. stearothermophilus	E. coli	Yeast	Human liver
B. stearothermophilus				
E. coli	59			
Yeast	39	37		
Human liver	48	41	43	

INTERRELATIONSHIPS BETWEEN THE Cu/Zn AND Mn SUPEROXIDE DISMUTASES

The synthesis of Cu/Zn and MnSOD is carried out by different genes. In humans, the gene for the Cu/ZnSOD is located on chromosome 21 (29) whilst the gene for the MnSOD is located on chromosome 6 (30). This indicates that the two proteins will have a different primary structure and a comparison of the amino acid sequence of human Cu/Zn and MnSODs shows this to be the case (Fig.4).

	1	2	3	4	5	6	7	8	9	10	11	12	13	14	15	16	17	18	19	20
CUZN SOD	ALA	THR	LYS	ALA	VAL	CYS	VAL	LEU	LYS	GLY	ASP	GLY	PRO	VAL	GLN	GLY	ILE	ILE	ASN	PHE
MN SOD	LYS	HIS	SER	LEU	PRO	ASP	LEU	PRO	TYR	ASP	TYR	GLY	ALA	LEU	GLU	PRO	HIS	ILE	ASN	ALA
	21	22	23	24	25	26	27	28	29	30	31	32	33	34	35	36	37	38	39	40
CUZN SOD	GLU	GLN	LYS	GLU	SER	ASN	GLY	PRO	VAL	LYS	VAL	TRP	GLY	SER	ILE	LYS	GLY	LEU	THR	GLU
MN SOD	GLN	ILE	MET	GLN	LEU	HIS	HIS	SER	LYS	HIS	HIS	ALA	ALA	TYR	VAL	ASN	ASN	LEU	ASN	VAL
	41	42	43	44	45	46	47	48	49	50	51	52	53	54	55	56	57	58	59	60
CUZN SOD	GLY	LEU	HIS	GLY	PHE	HIS	VAL	HIS	GLU	PHE	GLY	ASP	ASN	THR	ALA	GLY	CYS	THR	SER	ALA
MN SOD	THR	GLN	GLU	LYS	TYR	GLN	GLU	ALA	LEU	ALA	LYS	GLY	ASP	VAL	THR	ALA	GLN	ILE	ALA	LEU
	61	62	63	64	65	66	67	68	69	70	71	72	73	74	75	76	77	78	79	80
CUZN SOD	GLY	PRO	HIS	PHE	ASN	PRO	LEU	SER	ARG	LYS	HIS	GLY	GLY	PRO	LYS	ASP	GLU	GLU	ARG	HIS
MN SOD	GLN	PRO	ALA	LEU	LYS	PHE	ASN	GLY	GLY	GLY	HIS	ILE	ASN	HIS	SER	ILE	PHE	TRP	THR	ASN
	81	82	83	84	85	86	87	88	89	90	91	92	93	94	95	96	97	98	99	100
CUZN SOD	VAL	GLY	ASP	LEU	GLY	ASN	VAL	THR	ALA	ASP	LYS	ASP	GLY	VAL	ALA	ASP	VAL	SER	ILE	GLU
MN SOD	LEU	SER	PRO	ASN	GLY	GLY	GLY	GLN	PRO	GLN	GLY	GLU	LEU	LEU	GLU	ALA	ILE	LYS	ARG	ASP
	101	102	103	104	105	106	107	108	109	110	111	112	113	114	115	116	117	118	119	120
CUZN SOD	ASP	SER	VAL	ILE	SER	LEU	SER	GLY	ASP	HIS	CYS	ILE	ILE	GLY	ARG	THR	LEU	VAL	VAL	HIS
MN SOD	PHE	GLY	SER	PHE	ASP	LYS	PHE	LYS	GLN	LYS	LEU	THR	ALA	ALA	SER	VAL	GLY	VAL	GLN	GLY
	121	122	123	124	125	126	127	128	129	130	131	132	133	134	135	136	137	138	139	140
CUZN SOD	GLU	LYS	ALA	ASP	ASP	LEU	GLY	LYS	GLY	GLY	ASN	GLU	GLU	SER	THR	LYS	THR	GLY	ASN	ALA
MN SOD	SER	GLY	TRP	LEU	GLY	PHE	ASN	LYS	GLN	ARG	GLY	HIS	LEU	GLN	ILE	ALA	ALA	CYS	PRO	ASN
	141	142	143	144	145	146	147	148	149	150	151	152	153	154	155	156	157	158	159	160
CUZN SOD	GLY	SER	ARG	LEU	ALA	CYS	GLY	VAL	ILE	GLY	ILE	ALA	GLN							
MN SOD	GLN	ASP	PRO	LEU	GLN	GLY	THR	THR	GLY	LEU	ILE	PRO	LEU	LEU	GLY	ILE	ASP	VAL	TRP	GLU
	161	162	163	164	165	166	167	168	169	170	171	172	173	174	175	176	177	178	179	180
CUZN SOD																				
MN SOD	HIS	ALA	TYR	TYR	LEU	GLN	TYR	LYS	ASN	VAL	ARG	PRO	ASP	TYR	LEU	LYS	ALA	ILE	TRP	ASN
	181	182	183	184	185	186	187	188	189	190	191	192	193	194	195	196	197	198	199	200
CUZN SOD																				
MN SOD	VAL	ILE	ASN	TRP	GLU	ASN	VAL	THR	GLU	ARG	TYR	MET	ALA	CYS	LYS	LYS				

Fig. 4. Comparison of the Primary Structure of Human Cu/Zn and MnSODs.

When the two isozymes are aligned from the N-terminal there are only 15 identical residues, however when a computer alignment is carried out, 43 residues are found to align (Fig.5) and 10 of these are Gly residues. A comparison of the ligands to the Cu, Zn and Mn is given in Table 4.

	1	2	3	4	5	6	7	8	9	10	11	12	13	14	15	16	17	18	19	20
CUZN SOD	ALA	THR	LYS	-	ALA	VAL	CYS	VAL	LEU	LYS	GLY	-	-	-	ASP	-	GLY	-	-	-
MN SOD	-	-	LYS	HIS	SER	-	-	-	LEU	PRO	ASP	LEU	PRO	TYR	ASP	TYR	GLY	ALA	LEU	GLU
	21	22	23	24	25	26	27	28	29	30	31	32	33	34	35	36	37	38	39	40
CUZN SOD	PRO	-	VAL	GLN	GLY	ILE	-	ILE	ASN	PHE	GLU	GLN	-	LYS	GLU	SER	ASN	GLY	PRO	-
MN SOD	PRO	HIS	ILE	ASN	ALA	GLN	ILE	MET	GLN	LEU	HIS	HIS	SER	LYS	HIS	-	HIS	ALA	ALA	TYR
	41	42	43	44	45	46	47	48	49	50	51	52	53	54	55	56	57	58	59	60
CUZN SOD	VAL	LYS	-	VAL	TRP	GLY	SER	ILE	-	-	-	LYS	-	-	-	GLY	LEU	THR	GLU	GLY
MN SOD	VAL	ASN	ASN	LEU	-	-	ASN	VAL	THR	GLN	GLU	LYS	TYR	GLN	GLU	ALA	LEU	ALA	LYS	GLY
	61	62	63	64	65	66	67	68	69	70	71	72	73	74	75	76	77	78	79	80
CUZN SOD	-	-	-	-	-	-	-	LEU	HIS	GLY	-	-	-	PHE	-	-	-	-	HIS	VAL
MN SOD	ASP	VAL	THR	ALA	GLN	ILE	ALA	LEU	GLN	PRO	ALA	LEU	LYS	PHE	ASN	GLY	GLY	GLY	HIS	ILE
	81	82	83	84	85	86	87	88	89	90	91	92	93	94	95	96	97	98	99	100
CUZN SOD	-	HIS	GLU	-	PHE	-	-	-	-	-	-	-	GLY	ASP	ASN	THR	ALA	GLY	CYS	THR
MN SOD	ASN	HIS	SER	ILE	PHE	TRP	THR	ASN	LEU	SER	PRO	ASN	GLY	-	-	-	-	GLY	-	-
	101	102	103	104	105	106	107	108	109	110	111	112	113	114	115	116	117	118	119	120
CUZN SOD	SER	ALA	GLY	-	PRO	HIS	-	-	-	-	-	-	-	-	-	-	PHE	ASN	PRO	LEU
MN SOD	-	-	GLY	GLN	PRO	GLN	GLY	GLU	LEU	LEU	GLU	ALA	ILE	LYS	ARG	ASP	PHE	GLY	SER	PHE
	121	122	123	124	125	126	127	128	129	130	131	132	133	134	135	136	137	138	139	140
CUZN SOD	SER	ARG	-	LYS	HIS	-	-	-	-	-	-	-	GLY	-	-	GLY	PRO	-	-	-
MN SOD	ASP	LYS	PHE	LYS	GLN	LYS	LEU	THR	ALA	ALA	SER	VAL	GLY	VAL	GLN	GLY	SER	GLY	TRP	LEU
	141	142	143	144	145	146	147	148	149	150	151	152	153	154	155	156	157	158	159	160
CUZN SOD	-	-	-	LYS	ASP	GLU	GLU	ARG	HIS	VAL	GLY	ASP	LEU	GLY	ASN	VAL	THR	ALA	-	-
MN SOD	GLY	PHE	ASN	LYS	GLN	-	-	ARG	-	-	GLY	HIS	LEU	-	GLN	ILE	ALA	ALA	CYS	PRO
	161	162	163	164	165	166	167	168	169	170	171	172	173	174	175	176	177	178	179	180
CUZN SOD	ASP	LYS	ASP	-	-	-	GLY	VAL	ALA	ASP	VAL	SER	ILE	GLU	ASP	SER	VAL	ILE	SER	LEU
MN SOD	ASN	GLN	ASP	PRO	LEU	GLN	GLY	THR	THR	GLY	LEU	-	-	-	-	-	-	ILE	PRO	LEU
	181	182	183	184	185	186	187	188	189	190	191	192	193	194	195	196	197	198	199	200
CUZN SOD	SER	GLY	-	ASP	-	-	-	HIS	-	-	CYS	ILE	-	-	-	-	ILE	GLY	ARG	THR
MN SOD	LEU	GLY	ILE	ASP	VAL	TRP	GLU	HIS	ALA	TYR	TYR	LEU	GLN	TYR	LYS	ASN	VAL	-	ARG	PRO
	201	202	203	204	205	206	207	208	209	210	211	212	213	214	215	216	217	218	219	220
CUZN SOD	-	-	LEU	-	-	-	-	-	VAL	VAL	HIS	-	GLU	LYS	ALA	ASP	ASP	LEU	GLY	LYS
MN SOD	ASP	TYR	LEU	LYS	ALA	ILE	TRP	ASN	VAL	ILE	ASN	TRP	GLU	ASN	VAL	THR	GLU	-	-	-
	221	222	223	224	225	226	227	228	229	230	231	232	233	234	235	236	237	238	239	240
CUZN SOD	GLY	GLY	ASN	GLU	GLU	SER	THR	LYS	THR	GLY	ASN	ALA	GLY	SER	ARG	-	LEU	ALA	CYS	GLY
MN SOD	-	-	-	-	-	-	-	-	-	-	-	-	-	-	ARG	TYR	MET	ALA	CYS	-
	241	242	243	244	245	246	247	248	249	250										
CUZN SOD	VAL	ILE	GLY	ILE	ALA	GLN	-													
MN SOD	-	-	-	-	-	LYS	LYS													

Fig. 5. Computer Alignment of the Primary Structure of Human Cu/Zn and MnSODs.

Table 4 Comparison of the Ligands to the Cu, Zn and Mn

Copper	Zinc	Manganese
His 46 → 79 (His)	His 63 → 106 (His)	His 26 → 31 (Glu)
His 48 → 82 (His)	His 71 → 125 (Gln)	His 74 → 82 (His)
His 63 → 106 (His)	His 80 → 149 (—)	Asp 157 → 184 (Asp)
His 120 → 210 (Asn)	Asp 83 → 152 (His)	His 161 → 188 (His)

The corresponding amino acid is given in brackets

Histidine 82 is a ligand to both the Cu (His48 in the human Cu/ZnSOD) and the Mn (His72 in the human MnSOD). However, most interesting is the fact that the four histidine ligands in the Cu/ZnSOD align with three histidines and one aspartate in the MnSOD with correct sequence order. Three histidines and one aspartate are the ligands to the Mn (27) and recently the Cu has been found to displace the Mn from the MnSOD of B. stearothermophilus (31). The computer aligned homology of 43 residues out of 247 represents 17% homology. However when the predicted DNA sequence for the MnSOD is aligned with the known DNA sequence of the Cu/ZnSOD it appears that 46 residues have undergone a one base change. When this is taken into consideration the number of identical residues rises to 89 and the homology to 36%.

EVOLUTIONARY RELATIONSHIPS BETWEEN THE Cu/Zn and MnSODs

The topologies of the phylogenetic relationships between the SODs are shown in Fig.6. The topologies were calculated by the "present-day" ancestor approach (32). No attempt was made to assign branch length to the trees because of the inaccuracies involved in determining aligned nucleotide sequences from amino acid sequences. Analysis of the data produced topologies which were in agreement with the cytochrome c evolutionary tree and trees derived from classical taxonomy and paleontology. However the rate of evolutionary changes in terms of amino acid and nucleic acid substitution varied greatly between the organisms as recently reported (14).

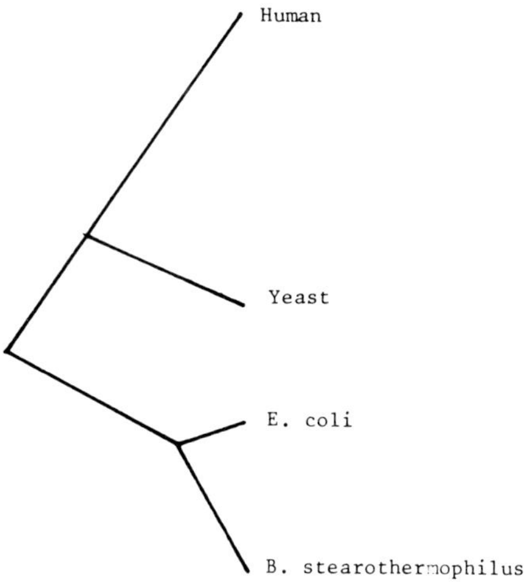

Evolutionary Tree Representation of Mn SODs

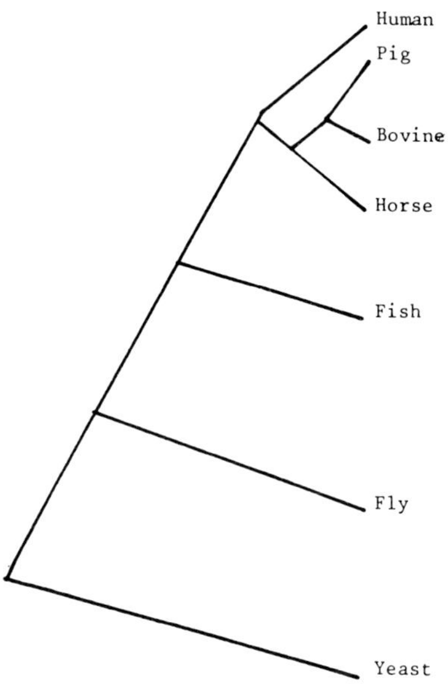

Evolutionary Tree Representation of Cu/Zn SODs

There has been considerable interest in recent years concerning the presence of both coding (exons) and non coding (introns) regions in genes. Gilbert (33) put forward the hypothesis that exons code for functional protein units. This was extended by Blake (34) who pointed out that exons might code for structural units as well. The observation of split gene structures led to the hypothesis that new proteins could evolve from ancestral proteins by gene shuffling (33,34). However the observation of the split gene structures solely in eukaryotes has been a puzzle. This posed the question whether introns evolved at the time of the prokaryote/eukaryote split or at some much earlier time. It would appear that the Cu/ZnSOD evolved at the time of the prokaryote/eukaryote split on the basis of its known phylogenetic distribution. The gene structure of human Cu/ZnSOD is composed of five exons and four introns (22). Each exon codes for a compact structural unit, three code for a ββ structure. A possible proposition might be that the protein evolved from gene shuffling and fusion of five different exons. Such a mechanism would account for the rapid evolution of the Cu/ZnSOD gene that would be required if it was developed late in evolutionary time. There is no evidence from a search of distant sequence relationships that the protein evolved from some other single protein.

REFERENCES

1. Parker MW, Schinina ME, Bossa F, Bannister JV (1984) Inorg Chim Acta 91:307-317

2. Puget K, Michelson AM (1974) Biochem Biophys Res Commun 58:830-838

3. Steinman HM (1982) J Biol Chem 257:8945-8952

4. Geller BL, Winge DR (1982) J Biol Chem 257:8945-8952

5. McCord JM, Boyle JA, Day EA, Rizzolo LJ, Salin ML (1977) In: Michelson AM, McCord JM, Fridovich I (eds) Superoxide and Superoxide Dismutase. Academic Press, New York, pp 129-138

6. Weisiger RA, Fridovich I (1973) J Biol Chem 248:3582-3592

7. Autor AP (1982) J Biol Chem 257:2713-2718

8. Williams RJP (1982) FEBS Lett 140:3-10

9. Bannister WH, Bannister JV, Searle AJF (1983) In: Cohen G, Greenwald R (eds) Oxy-Radicals and their Scavenging Systems: Molecular Aspects. Elsevier Biomedical, New York, pp 387-392

10. Barra D, Martini F, Bannister JV, Schinina ME, Rotilio G, Bannister WH, Bossa F (1980) FEBS Lett 121:215-218

11. Lerch K, Ammer D (1981) J Biol Chem 256:11545-11551

12. Steinman HM, Naik VR, Abernethy JL, Hill RL (1974) J Biol Chem 249:7376-7386

13. Schinina ME, Barra D, Simmaco M, Bassa F, Rotilio G (1985) FEBS Lett (In press)

14. Lee JM, Friedman DJ, Ayala FJ (1985) Proc Natl Acad Sci USA 82:824-828

15. Steinman HM (1980) J Biol Chem 255:6758-6765

16. Rocha H, Bannister WH, Bannister JV (1984) Eur J Biochem 145:477-484

17. Steffens GJ, Bannister JV, Bannister WH, Flohe L, Gunzler WA, Kim SA, Otting F (1983) Z Physiol Chem 364:675-690

18. Abe Y, Okazaki T, Shukuya R, Furuta H (1984) Comp Biochem Physiol 7718:125-130

19. Maeinowski D, Fridovich I (1979) Biochemistry 18:237-242

20. Martin JP, Fridovich I (1981) J Biol Chem 256:6020-6089

21. Bannister JV, Parker MW (1985) Proc Natl Acad Sci USA 82:149-152

22. Levanon D, Lieman-Hurwitz J, Dafni N, Wigderson M, Sherman L, Bernstein Y, Laver-Rudich Z, Danciger E, Stein O, Groner Y (1985) The EMBO Journal 4:77-84

23. Brock CJ, Walker JE (1980) Biochemistry 19:2873-2882

24. Steinman HM (1980) J Biol Chem 255:6758-6765

25. Ditlow C, Johansen JT, Martin BM, Ivendsen IB (1982) Carlsberg Res Commun 47:81-91

26. Barra D, Schinina ME, Simmaco M, Bannister JV, Bannister WH, Rotilio G, Bossa F (1984) J Biol Chem 259:12595-12601

27. Staelings WC, Pattridge KA, Strong RK, Ludwig ML (1984) J Bio Chem 259:10695-10699

28. Barra D, Schinina ME, Bossa F, Bannister JV (1985) FEBS Lett 179:329-331

29. Tan YH, Tischfield J, Ruddle FH (1973) J Exp Med 137:317-330

30. Creagan R, Tischfield J, Riciutti F, Ruddle FH (1973) Humangenetik 20:203-209

31. Bannister JV, Desideri A, Rotilio G (1985) FEBS Lett (In press)

32. Blanken RL, Klotz LC, Hinnebusch AG (1982) J Mol Evol 19:9-19

33. Gilbert W (1978) Nature 271:501

34. Blake CCF (1978) Nature 273:267

Superoxide and Superoxide Dismutase in Chemistry, Biology and Medicine; G. Rotilio, editor

SEQUENCING CONFIRMS NORMAL DIVERGENT EVOLUTION OF CU/ZN SOD

GERD STEFFENS*, ERNST FRANKUS*, KARIN HERING*, SUNG MAN A. KIM*, FRITZ ÖTTING*, EBERHARD SCHWERTNER*, KRYSTYNA PUGET**, ADOLF M. MICHELSON** and LEOPOLD FLOHE*

* Grünenthal GmbH, Center of Research, Zieglerstr. 6, D-5100 Aachen (FRG)
**Fondation Edmond de Rothschild, Institut de Biologie Physico-Chimique, 13, Rue Pierre et Marie Curie, F-75005 Paris (France)

INTRODUCTION

A spectacular step in the phylogenetic evolution of Cu/Zn superoxide dismutase (SOD), i.e. a natural gene transfer between a fish and its symbiotic bacterium, has been postulated (1), challenged (2), and re-emphasized (3). We here report on newly established SOD sequences fortifying our conviction that nothing unusual happened during the evolution of this enzyme.

METHODS

Cu/Zn SODs of pig, rat, yeast and cabbage were isolated by conventional chromatographic procedures and sequenced essentially as described (2,4).

RESULTS AND DISCUSSION

Two further vertebrate Cu/Zn SOD sequences, those of pig and rat, have been completely established. In pigs, an allotypic variant differing in one amino acid residue (Val replacing Leu) is detected. The sequence of the enzyme from *S. cerevisiae* (5) is confirmed. For the first time also sequence data on a plant Cu/Zn SOD are presented.

In agreement with existing data, the amino acid residues presumed to bind the metal ions are found to be conserved in all Cu/Zn SODs. Further identities, in particular the glycines at the turning points of the pleated sheets, the disulfide bridge, the C-terminal cluster and the functionally relevant Arg, prove that the SODs of vertebrates, bacteria, yeast and higher plants are derived from a common ancestor by divergent evolution (Fig. 1). The relatedness of the SODs, calculated from identities in homologous positions (Fig. 2), agrees perfectly with the taxonomical positions, i.e. phylogenetic distance of the corresponding species, as similarly observed in the cytochrome c superfamily (6).

The arguments again supporting a transfer of the SOD gene from pony fish

```
                   1                                   20
Pig (4)          Ac-A T K A V C V L K G D G P V Q G T I Y F E L K G E K - T
Cattle (7)       Ac-A T K A V C V L K G D G P V Q G T I H F E A K G D - - T
Man (4)          Ac-A T K A V C V L K G D G P V Q G I I N F E Q K E S N G P
Horse (8)        Ac-A L K A V C V L K G D G P V H G V I H F E Q Q Q E G G P
Rat              Ac-A M K A V C V L K G D G P V Q G V I H F E Q K A S G E P
Swordfish (3)       V L K A V C V L R G A G E T T G T V Y F E Q E G N A N A
Cabbage               A K C V A V L N S S E G V K G T I F F T H E G N G - A
Yeast (5)             V Q A V A V L K G D A G V S G V V K F E Q A S E S E P
P.leiognathi (2)      Q D L T V K M T D L Q T G K P V G T I E L S Q N K Y G

                      V                 40                 +   +
Pig                 V L V T G T I K G - L A E G D H G F H V H Q F G D N T Q
Cattle              V V V T G S I T G - L T E G D H G F H V H Q F G D N T Q
Man                 V K V W G S I K G - L T E G L H G F H V H E F G D N T A
Horse               V V L K G F I E G - L T K G D H G F H V H E F G D N T Q
Rat                 V V V S G Q I T G - L T E G E H G F H V H Q Y G D N T Q
Swordfish           V G K G I I L K G - L T P G E H G F H V H G F G D N T N
Cabbage             T T V T G T V S G - L R P G L H G F H V H A L G D N T N
Yeast               T T V S Y E I A G N S P N A E R G F H I H E F G D A T N
P.leiognathi        V V F T P E L A D - L T P G M H G F H I H Q N G - - - -

                          60                           +      80
Pig                 G C T S - - - - - - - - - - - - A G P H F N P - E S K K
Cattle              G C T S - - - - - - - - - - - - A G P H F N P - L S K K
Man                 G C T S - - - - - - - - - - - - A G P H F N P - L S R K
Horse               G C T T - - - - - - - - - - - - A G A H F N P - L S K K
Rat                 G C T T - - - - - - - - - - - - A G P H F N P - H S K K
Swordfish           G C I S - - - - - - - - - - - - A G P H F N P - A S K K
Cabbage             G C M S - - - - - - - - - - - - T G P H F N P - D G K T
Yeast               G C V S - - - - - - - - - - - - A G P H F N P - F K K T
P.leiognathi        S C A S S E K D G K V V L G G A A G G H Y D P E H T N K

                    +                 +   +         100
Pig                 H G G P K D Q E R H V G D L G N V T A G K D G V A T V Y
Cattle              H G G P K D E E R H V G D L G N V T A D K N G V A I V D
Man                 H G G P K D E E R H V G D L G N V T A D K D G V A D V S
Horse               H G G P K D E E R H V G D L G N V T A D E N G K A D V D
Rat                 H G G P A D E E R H V G D L G N V A A G K D G V A N V S
Swordfish           H A G P K D E D R H V G D L G N V T A D A N G V A K I D
Cabbage             H G A P E D A N R H A G D L G N I I V G D D G T A T F T
Yeast               H G A P T D E V R H V G D M G N V K T D E N G V A K G S
P.leiognathi        H G F P W T D D N H K G D L P A L F V S A N G L A T N P

                                120                          +        140
Pig                 I E D S V I A L S G D H S I I G R T M V V H E K P D D L
Cattle              I V D P L I S L S G E Y S I I G R T M V V H E K P D D L
Man                 I E D S V I S L S G D H C I I G R T L V V H E K A D D L
Horse               M K D S V I S L S G K H S I I G R T M V V H E K Q D D L
Rat                 I E D R V I S L S G E H S I I G R T M V V H E K Q D D L
Swordfish           I T D - K I S L T G P Y S I I G R T M V I H E K A D D L
Cabbage             I T D S Q I P L S G P N S I V G R A I V V H A D P D D L
Yeast               F K D S L I K L I G P T S V V G R S V V I H A G Q D D L
P.leiognathi        V L A P R L T L K - - - E L K G H A I M I H A G G D N H

                                                   160
Pig                 G R G G N E E S T K T G N A G S R L A C G V I G I T Q
Cattle              G R G G N E E S T K T G N A G S R L A C G V I G I A K
Man                 G K G G N E E S T K T G N A G S R L A C G V I G I A Q
Horse               G K G G N E E S T K T G N A G S R L A C G V I G I A P
Rat                 G K G G N E E S T K T G N A G S R L A C G V I G I A Q
Swordfish           G R G G N E E S L K T G N A G S R L A C G V I G T E -
Cabbage             G K G G H E L S L S T G N A G G R V A C G I I G L Q G
Yeast               G K G D T E E S L K T G N A G P R P A C G V I G L T N
P.leiognathi        - - - - S D M P K A L G G G G A R V A C G V I Q - - -
```

Fig. 1: SOD sequences. Boxes: conserved residues; (+) metal ligands.

	Pig	Cattle	Man	Horse	Rat	Swordfish	Cabbage	Yeast	P.leiognathi
Pig	152								
Cattle	130	151							
Man	128	125	153						
Horse	117	122	125	153					
Rat	125	127	130	123	153				
Swordfish	104	109	103	103	99	151			
Cabbage	87	81	79	76	79	80	151		
Yeast	81	82	85	87	87	83	77	153	
P.leiognathi	43	44	39	42	43	45	43	40	151

Fig.2: Identical amino acid residues in homologous positions of Cu/Zn SODs

to *P. leiognathi* (3) appear weak: 1. calculating the homology of arbitrarily chosen DNA sequences, instead of the amino acid sequences they are deduced from, weakens hard data by bias; 2. fitting, by molecular modelling, the amino acid sequence of the bacteriocuprein into the tertiary structure of an homologous vertebrate SOD and, thereby, concluding relatedness of the two proteins meets the definition of tautology. In brief therefore, the hypothesis that the symbiont *P. leiognathi* has acquired its Cu/Zn SOD gene from its teleost host appears less appealing than the idea that man could have incorporated the SOD gene while devouring yeast-spoiled grapes or *Sauerkraut*. The experimental data yet support an uncomplicated evolution of the Cu/Zn SOD family.

REFERENCES

1. Martin JP Jr., Fridovich I (1981) J Biol Chem 256: 6080-6089
2. Steffens GJ, Bannister JV, Bannister WH, Flohé L, Günzler WA, Kim SMA, Ötting F (1983) Hoppe-Seyler's Z Physiol Chem 364: 675-690
3. Bannister JV, Parker MW (1985) Proc Natl Acad Sci USA 82: 149-152
4. Hering K, Kim SMA, Michelson AM, Ötting F, Puget K, Steffens GJ, Flohé L (1985) Biol Chem Hoppe-Seyler 366: 435-445
5. Johansen JT, Overballe-Petersen C, Martin B, Hassemann V, Svendsen I (1979) Carlsberg Res Commun 44: 201-217
6. Dayhoff MO (1976) In: Atlas of Protein Sequence and Structure, Vol. 5, Suppl. 2, pp 25-35, Natl Biomed Res Found, Washington, D.C.
7. Steinman HM, Naik VR, Abernethy JL, Hill RL (1974) J Biol Chem 249: 7326-7338
8. Lerch K, Ammer D (1981) J Biol Chem 256: 11545-11551

Superoxide and Superoxide Dismutase in Chemistry,
Biology and Medicine; G. Rotilio, editor

STRUCTURE OF THE HUMAN Cu/Zn SOD GENE

ROBERT A. HALLEWELL, JOHN P. PUMA, GUY T. MULLENBACH, RICHARD C. NAJARIAN
Chiron Research Laboratories, Chiron Corporation, 4560 Horton Street, Emeryville, California (USA)

INTRODUCTION

The isolation, mapping and nucleotide sequence of a human Cu/Zn superoxide dismutase (HSOD) gene is reported. The location of the promoter, transcription start site(s), and polyA addition sites are described or discussed, particularly where they differ from the previously described HSOD gene (1). The location of the four introns and the splice donor and acceptor sequences are reported. The locations of the five exons in the crystal structure of HSOD are inferred by analogy with the known structure of bovine SOD (2). None of the exons appear to be a clearly defined functional or structural domain similar to those found in many other proteins (3). All but one of the exon-intron junctions occurs at the protein surface in agreement with the suggestion that addition of exons at the protein surface facilitates evolution without disrupting the existing structure and function of the protein (4).

MATERIALS AND METHODS

Isolation and Sequencing

The gene was isolated from a bacteriophage lambda library of human DNA (5,6) using radiolabelled cDNA (7) as a probe. After shotgun cloning restriction endonuclease digests of the bacteriophage DNA into M13 vectors (8) appropriate clones were isolated using radiolabelled synthetic oligonucleotides as probes (9) and sequenced by the chain termination method (10) using the probes as primers.

Mapping the Exons

A PstI map was obtained using the partial digestion method (6) using various radiolabelled synthetic DNA probes. The precise locations of the exons on the PstI map were obtained by mapping restriction sites which had been sequenced (see Fig. 1). Such sites were mapped within a particular PstI fragment by a combination of end labelling and Southern blotting (6) using various radiolabelled synthetic DNA probes.

RESULTS

The HSOD nucleotide sequence shown in Fig. 1 demonstrates most of the characteristics of genes from higher eukaryotes. The promoter region contains

```
GAATTCTGCC AACCAAATAA GAAACTCTAT ACTAAGGACT AAGAAAATTG CAGGGGAAGA AAAGGTAAGT CCCGGGATTG AGGTGTAGCG ACTTTCTATA   (100)
  EcoRI

CCCTCAGAAA ACTAAAAAAC AAGACAAAAA AATGAAAACT ACAAAAGCAT CCATCTTGGG GCGTCCCAAT TGCTGAGTAA CAAATGAGAC GCTGTGGCCA   (200)

                                                                                         T
AACTCACGTC ATAACTAATG ACATTTCTAG ACAAAGTGAC TTCAGATTTT CAAAGCGTAC CCTGTTTACA TCATTTTGCC AATTTCGCGT ACTGCAACCG   (300)

                                             G T
GCGGGCCACG CCCCCGTGAA AAGAAGGTTG TTTTCTCCAC ATTTCGGGGT TCTGGACGTT TCCCGGCTGC GGGGCGGGGG GAGTCTCCGG CGCACGCGGC   (400)

                                                                                         C
CCCTTGGCCC CGCCCCCAGT CATTCCCGGC CACTCGCGAC CCGAGGCTGC CGCAGGGGGC GGGCTGAGCG CGTGCGAGGC GATTGGTTTG GGGCCAGAGT   (500)
                                                                              Pst I

GGGCGAGGCG CGGAGGTCTG GCCTATAAAG TAGTCGCGGA GACGGGGTGC TGGTTTGCGT CGTAGTCTCC TGCAGCGTCT GGGGTTTCCG TTGCAGTCCT   (600)
                                                   mRNA start site

                                            1
                                            Met Ala Thr Lys Ala Val Cys Val Leu Lys Gly Asp Gly Pro Val Gln Gly Ile Ile Asn
CGGAACCAGG ACCTCGGCGT GGCCTAGCGA GTT ATG GCG ACG AAG GCC GTG TGC GTG CTG AAG GGC GAC GGC CCA GTG CAG GGC ATC ATC AAT (693)

                23
Phe Glu Gln Lys               C
TTC GAG CAG AAG GCAAGGGCTG GGACGG.....................CAGGTTAAGC AGCTTGCTGG AGGTTCACTG GCTAGAAAGT GGTCAGCCTG   (771)

 C
GGATTTGGAC ACAGATTTTT CCACTCCCAA GTCTGGCTGC TTTTTACTTC ACTGTGAGGG GTAAAGGTAA ATCAGCTGTT TTCTTTGTTC AGAAACTCTC   (871)
                                                                              PvuII

                            24
                            Glu Ser Asn Gly Pro Val Lys Val Trp Gly Ser Ile Lys Gly Leu Thr Glu Gly Leu His Gly Phe
TCCAACTTTG CACTTTTCTT AAAG GAA AGT AAT GGA CCA GTG AAG GTG TGG GGA AGC ATT AAA GGA CTG ACT GAA GGC CTG CAT GGA TTC (961)

                                    55
His Val His Glu Phe Gly Asp Asn Thr Ala
CAT GTT CAT GAG TTT GGA GAT AAT ACA GCA G GTCGGGTGTT ........................GTGTTTCTTT TTAGAATGTA TTTGGGAACT   (1032)
```

```
                                           56
                                           Gly Cys Thr Ser Ala Gly Pro His Phe Asn Pro Leu Ser Arg Lys His Gly Gly
TTAATTCATA ATTTAGCTTT TTTTTCTTCT TCTTATAAAT AG GC TGT ACC AGT GCA GGT CCT CAC TTT AAT CCT CTA TCC AGA AAA CAC GGT GGG (1127)

                 79
Pro Lys Asp Glu Glu Arg                       C       ▽    AC  ▽ T          |                          C
CCA AAG GAT GAA GAG AG GTAACAAGAT GCTTAACTCT TGTAATAATg gccGATCATG gTTCTGGAGT TCATATGGTA TACTACTTGT AAATATGTGC (1134)
                                              ^                     ^          NdeI                         ^
                                                         GG
TAAGATAATT CCGTGTTTCC CCCACCTTTG CTTTTGAACT TGCTGACTCA TCTAAA..........................GTACTTCTGA AATCAGGTGC  (1300)
                                                         ^
                                                                          |
AGCCCCATCT TTCTTCCCAG AGCATTAGTG TGTAGACGTG AAGCCTTGTT TGAAGAGCTG TATTTAGAAT GCCTAGCTAC TTGTTTGCAA ATTTGTGTCC  (1400)
                                                                          BsmI
                                                                                                 80
                                                                                    ▽  ▽       His Val Gly Asp
TACTCAGTCA AGTTTTAATT TAGCTCATGA ACTACCTTGA TGTTTAGTGg CATCAGCCCT AATCCATCTG ATGCTTTTTC ATTATTAG G CAT GTT GGA GAC (1501)

Leu Gly Asn Val Thr Ala Asp Lys Asp Gly Val Ala Asp Val Ser Ile Glu Asp Ser Val Ile Ser Leu Ser Gly Asp His Cys Ile
TTG GGC AAT GTG ACT GCT GAC AAA GAT GGT GTG GCC GAT GTG TCT ATT GAA GAT TCT GTG ATC TCA CTC TCA GGA GAC CAT TGC ATC  (1588)

                 118
Ile Gly Arg Thr Leu Val            ▽                                           G                 ▽
ATT GGC CGC ACA CTG GTG GTAAGTTTTC ATAAAAGGAT ATGCATAAAA CTTCTTCTAA CATACAGTCA TGTATCTTTT CACTTTGATT GTTAGTCGCG (1686)
                                                                               ^
AA                                   |                        G
GTTTCT..........................GAAAAAGCTT TGAGTAGTAG TTTCTACTTT TAAACTACTA AATATTAGTA TATCTCTCTA CTAGGATTAA (1762)
                                      HindIII
TGTTATTTTT CTAATATTAT GAGGTTCTTA AACATCTTTT GGGTATTGTT GGGAGGAGGT AGTGATTACT TGACAGCCCA AAGTTATCTT CTTAAAATTT   (1862)

       119
       Val His Glu Lys Ala Asp Asp Leu Gly Lys Gly Gly Asn Glu Glu Ser Thr Lys Thr Gly Asn Ala Gly Ser Arg Leu Ala
TTTACAG GTC CAT GAA AAA GCA GAT GAC TTG GGC AAA GGT GGA AAT GAA GAA AGT ACA AAG ACA GGA AAC GCT GGA AGT CGT TTG GCT  (1950)
```

```
                                    153
Cys Gly Val Ile Gly Ile Ala Gln STOP
TGT GGT GTA ATT GGG ATC GCC CAA TAA  ACATTCCCTT GGATGTAGTC TGAGGCCCCT TAACTCATCT GTTATCCTGC TAGCTGTAGA AATGTATCCT   (2047)

                              •PolyA
GATAAACATT AAACACTGTA ATCTTAAAAG TGTAATTGTG TGACTTTTTC AGAGTTGCTT TAAAGTACCT GTAGTGAGAA ACTGATTTAT GATCACTTGG  (2147)

AAGATTTGTA TAGTTTTATA AAACTCAGTT AAAATGTCTG TTTCAATGAC CTGTATTTTG CCAGACTTAA ATCACAGATG GGTATTAAAC TTGTCAGAAT  (2247)
                                                                                    •PolyA
TTCTTTGTCA TTCAAGCCTG TGAATAAAAA CCCTGTATGG CACTTATTAT GAGGCTATTA AAAGAATCCA AATTCAAACT AAATTAGCTC TGATACTTAT  (2347)

           A
TTATATAAAC TGCTTCAGTG GAACAGATTT AGTAATACTA ACAGTGATAG CATTTTATTT TGAAAGTGTT TTGAGACCAT CAAAATGCAT ACTTTAAAAC  (2447)

AGCAGGTCT (2456)
```

Fig. 1. Nucleotide sequence of the HSOD gene. Nucleotides underlined with arrows indicate direct or inverted repeat sequences. The CATT and TATAAA sequences in the promoter are boxed and the GGGCGG sequences are underlined, AATAAA-like sequences are semi-boxed. Discrepancies with the previously published sequence (1) are marked (∇ deletion, Λ insertion). Dinucleotides associated with splice donor (GT) or acceptor (AG) sites are underlined. Nucleotides which were not determined unambiguously are printed lower case.

TATAA and CATT sequences (beginning at nucleotides 421 and 524) which have been shown to be associated with transcription initiation (11). In addition the promoter region contains three 5' GGGCGG 3' boxes. In some cases such sequences appear to activate transcription by a mechanism that involves binding of one protomer of the cellular protein Sp1 for each GGGCGG box. No shared regulatory feature is evident among the diverse promoters with actual or potential Sp1 binding sites (12). Transcription initiation probably occurs at the G-residue at nucleotide 553. This location and nucleotide are in agreement both with the observed distance from the TATAA box (11) and with direct mapping studies (12).

A large number of discrepancies were observed between our nucleotide sequence and that reported by Groner et al. (1). Although both genes were isolated from the same library (5) it is possible that we have sequenced a different allele of the same gene (our gene corresponds to λA-2 of Groner et al. [1]) thus accounting for some differences. One quite important difference is the absence of a CATT box at nucleotide 480 in the sequence presented here which Groner et al. (1) suggest may have special significance because it is flanked by inverted repeat sequences. The promoter region was unusually difficult to sequence using the chain termination method (10) probably due to secondary structure effects caused by the various repeated seqences and required several primers and sequencing on both strands to resolve ambiguities. Fig. 2 demonstrates the effects of groups of dC-residues on sequencing gels and illustrates how the first band of a C-track can be overlooked. These C-residues at nucleotides 408 and 413 (see Fig. 1) were not observed in the report of Groner et al. (1).

It has been shown (13,7) that the SOD gene has multiple polyA-addition signals and that there are two predominant mRNA species which differ in size at their 3' ends. An as yet unresolved discrepancy exists regarding the location of the 3' end of the larger of the two RNA molecules. Mapping studies (13) have suggested that the 3' end is located at nucleotide 2288 whereas cDNA sequencing shows that an mRNA species exists with a 3' end at nucleotide 2322 (7). It is possible that the latter RNA species is rare and that most RNAs are processed at nucleotide 2288. Further studies are required to resolve this question.

The map of the SOD gene shown in Fig. 3 shows that the SOD promoter, coding sequence and 3' untranslated region are encoded by about 9.5 kb of DNA. The coding sequence is interrupted by four introns resulting in five SOD exons. The splice donor and acceptor dinucleotides agree with the consensus sequence (11) except for the GC dinucleotide at position 1606 which has been observed only twice previously (14).

It has been observed that exons quite often correspond to functional or structural domains of proteins. There is now some evidence (15) for the

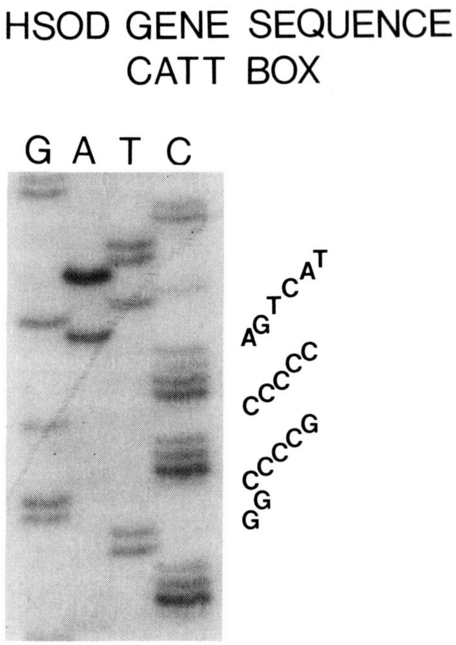

Fig. 2. Autoradiogram of sequencing gel showing the effect of groups of dC-residues on band intensities corresponding to dC-residues.

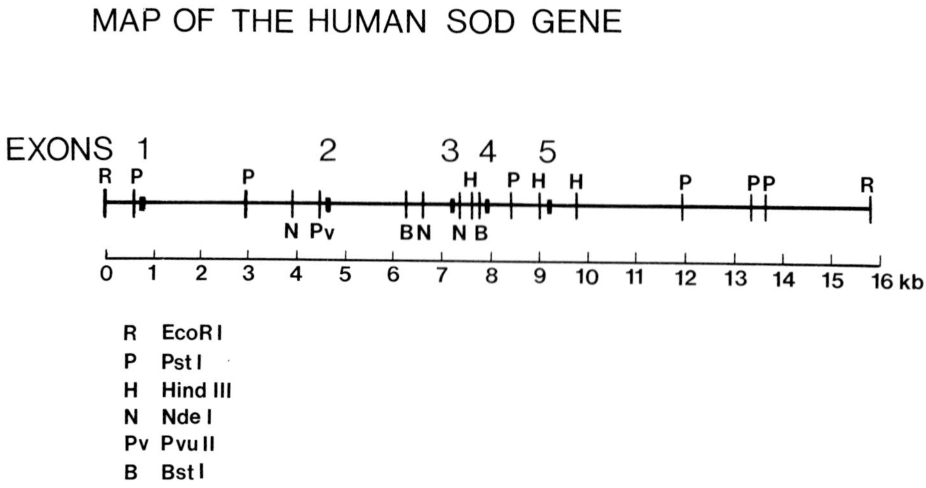

Fig. 3. Map of the HSOD gene. Distances are in kilobases (kb). Exons are indicated by solid rectangular boxes below the numbers. Note that the BstI map of the 5.5 kb PstI fragment is not complete.

suggestion (3) that an important function of introns is to facilitate protein evolution by addition, loss or substitution of exons. In addition, it has been observed that intron-exon junctions often occur at the protein surface (4). This may indicate that those exon changes which are selected during evolution most often occur at the protein surface so that the existing structure and function of the protein is not disrupted. Fig. 4 shows the locations of the five HSOD exons superimposed on the closely related bovine SOD crystal structure (2). None of the exons appears to correspond to a clearly defined structural or functional domain. Possible SOD domains might include metal binding, dimer binding and superoxide channeling. All intron-exon junctions except the last (at Val^{119}) occur at the protein surface suggesting that some of the exons may correspond to domains that modified SOD function during evolution. Further insight into the nature of exon-mediated Cu/Zn SOD evolution will probably be obtained as gene and protein structures of distantly related Cu/Zn SODs become available.

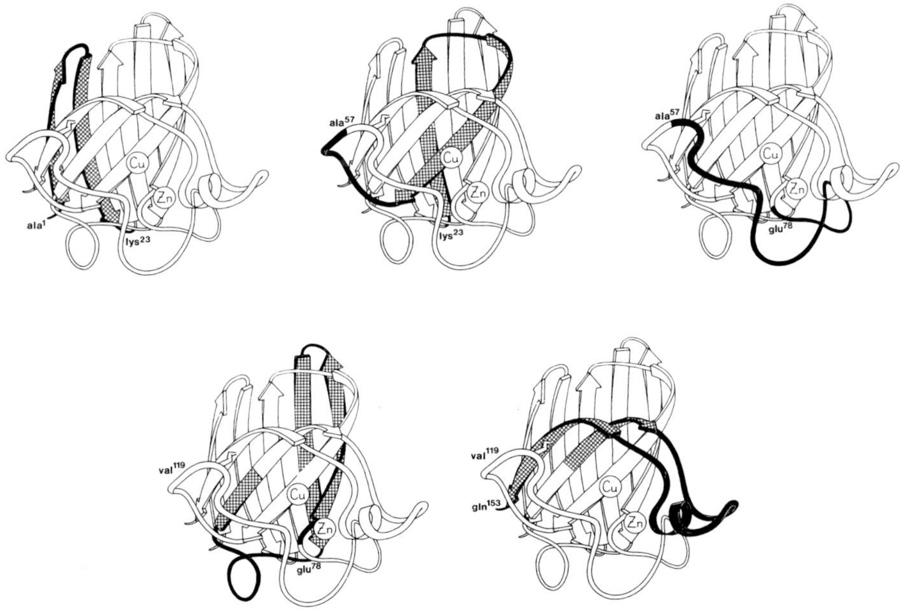

Fig. 4. Location of the five HSOD exons superimposed on the bovine SOD crystal structure.

REFERENCES

1. Levanon D, Lieman-Hurwitz J, Dafni N, Wigderson M, Sherman L, Bernstein Y, Laver-Rudich Z, Danciger E, Stein O, Groner Y (1985) The EMBO J 4:77-84

2. Tainer JA, Getzoff ED, Beem KM, Richardson JS, Richardson DC (1982) J Mol Biol 160:181-217

3. Gilbert W (1978) Nature 271:501

4. Craik CS, Rutter WJ, Fletterick R (1983) Science 220:1125-1129

5. Lawn RM, Fritsch EF, Parker RC, Blake G, Maniatis T (1978) Cell 15:1157-1174

6. Maniatis T, Fritsch EF, Sambrook J (1982) Molecular Cloning. Cold Spring Harbor Laboratory, Cold Spring Harbor, NY

7. Hallewell RA, Masiarz FR, Najarian RC, Puma JP, Quiroga MR, Randolph A, Sanchez-Pescador R, Scandella CJ, Smith B, Steimer KS, Mullenbach GT (1985) Nucl Acids Res 13:2017-2033

8. Messing J (1983) Meth Enzymol 101:20-78

9. Suggs SV, Wallace RB, Hirose T, Kawashima, EH, Itakura K (1981) Proc Natl Acad Sci USA 78:6613-6617

10. Sanger F, Nicklen S, Coulson AR (1977) Proc Natl Acad Sci USA 74:5463-5467

11. Breathnach R, Chambon P (1981) Ann Rev Biochem 50:349-383

12. Dynan WS, Tjian R (1985) Nature 316:774-778

13. Sherman L, Levanon D, Lieman-Hurwitz J, Dafni N, Groner Y (1985) Nucl Acid Res 12:9349-9366

14. Fischer HD, Dodgson JB, Hughes S and Engel JD (1984) Proc Natl Acad Sci USA 81:2733-2737

15. Rogers J (1985) Nature 315:458-459.

Superoxide and Superoxide Dismutase in Chemistry,
Biology and Medicine; G. Rotilio, editor

THE HUMAN Cu/Zn SUPEROXIDE DISMUTASE GENE FAMILY: ARCHITECTURE AND EXPRESSION OF THE CHROMOSOME 21-ENCODED FUNCTIONAL GENE AND ITS RELATED PROCESSED PSEUDOGENES

YORAM GRONER, JUDY LIEMAN-HURWITZ, NAOMI DAFNI, YAEL BERNSTEIN, LEVANA SHERMAN, DITSA LEVANON, EFRAT DANCIGER AND ORNA ELROY-STEIN
Department of Virology, The Weizmann Institute of Science, Rehovot, Israel 76100

Introduction

Superoxide dismutase (SOD:EC 1.15.1.1) is a "housekeeping" enzyme present in most aerobic organisms. It catalyzes the oxidation-reduction reaction: $O_2^{\dot{-}} + O_2^{\dot{-}} + 2H^+ \rightarrow H_2O_2 + O_2$ providing a defense against oxygen toxicity (reviewed in 1 & 2). Eukaryotic cells contain two distinct forms of SOD - a mitochondrial manganese-containing enzyme and a cytoplasmic copper/zinc-containing enzyme (SOD-1). The human SOD-1 (hSOD-1) is a dimer of 32 kd composed of two identical non-covalently linked subunits (3). The gene locus of hSOD-1 was assigned to chromosome 21 (4) and mapped at the chromosomal region 21q22 (reviewed in 5). Human chromosome 21 and in particular the region 21q22 - is involved in the most common genetic disorder known as trisomy 21, or Down's syndrome (5,6). The SOD-1 gene which resides at that region may be involved in the development of the syndrome. It is for this reason that we wished to isolate and study this gene.

Construction of a cDNA clone for hSOD-1

Poly(A)-containing RNA enriched for hSOD-1 mRNA was isolated and used to synthesize double-stranded complementary DNA (cDNA), which was inserted into the endonuclease PstI site of the plasmid pBR322. The chimeric molecules were used to transform *Escherichia coli*. Two clones containing hSOD-1 cDNA inserts were identified by their ability to hybridize with hSOD-1 mRNA. Each of these clones carried a 650 base-pair (bp.) insert, as was determined by restriction enzyme digestion and electron microscopic heteroduplex analysis (11). One clone, designated pS61-10, was sequenced (Fig. 1). It contains 459 nucleotides representing the entire coding region and 95 nucleotides of the 3'-untranslated region. The amino acid sequence predicted by the pS61-10 insert agrees with the published amino acid sequence of hSOD-1, with few exceptions (Table 1). Since the nucleotide sequence of the hSOD-1 gene was identical to that of the cDNA (see below) we assume that these differences reflect experimental errors rather than polymorphisms. Once the plasmid containing the hSOD-1 cDNA insert was identified, it was used as a probe for genomic DNA and RNA blots to obtain information about the hSOD-1 gene and the hSOD-1 mRNA. The results showed that the probe reacted with several genes in human cells and with two species of poly(A)-containing RNAs.

```
    1   2   3   4   5   6   7   8   9   10  11  12  13  14  15  16  17  18  19  20  21  22  23  24  25  26
    ALA THR LYS ALA VAL CYS VAL LEU LYS GLY ASP GLY PRO VAL GLN GLY ILE ILE ASN PHE GLU GLN LYS GLU SER ASN
G32 ATG,GCG,ACG,AAG,GCC,GTG,TGC,GTG,CTG,AAG,GGC,GAC,GGC,CCA,GTG,CAG,GGC,ATC,ATC,AAT,TTC,GAG,CAG,AAG,GAA,AGT,AAT,
         40         50              60             70           80              90            100           110

27  28  29  30  31  32  33  34  35  36  37  38  39  40  41  42  43  44  45  46  47  48  49  50  51  52  53
GLY PRO VAL LYS VAL TRP GLY SER ILE LYS GLY LEU THR GLU GLY LEU HIS GLY PHE HIS VAL HIS GLU PHE GLY ASP ASN
GGA,CCA,GTG,AAG,GTG,TGG,GGA,AGC,ATT,AAA,GGA,CTG,ACT,GAA,GGC,CTG,CAT,GGA,TTC,CAT,GTT,CAT,GAG,TTT,GGA,GAT,AAT,
      120           130           140             150            160           170            180           190

54  55  56  57  58  59  60  61  62  63  64  65  66  67  68  69  70  71  72  73  74  75  76  77  78  79  80
THR ALA GLY CYS THR SER ALA GLY PRO HIS PHE ASN PRO LEU SER ARG LYS HIS GLY GLY PRO LYS ASP GLU GLU ARG HIS
ACG,GCA,GGC,TGT,ACC,AGT,GCA,GGT,CCT,CAC,TTT,AAT,CCT,CTA,TCC,AGA,AAA,CAC,GGT,GGG,CCA,AAG,GAT,GAA,GAG,AGG,CAT,
     200              210         220          230                240          250          260            270

81  82  83  84  85  86  87  88  89  90  91  92  93  94  95  96  97  98  99  100 101 102 103 104 105 106 107
VAL GLY ASP LEU GLY ASN VAL THR ALA ASP LYS ASP GLY VAL ALA ASP VAL SER ILE GLU ASP SER VAL ILE SER LEU SER
GTT,GGA,GAC,TTG,GGC,AAT,GTG,ACT,GCT,GAC,AAA,GAT,GGT,GTG,GCC,GAT,GTG,TCT,ATT,GAA,GAT,TCT,GTG,ATC,TCA,CTC,TCA,
     280         290            300           310          320                330          340             350

108 109 110 111 112 113 114 115 116 117 118 119 120 121 122 123 124 125 126 127 128 129 130 131 132 133 134
GLY ASP HIS CYS ILE ILE GLY ARG THR LEU VAL VAL HIS GLU LYS ALA ASP ASP LEU GLY LYS GLY GLY ASN GLU GLU SER
GGA,GAC,CAT,TGC,ATC,ATT,GGC,CGC,ACA,CTG,GTG,GTC,CAT,GAA,AAA,GCA,GAT,GAC,TTG,GGC,AAA,GGT,GGA,AAT,GAA,GAA,AGT,
    360          370            380                 390           400           410           420           430

135 136 137 138 139 140 141 142 143 144 145 146 147 148 149 150 151 152 153
THR LYS THR GLY ASN ALA GLY SER ARG LEU ALA CYS GLY VAL ILE GLY ILE ALA GLN STOP
ACA,AAG,ACA,GGA,AAC,GCT,GGA,AGT,CGT,TTG,GCT,TGT,GGT,GTA,ATT,GGG,ATC,GCC,CAA, TAAACATTCCCTTGGATGTAGTCTGAGG
440             450        460          470                480           490           500         510         520

CCCCTTAACTCATCTGTTATCCTGCTAGCTGTAGAAATGTATCCTGATAAACATTAAACACTGTAATCTTAAAAAAAAAAAC17
     530         540         550         560         570            580         590          600
```

Fig. 1. DNA sequence of SOD-1 cDNA clone (7).

Table 1: Differences between the three published amino acid sequences of hSOD-1

	Residue No. as in Fig. 1:	11	17	26	49	52	53	92	98
a.	Nucleotide sequence of cDNA Sherman et al., (7), and gene Levanon et al., (8)	Asp	Ile	Asn	Glu	Asp	Asn	Asp	Ser
b.	Protein-Sequence-I Jabusch et al., (9)	[Asn]	Ile	Asn	Glu	Asp	Asn	[Asn]	Ser
c.	Protein Sequence-II Barra et al., (10)	Asp	[Ser]	[Asp]	[Gln]	[Asn]	[Asp]	Asp	[Val]

The Cu/Zn-SOD mRNAs

Two transcripts of 0.7 and 0.9 kilobases (Kb) were found in a variety of human cells (Fig. 2). The major 0.7 Kb species is approximately four times more abundant than the minor 0.9 Kb mRNA (7). These two species were also present in monkey cells, whereas mouse cells contained only a 0.7 Kb RNA. In the mouse/human hybrid line WAV4dF9-4a (4) that contains chromosome 21 as the only human chromosome, these two human RNAs were detected, indicating that both are encoded by this chromosome. The RNAs were found in poly(A)-containing polysomal RNA and were translated *in vitro* to immunoprecipitable hSOD-1 polypeptides, indicating that they are functional mRNAs. Isolation of the hSOD-1 gene revealed that the two mRNAs are transcribed from a unique hSOD-1 active

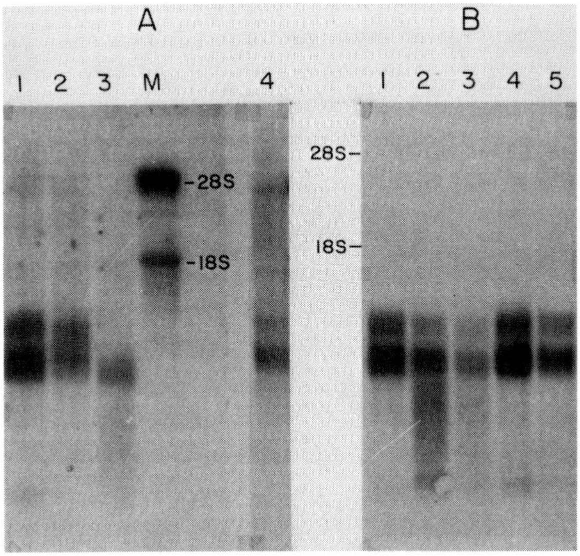

Fig. 2. Detection of hSOD-1 mRNAs. (A) RNA extracted from the following sources: lane 1 - SV80 human cells; lane 2 - COS monkey cells; lane 3 - A9 mouse cells; lane 4 - poly(A)-containing nuclear RNA. (B) RNA extracted from various human cell lines: lane 1 - GM-137; lane 2 - HL60; lane 3 - U937; lane 4 - SV80; lane 5 - FS-11. Analyses were carried out as described previously (7).

gene residing on chromosome 21, and that sequences at the 3'-untranslated region account for the difference in size between them (13)(see Fig. 3). Several lines of evidence support this conclusion. First, hybridization experiments with defined restriction fragments of hSOD-1 cDNA clone showed that both mRNAs contain sequences of the entire coding region plus the 95 nucleotides representing the 3'-untranslated region of the small hSOD-1 mRNA. Second, primer extension and S_1 mapping experiments with genomic DNA probes from the 5' and 3'-ends of the hSOD-1 gene indicated that the DNA sequences encoding the large transcripts are contiguous to those shared by the two mRNA species. Thirdly, only one region of human DNA with nucleotide sequences identical to the hSOD-1 cDNA was detected.

Anatomy of the chromosomal locus encoding the hSOD-1

The hSOD-1 gene locus, including more than 100 Kb of chromosomal DNA from the 21q22 region, was isolated and characterized (see Fig. 4). Human genomic libraries in lambda Ch4A were screened with the cloned hSOD-1 cDNA and several overlapping recombinant phages containing the whole hSOD-1 gene were isolated. Heteroduplex analysis and DNA sequencing (Fig. 3) reveals five rather small exons and four introns that interrupt the coding region. This gene is present

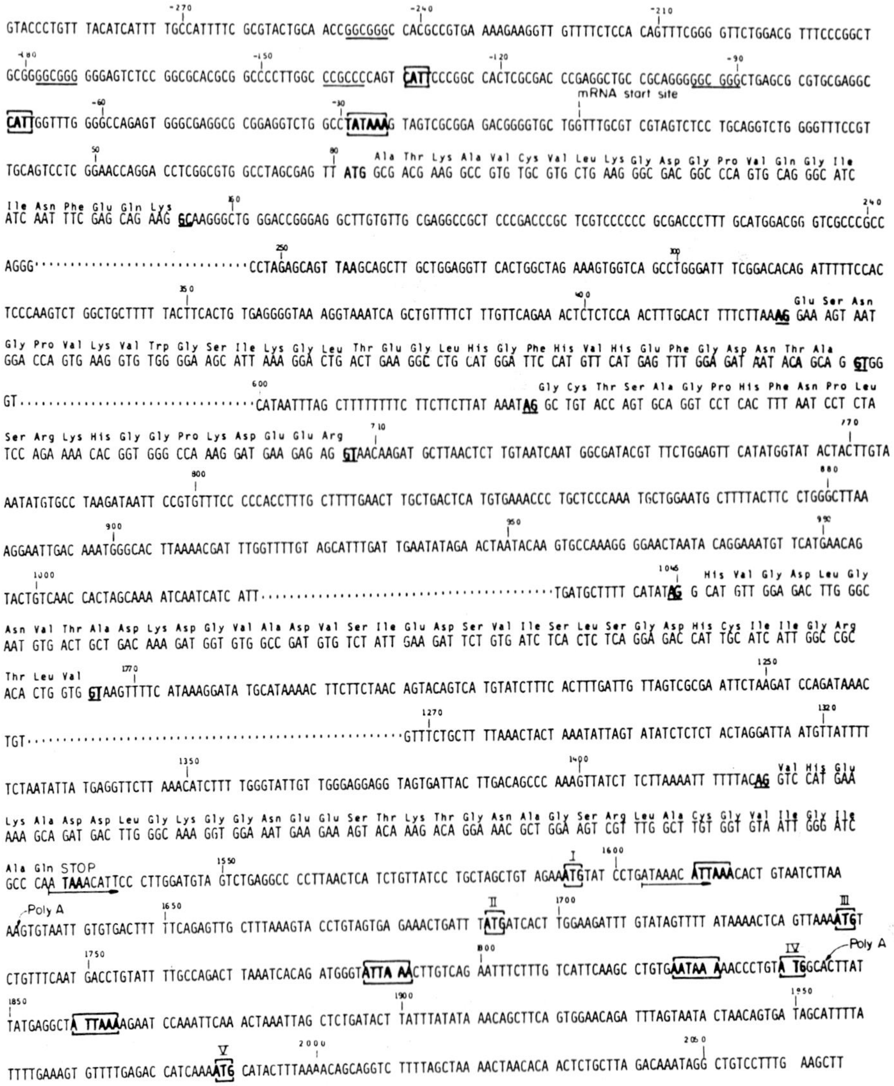

Fig. 3. Nucleotide and amino acid sequences of the hSOD-1 gene (8). The sequence of all coding regions and adjacent nucleotides are shown with 110 bases per line. The 'TATA', 'CAT' and polyadenylation sequences are boxed. The G-C elements and the splice junctions are underlined. The sites of initiation of transcription and poly(A)-tail are indicated.

as a single copy per haploid genome and the DNA regions present in the isolated phages (λA-2 and λB-1) are the only ones with nucleotide sequences identical to the hSOD-1 cDNA (Fig. 1). We therefore concluded that this region represents the unique hSOD-1 functional gene. The gene is approximately 11 Kb in length and is interrupted by four introns. In proportion to the sizes of the two mRNAs

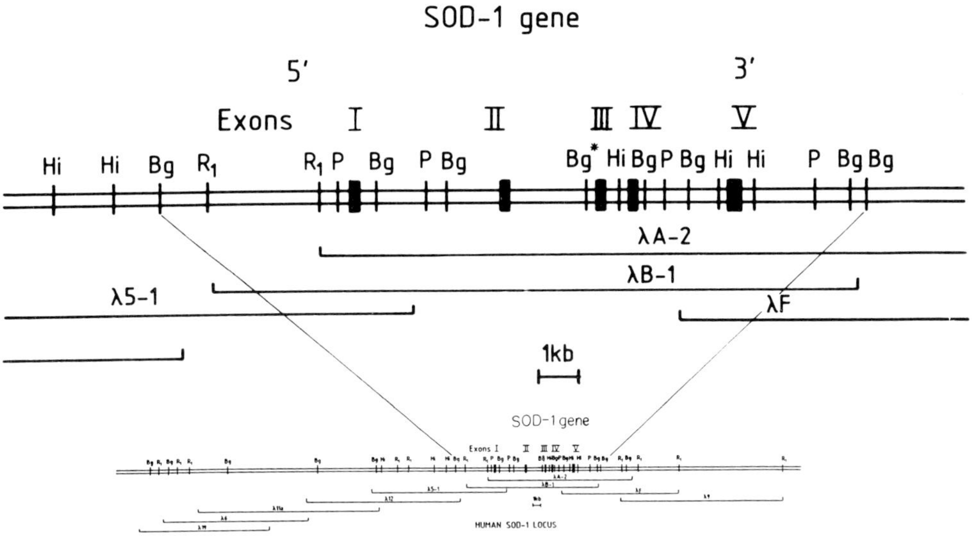

Fig. 4. Restriction map and organization of the hSOD-1 gene locus. Top line: genomic DNA from four overlapping λ clones with positions of four restriction enzymes R_1 = Eco RI, Bg = BglII, Hi = HindIII, P = PstI. The exons are the black rectangles numbered I-V. The polymorphism in the BglII site is indicated by an asterisk. Bottom line: molecular map of ∿100 kb of the hSOD-1 locus. The nine overlapping λ clones contain sequences of the gene and neighboring regions.

(0.7 and 0.9 Kb) this is a large gene, over 12 times the length of the longer mRNA species. In addition to this gene we have isolated four hSOD-1 related processed pseudogenes. Experiments with a genomic library of human chromosome 21 have indicated that the processed genes do not reside on this chromosome (see below). All the protein coding regions and part of the introns of the hSOD-1 gene were sequenced as well as 300 b.p. of the 5'-flanking region and 220 b.p. of the 3'-flanking region (Fig. 3). In the donor sequence of the first intron a T to C transition occurred and hence it deviates from the 5'GT....AG 3' consensus. The unusual 5'-GC donor site is not an artefact of the cloning procedure because it was detected in both λA-2 and λB-1 which represent the two alleles of the hSOD-1 gene. The two phages λA-2 and λB-1 contained the entire

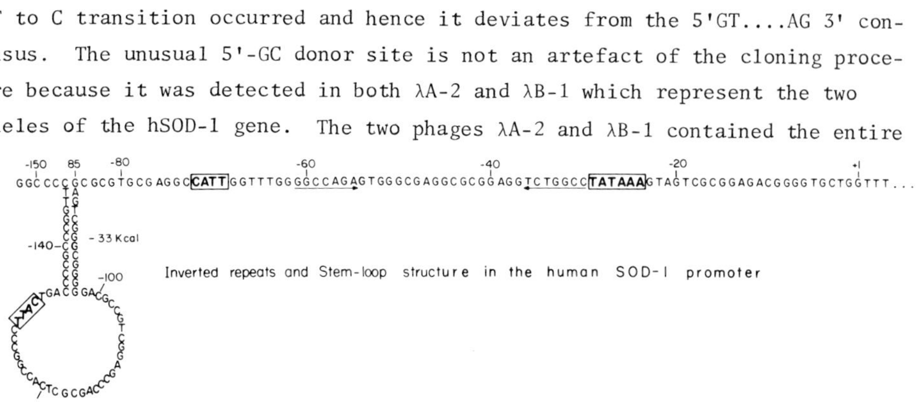

Fig. 5. Putative stem-loop structure located 5' to the 'TATA' and 'CAT' boxes of the hSOD-1 gene. The free energy per strand (-33 kcal/mol) is indicated.

hSOD-1 gene (Fig. 4). When hybridized to each other they formed a stable DNA duplex across their overlapping region. The restriction maps of that region are identical except for one BglII site (marked by an asterisk in Fig. 4), which is missing in λB-1. When DNA samples from ten unrelated individuals were digested with BglII and analyzed by Southern blot hybridization, three out of the ten contained this polymorphic BglII site. It was therefore concluded that the λA-2 and λB-1 contain the two alleles of the hSOD-1 gene and that the additional BglII site was created by an alteration in the nucleotide sequence which causes restriction fragment length polymorphism (RFLPs). At the 5' end of the gene there are the 'TATA' and 'CAT' promoter sequences as well as four copies of the -GGCGGG- hexanucleotide. Two of these -GC- elements are contained within a 13 nucleotide inverted repeat that can fold into a stem-loop structure (Fig. 5).

The hSOD-1 related pseudogenes

Earlier genomic blot experiments revealed several Eco RI fragments which showed homology with the hSOD-1 cDNA clone. The largest 18 Kb fragment harbors the functional gene described above. As for the other hSOD-1 related sequences, it was important to define their nature and location in order to determine whether we are dealing with multiple active genes residing on the same chromosome or, as described below, with a gene family containing one active gene, located on chromosome 21, and four pseudogenes that moved to other chromosomes. To isolate the sequences belonging to the other hSOD-1 homologous regions, two independently constructed human λCh4A libraries were screened with the cDNA probe.

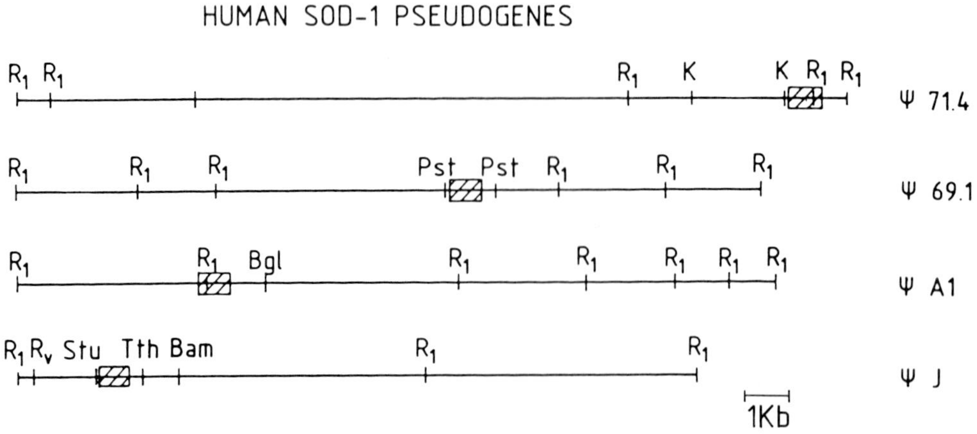

Fig. 6. Restriction enzyme maps of the four recombinant λ phages containing the hSOD-1 pseudogenes. The hatched boxes indicate the position of the region complementary to hSOD-1 RNA sequences. Bam = BamHI, Bgl = BglII, K = KpnI, Pst = PstI, R_1 = Eco RI, Rv = Eco Rv, Stu = StuI, Tth = Tth111I.

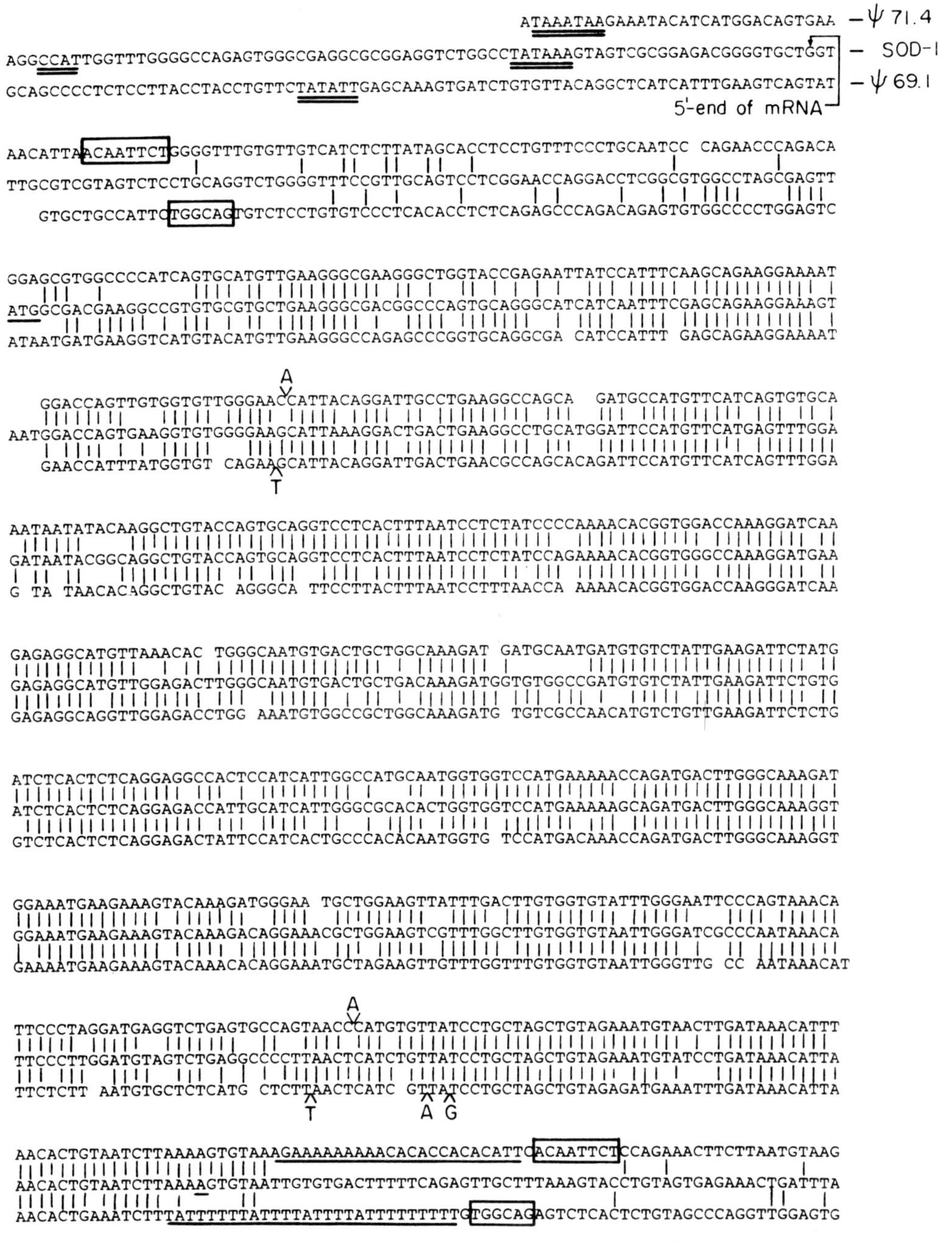

Fig. 7. Comparison of the DNA sequences of hSOD-1 gene and the two processed genes ψ71.4 and ψ69.1. The DNA sequences of the hSOD-1 (middle line), ψ71.4 (top line) and ψ69.1 (bottom line) were aligned to maximal homology. The vertical lines between the pseudogenes and the hSOD-1 indicate sequence identity. Small deletions are represented by gaps. The 5' and 3' direct repeats are indicated by boxes.

In each case, about 10^6 phages were screened, resulting in the isolation of four distinct recombinants (λ71.4, λ69.1, λA1, λJ) containing hSOD-1 related sequences (Fig. 6). Smaller DNA fragments containing hSOD-1 sequences were isolated by restriction enzymes digestion and blot hybridization. The relevant fragments were subcloned in pBR322 or pUC13 and the hSOD-1 related region identified and mapped either by heteroduplex analysis with the cDNA- or by complete sequencing (Fig. 7). All four regions are devoid of introns, which typifies pseudogenes of the processed type (14)(Fig. 6). Although the overall sequence homology of ψ71.4 and ψ69.1 to the SOD-1 gene was extensive they contain multiple genetic lesions, such as insertions, deletions and base substitutions, resulting in-frame termination codons, that preclude the translation of the normal hSOD-1 polypeptides. The 71.4 and 69.1 pseudogenes are also flanked by short direct repeat sequences - an additional hallmark of processed pseudogenes (Fig. 7). Experiments with a genomic library of human chromosome 21 and genomic blots of DNA from the mouse/hybrid line WAVR4dF9-4a which contains chromosome 21 as its only human chromosome (4) indicated that the processed pseudogenes do not reside on chromosome 21. Direct comparison of the nucleotide sequences within the coding region of the expressed hSOD-1 gene with those of the processed gene can provide an estimate of their divergence time (12). When base substitution in the pseudogenes were counted and the divergence value calculated it equalled - ∿17% for both pseudogenes - indicating that the separation between the functional and the pseudogenes occurred about 25 million years ago.

This work was supported by a Basic Research Grant No. 1-906 from the March of Dimes Birth Defects Foundation, by Biotechnology General Corp., (Israel), by the Weizmann Institute Leo and Julia Forchheimer Center of Molecular Genetics and by the Minerva Foundation, Munich, Germany.

REFERENCES

1. Fridovich I (1978) Science (Wash.) 201:875-880.
2. Fridovich I (1979) In: Eichhorn GI, Marzilli IG (eds) Advances in Inorganic Biochemistry, Elsevier, North Holland, NY, pp 67-90
3. Briggs RG, Fee JA (1978) Biochim Biophys Acta 537:86-99
4. Tan YH, Tischfield J, Ruddle FH (1973) J Exp Med 137:317-330
5. Summit RL (1981) In: de la Cruz FF, Gerald PS (eds) Trisomy 21 (Downs Syndrome) Research Perspective. University Park Press, Baltimore, pp 225-235
6. Lejeune JM, Gautier M, Turpin R (1959) C R Hebd Seances Acad Sci, Paris 248:1721-1722
7. Sherman L, Dafni N, Lieman-Hurwitz J, Groner Y (1983) Proc Natl Acad Sci USA 80:5465-5469
8. Levanon D, Lieman-Hurwitz J, Dafni N, Wigderson M, Sherman L, Bernstein Y, Laver-Rudich Z, Danciger E, Stein O, Groner, Y. (1985) EMBO J. 4:77-84

9. Jabusch JR, Farb DL, Kerschensteiner DA, Deutsch H (1980) Biochemistry (Wash.) 19:2310-2316

10. Barra D, Martini F, Bannister JV, Schinina ME, Rotilio G, Bannister WH, Bossa F (1980) FEBS Lett 120:53-56

11. Lieman-Hurwitz J, Dafni N, Lavie V, Groner Y (1982) Proc Natl Acad Sci USA 79:2808-2811

12. Perler F, Efstratiadis A, Lomedico P, Gilbert W, Kolodner R, Dodgson J (1980) Cell 20:555-566

13. Sherman L, Levanon D, Lieman-Hurwitz J, Dafni N, Groner Y (1984) Nucleic Acid Res 12:9349-9365

14. Jeffreys AJ, Harris S (1985) BioAssays 1:253-258

Superoxide and Superoxide Dismutase in Chemistry, Biology and Medicine; G. Rotilio, editor

COMPARISON OF HUMAN CU/ZN SUPEROXIDE DISMUTASE DERIVED FROM ERYTHROCYTES, RECOMBINANT E. COLI AND RECOMBINANT YEAST.

L. FLOHE*, S.-M. A. KIM*, F. ÖTTING*, D. SAUNDERS*, E. SCHWERTNER*, G.J. STEFFENS, R. BLACHER**, F. MASIARZ**, C. SCANDELLA**, and R. HALLEWELL**
* Grünenthal GmbH, Center of Research, Zieglerstr. 6, D-5100 Aachen/FRG
**Chiron Corp., 4560 Horton Street, Emeryville CA 94608/USA.

INTRODUCTION

The therapeutic potential (1) of superoxide dismutases (SOD) has stimulated various companies to produce an authentic human SOD (HSOD) by gene technology. While it has become state of the art to express heterologous amino acid sequences of considerable length in bacteria, it still represents a major challenge to obtain, by this methodology, a native mammalian enzyme, in particular if posttranslational processing is required. The cDNA (2-4) and genomic DNA (5) encoding human SOD have meanwhile been obtained and successful expression in microorganisms has been reported (3,4). We here compare the structure and function of Cu/Zn SOD obtained from recombinant *E. coli* (cHSOD) and *S. cerevisiae* (yHSOD) with that isolated from human erythrocytes (eHSOD).

METHODS

Crude recombinant cHSOD and yHSOD were obtained from Chiron Corp., Emeryville, Ca., and purified by chromatographic techniques. eHSOD was purified essentially as described (6). Specific activities and chromatographic behaviour of SODs were determined according to Flohé and Ötting (7). Tryptic peptides were obtained, separated and analysed as described by Steffens et al. (8).

The structure of the N-terminal tripeptide T1 of yHSOD was established by the fast atom bombardement (FAB) technique. 20 nMol of T1 was loaded in a glycerol matrix onto a metallic target and bombarded with accelerated argon atoms. Long-lived ion beams thus generated were mass analysed on a Finnigan mass spectrometer.

RESULTS

Every SOD sample investigated showed a specific activity in the cytochrome c reduction test (7) of $>$ 3,000 U/mg and a nearly molar content of Cu and Zn. In SDS gel electrophoresis, cHSDO, yHSDO and eHSOD were

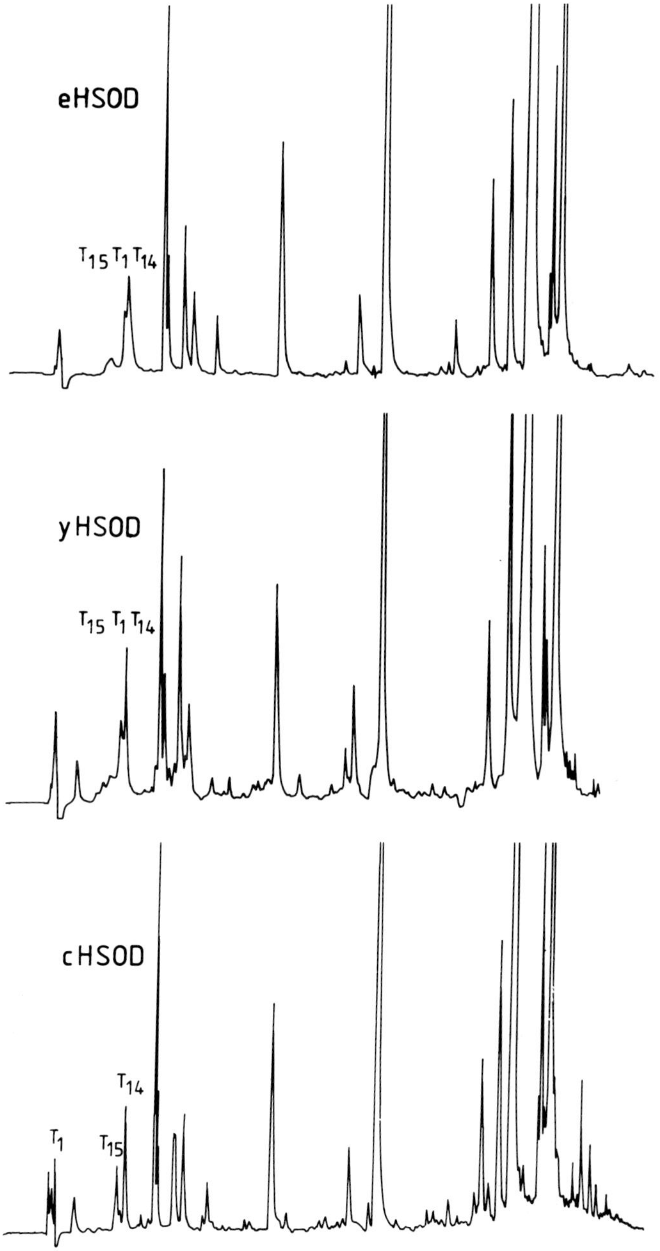

Fig. 1: HPLC finger prints of tryptic peptides derived from eHSOD, yHSOD and cHSOD. 10 - 15 nMol each were applied to V2A columns (25 cm x 7.2 mm) packed with Nucleosil 5-C18 and eluted with gradients of solutions A (0.05 % TFA) and B (0.025 % TFA) at 50°C and 1.6 ml/min as follows: Gradient A - B: 0 - 36 min 0 - 21.4 % B, 36 - 46 min 21.4 - 27 % B; 46 - 60 min 27 - 50 % B. Note that all peaks with the exception of T1 appear at identical retention times.

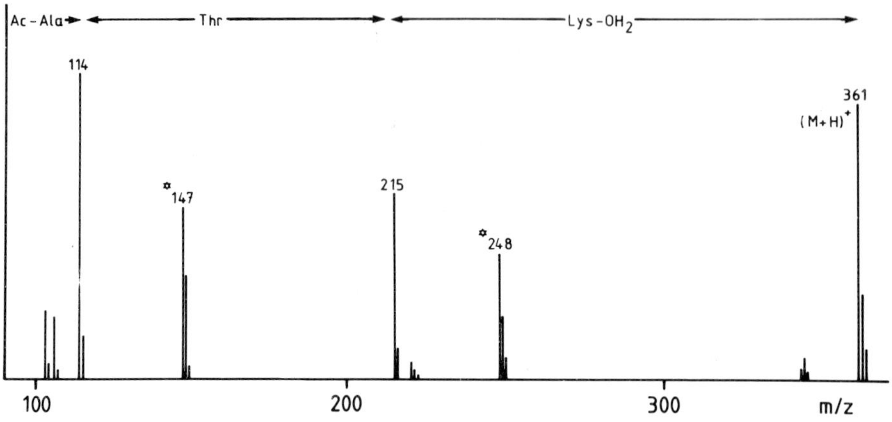

Fig. 2: Determination of the structure of the N-terminal tripeptide T1 isolated from a tryptic digest of yHSOD by FAB mass spectroscopy. The signals of the N-terminal fragments at m/z 114 and 215 were assigned to Ac-Ala and Ac-Ala-Thr, respectively. C-terminal fragments (*) were seen at m/z 147 and 248 assignable to H2-Lys-OH and H2-Thr-Lys-OH. The intense signal of the quasimolecular ion (M+H)+ at m/z 361 confirmed the molecular weight and sequence of T1.

indistinguishible. In the agarose gel electrophoresis (7) an identical pattern of charge isomers was obtained for yHSOD and eHSOD, although the relative intensities of the bands differed slightly, whereas cHSOD showed a multiplicity of active bands none of which corresponded to anyone of eHSOD.

In order to pinpoint the chemical difference between cHSOD and the other two preparations, the samples were subjected to tryptic cleavage and the resulting peptides separated by HPLC (Fig. 1). If slight variations due to variable efficacy of digestion are ignored, the difference between the samples appears restricted to the N-terminal tryptic peptide (T1). Further investigations after TLC purification of T1 revealed an identical amino acid composition. Carboxypeptidase treatment yielded lysine, threonine, and N-acetyl-alanine in case of eHSOD and yHSOD, but free alanine in case of cHSOD. The identity of N-acetyl-alanine thus obtained was established by conventional GC-MS. Similarly, N-terminal sequence analysis of the total proteins revealed a free N-terminus (Ala with some contamination by Met) in case of cHSOD, but a blocked terminus in yHSOD and eHSOD. Finally, the structure of the N-terminal tryptic peptide of yHSOD was determined by

FAB-MS (Fig. 2) and unambigously established as N-acetyl-alanyl-threonyl-lysine.

DISCUSSION

The results show that active recombinant HSOD can be obtained from *E. coli* and yeast. Relying on specific activity, we can conclude that correct protein folding forming an active center occurs similarly in both microorganisms. Expectedly, *E. coli* does not acetylate the N-terminus of HSOD which results in a different electrophoretic behaviour and, probably, due to N-terminal degradation, in a pronounced heterogeneity of active molecular species. These findings suggest that cHSOD is a less attractive candidate for the development of HSOD for clinical use. In contrast, yHSOD could be shown to be an authentic HSOD in all criteria investigated and thus represents the optimum tool to further explore the therapeutic potential of SOD in human diseases.

ACKNOWLEDGEMENTS

This work was supported by the *Bundesministerium für Forschung und Technologie* of the Federal Republic of Germany (Grant PTB 8482).

REFERENCES

1. Flohé L, Giertz H, Beckmann R (1985) In: Bonta IL, Bray MA Parnham MJ (eds) Handbook of Inflammation, Vol.5. Elsevier, pp 255-281

2. Sherman L, Dafni N, Lieman-Hurwitz J, Groner Y (1983) Proc Natl Acad Sci 80:5465-5469

3. Hallewell RA, Masiarz FR, Najarian RC, Puma JP, Quiroga MR, Randolph A, Sanchez-Pescador R., Scandella CJ, Smith B, Steiner KS, Mullenbach GT (1985) Nucleic Acids Res 13: 2017-2034

4. Hallewell RA, Tekamp-Olson P, Rosenberg S, Blacher R, Masiarz F Scandella C this issue

5. Hallewell RA, Puma JP, Mullenbach GT, Najarian RC this issue

6. Bannister JV, Bannister WH (1985) In: Packer L (ed), Methods in Enzymology vol. 105, pp 88-93

7. Flohé L, Ötting F (1985) In: Packer L (ed), Methods in Enzymology vol. 105, pp 93-104

8. Steffens GJ, Bannister JV, Bannister WH, Flohé L, Günzler WA, Kim SMA, Ötting, F (1983) Hoppe-Seyler's Z Physiol Chem 364: 675-690

© 1986 Elsevier Science Publishers B.V. (Biomedical Division)
Superoxide and Superoxide Dismutase in Chemistry,
Biology and Medicine; G. Rotilio, editor

DISTRIBUTION OF BACTERIOCUPREIN SUPEROXIDE DISMUTASES IN PROKARYOTES

HOWARD M. STEINMAN AND PAUL V. DUNLAP

Department of Biochemistry, Albert Einstein College of Medicine, 1300 Morris Park Avenue, Bronx, New York 10461 (USA), and Department of Microbiology, New York State College of Agriculture and Life Sciences, Cornell University, Ithaca, New York 14853 (USA)

INTRODUCTION

Although copper-zinc superoxide dismutases (CuZnSODs) are characteristic of eukaryotes, two bacterial species have been identified that contain CuZnSODs, termed bacteriocupreins: _Photobacterium leiognathi_, a marine, luminescent bacterium and light-organ symbiont of ponyfish (1) and _Caulobacter crescentus_, a fresh water bacterium that is neither luminescent nor a symbiont (2). Each of these bacteria also contains an FeSOD. Since the discovery of the first bacteriocuprein in _P. leiognathi_ there has been considerable interest in the mechanism by which that bacterium acquired an SOD of striking similarity to the CuZnSODs of eukaryotes, and whether bacteriocupreins are as sparsely distributed as they seem to be. This article reviews recent studies of bacteriocuprein distribution which are also pertinent to the question of bacteriocuprein origin.

VARIATION OF BACTERIOCUPREINS AMONG SYMBIOTIC STRAINS OF _Photobacterium leiognathi_ (3)

While the type strain of _P. leiognathi_, American Type Culture Collection 25521, a symbiont of _Leiognathus splendens_, and its bacteriocuprein have been characterized, no other strains of _P. leiognathi_ have been examined for bacteriocupreins. To characterize the distribution of bacteriocupreins in _P. leiognathi_, nine new symbiotic strains of _P. leiognathi_ were isolated from the light organs of six additional species of ponyfish. (4,5,6, See Table I) Sonic extracts of these strains were compared with the type strain by Western blotting, using a monospecific polyclonal antibody raised against the bacteriocuprein purified from the type strain (7), and by _in situ_ staining for SOD activity.

TABLE I

BACTERIOCUPREIN VARIANTS OF P. leiognathi

Strain	& Ponyfish	Ponyfish Source*	& Sex	Bacteriocuprein Mobility	& Activity	CRM*
Type	L. splendens	T	n.d.	fast	++	++
1	L. brevirostris	PI	n.d.	slow	++	+
2	L. equulus	PI	n.d.	slow	++	+
3	L. fasciatus	PI	n.d.	slow	++	+
4	L. nuchalis	J	n.d.		-	+
5	L. nuchalis	J	n.d.	fast	++	+
6	L. elongatus	J	male		-	-
7	L. elongatus	J	female	fast	++	++
8	L. rivulatus	J	male	fast	+	++
9	L. rivulatus	J	female		-	+

* T, Thailand, PI, Philippine Islands, J, Japan, n.d., not determined, CRM, immunologically cross-reactive material.

Only five of the newly isolated symbiotic strains contained bacteriocuprein activity comparable to that in the type strain, indicated by ++ in the next to the last column of Table I. Strain 8 had about 10% of the bacteriocuprein activity in the type strain and the remaining three strains had no detectable bacteriocuprein activity (less than 5-10% that in the type strain). A similar variation was observed in the amount of antigenic polypeptide of 16,000 daltons, the molecular weight of the bacteriocuprein subunit. The level of cross reactive material in strains 7 and 8 was comparable to that in the type strain. In the 6 additional strains, indicated by + in the last column of Table I, considerably less cross reactive material was detected, indicating lower levels of bacteriocupreins, or bacteriocupreins lacking the full set of determinants present in the immunizing antigen. In strain 6, no antigenic polypeptide was detected, even at loads ten times larger than that producing an intense zone with the type strain. The bacteriocuprein form of SOD is clearly not a fixed attribute of symbiotic P. leiognathi as a species, but varies from strain to strain. These observations, though drawn from a limited number of symbiotic strains of P. leiognathi, suggest that these bacteriocuprein variations might relate to differences in the species, sex, and geographical source of the ponyfish host.

The strain variation in *P. leiognathi* is pertinent to understanding the distribution of bacteriocupreins. Bacterial species have been principally screened for bacteriocupreins by activity assays. It is possible that species other than *P. leiognathi* exhibit a strain variation in bacteriocuprein activity like that demonstrated in Table I. If the particular strain selected as a representative of the species exhibited a low level of bacteriocuprein enzymatic activity, then its bacteriocuprein might be overlooked. Consequently, the bacteriocuprein form of SOD may be more widely distributed than presently perceived, because of strain variation in activity.

BACTERIOCUPREIN SUPEROXIDE DISMUTASES IN PSEUDOMONADS (7)

As part of an ongoing survey of prokaryotes for copper-based SOD activity, sonic extracts of nine strains representing seven species of the genus *Pseudomonas* were examined by *in situ* activity staining. Two species, *Pseudomonas diminuta* and *P. maltophilia*, contain a superoxide dismutase activity that is inhibited by cyanide, indicating that it is based on copper. These cyanide-sensitive activities are not attributable to a low molecular weight copper chelate with dismutase activity, because the apparent molecular weights of the activities are 31,000 to 41,000. Furthermore, the activities are not attributable to copper-containing proteins with trace SOD activity, because they are present in significant amounts in the extract. Indeed, in *P. diminuta*, the cyanide-sensitive activity is the major SOD activity in the extract. More definitive identification of the metal cofactor in these pseudomonad dismutases will be obtained from the future characterizations planned for these enzymes. However, the above data strongly suggest that *P. diminuta* and *P. maltophilia* contain bacteriocuprein SODs.

Martin and Fridovich proposed that the gene for bacteriocuprein of *P. leiognathi* was acquired by transfer of DNA encoding ponyfish CuZnSOD from the eukaryotic host fish to the prokaryotic symbiont (8, 9). For the three non-symbiotic strains that contain bacteriocuprein SODs, *C. crescentus* and the two pseudomonads, other explanations of their origin may be more valid, explanations that do not invoke an existing symbiotic association. Among the alternatives are: evolution of CuZnSODs independently in prokaryotes and eukaryotes, evolution in an ancestor common to both prokaryotes and eukaryotes, or origin in prokaryotes followed by gene transfer to eukaryotes.

ISOLATION OF THE BACTERIOCUPREIN GENE FROM *P. leiognathi* TYPE STRAIN (10)

To permit analysis of the bacteriocuprein phenomenon on the DNA level, complementing the above protein studies, the DNA encoding the bacteriocuprein of the type strain of *P. leiognathi* was cloned. Genomic DNA was restricted with Hind III and ligated into pUC12. Recombinants were screened by colony hybridization using a redundant, synthetic oligonucleotide probe derived from the amino sequence published for *P. leiognathi* bacteriocuprein, (11) spanning residues 83-88. Seven recombinants containing 3.6 kb inserts were thus identified, and the plasmid DNA from one of these was subcloned into pUC12 after digestion with BamH I. Three recombinants containing 1.8 kb inserts were identified by colony hybridizations using the above probe and a second probe derived from residues 133-137. All ten recombinant plasmids specifically hybridize with the two probes. In addition, sonic extracts of *E. coli* hosts containing these recombinant plasmids show a zone of superoxide dismutase activity, not found in control *E. coli* strains, that comigrates with the bacteriocuprein SOD in extracts of the *P. leiognathi* type strain and is sensitive to cyanide. Therefore, the cloned DNA contains the bacteriocuprein structural gene, and that gene is expressed in *E. coli* hosts. This isolated DNA, now being sequenced, will be particularly useful in further characterizing the bacteriocuprein variants in *P. leiognathi* and in determining the distribution of the bacteriocuprein gene or of homologous DNA in prokaryotes.

ACKNOWLEDGEMENTS

This investigation was supported by Grant GM 24527 from the National Institutes of Health and by American Cancer Society Institutional Grant IN-28-X (to H.M.S.). One of the redundant oligonucleotide mixtures was synthesized in the Oligonucleotide Facility of Albert Einstein College of Medicine.

REFERENCES

1. Puget K, Michelson AM (1974) Biochem Biophys Res Commun 58:830-838
2. Steinman HM (1982) J Biol Chem 257:10283-10293
3. Dunlap PV, Steinman HM (1985) J Bact, submitted
4. Dunlap PV (1984) Biol Bull 167:410-425
5. Dunlap PV (1985) Arch Microbiol 141:44-50
6. Dunlap PV, Fukasawa S. (1984) unpublished data
7. Steinman HM (1985) J Bact 162:1255-1260
8. Martin JP Jr, Fridovich I (1981) J Biol Chem 256:6080-6089
9. Bannister JV, Parker MW (1985) Proc Natl Acad Sci (USA) 82:149-152
10. Steinman HM, Furman AM (1985) unpublished data
11. Steffens G-J, Bannister JV, Bannister WH, Flohe L, Gunzler WA, Kim S-M, Otting F. (1983) Hoppe-Seyler's Z Physiol Chem 364:675-690

© 1986 Elsevier Science Publishers B.V. (Biomedical Division)
Superoxide and Superoxide Dismutase in Chemistry, Biology and Medicine; G. Rotilio, editor

REGULATION OF THE BIOSYNTHESIS OF SUPEROXIDE DISMUTASE IN PROCARYOTES

HOSNI M. HASSAN* AND CARMELLA S. MOODY**

*North Carolina State University, Raleigh, N.C. 27695-7624, and
**W. R. Grace & Co., Columbia, M.D. 21044

INTRODUCTION

Procaryotes exhibit the broadest range of diversity with respect to oxygen requirements and tolerance. This range extends from those organisms that survive and grow in the total lack of oxygen to those that are totally dependent on oxygen. This diversity makes them ideal systems to study the toxicity of oxygen and the regulation of superoxide dismutases that protect against the deleterious effects of oxygen.

Superoxide dismutases (SODs) are metalloenzymes found in organisms that are exposed to oxygen for transient or for long periods of time (1). Their sole function is to remove the superoxide anion (O_2^-) formed via the univalent reduction of oxygen, and thus protect the cells against oxygen toxicity (2). Most of the work that has led to a greater understanding of the regulation of superoxide dismutase and its role in protecting against oxygen toxicity has come from studies using the facultative anaerobe *Escherichia coli*. This review will be focused on what has been learned during the past decade about the biosynthesis of SOD in *E. coli*.

SUPEROXIDE DISMUTASE BIOSYNTHESIS IN ESCHERICHIA COLI.

Types of superoxide dismutase.

E. coli has three isoenzymic forms of superoxide dismutase. The manganese superoxide dismutase gene has been determined to be located at approximately 87 minutes on the *E. coli* K-12 map (3), and the iron superoxide dismutase has been cloned (4) and is thought to be located at approximately 36.5 min. The iron-containing form is present under both anaerobic and aerobic conditions and is, therefore, constitutive. The manganese-containing enzyme is absent under anaerobic conditions but is rapidly synthesized upon exposure to air (5). The hybrid form (6), which consists of one subunit each of the iron and manganese isoenzymes and has iron at the active site, is also synthesised in response to aerobic exposure. A recent report has indicated that the hybrid isoenzyme can have either manganese or iron as the catalytic metal (7). *Salmonella typhimurium* has the same isoenzymes and responds to oxygen exposures in a similar fashion as *E. coli* (8).

Superoxide dismutase biosynthesis in response to oxygen.

Derepression of the Mn-superoxide dismutase in *E. coli* occurs upon exposure to air after anaerobic growth. The derepression is rapid but the enzyme is not fully induced even after three hours of aerobic growth. When derepression of the enzyme under these conditions is prevented by inhibitors of protein biosyntheseis, the cells succumb to aerobic exposure (5). It has been proposed that the iron isoenzyme is constitutive to provide a balance between the need for a back-up defense against aerobic exposures and the energy expense required to maintain an adequate amount of enzyme necessary for protection during a completely aerobic existence (9). Increased synthesis of superoxide dismutase also occurs in response to concentrations of oxygen higher than 20% (10).

Superoxide dismutase biosynthesis in response to changes in specific growth rate.

The inducer of the MnSOD biosynthesis is not molecular oxygen, *per se* (11). This was initially determined when the levels of superoxide dismutase were determined to be lower when *E. coli* was grown at lower specific growth rates and increased when the specific growth rate was increased in a glucose limited chemostat culture grown at constant aeration. The varying levels of superoxide dismutase in response to changes in specific growth rate are explained by the higher rate of respiration and therefore the greater rate of superoxide production at higher specific growth rates than at lower growth rates.

Superoxide dismutase biosynthesis in response to varying carbon sources: fermentative versus oxidative metabolism.

Superoxide dismutase biosynthesis in *E. coli* has also been shown to vary with the carbon source being metabolized. In general, if *E. coli* is utilizing a fermentative carbon source, such as glucose, the SOD levels are much lower than when aerobically metabolized carbon sources such as succinate, lactate, or trypticase peptone, are utilized (12). It was concluded that there is greater superoxide production when the cells are utilizing an oxidative carbon source than when the cells are growing fermentatively (12) and that the increased levels of SOD under these conditions are a reckoned response to the increased superoxide production.

Superoxide dismutase biosynthesis in response to increased intracellular superoxide production by redox active compounds.

The final tug needed to remove the curtain of uncertainty that surrounded the identity of the inducer of MnSOD biosynthesis came when it was seen that, indeed, increasing intracellular superoxide production in medium containing glucose did result in increased MnSOD biosynthesis. The tools utilized to exacerbate intracellular O_2^- production were redox active compounds such as paraquat (12), pyocyanine, phenazine methosulfate, streptonigrin, juglone, plumbagin, menadione, methylene blue and azure C (13). All these compounds are capable of causing an increase in cyanide resistant respiration, an indication of intracellular O_2^- production, with no net change in respiration (14). Paraquat (PQ) was seen to be a superior inducer of MnSOD biosynthesis (14), however, all the compounds listed function in a similar manner to PQ. Inhibition of MnSOD induction was also seen in the presence of inhibitors of transcription but not of DNA replication. These results indicate that the induction of MnSOD is under transcriptional control and that O_2^- or a product formed from the interaction of O_2^- is the effector (13).

Genetic loss of superoxide dismutase and catalase activities: requirement for obligate anaerobiosis.

All the information presented so far has been circumstantial evidence that greatly supports the superoxide theory of oxygen toxicity and assesses the protective role played by superoxide dismutase. The validity of the theory would, however, be greatly strengthened if it were possible to obtain a mutant that has lost the ability to synthesize SOD and determine its ability to survive in the presence of oxygen. To accomplish this, *E. coli* mutants were isolated that were unable to synthesize either catalase or peroxidase but retained their ability to induce MnSOD as well as mutants that were incapable of synthesizing all three enzymes (15). All the mutants were found to be extremely sensitive to the lethality of oxygen. These data indicate that both O_2^- and H_2O_2 are important in mediating the toxicity of oxygen but that the rate of killing in the absence of MnSOD is at least 7 times greater than in the absence of catalase alone (15).

RECENT DEVELOPMENTS

Regulation of MnSOD is independent of the inducible DNA repair system (SOS).

Oxyradicals generated by paraquat have been shown to induce the SOS system in *E. coli* (16). This effect is oxygen dependent. Paraquat is also known to be toxic and mutagenic (8) and to induce MnSOD (13, 14). It is postulated that increased production of O_2^- by paraquat leads to DNA damage/mutation which in turn induces the SOS system as well as the MnSOD. Thus induction of SOS and SOD can be viewed as a defense mechanism against increased intracellular oxyradicals. These findings raised the possibility that the gene encoding the MnSOD is an SOS gene. This hypothesis was tested by studying the induction of MnSOD in several strains of *E. coli* with different mutations in *recA* and *lexA* (17). The data clearly demonstrate that the induction of SOS and MnSOD by oxyradicals are not co-regulated, and that they represent independent lines of defense against increased intracellular oxyradicals.

Induction of MnSOD by nalidixic acid.

Nalidixic acid was noted to cause a significant increase in MnSOD biosynthesis in *E. coli* (18). The stimulatory effect of nalidixic acid is not due to its ability to inhibit DNA gyrase nor to increase the intracellular flux of O_2^-, but rather due to its ability to chelate Fe^{2+} (18). This observation led to the finding of the role of iron in the regulation of MnSOD (19).

Role of iron in the regulation of MnSOD.

Addition of iron has a repressive effect on the synthesis of MnSOD under normal or inducing conditions (i.e. in presence of paraquat) (19). Addition of manganese to cultures already derepressed for MnSOD biosynthesis causes a further increase in the amount of active enzyme, however, this effect is also abolished by the addition of iron. Removal of metals from the growth medium by Chelex-100 also derepresses the synthesis of MnSOD but repletion of the medium with iron abolishes this effect. Chelators specific for Fe^{2+}, 2,2'-dipyridyl and 1,10-phenanthroline, cause a 5-to 7-fold increase in MnSOD. These results suggest that ferrous iron plays a regulatory role in the biosynthesis of MnSOD.

Induction of MnSOD in the absence of dioxygen.

The observation that MnSOD can be induced in *E. coli* by ferrous iron chelators, in the complete absence of dioxygen (19), led to the proposal that the regulatory repressor that controls the synthesis of MnSOD is an iron-containing protein. A schematic model is proposed to accommodate previous as well as present findings (Fig. 1). In this model, the control for MnSOD biosynthesis is envisioned as being a negative control operon, where the regulatory protein (RP) exists in either an active or an inactive state. The active repressor state has Fe^{2+} associated with it (RP-Fe^{2+}), while inactive repressor has either Fe^{3+} (RP-Fe^{3+}) or no iron (RP) associated with it. O_2^- plays a role in increasing biosynthesis of MnSOD by changing the valence of the iron associated with the repressor protein. When intracellular flux of O_2^- is increased by an increased oxidative metabolism or by redox-cycling compounds, it acts to oxidize the Fe^{2+} in the repressor protein. Other oxidants may also oxidize RP-Fe^{2+} to RP-Fe^{3+}. The RP-Fe^{3+}

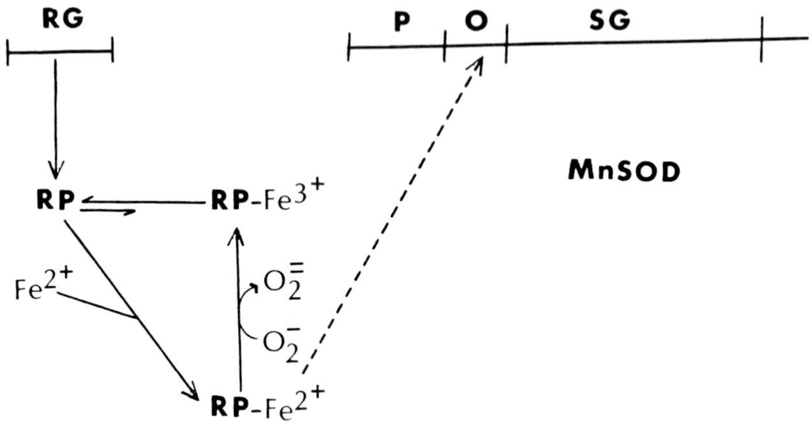

Fig. 1. Schematic model for the regulation of MnSOD by oxygen, superoxide radical, and iron chelators. RG, regulatory gene; P, promoter; O, operator; SG, structural gene (MnSOD); RP, apo-repressor protein (inactive); RP-Fe^{3+}, ferric repressor (inactive); FP-Fe^{2+}, ferrous repressor (active).

is an inactive form of the repressor, which may undergo a conformational change and lose the metal. The repressor is now inactive and MnSOD is synthesized. Under anaerobic contitions, where there is no superoxide or other oxidants to oxidize the Fe^{2+}, the repressor protein is always active and no MnSOD is made. However, the increased synthesis of MnSOD by chelators and by removal of iron from the medium under anaerobic conditions can be explained by Fe^{2+} being unavailable for binding with the repressor protein and therefore MnSOD is synthesized even in the absence of O_2, O_2^- or other oxidants. The increased synthesis of MnSOD, aerobically, in the iron-depleted medium or in the presence of chelators can also be explained by the formation of an inactive repressor protein lacking Fe^{2+} (RP).

Recent studies using ^{59}Fe demonstrated the presence of ^{59}Fe-protein under repressed conditions but not under derepressed conditions (unpublished results). Further studies on the regulatory gene are in progress.

Induction of superoxide dismutase in other procaryotes.

The effects of oxygen, paraquat and 2,2'-dipyridyl on the synthesis of SOD in various Gram-positive and Gram-negative bacteria were examined in order to determine whether or not the control of SOD biosynthesis in other procaryotes is similar to that of *E. coli* (20). Results indicate that the MnSODs from several Gram-positive bacteria, namely *Streptococcus faecalis*, *Staphylococcus epidermidis*, and *Listeria monocytogenes*, are non-inducible. Furthermore, FeSOD is inducible in two Gram-negative organisms, *Alcaligenes faecalis*, and *Pseudomonas aeruginosa*, which is different from the constitutive FeSOD in *E. coli*. Induction of MnSOD in other members of the Entrobacteriacae family is similar to that of *E. coli* (20, submitted for publication). These data clearly indicate that the regulatory mechanism(s) over MnSODs and FeSODs are not universal, and that the complete story has to await further investigation.

REFERENCES

1. Fridovich I (1978) Science 201:875-880
2. Hassan HM, Fridovich I (1980) In: Jacoby WB (ed) Enzymic Basis of Detoxication, Vol.1. Academic Press, New York, pp 311-332
3. Touati D (1983) J Bacteriol 155:1078-1087
4. Sakamoto H, Touati D (1984) J Bacteriol 159:418-420
5. Hassan HM, Fridovich I (1977) J Bacteriol 129:1574-1583
6. Dougherty HW, Sandowski SJ, Baker EE (1978) J Biol 253:5220-5223
7. Clare DA, Blum J. Fridovich I (1984) J Biol Chem 259:5932-5936
8. Moody CS, Hassan HM (1982) Proc Natl Acad Sci USA 79:2853-2859
9. Hassan HM (1981) In: Bloch K, Bolis L, Tosteson DC (eds) Membranes, Molecules, Toxins and Cells. John Wright, PSG Inc, Boston, pp 123-133
10. Gregory EM, Fridovich I (1973) J Bacteriol 114:543-548
11. Hassan HM, Fridovich I (1977) J Bacteriol 130:805-811
12. Hassan HM, Fridovich I (1977) J Bacteriol 132:505-510
13. Hassan HM, Fridovich I (1979) Arch Biochem Biophys 196:385-395
14. Hassan HM, Fridovich I (1977) J Biol Chem 252:7667-7672
15. Hassan HM, Fridovich I (1979) Rev Infect Dis 1:357-367
16. Brawn, K, Fridovich I (1985) J Biol Chem 260:922-925
17. Hancock LC, Hassan HM (1985) J Biol Chem (in press)
18. Hassan HM, Moody CS (1984) FEMS Microbiol Lett 25:233-236
19. Moody CS, Hassan HM (1984) J Biol Chem 259:12821-12825
20. Schiavone JR, Hassan HM (1985) Abst Am Soc Microbiol K67

Superoxide and Superoxide Dismutase in Chemistry, Biology and Medicine; G. Rotilio, editor

INDUCTION OF SUPEROXIDE DISMUTASES IN ESCHERICHIA COLI B BY METAL CHELATING AGENTS

SHIRLEY Y. R. PUGH[1] AND IRWIN FRIDOVICH
Duke University Medical Center, Durham, North Carolina 27710, U.S.A.

SUMMARY

The MnSOD[2] of E. coli B is induced by chelating agents, paraquat and Mn(II). The actions of Mn(II) and paraquat were synergistic and dependent upon the presence of O_2. Chelating agents induced the accumulation of the MnSOD, even in the absence of dioxygen, and simultaneously suppressed the FeSOD. A mechanism which incorporates autogenous repression by the apoenzymes, but not by the holoenzymes, is proposed to explain the control of the biosynthesis of these enzymes in E. coli.

E. coli produces two homodimeric SODs,[2] one of which contains iron (1-3) and the other manganese (4,5). Anaerobically-grown cells contain FeSOD, while aerobically-grown cells contain FeSOD, MnSOD and a hybrid SOD (6). Conditions which increased intracellular production of O_2^- induced biosynthesis of MnSOD (6-9). This suggested that FeSOD is constitutive in E. coli, whereas MnSOD is under repression control. Subsequent observations involving induction of MnSOD by iron starvation (10) or by enrichment of the medium with Mn(II) (11) suggested greater complexity in the control of SOD biosynthesis. Indeed the anaerobic induction of MnSOD biosynthesis by iron starvation or by chelating agents (10,12) indicates that O_2^- is not essential for MnSOD biosynthesis. We will now describe some of the pertinent data and consider alternative explanations.

MATERIALS AND METHODS

E. coli B B_{12}^- (ATCC 29682) were grown in VB (13) or TSY media (11). After 16 hr at 37°C, cells were washed three times in 5 mM K_{pi}, 1.0 mM EDTA, pH 7.8, at 4°C and were then suspended in ~1.0 ml of this buffer. The washed cells were lysed by two passages through a French Pressure Cell at 20,000 lb/in^2. Lysates were clarified at 13,000 x g for 10 min at 4°C. SOD was assayed as previously described (14). Protein was assayed colorimetrically (15). Cell suspensions were clarified in 0.33 M NaOH at 90°C for 90 min, prior to assay

---

1 Present address: A.E.R.E., Harwell, Didcot, Oxfordshire, England OX11 ORA.
2 Abbreviations: SOD, superoxide dismutase; MnSOD, manganese-containing superoxide dismutase; FeSOD, iron-containing superoxide dismutase; TSY, trypticase-soy-yeast extract; VB, Vogel Bonner, PQ, paraquat; 8-HQ, 8-hydroxyquinoline; K_{pi}, potassium phosphate; EDTA, ethylenediaminetetraacetate; DTPA, diethylenetriaminepentaacetate.

for protein content. In all cases bovine serum albumin was used as the protein standard.

SOD electromorphs were separated on 7% polyacrylamide gels (16) which were stained for activity (17). Linear scanning densitometry at 560 nm was used to estimate the relative amount of each electromorph.

RESULTS

Induction by Chelating Agents

Addition of 8-HQ, EDTA or of o-phenanthroline, to cells growing in VB or TSY media, caused first an increase and then a decrease in SOD content, as shown in fig. 1. DTPA or α,α'-dipyridyl were similarly active; whereas the non-

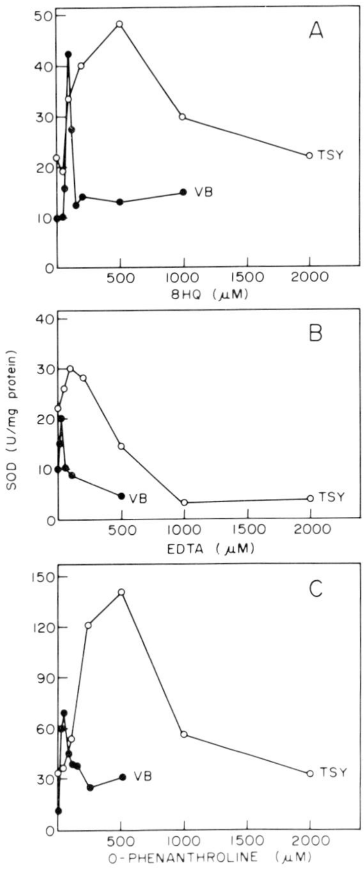

Fig. 1. Induction of SOD in aerobic TSY and VB media. 8-HQ (A), EDTA (B), or o-phenanthroline (C) was added to VB (●) or TSY (O) medium at the time of inoculation with *E. coli* B. Cells were harvested in the stationary phase of growth (after 16 h at 37°C), lysed, extracted, and assayed for total SOD and protein content. The specific activity of SOD in the extracts is here graphed as a function of the concentration of the chelating agent in the medium.

chelating analogue m-phenanthroline had no effect. The range of chelator concentration over which increased SOD content was seen was greater in TSY, then in VB medium, possibly due to the metal-buffering effect of the many natural chelating agents present in the complex TSY medium.

Fe(II) or Mn(II), added to VB medium to 100 μM, caused an approximate doubling of the SOD content of E. coli and when present simultaneously their effects were additive. PQ, at 10 μM in VB medium, only modestly increased the SOD content of cells grown in this medium, but exhibited a marked synergism with Mn(III), as shown in fig. 2.

Impressive inductions were achieved when metals, PQ and chelating agents were used in combination. Thus, E. coli grown in VB enriched with 100 μM Fe(II),

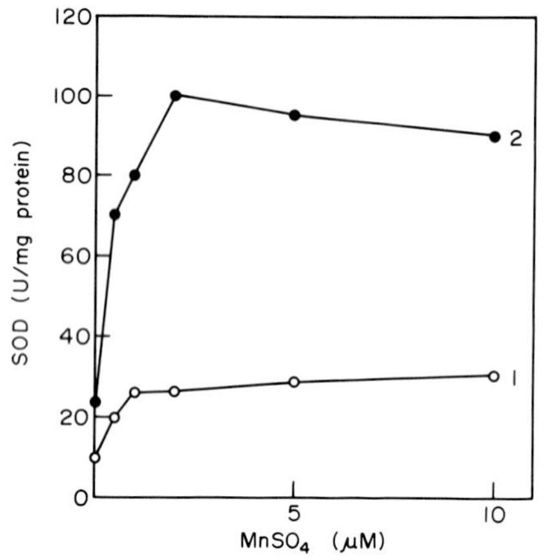

Fig. 2. Effect of PQ on SOD induction by Mn(II). E. coli B cells were grown in aerobic VB medium containing various concentrations of Mn(II) with (●) and without (O) 10 μM PQ.

100 μM Mn(II), 10 μM PQ and 100 μM 8-HQ, yielded extracts containing 320 units SOD/mg protein; which represents a 30-fold increase over the SOD content of cells grown in the absence of these additives. Cells inoculated directly into such supplemented medium did not grow well, giving $A_{600\ nm}$ of 0.22 after 12 hrs at 37°C, which is to be compared with $A_{600\ nm}$ = 1.2 seen with the simple VB medium. In contrast, when these compounds were added to mid-log cultures, they caused a profound increase in SOD content; without noticeable growth inhibition. This is shown in fig. 3.

Anaerobic Induction of MnSOD

E. coli grown in anaerobic TSY media containing progressively higher contents of 8-HQ exhibited first decreased and then increased SOD content. Polyacrylamide gel electrophoresis of extracts of these cells, followed by activity-staining demonstrated anaerobic induction of MnSOD by 8-HQ, as shown in fig. 4. Thus, gel 1 exhibited only FeSOD as would be expected for anaerobically grown cells; whereas gels II and III, representing cells grown in 0.5 and 1.0 mM 8-HQ respectively, exhibited more MnSOD than FeSOD. Mn(II) at 100 μM, without or

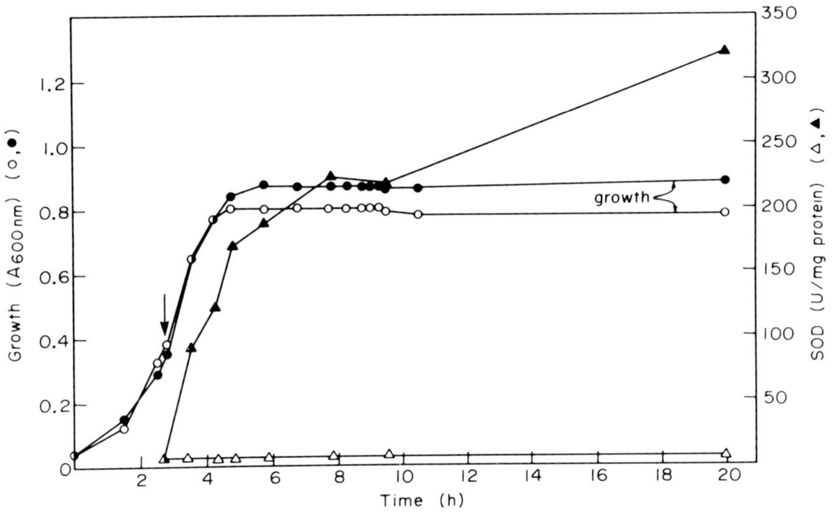

Fig. 3. Effect of Mn(II), Fe(II), 8-HQ, and PQ on growth and SOD content of *E. coli* B. Cells were grown aerobically in duplicate in VB medium. After 165 min of growth, 100 μM Mn(II), 100 μM Fe(II), 100 μM 8-HQ, and 10 μM PQ were added one culture. Growth was monitored by absorbance, and samples were taken at intervals for extraction and assay of specific SOD activity. Symbols: O,Δ, no additions to medium; ●,▲, with additions (components were added at the time point indicated by the arrows).

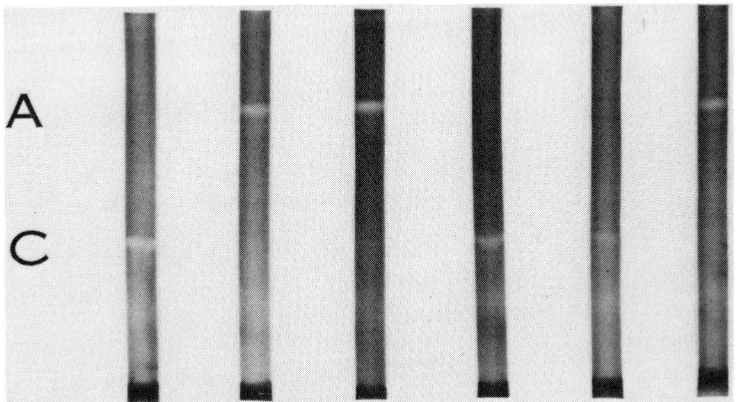

Fig. 4. Effect of Mn(II) and 8-HQ on SOD isozymes in anaerobically grown *E. coli* B cells. The bacteria were anaerobically grown in TSY medium modified as indicated below. Soluble extracts were prepared and analyzed for protein. Samples containing 50 μg of protein were applied to each gel, and after electrophoresis, the gels were stained for SOD activity. A, MnSOD band; C, FeSOD band. Gels: I, no additions; II, 500 μM 8-HQ; III, 1,000 μM 8-HQ; IV, 100 μM Mn(II); V, 100 μM Mn(II) plus 100 μM 8-HQ; VI, 100 μM Mn(II) plus 500 μM 8-HQ.

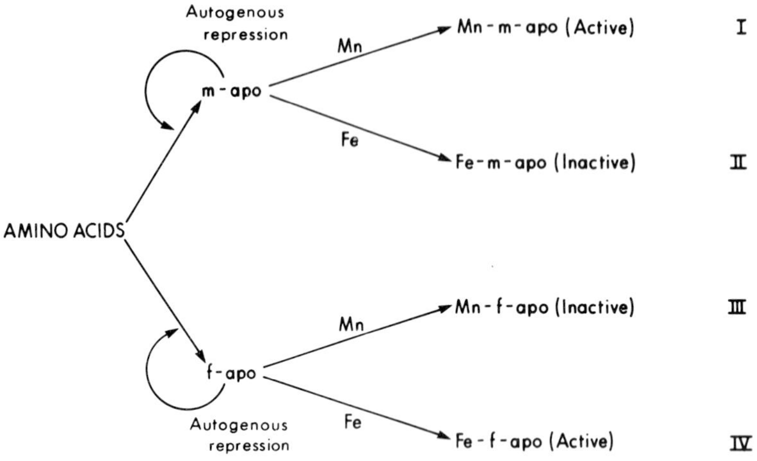

$$Mn(II) + O_2^- + 2H^+ \rightleftharpoons Mn(III) + H_2O_2$$
$$Fe(III) + O_2^- \rightleftharpoons Fe(II) + O_2$$

Fig. 5. Effects of metals on the biosynthesis of SODs--a proposal. Abbreviations: m-apo, apoprotein of MnSOD; f-apo, apoprotein of FeSOD. The valences of the metals are not shown. It is assumed that Fe(II) binds to both m-apo and f-apo more avidly than Mn(II), but not more avidly than Mn(III). The reactions illustrated at the bottom indicate why O_2^- increases the availability of Mn(III).

with equimolar 8-HQ, did not cause induction of MnSOD (gels IV and V). When 8-HQ was present in a 5-fold molar excess over Mn(II) MnSOD was again evident (gel VI).

DISCUSSION

Induction of MnSOD by nutritional restriction of iron or by chelating agents, can be explained on the basis of a repressor which is effective only when it contains bound iron (10). However, the biphasic induction seen when a range of chelator concentrations were examined (fig. 1) the synergism between Mn(II) and PQ (fig. 2) and the dioxygen-dependence of the induction of MnSOD by Mn(II) (11) are not readily accommodated by this proposal. Fig. 5 schematically presents an alternative hypothesis.

We propose that the apoSODs act as autogenous repressors; whereas the active holoenzymes do not. This provides for the very impressive inductions which can be seen under appropriate condition, while preventing futile overproduction of apoenzymes, which cannot be completed due to shortages of the necessary metals. We propose that manganese and iron compete for occupancy of the active sites of the apoenzymes, but that only manganese can impart activity to the apoMnSOD and only iron to the apoFeSOD. This proposal is in accord with the results of studies of metal removal and replacement performed with the isolated enzymes (18-23). We further propose that these metals are bound in the following order of affinities: Mn(III)>Fe(II)>Mn(II)>>0. Mn(II) is oxidized to Mn(III) by O_2^- (24-27) and in Staphylococcus aureus imposition of aerobic conditions caused oxidation of cellular Mn(II) to Mn(III) (28). In contrast Fe(III), whether in the form of an EDTA complex or in various methemes, is reduced by O_2^-. Hence, iron will be Fe(II), aerobically or anaerobically, whereas manganese will be Mn(II) or Mn(III), depending on whether or not O_2^- is being produced.

This proposal can account for the observed inductions of SOD in E. coli, as follows: a. Adding Mn(II) accelerates the conversion of apoMnSOD to the corresponding holoenzyme. This lowers the concentration of the repressing apoenzyme and so derepresses biosynthesis of this apoenzyme; b. Induction of MnSOD by Mn(II) was dioxygen-dependent, because Mn(II) cannot compete with Fe(II) and must be converted to Mn(III) by O_2^- before it can do so. c. PQ and Mn(II) were synergistic because PQ increased the intracellular flux of O_2^-, which converted a greater fraction of the total manganese into Mn(III), which competes favorably with Fe(II) for the apoMnSOD; d. Chelating agents which exhibit relatively greater affinity for iron, than for manganese, diminish the extent to which Fe(II) competes with Mn(III) or Mn(II) for the apoMnSOD and thus increase the amount of active MnSOD produced. Indeed, the chelating agents so far diminished the availability of Fe(II) that Mn(II), present anaerobically, could compete favorably with it. Hence chelating agents allowed MnSOD to be made anaerobically and simultaneously decreased cell content of FeSOD.

The proposal in fig. 5 incorporates autogenous repression by the apoSODs, but not by the metal substituted SODs. A further refinement would be to suppose that apoSODs and incorrectly substituted metallo-SODs all acted as autogenous repressors, while the active holoenzymes did not. This could, if correct, prevent the accumulation of inactive, incorrectly-substituted SODs such as Fe-m-apo and Mn-f-apo. It will now be necessary to subject the several aspects of this proposal to experimental testing.

ACKNOWLEDGEMENTS

This work was supported by Public Health Service research grants GM-10287 from the National Institute of General Medical Sciences, and BRSG S07 RR05405 from the Biomedical Research Support Grant Program, Division of Research

Resources, National Institutes of Health, and by research contract DAAG29-82-K-0020 from the Department of the Army, Research Triangle Park, N.C.

REFERENCES

1. Yost FJ Jr, Fridovich I (1973) J Biol Chem 248:4905
2. Slykehouse TO, Fee JA (1976) J Biol Chem 251:5472
3. Stallings WC, Pattridge KA, Strong, RK, Ludwig ML (1984) J Biol Chem 259:10695
4. Keele BB Jr, McCord JM, Fridovich I (1970) J Biol Chem 245:6176
5. Steinman HM (1978) J Biol Chem 253:8708
6. Hassan HM, Fridovich I (1977) J Biol Chem 253:8143
7. Hassan HM, Fridovich I (1976) J Bacteriol 129:1574
8. Hassan HM, Fridovich I (1977) J Biol Chem 252:7667
9. Hassan HM, Fridovich I (1979) Arch Biochem Biophys 196:385
10. Moody CS, Hassan HM (1984) J Biol Chem 259:12821
11. Pugh SYR, DiGuiseppi JL, Fridovich I (1984) J Bacteriol 160:137
12. Pugh SYR, Fridovich I (1985) J Bacteriol 162:196
13. Vogel HJ, Bonner DM (1956) J Biol Chem 218:97
14. McCord JM, Fridovich I (1969) J Biol Chem 244:6049
15. Lowry OH, Rosenbrough NJ, Farr AL, Randall RJ (1951) J Biol Chem 193:265
16. Davis BJ (1964) Ann NY Acad Sci 121:404
17. Beauchamp C, Fridovich I (1971) Anal Biochem 44:276
18. Puget K, Lavelle F, Michelson AM (1977) In: Michelson AM, McCord JM, Fridovich I (eds) Superoxide and Superoxide Dismutases. Academic, New York, London, p. 139.
19. Ose DE, Fridovich I (1979) Arch Biochem Biophys 194:360
20. Brock CJ, Harris JI (1977) Biochem Soc Trans 5:1537
21. Brock CJ, Walker JE (1980) Dev Biochem 11A:237
22. Yamakura F, Suzuki K (1980) J Biochem 88:191
23. Kirby T, Blum J, Kahane I, Fridovich I (1980) Arch Biochem Biophys 201:551
24. Homann PH (1965) Biochemistry 4:1902
25. Kono Y, Takahashi MA, Asada K (1976) Arch Biochem Biophys 174:454
26. Archibald FS, Fridovich I (1982) Arch Biochem Biophys 214:452
27. Cabelli DE and Bielski BHJ (1984) J Phys Chem 88:3111
28. Ezra FS, Lucas DS, Russell AF (1984) Biochim Biophys Acta 803:90
29. Ilan YA, Czapski G (1977) Biochim Biophys Acta 498:386
30. Sutton HC, Roberts PH, Winterbourn CC (1976) Biochem J 155:503
31. Lynch RE, Thomas JE, Lee GR (1977) Biochemistry 16:4563
32. Ferradini C, Foos J, Gilles L, Haristoy D, Pucheault J (1978) Photochem Photobiol 28:851
33. McCord JM, Fridovich I (1968) J Biol Chem 243:5753

Superoxide and Superoxide Dismutase in Chemistry,
Biology and Medicine; G. Rotilio, editor

SUPEROXIDE DISMUTASE MUTANTS OF *ESCHERICHIA COLI*.

DANIELE TOUATI and ANTOINE CARLIOZ
Génétique et Membranes, Institut Jacques Monod, CNRS, Université Paris VII, 2 Place Jussieu, 75251 Paris cedex 05, France.

INTRODUCTION

Superoxide dismutase is the first line of defense against oxygen toxicity; it has been found in almost all organisms exposed to oxygen. Therefore, the question arises of the possibility of life without superoxide dismutase.

E. coli, a facultative anaerobic bacterium contains two superoxide dismutases (1,2), an iron enzyme (FeSOD) and a manganese enzyme (MnSOD); a hybrid (3) enzyme is also found in the cells. The iron isoenzyme is expressed at the same rate in aerobic and anaerobic conditions; the manganese enzyme is induced by oxygen or superoxide radicals (4).
We previously described the cloning of the MnSOD gene (*sodA*, 5) and the FeSOD gene (*sodB*, 6). Using these clones, we attempted to obtain mutants to study the role and regulation of the SODs. The isolation and characterisation of such mutants and preliminary studies of their physiology and regulation are described.

CONSTRUCTION OF IRON OR MANGANESE SUPEROXIDE DISMUTASE MUTANTS

Rationale

- Insertion of a Mu transposon in a plasmid carrying a *sod* gene, as described (7) and screening for insertions in the *sod* gene.
- Replacement of chromosomal DNA by the segment carrying the mutated gene done in a *polA* strain as described (8)
- The mutated gene is then transduced for further studies in an appropriate strain, using Mu transposon markers for selection.

Material

The plasmids used both derived from pB R322.
pDT1-5 (5) carries the *sodA* gene on a 4.8 kb bacterial DNA fragment and the *bla* gene (Amp^R) on the vector; production of SOD in the pDT1-5 bearing clones is about 4 fold that in wild type.
pHS1-7, a sub clone of pHS1-0 (6), carries the *sodB* gene on a 10 kb bacterial DNA fragment and the *bla* gene (Amp^R) on the vector. Production of SOD in pHS1-7 bearing clones is about 6 fold that in wild type.

The transposons : two mini-Mus were used in which most of the Mu functions are defective; they can transpose only if complemented by wild type Mu; they carry the Tn9 transposon (Cm^R) and either the neomycine phosphotransferase (NPT) structural gene (9) or lactose operon structural genes (unpublished, derived from the former). Upon insertion they are able to create gene fusions.

Construction.

Insertions into plasmids were selected by Cm^R. Among these, insertions into *sod* gene were selected as clones in which SOD level was back to the wild type level (immunoprecipitation, SOD measurements).

A fusion between the *sodB* and the NPT gene (*sodB*::Km) and a fusion between the *sodA* and the β- galactosidase gene (*sodA*::*lac*) were selected for the studies described below. The Tn9 transposon was deleted from the *sodB*::Km fusion.

Mutations were introduced onto the bacterial chromosome by allele exchange (8) and transduced into a prototrophic strain deleted for the lactose operon.

CHARACTERISATION OF SOD A AND SOD B MUTANTS

A *sodA* mutant candidate (Cm^R, Lac^+, Amp^S) and *sodB* mutant candidate (Km^R, Amp^S) were assayed for SOD activity. No MnSOD activity was found in the *sodA* mutant, no FeSOD was found in *sodB* mutants, as shown by activity revelation on an electrophoresis gel (Fig 1) or by "western-blotting" (not shown). Levels of MnSOD or FeSOD in *sodB* or in *sodA* were as in wild type.

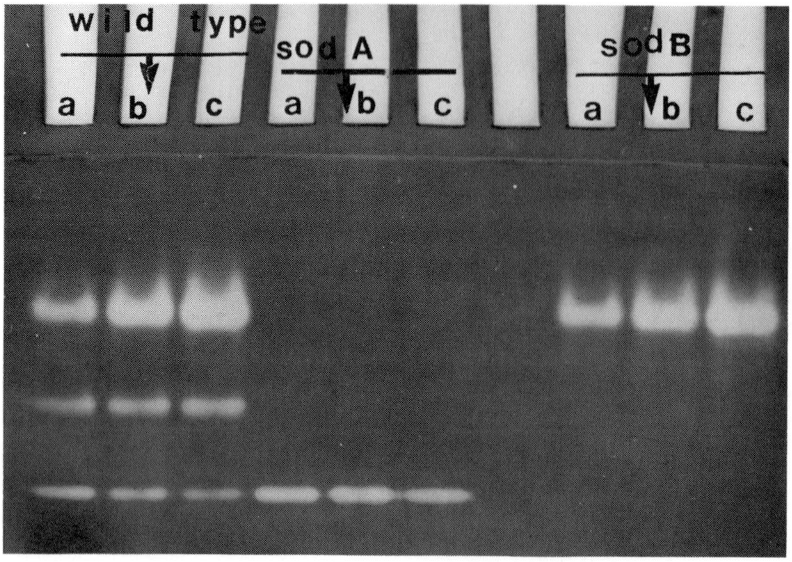

Fig. 1. SOD activity in *E. coli* mutants.
Paraquat : lane a : 0M; lane b : 5 x 10^{-6}M; lane c : 5 x 10^{-5}M.

In strains grown anaerobically no MnSOD is observed; only FeSOD is present in wild type and in *sodA* mutant; the *sodB* mutant has no detectable level of superoxide dismutase activity.

PHYSIOLOGICAL PROPERTIES OF SOD A AND SOD B MUTANTS

- growth in rich medium in the presence of paraquat

Growth in LB (rich) medium was assayed in the presence of paraquat (10^{-4}M, 5 x 10^{-5}M, 5 x 10^{-6}M). No difference was observed between the *sodB* mutant and wild type; thus lack of FeSOD does not affect growth in rich medium. A slight increase of sensitivity was observed in the *sodA* mutant.

- growth in minimal medium in the presence of paraquat (Fig. 2).

In minimal medium the absence of FeSOD and, even more, the absence of MnSOD drastically affect growth.

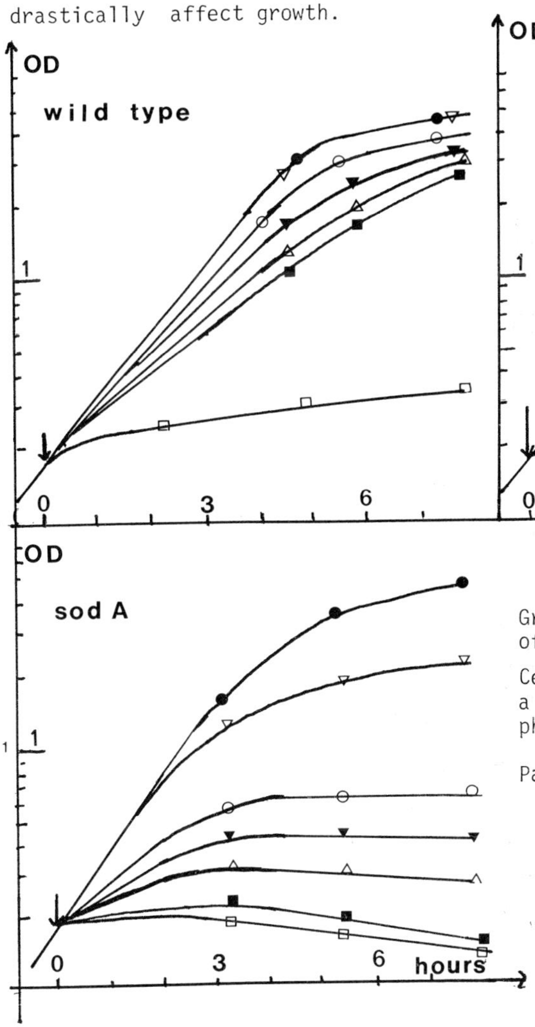

Fig.2.

Growth in minimal medium in presence of paraquat.

Cells were inoculated at OD 0.03 from a preculture stopped in exponential phase.Paraquat was added at OD about 0.2

Paraquat: ● 0M, ▽ 10^{-7}M, ○ 3 10^{-7}M, ▼ 10^{-6}M, △ 3 10^{-6}M, ■ 10^{-5}M, □ 3 10^{-5}M.

REGULATION OF MnSOD SYNTHESIS

Induction of β-galactosidase in the *sodA::lac* fusion was assayed in various conditions. In Fig. 3 is represented the induction of β-galactosidase in rich medium in the presence of paraquat. The induction observed demonstrates that β-galactosidase is effectively under the same control as MnSOD. The degradation observed presumably results from a peculiar sensitivity to paraquat of the MnSOD-β-galactosidase hybrid protein and is under study.

In Fig. 4 is represented the induction of β-galactosidase of the MnSOD overproducer plasmid. The weaker induction observed in the presence of excess MnSOD demonstrates that MnSOD acts on its own induction. Similarly, almost no increase of MnSOD was observed in a wild type strain carrying an MnSOD overproducer when grown in the presence of paraquat (x 1.5 for 5.10^{-5}M paraquat).
On the other hand, no effect of FeSOD overproduction on β galactosidase induction is observed.

Fig.3. Fig.4.

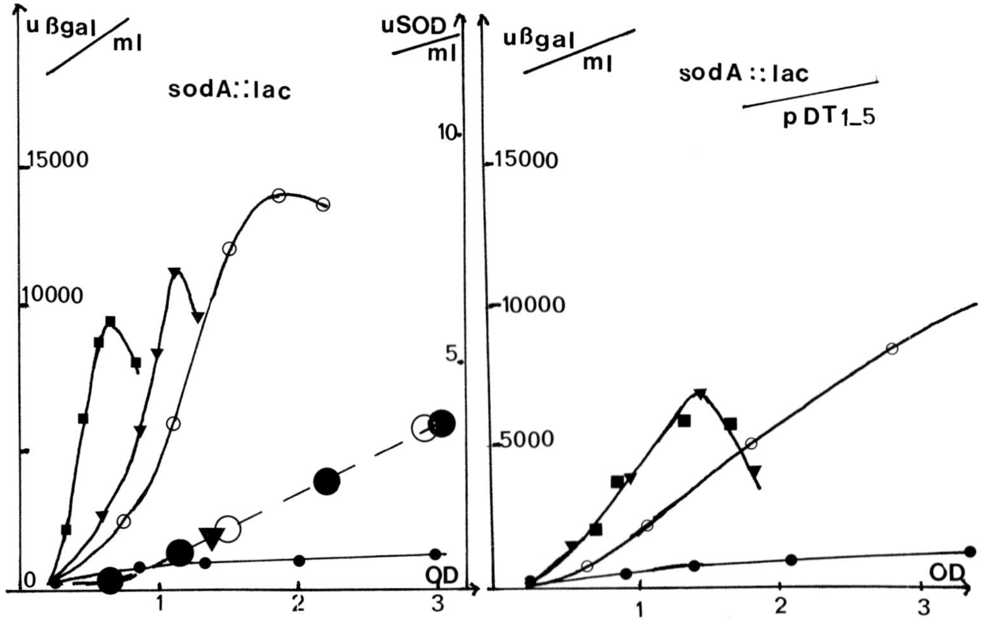

Figs. 3 and 4. β-galactosidase induction by paraquat in a *sodA::lac* fusion in absence (Fig. 3) or in presence (Fig. 4) of a MnSOD overproducer plasmid (pDT1-5). Paraquat ● 0M, ○ 5.10^{-6}M, ▼ 5.10^{-5}M, ■ 10^{-4}M. Large symbols in Fig. 3 represent SOD measurements (FeSOD).

CONSTRUCTION, CHARACTERISATION AND PHYSIOLOGY OF A SOD A SOD B DOUBLE MUTANT.

Construction of a double mutant was attempted by transduction of the *sodB*::Km fusion into a *sodA* mutant, selecting for Km^R. Plating immediately after transduction was done in anaerobic and aerobic conditions. Transductants (Cm^R Km^R) were obtained in both conditions and after further study were shown to be identical.

The presumptive *sodA sodB* mutant was assayed for superoxide dismutase activity in crude extracts by revelation of SOD activity on electrophoretic gels, total SOD activity measurements, and "western-blotting". No SOD activity could be detected by any of these methods.

The physiological properties of the *sodA sodB* mutant, devoid of superoxide dismutase, are summarized below.

Grown aerobically in rich medium :

- it has the same level of catalase as the wild type
- it shows slower growth and a very high sensitivity to superoxide radicals and hydrogen peroxide.

It does not grow aerobically on glucose minimal medium, this growth is restored either - (i) in anaerobiosis, (ii) by adding the 20 amino acids, (iii) by a MnSOD overproducer plasmid, (iv) by a FeSOD overproducer plasmid.

DISCUSSION

We have shown that :

- MnSOD acts on its own biosynthesis (i) very weak induction of MnSOD by paraquat in a wild type strain carrying a MnSOD overproducer plasmid, (ii) lower induction of β-galactosidase in a *sodA::lac* fusion in the presence of a MnSOD overproducer plasmid.
- FeSOD does not act on MnSOD induction (i) same level of MnSOD in *sodB* mutants and in wild type, (ii) same induction of β-galactosidase in a *sodA::lac* fusion in the presence of a FeSOD overproducer plasmid, (iii) no superoxide dismutase activity in *sodB* mutant grown anaerobically.
- MnSOD does not act on FeSOD expression - same level of FeSOD in *sodA* mutants and in wild type.

Recently various models have been proposed for MnSOD and FeSOD regulation (10,11) to account for the effects of metal salts on SOD biosynthesis. An iron protein was proposed by Hassan and Moody (10) to regulate MnSOD biosynthesis. Such a model does not easily explain the observed effect of MnSOD on its biosynthesis. Pugh and Fridovich (11) have proposed as a regulation model an autogenous repression by the apoenzyme. If this model is compatible with the effect of MnSOD on its biosynthesis, it would hardly explain the induction of β-galactosidase by superoxide radicals in the MnSOD-β-galactosidase protein fusion. The study of various fusions and the isolation of regulation mutants should allow progress in the understanding of these mechanisms.

Physiological studies of the double mutant are full of promise and should help to determine the type of biological damage and the targets of superoxide radicals (or derivatives). The inability to grow in glucose minimal medium is in agreement with previous work (12,13), showing that certain aminoacids (or precursors) were targets for oxygen toxicity.
The high hydrogen peroxide sensitivity supports the idea that the main toxicity of the superoxide radical would not be a direct effect, but more likely would result from the formation with H_2O_2 of very reactive hydroxyl radicals (14).

Since we have been able to isolate a double mutant, completely devoid of superoxide dismutase activity, we conclude that superoxide dismutase is not essential for aerobic life.
However, the extreme sensitivity of this mutant to oxygen radicals and hydrogen peroxide and its inability to grow on minimal medium, would rapidly counterselect such a mutant in the course of evolution.

REFERENCES

1. Yost FJ (1973) J Biol Chem 248:4905.
2. Keele BB Jr, McCord JM, Fridovic I (1970) J Biol Chem 245:6176.
3. Dougherty HW, Sadowski SJ, Baker EE (1978) J Biol Chem 253:5220.
4. Hassan HM, Fridovich I (1977) J Biol Chem 252:7667.
5. Touati D (1983) J Bacteriol 155:1978.
6. Sakamoto H, Touati D (1984) J Bacteriol 159:418-420.
7. Castilho BA, Olfson P, Casadaban MJ (1984) J Bacteriol 158:488.
8. Gutterson NI, Koshland DE Jr (1983) Proc Natl Acad Sci USA 80:4894.
9. Ratet P, Richaud F (1985) Submitted for publication.
10. Moody C, Hassan H (1984) J Biol Chem 259:12821.
11. Pugh SYR, Fridovich I (1985) J Bacteriol 162:196.
12. Brown OR, Seither RL (1983) Fund Appl Toxicol 3:209.
13. Fee JA, Lees AC, Block PL, Neidhardt FC (1979) In : Caughey WS (eds) Biochemical and Clinical Aspects of Oxygen. Academic Press, New York, pp. 639-641.
14. Halliwell B, Gutteridge JMC (1984) Biochem J 219:1.

© 1986 Elsevier Science Publishers B.V. (Biomedical Division)
Superoxide and Superoxide Dismutase in Chemistry,
Biology and Medicine; G. Rotilio, editor

PARAQUAT-MEDIATED SELECTION FOR MUTATIONS IN A CLONED MANGANESE-SUPEROXIDE DISMUTASE GENE

CRAIG A. BLOCH and FREDERICK M. AUSUBEL

Department of Molecular Biology, Massachusetts General Hospital and Department of Genetics, Harvard Medical School, Boston, Massachusetts 02114, U.S.A.

Introduction:

Paraquat (methyl viologen) is a bipyridyl compound believed to uncouple electron transport from oxidative phosphorylation by undergoing redox cycling within the cell (1). This model for toxicity predicts that paraquat (Pq) would decrease the fraction of reducing equivalents available for divalent electron transport and in the process would increase the specific production of superoxide per equivalent of electrons effectively utilized in the divalent pathway. In wild-type *E. coli*, while both Manganese-superoxide dismutase (Mn-SOD) and catalase are induced on exposure to 0.5 mM Pq, growth rate is not appreciably affected (2). The cloning of the *E. coli* Mn-SOD gene (*sodA*) by D. Touati (3), has made possible the examination of dioxygen's effects on a cell which, in response to a common positive regulatory stimulus, might be expected to overproduce Mn-SOD, relative to catalase and any other gene products coordinately involved in adaptation to oxidant stress. Strains transformed with a multicopy *sodA* plasmid (pDT1.5) have already been shown to grow normally in standard shake cultures in air (*op.cit.*) but we wondered whether such would be the case under conditions which augment oxygen toxicity.

Results:

In developing a rapid method to screen for the presence of the Mn-SOD gene on a multicopy plasmid, we examined the growth of *E. coli* pDT1.5 transformants on Pq. Scored at 24 hours, LB agar plates containing 0.2 mM Pq and 50 μg/ml ampicillin (Ap) which had been inoculated with <200 CFU/plate of DK1/pDT1.5, had no growth or only barely perceptible pinpoint colonies, whereas in the vector carrying positive control (DK1/pHC79) colonies grew to essentially normal size (results not shown). Both DK1/pDT1.5 and DK1/pHC79 grew normally on LB agar containing Ap and no Pq.

We next tested whether we could identify the *sodA* gene by taking advantage of the fact that Pq inhibits the growth of DK1/pDT1.5 on solid medium; we generated mutations in pDT1.5 using transposon Tn*5* insertional mutagenesis (4) and then screened the clones transformed with the Tn*5* mutated plasmids for those with growth on 0.5 mM Pq. The results are shown in Fig. 1. Of the 144 pDT1.5::Tn*5* inserts tested, 24 grew on Pq as the DK1/pHC79 control strain did, yet retained pDT1.5. The insertion sites of 140 of the 144 Tn*5* insertions were determined by restriction endo-nuclease analysis (mutants 55, 57, 59

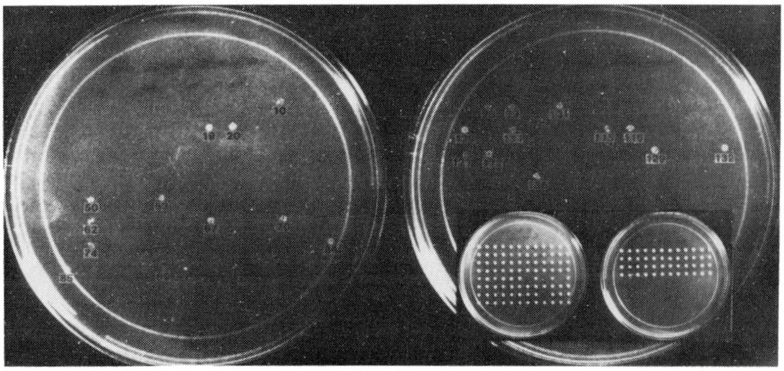

Figure 1: Selection of pDT1.5::Tn*5* mutations on Pq. Saturated cultures in LB broth with Ap and Km selection were diluted $5x10^2$ before spotting onto LB agar containing 50 μg/ml Ap, 50 μg/ml Km and 0.5 mM Pq. The plates were incubated at 37°C for 24 hours and then scored.

and 81 were lost in storage). Those mapping to the plasmid insert are depicted in Fig. 2. Twenty of the 24 Tn*5* inserts which allowed growth on Pq were contiguous and mapped to a 0.6 kb region on the pDT1.5 insert. The remaining four (62, 115, 122 and 136) all mapped to the pDT1.5 vector. None of the 116 Tn*5* inserts not allowing growth on Pq mapped to this 0.6 kb segment. The size of the zone delimited by the 20 contiguous Tn*5* inserts which permit growth on Pq, is close to a reasonable predicted size for the *sodA* transcriptional unit of 0.7 kb (0.62kb coding region for a 205 amino acid structural gene and a 0.1 kb promoter region).

All clones grew normally on control plates containing Ap and kanamycin (Km) selection but no Pq (Fig. 1, inset.) and all clones grew normally on plates containing 0.5 mM Pq in addition to antibiotic selection, in the absence of oxygen (results identical to Fig. 1 inset). Thus Pq had no effect on growth under anaerobic conditions.

The specific activity of SOD in supernatant extracts are depicted for twelve Tn*5* insertions in Fig. 2, inset. Six Tn*5* insertions (numbers 1, 9, 24, 32, 96, and 139) which mapped to selected sites outside of the 0.6 kb presumptive *sodA* gene, all expressed high levels of SOD activity equal to or greater than that of the non-mutated DK1/pDT1.5. In contrast, SOD activity in each of the strains tested which carried Tn*5* inserts allowing growth on Pq (insertions 10, 19, 20, 53, 70 and 116) was approximately equal to that in the negative control, DK1/pHC79.

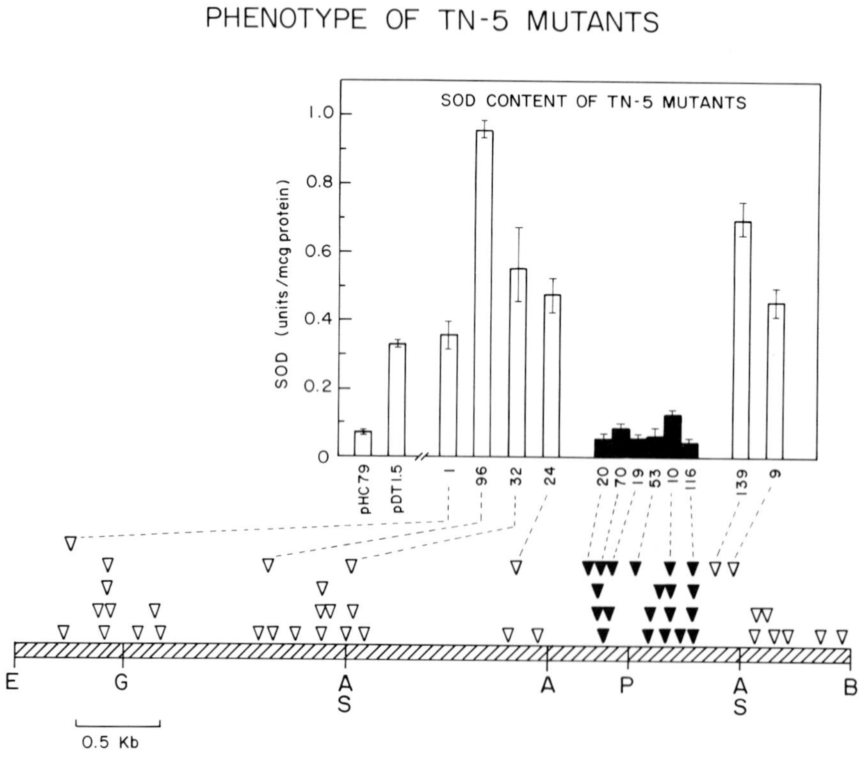

Figure 2: Phenotype of pDT1.5::Tn*5* insertion mutations and SOD levels in transformed *E. coli*. The insertion sites of 20 of the 24 Tn*5* mutations affecting growth on Pq mapped to the pDT1.5 insert and are depicted as shaded wedges. The specific activities for SOD of six clones bearing these inserts are shown in the bar graphs above each site. The insert site of the mutations not affecting growth on Pq are depicted as unshaded wedges with the SOD specific activities for six of them shown above each site. Also shown are the specific activities of non-mutagenized DK1/pHC79 and DK1/pDT1.5.

Discussion:

We have shown that Pq exerts a selection against *E. coli* bearing extra copies of pDT1.5. In addition we have demonstrated that the effect is reversed by mutations within a 0.6 kb segment on pDT1.5 and that these mutations also cause, under standard conditions (in the absence of Pq),a five-fold or greater reduction in total SOD expression by the pDT1.5 transformant. We have not provided evidence concerning SOD expression during exposure to Pq, however. Therefore, our data demonstrate that the Pq effect is dependent on the function associated with a specific 0.6 kb locus, and only strongly suggest that it is mediated by an increased intracellular specific activity of Mn-SOD. They also suggest that our 0.6 kb segment represents, in fact, the location and minimal size of the *sodA* transcriptional unit.

The striking sensitivity to Pq of a strain which overproduces SOD might at first seem paradoxical. Our data might be rationalized if it is assumed that the response to such an oxidative stress requires the coordinated activity of several enzymes. Hassan and Fridovich (2) have found that wild-type *E. coli*, upon exposure to the same concentration of Pq that we used, induces both SOD and catalase, and maintains normal growth. Christman, *et al.* have recently shown the existence of a positive regulatory element (*oxyR*) controlling expression of an anti-oxidant regulon in Salmonella (5). The genes belonging to the regulon which they identified include those coding for Mn-SOD, catalase, peroxidase, glutathione reductase, and five other genes.

In cells bearing extra-chromosomal copies of the Mn-SOD gene on a multicopy plasmid, a reasonable expectation would be for any positive stimulus to an anti-oxidant regulon in *E. coli* to yield a higher ratio of Mn-SOD to other regulon products than in wild-type. The biochemical consequences of such an imbalance in regulon expression might be an escape from the detoxification pathway of harmful reactive by-products. These would be allowed to form through the increased occurance of side-reactions which only become significant when the proper ratios of intermediates are derranged by an imbalance in enzyme expression. For instance, were SOD activity to be disproportionately greater than that of catalase (and/or peroxidase), a larger steady-state pool of hydrogen peroxide would be expected. It has been postulated that in the presence of high concentrations of both superoxide and hydrogen peroxide , the two may react to form hydroxyl radical (6).

The selection exerted by Pq against cloned Mn-SOD in *E. coli* may be generally applicable to the isolation of Mn-SOD genes in related species, and a greater understanding of the mechanism of Pq toxicity in this case may provide insight on the mechanism by which biological systems protect themselves from the toxicity of oxygen.

Acknowledgements:

This work was supported by a grant from Hoechst, AG. We wish to thank D. Touati for her gift of pDT1.5. We also thank N. Olszewski, K. Dunn, D. Marvel and J. Badwey for helpful criticism and discussion.

References:

1. Bus JS, Cagen SZ, Olgaard M, Gibson JE (1976) Toxicology and Appl Pharmacology **35**:501-513
2. Hassan HM and Fridovich I (1977) J Biol Chem **252**:7667-7672
3. Touati D (1983) J Bacteriol **155**:1078-1087
4. de Bruijn FJ, Lupski JR (1984) Gene **27**:131-149
5. Christman MF, Morgan RW, Jacobson FS, Ames BN (1985) Cell **41**:753-762
6. Beauchamp C, Fridovich F (1970) J Biol Chem **245**:4641-4646

Superoxide and Superoxide Dismutase in Chemistry, Biology and Medicine; G. Rotilio, editor

SUPEROXIDE DISMUTASE AND THE PHYSIOLOGICAL RESPONSE TO COPPER IN S. CEREVISIAE

M. A. GRECO, W. MAGNER, D. HRAB, AND D. KOSMAN
Department of Biochemistry, SUNY, Buffalo, NY 14214 USA

INTRODUCTION

Organisms have highly integrated, adaptive responses to stress. As is true of most nutrients, either a deficiency or an excess of copper results in cellular stress. Characterization of the molecular details of such responses and metabolism requires an organism amenable to physiologic, biochemical, and genetic approaches. Clearly, S. cerevisiae fulfils this requirement.

We have investigated the response of a wild-type, $CUP1^R$ strain of S. cerevisiae to the extremes of medium Cu(II) concentrations. Our objective was to establish additional biochemical markers of cellular copper status as well as incipient copper toxicity. In particular, we focused on the expression of SOD-1 in relation to copper level since the level of this copper enzyme does vary with nutrient Cu in other organisms (1), and has been implicated in the defense against Cu-related cellular stress (2, 3).

MATERIALS AND METHODS

The diploid yeast strain used, 435 (A52-2A, F. Sherman), was MATa, trp1-2. Cells were grown in either a standard YNBBD or a synthetic medium prepared with AnalaR (BDH) reagents. In the latter, residual Cu was below detectable limits (50 nM, flameless atomic absorption spectrophotometry). Cell extracts and mitochondria were prepared by homogenization in the presence of glass beads (0.3-0.45 mm). Superoxide dismutase, catalase, glyceraldehyde-3-phosphate dehydrogenase, cytochrome oxidase, and succinic dehydrogenase were assayed using standard procedures. Protein determinations and labelling of protein and RNA followed standard procedures.

RESULTS

Cellular response to Cu-deficiency: superoxide dismutases. The strain of S. cerevisiae used exhibited the resistance to copper characteristic of the $CUP1^R$, multi-copy copper thionein genotype. Although growth of this strain appeared normal when Cu < 50 nM (data not shown), another index of cellular Cu deficiency was evaluated, namely the activity of the Cu/Zn superoxide dismutase, SOD-1. In fact, this enzyme activity did decrease as available medium Cu decreased. However, the total cellular dismutase activity remained essentially unchanged due to a compensating increase in the activity attributable to the Mn-containing SOD, SOD-2 (Table 1). Thus, as demonstrated in other organisms, SOD-1 activity can reflect cellular Cu status and showed this yeast to be Cu-deficient despite its apparent normal growth.

TABLE I

SOD ACTIVITIES IN CU-DEFICIENT AND NORMAL YEAST

Media Cu, nM	Total SOD*	SOD-1	SOD-2
<50**	55	22 (40)	33 (60)
150***	60	55 (92)	5 (8)

* Activities expressed as units/mg soluble protein in mid-log cells

** Value represents detectable limit of flameless atomic absorption

*** Value represents normal media without added Cu

<u>Cellular response to incipient Cu toxicity: superoxide dismutases and cellular protein</u>. Of interest was whether increases in medium Cu above normal levels increased SOD-1 levels further. In fact, addition of 10 μM ⩽ Cu ⩽ 250 μM to aerobic cultures in early- to mid-log growth caused an immediate ($t_{1/2}$ ~ 20 min) 2-fold increase in the amount of cellular SOD-1 activity. A corresponding increase SOD-1 specific activity was also noted. These increases were inhibited by cycloheximide (Fig. 1A, B).

In contrast, when the added metal resulted in medium containing > 200 μM Cu, the <u>specific activity</u> (units/mg soluble protein) increased 5-6 fold (Fig. 1B);

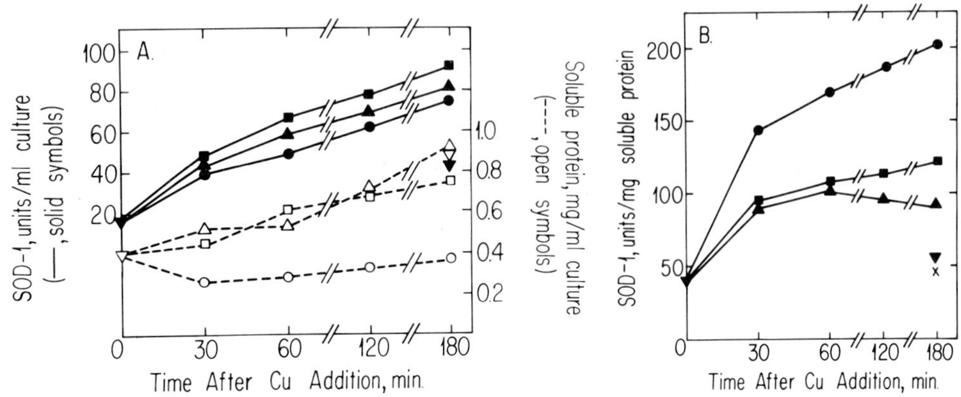

Figure 1. Cu effects on SOD-1 and soluble protein. (A) Total SOD-1 activity (solid symbols) and soluble protein (open symbols) are monitored following addition of Cu at 0.1 (▲ and △), 0.25 (■ and □), and 2 mM (● and ○). Controls are given by ▼ and ▽ (~ 0.2 μM Cu). (B) SOD-1 specific activity is monitored following addition of Cu; concentrations and symbols are as in (A). Value represented by X is after treatment with 0.25 mM Cu in the presence of 0.1 mg/ml cycloheximide.

this large difference was due to a decrease in the measureable soluble cellular protein (Fig. 1A). Thus, while Cu approaching cytotoxic levels appears to increase SOD-1 activity, at the same time it induces a precipitant decline in cellular protein. Soluble protein declines from 32% of total cellular protein in control cells to 6%. Clearly, however, this relative susceptibility is not shared by SOD-1 since its specific activity increased 3-fold more than does the total cellular SOD-1 in cells grown in 2 mM Cu.

Cellular response to incipient Cu toxicity: RNA and protein synthesis. The sensitivity of soluble proteins to Cu was reflected in the effects of Cu on RNA and protein synthesis. Both were inhibited by Cu in a dose- and time-dependent fashion (Fig. 2A, B). RNA synthesis was somewhat more sensitive. Note that these effects occurred at Cu levels at which cellular growth, as monitored by turbidity, appeared normal. In contrast, as measured by total soluble protein, growth was clearly impaired (Fig. 1A). This correlated with the combined effects of Cu on protein stability and synthesis. Again, note that SOD-1 levels increase despite the fact that, overall, the cellular complement of protein decreases and protein (and RNA) synthesis is inhibited. Autoradiographs of gels of *in vivo* ^{35}S-labelled SOD-1 immune precipitates confirmed this observation.

O_2 dependence of Cu toxicity. Analysis of the effects of Cu on both protein synthesis and the level of SOD were evaluated in anaerobic cultures, as well. The inhibition of protein synthesis was unaltered from that illustrated in Figure 2A. In contrast, no induction by Cu of SOD was observed in the absence of O_2.

Effects of Cu on other marker enzymes. Cu had differential effects on four other enzymes selected as markers (Table II). The decrease in GPD reflects the effect of Cu on soluble protein; GPD is a major cytosolic enzyme in yeast. In contrast, total catalase activity increases as does its specific activity. As markers of mitochondrial integrity, succinic dehydrogenase and cytochrome oxidase differ in their response to Cu. Understandably, Cu-deficient cells

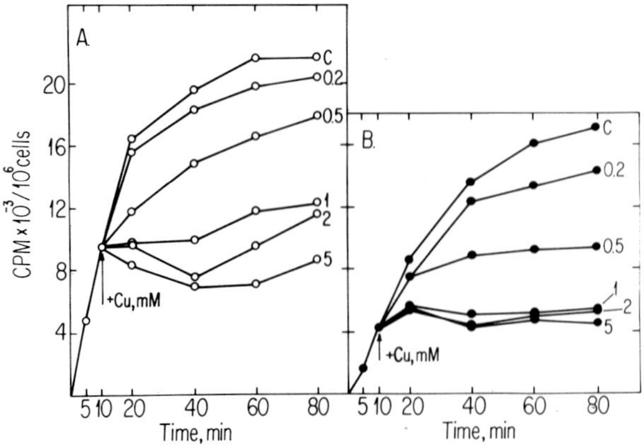

Figure 2. Cu effects on protein and RNA synthesis. In (A) ^{3}H-leu incorporation is followed in control and Cu-treated cells at metal concentration as noted. In (B), RNA synthesis is monitored by ^{3}H-uracil incorporation.

exhibit a decrease in the activity of the latter enzyme, an activity which is also depressed by cytotoxic metal levels. This response appears unrelated to a non-specific perturbation of mitochondrial function since succinic dehydrogenase activity increases at such levels. The response to Cu of total cellular respiration correlates with the decrease in oxidase activity (data not shown).

TABLE II

CU DOSAGE EFFECTS ON SELECTED YEAST ENZYMES*

Media Cu, mM	GPD	Catalase	Succinic Dehydrogenase	Cytochrome Oxidase
$< 5 \times 10^{-5}$	109	60	138	54
1×10^{-3}	100	100	100	100
0.5	3	184	154	84
1.0	3	140	174	25

* Expressed as % of control, 1 μM Cu

CONCLUSION

Cu inhibits protein and RNA synthesis in yeast. This inhibition is not O_2-dependent. This inhibition is not accompanied by significant effect on the protein complement of the cell: protein turnover is not accelerated nor is SOD-1 induced. These latter two phenomena are observed only when O_2 is present. Reasonably, SOD-1 is not induced directly by Cu, nor by "signals" associated with inhibition of macromolecular synthesis. Rather, this induction - and that of catalase - is related to a sequence of events requiring both Cu and O_2, one which, at least in part, involves a significant increase in protein turnover. We hope to understand the chemistry and biochemistry of this metal- and O_2-dependent cellular stress response. (Supported by NIH AM-19708).

REFERENCES

1. Shatzman, A.R., Kosman, D.J. (1978) **Biochim. Biophys. Acta. 544:** 163-179 and references therein.
2. Gregory, E.M., Goscin, S.A., Fridovich, I. (1974) **J. Bacteriol. 117:** 456-460.
3. Naiki, N. (1980) **Plant Cell Physiol. 21:** 775-783.
4. Hamer, D.H., Thiele, D.J., Lemontt, J.F. (1985) **Science 228:** 685-690 and references therein.

Superoxide and Superoxide Dismutase in Chemistry, Biology and Medicine; G. Rotilio, editor

EFFECTS OF CYTOCHROME-c DEFICIENCY, PARAQUAT, AND COPPER ON THE BIOSYNTHESIS OF COPPER-ZINC SUPEROXIDE DISMUTASE AND CATALASE IN *SACCHAROMYCES CEREVISIAE*.

HOSNI M. HASSAN AND FANG-JEN LEE
North Carolina State University, Raleigh, N.C. 27695-7624 USA

INTRODUCTION

The application of superoxide dismutase and catalase as antioxidants (1-3), and as anti-inflammatory and radioprotective agents (4-6) has prompted its commercial production from bovine blood and from genetically engineered microorganisms (7). Previous studies in *E. coli* and other procaryotes (8-14) have demonstrated that increasing the intracellular flux of O_2^-, via cyclic oxidation-reduction of several redox active compounds, induces the synthesis of the manganese-containing SOD (MnSOD). The objective of this study was to determine the optimum conditions for maximal biosynthesis of superoxide dismutase in *S. cerevisiae*. We demonstrate that both paraquat and copper increase the specific activity of CuZnSOD in *S. cerevisiae*, while having no effect on catalase.

MATERIALS AND METHODS

Organisms. *Saccharomyces cerevisiae* var. *ellipsoideus* ATCC 18790 (wild type) and ATCC 18789 (cytochrome *c* deficient mutant) were obtained from the American Type Culture Collection.

Media and growth conditions. The composition of the growth medium was (per liter): $MgSO_4 \cdot 7H_2O$, 0.5 g; KH_2PO_4, 1.0 g; $(NH_4)_2 SO_4$, 5.0 g; $CaCl_2$, 0.1 g; NaCl, 0.1 g; yeast extract, 7.0 g (BBL). Growth conditions were the same as reported before (15).

Cell-free extracts preparation and assays. Cells were harvested, and dialyzed cell-free extracts were prepared as previously described (15). Superoxide dismutase was assayed according to McCord and Fridovich (16), modified by adding 10^{-5} M CN^- to inhibit cytochrome oxidase as described (15). Catalase was assayed as described before (15).

RESULTS AND DISCUSSION

Effect of oxygen on superoxide dismutase and catalase biosynthesis. Data in Table 1 clearly show that both strains responded to the presence of molecular oxygen by increasing their intracellular levels of both SOD and catalase. The level of SOD in the mutant cells exposed to oxygen was about twice that seen in the wild type. This might indicate that the intracellular flux of O_2^- in the mutant cells is significantly higher than that experienced in the wild type cells. On the other hand, catalase levels were much higher in the wild type than in the mutant cells.

TABLE 1

EFFECT OF OXYGEN ON SUPEROXIDE DISMUTASE AND CATALASE BIOSYNTHESIS

GROWTH	WILD TYPE		MUTANT	
CONDITIONS	SOD	CATALASE	SOD	CATALASE
	(U/mg protein)			
AEROBIC	11.8 ± 2.0	17.4 ± 1.0	22.4 ± 5.2	6.7 ± 2.0
ANAEROBIC	0.88	0.1	0.93	0.2

Effect of paraquat on the kinetics of superoxide dismtase and catalase biosynthesis. Previous studies (11-14) have shown that increasing the intracellular flux of O_2^- increases SOD biosynthesis in *E. coli*. Data presented in Figure 1 strongly support this thesis and show that the same is true in a simple eucaryotic organism. Maximum induction of SOD by PQ^{2+} (2.5 mM) in both strains was observed at late log phase of growth. SOD was induced to a greater extent in the mutant than in the wild type (Fig. 1). The presence of 2.5 mM PQ^{2+} did not have significant effect on catalase and peroxidase biosynthesis (data not shown).

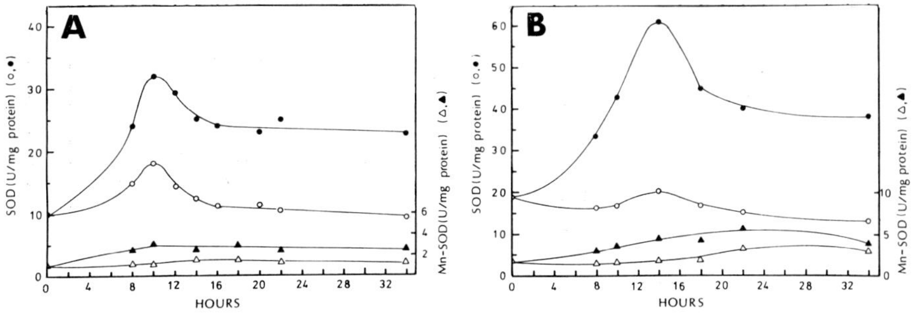

Fig. 1. Effect of paraquat on the kinetics of superoxide dismutase biosynthesis in *S. cerevisiae*. (A) wild type, (B) mutant, no paraquat (open symbols), 2.5 mM PQ^{2+} (closed symbols).

Effect of paraquat concentration on SOD and catalase biosynthesis. Data presented in Figure 2. clearly indicate that the addition of PQ^{2+} to the glucose-rich medium (GRM) or glucose-limited medium (GLM) induced the synthesis of total SOD as well as MnSOD in both the mutant and the wild type strains. In GRM, total SOD biosynthesis in the mutant strain was more responsive to the presence of increasing concentrations of PQ^{2+} than in the wild type strain (Fig. 2A). However, in GLM both the mutant and the wild type strains responded in a similar fashion to paraquat up to 1 mM. Raising $[PQ^{2+}]$ above this resulted in a decline in the total SOD content of the wild type whereas the mutant increased its SOD levels up to 2.5 mM PQ^{2+} (Fig. 2B). On the other hand, the biosynthesis of catalase in the mutant was not affected by the concentration of paraquat added to either GRM or GLM (data not shown), but increasing the concentration of paraquat in either type of medium caused a decline in the catalase biosynthesis in the wild type strain.

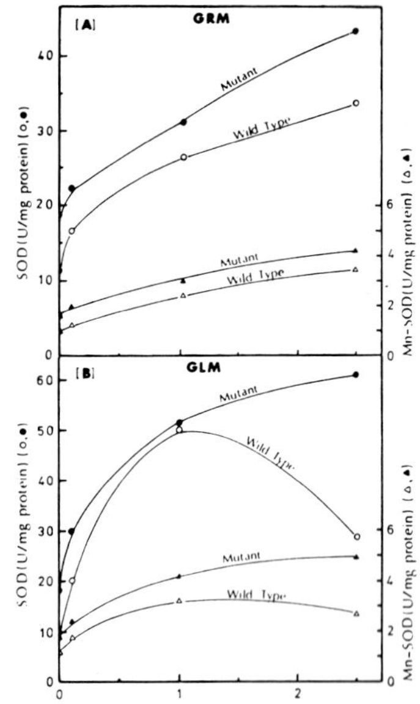

Fig. 2. Effect of $[PQ2^{+}]$ on the biosynthesis of SOD.

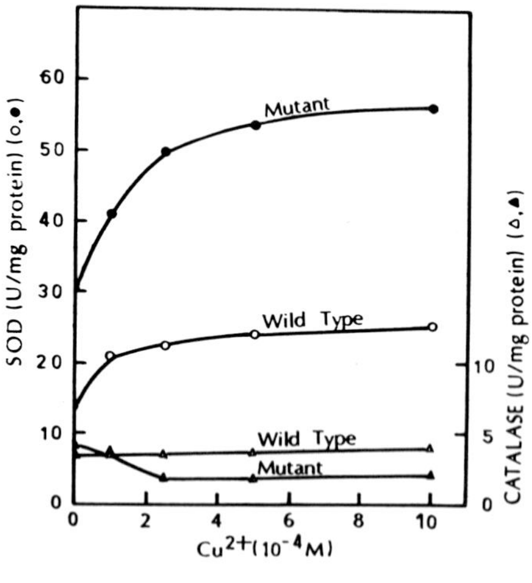

Fig. 3. Effect of $[Cu2^{+}]$ on the levels of SOD and catalase.

Effect of copper on the biosynthesis of SOD and catalase. Because copper is bound in the active center of CuZnSOD it appeared possible that addition of excess Cu^{2+} to the growth medium might increase the biosynthesis of this enzyme in the test organisms. Thus, the two strains of S. cerevisiae were grown aerobically in medium containing different concentrations of $CuSO_4$. Extensive dialysis of the cell-free extracts was performed to insure the complete removal of free copper that could interfere with the SOD assay. Figure 3 shows that added Cu^{2+} increased SOD by 80 - 100% in both strains of S. cerevisiae, while having no significant effect on catalase activity. Furthermore, de novo protein biosynthesis was required since the stimulatory effects of PQ^{2+} or Cu^{2+} on the biosynthesis of CuZnSOD were inhibited by the addition of cycloheximide (2.5 ug/ml).

In the presence of both PQ^{2+} and Cu^{2+}, both the mutant and the wild type strains produced more SOD than in the presence of either PQ^{2+} or Cu^{2+} alone. However, their effects were neither additive nor synergistic (data not shown).

ACKNOWLEDGEMENT

This work was supported in part by grant PCM-8213853 from the National Science Foundation.

REFERENCES

1. Michelson AM (1977) US Patent 4,029,819
2. Michelson AM, Monod J (1975) US Patent 3,920,521
3. Hicks CL (1980) J Dairy Sci 63:1199-1204
4. McCord JM (1974) Science 185:529-531
5. McCord JM, English D (1981) In: Holcember JS (eds) Enzymes as drugs. Jone W & S Inc New York, pp 353-365
6. Petkau A (1978) Photochem Photobiol 28:765-774
7. Hallewell RA, Masiarz FR, Najarian RC, Puma JP, Quiroga MR, Randolph A, Sanchez-Pescador R, Sacndella CJ, Smith B, Steimer KS, Mullenbach (1985) Nucleic Acids Res 6:2017-2034
8. Hassan HM, Fridovich I (1978) J Biol Chem 253:8143-8148
9. Moody CS, Hassan HM (1982) PNAS USA 79:2855-2859
10. Hassan HM, Fridovich I (1977) J Bacteriol 130:805-811
11. Hassan HM, Fridovich I (1977) J Bacteriol 132:505-510
12. Hassan HM, Fridovich I (1977) J Biol Chem 252:7667-7672
13. Hassan HM, Fridovich I (1979) Arch Biochem Biophys 196:385-395
14. Hassan HM, Fridovich I (1979) J Biol Chem 254:10846-10854
15. Lee FJ, Hassan HM (1985) J Biotech (submitted)
16. McCord JM, Fridovich I (1969) J Biol Chem 244:6049-6055

Superoxide and Superoxide Dismutase in Chemistry, Biology and Medicine; G. Rotilio, editor

HEPATIC COPPER AND SUPEROXIDE DISMUTASE ACTIVITY IN TOXIC MILK MUTANT MICE

HAROLD RAUCH, DAMIAN DUPUY, RICHARD J STOCKERT & IRMIN STERNLIEB
Univ of Massachusetts, Amherst MA and A Einstein College of Medicine, Bronx, NY

INTRODUCTION

Toxic milk (tx) is an autosomal recessive mutation which affects copper metabolism in mice. Homozygotes amass large amounts of copper in livers while other tissues become somewhat deficient. In particular, fetuses carried by mutant females and the milk they secrete are extremely low in copper, a condition which leads to severe neonatal copper deficiency and ultimately to infant death (1). Concomitant to high hepatic copper concentration, mutants show reduced ceruloplasmin and superoxide dismutase (SOD) activities. We report here on our investigation of the distribution of copper in the livers of mutant mice and its relationship to SOD activity.

MATERIAL AND METHODS

Analyses were performed on young normal (+) and mutant (tx) adults 6 to 10 weeks of age. Livers were excised, homogenized in isotonic saline and centrifuged at 100000g to give supernatants which were resolved by gel filtration on Sephadex G-75. Pooled copper-containing fractions were treated with chloroform/ethanol followed by acetone precipitation (45-75%). The recovered material was further characterized by high performance ion-exchange chromatography using an LKB Ultrapor DEAE column; for elution a linear gradient to .2M phosphate in 60 minutes was applied. Copper concentrations were determined by atomic absorption spectrophotometry. SOD activity was assayed according to Marklund and Marklund (2).

RESULTS AND DISCUSSION

Collected data are presented in Table I. It is apparent that in both genotypes most hepatic copper is cytosolic. When supernatants are resolved on Sephadex G-75, copper is distributed among three peaks: Peak I corresponds to the void volume and represents proteins with molecular weights (MW) of 50000 and above; Peak II corresponds to proteins in the 30-35000 MW range; Peak III proteins are in the 6-10000 range. Most copper appears in Peak II in normals while much copper is found in both II and III in mutants. In both genotypes, SOD activity is restricted to Peak II fractions.

An SOD-enriched fraction, SA75, was prepared by organic solvent precipitation. While the mutant preparation has 35 times as much copper as the normal it exhibits less than two-thirds the SOD activity. In the process of SOD-enrichment there is to be expected a shift in the Cu:Zn ratio toward the 1:1 characteristic of normal CuZnSOD. This shift is seen in the preparation from normal animals, but not in mutants where the ratio remains 2.5:1. This suggests the presence in the mutant fraction of an aberrant SOD molecule having a greater Cu/Zn ratio, or the presence of cuproprotein other than SOD.

TABLE I

	Copper Concentration[a]		SOD Activity[c]		Cu/Zn Ratio	
	+	TX	+	TX	+	TX
HOMOGENATE	38.6	1086			0.2	2.9
SUPERNATANT	37.5	1012	1.7	0.6	0.4	2.6
SEPHADEX PEAK I	0.1	46				
PEAK II	22.4	246				
	(1.3	9.8)[b]	42.0	23.8		
PEAK III	TRACE	385				
SA75	(0.6	21.0)[b]	52.0	30.0	1.5	2.5

a. ug Cu/ g Liver
b. ug Cu/ mg Protein
c. units/ mg Protein

When SA75 fractions were further characterized by ion-exchange chromatography, it became apparent that retention time with respect to copper content and SOD activity differs in the two genotypes (Figure 1). In normals, SOD activity, peak copper and protein concentrations are found in the same fraction at .1M in the gradient. In mutants however, the copper and protein peak showing SOD activity occurs at .053M; in addition there is a second copper peak appearing at .066M but which is devoid of SOD activity. An SA75 preparation from a heterozygous mouse (+/tx) gave copper peaks at .053M and .066M as in mutants as well as one at .112M as in normals; SOD activity was exhibited both at .054M and .112M but not at .066M. These observations point to variation in the SOD molecule which correlates with genotype; the appearance of both types in the heterozygous preparation is especially telling in this regard.

The excess copper accumulated in the liver of the toxic milk mouse is associated primarily with proteins in two MW classes, the heavier corresponding to the range

FIGURE 1

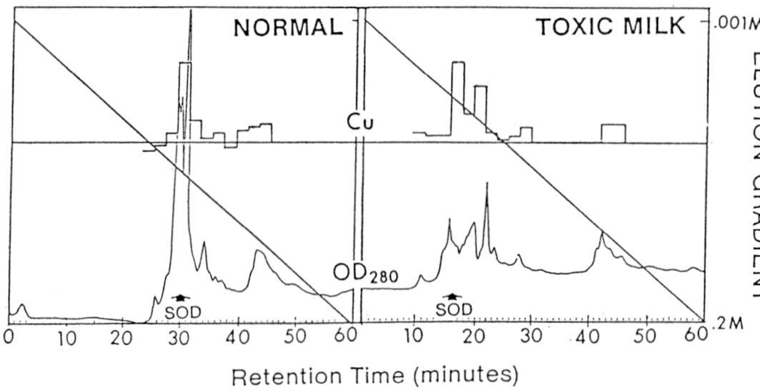

which includes SOD. Despite much copper and protein in this peak, there is markedly reduced enzyme activity. This could be due to qualitative and/or quantitative differences in the SOD molecules produced in mutant liver. The behavior exhibited on the ion-exchange column suggests at least qualitative changes in this SOD. This is supported by the fact that SOD from normals and mutants exhibit different mobilities on polyacrylamide gel electrophoresis. The cuproproteins eluting at .066M are of interest since they account for much of the copper in this peak. Whether they represent SOD molecules so modified as to have lost enzymatic activity or entirely novel proteins has yet to be determined. Finally the question remains as to whether the SOD and the other cuproprotein are responsible for or are the result of the hepatic copper accumulation. In investigating this question, the copper proteins in the lower MW peak, eg metallothioneins, cannot be overlooked. Overall, this mutation provides an exceptional tool for probing copper homoeostasis.

ACKNOWLEDGEMENT

This study has been supported in part by the following grants: BRSG RRO 7048 (U Massachusetts); NIAMDD AM017702 and AM-32972 (A Einstein COM).

REFERENCES

1. Rauch H (1983) J Hered 74:141
2. Marklund S, Marklund G (1974) Eur J Biochem 47:469

Superoxide and Superoxide Dismutase in Chemistry,
Biology and Medicine; G. Rotilio, editor

INTRAUTERINE MATURATION OF LUNG ANTIOXIDANT ENZYMATIC DEFENCE IN THE RAT

EVA GERDIN, ULF J. ERIKSSON, OLOF TYDÉN & BENGT GERDIN

Department of Obstetrics and Gynecology and Department of Surgery, University Hospital, Uppsala, Department of Medical Cell Biology, Biomedicum, University of Uppsala, and Department of Experimental Medicine, Pharmacia AB, Uppsala (Sweden)

INTRODUCTION

Toxic effects of oxygen are reported to contribute to respiratory disorders in the newborn and the premature infant. The lungs of the newborn are exposed to five times the intrauterine oxygen tension. It is still not clear to what extent the late fetal development involves a biological preparation for this relative hyperoxia. To elucidate this question we have determined the activities of the antioxidative enzymes glutathione peroxidase (GSH-Px), catalase (CAT) and superoxide dismutase (SOD) in perinatal rat lung.

METHODS

Lung and blood from fetuses of Sprague-Dawley rats of gestational ages 16-22 days (pregnancy term) and from two-day-old newborn rats were studied. The lungs were trimmed, washed in saline, diced with scissors, sonicated in a volume of 1.2 ml of phosphate-buffered saline and centrifuged. GSH-Px activity was assayed with a slight modification of the method of Günzler et al. (1) at 37°C, using t-butyl hydroperoxide as substrate. CAT activity was measured with a Clark oxygen electrode as described by Del Rio et al. (2). For determining SOD activity the pyrogallol method of Marklund and Marklund (3) was used. MnSOD activity was assayed in some of the samples after inhibition of CuZnSOD with 3 mM KCN. The very small amount of blood contaminating the lung homogenates was measured by the method of Marklund (4), utilizing the peroxidase activity of hemoglobin. The DNA content of the lungs was determined by a fluorescence method. The results are given as mean ± S.E. of 5 litters.

RESULTS

In spite of careful mechanical washing of the lungs, 10-30 % of the GSH-Px (Fig. 1) and CAT (Fig. 2) in the lung homogenate originated from contaminating blood. The SOD activity that derived from the blood was found to be less than 2 %, and is therefore not corrected for.

GSH-Px increased only slightly during early fetal life, but to a greater extent during the last three days before birth.

The increase in lung GSH-Px activity appeared to be more pronounced when the data were expressed per mg DNA (Fig. 3). The CAT activity showed only a minor in-

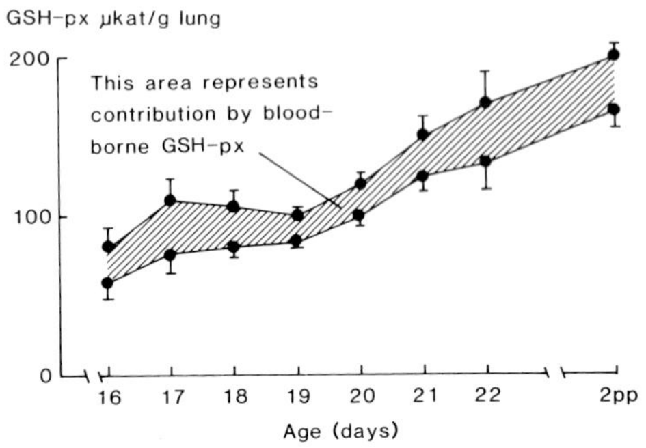

Fig. 1 GSH-Px activity (µkat/g lung). Upper curve = total activity. Lower curve = net activity after subtraction of blood-borne GSH-Px.

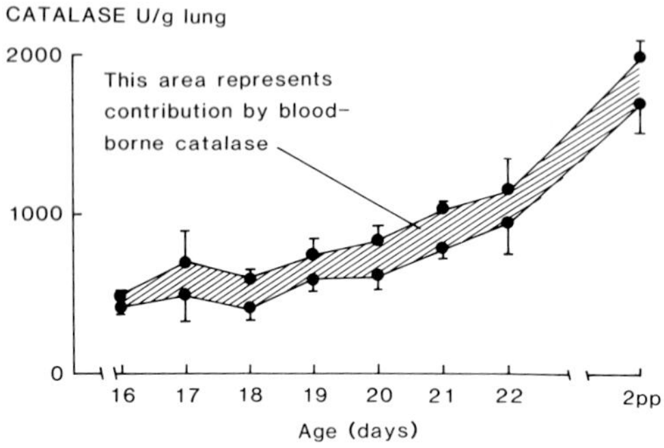

Fig. 2. CAT activity (U/g lung). Upper curve = total activity. Lower curve = net activity after subtraction of blood-borne CAT.

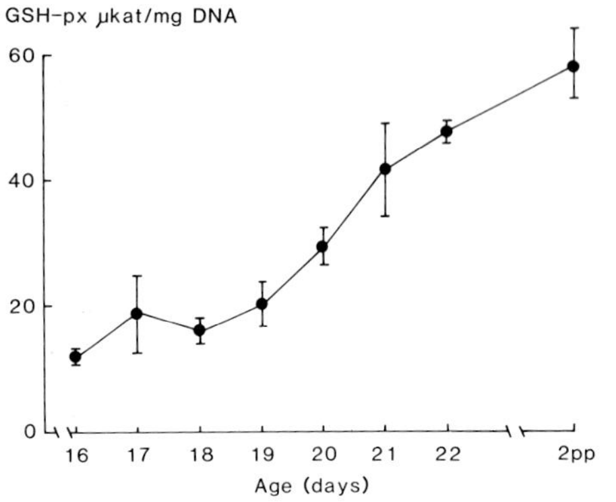

Fig. 3. GSH-Px activity after subtraction of blood-borne GSH-Px (µkat/mg DNA).

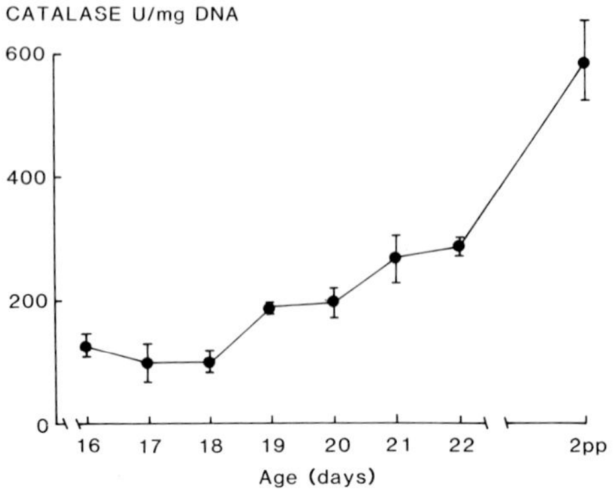

Fig. 4. CAT activity after subtraction of blood-borne CAT (U/mg DNA).

crease during intrauterine life, similar to that in GSH-Px (Figs. 2 and 4). The main acceleration of CAT activity occurred after birth. The total SOD activity did not change significantly in late pregnancy and the activity of CuZnSOD was markedly greater than that of MnSOD (Fig. 5). When expressed per mg DNA, however, the SOD activity showed a slight increase between gestational day 19 and day 2 post-partum (Fig. 6).

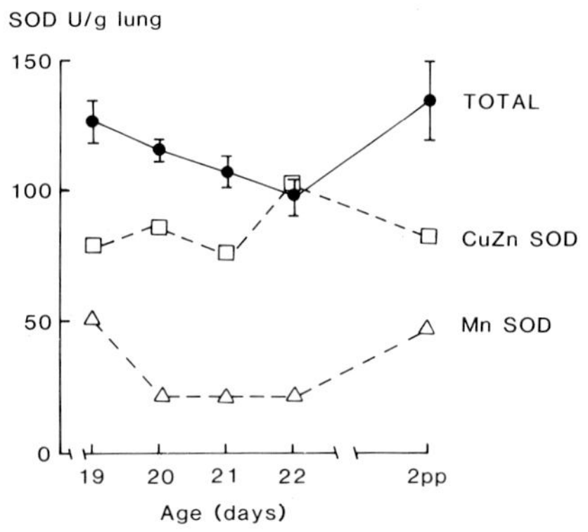

Fig. 5. SOD activity (U/g lung). CuZnSOD and MnSOD activity was measured in one lung for each gestational day.

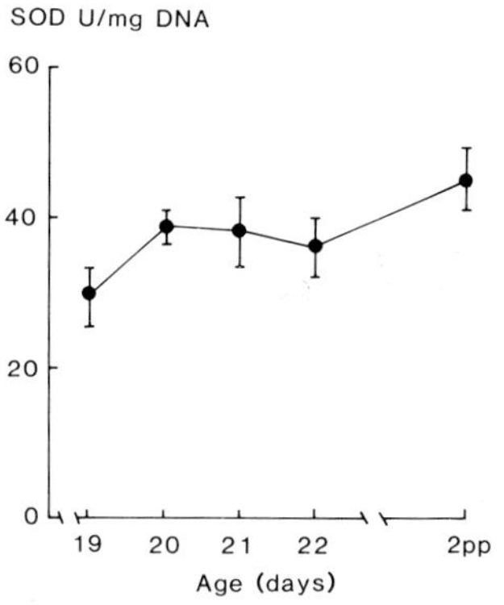

Fig. 6. SOD activity (U/mg DNA).

DISCUSSION

The three antioxidant enzymes studied each displayed characteristic developmental patterns. GSH-Px increased mainly in late fetal life; the greatest rise in CAT occurred immediately after birth, whereas SOD showed a very modest increase with time. The results differ from those of Tanswell and Freeman (5), who found no change in perinatal CAT and reported that the major part of lung SOD activity consisted in MnSOD. Furthermore, the increase in SOD was more pronounced in the studies by Frank and collaborators in the perinatal rabbit lung (6) and perinatal rat lung (7). No systematic investigation on this matter has yet been performed in man.

The present study represents an experimental approach to the investigation of an important aspect of fetal and neonatal pulmonary maturation. Before a complete understanding of this process can be reached, however, it is necessary to clarify the question how the demonstrated changes in enzymatic activities are related to other morphological and biochemical events in the developing fetal lung.

REFERENCES

1. Günzler WA, Kremers H, Flohé L (1974) Z Klin Chem Klin Biochem 12:444-448
2. Del Rio LA, Ortega MG, Lopez AL, Gorgé JL (1977) Anal Biochem 80:409-415
3. Marklund S, Marklund G (1974) Eur J Biochem 47:469-474
4. Marklund S (1979) Clin Chim Acta 92:229-234
5. Tanswell AK, Freeman BA (1984) Pediatr Res 18:584-587
6. Frank L, Groseclose EE (1984) Pediatr Res 18:240-244
7. Frank L, Lewis PL, Sosenko IRS (1985) Pediatrics 75:569-574

Superoxide and Superoxide Dismutase in Chemistry,
Biology and Medicine; G. Rotilio, editor

DISTRIBUTION OF SUPEROXIDE DISMUTASE, GLUTATHIONE PEROXIDASE AND CATALASE IN DEVELOPING RAT BRAIN

ROLANDO F. DEL MAESTRO AND WARREN MCDONALD

Brain Research Laboratory, Victoria Hospital, University of Western Ontario, London, Ontario, N6A 4G5, Canada

INTRODUCTION

The concentration of individual scavenging enzymes has been assessed in cerebral tissues as a function of age in a variety of species (1-4). A consistent developmental pattern has not been observed (4). These enzymes may function as a scavenging system and the level of any individual enzyme may not be as useful as the assessment of a number of enzymes in the specimens from various cerebral sites in the same animal. Cerebral maturation is associated with increasing aerobic metabolism (5) and it was hypothesized that a developmental pattern of scavenging enzymes would be observed during brain maturation. The developmental pattern of copper- and zinc-containing superoxide dismutase (CuZnSOD), manganese-containing superoxide dismutase (MnSOD), glutathione peroxidase (GSH-Px) and catalase (CAT) was assessed in developing rat brain. The enzymes studied were assayed in different brain regions (cerebral cortex, striatum, cerebellum and brainstem) and all enzyme values were assayed in each specimen with correction for erythrocyte contamination.

MATERIALS AND METHODS

Wistar rats of both sexes and of various ages were used. Prenatal animals (19 days gestational age) were obtained by caesarean section and blood and brain samples pooled from three animals to obtain each individual specimen. For the other time periods studied, 2, 10, 19, 77 and 280 days, brain regions and each animal's corresponding blood sample were used.

Enzyme Assays

Superoxide dismutases. The pyrogallol autoxidation method was used to determine these enzymes (6,7). Cyanide insensitive SOD which was considered to represent MnSOD was assayed by adding 1mM sodium cyanide to the supernatant.

Glutathione peroxidase. The method of Gunzler *et al*. (8) was employed to measure this enzyme.

Catalase. This enzyme was determined immediately after obtaining the supernatant using a Clark electrode (9).

Correction for erythrocyte contamination. Although care was taken to obtain specimens contaminated with as little blood as possible, it was necessary to compensate for enzyme activity contributed by erythrocytes. The hemoglobin (10) and enzyme content of each tissue homogenate and its corresponding blood sample were determined and appropriate correction made.

RESULTS

Table I reports the changes observed in the enzymes studied in cerebral cortex. Up to 10 days, total SOD activity was stable and then a progressive increase in total SOD was seen which stabilized after 11 weeks in all regions studied. The increase in total SOD activity was accounted for by a progressive increase in CuZnSOD. The GSH-Px activity fell in all brain areas studied after birth and increased after 10 days reaching a maximum value at 11 weeks. A progressive fall in CAT activity was seen in all brain regions studied with very low levels of CAT activity seen at 11 and 40 weeks of age.

TABLE I

LEVEL OF SCAVENGING ENZYMES IN CEREBRAL CORTEX DURING RAT DEVELOPMENT (N = 6 AT EACH TIME PERIOD AND VALUES ARE MEANS ± SEM)

Age	Superoxide Dismutase ng/mg protein			Glutathione Peroxidase m Units/ng protein	Catalase Units/ng protein
	CuZn	Mn	Total		
Prenates	102± 15	401±22	503± 14	41± 3	1.9±0.1
2 days	106± 14	409±31	515± 25	37± 3	1.7±0.1
10 days	146± 28	336±15	481± 28	19± 2	0.8±0.1
19 days	340± 31	367±23	707± 39	32± 6	0.7±0.1
77 days	461± 88	544±29	1005± 93	75±10	0.1±0.03
280 days	619±147	543±42	1161±178	51±13	0.4±0.1

DISCUSSION

In the Wistar rat, neonatal cerebral development is characterized by increasing CuZnSOD activity, a decrease in CAT and, after an initial decrease, increasing GSH-Px activity. All brain regions assessed underwent a similar pattern. The results of this study are consistent with the concept of increasing SOD activity in cerebral tissue during development (3,4). Parallel changes do not occur in erythrocytes (results not shown) and cerebral tissue since erythrocyte SOD activity was stable throughout the study period. The reason and mechanism for the increase seen in SOD activity in cerebral tissue remains unknown. Parallel changes in postnatal GSH-Px activity was seen in erythrocyte and cerebral tissue. The initial fall in GSH-Px activity may be related to decreased nutritional availability of selenium during early development (11,12). The marked decrease in CAT activity associated with postnatal development was not associated with similar changes in erythrocyte CAT activity. The rise in GSH-Px activity may be a response to a need for further enzymatic capacity to deal with the production of H_2O_2 possibly associated with increased SOD activity.

The ontogenetic pattern of these enzymes during cerebral development is rather complex. An analysis of subcellular localization of these enzymes and their possible relationships to oxidative metabolism may clarify the role of these enzymes in cerebral metabolism.

ACKNOWLEDGEMENTS

This project was supported by the Medical Research Council of Canada Grant No. MA-8104 and the Brain Research Fund Foundation.

REFERENCES

1. Kellogg EW III and Fridovich I (1976) J Geronatol 31:405-408
2. Reiss U and Gershon D (1976) Biochem Biophys Res Commun 73:255-262
3. Mavelli L, Mondovi B, Federico R and Rotilio G (1978) J Neurochem 31:361-364
4. Mavelli L, Rigo A, Federico R, Ciriolo MR and Rotilio G (1982) Biochem J 204:535-540
5. McIlwain H and Bachelard S (1971) In: McIlwain H, Bachelard S (eds) Biochemistry and the Central Nervous System. Churchill Livingstone, London, pp 406-444
6. Marklund S and Marklund G (1974) Eur J Biochem 47:469-474
7. Del Maestro RF, McDonald W and Anderson R (1983) In: Greenwald RA, Cohen G (eds) Oxy Radicals and Their Scavenger Systems Vol. II. Elsevier Science, New York, pp 28-35
8. Gunzler WA, Kramers H and Flohé L (1974) Z Klin Chem Klin Biochem 12:444-448
9. Del Rio LA, Ortega MG, Lopez AL and Gorge JL (1977) Anal Biochem 80:409-415
10. Harboe M (1959) Scandinav J Clin & Lab Investigation II:66-70
11. Prohaska JR and Ganther HE (1976) J Neurochem 27:1379-1387
12. Butler JA, Whanger PD and Tripp MJ (1982) Am J Clin Nutr 36:15-23

Superoxide and Superoxide Dismutase in Chemistry,
Biology and Medicine; G. Rotilio, editor

ECDYSTERONE ENHANCES OXYGEN CELLULAR CONSUMPTION AND SUPEROXIDE DISMUTASE ACTIVITY IN DROSOPHILA AEROBIC CELLS.

M. ROPP *, R. CALVAYRAC°, M. BEST-BELPOMME *

* U.A. CNRS 1135, 7 quai St-Bernard 75005 Paris - France
° Paris VII, 4, place Jussieu 75005, Paris - France

Drosophila cells cultured in vitro were shown to respond specifically to the steroid hormone known to play a key role in the development of insects : the 20-hydroxyecdysone (ecdysterone) (1). Indeed the addition of ecdysterone into the culture medium provokes: morphological changes (2) with an increase of cyclic GMP (3), arrest of division (4), induction of polypeptides (5) and induction of enzymatic activities: acetylcholinesterase (6), β galactosidase (7), catalase (8). Knowing that Drosophila cells are aerobic (9) and that catalase is one of the enzyme which protects the aerobic cells against the cellular damages caused by the oxygen reduced species, (10, 11) we were led to measure, after hormonal stimulation, the superoxide dismutase activity and the cellular consumption of oxygen.

Materials and Methods. Cells and hormonal treatment : The clone 8-9K was chosen. These embryonic Drosophila cells derived from the Kc line (12). Cells were grown at 23°C in D22 medium supplemented with 5% fetal calf serum (12). Ecdysterone was added into the growth medium 2 days after the subculture.

Measure of oxygen cellular consumption : The cells were gently detached from the bottom of the flask, centrifugated at 400g for 10 min., resuspended in a fresch medium saturated in O2. The O2 concentration in the suspension (s in µM) was measured at 25°C, in a closed Clark electrode. Since Drosophila cells consummed oxygen, the O2 measurement was continuous by monitored untill its concentration reached practically zero. The rate of the cellular O2 consumption (v) was calculated from the slopes of the decreasing (O2) curve and expressed in fentomoles. min $^{-1}$. cell. $^{-1}$. For each determination 1 - 2. 10^8 cells were used.

Homogeneization of cells. The cells, detached from the bottom of the flasks were washed in an isotonic solution and homogenized at 0°C with a glass - Teflon Potter in a Tris-Triton buffer (0.01 M Tris. pH 7, 5mM $MgCl_2$, 1 M NaCl, 1% (V/V) Triton X-100). The homogenate was then centrifugated (20.000g, 20 min, 4°C). Protein concentration and superoxide dismutase activity were measured in the supernatant (cellular extract).

Assay of superoxide dismutase : In order to measure the superoxide dismutase activity in the cellular extracts, the luminol method (13) was chosen since it the most accurate technic in cellular crude extracts (14). 100 µl of A (0.23 M glycine buffer pH9, 0.1mM luminol, 4mM E.D.T.A.) were mixed with 100 µl of B (0.77mM hypoxanthine, 0.23M glycine buffer pH9). Each cellular extract was diluted with the Tris-Triton buffer 1, 2, 4, 8, 16 and 32 fold. The superoxide dismutase activity was so measured in function of the protein concentration for a given cellular extract. For this purpose 5µl of each dilution were added to the A+B mixture and the reaction was initiated by adding 10 µl of xanthine oxidase (Sigma : 0.9 units/ml). The luminescence was continuously recorded with a nucleotimètre C.L.V., France) and its maximum, reached after about 75 seconds, was measured in arbitrary Relative Light Unit (R.L.U.). This value was normalized by dividing it by the R L Us of the "blanks" (i.e. no protein but the same Tris-Triton buffer final concentration). We so obtained y ($0 < y < 1$) wich decreased when the quantity of proteins (x, in mg) in the assay mixture increased. Since y was not a linear function of x we linearized the results by plotting $\frac{y}{1-y}$ in function of $\frac{y}{5x}$ .We verified the validity of this representation by using data published (14). Thus, we obtained a linear representation, each point representing an independant assay, y warying with x. For a value of $\frac{y}{1-y}=1$ on the ordinates (i.e. y = 0,5), a precise interpolation gave directly the specific superoxide dismutase activity expressed in luminol units (15).

Results : Ecdysterone enhances oxygen cellular uptake : For studying the kinetic parameters of the cellular oxygen uptake, (i.e. the maximal rate Vm of O2 consumption and Km for O2), we decided to represent $\frac{v}{s}$ in function of v. This representation was chosen because it allows a better precision and an easy correction of the contribution of a linear increase of v in function of s (16) at least due to the O2 autoconsumption of the electrode . As seen in the Fig. 1 a straight line is indeed obtained. In untreated cells the kinetic parameters of the oxygen consumption were found to be 0,20 ± 0,06 fentomoles $O2.mn^{-1}.cell^{-1}$ for Vm and 4 ± 1µM for Km (9). When cells are treated by ecdysterone the same kind of measurement showed an increase of the Vm wich is lenght of hormone treatment and steroid concentration dependant (table 1). In all these cases, the Km remains constant even if the Vm is multiplied by at least 3 fold. The cellular oxygen uptake can be greatly diminished with the addition of 1mM KCN. Half effect corresponds to a KCN concentration of about 17 µM in untreated cells and becomes higher about 50 µM in hormone treated cells (0,5 µM, 4 days).

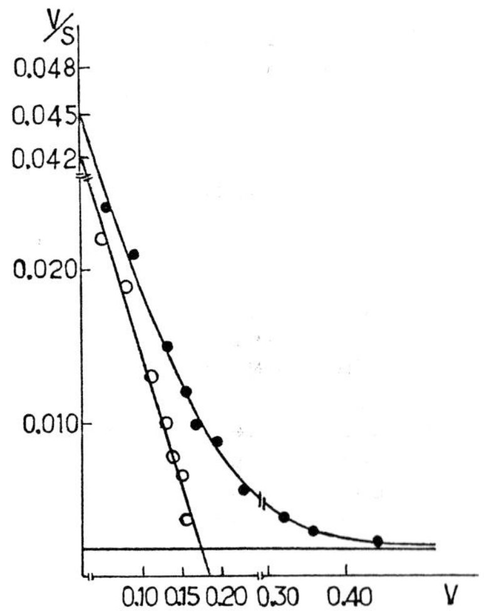

Figure 1: Maximal rate of cellular consumption V_M and K_m for O_2.

Table 1 : Comparative values of V_M in untreated and treated cells.

Treatment \ Time after treatment	2 days	4 days
Control cells	0,28	0,18
Ecdysterone 0,25 µM	0,4	0,4
Ecdysterone 0,5 µM	0,74	0,66

Table 2 : Superoxide dismutase activity

Treatment \ Time after treatment	2 days	4 days
Control cells	6,8	12
Ecdysterone 0,025 µM	10,6	21,3
0,1 µM	12,5	26,5
0,5 µM	13,7	27,5

Ecdysterone enhances superoxide dismutase activity : Since we observed that the hormone increases the cellular O2 uptake, we decided to measure superoxide dismutase activity in untreated and steroid treated cells. This activity is indeed known as a fundamental cellular defense against damages caused by the first product of oxygen reduction : the superoxyde ion. Superoxide dismutase activity is increased by the hormone treatment (table 2). This increase is dependant of the time and concentration of the hormonal treatment. When data of table 1 are compared with those of table 2, a parallelism between the increases of Vm and of superoxide dismutase activity can be observed. Sephadex G-100 chromatographies (not shown) have revealed that superoxide dismutase activities are splitted into two peaks having apparent molecular weights of about 80K daltons (25% of the total activity) and about 35K daltons. The first one seems to correspond to the mitochondrial activity in respect of its M.W. already known in Ceratitis Capitata (17). Assaying fractions, we observe that catalase activity, which is excluded in this type of chromatography, does not interfere with the luminol assay.

Discussion : It is known that mitochondria consume (reduce) the great majority of the O2 entering in a cell (10). Consequently the most simple hypothesis is that the increase of the oxygen uptake in the hormone treatment cells is due to an activation of the mitochondrial properties of reducing oxygen. For instance, after the steroid action the cytochrome C oxidase activity could be increased. Several arguments are in favour of this suggestion, namely the value of Km remaining constant, the cyanide effect, and indeed preliminar data of our group showing that the cytochrome C oxidase activity is increased in hormone treated cells. Other explanations should be offered . For example the steroid treatment could have a decoupling effect. This is unlekely since the ATP pool we measured in control and hormone treated cells remain constant and moreover tends to increase at the fourth day of a 0,1 μM ecdysterone treatment.

In conclusion, in the context of the scavenging of the oxygen reduced species which are also considered as the possible inducers of the so-called heat shock proteins (9), it is interesting to observe that a steroid can induce in target cells an increase of catalase (8), superoxide dismutase activities accompanied by the enhancement of the O_2 cellular uptake.

1. Best-Belpomme,M. Courgeon,A.M.(1980)In progress in ecdysone research, p.379-382, North Holland Biomedical Press.
2. Courgeon,A.M. (1972) Exp.Cell.Res. 74, 327-336
3. Courgeon,A.M.,Caila,H.L. (1981) Exp.Cell.Res. 133, 15-22
4. Courgeon,A.M. (1972) Nature New.Biol. 238, 250-251
5. Couderc,J.L., Dastugue,B. (1980) Bioch.Bioph.Res.Com.97, 173-181
6. Best-Belpomme,M.,Courgeon,A.M. (1977) FEBS Letters 82, 345-347
7. Best-Belpomme,M.,Courgeon,A.M.,Rambach,A. (1978) PNAS USA 75, 6102-6106
8. Best-Belpomme,M.,Ropp,M.(1982) Eur.J.Biochem. 121, 349-355
9. Ropp,M.,Courgeon,A.M.,Calvayrac,R. and Best-Belpomme,M. (1983) Canadian J.of Bioch. and Cell Biol. 61, 456-461
10. Chance,B.,Sies,H.,Boveris,A. (1979) Physiol. Rev. 59, 527-604
11. Fridovitch,I. (1976) in Free radical in biology (A.Pnyon ed.) Vol.1, p.239-271, Academic press New-York
12. Echalier,G.,Ohanessian,A. (1970) In vitro 6, 162-172
13. Puget,K.,Michelson,A. (1974) Biochimie 56, 1255-1261
14. Bensinger,R.,Johnson,C. (1981) Anal.Bioch. 116, 142-145
15. Lee,Y.M.,Ayala,F.,Hara,P. (1981) J.of Biol.Chem.256, 8506-8509
16. Ropp,M. Thèse 3ème Cycle
17. Michelson,A. Pers.Com.

Superoxide and Superoxide Dismutase in Chemistry, Biology and Medicine; G. Rotilio, editor

INDUCTION OF SUPEROXIDE DISMUTASES IN LEMON LEAVES BY INFILTRATION WITH IRON(II).

FRANCISCA SEVILLA*, ELADIO HELLIN*, CARLOS F. ALCARAZ* and LUIS A. DEL RIO**

*CSIC. Centro de Edafología y Biología Aplicada del Segura. P.O. Box 195. 30003-Murcia (Spain).
**CSIC. Estación Experimental del Zaidín. P.O. Box 419. 18008-Granada (Spain).

INTRODUCTION

Recently we found that leaf extracts from lemon trees contained at least five electrophoretically distinct superoxide dismutase isozymes: two Cu-Zn-SODs, one Mn-SOD and, interestingly, two Fe-SODs (1), a type of superoxide dismutase mainly found in procaryotic organisms and which is only present in the plant families Gingkoaceae, Nymphacea, and Cruciferae (2), and in some eucaryotic algae (3,4).

The metalloenzyme system superoxide dismutase containing four different metals, could be useful for the appraisal of metal imbalances in citrus plants, particularly iron deficiencies, and in this communication we further explore this possibility by studying the induction of SOD activity in lemon leaves by vacuum-infiltration with iron(II) salts.

MATERIAL AND METHODS

Leaves (8-10 months-old) were sampled from lemon trees (*Citrus limonum* R.) grown by strip irrigation under field conditions. Leaves from Fe-deficient trees were vacuum-infiltrated with 5.4mM $FeSO_4$ and with distilled water (controls) for 3 min, and incubated in the dark at 25°C for 24h, in the presence of 0.18mM $FeSO_4$. Leaves were then illuminated (4500 LUX) for 24-48h in a growth chamber under continuous aeration.

Leaf extracts from lemon leaves were prepared as described earlier (1). Total SOD activity was estimated by a photochemical method (5). SOD isozymes were individualized by isoelectric focusing on 7.5% polyacrilamide gels containing ampholytes (a mixture of pH ranges 4.2-4.9 and 4.5-5.4) purchased from Pharmacia. SOD activity in gels was localized by the photochemical method (6) and quantitated by linear scanning densitometry. Peroxidase isozymes were detected by activity staining with o-dianisidine (7). Metal concentration of lemon leaves was determined by atomic absorption spectrophotometry and proteins were estimated according to the Bradford procedure (8).

RESULTS AND DISCUSSION

Isoelectric focusing of crude extracts from lemon leaves revealed the existence

Abreviations: SOD, superoxide dismutase

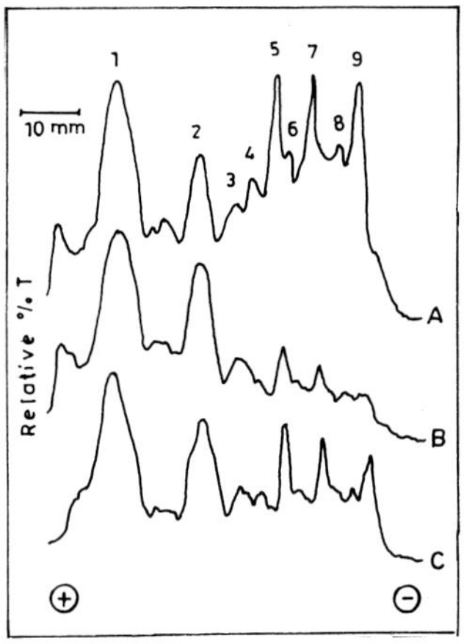

Fig. 1.- Densitometric scans of SODs from lemon leaves infiltrated with Fe(II). A: Full-nutrient leaves; B: Fe-deficient leaves infiltrated with H_2O and illuminated for 24h (control); C: Fe-deficient leaves infiltrated with 5.4mM $FeSO_4$ and illuminated for 24h.

of heterogeneity in the SOD activity bands previously separated by polyacrylamide gel electrophoresis (1), with four additional SOD isozymes being detected. Leaf extracts contained: four Cu,Zn-SODs (bands 1-4), three Fe-SODs (bands 5-7), and two Mn-SODs (bands 8 and 9). The existence of charge variants of Mn-SOD and Fe-SOD has also been described recently in *E. coli* using isoelectric focusing (9). The isoelectric points determined for the different lemon SODs were: 1, 4.52; 2, 4.74; 3, 4.84; 4, 4.89; 5, 4.95; 6, 4.98; 7, 5.05; 8, 5.13; and 9, 5.18. The staining for peroxidase activity of parallel gels showed the presence of two isoperoxidases, one partially overlapping SOD band 1 and the other between this band and band 2.

Vacuum infiltration of Fe-deficient leaves with 5.4mM $FeSO_4$ brought about the induction of Fe-SOD isozymes 5 and 7 and, to a lesser extent, of Mn-SOD isozymes 8 and 9 (Fig. 1). This effect was intensified after 48h illumination (Table 1). The concentration of iron in Fe-deficient tissues after Fe(II) infiltration rised considerably, and was even higher than levels found in normal leaves.

Illumination for 48h under continuous aeration of Fe-deficient leaves previously vacuum-infiltrated with water (control), showed an interesting increase of CN-resistant SODs (Fe- and Mn-containing), particularly the Mn-SODs (bands 8 and 9) and the most cathodic Fe-SOD (band 7) (Fig. 2).

Results found showed the induction by Fe(II) of Fe-SOD isozymes in iron-deprived lemon leaves. The induction of Fe-SOD by Fe(II) has been reported recently in *E. coli* (10) and a biosynthesis mechanism based on the autogenous repression by the apoSODs has been proposed (11). However, the induction of Mn-SODs and Fe-SOD by the action of light under strong aeration suggests an induction mechanism mediated by activated oxygen species generated by photoxidative processes (12).

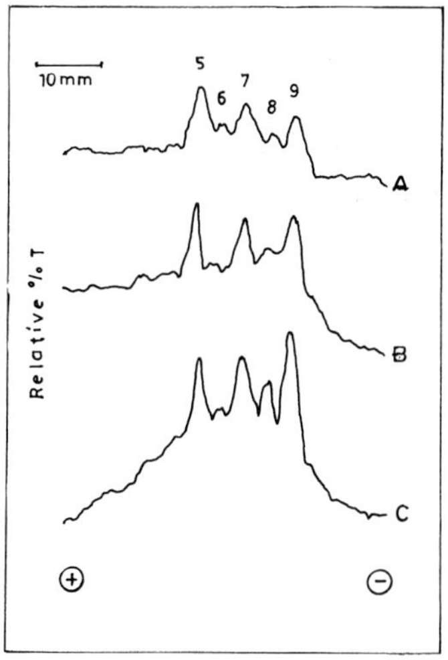

Fig. 2.- Densitometric scans of CN-resitant SODs from control Fe-deficient leaves illuminated under intense aeration. A: leaves kept under dark; B: leaves illuminated for 24h; C: for 48h.

Table 1.- SOD activity (U/ml) of Fe-deficient lemon leaves after vacuum-infiltration with distilled water and Fe(II) salts and illumination.

Illumination time	Infiltrating reagent	SOD activity bands (nº)								
		1	2	3	4	5	6	7	8	9
24h	H_2O	9.73	8.50	2.70	0.63	2.96	1.31	2.81	1.36	2.09
24h	5.4mM Fe(II)	10.28	8.01	2.29	1.95	4.86	2.57	5.93	2.97	4.91
48h	H_2O	10.78	6.21	2.32	2.14	3.06	1.60	3.26	4.71	5.81
48h	5.4mM Fe(II)	13.22	7.96	2.22	2.00	5.48	2.33	8.02	4.81	6.89

Values are the average of three different samples.

REFERENCES

1. Sevilla, F.; del Río, L.A. and Hellín, E. (1984). *J. Plant Physiol.*, *116*: 381-7
2. Bridges, S.M. and Salin, M.L. (1981). *Plant Physiol.*, *68*: 275-8.
3. Lengfelder, E. and Elstner, E.F. (1979). *Z. Naturforsch.*, *34*: 374-80.
4. Kanematsu, S. and Asada, K. (1979). *Arch. Biochem. Biophys.*, *195*: 535-45.
5. Giannopolitis, C.N. and Ries, S.K. (1977). *Plant Physiol.*, *59*: 309-14.
6. Beauchamp, C.O. and Fridovich, I. (1971). *Anal. Biochem.*, *44*: 276-87.
7. García, J.E.; Gómez, M.; Yáñez, J.; López-Gorgé, J. and del Río, L.A. (1982). *Agrochimica*, *26*: 477-86.
8. Bradford, M.M. (1976). *Anal. Biochem.*, *72*: 248-54.
9. Clare, D.A.; Blum, J. and Fridovich, I. (1984). *J. Biol. Chem.*, *259*: 5932-6.
10. Pugh, S.Y.R.; di Giuseppi, J.L. and Fridovich, I. (1984). *J. Bacteriol.*, *160*: 137-42.
11. Pugh, S.Y.R. and Fridovich, I. (1985). *J. Bacteriol.*, *162*: 196-202.
12. Elstner, E.F. (1982). *Ann. Rev. Plant Physiol.*, *33*: 73-96.

PRESENCE OF SUPEROXIDE DISMUTASE ACTIVITY IN GLYOXYSOMES FROM WATERMELON, CITRULLUS VULGARIS

LUISA Mª SANDALIO AND LUIS A. DEL RIO
Unidad de Bioquímica Vegetal, Estación Experimental del Zaidín, C.S.I.C., Aptdo. 419, 18008 Granada (Spain)

INTRODUCTION

Peroxisomes are subcellular "respiratory" organelles containing as basic enzymatic constituents catalase and at least one H_2O_2-producing flavin oxidase, and have different metabolic pathways depending upon the tissue (1). In a previous work, by applying immunocytochemical techniques to protoplasts isolated from pea leaves (Pisum sativum L), the presence of manganese-containing superoxide dismutase* in pea peroxisomes was demonstrated (2). Since glyoxysomes are functionally specialized peroxisomes occurring in the storage tissues of oilseeds (1), we decided to investigate the presence of SOD activity in these organelles. This communication reports the SODs present in watermelon cotyledons, their molecular weights and subcellular localization, with the conclusion of the existence of a Cu-Zn/SOD in glyoxysomes.

MATERIALS AND METHODS

Watermelon seeds (Citrullus vulgaris Schrad., var. sugar baby) were germinated under dark conditions, essentially as described by Hock (3). Four-days-old cotyledons (75 pairs) were carefully homogenized in 75 ml of a medium containing 0.25 M sucrose, 50 mM Hepes buffer (pH 7.5), 0.1% BSA, 10 mM KCl, 1 mM $MgCl_2$, 1 mM Na_2-EDTA and 10 mM DTT. All operations were performed at 0-4°C. The homogenate was centrifuged at 270 g for 10 min and then at 12,000 g for 15 min. The resulting pellet was gently resuspended in 2.5 ml of extraction medium and centrifuged on a discontinuous sucrose gradient (20-57%; w/w) at 83,000 g for 4 h. Glyoxysomes and mitochondria were identified in the gradient fractions (ca. 1.2 ml) by using catalase (4) and cytochrome oxidase (5), respectively, as enzyme markers. Total SOD activity was determined as described by Elstner and Heupel (6) and proteins by the Bradford procedure (7). Gel electrophoresis was performed on 10% acrylamide gels (8) and SOD activity was localized by the photochemical method (9). Activity bands were recorded by gel densitometry. The molecular weight of SODs was determined by thin-layer electrophoresis at different polyacrylamide-gel concentrations (5-10%)(10).

---

*Abbreviations: SOD, superoxide dismutase; Cu-Zn/SOD, cuprozinc superoxide dismutase; Mn-SOD, manganese superoxide dismutase.

RESULTS AND DISCUSSION

Watermelon cotyledons contained at least four electrophoretically distinct SODs which were identified, by using 2 mM CN^- and 2 mM H_2O_2, as two manganese- and two copperzinc-containing SODs (Fig. 1).

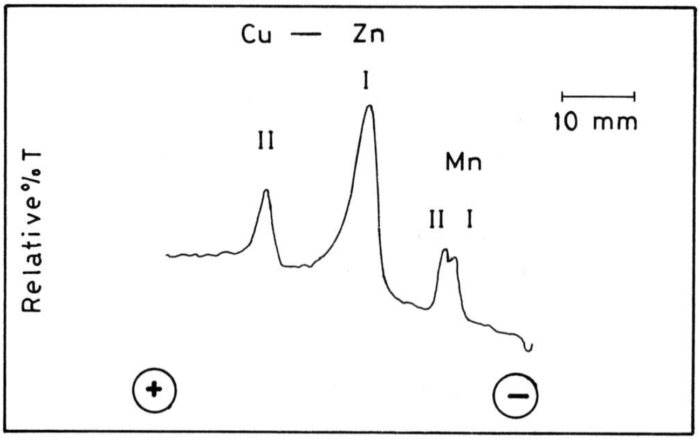

Fig. 1. Densitometric scan of SOD activity-stained gel electrophoretogram of extracts from watermelon cotyledons.

The molecular weights of SOD isozymes were about 76,000 and 88,000 for Mn-SODs I and II, and 39,000 and 32,000 for Cu-Zn/SODs I and II, respectively. The Ferguson plots obtained (10) indicated that Cu-Zn/SODs I and II could be charge isomers, varying mainly in their net charge.

Sucrose gradient centrifugation of homogenates from watermelon cotyledons provided a neat separation of glyoxysomes (fractions 4-8) from mitochondria (fractions 16-20) (Fig. 2a), and two main peaks of SOD activity were found to coincide with those two subcellular particles (Fig. 2b). The SODs present in the isolated cell organelles were identified by polyacrylamide gel electrophoresis, all of the fractions defining each peak being analyzed. Purified glyoxysomes contained preponderatingly Cu-Zn/SOD II isozyme and, to a lesser extent, a Mn-SOD isozyme. The bulk of Mn-SOD activity was present in mitochondria, whereas Cu-Zn/SOD I isozyme apparently was distributed between mitochondria and the cytosol.

The presence of SOD activity in watermelon glyoxysomes, a type of specialized plant peroxisomes, is in accordance with previous results reporting the localization of SOD activity in pea leaf peroxisomes (2) and provides an additional evidence to hypothesize a role for superoxide dismutases in the metabolism of peroxisomes.

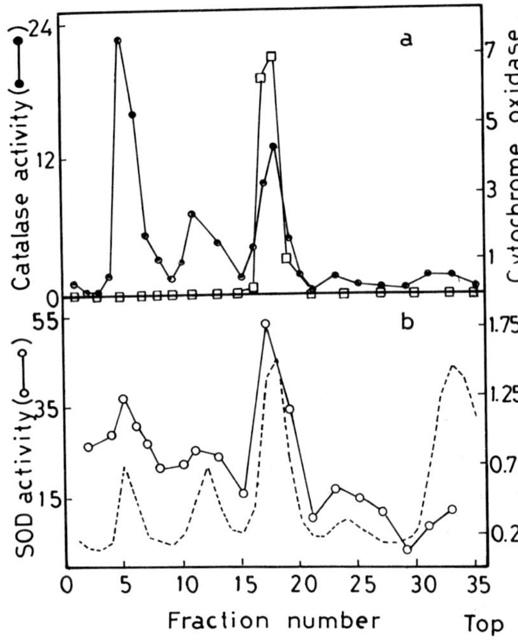

Fig. 2. Sucrose density gradient separation of crude particulate pellet from watermelon cotyledons. Fractions were eluted from the bottom of the tubes.

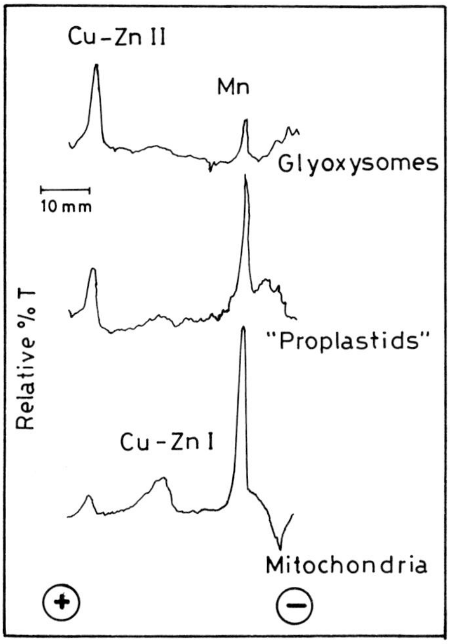

Fig. 3. Distribution of SOD isozymes in subcellular fractions isolated from watermelon cotyledons. Analyzed fractions corresponded to the maximum of each organelle peak.

ACKNOWLEDGEMENTS

LAR would like to thank Prof. Bertold Hock, Institute of Botany, Technical University of Munich, FRG, for his kind help in the initial stages of this project and his encouraging advices. This work was supported by research grant # 2511/83 from the CAICYT (Spain).

REFERENCES

1. Huang AHC, Trelease RN and Moore TS,Jr (1983) Plant Peroxisomes. Academic Press, New York
2. Del Río LA, Lyon DS, Olah I, Glick B and Salin ML (1983) Planta 158: 216-224
3. Hock B (1969) Planta 85: 340-350
4. Del Río LA, Gómez M, Leal A and López J (1977) Anal Biochem 80: 409-415
5. Schnarrenberger CA, Oeser A and Tolbert NE (1971) Plant Physiol 48: 566-574
6. Elstner EF and Heupel A (1976) Anal Biochem 70: 616-620
7. Bradford MM (1976) Anal Biochem 72: 248-254
8. Davis BJ (1964) Ann N Y Acad Sci 121: 404-427
9. Weisiger RA and Fridovich I (1973) J Biol Chem 248: 3582-3592
10. Hedrick JL and Smith AJ (1968) Arch Biochem Biophys 126: 155-164

Superoxide and Superoxide Dismutase in Chemistry, Biology and Medicine; G. Rotilio, editor

GENERATION OF SUPEROXIDE ANION AND SUPEROXIDE DISMUTASE ACTIVITY IN TOMATO PLANTS INFECTED BY NEMATODES

GIUSEPPE ZACHEO AND TERESA BLEVE ZACHEO
Istituto di Nematologia Agraria, C.N.R., Via G. Amendola 165/A, 70126 Bari, Italy.

INTRODUCTION

Tomato plant cells resistant to the root-knot nematode, *Meloidogyne incognita*, react hypersensitively within about 24 h to nematode invasion[1,2]. Hypersensitive response of host cells to nematode attack results in cell browning and is assumed to be an early event associated with the resistance.

Hypersensitive cell death has been found associated with elicitation of physiological and biochemical reactions in the tissues neighbouring the dead cells.

During the invasion of resistant, but not of susceptible tomato cultivars by *M. incognita*, superoxide dismutases (SOD) decrease and peroxidases as well as the rate of a non cytochrome respiratory pathway increase, while the course of the test based on reduction of extracellular cytochrome c also accelerated[3,4,5,6]. However it still remains to be elucidated why nematode feeding induces cell death in the resistant root tissue and giant cells in the susceptible one. Preliminary experiments suggest that superoxides, which are toxic and destructive to living cells, may be involved in the occurrence of hypersensitive cell death[6]. The present data describe the observations on the reduction of nitroblue tetrazolium (NBT) and changes in superoxide dismutase activity in susceptible and resistant tomato roots, attacked by *M. incognita*.

MATERIAL AND METHODS

Seedlings of tomato cultivars Roma VF, Early Pak and Red Stone (susceptible) and VFN8, Piersol and Ronita (resistant) were germinated in sterilized quartz sand. Uniformly germinated seeds were transferred to 3 cm clay pots containing quartz sand and divided in two groups, one of which, uninfested, was used as a control whereas the other was immediately inoculated with larvae of *M. incognita*. Four days after inoculation, healthy and infested roots were removed and tested. SOD was assayed by the xanthine oxidase cytochrome c method[7] with 0.02 mM KCN added to suppress cytochrome c oxidase.

The detection of $O_2^{\overline{\cdot}}$ was based on its ability to reduce nitroblue tetrazolium (NBT)[8]. For microscopic study, infested roots were immersed in a solution containing 0.05% NBT in 10 mM potassium phosphate buffer or in buffer alone for 1 h. Roots were fixed in 3% glutaraldehyde-formaldehyde in 0.05 M cacodylate buffer for 2 h, dehydrated and embedded in Spurr's medium. Semithin sections, either unstained or stained with toluidine blue, were observed under a light microscope.

RESULTS

The solution containing NBT, which was originally pale yellow, became bluish after the tomato roots were immersed and heated in it. This indicated that NBT had been reduced to formazan. The susceptible roots infested with *M. incognita* showed limited evidence of formazan formation outside the tissue, compared with the uninfested control. The resistant roots showed a marked activity in reducing the NBT (Table I).

TABLE I

REDUCTION OF NBT AND SOD ACTIVITY IN ROOTS OF TOMATO CULTIVARS SUSCEPTIBLE AND RESISTANT TO *M. INCOGNITA*.

NBT reduction is expressed as OD_{580} dry weight per hour; SOD activity is expressed as U/mg of proteins

	NBT reduction			SOD		
Cultivar	Not infested	Infested	Significance (LSD)	Not infested	Infested	Significance (LSD)
Roma VF	427	161	0.01	13	39	0.01
Early Pak	242	149	0.01	12	23	0.01
Red Stone	312	191	0.01	6.52	18.62	0.01
VFN8	235	395	0.01	8.20	8.20	N.S.
Ronita	489	619	0.01	12	10.57	N.S.
Piersol	356	556	0.01	9.70	8.70	N.S.

SOD activity following nematode attack increased considerably, i.e., three times over the uninfested control, in susceptible tomato roots, whereas decreased slightly in roots of the resistant cultivars (Table I). An inverse correlation was observed between changes in SOD and NBT reduction. Whenever NBT reduction increased, SOD activity decreased (resistant plants) and when NBT reduction decreased, SOD activity increased (susceptible plants). The reduction of NBT was stimulated by adding NADPH or NADH to the reaction mixture containing NBT. The effect of the chemicals, KCN and SHAM (salicylhydroxamic acid), which are known to inhibit cyanide sensitive (cytochrome oxidase) and cyanide-resistant respiration (alternative oxidase) respectively, was tested. Addition of KCN in the reaction system increased the ability of roots in reducing NBT; when SHAM was added, the reduction of NBT was lowered (Table II). As the reduction of NBT seems to be mediated by superoxide, which depends on NADPH oxidation, it is assumed that a terminal oxidase may be involved in the reaction system and the increased SOD activity in susceptible cultivars contributes to lower the amount of superoxides. In resistant tomato roots, a localized accumulation of blue

TABLE II

EFFECT OF SOME COMPOUNDS ON THE REDUCTION OF NITROBLUE TETRAZOLIUM (NBT) BY ROOTS OF TOMATO, ROMA VF (SUSCEPTIBLE) AND VFN8 (RESISTANT), INFESTED AND NOT WITH *M. INCOGNITA*.

Roots were immersed in 0.01 M potassium phosphate buffer pH 7.8 containing 0.05% NBT and 0.1 mM EDTA. The reducing activity is expressed as OD_{580} dry weight per hour.

	Roma VF			VFN8		
Treatment	Not infested	Infested	Significance (LSD)	Not infested	Infested	Significance (LSD)
Buffer	183	146	0.01	118	196	0.01
10 μM NADPH	235	204	0.01	140	257	0.01
10 μM NADH	245	196	0.01	217	358	0.01
0.4 mM KCN	253	206	0.01	192	309	0.01
1 mM SHAM	85	96	N.S.	59	102	0.01

pigment (formazan) was found around the feeding site of nematodes (Fig. 1).

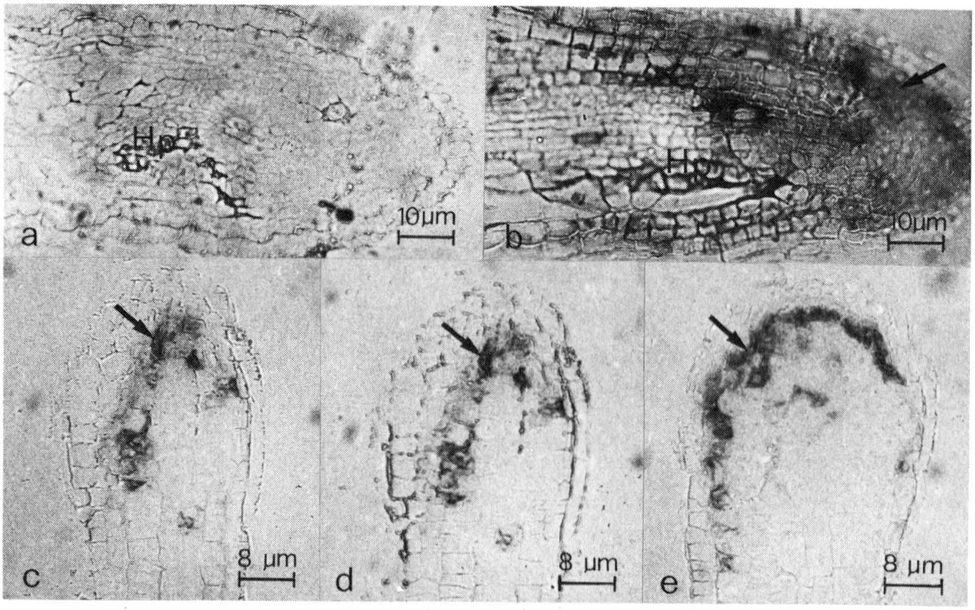

Fig. 1. Localized activation of NBT reduction in resistant tomato root infested by *M. incognita*. a) Section of an untreated root tip. b) Section of a root tip treated with NBT, where active reducing sites of NBT (arrow) and hypersensitively dead cells (Hp) are present. c, d, e) Unstained serial sections of the same root tip, treated with NBT.

The formazan was typically seen in the cells not directly affected by nematode feeding. In dead cells, fed up by nematodes, the blue pigment was scattered in a granulated form. Susceptible roots showed no localized accumulation of formazan in the tissues.

REFERENCES

1. Paulson RE, Webster JM (1972) Ultrastructure of the hypersensitive reaction in roots of tomato, *Lycopersicon esculentum* L., to infection by the root-knot nematode *Meloidogyne incognita*. Physiol Plant Pathol 2: 227-234.
2. Bleve Zacheo T, Zacheo G, Melillo MT, Lamberti F (1982) Ultrastructural aspects of the hypersensitive reaction in tomato root cells resistant to *Meloidogyne incognita*. Nematol medit 10: 81-90.
3. Zacheo G, Bleve Zacheo T, Lamberti F (1982) Role of peroxidase and superoxide dismutase activity in resistant and susceptible tomato cultivars infested by *Meloidogyne incognita*. Nematol medit 10: 75-80.
4. Zacheo G, Arrigoni Liso R, Bleve Zacheo T, Lamberti F, Arrigoni O (1983) Mitochondrial peroxidase and superoxide dismutase activities during the infection by *Meloidogyne incognita* of susceptible and resistant tomato plants. Nematol medit 11: 107-114.
5. Zacheo G, Bleve Zacheo T (1984) Superoxide anion generation during the hypersensitive response in plants infested by nematodes. Phytopath medit XXIII: in press.
6. Zacheo G, Bleve Zacheo T (1985) Cyanide-resistant respiration in tomato roots infested by *Meloidogyne incognita*. Nematol medit 13: in press.
7. Blum J, Fridovich I (1983) Superoxide, hydrogen peroxide and oxygen toxicity in two free-living nematode species. Archiv Biochem Biophys 222: 35-43.
8. Doke N (1983) Involvement of superoxide anion generation in the hypersensitive response of potato tuber tissues to infection with an incompatible race of *Phytophthora infestans* and to the hyphal wall components. Physiol Plant Pathol 23: 345-357.

CELLULAR AND PHYSIOLOGICAL ASPECTS OF OXY-RADICALS AND THEIR RELATED ENZYMES

Superoxide and Superoxide Dismutase in Chemistry,
Biology and Medicine; G. Rotilio, editor

DETECTION OF SUPEROXIDE AS A CONSEQUENCE OF CELLULAR DRUG METABOLISM

GERALD M. ROSEN, BRUCE A. FREEMAN AND CHARLES M. MANSBACH, II.

Departments of Pharmacology and Medicine, Duke University Medical Center, Durham, North Carolina 27710 (USA)

INTRODUCTION

The most difficult obstacle to the study of biologically generated free radicals has been the identification of these highly reactive intermediates. One method that has been used successfully is spin trapping (1). This technique consists of using a nitrone or nitroso compound to "trap" the initial unstable free radical as a "long-lived" nitroxide that can be observed using electron spin resonance (ESR) spectrometric procedures (2).

Spin trapping was used to study the formation of superoxide arising as the result of cellular metabolism of either the quinone, menadione or the nitro-containing compound, nitrazepam. Two different cell types were employed: (1) cultured endothelial cells isolated from the porcine thoracic aorta, and (2) isolated enterocytes from the rat intestine.

MATERIALS AND METHODS

General Comments. Menadione (2-methyl-1,4-naphthoquinone), N,N-bis(2-[bis(carboxymethyl)amino]ethyl)glycinediethylenetriaminepentaacetic acid (DETAPAC), bovine erythrocyte Cu/Zn superoxide dismutase, glutathione, glutathione peroxidase and NADPH were purchased from Sigma Chemical Company. Nitrazepam (1,3-dihydro-7-nitro-5-phenyl-2H-1,4-benzodiazepin-2-one) was generously provided by Hoffmann-LaRoche. Chelex 100 ion-exchange resin was obtained from Bio-Rad. The spin trap 5,5-dimethyl-1-pyrroline-1-oxide (DMPO) was synthesized according to the procedure of Bonnett, et al. (3). Unless stated otherwise, all buffers were passed through a Chelex 100 ion exchange column to remove divalent metal ion impurities as described by Poyer and McCay (4). ESR spectra were recorded on a Varian Associates Model E-9 spectrometer. NADPH-cytochrome P-450 reductase was purified from rat hepatic microsomes according to the procedure of Yasukochi and Masters (5).

Cell Preparation. Porcine thoracic aorta endothelial cells were grown according to published procedures (6). Rat enterocytes were isolated as described previously (7).

Biochemical Analyses. Menadione (0.1 mM in 0.14 M DMSO) or nitrazepam (1 mM in 0.14 M DMSO) was incubated with purified NADPH-cytochrome P-450 reductase (10 μg/ml) and NADPH (250 μ M), which, in the presence of succinoylated cytochrome c was used to determine the rate of superoxide generation (8). Under these conditions, 20 μ M/min of superoxide was produced. Preliminary experiments had indicated that this rate of superoxide generation was satisfactory, these drug concentrations were used for all future cell suspension studies. Extracellular superoxide generation was measured as the SOD-inhibitable reduction of succinoylated cytochrome c (9). NADPH-Cytochrome P-450 reductase, DT-diaphorase, and glutathione peroxidase activities were determined as described (10).

Spin Trapping. Experiments were designed to detect menadione- or nitrazepam-stimulated superoxide generation by cells. A typical reaction mixture contained cells (0.4 - 1 million), menadione (0.1 mM in 0.14 M DMSO), nitrazepam (1 mM in 0.14 DMSO) or DMSO (0.14 M), DMPO (0.1 M), DETAPAC (1 mM), and sufficient buffer to bring the final volume to 0.5 ml.

RESULTS AND DISCUSSION

Incubation of a suspension of cultured endothelial cells with menadione in the presence of DMPO resulted in the formation of paramagnetic species (Fig. 1). In the absence of menadione, no ESR spectrum was observed. Scan B in Fig. 1, obtained 10 min after scan A was initiated, points to the existence of three distinct products: DMPO-CH_3 (peak 1), DMPO-OH (peak 2), and DMPO-OOH (peak 3). Identification of these spin trapped adducts was based on a number of factors including the published ESR spectra of these nitroxides (1,11). Inclusion of SOD (20 μg/ml) in the experiment gave the ESR spectrum shown in Fig. 2, which is characteristic of DMPO-OH (11). These data point to the fact that significant quantities of superoxide was spin trapped, extracellularly. The source of this superoxide is of interest. Recently, Thor, et al. (12) had suggested that during menadione metabolism by hepatocytes, superoxide was generated intracellularly, which diffused into extracellular spaces. Our experiments under various experimental conditions demonstrated that extracellular superoxide is the result of endotheliar metabolism of menadione to menadiol which diffuses across the cell membrane. Reaction with menadione gives the semiquinone. In the presence of oxygen, superoxide is generated (13).

The ESR spectrum of DMPO-OH (Fig. 2) does not, without additional evidence, prove that hydroxyl radical was spin trapped. It merely defines the stability of other spin trapped adducts. We have previously shown that DMPO-OOH is unstable and rapidly decomposes into several species one of which is DMPO-OH (14). More recently, we have found that glutathione peroxidase will rapidly reduce DMPO-OOH to DMPO-OH (13). For this reason, we measured glutathione peroxidase activity of endothelial cells and found it to be 40 nmoles/min/mg protein. Although this is not unequivocal proof that Fig. 2 is the result of intracellular spin trapping of superoxide followed by rapid glutathione peroxidase reduction of DMPO-OOH to DMPO-OH, the evidence gathered points to this conclusion.

We have found that nitrazepam can be reduced by NADPH-cytochrome P-450 reductase to its nitro anion free radical and that, in the presence of oxygen,

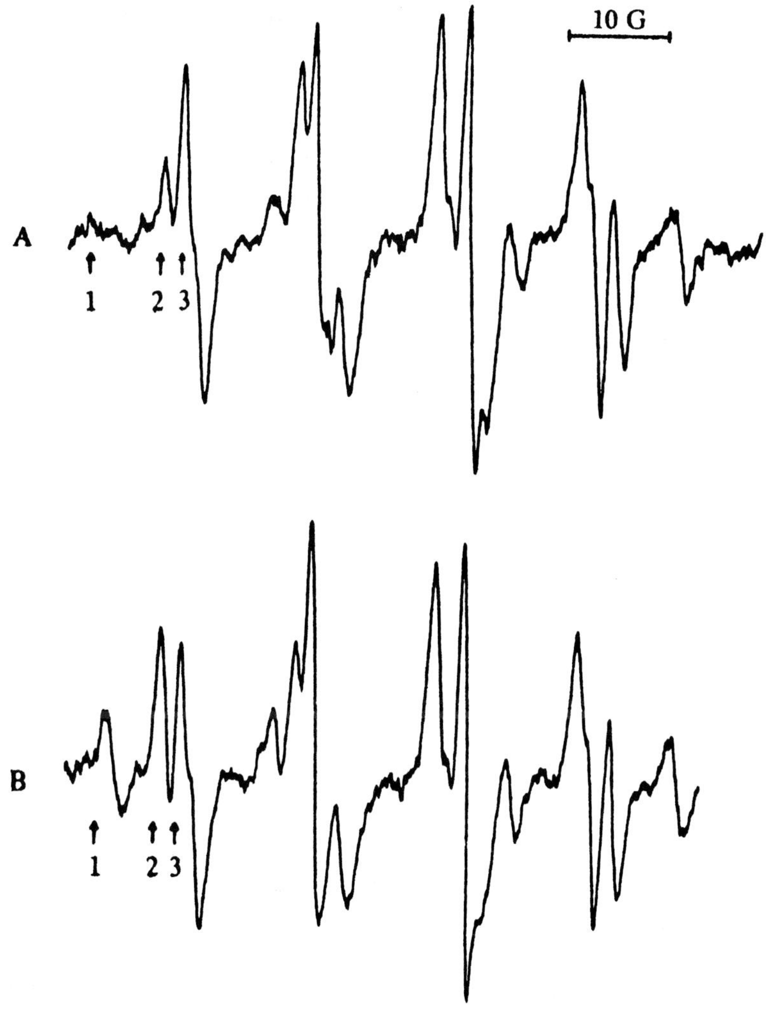

Fig. 1. ESR spectra obtained when endothelial cells were incubated with menadione in the presence of DMPO. Of interest is the observation that this spectrum is a composite of three spin-trapped adducts: 2,2,5-trimethylpyrrolidinyloxyl (DMPO-CH3, peak 1), 2,2-dimethyl-5-hydroxy-1-pyrrolidinyloxyl (DMPO-OH, peak 2), and 2,2-dimethyl-5-hydroperoxy-1-pyrrolidinyloxyl (DMPO-OOH, peak 3). The spectra were obtained under identical conditiond except that scan B was initiated 10 min after the reaction was started.

superoxide is generated (15). We felt that this compound would be an idea probe to generate intracellular superoxide, as a result of cellular metabolism, since no source of extracellular superoxide can be obtained under these experimental conditions. Incubation of endothelial cells with nitrazepam and DMPO resulted in an ESR spectrum identical to that observed when menadione was added to endothelial cells in the presence of DMPO and extracellular superoxide dismutase (Fig. 2). The addition of SOD (20 μg/ml) to the nitrazepam-endothelial cell experiment in the presence of the spin trap did not alter the ESR spectrum. In the absence of nitrazepam, no ESR spectrum was observed. These results, in combination with those obtained using menadione, demonstrate that spin trapping techniques can detect intracellularly generated superoxide.

Incubation of enterocytes with menadione and DMPO resulted in the formation of a paramagnetic species (Fig. 3A), which was not inhibited in the presence of SOD. In the absence of menadione, no ESR spectrum was observed (Fig. 3B). The spectrum depicted in Fig. 3A is characteristic of DMPO-OH (11), and is identical to that ESR spectrum recorded when endothelial cells are incubated with menadione and DMPO in the presence of SOD (Fig. 2). As discussed earlier, the appearance of the ESR spectrum for DMPO-OH does not prove that hydroxyl radical was spin trapped. As in the case of endothelial cells, all evidence gathered points to the fact that only superoxide is generated during the metabolism of menadione by enterocytes (10). The lack of stability of DMPO-OOH may be the result of rapid bioreduction by glutathione peroxidase. For this reason, we determined the glutathione peroxidase activity of enterocytes and found it to be 20 nmoles/min/mg protein.

Finally, our results raise the question as to why only intracellular superoxide was spin trapped. One possible explanation is that the rate of superoxide generation extracellularly is too slow to allow DMPO to realistically compete with other reactions (e.g., self dismutation of superoxide)(16). To test this theory, we directly measured the extracellular production of superoxide by incubating enterocytes with menadione in the presence of succinoylated cytochrome c. We found that the rate of extracellular superoxide was only 70 nM/min/million cells, apparently too slow to be measured by spin trapping methods.

Since we have earlier demonstrated the spin trapping of extracellular superoxide during the metabolism of menadione by endothelial cells, we questioned why this was not the case for enterocytes. We found, using homogenates of enterocytes, that the reduction of menadione via a one-electron route (NADPH-cytochrome P-450 reductase) was 69 nmoles/min/mg protein, while for the two-electron pathway (DT-diaphorase), the reduction was only 34 nmoles/min/mg protein. This three-fold difference in reductive pathways could account for the low level of extracellular superoxide produced. As with

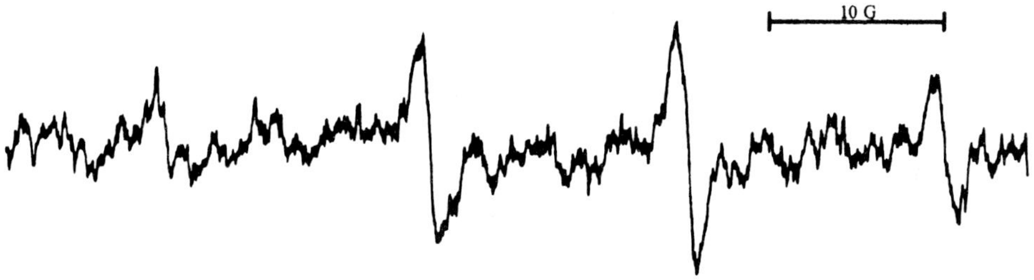

Fig. 2. ESR spectrum obtained when endothelial cells were incubated with menadione in the presence of DMPO and SOD. Hyperifine coupling constants: A_N = A_H = 14.9 G.

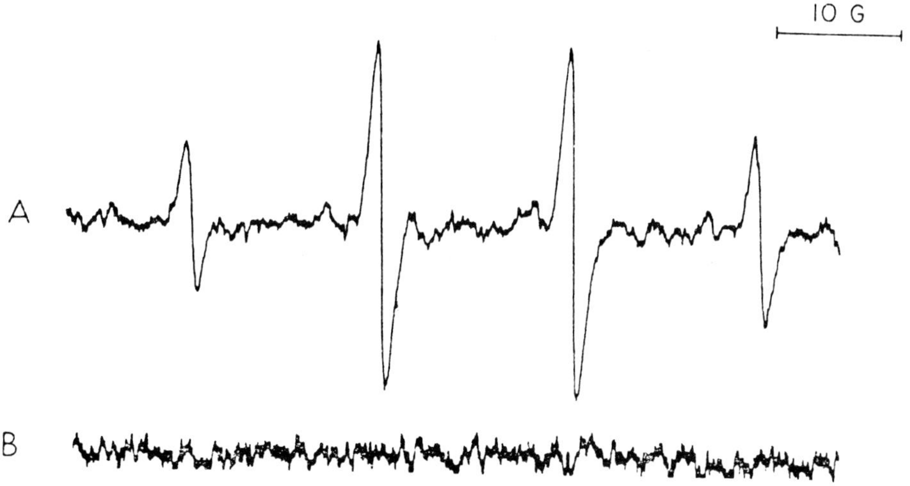

Fig. 3. ESR spectra obtained when enterocytes were incubated with DMPO in the presence (scan A) and absence (scan B) of menadione. Identical results were obtained when nitrazepam was substituted for menadione. Hyperfine coupling constants: A_N = A_H = 14.9 G.

endothelial cells, enterocytes incubated with nitrazepam gave only intracellular superoxide as detected by spin trapping procedures (Fig. 3). Under these experimental conditions, we were unable to detect any extracellular superoxide even using the ferricytochrome c reduction procedure.

We have demonstrated that endothelial cell reductases can univalently reduce menadione to menadione semiquinone, which can then transfer an electron to oxygen, yielding superoxide. Cellular DT-diaphorase can divalently reduce manadione to menadiol, which can diffuse across the plasma membrane and react with menadione, forming menadione semiquinone by disproportionation. Reduction of oxygen gives superoxide. In this manner, superoxide can be generated and detected both intra- and extracellularly using spin trapping methods. For enterocytes, the univalent reductive pathway predominates, and as such only intracellular superoxide was observed, as measured by spin trapping procedures.

REFERENCES

1. Finkelstein E, Rosen GM, Rauckman EJ (1980) Arch Biochem Biophys 200:1-16

2. Janzen EG (1980) In: Pryor WA (ed) Free Radical in Biology. Vol 4, Academic Press, New York, pp 115-154

3. Bonnett R, Brown RFC, Clark VM, Sutherland IO, Todd A (1959) J Chem Soc 2094-2101

4. Poyer JL, McCay PB (1971) J Biol Chem 246:263-269

5. Yasukochi Y, Masters BSS (1976) J Biol Chem 251:5337-5344

6. Freeman BA, Young SL, Crapo JD (1983) J Biol Chem 258:12534-12542

7. Hartman F, Owen R, Bissell DM (1982) Am J Physiol 242:G147-G155

8. Finkelstein E, Rosen GM, Patton SE, Cohen MS, Rauckman EJ (1981) Biochem Biophys Res Commun 102:1008-1015

9. Kuthan H, Ullrich V, Estabrook RW (1982) Biochem J 203: 551-558

10. Mansbach CM, II, Rosen GM, Rahn CA, Strauss KE (1985) Mol Pharmacol (submitted for publication)

11. Finkelstein E, Rosen GM, Rauckman EJ, Paxton J (1979) Mol Pharmacol 16: 676-685

12. Thor H, Smith MT, Hartzell P, Bellomo G, Jewel SA, Orrenius S (1982) J Biol Chem 257:12419-12425

13. Rosen GM, Freeman BA (1984) Proc Natl Acad Sci USA 81:7269-7273

14. Finkelstein E, Rosen GM, Rauckman EJ (1982) Mol Pharmacol 21:262-265

15. Rosen GM,Rauckman EJ, Wilson RL, Jr, Tschanz C (1984) Xenobiotica 14:785-794

16. Finkelstein E, Rosen GM, Rauckman EJ (1980) J Amer Chem Soc 102:4994-4999

Superoxide and Superoxide Dismutase in Chemistry,
Biology and Medicine; G. Rotilio, editor

INTRACELLULAR HYDROGEN PEROXIDE STIMULATES LOW LEVEL CHEMILUMINESCENCE

JULIO F. TURRENS, CECILIA GIULIVI and ALBERTO BOVERIS

Instituto de Química y Fisicoquímica Biológicas, Facultad de Farmacia y Bioquímica, Universidad de Buenos Aires. Junín 956, 1113-Buenos Aires (Argentina).

INTRODUCTION

The emission of low level chemiluminescence appears to be the consequence of free radical generation and correlates with lipid peroxidation (1-3). Different pieces of evidence point to both singlet oxygen and excited carbonyls as the main source of light emission (1,4,5). Several biological systems including intact liver (1,3), isolated hepatocytes (2), subcellular fractions (6) and isolated enzymatic systems (7) are effective sources of chemiluminescence. In this study we report the effect of increasing the physiological steady state concentration of H_2O_2 on the low level chemiluminescence of both intact hepatocytes and liver homogenates.

MATERIALS AND METHODS

Hepatocytes were isolated by liver perfusion with collagenase (viability > 95%). Cells and homogenates were suspended in 6 mM glucose, 135 mM NaCl and 10 mM Tris-HCl (pH 7.4). Chemiluminescence was measured in a Johnson Research Foundation Photon Counter (1), and expressed as counts per second (cps).

RESULTS AND DISCUSSION

The intracellular concentration of H_2O_2 was increased by inhibiting its metabolism through the enzymes catalase and glutathione peroxidase. For this purpose we used 3-amino-1,2,4-triazol (ATZ) and iodoacetamide (IAM). The first drug inhibits catalase reacting with the enzyme-substrate Compound I. The latter reacts with

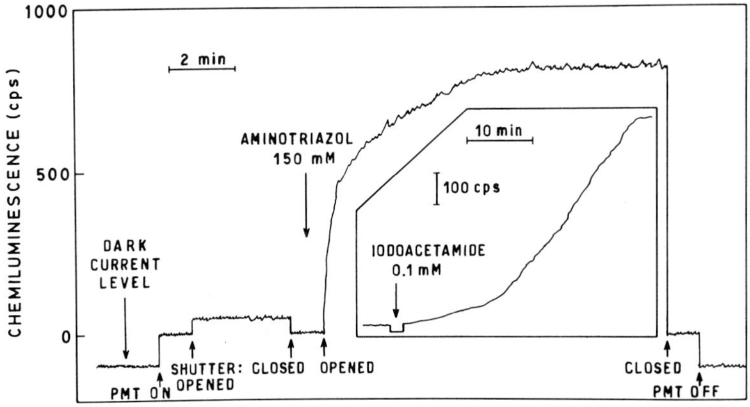

Figure 1: Low level chemiluminescence induced by ATZ or IAM addition. When ATZ was used, osmolarity was kept constant by varying NaCl concentration. PMT means photomultiplier. Protein concentration: 1 mg/ml.

reduced glutathione, thus inhibiting glutathione peroxidase activity through elimination of one substrate.

Spontaneous chemiluminescence from isolated hepatocytes (25 ± 1 cps/mg protein) was increased up to 20-fold after addition of either 150 mM ATZ or 0.1 mM IAM (Figure 1). When catalase was inhibited, chemiluminescence reached a maximum 3 to 5 min after addition of ATZ. Iodoacetamide-dependent chemiluminescence showed a 10 to 20 min lag phase before increasing steadily during 20 to 30 min. The difference in the response time may reflect different rates of reaction between inhibitors and substrates and also the fact that ATZ inhibits an enzyme whose activity directly regulates the H_2O_2 steady state whereas glutathione peroxidase activity will only be affected when reduced glutathione concentrations fall underneath the Km value.

Aminotriazol enhanced chemiluminescence in a dose dependent manner when added to both intact hepatocytes and liver homogenates (Figure 2A). The increase in chemiluminescence correlated with catalase inhibition. The concentration of ATZ that produced 50% of the maximal chemiluminescence (32.5 mM) was very close to the experimentally determined Ki for catalase (33.2 mM). Similar effects on chemiluminescence by hepatocytes and liver homogenates were found upon IAM addition. The maximal effect corresponded to IAM concentrations close to 150 nmol / mg protein which are similar to the intracellular glutathione concentration (Figure 2B and Ref. 8). Iodoacetamide-dependent chemiluminescence could be reverted upon addition of 2-mercaptoethanol (Table 1) confirming that the increase in chemiluminescence was due to reduced glutathione depletion. Reduced glutathione is effective in keeping a low rate of free radical production not only as a glutathione peroxidase cofactor but as a free radical scavenger (9). When both ATZ and IAM were present together, further addition of 5 mM 2-mercaptoethanol

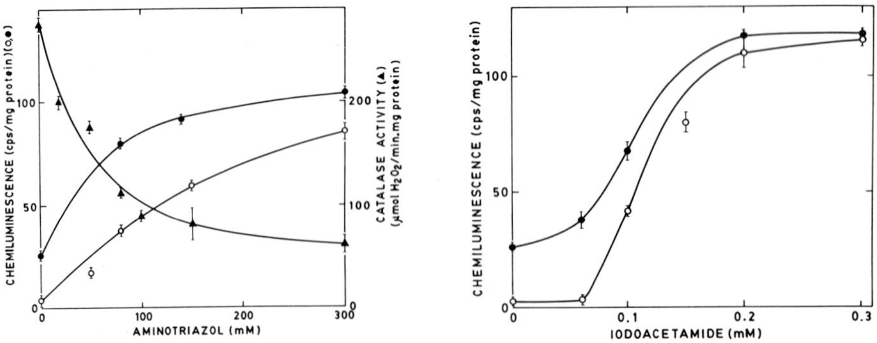

Figure 2: Effect of ATZ and IAM on low level chemiluminescence by intact hepatocytes (o) and liver homogenates (●). Fig. 2A also shows the ATZ-dependent inhibition of catalase activity (▲).

TABLE I

2-MERCAPTOETHANOL INHIBITED IAM-DEPENDENT CHEMILUMINESCENCE OF HEPATOCYTES

Values of chemiluminescence were taken 30 minutes after addition of IAM.

Iodoacetamide (mM)	Chemiluminescence (cps / mg protein)	Chemiluminescence after addition of 2-mercaptoethanol (cps / mg protein)
0	25	24
0.05	50	30
0.10	65	25
0.20	103	32
0.30	120	37

inhibited chemiluminescence in 40%, suggesting that there are two different sources responsible for maintaining an intracellular steady state concentration of H_2O_2: the peroxisomes and a second one composed by mitochondria and cytosolic enzymes (10).

From these results it can be concluded that intracellular H_2O_2 is one of the main species leading to the generation of free radicals reasponsible for lipid peroxidation.

ACKNOWLEDGEMENTS

We would like to acknowledge Dr. J.M. Dellacha and Mr. E.J. Robbetto for providing the hepatocytes used in these studies. This work was supported by grants from CONICET and SECYT.

REFERENCES

1. Boveris A, Cadenas E, Reiter R, Filipkowski M, Nakase M, Chance B (1980) Proc. Natl. Acad. Sci. USA 77: 347-351.
2. Cadenas E, Wefers H, Sies H (1981) Eur. J. Biochem. 119:531-536
3. Videla LA, Fraga CG, Koch OR, Boveris A (1983) Biochem. Pharmacol. 32:2822-25
4. Cadenas E, Sies H (1982) Eur. J. Biochem. 124: 349-356
5. Cadenas E, Sies H, Campa A, Cilento G (1984) Photochem. Photobiol. 40:661-666
6. Boveris A, Fraga CG, Varsavsky AI, Koch OR (1983) Arch. Biochem Biophys. 227: 534-541.
7. Boveris A, Cadenas E, Chance B (1980) Photobiochem. Photobiophys. 1:175-182
8. Reed DJ, Beatty PW (1978) In: Sies H, Wendel A, (eds) Functions of Glutathione in Liver and Kidney. Springer-Verlag,Berlin Heidelberg. pp. 13-21.
9. Kosower NS, Kosower EM (1976) In: Pryor WA (ed), Free Radicals in Biology. Academic Press, New York. Vol. 2, pp. 55-84.
10. Boveris A, Oshino N, Chance B (1972) Biochem. J. 128:617-630.

Superoxide and Superoxide Dismutase in Chemistry,
Biology and Medicine; G. Rotilio, editor

DNA STRAND BREAKS IN CULTURED ENDOTHELIAL CELLS INDUCED BY REACTIVE OXYGEN SPECIES: IMPLICATIONS FOR TOXIC MECHANISM

ANNE P. AUTOR AND ROBERT L. THIES

University of British Columbia, Pulmonary Research Laboratory and Department of Pathology; St. Paul's Hospital, 1081 Burrard St., Vancouver, B.C., Canada V6Z 1Y6 and Department of Pharmacology, The University of Iowa, Iowa City IA USA 52242

INTRODUCTION

Reactive oxygen species ($O_2^{-\cdot}$, H_2O_2, $HO^{\cdot}$) generated by a variety of means are known to cause tissue injury[1]. In vivo, $O_2^{-\cdot}$ and H_2O_2 are produced, for example as by-products of normal metabolism, during the metabolism of drugs capable of redox cycling, (eg. quinoid antibiotics) or by activated phagocytic cells. As a result of the catalytic action of iron, $O_2^{-\cdot}$ and H_2O_2 can produce the highly reactive hydroxyl radical, $HO^{\cdot}$[2]. The presence of reactive oxygen species renders a number of macromolecular components of the cell vulnerable to structural damage. Evidence has now accumulated indicating that reactive oxygen can induce single strand breaks in DNA[3]. Reactive oxygen species have also been implicated in the decrease of NADPH reported as a decrease of the NADPH/NADP ratio in oxidatively stressed cells[4]. Since cellular transhydrogenases maintain a balance of NADH and NADPH, oxidant stress can significantly decrease both forms of reduced pyridine nucleotides in the cell with a concomitant increase in oxidized pyridine nucleotides, thus affecting many synthetic, regulatory and reparative processes.

In our studies, isolated bovine pulmonary endothelial cells grown to confluence were used as the cellular target for oxidative stress. These cells were chosen because it is known that the pulmonary endothelium is the site of lung oxidant injury[5]. Dihydroxyfumarate (DHF) which produces reactive oxygen species during its oxidation to diketosuccinate was added to the suspending medium at the source of reactive oxygen. In previous studies using this system we observed a rapid decrease in the endothelial cell content not only of NADH but also of the total pyridine nucleotides[6]. The major pathway by which NAD^+ is catabolized is via poly (ADP-ribose) synthetase which catalyses the transfer of the ADP-ribose moiety from NAD^+ to nuclear proteins[7]. Normally at a low level of activity in cells, poly(ADP-ribose) synthetase is stimulated by cellular DNA specifically damaged by strand scission[8]. In order to determine whether pyridine nucleotide depletion is the result of a DNA strand break-induced enhancement of poly (ADP ribose) synthetase activity in oxidatively stressed cells we initiated a study to determine the effect of DHF on endothelial cell DNA in situ. The ability of superoxide dismutase, catalase, the iron chelator, deferoxamine and 3-aminobenzamide, an inhibitor of poly(ADP ribose) synthetase, to protect against DHF induced DNA strand breaks was also determined.

MATERIALS AND METHODS

Endothelial Cells

Pulmonary artery endothelial cells were isolated for culture by collagenase treatment of bovine arteries[9,10] and grown to confluence in 25 cm^2 tissue culture flasks using Medium 199 containing 20% fetal calf serum.

Endothelial cell DNA was radiolabelled by incubating the cells at 37° 24 hours prior to confluence in fresh growth medium containing 0.05 μCi/ml of ^{14}C-methyl thymidine (Amersham, Oakville Ontario). After confluence was achieved, cells were washed with Hanks buffered salt solution (HBSS) and incubated for an additional 12 hours in growth medium without radiolabelled thymidine.

Dihydroxyfumarate

Stock solutions of dihydroxyfumarate (DHF) were freshly prepared before use by dissolving 20 mg DHF in 1 ml of nitrogen-sparged 0.1N NaOH at 60° followed by dilution to a final volume of 5 ml with nitrogen-sparged distilled water. The pH of the DHF solution was adjusted to 6.8 (approximately) with 30 μl 3N NaOH. The final concentration of the stock solution was 4 mg/ml.

Incubation of endothelial cells

Twelve hours post-confluence the ^{14}c-thymidine-labeled endothelial cells were removed from the surface of the culture flasks by trypsinization. The suspended cells were washed three times and resuspended in HBSS to a density of 100,000 cells per ml. An aliquot containing approximately 5 x 10^4 cells was transferred to plastic tubes. Cells were preincubated for 30 minutes at 37° in an atmosphere of O_2:CO_2, 95%:5%. Yeast Cu/Zn superoxide dismutase, catalase and/or deferoxamine were added immediately prior to the addition of an aliquot of DHF stock solution. The DHF concentration was 100 μg in a final incubation volume of 1 ml. The suspended cells were incubated at 37° with constant agitation in an atmosphere of O_2:CO_2, 95%:5% for 30 minutes. After incubation the cell suspension was immediately chilled in ice, centrifuged and washed 3 times with cold HBSS at 4° to remove DHF and the various added agents.

Analysis of DNA

Endothelial cell DNA was evaluated by determining single strand breaks in cellular DNA before and after exposure to reactive oxygen species and specified protective agents. Single strand breaks in the DNA were determined by the alkaline unwinding technique by Rydberg[11]. Briefly, cells are exposed to an alkaline, high salt solution for a preset time in order to lyse the cells and unwind the DNA. The cell lysate is neutralized and sonicated. The fraction of

DNA pieces isolated as single stranded DNA after alkali-high salt treatment is proportional to the original number of single strand breaks in the cellular DNA. Data are expressed as the percentage remaining double stranded DNA compared with total DNA and can be expected to decrease as DNA strand scission occurs.

TABLE I

RELATIVE DNA STRAND BREAKS IN PULMONARY ARTERY ENDOTHELIAL CELLS EXPOSED TO OXYGEN RADICAL FLUX WITH DIHYDROXYFUMARATE: EFFECT OF SUPEROXIDE DISMUTASE, CATALASE, DEFEROXAMINE, OR 3-AMINOBENZAMIDE

Additions	n	Dihyroxyfumarate (DHF) (100 μg/ml)*	Double Stranded DNA** (% of total DNA)***
-	9	-	91 ± 1
-	9	+	33 ± 1
superoxide dismutase (SOD)			
(19 U/ml)	3	+	85 ± 1
(38 U/ml)	4	+	76 ± 3
(75 U/ml)	7	+	72 ± 2
(150 U/ml)	7	+	67 ± 2
(300 U/ml)	2	+	62
(600 U/ml)	7	+	38 ± 1
catalase (100 /ml)	5	+	86 ± 1
deferoxamine (DFX) (0.015 mM)	3	+	34 ± 1
(0.15 mM)	3	+	35 ± 1
(1.5 mM)	8	+	49 ± 2
(4.5 mM)	3	+	45 ± 6
SOD (75 /ml) + DFX (1.5 mM)	3	+	86 ± 1
(150 /ml) + DFX (1.5 mM)	3	+	90 ± 1
(600 /ml) + DFX (1.5 mM)	3	+	87 ± 1
3-aminobenzamide (1 mM)	3	+	32 ± 3

* Cells were incubated with DHF/added agents for 30 minutes at 37°.
The superoxide dismutase inhibitable rate of $O_2^{-\cdot}$ generation by 100 μg/ml DHF at 37° in HBSS = 3.6 ± 0.6 nmol/min.

** Calculated as follows: [double stranded DNA ÷ (double stranded DNA + single stranded DNA)] x 100

*** Data are expressed as the mean ± SEM

RESULTS AND DISCUSSION

It is clear that the DNA of pulmonary endothelial cells is damaged in situ when the cultured cells are exposed to an extracellular source of reactive oxygen (Table 1). After 30 minutes exposure to reactive oxygen the percentage of double stranded DNA diminished by approximately 60%; directly proportional to the number of single strand breaks induced in the nuclear DNA.

Several years ago it was reported that strand breakage could be induced in isolated supercoiled DNA by the action of the xanthine oxidase-xanthine system. Both superoxide dismutase and catalase protected against the DNA damage leading the authors to conclude that HO˙ was the damaging agent[12].

Recently several articles have appeared implicating oxygen free radicals in mutagenesis and DNA damage in intact cells [3,13]. DNA strand breakage and sister chromatid exchange has been observed in several systems including lymphocytes, neutrophils and cultured Chinese hamster ovary cells[14-17].

Superoxide dismutase and catalase when added extracellularly in our system protected cultured endothelial cells against DHF-induced strand breakage (Table I). Partial protection was provided by the iron chelator deferoxamine. These results are consistant with the interpretation that HO˙ produced by the reaction of $O_2^{-\cdot}$, H_2O_2 catalyzed by iron is the ultimate damaging agent.

The protective effect of superoxide dismutase diminished, however, with increasing concentrations of the enzyme. Addition of deferoxamine with these higher concentrations of superoxide dismutase restored the complete protection against oxygen radical induced DNA strand scission. These results suggest that in this system increasing concentrations of superoxide dismutase produced a higher level of H_2O_2 capable of being reduced to HO˙. Deferoxamine could prevent this reaction by removing catalytic iron thus preventing DNA damage. Cu/Zn superoxide dismutase from both bovine and yeast sources showed the same effects.

The diabetogenic agent alloxan also causes DNA strand breaks in rat pancreatic islets[18]. Furthermore, stimulation of nuclear poly(ADP ribose) synthetase and depletion of intracellular NAD was reported in this system. Superoxide dismutase, catalase and inhibitors of poly(ADP ribose) synthetase prevented the alloxan-induced inhibition of proinsulin synthesis. Catalase and superoxide dismutase but not poly(ADP ribose) synthetase inhibitors protected against DNA strand breaks. All three agents prevented alloxan induced depletion of NAD.

In our system the poly(ADP ribose) synthetase inhibitor 3-aminobenzamide gave partial protection against DHF induced depletion of intracellular NAD in pulmonary endothelial cells but was ineffective in preventing DNA strand breaks[6]. It is possible to suggest, therefore, that the NAD depletion observed in pulmonary endothelial cells exposed to reactive oxygen may be the result of the catalytic activity of poly(ADP ribose) synthetase activated by damaged DNA in these cells. Substantial depletion of pyridine nucleotides could compromise repair capacity in these cells ultimately leading to cell lysis.

REFERENCES

1. Autor AP (ed) (1982) Pathology of Oxygen. Academic Press, New York.
2. Aust SD, Morehouse LA, Thomas CE (1985) J Free Radical Biol Med 1:3-25.
3. Cerutti PA (1985) Science 227:375-381.
4. Chance B, Boveris A (1980) In: Robin ED (ed) Extrapulmonary Manifestations of Respiratory Disease. Marcell Dekker, New York, pp. 185
5. Till GO, Johnson KI, Kunkel R, Ward PA (1982) J Clin Invest 69:1126-1133.
6. Thies RL, Autor AP In preparation.
7. Olivera BM, Ferro AM (1982) In: Hayaishi O, Ueda K (ed) ADP-Ribosylation Reactions. Academic Press, New York pp. 19-40.
8. Berger N (1985) Radiat Res 101:4-15.
9. Booyse FM, Sedlak BJ, Rafelson ME (1975) Thrombos Diathes Haemorrh (Stattg) 34:825-839.
10. Autor AP, Bonham AC, Thies RL (1984) J Tox Envir Health 13:387-395.
11. Rydberg B (1975) Radiat Res 61:274-287.
12. Brawn K and Fridovich I (1980) Acta Physiol Scand Suppl 492:9-18.
13. Weitzman SA, Weitberg AB, Clark EP, Stossel TP (1985) Science 227:1231-1233.
14. Morgan AR, Cone RL, Elgert TM (1976) Nucl Acid Res 3:1139-1149.
15. Emerit I, Cerutti PA (1982) Proc Soc Acad Sci 79:7509-7513.
16. Phillips BJ, James TEB, Anderson D (1984) Muta Res 126:265-271.
17. Birnboim HC (1982) Science 215:1247-1249.
18. Uchigata Y, Yamamoto H, Kawamura A, Okamoto H (1982) J Biol Chem 257:6084-6088.

BACTERICIDAL EFFECT OF H_2O_2 AND DNA DAMAGE IN xthA MUTANTS OF E. COLI

J. ARONOVITCH*, A. SAMUNI, D. GODINGER, M. GREENBAUM AND G. CZAPSKI**
Department of Molecular Biology*, School of Medicine and Department of Physical Chemistry**, Hebrew University, Jerusalem, ISRAEL.

INTRODUCTION

Hydrogen peroxide is a normal metabolite in aerobic cells. Although not a very toxic species on its own it may react with transition metals or metal chelates to produce radicals that have been shown to damage membrane lipids and nucleic acids. E. coli cells bearing functional DNA repair genes are resistant to millimolar concentrations of H_2O_2. Endonuclease III deficient (xthA) and recA mutants have been shown to be hypersensitive to H_2O_2 (1,2). We have used these mutants to investigate the role of endogenous iron and copper in promoting the DNA damaging and bactericidal effects of H_2O_2 on growing bacterial cells.

MATERIAL AND METHODS

Wild type E. coli KL-16 and its xthA and recA mutants were grown in a glucose-mineral medium supplemented with 1% Casamino acids. Log phase cells in growth medium or washed cells suspended in 1mM phophate buffer pH 7.4 were treated with H_2O_2 (0,1-5mM); H_2O_2+1,10 phenanthroline (OP) (4-10μM), or Cu^{2+} (1-10μM)+Ascorbic acid (0.1mM). Surviving cells were counted on L-agar plates.

RESULTS AND DISCUSSION

We have observed that washed xthA cells do not differ much from wild type cells in their sensitivity to several oxygen radical generating systems i.e. H_2O_2, H_2O_2-OP, copper-ascorbic acid, UV and γ-radiation. The H_2O_2 hypersensitivity of xthA mutants was demonstrable only when H_2O_2 was added to log phase growing cells. The rate of killing increased with the ionic strength of the growth medium and with the growth rate of the culture and was markedly enhanced by anoxia (Table 1). Inhibition of protein, RNA or DNA synthesis with antibiotics had little effect on the killing rate constant. Treatment of growing cultures with H_2O_2 caused rapid depletion of the ATP pools of the cells. Removal of the H_2O_2 by catalase was followed by rapid replenishment of the ATP pools. In washed cells the decrease in the ATP level was accompanied by leakage of K^+, we therefore assume that the depletion of ATP is a result of membrane damage which activates the membrane ATPase. The time course of ATP depletion in wild type, xthA and recA cells was similar, it seems therefore that this effect of H_2O_2 does not contribute to the hypersensitivity of xthA mutants to H_2O_2.

TABLE 1. SUSCEPTIBILITY OF xthA AND recA MUTANTS (K,min^{-1}) TO VARIOUS BACTERICIDAL SYSTEMS; GROWING (G) AND WASHED (W) CELLS.

Bactericidal system:	H_2O_2	H_2O_2 anoxia	H_2O_2	H_2O_2-OP	H_2O_2-OP anoxia	H_2O_2-OP	Cu^{2+} ascorb.	UV	γ-rad.
H_2O_2 mM:	5	5	5	5	0.5	0.5	0.1	–	–
	G	G	W	G	W	W	W	W	G & W
KL-16	<0.1	0.4	<0.1	2.4	1.9	0.7	6.7	1.7	0.2
KL-16 xthA	1.15	3.4	<0.2	3.6	3.6	1.6	6.5	1.8	0.3
KL-16 recA	1.64	4.1	<0.2	5.7	4.8	1.5	7.4	34	1.4

The key role of transition metal ions in the toxicity of H_2O_2 is well documented. However, additon of various chelating agents, such as EDTA, DTPA and Desferrioxamine to growing xthA cells just before the introduction of H_2O_2 had only a slight protective effect. In contrast, preincubation of these cells with chelators, including OP, followed by washing and resuspension in growth medium had a marked protective effect. To test the premise that the intrinsic transition metal concentration determines the toxicity of H_2O_2 we used a metal poor medium. Cells were grown in this medium in the presence of DTPA or metal salts, washed, resuspended in the basal medium and then challenged with H_2O_2. Results showed that killing by H_2O_2 was proportional to the intrinsic iron content of the cells, whereas the killing of resting cells by copper-ascorbic acid was proportional to the amount of Cu^{2+} adsorbed to the cells. Washing the cells with 2% NaCl released part of the adsorbed Cu^{2+} and lowered the killing rate constant. Table 2 shows the effect of the growth medium on the sensitivity to H_2O_2 and copper-ascorbic acid.

TABLE 2. EFFECT OF GROWTH MEDIUM ON KILLING RATE K,(min^{-1}) OF xthA CELLS.

Killing by	Additions to growth medium				
	--	DTPA	DSFA	Cu^{2+}	Fe^{3+}
H_2O_2 (5mM)	0.6	0.36	0.3	0.58	0.9
Cu^{2+}(4μM),Ascorbate(0.1mM)	0.8	0.56	--	1.2	0.66

Treatment of growing cultures of wild type or mutant cells with non-lethal concentrations of H_2O_2 caused an appreciable number of single strand breaks (SSB). With bactericidal concentrations we found a correlation between the killing effect of H_2O_2 or H_2O_2-OP and the extent of DNA degradation (SSB and DSB) the hypersensitivity of xthA and recA cells to the bactericidal effect of H_2O_2 correlated with an increase in the number of SSB (Fig. 1). Since recA cells are also hypersensitive to H_2O_2 and a great part of the SSB is repairable upon incubation in growth medium, it follows that the repair of the majority of the lesions does not involve exonuclease III which explains the almost normal sensitivity of xthA cells to γ-radiation. We are currently investigating the reasons for the relative insensitivity of washed cells to H_2O_2 and why the xthA phenotype is expressed only in growing cells.

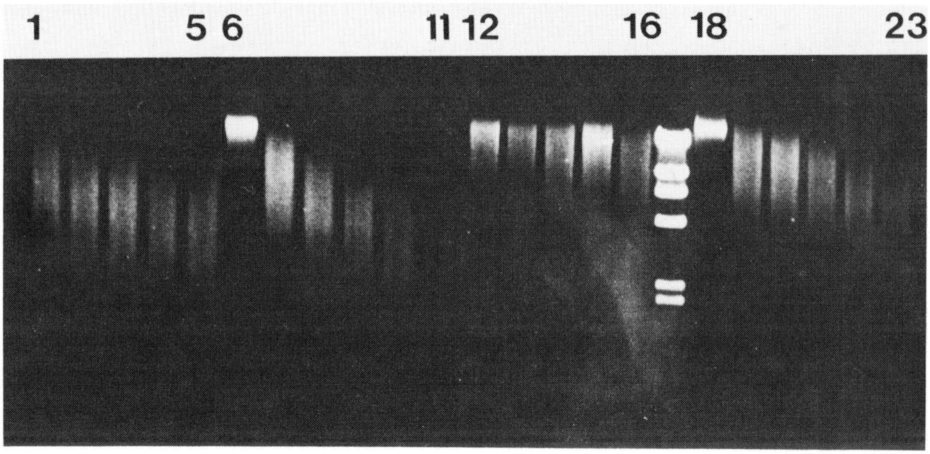

FIG. 1

	KL-16 xthA						KL-16					
	DNA CLEAVAGE (SSB) BY H_2O_2 (5mM) IN GROWING CELLS											
Lane No.	1	2	3	4	5	6	12	13	14	15	16	18
Reaction time (min)	0.25	0.5	0.75	1	1.5	0	2	4	6	8	16	0
Survival %	73	59	46	37	23	100	96	86	64	56	34	100
	DNA CLEAVAGE (SSB) BY H_2O_2 (5mM) + OP (0.02mM) IN GROWING CELLS.											
Lane No.	7	8	9	10	11	6	19	20	21	22	23	18
Reaction time (min)	0.15	0.3	0.5	0.75	1	0	0.5	1	2	3	4	0
Survival %	87	61	29	9	3	100	100	92	61	37	19	100

ACKNOWLEDGEMENT

This investigation was supported by grant #1409 of the council of Tobacco Research and by GSF Munich.

REFERENCES

1. Hartman PS, Eisenstark, A (1978) J Bact 133:769-774
2. Demple B, Halbrook J, Linn S (1983) J Bact 153:1079-1082

Superoxide and Superoxide Dismutase in Chemistry, Biology and Medicine; G. Rotilio, editor

IN VIVO DEGRADATION OF BACTERIAL DNA BY H_2O_2 AND O-PHENANTHROLINE

J. ARONOVITCH*, A. SAMUNI*, D. GODINGER* AND G. CZAPSKI**
Department of Molecular Biology*, School of Medicine and Department of Physical Chemistry**, Hebrew University, Jerusalem, ISRAEL.

INTRODUCTION

Cell damaging processes caused by active oxygen species are often mediated by H_2O_2 and transition metals and may therefore be modified by metal chelators. 1,10-phenanthroline (OP) and its derivatives which form stable chelates with transition metals inhibit the proliferation of various procaryotic and eucaryotic cells. These effects were ascribed to the chelation of enzyme-bound metal and depression of respiration (1,2). Recently it was shown that the OP-cuprous ion complex $[(OP)_2Cu^+]$ cleaves DNA in solution in a reaction requiring H_2O_2 and a reducing agent (3-5). However, OP which is toxic to cells in culture became non toxic on the addition of copper, zinc or iron (6). Moreover, OP prevented the formation of DNA breaks and the killing of mouse fibroblasts exposed to H_2O_2 or to a superoxide generating system (7-9). We tested therefore the effects of OP on bacterial cells the DNA of which may be more accesible to OP. We tried to determine whether OP enhances DNA fragmentation by H_2O_2 in vivo and if there is a correlation between the extent of DNA damage and the bactericidal effect of OP and what is the role of intrinsic metal ions in this system.

MATERIAL AND METHODS

Washed E. coli KL-16 cells suspended in phosphate buffer pH 7.4 (1mM) were treated with H_2O_2 (0.5-2 mM) and OP or OP-Cu^{2+} complexes (2-20 μM) at 37°C. Surviving cells were counted on L-agar plates. The DNA of the cells was extracted and analysed by conventional agarose gel electrophoresis methods.

RESULTS AND DISCUSSION

Micromolar concentrations of OP added to non bactericidal concentrations of H_2O_2 caused rapid killing of washed E. coli cells (Fig. 1). The rate of killing increased with the concentration of the reactants and decreased with the increase in ionic strength of the incubation mixture. Incubation of the cells with metal chelating compounds (EDTA, DTPA, desferrioxamine, OP) prior to washing, protected them against killing by subsequent addition of OP and H_2O_2. The treatment with chelators had only a small effect on the killing rate by OP-Cu^{2+} complexes. Cells grown in the presence of Cu^{2+} or Fe^{3+} salts were more sensitive to OP than cells grown in normal media. These observations supported the contention that the rate of killing by OP was dependent on the intrinsic concentration of metal ions. By determining the killing rate constants of mixtures of OP-Cu^{2+} complexes containing varying ratios of the complexes $OPCu^{2+}$, $(OP)_2Cu^{2+}$ and $(OP)_3Cu^{2+}$, we concluded that the active species is the $(OP)_2Cu^{2+}$ complex. Hydroxyl radical scavengers had no effect on the rate of killing.

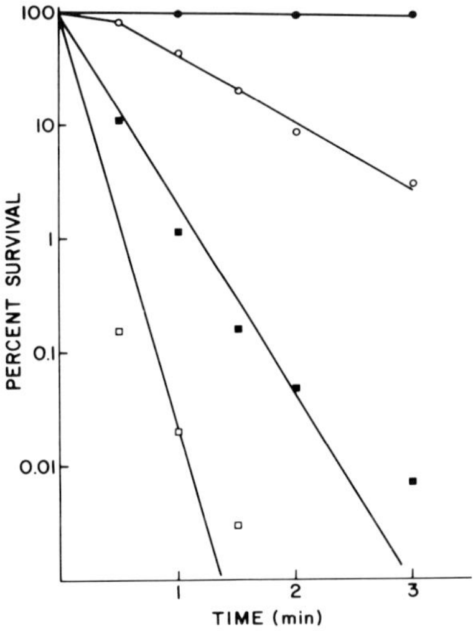

Fig. 1

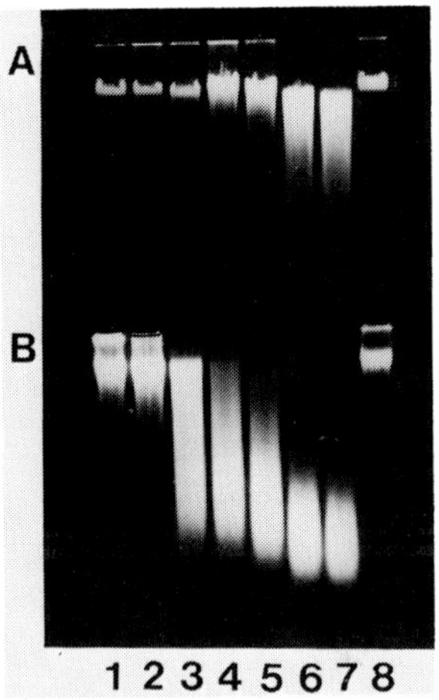

Fig. 2.

Fig. 1. Bactericidal effect of OP. ●---● H_2O_2 0.5mM; O---O H_2O_2+OP 5μM; ■---■ H_2O_2+OP:Cu^{2+} complex (20:1); □---□ H_2O_2+OP:Cu^{2+} complex under N_2.

Fig. 2. DSB (A) and SSB (B) in DNA extracted from H_2O_2+OP treated cells; Lane 1: H_2O_2 2mM; Lanes 2-7: H_2O_2+OP; 1,2,4,6,8,10 μM; Lane 8: control.

Loss of viability following treatment with copper and ascorbic acid is accompanied by manifestations of extensive membrane damage including loss of active transport, leakage of K^+ and various small molecules and depletion of ATP. In contrast we observed very little membrane damage in cells treated with H_2O_2 and OP. We tested therefore the effect of H_2O_2 with OP on the bacterial DNA. Although agarose gel electrophoresis gives mainly qualitative information on DNA fragmentation, it was clear that OP enhanced the production of both single (SSB) and double (DSB) strand breaks. The decrease in the number of surviving cells and the extent of DNA degradation were directly related to the time of exposure and to the concentration of H_2O_2 and (OP) (Fig. 2). H_2O_2 by itself ($2x10^{-3}$M) produced very little SSB. Addition of OP under conditions allowing 90% survival, caused extensive nicking of the bacterial DNA. Most of these SSB were repairable upon incubation of the cells in growth medium for 5 minutes. Double strand breaks were observable only at 10% survival or lower and there was a good correlation between the number of DBS and survival down to the level of 0.01% survival. It was shown that the copper-OP catalyzed oxidation of various biological reductants by H_2O_2 produces OH$^\bullet$ without superoxide radical being formed as an intermediate (10). We have tested the effect of anoxia on the rate of killing and the extent of DNA degradation and found that both processes are enhanced by removal of oxygen (Fig. 1) indicating that the reduction of the DNA-$(OP)_2Cu^+$ ternary complex may be the rate limiting step in vivo.

SUMMARY

1:10-Orthophenanthroline enhances killing of E. coli cells by H_2O_2 and the in vivo fragmentation of the bacterial DNA. There is a good correlation between the bactericidal effect and the number of single and double strand breaks produced in the bacterial DNA. The results suggest that intrinsic metal ions are involved in the bactericidal effect of OP and superoxide radical is probably not an obligatory intermediate.

ACKNOWLEDGMENT

This investigation was supported by grant #1409 of the council of Tobacco Research and GSF Munich.

REFERENCES

1. Shulman A, Dwyer FP (1964) In: Dwyer FP, Mellor DP (eds) Chelating Agents and Metal Chelates. Academic Press, New York, pp 383-433.
2. Smit H, VanDerGoot H, Nauta WT, Timmerman H, DeBolster MWG, Jochensen AG, Stouthamer AH, Vis RD (1981) Antimicrob. Agents and Chemother. 20:455-462.
3. Sigman DS, Graham DR, D'Aurora V, Stern AM (1979) J. Biol. Chem. 254: 12269-12272.
4. Doenry KM, Que BG, So AG (1980) Biochem. Biophys. Res. Commun. 93:264-270.
5. Gutteridge JMC, Halliwell B (1982) Biochem. Pharmacol. 31: 2801-2805.
6. Falchuk KH, Krishan A (1977) Cancer Res. 37: 2050-2056.
7. Mello Filho AC, Meneghini R (1984) Biochim. Biophys. Acta 781: 56-63.
8. Mello Filho AC, Hoffmann ME, Meneghini R (1984) Biochem. J. 218: 273-275.
9. Mello Filho AC, Meneghini R (1985) Biochim. Biophys. Acta in press.
10. Florence TM (1984) J. Inorganic Biochem. 22: 221-230.

Superoxide and Superoxide Dismutase in Chemistry,
Biology and Medicine; G. Rotilio, editor

PARAQUAT TOXICITY IS MEDIATED BY IRON OR COPPER: THE TARGET ORGANELLE IN E. COLI IS THE CYTOPLASMIC MEMBRANE

RON KOHEN, PNINA KORBASHI AND MORDECHAI CHEVION
Department of Cellular Biochemistry, The Hebrew University,
Jerusalem 91010, Israel

INTRODUCTION

Paraquat (PQ^{+2}) is a widely used herbicide, that is metabolized in various cells to yield the paraquat monocation radical ($PQ^{+\bullet}$) (1), that in time, reacts very fast with oxygen ($7.7x10^{8}M^{-1}sec^{-1}$) (2) producing the superoxide radical ($O_2^{\bar{\bullet}}$). Considerable biological understanding of the mechanism of paraquat toxicity in bacterial cells has been gained by Fridovich and co-workers (3,4,5). They showed that bacterial killing requires molecular oxygen and carbon source and that the induction of cellular SOD led to a considerable protection.

RESULTS AND DISCUSSION

We have examined the involvement of the transition metals iron and copper in paraquat toxicity both in bacterial cells and in mammalian model. With bacterial cells we have carried out a broad series of experiments in which adventitious metals or chelators were used, aerobic and anaerobic conditions were employed, free radical scavengers were introduced, and the endogenous SOD and catalases were induced. All these experiments pointed to the dramatic enhancement (over 100 fold) of the rate of bacterial inactivation by copper and iron and the total protection by chelators (6).

The metal-mediated site specific Haber-Weiss reaction is the suggested mechanism of paraquat toxicity (see scheme).

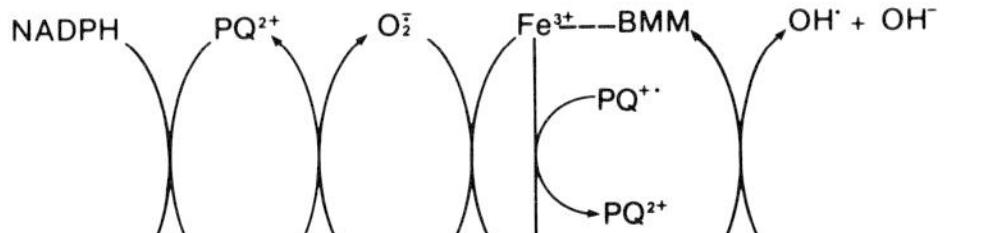

BMM = BIOLOGICAL MACRO MOLECULES

Paraquat radical reacts very fast with oxygen to generate O_2^{-}. It can also directly reduce biological complexes of transition metal ions (7,8). The superoxide radical thus formed can dismute, enzymatically or spontaneously, to produce hydrogen peroxide. Subsequently, the reduced metal complexes can react with hydrogen peroxide to produce hydroxyl radical in a site specific fashion which may be responsible for the biological damage.

Following four different paramaters in mice we could show the validity of the involvement of transition metals in paraquat intoxicated mammals (9). We used the survival of the animals, loss of weight, loss of activity of lung acetylcholine esterase and the water content of the lungs. Chelators therapy provided a significant protection (up to 70% (10)), in all four parameters, while loading the animals with transition metals markedly exacerbated the poisoning.

We have further investigated paraquat poisoning in the E. coli B cells trying to identify the target organella. Using agarose gel electrophoresis in neutral or alkaline condition, we identified *in vivo* only single strand DNA breaks, that could not account for the massive bacteria killing (3 log units after 17 min).

TABLE I

THE ROLE OF COPPER IN PARAQUAT-INDUCED DAMAGE TO E. COLI MEMBRANE

Cells were incubated at 37° for 15 min in a reaction mixture containing phosphate buffer (1mM, pH 7.4), glucose (0.5% w/v), paraquat, copper and other additions as noted. Subsequently, cellular ATP content was determined by the luciferin-luciferase method (11), retention of leucine by the cells was quantitated by counting the radioactivity measurements, and the cellular level of potassium was recorded by atomic absorption*.

Parameter	Cells/ml	PQ [mM]	Cu [μM]	Survival %	Cell content %	Remarks
ATP	$7x10^7$	0.25	5	0.01	5	
	$7x10^7$	0.25	5	91	84	Reaction carried out at 0°C
	$1x10^8$	0.25	0.6	12	6	
	$1x10^8$	0.25	-	82	98	+ Detapac [0.1mM]
	$1x10^8$		0.6	70	100	
	$1x10^8$	0.25	5	66	115	Anaerobic
	$1x10^8$	0.25	5	0.1	2.4	Anaerobic + H_2O_2 [0.1mM]
	$1x10^8$		5	101	100	
K^+	$8x10^8$	2.0	5	0.5	31	
	$8x10^8$	2.0		64	81	+ Detapac [0.1mM]
	$8x10^8$		5	114	80	
^{14}C-Leucine	$2x10^8$			100	100	
	$2x10^8$	0.5	5	0.1	1	
	$2x10^8$	0.5		95	120	+ Detapac [0.1mM]
	$2x10^8$		5	100	135	

* personal communication - J. Aronovitch

Damage to the cytoplasmic membrane was studied by following the changes of cellular level of three parameters as a function of exposure to paraquat: the content of cellular ATP and potassium (11,12) and the retention of radioactive leucine (table 1). From this table it is clear that exposure of E. coli cells to paraquat and a chelator where transition metals are removed does not cause any damage to the membrane and does not induce any bacterial killing. In contrast, when traces of copper are present in the system there is a dramatic fall in all three parameters indicating severe membrane damage that parallels or even preceded the cellular inactivation of the bacteria. Further substantiation for membrane damage can be seen from the electron microscopy studies that yield results similar to those reported by Hassan and Fridovich (13).

REFERENCES
1. Sharp CWM, Ottolenghi A, Posner HS (1972) Toxicol Appl Pharmacol 22:241
2. Farrington JA, Land ME, Fletcher K (1973) Biochim Biophys Acta 314:372
3. Hassan M, Fridovich I (1978) J Biol Chem 253:8148
4. Hassan M, Fridovich I (1977) J Biol Chem 252:7667
5. Hassan M, Fridovich I (1979) J Biol Chem 254:10846
6. Kohen R, Chevion M (1985) Free Radical Res Commun 1, in press
7. Sutton HC, Winterbourn CC (1984) Arch Biochem Biophys 235:106
8. Winterbourn CC, Sutton HC (1984) Arch Biochem Biophys 235:116
9. Kohen R, Chevion M (1985) Biochem Pharmacol 34:1841
10. Kohen R, Chevion M (1985) In:Karlin KD. Zubrita J (eds) Biochemical and Inorganic Aspects of Copper Coordination. Adenine Press, in press
11. Cole HA, Wimpenny JWT, Hughes PE (1967) Biochim Biophys Acta 143:445
12. Lundin A, Thore A (1975) Appl Microbiol 30:713
13. Hassan HM, Fridovich I (1981) Can J Physiol Pharmacol 60:1367

ACKNOWLEDGEMENT: This study was supported by a grant from the United States - Israel Binational Science Foundation.

Superoxide and Superoxide Dismutase in Chemistry, Biology and Medicine; G. Rotilio, editor

PARAQUAT KILLS CHINESE HAMSTER OVARY CELLS BY FORMING SUPEROXIDE

BAGLEY, A, KRALL, J, AND LYNCH, RE

Department of Pathology, University of Utah Medical School, Salt Lake City, Utah 84132, USA

INTRODUCTION

The ability to introduce enzymes directly into cultured cells allows one to use the specificity of an enzyme for its substrate to determine the intracellular modes of action of a drug. One such drug whose mode of action is of great interest is paraquat, which engenders superoxide as a means of killing E. coli (1). To determine whether paraquat injures cultured mammalian cells by the formation of superoxide we have introduced superoxide dismutase into Chinese hamster ovary cells to determine whether cells so treated are protected from the cytotoxicity of paraquat.

MATERIALS AND METHODS

Chinese hamster ovary K1 cells were grown in F-12 with 5/5% donor/fetal calf serum at 37 degrees in a CO_2 incubator. Enzymes were introduced into cultured cells by scrape-loading as described (2) except that scraping was done at 4 degrees. Superoxide dismutase was inactivated by hydrogen peroxide at pH 10.2 in 50 mM carbonate buffer (3) at 37 degrees. The toxicity of paraquat was assessed by exposure of cells at a density of 200 cells/60 mm dish in F-12 with 5/5% donor/fetal calf serum to the indicated concentrations of paraquat at 37 degrees in 94.5% air/5.5% CO_2 for 24 hours with constant gentle rocking. At the end of 24 hours the medium was replaced and the dishes allowed to incubate stationarily without paraquat until colonies were visible.

RESULTS

Exposure of Chinese hamster ovary cells to paraquat for 24 hours produced a dose-dependent loss of the ability to form colonies. Addition of either superoxide dismutase or catalase (40 ug/ml) to the medium did not prevent the toxicity of paraquat. However, when superoxide dismutase was loaded into the cells by scrape-loading (2), the toxicity of paraquat was decreased. The percent survival in 80 uM paraquat of cells loaded with superoxide dismutase (55.8% ± 2.8%; mean ± standard deviation) or with superoxide dismutase and catalase (46.6% ± 6/4%) was greater than the survival of cells which were loaded without added enzymes (25.52% ± 6.1%) or of cells loaded with catalase alone (29.0% ± 5.4%).

The dependency of superoxide dismutase on its enzymic activity for this effect was shown by a comparison of the effects of native vs. peroxide-inactivated enzyme on the toxicity of paraquat after scrape-loading into Chinese hamster ovary cells. At 100 uM paraquat the survival of cells loaded with superoxide dismutase (78.7% ± 3.9%) was greater than that of cells loaded without added enzymes (34.8% ± 2.0%) or of cells loaded with peroxide-inactivated superoxide dismutase (33.7% ± 4.4%). Neither dimethylsulfoxide 50 mM nor desferrioxamine 50 uM reduced the toxicity of paraquat for Chinese hamster ovary cells, suggesting that a metal-catalyzed formation of the hydroxyl radical (4-6) may not cause this cytotoxic effect of paraquat.

These data suggest that the toxicity of paraquat for these cultured mammalian cells is mediated at least partly by superoxide anion. The lack of inhibition by catalase, by desferrioxamine, or by dimethylsulfoxide suggests that the formation of the cytocidal species may not be dependent on hydrogen peroxide and transition metals. However, Chinese hamster ovary cells are rich in catalase (7). They may therefore scavenge hydrogen peroxide so efficiently that any increment in the activity of catalase produced by scrape-loading is insignificant compared with the activity already present.

REFERENCES

1. Hassan, HM & Fridovich, I (1975) J Biol Chem 253:8143-8148
2. McNeil PL, Murphy, RF, Lanni, F, Taylor, DL (1984) J Cell Biol 89:1556-1564
3. Hodgson, EK & Fridovich, I (1975) Biochemistry 14:5294-5299
4. Fenton, HFH (1894) J Chem Soc 65:899-910
5. Haber, F & Weiss, J (1934) Proc R Soc London, Ser A 147:332-351
6. Weinstein, J & Bielski, BHJ (1979) J Am Chem Soc 101:58-62
7. Lynch, RE, Unpublished observations

Superoxide and Superoxide Dismutase in Chemistry, Biology and Medicine; G. Rotilio, editor

AN ALTERNATE DEFENSE MECHANISM AGAINST PARAQUAT TOXICITY IN *ESCHERICHIA COLI*.

HOSNI M. HASSAN AND SU-MEI KAO
North Carolina State University, Raleigh, N.C. 27695-7624 USA

INTRODUCTION

Paraquat is toxic, mutagenic and causes the induction of MnSOD in *E. coli* (1-8). Recently, we have isolated mutants of *E. coli* that were about 50-fold more resistant to PQ^{2+} than the parent strain. We originally postulated that these PQ^{2+}-resistant mutants might be overproducers of MnSOD or constitutive in its biosynthesis; but this was not the case (9). The biochemical basis for an alternate defense mechanism against PQ^{2+} toxicity was explored in one of the mutants. The data indicate that the mutant is impaired in its ability to actively transport PQ^{2+}.

MATERIALS AND METHODS

Bacterial strains. Escherichia coli B (B12) ATCC 29682 and a PQ^{2+}-resistant strain AI-11 (mutant) were used (9).

Growth conditions and assays. Growth media and the preparation of dialyzed cell free extracts were as described (9,10). Superoxide dismutase was assayed according to (11). G6PD was assayed by following the reduction of NADP at 340 nm (12). PQ^{2+}-dependent NADPH-diaphorase was estimated from the rate of PQ^{2+}-dependent NADPH oxidation measured at 340mm(10). The rate of extracellular superoxide radical generation was determined according to (5).

RESULTS AND DISCUSSION.

Effects of [PQ^{2+}] on Growth and SOD Biosynthesis. The wild type and the mutant strain were exposed to a wide range of [PQ^{2+}] in a TSY medium. The mutant cells were about 50-fold more resistant to PQ^{2+} than the wild type cells (Table 1). PQ^{2+} (0.1 mM) elicited a 10-fold increase in the MnSOD of the wild type, while 10 mM of PQ^{2+} was required to elicit the same effect in the mutant. These results clearly indicate that the mutant must have some alternate mechanism(s) that allows it to grow in the presence of high concentrations of PQ^{2+} without proportionate increases in the defensive enzyme, MnSOD. This resistance to PQ^{2+} was not due to changes in hydroperoxidases (9).

TABLE 1

EFFECTS OF [PQ^{2+}] ON GROWTH AND ON THE SYNTHESIS OF SOD

STRAIN	PQ^{2+}(mM)	GENERATION TIME (min)	SOD TOTAL	SOD MnSOD
			(U/mg protein)	
29682	0	29	25.9	7.6
	0.1	57	84.5	71.6
	1.0	69	131.2	120.9
AI-11	0	30	17.1	6.7
	1.0	44	39.1	31.3
	10.0	73	62.2	54.5
	50.0	87	210.4	197.5

Effects of [PQ^{2+}] on Glucose-6-Phosphate Dehydrogenase and NADPH-:PQ^{2+}-diaphorase. In *E. coli*, the generation of $PQ^{\dot{+}}$ is catalyzed by a diaphorase at the expense of NADPH (5). Thus, the rate of $PQ^{\dot{+}}$ formation is directly related to the activity of the NADPH:PQ^{2+}-diaphorase (diaphorase) and the supply of NADPH, which is provided by the activity of G6PD. When grown in the absence of PQ^{2+}, the specific activities of these enzymes in the wild type and the mutant were comparable. On the other hand, growth in the presence of 1 mM PQ^{2+} markedly increased the activities of these enzymes in the wild type as compared to the mutant. Thus, 1 mM PQ^{2+}increased the activity of diaphorase by 3.5- and 1.5-fold, and the activity of G6PD by 10- and 4-fold in the wild type and the mutant, respectively. However, growing the mutant in the presence of 50mM PQ^{2+} resulted in further increase in the specific activities of these two enzymes to a level comparable to that of the wild type grown in the presence of 1mM PQ^{2+} (Table 2).

TABLE 2

EFFECT OF [PQ^{2+}] ON THE SYNTHESIS OF G6PD AND DIAPHORASE

STRAIN	PQ^{2+}(mM)	G6PD (U/mg protein)	DIAPHORASE (U/mg protein)
29682	0	0.22	0.04
	1	2.22	0.14
AI-11	0	0.20	0.05
	1	0.71	0.08
	50	1.59	0.17

Effects of [PQ^{2+}] on Intracellular O_2^- Generation. Both the wild type and the mutant were tested for their rate of oxygen consumption in the presence of various [PQ^{2+}] with and without 1 mM cyanide. Both strains gave biphasic reciprocal plots indicating apparent high- and low-affinity (1/Km) behavior towards PQ^{2+} (Table 3). The results indicate that the intracellular flux of the PQ^{2+}-mediated O_2^- was significantly lower in the mutant than in the wild type. Furthermore, the rate of efflux of $PQ^{\dot{+}}$ from the mutant was similar to that in the wild type (10).

TABLE 3

EFFECT OF [PQ^{2+}] ON CYANIDE RESISTANT RESPIRATION*

STRAIN	APPARENT Km (uM)	APPARENT Vmax (%)
29682	6.6	25
	114	96
AI-11	127	25
	1005	96

* Paraquat-mediated cyanide resistant respiration was determined at 37°C using a Clark type electrode. The rate of oxygen uptake by cell suspension ($O.D._{600}$=1.0) was determined in the absence and in the presence of 1.0 mM cyanide. Increases (%) in the rate of CNRR were determined by adding different concentrations of PQ^{2+} to the reaction mixtures, after a linear rate in presence of CN^- was established. Reciprocals of the CNRR% were plotted against reciprocals of the external concentrations of PQ^{2+} and the apparent Km and Vmax were calculated.

Cross Resistance to Other Redox Active Compounds. If resistance to PQ^{2+} is due to reduced levels of G6PD, diaphorase, or intracellular pO_2 then the mutant should also be resistant to other redox active compounds (5,8). We compared the sensitivity of the wild type and the mutant toward PQ^{2+}, juglone, and plumbagin and found that resistance to PQ^{2+} was not accompanied by cross resistance toward the other redox compounds (10,data not shown). This led us to conclude that the mechanism of resistance must be due to a decreased rate of PQ^{2+} uptake in the mutant strain.

Kinetics of PQ^{2+} Uptake. The kinetics of ^{14}C-PQ^{2+} uptake in the mutant and the wild type were markedly different (Fig. 1). The wild type strain accumulated PQ^{2+} by active uptake against the concentration gradient, while the mutant seemed to accumulate PQ^{2+} passively. It is possible that the mutant has lost the

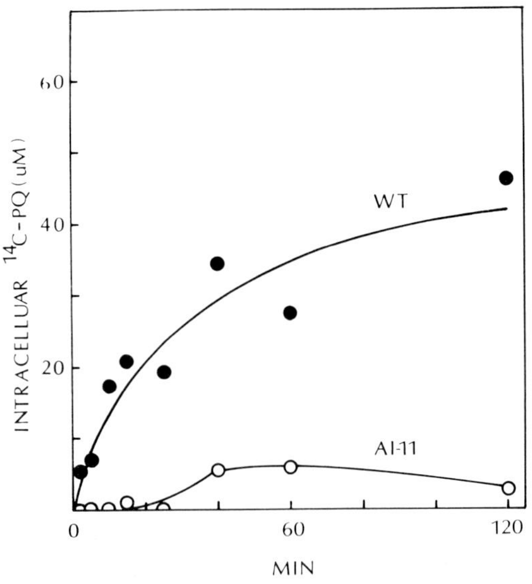

Fig. 1. Kinetics of ^{14}C-paraquat uptake.

energy-dependent PQ^{2+} uptake capacity, presumably by the alteration of some membrane-bound transport protein. The differences between the cytoplasmic membrane proteins of the wild type and the mutant and the nature of PQ^{2+} transporter require further studies.

ACKNOWLEDGEMENT

This work was supported in part by grant PCM-8213853 from the National Science Foundation.

REFERENCES

1. Fridovich I, Hassan, HM (1979) Trends Biochem. Sci. 4:113-115
2. Montgomery MR (1977) Res Commun Chem Pathol Pharmocol 16:155-158
3. Hassan HM, Fridovich I (1977) J Biol Chem 252:7667-7672
4. Hassan HM, Fridovich I (1978) J Biol Chem 253:8143-8148
5. Hassan HM, Fridovich I (1979) J Biol Chem 254:10846-10852
6. Hassan HM, Moody CS (1982) Can J Physiol Pharma 60:1367-1373
7. Moody CS, Hassan HM (1982) PNAS USA 79:2855-2859
8. Hassan HM (1984) Methods in Enzymology 105:523-532
9. Kao SM, Hassan HM (1985) FEMS Microbiol Lett 28:93-97
10. Kao SM, Hassan HM (1985) J Biol Chem (In press)
11. McCord JM, Fridovich I (1969) J Biol Chem 244:6049-6055
12. Anderson WB, Nordlie RC (1968) Biochem 7:1479-1485

© 1986 Elsevier Science Publishers B.V. (Biomedical Divison)
Superoxide and Superoxide Dismutase in Chemistry, Biology and Medicine; G. Rotilio, editor

THE EFFECT OF PARAQUAT ON THE PEROXIDE METABOLISM ENZYMES AND LIPID PEROXIDATION IN ERYTHROCYTES OF CARP /CYPRINUS CARPIO/. II

TERESA GABRYELAK
Institute of Biochemistry and Biophysics, Department of Biophysics
University of Łódź, Poland

INTRODUCTION

Paraquat /1,1'-dimethyl-4,4'-dipyridylium dichloride/ is a widely used herbicyde. Gramoxone which has 25 percent paraquat dichloride and 75 percent water is available in Poland.

The redox reactions involving paraquat /PQ/ in the presence of oxygen produce the superoxide radical /O_2^-/ an agent strongly influencing the function and the structure of important cellular molecules. For example, its action on polyunsaturated cell membrane lipids after dehydrogenation initiates lipid peroxidation, a process damaging the integrity of cell membrane [1].

It is well-known fact that all organisms posses a system providing defense against O_2^- toxicity. This system includes superoxide dismutase /SOD/, catalase /C/ and peroxidase /P/ among others. SOD scavenges superoxide radicals and the hydrogen peroxide generated as a result of this reaction is eliminated by catalase and peroxidase. Thus by measuring the peroxide metabolism enzyme activities, we tested the effect of the herbicide on freshwater fish. Assuming that PQ influences the intracellular flux of superoxide radical anion[2], it should cause changes in the antioxidant enzyme activities.

MATERIAL AND METHODS

Carps /600-800g/ were acclimated for a minimum of 7 days in aquarium water at a temperature of 18 $\pm$ 1°C. The exposures to 10ppm PQ were 4, 12, 48 and 96 hr in in vivo experiments. The in vitro experiments were carried out with erythrocytes incubated for 30 min in the presence of 4, 10, 20ppm PQ. Blood was taken in heparin from the tail vein. After removal of serum the red cells were washed with 0.6% NaCl and then lysed by addition of 1 vol of distilled water. DNase /50μg/ml/ was added to the lysate which was incubated for 2 hr at room temperature and frozen overnight at -20°C. After centrifugation the hemoglobin concentration was estimated by the method of Drabkin[3]. The SOD activity was measured by the method of Misra and Fridovich[4]. The catalase activity was determined at 240 nm according to the method of Beers and Sizer[5]. Peroxidase activity was measured spectrophotometrically by the guaiacol method of George as in Colowik and Kaplan[6]. The glutathione peroxidase /GSH-Px/ activities were determined with cumene-OOH as substrate by a combination of methods of Chiu et al.[7] and of Sedlak et al.[8]. The extent of lipid peroxidation /LP/ was determined by the method of Placer et al.[9]. All the experiments were carried out at 25°C.

RESULTS AND DISCUSSION

The results obtained demonstrate differences between in vivo and in vitro experiments as well as between preexposed and exposed animals. The toxic effect of PQ in vivo was expressed in our studies by an increase in the level of malondialdehyde /MDA/, one of the main products of degradation of polyunsaturated faty acids and an increase in the activity of the GSH-Px or inhibition of enzymes SOD, C and P as shown in Fig. 1.

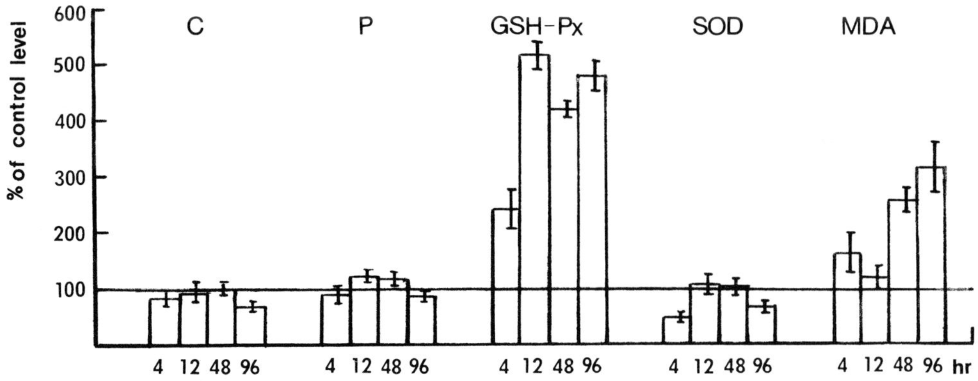

Fig. 1. Changes of C, P, GSH-Px and SOD activity and MDA level in erythrocytes of carps incubated for 4,12,48 and 96 hr in vivo in a medium containing 10ppm PQ. Mean number of animals studied: 4-5.

It is possible that C, P and SOD protective enzymes are inactivated in the course of increased O_2^- production and peroxidation of the membrane lipid may occur. In contrast, GSH-Px activity was found to be the most sensitive index of PQ intoxication. The mechanism of GSH-Px action under these conditions is hard to explain at present. Possibly, the abrupt increase in the level of hydrogen peroxide as suggested by Combs et al.[10].

In erythrocytes incubated in the presence of 4, 10, 20ppm PQ during 30 min, the MDA level and activities of C, P and GSH-Px were not changed but SOD activity was increased by about 200% at concentration of 4ppm PQ /Fig. 2./.

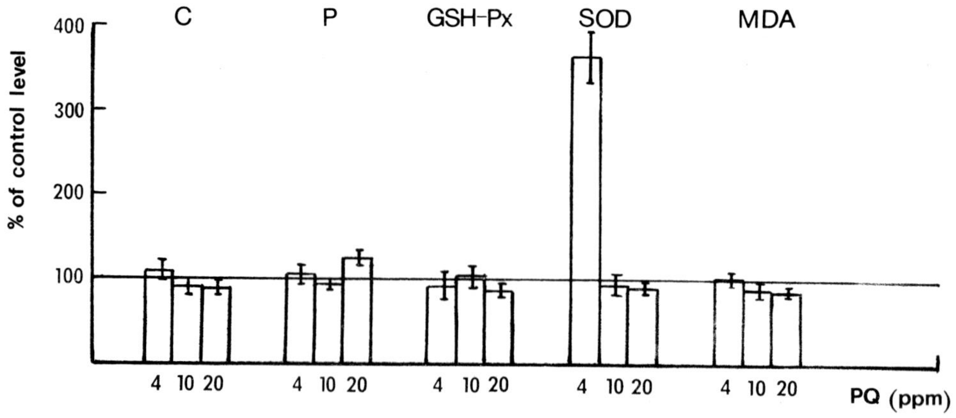

Fig. 2. Changes of P, C, GSH-Px, SOD activities and MDA level in carp erythrocytes. The cells were incubated in vitro for 30 min in medium containing 4,10,20ppm PQ. Mean number of individuals: 9.

The so high increase in SOD activity may be due to an elevation in the concentration of its substrate, O_2^-, which is generated by PQ, at a concentration as low as 4ppm. On the other hand, at higher PQ concentration activity of the enzyme decreased, conceivebly due to inactivation.

REFERENCES

1. Bus JS, Aust SD, Gibson JE /1975/ Res Commun Chem Pathol Pharmacol 11:31-38
2. Hassan HM, Fridovich I /1979/ J Biol Chem 254:10846-10852
3. Drabkin DL /1946/ J Biol Chem 164:703-723
4. Misra HP, Fridovich I /1972/ J Biol Chem 247:3170-3175
5. Beers RJ, Sizer IW /1952/ J Biol Chem 195:133-140
6. Colowick SP, Kaplan NO /1955/ In: Methods in Enzymology, Academic Press, New York, pp 764-775
7. Chiu DTY, Stultes FH, Tappel AL /1976/ Biochim Biophys Acta 445:558-566
8. Sedlak I, Lindsay RH /1968/ Analyt Biochem 25:192-205
9. Placer ZA, Cushman L, Johnson BC /1966/ Analyt Biochem 16:359-364
10. Combs GF, Peterson FJ /1983/ J Nutr 113:538-545

Superoxide and Superoxide Dismutase in Chemistry,
Biology and Medicine; G. Rotilio, editor

DIQUAT METABOLISM AND CYTOTOXICITY IN ISOLATED HEPATOCYTES

LUISA ROSSI*, LENA EKLÖW-LÅSTBOM, HJÖRDIS THOR and STEN ORRENIUS
Department of Toxicology, Karolinska Institutet, Stockholm, Sweden, and *Department of Biology, II University of Rome, Rome, Italy.

INTRODUCTION

Diquat, a bipyridilium herbicide related to paraquat, has been shown to be toxic to various animal species, probably as a result of the oxidative stress associated with its intracellular redox cycling. In the liver, diquat is reduced by hepatocellular electron transfer systems (1) to the diquat radical which can subsequently react with O_2; the $O_2^{\dot{-}}$ formed in this reaction is converted to H_2O_2 by dismutation. As demonstrated (2), the H_2O_2 produced in the cytosol of hepatocytes is mainly metabolized by glutathione peroxidase, with the concomitant oxidation of glutathione (GSH) to glutathione disulfide (GSSG); GSH is then regenerated by the action of GSSG reductase at the expense of NADPH. In addition, the non-enzymatic, iron-catalyzed reduction of H_2O_2 by $O_2^{\dot{-}}$ (Haber-Weiss reaction) can give rise to formation of the hydroxyl radical ($^{\bullet}OH$), an extremely reactive oxygen species. Both NADPH oxidation (3) and lipid peroxidation (LP) initiated by $^{\bullet}OH$ (4) have been suggested as possible mechanisms of the cytotoxicity induced by bipyridilium compounds.

We have investigated the mechanism of diquat-induced toxicity in isolated rat hepatocytes, with particular regard to the importance of the intracellular GSH level in protecting the cells against the oxidative challenge produced by diquat.

MATERIALS AND METHODS

Diquat dibromide was a gift of Dr L.L. Smith, I.C.I., England.

Hepatocytes were isolated from phenobarbital-treated male Sprague-Dawley rats and incubated as in (5). Some animals received diethylmaleate (DEM), 4 nmol/kg i.p., one hour before hepatocyte isolation to deplete intracellular GSH (6). Loss of cell viability was tested by Trypan blue uptake (5). To inhibit GSSG reductase activity, some cell suspensions were pretreated with 1,3-bis(2-chloroethyl)-1-nitrosourea (BCNU) (7).

Diquat (2 mM) was added directly to the cell suspension and aliquots were removed for spectrophotometric analysis of GSH

(2) and $NADPH/NADP^+$ (7) levels. LP was estimated by malondialdehyde (MDA) production by the thiobarbituric acid method.

RESULTS

Table I shows that incubation of hepatocytes with 2 mM diquat caused loss of cell viability when GSSG reductase activity was inhibited by BCNU-pretreatment or when the initial level of GSH had been decreased to 20% of control level by DEM-treatment of the rats. Thus, cell damage by diquat was only apparent when the GSH supply was compromised. BCNU-treatment was particularly effective, as 95% cell death occurred after only 2 h incubation with diquat, while 75% cell death was observed after 4 h in hepatocytes from DEM-treated rats.

TABLE I

EFFECTS OF VARIOUS INTERVENTIONS WHICH ALTER GSH CONTENT ON THE TOXICITY OF DIQUAT (2 mM) IN HEPATOCYTES

	Incubation time, h	GSH, % of zero time	Trypan blue uptake %
Hepatocytes no intervention	4	75	25
BCNU-treated	2	6	95
Hepatocytes from DEM-treated rats	4	20	75

As shown in Table II, addition of diquat to hepatocytes from DEM-treated rats caused extensive LP and a decrease in the $NADPH/NADP^+$ redox ratio; the observed decrease in GSH level could be due at least in part to LP, as lipid hydroperoxides are substrates of GSH peroxidase.

The presence of the iron chelator desferrioxamine (DFO) during incubation with diquat abolished LP, partially restored the GSH level and prevented cell death, without affecting the change in $NADPH/NADP^+$ redox level. These results suggest that diquat toxicity was dependent upon iron-catalyzed $^{\bullet}OH$ formation.

When the $^{\bullet}OH$-scavenger α-keto-γ-methiol-butyric acid (KMBA) was added to the incubation of hepatocytes from DEM-treated rats with diquat, partial protection of cell viability (25%)

TABLE II

THE PROTECTION BY VARIOUS AGENTS OF DIQUAT-INDUCED ALTERATIONS IN HEPATOCYTES FROM DEM-TREATED RATS*

Addition	GSH, nmol/ 10^6 cells**	MDA, nmol/ 10^6 cells	Trypan blue uptake, %	NADPH/ $\lvert NADPH+NADP^+ \rvert$
None	30	5	25	0.85
Diquat, 2 mM	4	25	75	0.50
+ DFO, 0.25 mM	14	4	25	0.50
+ KMBA, 1 mM	-	20	50	-
+ α-Tocopherol, 1 mM	-	3	40	-

* Values observed after 4 hours incubation are reported.
** Starting GSH value: 10 nmol/10^6 cells.

was observed. However, LP was not affected. In contrast, the antioxidant α-tocopherol completely abolished LP, with only 50% protection of cell viability. Thus, it appears that LP *per se* was not entirely responsible for the cytotoxicity produced by diquat under the present experimental conditions.

We conclude that the production of ˙OH by the redox cycling of diquat may be a critical event in the cytotoxicity of this agent. LP occurs as a consequence of ˙OH generation which, however, appears to be less critical for loss of cell viability than the disruption of intracellular thiol homeostasis which depends only in part on GSH consumption due to the metabolism of lipid hydroperoxides by GSH peroxidase.

REFERENCES

1. Gage JC (1968) Biochem J 109:757-761
2. Jones DP, Eklöw L, Thor H and Orrenius S (1981) Arch Biochem Biophys 210:505-516
3. Rose MF, Smith LL and Wyatt I (1976) Biochem Pharmacol 25:1763-1767
4. Bus JS, Aust SD and Gibson JE (1974) Biochem Biophys Res Commun 58:749-755
5. Moldéus P, Högberg J and Orrenius S (1978) Methods Enzymol 52:60-71
6. Boyland E and Chasseaud LF (1967) Biochem J 104:95-115
7. Eklöw L, Moldéus P and Orrenius S (1984) Eur J Biochem 138:459-463

© 1986 Elsevier Science Publishers B.V. (Biomedical Division)
Superoxide and Superoxide Dismutase in Chemistry,
Biology and Medicine; G. Rotilio, editor

DEGRADATION BY ENDOTHELIAL CELLS OF H_2O_2 PRODUCED BY ACTIVATED NEUTROPHILS

DOBRINA A., G.ZABUCCHI and P.PATRIARCA
Institute of General Pathology, University of Trieste, via Fleming 22,34127 Italy

Endothelial cells (EC) may be exposed, at sites of inflammation, to hydrogen peroxide and other products of oxygen reduction which are released by activated neutrophilic leukocytes (PMN). In the present study we show that bovine microvascular EC[1] are able to degrade the H_2O_2 produced by activated bovine neutrophils[2] and describe the biochemical pathways of H_2O_2 degradation.

Assays of H_2O_2 degrading enzymes on EC homogenates (Table I) showed that these cells have both glutathione peroxidase and glutathione reductase activities, a high catalase activity but no peroxidase activity (guaiacol method). The activity of the same enzymes in PMN, a cell which is notoriously involved in H_2O_2 processing, is also shown for comparison in the table.

EC were exposed for ten min to H_2O_2 continuously generated by phorbol-12myristate-13acetate (PMA)-activated PMN and the amount of H_2O_2 that accumulated in the medium was measured fluorimetrically[3]. In the presence of EC less H_2O_2 was recovered than in their absence, indicating H_2O_2 degradation by EC. As shown in Fig. 1, the peroxide degraded by EC increased as a function of the rate of H_2O_2 production, until a plateau was reached at a level of 4.0±1.2(SD)nmol H_2O_2 degraded/10^6 EC/min. Since the average H_2O_2 production rate by neutrophils was 1.3±0.5 (SD)nmol/10^6 PMN/min, it was calculated that a single EC could degrade the H_2O_2 continuously released by about three maximally stimulated PMN. When the experiments were performed in the presence of the catalase inhibitor NaN_3, H_2O_2 degradation by EC was strongly lowered at relatively high H_2O_2 concentratios but was slightly affected, or not affected at all, at low H_2O_2 concentrations, suggesting that at least two mechanisms, one NaN_3-sensitive (catalase) and the other NaN_3-

TABLE I - ENZYME ACTIVITIES IN BOVINE MICROVASCULAR ENDOTHELIAL CELLS AND PMN

Catalase:μmol H_2O_2/min. Glutathione peroxidase and glutathione reductase:nmol NADPH/min. Peroxidase:nmol tetraguaiacol/min. Protein:μgrams. All values are referred to 10^6 EC. The means ± SEM of (n) experiments are reported. Assays were performed as reported elsewhere[3].

	Endothelial cells		Blood granulocytes	
Catalase	3.00±0.57	(7)	0.27±0.02	(4)
Glutathione peroxidase	2.47±0.12	(7)	1.63±0.25	(5)
Glutathione reductase	2.98±0.14	(7)	5.16±0.24	(5)
Peroxidase	< 0.1	(5)	44.1 ±8.6	(5)
Protein	104.1 ±5.0	(17)	91.0 ±4.8	(6)

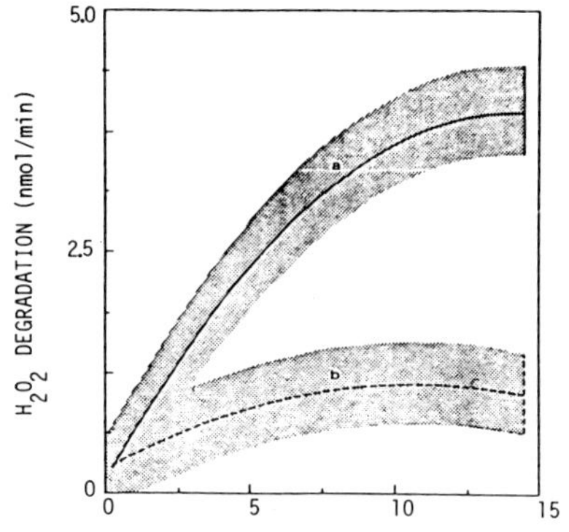

1

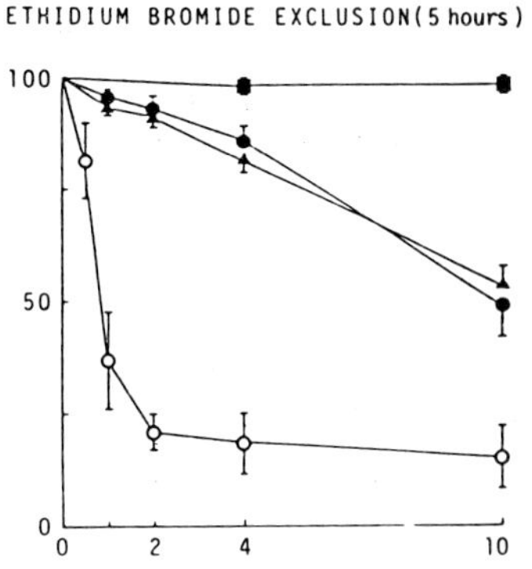

2

Fig. 1. Degradation by endothelial cells of H_2O_2 produced by activated PMN. EC (10^6)were incubated with different numbers of PMA (0.1 µg/ml)-stimulated neutrophils (0.25-10x10^6 cells) in the absence (a) or presence (b) of NaN_3 (2mM). The values of H_2O_2 degraded by EC were calculated from the difference of H_2O_2 measured in samples without and with EC. Curves represent multiple linear regression equations ± mean square deviations[4], as computed on 40 to 50 values from six experiments.

Fig. 2. Ethidium bromide exclusion by endothelial cells incubated for five hours with PMA (0.1 µg/ml)-stimulated PMN. 10^6 EC were incubated in 1 ml Krebs-Ringer-phosphate (pH 7.4) containing 5.5mM glucose with different numbers of PMN (0.5-10x10^6) and in the absence (●) or presence of 2mM NaN_3 (○), 40 µg/ml superoxide dismutase (▲) or 20 µg/ml catalase (■). The results are expressed as percentage of control values, that is values of samples incubated in the absence of PMN. Values are means ± 1SE of nine experiments.

insensitive were involved. The following evidence indicated that the glutathione redox cycle was operative in these cells and was the only NaN_3-insensitive pathway in H_2O_2 degradation, 1) EC exposed to H_2O_2 showed a dramatic and specific increase in the hexose monophosphate shunt (HMPS) activity (the production of $^{14}CO_2$ from [1-^{14}C]-glucose[5] was increased more than 25 fold over basal levels), as it has already been shown in other cell types that dispose off the H_2O_2 through the glutathione cycle, 2) the ratio between the amount of H_2O_2 degraded and the CO_2 produced by the cells in the presence of NaN_3 was 2:1, as theoretically expected if all the H_2O_2 degraded went through the glutathione cycle coupled with the HMPS, according to the following equations: $H_2O_2+2GSH \rightleftharpoons H_2O+GSSG$; $GSSG+NADPH+H^+ \rightleftharpoons 2GSH+NADP^+$; $2NADP^+ + glucose\text{-}6P \rightleftharpoons 2NADPH+H^+ + ribulose\text{-}5P+CO_2$, and 3) the ratio oxidized glutathione/reduced glutathione (GSSG/GSH) did not significantly increase over basal levels (0.02±0.01) in cells exposed to H_2O_2 in the presence of NaN_3 until some of the peroxide began to escape degradation (GSH and GSSG were measured fluorimetrically, as described by Voetman et al[6]. The total glutathione content of the cells was 14.0±1.6(SD)nmol GSH/10^7 cells,n=7). By combining the data on the total amount of H_2O_2 degraded and the amount of CO_2 produced through the HMPS it was possible to calculate the relative contribution of catalase and of the glutathione redox cycle in H_2O_2 degradation by EC. It was seen that degradation by catalase was predominant at H_2O_2 generation rates above 0.6 nmol/min/10^6 cells while the glutathione redox cycle was the most effective at lower H_2O_2 generation rates.

It is worth of note that the viability of EC (as determined by the ethidium bromide exclusion test[7]) was not appreciably affected by exposure of the cells to H_2O_2 generated by PMA-activated PMN at rates within the degradative potential of EC, that is at PMN:EC ratios between 2:1 and 4:1. At higher PMN:EC ratios instead, EC viability began to decline (Fig. 2). Addition of catalase to the incubation medium prevented cell damage while the addition of superoxide dismutase was ineffective in this respect, indicating that the main damaging agent for EC was H_2O_2. When EC catalase was poisoned with NaN_3, loss of EC viability was seen at H_2O_2 generation rates much lower than in the absence of the inhibitor. It is concluded that endothelial cells can protect themselves from PMN-induced oxidative injury in a well defined range of environmental H_2O_2 concentrations by actively degrading the peroxide. These results are in agreement with the observation that local accumulation of PMN is often not accompanied by any evidence of vascular damage[8].

ACKNOWLEDGEMENTS

Supported by grants from CNR (84.02024.52) and MPI (40% and 60%) of Italy.

REFERENCES

1. Dobrina A, Soranzo MR, Rossi F (1983) Cell Tissue Res 232:579
2. Gennaro R, Schneider C, deNicola G, Romeo D (1978) Proc Soc Exp Biol Med 157: 342
3. Rossi F, Bellavite P, Dobrina A, Dri P, Zabucchi G (1980) In: van Furth R (ed) Mononuclear phagocytes: Functional aspects. Martinus Nijhoff, The Hague, pp 1187-1213
4. Ostle B (1954) in: Ostle B (ed) Basic concepts and techniques for research workers. The Jowa State College Press, Arnes, Jowa, pp 202-231
5. Dobrina A, Rossi F (1983) Biochim Biophys Acta 762:295
6. Voetman AA, Loos JA, Roos D (1980) Blood 55:741
7. Voetman AA, Roos D (1980) Blood 56:846
8. Shaw JO, Henson PM (1982) Am J Pathol 108:17.

Superoxide and Superoxide Dismutase in Chemistry,
Biology and Medicine; G. Rotilio, editor

RELATIONSHIPS BETWEEN HYDROGEN PEROXIDE AND CALCIUM IN THE AGGREGATION OF HUMAN PLATELETS.

DOMENICO DEL PRINCIPE, ADRIANA MENICHELLI, WALTER DE MATTEIS, MARIA LUISA DI CORPO, STEFANO DI GIULIO, ANNA MARIA PENTASSUGLIO, and ALESSANDRO FINAZZI-AGRO'.

Dept.of Pediatrics, University of Rome "La Sapienza", and Dept.of Experimental Medicine and Biochemical Sciences, II University of Rome, Rome, Italy.

INTRODUCTION

Hydrogen peroxide produced by phagocytic cells during inflammatory reactions as such, or associated with myeloperoxidase system, induces platelet activation(1). We suggest that H_2O_2 has a role in platelet function on the basis of the following evidences. Platelets carry a membrane-bound H_2O_2 generating system(2) which releases minute amounts of H_2O_2 after stimulation with latex or opsonized-zymosan(3). The addition of hydrogen peroxide enhances the aggregation induced by several stimuli, except by the calcium ionophore A23187(4). The external catalase did not affect the aggregation induced by A23187 nor by soluble stimuli, but strongly reduced the aggregation elicited by particulate stimuli. In this paper we showed that H_2O_2 may act, at least in the case of thrombin-induced platelet aggregation, through an increase of platelet sensitivity to Ca^{2+}.

MATERIALS AND METHODS

Platelets were prepared from blood obtained from healthy donors, not taking any drugs for at least 20 days, after informed consent. Platelet preparation and activation was carried out as previously described(3). The agonist was thrombin(0.05 U/ml) in the presence or absence of external calcium. Platelet aggregation was monitored by the change in light transmission at 650 nm in a standard platelet aggregometer, according to the method of Born(1962).

RESULTS AND DISCUSSION

The addition of peroxide increased the aggregation induced by thrombin, ADP, opsonized-zymosan and collagen, whereas A23187-induced aggregation was inhibited(4).The aggregation induced by thrombin was dependent on the concentration of calcium in the medium.The addition of H_2O_2 reduced the amount of Ca^{2+} needed for the aggregation (Figs.1 and 2).At higher Ca^{2+} concentration($\geq 1x10^{-5}$)no more effect of H_2O_2 was observed.The figures illustrate the data obtained from 1 experiment,but are representative of all experiments done(n=15).

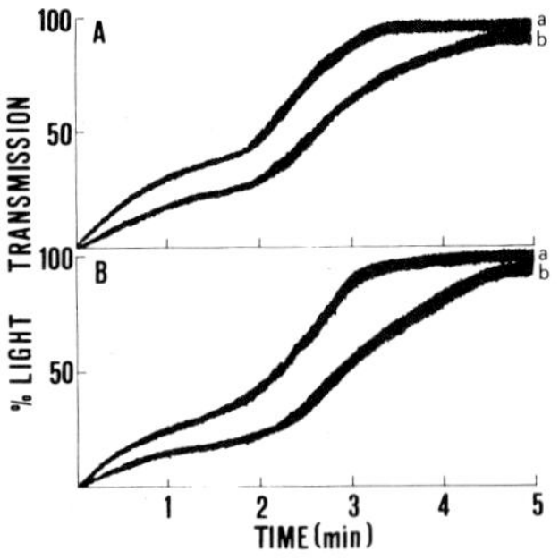

Fig.1.Thrombin-induced aggregation of platelets in the presence(a)or absence(b)of hydrogen peroxide.In the panel A the external calcium concentration was 10^{-7}M,in B 10^{-6}M.

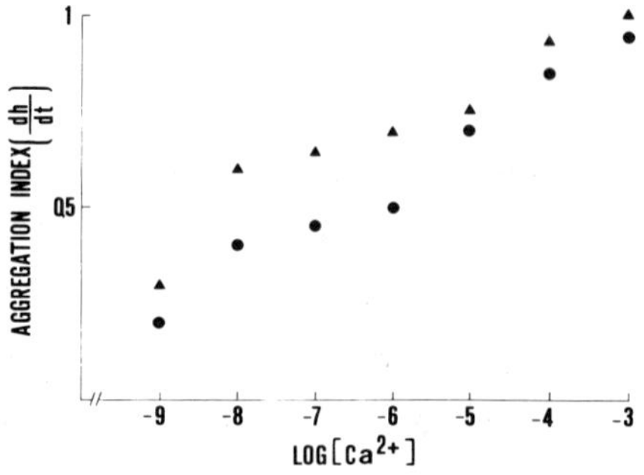

Fig.2.Ca^{2+} sensitivity of platelets in the presence(▲)or absence(●)of H_2O_2.dh=distance from the peak of maximum aggregation to baseline;dt=lenght from the point of the addition of agonist to interception with dh.

These experiments show that H_2O_2 produced in the course of platelet activation may have a role in the transduction of external signals.The different sensitivity to external catalase(4)may indicate that different transduction mechanisms of the external stimuli may be operating.Nishizuka suggested that calcium sensitivity modulation,i.e.,the synergistic role of calcium mobilization and protein kinase activation is crucial for the transduction events(5).The calcium sensitivity modulation may exert different control on different activation process induced by most agonists.The data reported here suggest that H_2O_2 could modify the platelet responses by regulating the calcium sensitivity of platelets.Using platelets made permeable by exposure to an intense electric field,it has been found that thrombin and phorbol esters may modulate the calcium sensitivity of platelet secretion(6).A clear correlation between H_2O_2 and calcium fluxes in mitochondria,microsomal vescicles and plasma membrane has been established.Hydrogen peroxide is endogenously produced during the activation of platelets,as demonstrated by the inactivation of intraplatelet catalase incubated with aminotriazole(4).

A physiological relevance of calcium mobilization induced by the peroxide in human platelets could be suggested.

ACKNOWLEDGEMENTS

This research was in part supported by Ministero della Pubblica Istruzione, Special Project Coagulazione intravascolare disseminata, and By Consiglio Nazionale delle Ricerche, Special Project Oncologia.

REFERENCES

1. Clark RA,and Klebanoff SJ(1980)J Immunol 124:399-405

2. Del Principe D,Mancuso G,Menichelli A,Gabriotti M,Cosmi EV,Gherardi G(1980)Biol Cell 38:135-140

3. Finazzi-Agrò A,Menichelli A,Persiani M,Biancini G,and Del Principe D(1982)Biochim Biophys Acta 718:21-25

4. Del Principe D,Menichelli A,De Matteis W,Di Corpo ML,Di Giulio S and Finazzi-Agrò A(1985)FEBS Lett 185:142-146

5. Nishizuka Y(1984)Trends Biochem Sci 9:163-166

6. Knight DE and Scrutton MC(1984)Nature 309:66-68

Superoxide and Superoxide Dismutase in Chemistry,
Biology and Medicine; G. Rotilio, editor

PHOTOCHEMICAL INACTIVATION OF MACROPHAGE PROGENITOR CELLS IN MOUSE BONE MARROW.

A. PETKAU AND W.S. CHELACK

Medical Biophysics Branch, Atomic Energy of Canada Limited Research Company, Whiteshell Nuclear Research Establishment, Pinawa, Manitoba, Canada, ROE 1LO

INTRODUCTION

The sensitivity of macrophage progenitor cells in mouse bone marrow to superoxide radicals ($O_2^{\cdot-}$) was previously demonstrated by both photochemical and radiation chemical methods using superoxide dismutase (SOD) as probe (1-3). The photochemical generator of $O_2^{\cdot-}$ involved the reduction of 2.4 μM riboflavin by 10 mM methionine in the presence of light (4-8). The excited riboflavin also reacted with methionine to form a semiquinone radical whose further reduction was associated with the sequential decarboxylation and deamination of the methionine radical to methional (4-8). The degradation of methional to methyl disulfide, ethylene and formic acid is catalyzed either by triplet state riboflavin, $O_2^{\cdot-}$ or its protonated form $HO_2^{\cdot}$ (4-7). In a minor pathway ($\leq$20%), methional may be degraded to methionine sulfoxide (7,8). The in vitro toxicity of these photoproducts to macrophage progenitor cells in mouse bone marrow was tested. The results indicated that only methional was toxic at the concentration elaborated by the photochemical generator of superoxide. SOD had no inhibitory effect on this toxicity.

MATERIALS AND METHODS

The sensitivity of the macrophage progenitor cells in phosphate buffered saline (PBS) at pH 7.4 to methional, methyl disulfide, ethylene, methionine sulfoxide and formic acid was tested by treating the washed cells (1-3) with different concentrations of the compounds for up to 60 min at room temperature. After the treatments, the cells were plated and cultured as previously described (1,2). Macrophage colonies containing more than 25 cells were counted after 7 days (1-3).

RESULTS AND DISCUSSION

Calculations indicated that a 10-min illumination of the photochemical generator of $O_2^{\cdot-}$ in PBS would elaborate ~ 0.3 mM methional. The effect on survival of exposing the macrophage progenitor cells for varying periods to 0.32 mM methional is shown in Fig. 1. SOD and catalase had no effect on the toxicity of methional to these cells (data not shown). Products derived from methional, such as methyl disulfide, formic acid and ethylene, were not toxic to the

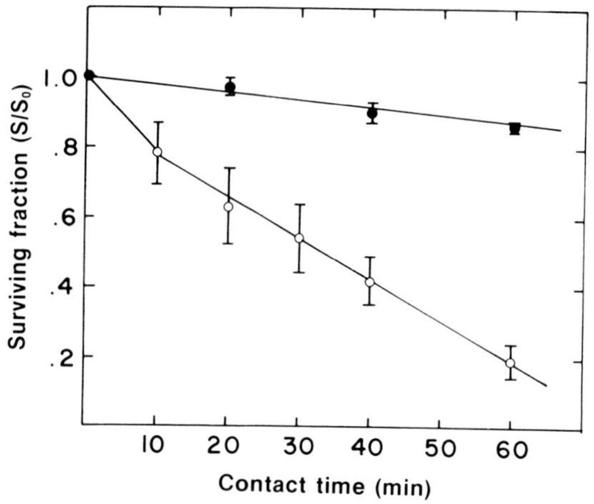

Fig. 1. Surviving fraction of macrophage progenitor cells vs contact time with 0.32 mM methional in PBS at room temperature. Control in PBS (.), 0.32 mM methional in PBS (o). Error bars ± 1 standard deviation.

cells at the concentrations expected after 10 min of illumination (Table). For methyl disulfide and formic acid, the expected concentrations were less than 0.7 and 2.2 mM, respectively. At concentrations greater than 0.7 mM, methyl disulfide inactivated the progenitor cells to an increasing extent (Table). Inactivation of the cells by formic acid at 22 mM (Table) or more appears to be a pH effect as 43 mM of formic acid dropped the pH of the PBS from 7.4 to 7.15. The concentration of ethylene was not measured directly because of its limited solubility in aqueous solution. An indirect method was used. It was found that illumination of the photochemical generator in PBS for 10 min reduced the pO_2 to 7.5, a reduction that could be reproduced by bubbling a solution of PBS with ethylene. The concentration of ethylene required to give a pO_2 of 7.5 in PBS had no effect on the survival of macrophage progenitor cells (Table). Therefore, it was concluded that the amount of ethylene elaborated by a 10-min illumination of the photochemical generator had no effect on the colony forming capacity of the macrophage progenitor cells. Similarly, methionine sulfoxide was not toxic to the progenitor cells, even at concentrations 100-fold higher than the one expected of 0.06 mM after a 10 min illumination (Table). Thus, aside from the $O_2^{\bar{\cdot}}$ mediated toxicity previously reported and inhibited by SOD (1,2), methional produced during the photochemical generation of $O_2^{\bar{\cdot}}$ from riboflavin and methionine has also now been shown to be toxic to macrophage progenitor cells. If methional has a similar toxicity on other cells, its use as an hydroxyl radical ($^{\cdot}OH$) scavenger could lead to an underestimate of the role of $^{\cdot}OH$ in cytotoxicity (9,10).

TABLE: TOXICITY OF THE PRODUCTS OF THE PHOTOCHEMICAL GENERATOR OF $O_2^{\cdot-}$ TO MACROPHAGE PROGENITOR CELLS AFTER A 40 MINUTE EXPOSURE AT ROOM TEMPERATURE.

Product	Concentration (mM)	Surviving Fraction (S/S_o)
Methyl disulfide	0.7	1.00
	1.8	0.89
	3.5	0.49
	17.5	0.04
	35	0
Formic acid	2.2-8.7	0.98
	22	0.28
	43	0
Ethylene	-*	1.00
Methional	0.3	0.60
	0.6	0.11
	1.6	0
Methionine sulfoxide	0.06-6.0	0.96-1.03

*- Not measured directly. See text.

REFERENCES

1. Petkau A, Chelack WS (1983) In: Greenwald RA, Cohen G (eds) Oxy Radicals and their Scavenger Systems, Vol 2: Cellular and Medical Aspects. Elsevier, New York, pp 167-172
2. Chelack WS, Petkau A (1983) Int J Radiat Biol 44:523-529
3. Petkau A, Chelack WS (1984) Biochem Biophys Res Commun 119:1089-1095
4. Massey V, Strickland S, Mayhew SG, Howell LG, Engel PC, Mathews RG, Schuman M, Sullivan PA (1969) Biochem Biophys Res Commun 36:891-897.
5. Ballou D, Palmer G, Massey V (1969) Biochem Biophys Res Commun 36:898-904
6. Moore WM, Spence JT, Raymond FA, Colson SD (1963) J Amer Chem Soc 85:3367-3372
7. Yang SF, Ku HS, Pratt HK (1967) J Biol Chem 242:5274-5280
8. Spikes JD, MacKnight ML (1970) Annals NY Acad Sci 171:149-162
9. Beauchamp C, Fridovich I (1970) J Biol Chem 245:4641-4646
10. Cohen G, Heikkila RE (1977) In: Michelson AM, McCord JM, Fridovich I (eds) Superoxide Dismutases. Academic Press, New York, pp 351-365

© 1986 Elsevier Science Publishers B.V. (Biomedical Division)
Superoxide and Superoxide Dismutase in Chemistry,
Biology and Medicine; G. Rotilio, editor

SUPEROXIDE FORMATION ACCOUNTS FOR ALMOST ALL EXTRA OXYGEN CONSUMPTION UPON STIMULATION OF LEUKOCYTES BY PMA

GERRIET LOSCHEN
Center of Research, Grünenthal GmbH, Zieglerstraße 6, D-5100 Aachen (FRG)

INTRODUCTION

Phagocytosis by granulocytes (PMNL) is accompanied by a burst of O_2 consumption, and a formation of $^{\bullet}O_2^-$ and H_2O_2. Though it is generally believed that this is mainly due to $^{\bullet}O_2^-$ formation followed by dismutation, this assumption could never be convincingly substantiated by independent measurements of O_2 consumption, and formation of $^{\bullet}O_2^-$ and H_2O_2 (1 - 4). The stoichiometry of the presumed mechanism was therefore investigated.

MATERIAL AND METHODS

2,2'-azino-di-(3-ethyl-benzothiazoline-(6)-sulfonic acid) (ABTS) was purchased from Boehringer, Mannheim, FRG, and epinephrine (adrenalin) from Sigma, St. Louis, Mi., U.S.A.. Bovine Cu/Zn-SOD was a commercial product of Grünenthal GmbH, Stolberg, FRG. PMNL were isolated according to (5). O_2 consumption of PMNL stimulated by phorbol myristyl acetate (PMA) was monitored by a Clark electrode. Stock solutions of the stable $ABTS^{\bullet +}$ radical were prepared by mixing 10 mM H_2O_2 and 20 mM ABTS in the presence of horse radish peroxidase (10 µg/ml) immediately before use.

THEORY

Four oxygen consuming reactions have to be considered in stimulated leukocytes: 1. mitochondrial respiration, 2. biosynthesis of eicosanoids, 3. bivalent oxygen reduction to H_2O_2, 4. monovalent oxygen reduction to $^{\bullet}O_2^-$ followed by dismutation to O_2 and H_2O_2. If the extra oxygen consumption upon stimulation of PMNL is solely due to a monovalent reduction of n O_2 to n $^{\bullet}O_2^-$,50 % of the O_2 consumed is recovered by dismutation (Fig. 1). The rate of $^{\bullet}O_2^-$ formation thus should be 200 % of the rate of extra O_2 consumption. Reduction of the $^{\bullet}O_2^-$ should prevent recovery of O_2 by dismutation and the rate of the O_2 consumption should be doubled. Re-oxidation of $^{\bullet}O_2^-$ to O_2 should recover all of the monovalently reduced O_2, seemingly blocking the extra oxygen consumption. SOD increases the speed of dismutation without affecting stoichiometry unless it is disturbed by agents trapping $^{\bullet}O_2^-$. Catalase should diminish the rate of O_2 consumption to 50 %, but only if H_2O_2 were completely accumulated.

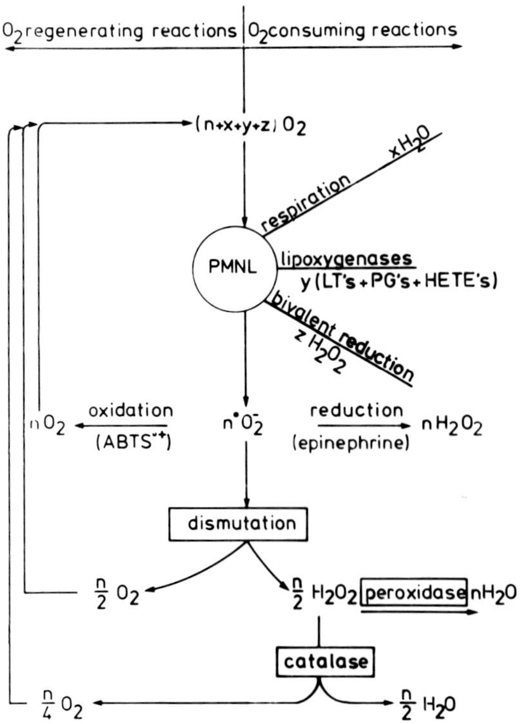

Fig. 1 Theoretical considerations of the stoichiometry of the respiratory burst of stimulated leukocytes.

RESULTS AND DISCUSSION

Fig. 2A shows that the PMA-stimulated (5 µg/ml) O_2 consumption is roughly doubled by addition by 300 µM epinephrine. This indicates that epinephrine is oxidized by $^{\bullet}O_2^-$ thereby preventing O_2 formation by $^{\bullet}O_2^-$ dismutation (Fig. 1). As expected, excess SOD (80 µg/ml) interfers with the epinephrine oxidation and almost restores the original rate of stimulated O_2 consumption.

Inversely, stimulated O_2 consumption is practically abolished by addition of ABTS$^{\cdot+}$ (400 µM) (Fig. 2B), since it reoxidizes $^{\bullet}O_2^-$ to O_2. Under these conditions H_2O_2 is not measurable. Again excess SOD (80 µg/ml) prevents $^{\bullet}O_2^-$ from reacting with ABTS$^{\cdot+}$ and the rate of O_2 consumption approaches that of the stimulated PMNL in the absence of any scavenger.

From these experiments it can be concluded that 1. no measurable H_2O_2 is formed by bivalent reduction of oxygen; 2. only a minor percentage (<10%) of the extra oxygen consumption of PMA-stimulated PMNL is used for other oxidative pathways, e.g. eicosanoid biosynthesis, as is also evident from lack of additional eicosanoids detected upon PMA stimulation. Monovalent O_2 reduction thus is the major route of O_2 consumption by PMNL under the conditions investigated.

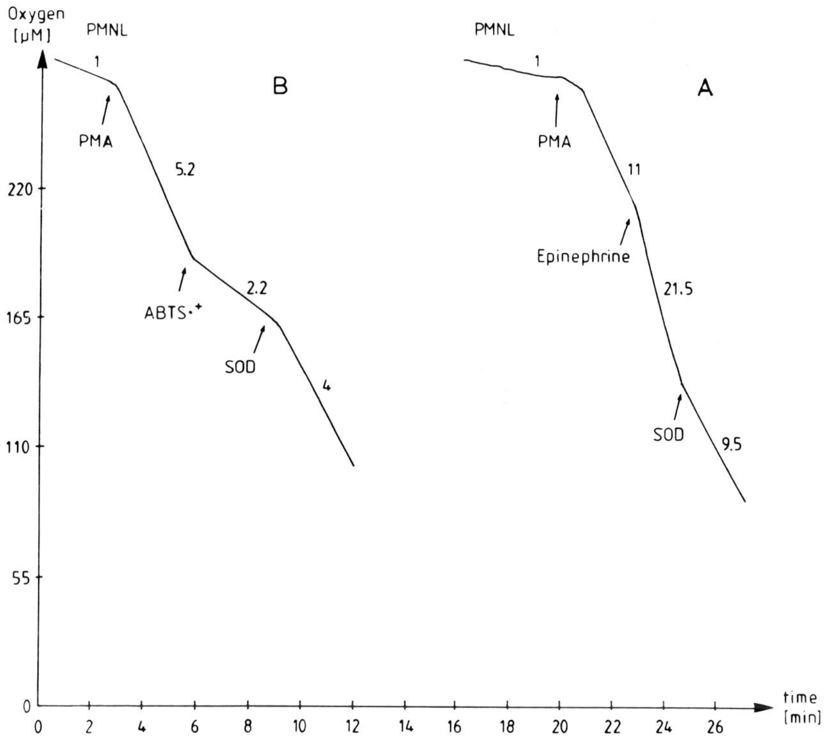

Fig. 2 Modulation of the apparent oxygen consumption of PMNL stimulated with PMA. Number above slopes indicate stimulation factors as compared to basal O_2 consumption (= 1).

ACKNOWLEDGEMENTS

This work was supported by the Bundesministerium für Forschung und Technologie (Grant PTB 8377 and 03 85696).

REFERENCES

1. Drath DB, Karnovsky ML (1975) J Exp Med 141:257 - 262.

2. Root RK, Metcalf J, Oshino N, Chance B (1975) J Clin Invest 55:945 -955.

3. Weening RS, Wever R, Roos D (1975) J Lab Clin Med 85:245 - 252.

4. Markert M, Allaz MJ, Frei J (1980) Febs Letters 113:225 - 230.

5. Pertoft H, Johnsson A, Wärmegard B, Seljelid R (1980) J Immunol Meth 33:221 - 229.

© 1986 Elsevier Science Publishers B.V. (Biomedical Division)
Superoxide and Superoxide Dismutase in Chemistry,
Biology and Medicine; G. Rotilio, editor

EVIDENCE FOR INVOLVEMENT OF OXYGEN RADICALS IN PMNL CHEMOTAXIS

LEOPOLD FLOHE, WOLFGANG A. GÜNZLER, GERRIET LOSCHEN, WOLFGANG MARTIN, BEATE MEIER*

Grünenthal GmbH, Center of Research, Zieglerstr. 6, D-5100 Aachen, FRG
*Kernforschungsanlage, D-5170 Jülich, FRG

INTRODUCTION

The migratory response of polymorphonuclear leukocytes (PMNL) to chemoattractants and the ability to release superoxide anions ($\cdot O_2^-$) have been considered independent processes of the host defense phenomenon. It is, however, well established that all relevant chemotactic factors, e.g. N-formyl-methionyl-leucyl-phenylalanine (FMLP) (1), leukotriene B_4 (LTB_4; 5S,12R-dihydroxy-6,14-cis-8,10-trans-eicosatetraenoic acid) (2), w-OH-LTB_4 (3), the complement fragment C5a (4) and the platelet activating factor (PAF-acether; PAF) (5) induce a small but consistent $\cdot O_2^-$ release. This intriguing parallelism prompted us to investigate the effect of superoxide dismutase (SOD) on the chemotactic response of PMNL towards these chemoattractants.

MATERIAL AND METHODS

FMLP was obtained from Sigma (St. Louis, Mo., USA). Purified C5a was kindly provided by T.E. Hugli (Scribbs Clinic, La Jolla, Ca., USA). Authentic LTB_4 was a gift from E.J. Corey (Boston, Ma., USA). PAF-acether (1-alkyl-2-acetyl-sn-glycero-3-phosphocholine) was a gift of Nattermann Chemie GmbH, Cologne, FRG. w-OH-LTB_4 was prepared biosynthetically from arachidonic acid and PMNL in the presence of the ionophor A23187 and isolated by reverse phase HPLC according to Powell (6).

Bovine erythrocyte Cu/Zn-SOD of at least 99.9 % purity was a commercial product of Grünenthal GmbH, Stolberg, FRG. Its specific activity was ca. 3,000 U/mg as measured by the cytochrome c reduction test (7). The corresponding apoSOD was prepared according to McCord and Fridovich (8). Fe-SOD and Mn-SOD were isolated from Propionibacterium shermanii (9). Catalase was purchased from Boehringer, Mannheim, FRG.

PMNL were isolated from venous blood of healthy volunteers after dextran sedimentation by centrifugation on a discontinuous Percoll gradient (Pharmacia, Freiburg i.Br., FRG) essentially as described by Pertoft et al. (10). Blind well chemotaxis chambers were used to investigate directed

PMNL locomotion through porous membrane filters according to Boyden (11). PMNL suspended at a concentration of 2 x 10^6/ml in TCM medium (120 mM NaCl, 25 mM Tris-HCl, 4 mM KCl, 0.6 mM $CaCl_2$, 1 mM $MgCl_2$, pH 7.2) were placed in the upper compartment of the Boyden chamber. The lower compartments were filled with TCM medium alone or FMLP, LTB_4, w-OH-LTB_4 C5a or PAF-acether of various concentrations in TCM medium. The compartments were separated by PVPF polycarbonate membrane filters from Nucleopore (Pleasanton, Ca., USA) of 10 µm thickness perforated with 3 µm pores.

Results were expressed as chemotactic index (CI) meaning the ratio between stimulated and unstimulated migration, as measured by counting the cells appearing at the lower surface of the filters after 45 min at 37°C.

Chemotaxis under agarose was performed essentially as described by Chenoweth et al. (12). Cu/Zn-SOD was embedded into the agarose gel by adding various concentrations to the warm agarose/gelatine mixture before solidification. Cellular movements were evaluated by light microscopy at 50-fold magnification by means of a calibrated ocular micrometer and expressed as "chemotactic differential" (CD), i.e. the difference between directed migration and random migration in µm.

RESULTS

The LTB_4-induced chemotaxis of PMNL was inhibited dose-dependently by Cu/Zn-SOD, Fe-SOD and Mn-SOD (Fig. 1) at concentrations between 10 to 50 µg/ml. This effect was neither seen in the presence of the apo form of Cu/Zn-SOD (Fig. 1) nor in the presence of unfractionated serum protein (2). Leukotaxis induced by other chemoattractants (FMLP, PAF, w-OH-LTB_4 or C5a) was also inhibited by the Cu/Zn-SOD (Fig. 2).

Since chemotaxis measured in blind well chambers does not clearly allow to distinguish between inhibition of directed migration and stimulation of random migration, the effect of Cu/Zn-SOD on PMNL chemotaxis was investigated by the "under agarose technique" (12). As shown in Fig. 3, Cu/Zn-SOD also inhibits PMNL chemotaxis under agarose dose-dependently when embedded into the agarose gel.

Inhibition of PMNL chemotaxis by superoxide dismutases implies a direct or indirect involvement of $^{\bullet}O_2^-$ in the chemotactic response. In order to further clarify this problem the effect of catalase on PMNL chemotaxis was tested (Fig. 4). Catalase, too, inhibits dose-dependently the chemotactic

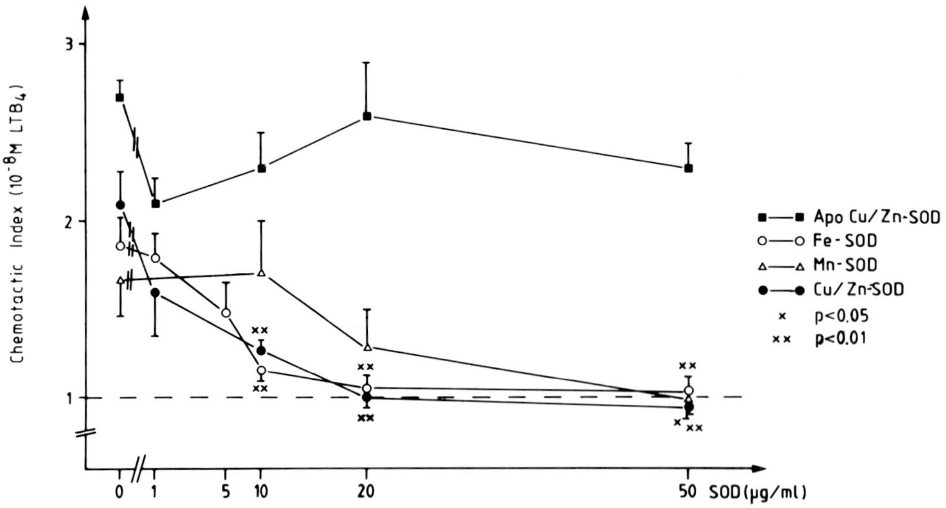

Fig. 1: Effect of various superoxide dismutases on leukotriene B_4-induced PMNL chemotaxis. Chemotaxis induced by 10 nM LTB_4 was tested in blind well chambers in the presence of various superoxide dismutases added to the cell suspension in the upper compartment. Values represent means ± SEM of n = 5; *p < 0.05; **p < 0.01.

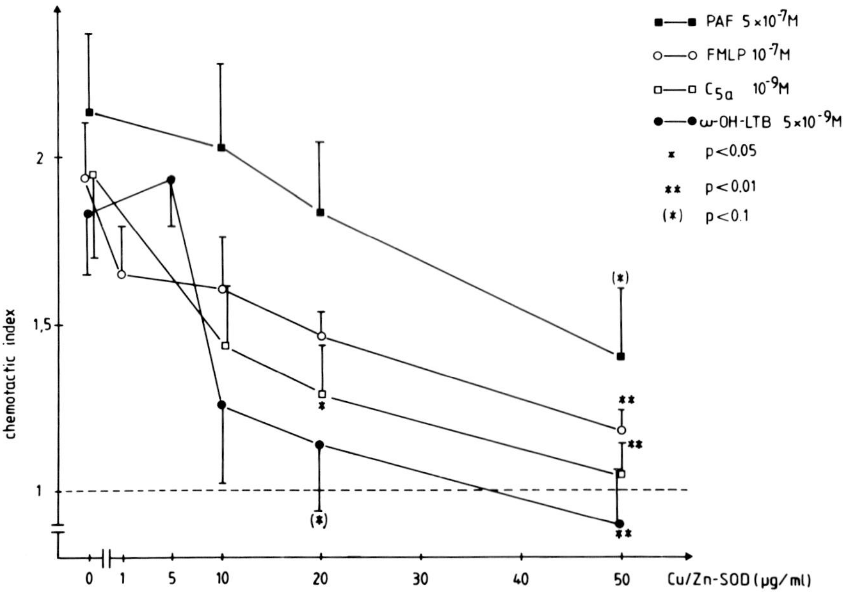

Fig. 2: Effect of Cu/Zn-SOD on PMNL chemotaxis induced by various chemoattractants. PMNL chemotaxis was performed in blind well chambers. Various concentrations of Cu/Zn-SOD were added to the cell suspension in the upper compartment, while the chemoattractants were placed in the lower compartment. Values represent means ± SEM of n = 5; (*) p < 0.1; *p < 0.05; **p < 0.01.

response of PMNL, as exemplified with FMLP, LTB_4 and C5a. This observation reveals a contribution of H_2O_2 to the chemotactic process and, therefore, an indirect involvement of $\cdot O_2^-$.

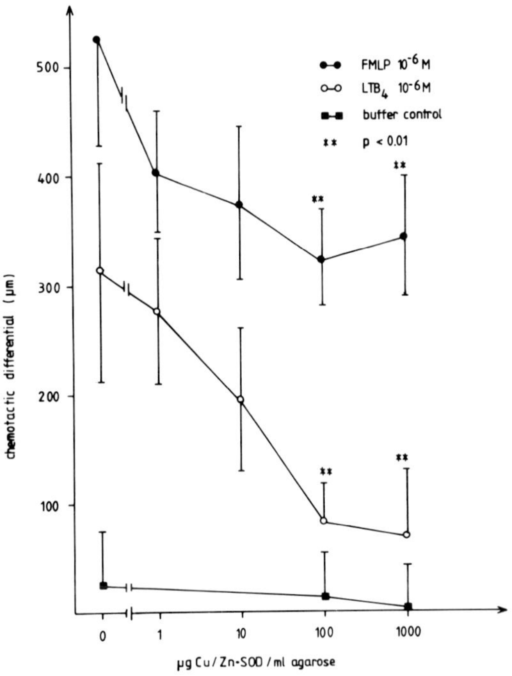

Fig. 3: Influence of Cu/Zn-SOD on PMNL chemotaxis under agarose. FMLP (1 µM) and LTB_4 (1 µM) were used as chemoattractants, buffer as control for random migration. Various concentrations of Cu/Zn-SOD were embedded into the agarose gels. Values represent means ± SEM of n = 6; $^{**}p < 0.01$; $^{*}p < 0.05$).

DISCUSSION

Inhibition of the directed migration of PMNL towards the chemoattractants by SOD suggests an involvement of $\cdot O_2^-$ radicals in the migratory response. A similar effect of SOD in endotoxin-induced leukotaxis was observed by Rosenfeld et al. (13). This assumption is further supported by the detection of $\cdot O_2^-$ radicals induced by these chemoattractants near the optimum of their chemotactic activity (LTB_4: 10 nM; FMPL: 100 nM; PAF-acether:500 nM; C5a: 10 nM; w-OH-LTB_4: 5 nM). No such $\cdot O_2^-$ release could be observed by the two chemotactically inactive isomers of LTB_4, the 6-trans-LTB_4 and 6-trans-12-epi-LTB_4 (2). w-OH-LTB_4, however, which shows nearly the same chemotactic activity as LTB_4 again stimulates $\cdot O_2^-$ release in PMNL at the chemotactically active concentration of 10 nM (3). Admittedly, any quantitative relationship between the efficacy in the induction of $^{\bullet}O_2^-$ release and chemotactic potency does obviously not exist. Also the threshold concentrations of the chemoattractants appear to be slightly lower for chemotaxis than for the induction of $^{\bullet}O_2^-$ release, even if measured by means of the sensitive luminol-enhanced luminescence. We are, however, inclined to attribute this small discrepancy in threshold dosages (at best half an order of magnitude) to the difficulties in detecting very low rates of $^{\bullet}O_2$ production in a complex cellular system, and we suggest that just a minute amount of $^{\bullet}O_2^-$ formed locally upon contact with a chemotattractant is the prerequisite for directed migration.

The observation that also catalase inhibits the leukotactic response of PMNL rules out a direct involvement of $^{\bullet}O_2^-$ radicals in the leukotactic response. Rather some product resulting from the interaction of $^{\bullet}O_2$ and H_2O_2 mediates a redox change in the cellular membrane. Furthermore, as not all stimuli of the resporatory burst of PMNL are chemoattractants, neither H_2O_2 nor $^{\bullet}O_2^-$ *per se* can be regarded as direct mediators of leukotaxis. $^{\bullet}O_2^-$, however, may play a crucial role in directed migration of PMNL by initiating some free radical chain leading to a localized alteration of the cell membrane which might be a prerequisite for the polarization of the cell.

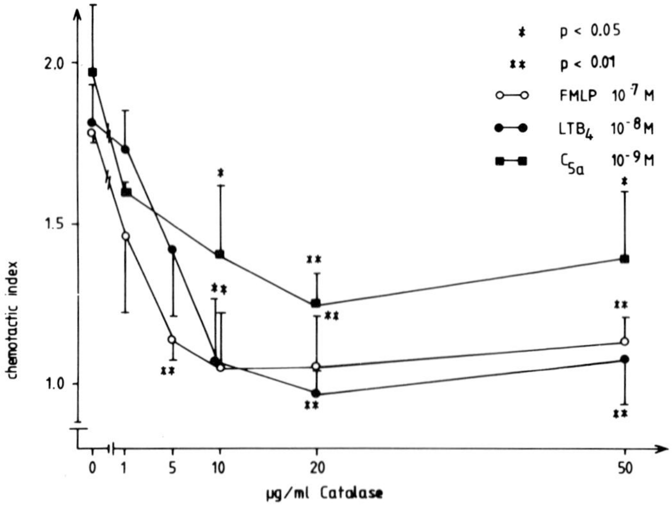

Fig. 4: Inhibition of PMNL chemotaxis by catalase.
PMNL were chemotactically stimulated in blind well chambers by FMLP (100 nM), LTB_4 (10 nM;) and C5a (1 nM) as chemoattractants. Various concentrations of catalase were added to the cellular compartments. Values represent means $\pm$ SEM of n = 4; **p < 0.01, *p < 0.05.

Evidently, the precise mechanism of chemotaxis inhibition by SOD remains to be worked out. It should also be mentioned that this observation substantiated here for five chemoattractants and, in part, with two test systems differing substantially, could not be reproduced in Boyden chamber experiments using thick membranes consisting of cellulose nitrate (Kownatzki, personal communication). Since it appears unclear which of the *in vitro* methods of chemotaxis mimics which cellular behaviour *in vivo*, the effects described here have to await corresponding *in vivo* experiments, before any definite conclusion as to their biological relevance can be drawn. It is our feeling, however, that our findings add further support to the concept that $^{\bullet}O_2^-$ plays a crucial role in initiating and, by cell recruitment, sustaining inflammatory processes (14).

ACKNOWLEDGEMENTS

This work was supported by grant PTB 8377 and 0385696 of the *Bundesministerium für Forschung und Technologie* of the Federal Republic of Germany.

REFERENCES

1. Kitagawa S, Takaku F (1981) FEBS Letters 128: 5-8
2. Flohé L, Martin W, Loschen G, Günzler WA (1984) Life Chem Rep 2: 318-324
3. Martin W, Loschen G, Flohé L (1985) Joint Meeting of the Belgian, Dutch and German Pharmacological and Toxicological Societies, Aachen, Sept. 23-26, Abstract
4. Webster RO, Hong SR, Johnston RB Jr., Henson PM (1980) Immunopharmacol 2: 201-219
5. Hartung HP, Parnham MJ, Winkelmann J, Englberger W, Hadding U (1983) Int J Immunopharmac 5: 115-121
6. Powell WS (1984) J Biol Chem 259: 3082-3089
7. Flohé L, Ötting F (1984) In: Packer L (ed) Oxygen Radicals in Biological Systems, Methods in Enzymology, Vol. 105, Academic Press, Inc., pp 93-104
8. McCord JM, Fridovich I (1969) J Biol Chem 244: 6049-6055
9. Meier B, Barra D, Bossa F, Calabrese L, Rotilio G (1982) J Biol Chem 257: 13977-13980
10. Pertoft H, Johnsson A, Wärmegard B, Seljelid R (1980) J Immunol Meth 33: 221-229
11. Boyden S (1962) J Exp Med 115: 453-466
12. Chenoweth DE, Rowe JG, Hugli TE (1979) J Immunol Meth 25: 337-353
13. Rosenfeld W, Khan AJ, Jhaveri R, Evans HE (1983) Clin Res 31: 454A
14. Flohé L, Giertz H, Beckmann R (1985) In: Bonta IL, Bray MA, Parnham MJ (eds) Handbook of Inflammation, Vol. 5. Elsevier, Amsterdam, pp 255-281

Superoxide and Superoxide Dismutase in Chemistry, Biology and Medicine; G. Rotilio, editor

INVESTIGATIONS ON THE HYPOTHETICAL ROLE OF GLUTATHIONE PEROXIDASE IN LEUKOTRIENE FORMATION BY RAT WHITE BLOOD CELLS

WOLFGANG A. GÜNZLER, MATTHIAS LEMKE, GERRIET LOSCHEN, WOLFGANG MARTIN, WOLFGANG H. SÖHNGEN, LEOPOLD FLOHÉ
Center of Research, Grünenthal GmbH, Zieglerstraße 6, D-5100 Aachen (FRG)

INTRODUCTION

Leukotrienes are formed in white blood cells by the lipoxygenase pathway from arachidonic acid (AA) with 5-hydroperoxy-6,8,11,14-eicosatetraenoic acid (5-HPETE) as an essential intermediate. Due to its low selectivity for hydroperoxide substrates (1), selenium-dependent glutathione peroxidase (GSH-Px) should be able to reduce 5-HPETE to 5-hydroxy-6,8,11,14-eicosatetraenoic acid (5-HETE) thus preventing formation of bioactive leukotrienes, e.g. the leukotaxin LTB_4. Evidence has been provided that in platelets impairment of GSH-Px activity results in an altered product pattern of the lipoxygenase pathway (2,3). We extended these investigations to two types of rat leukocytes, basophil leukemia cells (RBL-1) and polymorphonuclear leukocytes (PMNL).

MATERIAL AND METHODS

Cells

RBL-1 cells were grown in calf serum-enriched medium, used plain or supplemented with sodium selenite up to 5 µg selenium/l.

PMNL from female rats which had been kept on selenium-deficient or on adequate diets containing 0.01 and 0.22 mg selenium/kg, respectively, were elicited by i.p. injection of 10 % glycogen, collected by peritoneal lavage and isolated by density gradient centrifugation.

Assays of AA metabolization

800 g supernatants of sonified RBL-1 cells were incubated with 20 µM AA, 3 mM GSH and 2 mM $CaCl_2$ at 37°C for 30 min. Products were pre-purified on C_{18} silica gel and separated by HPLC on Nucleosil $5C_{18}$. Peaks of LTB_4 isomers designated I-IV in order of elution and of 5-HETE were quantified from peak areas monitored at 280 and 237 nm, respectively.

Intact viable (> 98 % Trypan blue exclusion) PMNL were incubated with 40 µM ^{14}C-AA, 5 mM GSH, 2 mM $CaCl_2$ and 20 µM ionophore A 23 187 for 15 min at 37°C. LTB_4 and 5-HETE were separated by TLC and quantified by radioscanning. LTB_4 isomers were determined in pre-purified samples by HPLC as described.

Determination of GSH-Px

Activity of GSH-Px was determined in cell homogenates with H_2O_2 as substrate (4) and refered to protein concentration.

RESULTS AND DISCUSSION

In RBL-1 cells supplementation of culture medium with sodium selenite resulted in up to 3-fold increased activities of selenium-dependent GSH-Px (Fig.1). In homogenates of RBL-cells grown on selenium-supplemented media the ratio LTB_4/5-HETE was found to be significantly lower than the corresponding ratio obtained with control cells (Fig. 1). Similar results were obtained when ratios were calculated from all LTB_4 isomers (I-IV) or from the biologically most potent LTB_4 isomer III alone. The inverse relationship between GSH-Px activity and LTB_4/ 5-HETE ratios gives support to the hypothesis that endogenous GSH-Px of RBL-1 cells is able to reduce 5-HPETE arising from the 5-lipoxygenase pathway.

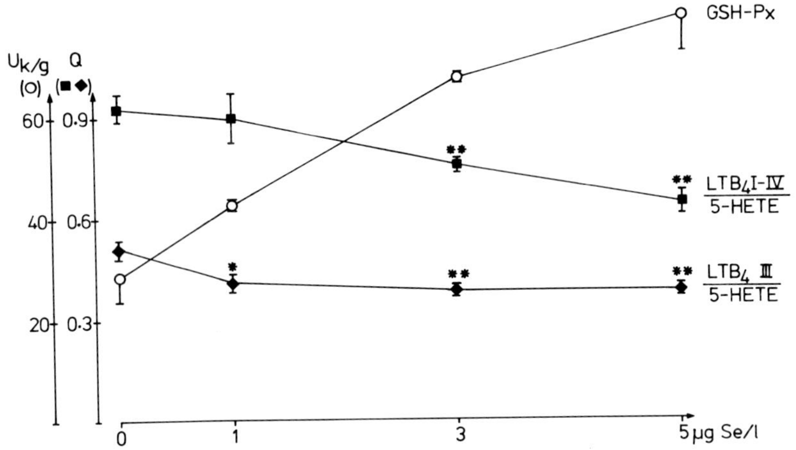

Fig.1. Activity of GSH-Px (U_k/g) and ratios LTB_4/5-HETE (Q) formed from AA by broken up RBL-1 cells grown in selenite-supplemented media. Means and SD, N=3. Significance of Q values from selenium-supplemented cells versus unsupplemented controls; * $p < 0.05$, ** $p < 0.01$.

However, in intact viable rat PMNL no significant alterations of LTB_4/5-HETE ratios were observed, although the activity of GSH-Px in PMNL had been reduced to less than 50 % of controls by selenium deficiency (Fig. 2).

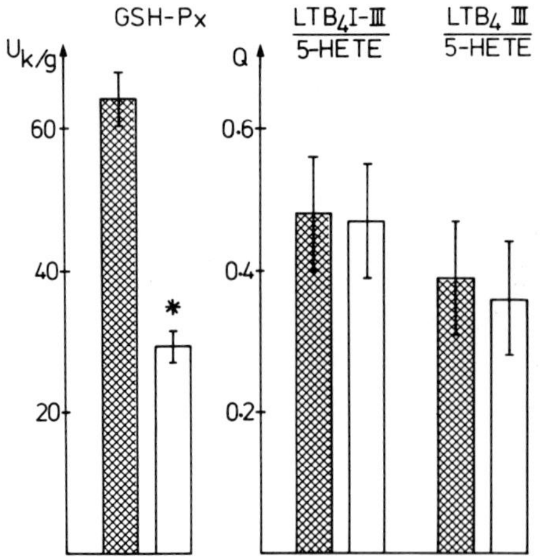

Fig.2. Activity of GSH-Px (U_k/g) and ratios LTB_4/5-HETE (Q) formed from AA by viable PMNL obtained from selenium-adequate (▩) and selenium-deprived (▭) rats. Means and S.D., N=15. Significance of reduction of GSH-Px activity: * $p < 0.002$.

It is concluded that, in principal, GSH-Px is able to reduce 5-HPETE and to interfere with leukotriene biosynthesis as demonstrated in homogenates of RBL-1 cells and earlier (2,3) in platelets. However, even substantial loss of GSH-Px activity appears not to affect leukotriene biosynthesis provided cellular compartimentation, as in vital PMNL, is intact. Any generalization as to the involvement of GSH-Px in the metabolism of hydroperoxides in vivo should therefore be considered with caution.

ACKNOWLEDGEMENT

This work was supported by grants PTB 8377 and 0385696 of the Bundesministerium für Forschung und Technologie (FRG).

REFERENCES

1. Flohé L, Günzler WA, Loschen G (1979) In: Kharash N (ed) Trace Metals in Health and Disease. Raven Press, New York pp 263 - 285.
2. Bryant RW, Bailey JM (1980) Biochem Biophys Res. Commun 92:268 - 276
3. Bryant RW, Simon TC, Bailey JM (1982) J Biol Chem 257: 14 937 - 14 943
4. Flohé L, Günzler WA (1984) Meth Enzymol 105:114 - 121

© 1986 Elsevier Science Publishers B.V. (Biomedical Division)
Superoxide and Superoxide Dismutase in Chemistry,
Biology and Medicine; G. Rotilio, editor

ACTIVATION OF GRANULOCYTE OXYGEN RADICAL PRODUCTION BY A MONONUCLEAR CELL-DERIVED CYTOKINE, GRANULOCYTE CHEMILUMINESCENCE INDUCER (GCI)

F.-E. MALY, A. KAPP* AND A.L. DE WECK
Institut für klinische Immunologie, Inselspital Bern and *Department of Dermatology, University of Freiburg

INTRODUCTION

In inflammatory processes, the importance of granulocyte-derived reactive oxygen species is subject of increasing interest. We investigated whether mononuclear cells can produce soluble mediators, different from immunoglobulins and complement components, which in turn will activate granulocytes to produce free radicals. Using lucigenin-dependent chemiluminescence to monitor oxygen radical production, we found that mononuclear cells can in fact produce a cytokine-like mediator(s), which stimulates production of oxygen radicals by granulocytes (granulocyte chemiluminescence inducer, GCI). This finding could be confirmed by cytofluorographic measurement of fluorescent oxidation products of 2',7'-Dichlorofluorescein-Diacetate (Maly et al., in preparation).

MATERIALS AND METHODS

GCI-containing supernatant: Mononuclear cells were isolated from blood by ficoll gradient and cultured in MEM with BSA (100 ug/ml). To stimulate mediator production, LPS (s. abortus equi) was added (500 ng/ml). After 24 hrs, supernatant was collected and stored at -20°C.

Granulocyte chemiluminescence: $1x10^6$ in MEM with 1 mM Lucigenin granulocytes, isolated by ficoll gradient and hypotonic lysis of erythrocytes with a purity of >95% and viability >98%, were placed in a luminometer (Berthold, West Germany) and test samples added.

HPLC gel filtration was used to estimate molecular weight of GCI. Monocyte-depleted lymphocytes and lymphocyte-depleted monocytes were prepared by counterflow elutriation.

Composition of cell preparations was analyzed by cytofluorographic analysis using monoclonal antibodies.

Interleukin 1-activity was measured in a thymocyte costimulator assay.

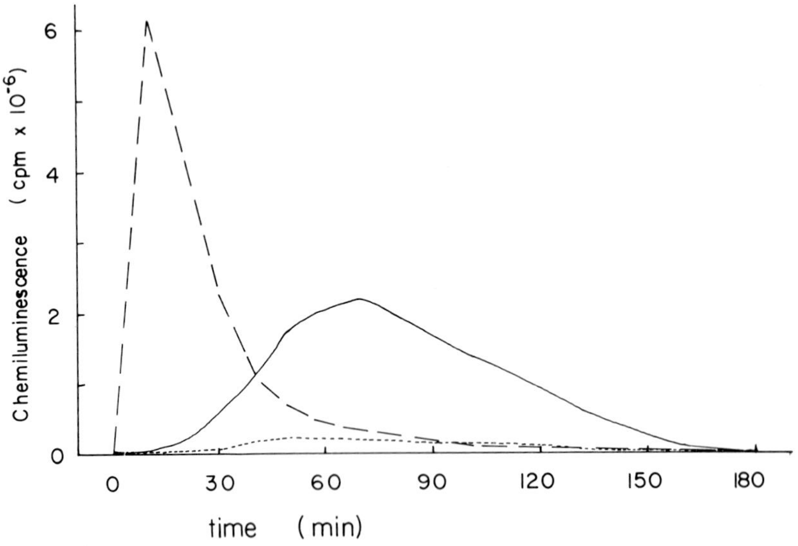

Figure 1 Granulocytes were stimulated with supernatant from LPS-stimulated (solid line) or unstimulated mononuclear cells (pointed line) at a final dilution of 1:2. Phorbol myristate acetate (100 ng/ml) served as a positive control (dashed line).

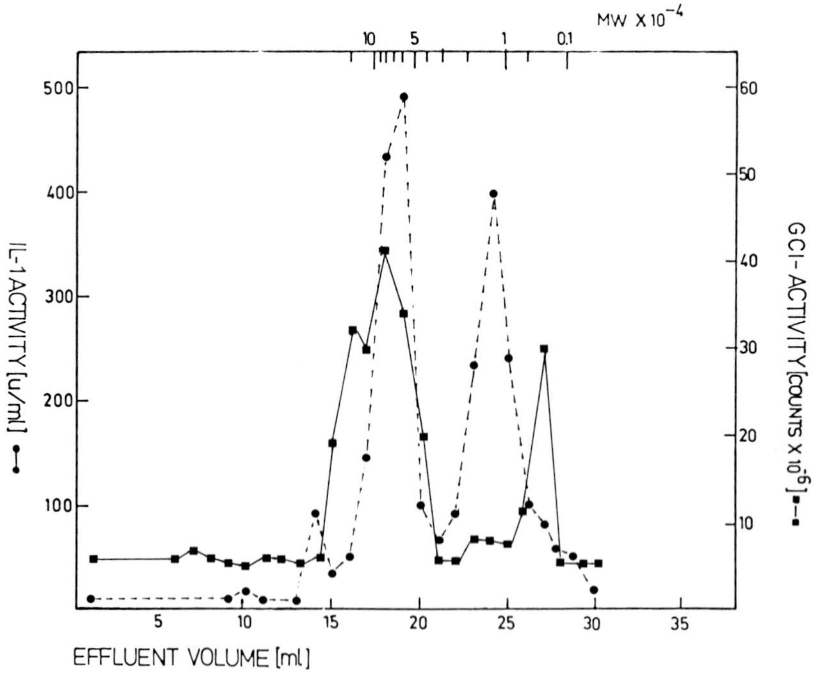

Figure 2 HPLC-fractionation on TSK 125 of LPS-induced mononuclear cell supernatant. Fractions were tested for GCI-activity (solid line) and interleukin 1-activity (broken line).

RESULTS

Supernatant of LPS-stimulated mononuclear cells strongly induced granulocyte chemiluminescence, which was inhibited to over 90% by superoxide dismutase and clearly differed from the response to supernatant of not stimulated mononuclear cells or LPS alone (figure 1). From experiments with practically pure lymphocytes and monocytes, obtained by counterflow elutriation and characterized by cytofluorography using monoclonal antibodies, monocytes appeared to be sufficient for production of GCI-activity and therefore probably represent the cellular origin or GCI. The active molecule appeared quite stable and to be of peptide nature, according to its sensitivity to trypsin and to inhibition of its production by cycloheximide (table 1). Upon HPLC, GCI eluted with a molecular weight of 60-70 kda and could also be separated from interleukin 1-activity, the best characterized monocyte-derived cytokine (figure 2).

TABLE 1 Biochemical properties of GCI

treatment	residual activity (%)
untreated GCI-containing supernatant	100
56°C, 60 min	42
100°C, 60 min	10
dialysis	80
trypsin treatment	41
ph change (2-12)	90
cycloheximide (20 ug/ml) during culture	20
monocyte depletion before culture	5
lymphocyte depletion before culture	98

CONCLUSIONS

From the data presented it is evident that mononuclear cells can produce a mediator(s) (GCI), which in turn will activate granulocytes to produce oxygen radicals, as evidenced by chemiluminescence. This mediator appears to be a peptide produced by monocytes and seems to be different from immunoglobulins, complement components and from another major monocyte product, interleukin 1. It should be mentioned that not only LPS, but also other mitogens and silica particles can stimulate GCI-production by mononuclear cells.

Therefore, stimulation of granulocyte oxygen radical production can in fact be achieved without need for immune complexes, complement components or phagocytosis, but by a cytokine-like mediator derived from monocytes. We suggest that also in vivo, such cytokine-mediated enhancement of free radical fluxes may occur and be of pathophysiological importance.

Superoxide and Superoxide Dismutase in Chemistry, Biology and Medicine; G. Rotilio, editor

STIMULATED SUPEROXIDE ION FLUXES IN HUMAN ERYTHROCYTES.

MARINA SCARPA*, PAOLO VIGLINO+ and ADELIO RIGO*.
*Institute of Biological Chemistry, University of Padova, Padova, Italy.
+Department of Spectroscopy, University of Venice, Venice, Italy.

INTRODUCTION

In previous papers we have shown that O_2^- is continuously generated in human erythrocytes (RBC) (1). In healthy individuals the rate of O_2^- generation ($R_\circ$), was found to be, in average, 2×10^{-8} M sec^{-1} while it resulted more than doubled in Fanconi anemia patients (2). In this communication we report on the effects of two quinone containing drugs, menadione and adriamycin, on the rate of O_2^- generation and we discuss their possible reactions.

RESULTS AND DISCUSSION

The methods used were as previously described (1). The approach to the steady-state $(ECu^{++})_s$ of the oxidized superoxide dismutase (SOD) added to a lysate of packed human RBC in the presence and in the absence of menadione or adriamycin was first-order in the oxidized SOD. From the slope of the plots $\ln \{([ECu^{++}]_\circ - [ECu^{++}]_s)/([ECu^{++}]_\circ - [ECu^{++}])\}$ vs time, the kinetic rate constant, k, of the attainment of the steady-state were calculated (see Table 1). The presence of menadione produced a strong increase of the k value, while this increase was smaller in the presence of adriamycin and it was observed only when the drug was incubated with Fe^{3+} or with plasma. On the basis of a general kinetic study the increase of the rate of O_2^- generation, in the presence of compounds containing a quinone structure, indicates that only O_2^- is directly involved in the reduction and oxidation of SOD, thus excluding the direct reaction of the copper ion with drugs or with their products. This behaviour is in accordance with the mono-electronic reduction of the quinone group (Q) to semiquinone (Q°):

1) $Q + e \longrightarrow Q^\circ$ followed by the reaction:

2) $Q^\circ + O_2 \rightleftharpoons Q + O_2^-$ see ref. 3)

TABLE 1

O_2^- GENERATION IN RBC IN THE PRESENCE OF MENADIONE AND ADRIAMYCIN.
The added bovine SOD was 2×10^{-5} M.

Drug	Conc. $M \times 10^5$	k $s^{-1} \times 10^3$	$(ECu^{++})_s$ %	$Ro + R_2$ $Ms^{-1} \times 10^8$
Menadione	---	0.7	59	1.4
Menadione	0.5	3.8	56	7.6
Menadione	1.0	9.0	56	18
Adriamycin	---	0.8	57	1.6
Adriamycin	4.0	0.8	60	1.6
Adriamycin	0.8	1.3*	57	2.6
Adriamycin	4.0	3.6*	55	7.2
Adriamycin	4.0	2.4**	56	4.8

* Fe^{3+} was added; $[Fe^{3+}]/[Adriamycin] = 1$
** Adriamycin was incubated with the plasma of the same donor for 5 min.

Therefore the removing of O_2^- by the reaction:

$$3)\qquad 2O_2^- + H^+ \xrightarrow[SOD]{k_3} O_2 + H_2O_2$$

could perturb the equilibrium 2. As a consequence it must be evaluated if the increase of the rate of SOD reduction in the presence of Q reflects a real increase of the O_2^- flux in RBC or it is due to the perturbation of the equilibrium 2 by the added SOD. Then taking into account the possible reactions between the radical Q° and other species N_i present in RBC:

$$4)\qquad N_i + Q° \xrightarrow{k_4} \text{products}$$

considering eqn. 1-3 and applying the steady-state approximation for the radicals Q° and O_2^-, the concentration of O_2^- at the steady-state results:

$$5)\quad [O_2^-]_{ss} = \left\{Ro + R_2/(1 + k_4\,[N_i]/k_2\,[O_2])\right\} / \left\{k_3\,[SOD] + k_{-2}\,[Q]\left[1 + -1/(1 + k_4\,[N_i]\,/k_2\,[O_2])\right]\right\}$$

where R_2 is the rate of Q° formation and R_o is the rate of O_2^- generation in RBC when no drug is present.

If 6) $k_2\,[O_2] >> k_4\,[N_i]$ eqn. 5 becomes:

$$7)\qquad [O_2^-]_s = (Ro + R_2)/k_3\,[SOD]$$

If condition 6 is not fulfilled the O_2^- flux, corresponding to Ro + R_2, calculated from eqn. 7 is underestimated. Since in human RBC k_3 [SOD] = 2.3 x 10^9 x 2 x 10^{-6} = 5 x 10^3 sec^{-1} it usually results:

8) k_3 [SOD] >> k_{-2} [Q]

and therefore, the addition of bovine SOD does not perturb the O_2^- flux in lysates.

In the case of menadione utilizing the kinetic rate constants calculated from the data of Winterbourn et al. (4, 5) we verified that conditions 6 and 8 hold and therefore the O_2^- fluxes (Ro + R_2) calculated according to eqn. 7, see Table 1, are correct. The rate of O_2^- generation, so calculated, corresponds, within 30%, to the flux of the menadione semiquinone generated in the following reaction (4):

9) Hb O_2 + Q ⟶ Hb^{3+} + Q° + O_2

thus indicating that the oxidation of Hb by menadione is the main source of O_2^-. Similar considerations can be applied to adriamycin, however, in this case, the rate of O_2^- generation calculated from eqn. 7 may be underestimated since we don't know if condition 6 is met.

In conclusion, the increased flux of O_2^- observed in RBC in the presence of quinone containing drugs, though controlled by SOD, determines an increase of the steady-state concentration of O_2^- which, according to eqns. 5 and 7, is approximately proportional to the rate of O_2^- generation.

REFERENCES

1. Scarpa M, Viglino P, Contri D, Rigo A (1984) J Biol Chem 259: 10657-10659.

2. Scarpa M, Rigo A, Momo F, Isacchi G, Novelli G and Dallapiccola B (1985) Biochem Biophys Res Commun 130: 127-132.

3. Patel K and Willson R (1973) J Chem Soc Faraday Trans 1: 814-825.

4. Winterbourn C, French J and Claridge R (1979) Biochem J 179: 665-673.

5. Winterbourn C, French J and Claridge R (1978) FEBS Lett 94: 269-272.

Superoxide and Superoxide Dismutase in Chemistry,
Biology and Medicine; G. Rotilio, editor

ERYTHROCYTE ANTIOXIDANTS AS SCAVENGERS AND/OR INDICATORS OF OXYGEN METABOLITES IN BIOLOGICAL SYSTEMS

KAREN M. TOTH, CONNIE J. BEEHLER, STEVE S. PARRY, MILENA M. HALEK, ELAINE M. BERGER AND JOHN E. REPINE

Departments of Medicine and Pediatrics and the Webb-Waring Lung Institute, University of Colorado Health Sciences Center, Box C322, 4200 East 9th Avenue, Denver, Colorado, 80262 (USA)

INTRODUCTION

Previous studies have usually focused on the ability of erythrocyte (RBC) antioxidants to protect erythrocytes from oxidant injury (1). However, RBC antioxidants might also protect other tissues against O_2 metabolite induced injury. RBC contain large amounts of antioxidants, are abundant at sites of inflammation and injury where O_2 metabolites are being generated, and can decrease O_2 metabolite concentrations *in vitro* (2). In addition, changes in RBC antioxidants might reflect O_2 metabolite concentrations and thereby predict injury (3).

Based on these findings, we hypothesized that RBC antioxidants might serve as scavengers and/or indicators of O_2 metabolites in biological systems. Our preliminary results show that 1) RBC can prevent H_2O_2 mediated tissue injury *in vitro* and 2) that RBC from human cigarette smokers contain more glutathione (GSH) and catalase than RBC from non-smokers.

MATERIALS AND METHODS

The methods used have been reported in detail previously (4). RBC glutathione, catalase and glutathione peroxidase levels were measured using conventional assays (5-9).

RESULTS

Addition to perfusates of increasing concentrations of washed human RBC progressively decreased lung weight gains and lung lavage albumin concentration increases seen in isolated lungs perfused with hypoxanthine (HX) and xanthine oxidase (XO). The mechanism of RBC protection against HX and XO induced lung injury appeared to involve scavenging of hydrogen peroxide (H_2O_2) by intracellular RBC antioxidants. First, addition of glutaraldehyde prefixed RBC which did not release their intracellular contents when treated with H_2O_2 _in vitro_ were equally as effective as untreated RBC in preventing injury in isolated lungs perfused with HX and XO. Second, addition of diethyldithiocarbamate (DDC, an inhibitor of SOD) treated RBC also provided protection. The same pattern of inhibition was seen when RBC were added to cultured lung endothelial cells (EC) which were damaged (LDH release) by addition of H_2O_2 or H_2O_2 dependent oxidation of reduced cytochrome c _in vitro_ (4). Glutathione and catalase activities were increased in RBC obtained from human cigarette smokers (>1 pack/day) (Table I).

TABLE I

ANTIOXIDANTS AND ENDOTHELIAL PROTECTION BY RBC FROM HUMAN CIGARETTE SMOKERS

Test Subjects	RBC catalase activity (units/ml of packed RBC)	RBC GSH activity (uM/gm Hb)	RBC protection of EC from H_2O_2 § (%)
Non-smokers	64,242±1280 (14)*	5.0±1.8 (14)	24±1.8 (14)
Cigarette smokers	78,242±2580 (14)†	6.4±2.3 (14)†	53±2.1 (14)†

* Mean ± standard error (number of determinations).

† Value significantly ($p<0.01$) different from value for non-smokers.

§ 100- (LDH from EC + RBC + H_2O_2/LDH from EC + H_2O_2) x 100%

DISCUSSION

The ability of RBC to decrease O_2 metabolite concentrations *in vitro* and O_2 metabolite induced injury to other tissues is very striking (4). Addition of as little as 1% RBC consistently and completely prevented H_2O_2 mediated injury to isolated perfused lungs and endothelial cells in culture. Additional studies indicate that the mechanism of protection by RBC involves intracellular scavenging of O_2 metabolites by intact RBC. The results are similar to recent work which shows that insufflation of RBC protects rats against hyperoxia induced acute edematous lung injury - again by mechanisms that appear to involve scavenging by intracellular antioxidant (glutathione) activities (10). Although experiments in these *in vitro* models suggest that RBC can protect other tissues, it has been difficult to design experiments which definitely prove that RBC can scavenge O_2 metabolites in animals or human subjects. Furthermore, one must reconcile that RBC antioxidants might not be effective in more complete biological systems because they may not reach key intracellular locations or tight junctions where neutrophils are sticking to endothelial cells. The possibility that O_2 metabolites might preferentially react with ubiquitous other compounds rather than RBC antioxidants is also raised as a consideration.

Even if RBC antioxidants do not confer some protection by scavenging O_2 metabolites, they may reflect O_2 metabolite concentrations and/or predict susceptibility to O_2 metabolite induced injury. Increases in GSH and catalase were found in RBC from human cigarette smokers. These increases most likely are a result of exposure of RBC to oxidants present in cigarette smoke and/or generated by activated phagocytes (neutrophils or macrophages) in cigarette smokers. In addition, because considerable variation exists in the levels of RBC antioxidants among individual cigarette smokers, RBC antioxidant levels might predict individual susceptibility to cigarette smoke induced diseases, such as vascular injury (atherosclerosis), lung injury (emphysema) and/or lung cancer. The latter might help explain why only some cigarette smokers develop these diseases while other cigarette smokers remain unexplainably healthy. This oversimplification ignores the complex interaction which must exist between the oxidants generated and increases in antioxidant defenses in these disease processes (11). Obviously, the "net" or excess production of O_2 metabolites is the critical determinant. Regardless, RBC antioxidants provide a new area for investigating the contribution of O_2 metabolites to important diseases.

REFERENCES

1. Weiss SJ (1982)J Biol Chem 257:2947-2953
2. Test ST, Weiss SJ (1984) J Biol Chem 259:399-405
3. Frank L (1982) J Appl Physiol 53(2):475-479
4. Toth KM, Clifford DP, Berger EM, White CW, Repine JE (1984) J Clin Invest 74:292-295
5. Beers RF, Sizer JW (1952) J Biol Chem 195:133
6. Oshino N, Chance B, Sies H, Bucker T (1973) Arch Biochem Biophys 154:117-131
7. Nicholls P (1965) Biochem Biophys Acta 99:286-297
8. Lauterburg BH, Smith CV, Hughes H, Mitchell JR (1984) J Clin Invest 73:124-133
9. Griffith OW (1981) J Biol Chem 2564900-4904
10. vanAsbeck S, Hoidal J, Schwartz B, Vercellotti G, Moldow C, Jacob H (1984) Clin Res 32:563A
11. Hoidal JR, Fox RB, LeMarbe PA, Perri R, Repine JE (1981) Am Rev Respir Dis 123:85-89

Superoxide and Superoxide Dismutase in Chemistry,
Biology and Medicine; G. Rotilio, editor

METHEMOGLOBIN REDUCTION BY DIVICINE: A HEXOSE MONOPHOSPHATE SHUNT-DEPENDENT PROCESS PROTECTING FROM ERYTHROCYTE-DAMAGING AUTOXIDATION MECHANISMS

A. De Flora, U. Benatti, L. Guida, M. Grasso, M. Tonetti, and C.C. Winterbourn*

Institute of Biochemistry, University of Genoa, Genoa, Italy, and *Department of Pathology, Clinical School of Medicine, Christchurch Hospital, Christchurch, New Zealand.

Favism is an acute hemolytic anemia affecting a number of genetically G6PD-deficient individuals. Two pyrimidine aglycones, divicine (DH_2) and isouramil, that are stored in fava beans in the form of the corresponding ß-glucosides (vicine and convicine, respectively), are among the toxic exogenous agents responsible for oxidative hemolysis (1, 2). Redox cycling, involving autoxidation of both native aglycones and production of H_2O_2, followed by reduction of the quinonic species by cellular metabolites, is thought to determine irreversible erythrocyte (RBC) damage.

Evidence is here presented that DH_2 can undergo an alternative oxidation pathway, by reducing MetHb stoichiometrically. This non-enzymatic process is dependent on intracellular hexose monophosphate shunt (HMS) activity supplying NADPH and GSH for the continuous reduction of the quinone (D_{ox}) and the consequent electron transfer from DH_2 to MetHb. G6PD-deficient RBC, whose HMS activity is inadequate to keep DH_2 reduced, are less competent than normal cells in reducing MetHb, this resulting in uncontrolled formation of highly reactive species including ferrylhemoglobin (3).

MATERIALS AND METHODS

Normal human RBC and cells with the Mediterranean type of G6PD deficiency were obtained from healthy male subjects. Preparation of DH_2 and D_{ox}, nitrite treatment of RBC to obtain 100% MetHb, experimental conditions for incubation of both MetHb-containing RBC and crystallized MetHb with either DH_2 or D_{ox} (all carried out under nitrogen), have been described elsewhere (4).

RESULTS AND DISCUSSION

Anaerobic incubation of MetHb-containing normal (A) and G6PD-deficient (B) RBC with DH_2 or D_{ox} resulted in a progressive MetHb reduction (Fig 1). Prior removal of glucose yielded kinetic patterns and a stoichiometry (1.5-2.5 moles of heme/mole of DH_2) that were not dissimilar in the two RBC types. In the presence of glucose, however, MetHb reduction was consistently enhanced in the normal RBC only. Thus, the DH_2-dependent MetHb reduction is stimulated by a glucose-consuming pathway and this is identifiable with the HMS which is severely defective in the G6PD-deficient RBC. Accordingly, the HMS activity was markedly increased in the normal, but not in the G6PD-deficient RBC, when challenged with DH_2.

Experiments with purified MetHb and with dialysed, MetHb-containing hemolysates showed that NADPH and GSH were the cellular metabolites mediating the glucose effect. Thus, in both systems, although NADPH, NADH and GSH alone did not reduce MetHb, each of the three compounds increased the degree of DH_2-dependent MetHb reduction and also enabled D_{ox}, which was ineffective *per se*, to give a comparable amount of reduction (Table I). These findings imply a

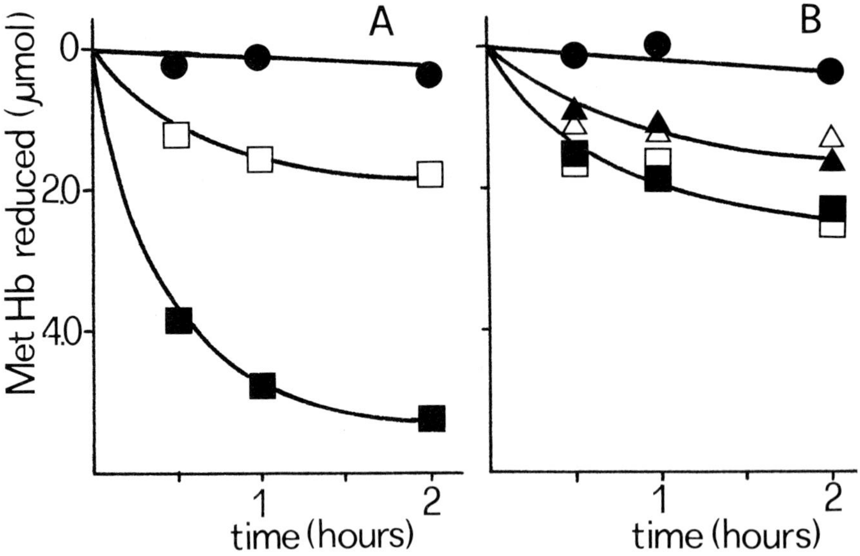

Fig. 1. Reduction of MetHb in normal and G6PD-deficient RBC incubated with DH_2 and D_{ox}.

Normal (A) and G6PD-deficient (B) MetHb-containing RBC were incubated under nitrogen at 37°C in PBS in the presence of 1 mM glucose (●), 1 mM DH_2 and 1 mM glucose (■), 1 mM DH_2 and 1 mM glucose 6-P (□), 1 mM D_{ox} and 1 mM glucose (▲), 1 mM D_{ox} and 1 mM glucose 6-P (△). The final hematocrit was 30%. The values of intracellular HMS activity, expressed as nanomoles of $^{14}CO_2$ released from ^{14}C-1-glucose per hour per ml of RBC incubated under nitrogen (5), were the following: 60 (no DH_2) and 535 (with DH_2) for the normal RBC, and 29 (no DH_2) and 45 (with DH_2) for the G6PD-deficient cells.

two-step electron transfer started by D_{ox} and either NADPH, NADH or GSH, with DH_2 as the final MetHb reductant. In spite of NADH having a comparable effect to NADPH or GSH, the results obtained with the G6PD-deficient RBC (Fig. 1) seem to rule out the involvement of metabolic processes other than the HMS pathway in perpetuating the DH_2-dependent MetHb reduction.

Several lines of evidence tend to exclude an enzymatic mechanism. a) No quantitative difference was observed with different compounds as the primary reducing agents (Table I); b) different amounts of hemolysate produced the same degree of DH_2-dependent MetHb reduction; c) there appears to be no involvement of NADPH diaphorase (6). These data, albeit negative, are important in view of the features of acute hemolysis in favic patients, which is totally unpredictable and certainly related to some still undefined individual variability: accordingly, such variability is not inherent to any enzyme system involved in the DH_2-dependent MetHb reduction.

Since it competes with DH_2 autoxidation and with its detrimental consequences (for instance, formation of ferrylhemoglobin (3)), MetHb reduction by

Table I. Reduction of purified Methemoglobin by Divicine.

Additions*	MetHb reduced in 30 min (umol)	
	by DH_2	by D_{ox}
None	0.56	0.20
NADPH	0.81	0.60
NADH	0.81	0.60
GSH	0.88	0.69

*Final concentrations were: 0.5 mM (DH_2 and D_{ox}); 2 mM (NADPH, NADH and GSH); 1 mM (MetHb). No reduction by GSH, NADH or NADPH alone detectable.

DH_2 may protect RBC considerably against oxidant damage by ingestion of fava beans. Its dependence on HMS activity can explain, at least in part, the much higher susceptibility to DH_2 of G6PD-deficient than of normal RBC.

ACKNOWLEDGEMENTS

Supported in part by the Special C.N.R. Project "Ingegneria Genetica e Basi Molecolari delle Malattie Ereditarie", by the Italian Ministry of Education and by the M.R.C. of New Zealand.

REFERENCES

1. Mager, J., Glaser, G., Razin, A., Izak, G., Bien, S., and Noam, M. (1965) Biochem. Biophys. Res. Commun. 20: 235-240
2. Chevion, M., Navok, T., Glaser, G., and Mager, J. (1982) Eur. J. Biochem. 127: 405-409
3. Winterbourn, C.C. (1985) Environ. Health Pers., in press
4. Benatti, U., Guida, L., Grasso, M., Tonetti, M., De Flora, A., and Winterbourn, C.C. (1985) Arch. Biochem. Biophys., in press
5. De Flora, A., Benatti, U., Morelli, A., and Guida, L. (1983) Biochem. Int. 7: 281-289
6. Lind, C., Hochstein, P., and Ernster, L. (1982) Arch. Biochem. Biophys. 216: 178-185

© 1986 Elsevier Science Publishers B.V. (Biomedical Division)
Superoxide and Superoxide Dismutase in Chemistry,
Biology and Medicine; G. Rotilio, editor

ROLE OF GLUTATHIONE REDUCTASE AND METHEMOGLOBIN REDUCTASE IN THE PROTECTION OF ERYTHROCYTES OF DIFFERENT AGE FROM OXIDATIVE STRESS

MASSIMO D'AQUINO AND SANCIA GAETANI

Istituto Nazionale della Nutrizione, Via Ardeatina 546, Roma

INTRODUCTION

Glutathione reductase (GR) and methemoglobin reductase (MHbR) are flavin enzymes involved in the protection of the erythrocyte from oxidative insults. Moreover the activity of both enzymes declines with the aging of the erythrocyte (1,2). The aim of the present work was to understand the role GR and of MHbR in the protection of the erythrocyte from the oxidative stress of the natural oxidative compound divicine (3). To obtain cells with different amounts of the two enzymes, erythrocytes from rats riboflavin deficient (RD) and of different age were used.

MATERIAL AND METHODS

Animals

Male albino Sprague-Dawley rats (150g) were fed a Riboflavin deficient or supplemented diet for 3 or more months.

Age dependent fractionation

Uncoagulated blood taken by heart puncture was fractionated according to Bennett et al. (4) on Percoll-phosphate saline buffer (PBS). Three fractions from control erythrocytes (C) were collected from the gradient (young, intermediate, old) according to their density (Table I). Equivalent fractions of the RD cells were collected on the basis of the distance migrated. Cells in each fraction were quantitated by Hb determination (5).

Preparation of divicine and treatment of blood

Vicine dissolved in PBS was treated with ß-glycosidase as previously described (3). The erythrocyte fractions were incubated without or with 2.5mM or 5.0mM divicine prepared from vicine for 1h at 37°C in the absence of glucose.

GSH and Methemoglobin determination

GSH was assayed by the method of Beutler (6) and MetHb by that of Evelyn et al. (7).

Enzyme assay

The activity of hexokinase (HK), piruvate kinase (PK) and glucose-6-phosphate dehydrogenase (G6PD) were assayed according to Beutler (8); GR activity by the method of Tillotson (9) and MHbR according to Hegesh (10).

The results of all the tables represent the mean value ± SD.

RESULTS AND DISCUSSION

Table I shows the Hb% of total blood in the three age fractions of C and RD erythrocytes.

TABLE I

Erythrocytes	% Erythrocytes		
	Young	Intermediate	Old
C	82.1 ± 5.41	12.7 ± 1.30	5.2 ± 0.67
RD	42.4 ± 3.22	32.9 ± 2.72	19.8 ± 1.83

It can be seen that riboflavin deficiency brings about an acceleration of the erythrocyte cycle with a shift of the cells towards older fractions. However when the deficiency becomes more severe (data not shown) only young cells and retyculocytes are present in the blood.

In table II enzyme activities in unfractionated blood and in cells of different age are reported. HX, G6PD and PK activities decline with increasing age to the same extent in C and in RD cells with the exception of the PK activity in old cells where the decrease in RD erythrocytes is more pronounced than in C cells. MHbR activity also decreases with the age of the cells; however the absolute values are in RD erythrocytes lower in all the age fractions than in the correspondent fractions of C cells.

TABLE II

Enzyme	Erythr.	Activity (IU)			
		Unfractionated	Young	Intermediate	Old
EK	C	0.72±0.21	0.92±0.34	0.74±0.27	0.66±0.18
	RD	0.68±0.17	0.84±0.29	0.59±0.23	0.59±0.23
G6PD	C	7.9 ±1.42	8.7 ±1.43	5.6 ±1.27	4.0 ±1.10
	RD	7.5 ±1.41	7.6 ±1.61	4.4 ±1.11	3.2 ±1.21
PK	C	4.9 ±1.72	4.6 ±1.50	3.4 ±0.51	1.8 ±0.09
	RD	4.3 ±1.31	4.4 ±1.39	2.3 ±0.41	0.9 ±0.06
MHbR	C	3.9 ±0.41	5.5 ±0.47	3.5 ±0.38	1.8 ±0.21
	RD	2.9 ±0.27	3.0 ±0.33	2.2 ±0.51	1.1 ±0.26

GR activity (Table III) without FAD addition is much lower in all age fractions of RD erythrocytes than in the respective fractions of the C cells. When FAD is added only old cells have a lower activity in RD than in C erythrocytes. AC is the ratio of the enzyme activity + FAD over the activity - FAD.

TABLE III

Erythrocytes		GR Activity (IU) Unfractionated	Young	Intermediate	Old
C	- FAD	6.7 ±1.52	7.5 ±1.63	6.3 ±1.45	5.2 ±1.48
	+ FAD	7.1 ±1.60	7.9 ±1.70	6.8 ±1.57	6.2 ±1.48
	AC	1.0 ±0.04	1.0 ±0.05	1.0 ±0.04	1.1 ±0.07
RD	- FAD	2.5 ±0.20	2.7 ±0.30	1.2 ±0.70	0.7 ±0.10
	+ FAD	6.3 ±1.30	7.4 ±1.69	5.8 ±1.48	4.2 ±1.40
	AC	2.9 ±0.15	2.9 ±0.10	4.8 ±0.30	6.3 ±1.20

MetHb as a % of total Hb was 3 in C and in RD cells of all ages. However the treatment with 2.5mM divicine caused an increase from 7 (young) to 13 (old) in C and from 11 to 19 in RD cells respectively. With 5.0mM divicine the effect was more pronounced: from 14 to 25 in C and from 25 to 34 in RD.

GSH of young and intermediate RD erythrocytes was more sensitive to the effect of divicine (5.0mM) than GSH of their respective fractions of C erythrocytes. In the old cells of both groups of rats the amount of GSH under the effect of divicine became undetectable, probably as a result of the low G6PD activity.

ACKNOWLEDGEMENTS

This work was supported by grants from the Progetto Finalizzato "Ingegneria genetica e basi molecolari delle malattie ereditarie" CNR, Roma.

REFERENCES

1. Powers J.H. and Thurnham D.I. (1981) Brit. J. Nutr. 46:257
2. Matsuki T., Tamura M., Takeshita M., Yoneyama Y. (1981) Bioch. J. 194:327
3. D'Aquino M., Gaetani S. and Spadoni M.A. (1983) Biochem. Biophys. Acta 731:161
4. Bennet G.D. and Kay M.M.B. (1981) Exp. Hematol. 9:297
5. Van Kampen E.J. and Zijlstra W.G. (1961) Clin. Chim. Acta 6:538
6. Beutler E., Duron O. and Kelley B.M. (1963) J. Lab. Clin. Med. 61:882
7. Evelyn K.A. and Malloy H.T. (1938) J. Biol. Chem. 126:655
8. Beutler E. (1971) Red cell metabolism. Grune and Stratton Inc. N.Y.
9. Tillotson J.A. and Sauberlich H.E. (1971) J. Nutr. 101:1459
10. Hegesh E., Calmanovici N. and Avron M. (1968) J. Lab. Clin. Med. 72:339

REGULATION OF GLUCOSE-6-PHOSPHATE DEHYDROGENASE UNDER OXIDATIVE STRESS

REGINA BRIGELIUS

Univ. Düsseldorf, Dept.Physiol.Chem. I, D-4000 Düsseldorf, FRG

INTRODUCTION

Direct regulation of Glucose-6-phosphate dehydrogenase (G6PDH) by GSSG has been deduced from the following observations: reversal by GSSG of NADPH inhibition of G6PDH in rat liver cytosol despite seemingly blocked glutathione reductase (GR) by $ZnCl_2$ (1); persistent G6PDH activation associated with increase in glutathione mixed disulfides in rat livers perfused with t-butyl hydroperoxide (tBOOH) (2). Since any contribution of GSSG to G6PDH regulation has been questioned (3), we designed experiments to decide whether alterations of the $NADPH/NADP^+$ ratio alone can account for changes in G6PDH activity.

METHODS

Rat liver cytosol was prepared either in 0.25 M sucrose or Krebs-Henseleit medium. Each 5 ml of these preparations were incubated with 5 ml 0.1 M Tris buffer, pH 8.0, containing 9 mM $MgCl_2$ and 60 µM $NADP^+$ and with 830 µl buffered 5 mM GSSG solution at 32°C. Optionally, $ZnCl_2$ (33 µM final concentration) was added to block GR as suggested (1). At given times of preincubation 50 µl aliquots were added to 630 µl test solution (50 mM Tris, pH 7.4, 3 mM $MgCl_2$, 17.5 µM $NADP^+$, 85 µM NADPH) and completion of GR-dependent NADPH-consumption was monitored at 366-436 nm. When a stable absorption was established (3 min), the G6PDH reaction was started by 20 µl 35 mM glucose-6-phosphate (G-6-P). At corresponding preincubation times 500 µl aliquots were mixed with 500 µl 1 N perchloric acid containing 6.25 mg N-ethyl maleimid/ml to estimate the final GSSG content (4) at the end of the preincubation. $NADP^+$ and NADPH at the starting time of the G6PDH reaction were calculated from the absorption readings after termination of the GR reaction.

RESULTS

Preincubation of rat liver cytosol resulted in a time-dependent loss of added GSSG. Before measurement of G6PDH activity the remaining GSSG was reacted with endogeneous GR in the NADPH-containing test system. GR-reaction took place irrespective of the presence of Zn^{2+}. G6PDH activity at defined $NADPH/NADP^+$ ratios (at zero concentrations of GSSG) could then be determined.

*Present address: Grünenthal GmbH, Zieglerstr. 6, D-5100 Aachen, FRG

This experimental design allowed to correlate G6PDH activity in rat liver cytosol with the GSSG content (at the end of preincubation) and, further with the actual NADPH/NADP$^+$ ratio at the start of the G6PDH reaction.

For reasons not to be discussed here, the time required to establish the equilibrium between GSSG and the constituents of the cytosol differed substantially in the different media. In sucrose medium GSSG dropped to zero within

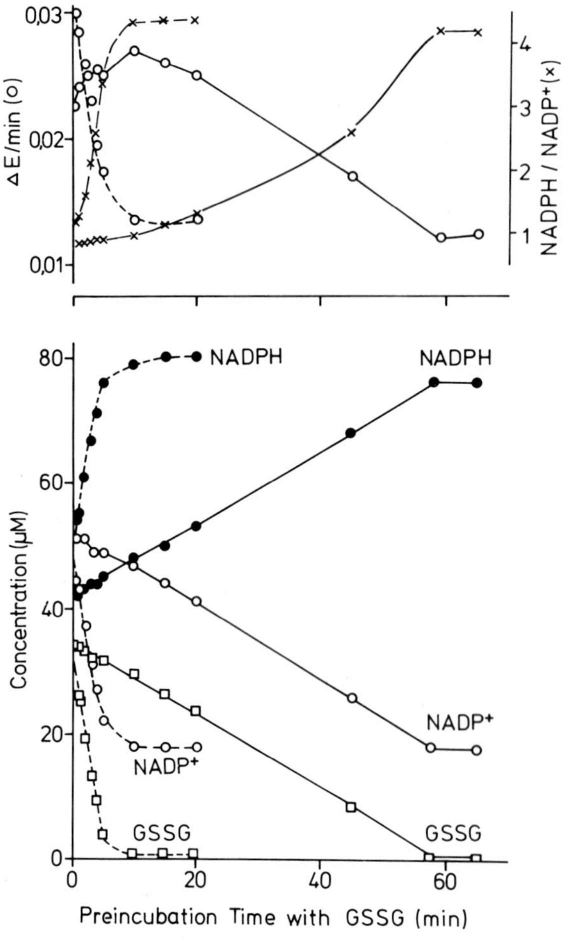

G6PDH activity after preincubation with GSSG in rat liver cytosol in relation to different NADPH/NADP$^+$ ratios (upper part), and NADPH (●), NADP$^+$ (o) and pre-test GSSG (□) levels (lower part). Solid lines: cytosol prepared in Krebs-Henseleit medium; broken lines: cytosol prepared in 0.25 M sucrose medium and preincubated with 33 µM $ZnCl_2$.

5 min whereas it needed 1 h to be consumed in Krebs-Henseleit medium. The residual levels of GSSG after the preincubation determined the final NADPH/ $NADP^+$ ratio in the G6PDH test system.

Evidently GSSG hat little chance to interact with G6PDH within the limited time of stability during the preincubation in the sucrose medium. Correspondingly, the G6PDH activity appeared unaffected by GSSG, but just inversely related to the NADPH/$NADP^+$ ratio. In the Krebs-Henseleit medium however, after up to 20 min preincubation with GSSG, G6PDH activity increased with the NADPH/$NADP^+$ ratio, whereas the anticipated inverse relationship to the NADPH/ $NADP^+$ ratio was only observed at lower final GSSG concentration.

DISCUSSION

The results show that activation of G6PDH by GSSG a) requires relatively high concentration (around 0.3 mM), b) depends on exposure to such GSSG concentrations for several minutes and c) persists for at least 20 minutes. Without prolongued exposure to high GSSG concentration, G6PDH activity appears dominated by the NADPH/$NADP^+$ ratio. The time dependency of the GSSG effect suggests a covalent regulation, probably a mixed disulfide formation (2). The covalent character is also supported by the fact that G6PDH activity of rat livers is persistently increased after perfusion with tBOOH (2), an effect attributable to acceleration of V_{max} for both substrates $NADP^+$ and G-6-P, but not, as suggested earlier (1), to a change in K_i for NADPH (unpublished results). The mechanism of this covalent regulation however, could not be clarified, since partially purified G6PDH was not activated by GSSG, though by thioltransferase presumed to catalyse disulfide exchange, but also in the absence of GSSG (unpublished observations).

In conclusion, under severe oxidative stress (peroxide perfusion or exposure to GSSG in vitro), rat liver G6PDH is definitely increased apparently by covalent regulation, whereas under a wide range of conditions its activity is determined by NADPH/$NADP^+$ ratio.

Supported by DFG grant Br 778/1-2

REFERENCES

(1) Eggleston, L.V. and Krebs, H.A. (1974) Biochem. J. 138, 425-435

(2) Brigelius, R. (1985). In: "Oxidative Stress", (H. Sies, ed.) pp. 243-272, Acad. Press, London.

(3) Levy, H.R. and Christoff, M. (1983) Biochem. J. 214, 959-965

(4) Akerboom, T.P.M. and Sies, H. (1981) Methods in Enzymol. 77, 373-382

Superoxide and Superoxide Dismutase in Chemistry, Biology and Medicine; G. Rotilio, editor

THE EFFECT OF POST-DECAPITATIVE ISCHEMIA ON SUCCINATE-SUPPORTED HYDROGEN PEROXIDE RELEASE BY BRAIN MITOCHONDRIA

MARIA CINO[+] and ROLANDO F. DEL MAESTRO[++]

Brain Research Laboratory, Victoria Hospital, Depts. of Biochemistry[+] and Clinical Neurological Sciences[+], University of Western Ontario, London, Ontario, N6A 4G5, Canada

INTRODUCTION

Under aerobic conditions, isolated mitochondria from a variety of tissues are effective sources of H_2O_2 (1). It has been shown that O_2^- is the essential precursor of H_2O_2 in mitochondria (2,3) and that the generation of H_2O_2 is related to the energy state of these organelles (4). There are two major sites of free radical generation in the electron transport chain, the ubiquinone-cytochrome b (5) and the NADH dehydrogenase regions (6,7). Despite the toxic potential of H_2O_2 and relatively low levels of free radical-scavenging enzymes in brain (8,9), the production of H_2O_2 by brain mitochondria remains controversial (10,11). Hence, one of the main goals of this study was to ascertain whether or not the electron transport chain of brain mitochondria produces H_2O_2.

It has been suggested that during ischemia electron transport components dissociate resulting in the accumulation of ubisemiquinone, which upon reoxygenation autoxidizes generating O_2^- (12,13). We tested the hypothesis that mitochondria damaged during ischemia generate increased amounts of O_2^- and H_2O_2 upon reoxygenation. The incubation of isolated mitochondria (following complete brain ischemia) in buffer, in which oxygen and substrates are not limiting, provides a crude *in vitro* model of the effect of post-ischemic reperfusion on the mitochondria.

MATERIALS AND METHODS

Model of Complete Ischemia

Female Wistar rats, weighing 150-175 g were sacrificed by decapitation. Brain mitochondria were prepared by the method described by Rehncrona *et al.* (14) either immediately following decapitation (control fraction) or following a 15 minute incubation period in nitrogen-saturated 0.9% (w/v) saline bath at 37°C (post-ischemic fraction).

Measurement of Extramitochondrial Hydrogen Peroxide Release

Hydrogen peroxide generation by control and post-ischemic mitochondrial fractions was measured by monitoring the formation of Compound II at 417-402nm ($E_{417-402} = 50\ mM^{-1}\cdot cm^{-1}$) in an Aminco DW-2a dual wavelength spectrophotometer as described by Turrens *et al.* (15). The reaction was initiated by the addition of 6mM succinate to a cuvette containing 0.25 mg protein$\cdot ml^{-1}$ and 0.5μM horseradish peroxidase in 145mM KCl, 5mM potassium phosphate, 3mM $MgCl_2$, 0.1mM EGTA and 30mM HEPES, pH 7.4. The effect of respiratory inhibitors (0.2μM rotenone and 0.2μM antimycin A) on H_2O_2 release by brain mitochondria was also investigated.

RESULTS

Figure 1a shows a spectrophotometric tracing of Compound II formation by control and post-ischemic mitochondrial fractions. Figure 1b shows the effect of respiratory inhibitors on Compound II formation by control mitochondria.

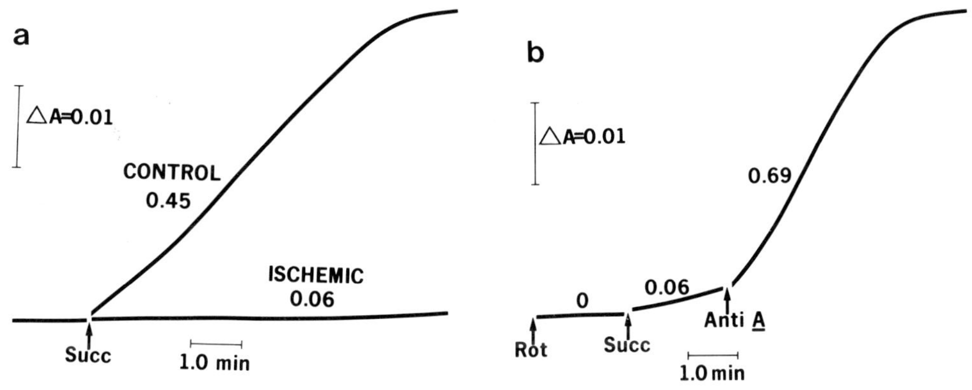

Figure 1. (a) The effect of complete ischemia on H_2O_2 generation by isolated brain mitochondria. (b) The effect of respiratory inhibitors on succinate-supported H_2O_2 release by brain mitochondria. Abbreviations: Succ, succinate; Rot, rotenone; Anti A, antimycin A. The numbers above the traces indicate nmole H_2O_2 per min per mg protein. The above patterns were characteristic of 7 different mitochondrial preparations.

Succinate supported extramitochondrial H_2O_2 release in the absence of respiratory inhibitors in control mitochondrial fractions. Upon reoxygenation following 15 minutes of ischemia, however, the ability of brain mitochondria to generate H_2O_2 decreased greater than 87% (Fig. 1a). Preincubation of brain mitochondria with rotenone largely prevented the stimulation of H_2O_2 release by succinate; however, the subsequent addition of antimycin A to either control (Fig. 1b) or post-ischemic preparations (results not shown) led to the equivalent stimulation of H_2O_2.

DISCUSSION

In this study we have demonstrated that brain mitochondria can generate H_2O_2 in the presence of succinate and that under these conditions this activity is rotenone-sensitive. It has been shown that succinate, in the absence of ADP, causes the reduction of pyridine nucleotides by reversed electron transfer from ubiquinone to NADH dehydrogenase (16). Since rotenone blocks the transfer of electrons from NADH dehydrogenase to ubiquinone, it appears that most of the succinate-supported H_2O_2 generation in brain mitochondria occurs via the transfer of reducing equivalents by succinate to the NADH dehydrogenase portion of the chain. The enhancement of extramitochondrial H_2O_2 release by antimycin A indicates that under conditions where it is maximally reduced, the ubiquinone-cytochrome b region is also a source of H_2O_2 in brain mitochondria.

We have further demonstrated that mitochondria damaged during ischemia release H_2O_2 at significantly lower rates than control mitochondrial fractions upon reoxygenation, and by inference, upon post-ischemic reperfusion. Thus, the hypothesis that mitochondrial free radical generation is enhanced following ischemia is not supported. These results are not consistent with the hypothesis that disruption of the electron transport chain during ischemia leads to enhanced autoxidation upon post-ischemic reperfusion (11). The rate of succinate-supported H_2O_2 release by post-ischemic mitochondrial fractions is comparable to that by rotenone-inhibited control mitochondrial fractions. Since a major portion of succinate-supported H_2O_2 release is due to the NADH dehydrogenase portion of the electron transport chain in brain mitochondria, it appears that the decreased ability of the post-ischemic mitochondrial fractions to generate H_2O_2 may be related to an impairment of reversed electron transfer.

ACKNOWLEDGEMENTS

This study was supported by the Ontario Heart Foundation Grant No. 3-26 and the Brain Research Fund Foundation.

REFERENCES

1. Chance B, Sies H and Boveris A (1979) Phys Rev 59:527-605
2. Boveris A and Cadenas E (1975) FEBS Lett 54:311-315
3. Dionisi O, Galeotti T, Terranova T and Azzi A (1975) Biochim Biophys Acta 403:292-301
4. Boveris A and Cadenas E (1982) In: Oberley W (ed) Superoxide Dismutase. CRC Press, Florida, pp 15-30
5. Cadenas E, Boveris A, Ragan CI and Stoppani AOM (1977) Arch Biochem Biophys 180:248-257.
6. Takeshige K and Minakami S (1979) Biochem J 180:129-135
7. Turrens JF and Boveris A (1980) Biochem J 191:421-427
9. Sinet PM, Heikkila RE and Cohen G (1980) J Neurochem 34:1421-1428
10. Sorgato MC, Sartorelli I, Loschen G and Azzi A (1974) FEBS Lett 45:92-95
11. Forman HJ and Kennedy J (1976) Arch Biochem Biophys 173:219-224
12. Demopoulos HB, Flamm ES, Pietronigro DD and Seligman ML (1980) Acta Scand 492(Suppl):91-120
13. Braquet P, De Feudis FV, Deby C and Braquet M (1984) In: Bes A, Braquet P, Paoletti R and Siesjo BK (eds) Cerebral Ischemia. Elsevier Science, Amsterdam, pp 265-281
14. Rehncrona S, Mela L and Siesjo BK (1979) Stroke 10:437-446
15. Turrens JF, Freeman BA and Crapo JD (1982) Arch Biochem Biophys 217:411-421
16. Chance B and Hollunger G (1957) Fed Proc 16(abstr):163

ASSESSMENT OF LIPID AND PROTEIN OXIDATION IN BRAIN MITOCHONDRIA FOLLOWING *IN VITRO* EXPOSURE TO FREE RADICAL-GENERATING SYSTEMS AND POST-DECAPITATIVE ISCHEMIA

MARIA CINO[+], CHRISTOPHER H. VIERECK[+] AND ROLANDO F. DEL MAESTRO[++]

Brain Research Laboratory, Victoria Hospital, Depts. of Biochemistry[+] and Clinical Neurological Sciences[++], University of Western Ontario, London, Ontario, N6A 4G5, Canada

INTRODUCTION

Post-ischemic mitochondrial fractions are characterized by decreases in the rate of state 3 respiration (1,2), cytochrome *c* oxidase activity (2) and phospholipid levels and increases in free fatty acids (3,4). These alterations may involve free radical damage to mitochondrial membranes since these contain (i) large amounts of polyunsaturated fatty acids which are easily peroxidizable (5), (ii) protein sulfhydryl groups which upon oxidation may form inter-/intra-disulfide linkages possibly leading to enzyme inactivation (6) and (iii) electron transport components in close proximity with the lipids of the inner membrane (7). Furthermore, certain areas of the brain have high levels of total iron (8), which is a primary catalyst in oxidative reactions (9). Based on the hypothesis that mitochondrial free radical generation increases during ischemia and post-ischemic reperfusion (5), the extent of oxidative injury to mitochondrial membranes following *in vitro* exposure to free radical-generating systems and post-decapitative ischemia was investigated.

MATERIALS AND METHODS

Model of Complete Ischemia

Female Wistar rats, weighing 150-175 g were sacrificed by decapitation. Brain mitochondria were prepared as described by Rehncrona *et al.* (10) either immediately following decapitation (control fraction) or following a 15 or 30 minute incubation period in nitrogen-saturated 0.9% (w/v) saline bath at 37^oC (post-ischemic fraction).

Quantitation of Lipid Peroxidation Levels and Sulfhydryl Group Content

Free radical-induced peroxidation and sulfhydryl group oxidation. Brain mitochondria (6-8 mg protein·ml^{-1}) were incubated with 0.3mM Fe_2SO_4, 0.1mM ADP and 0.8mM ascorbate (11) or with 0.3mM Fe_2SO_4, 110μM hypoxanthine and 1 mg·ml^{-1} xanthine oxidase (12) for 1 hr at 37^oC (oxidized fraction).

Lipid extraction and measurement of lipid peroxidation. Total mitochondrial lipids were extracted with chloroform : methanol (1:1,v/v) (13). The extent of lipid peroxidation in these extracts was assessed by measuring the level of diene conjugates (14) and lipid-soluble fluorescent products (15).

Protein oxidation. Protein oxidation was assessed by measuring the sulfhydryl group content (16) in mitochondrial fractions.

RESULTS

Figure 1 shows the levels of lipid peroxidation products in brain mitochondrial fractions under the indicated conditions. Following in vitro exposure to (Fe^{2+}/ADP)-ascorbate and Fe^{2+}-(hypoxanthine/xanthine oxidase) free radical-generating systems, the levels of diene conjugate and lipid-soluble fluorescent products increased 2- and 3.5-fold respectively. There were, however, no significant differences in the accumulation of these products in control and post-ischemic mitochondrial fractions. Figure 2 shows the sulfhydryl

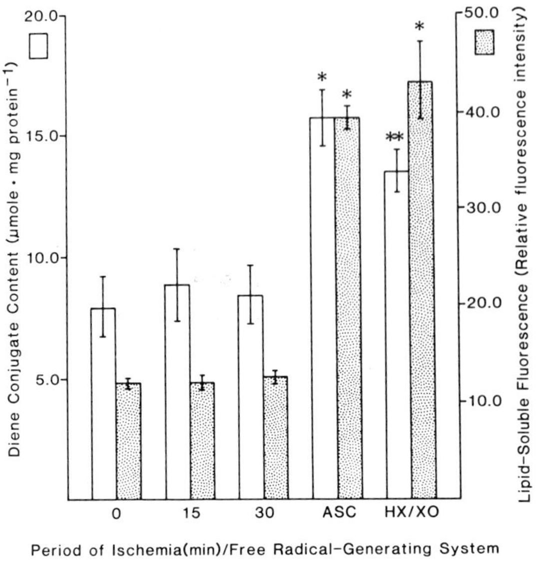

Fig. 1. Lipid peroxidation levels in brain mitochondria following post-decapitative ischemia and in vitro exposure to free radical-generating systems. Abbreviations: ASC, ascorbate; HX/XO, hypoxanthine/xanthine oxidase. Results represent mean values ± S.E.M. of 6 preparations.

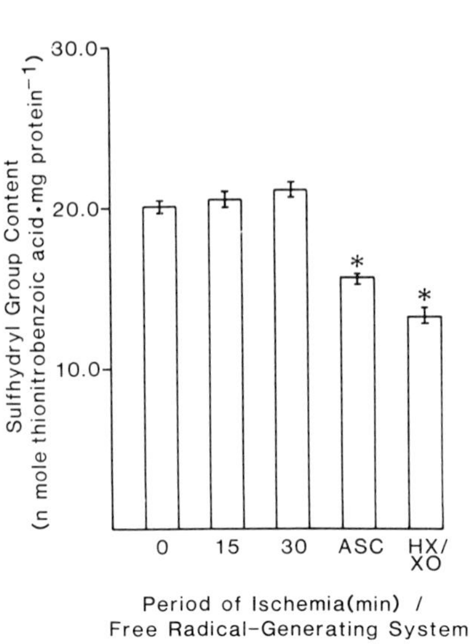

Fig. 2. Sulfhydryl group content in brain mitochondria following post-decapitative ischemia and in vitro exposure to free radical-generating systems. Results represent mean values + S.E.M. (0,15,30 min, n=5; ASC, HX/XO, n=4)

group content in control, post-ischemic and oxidized mitochondrial fractions. There was a 20-35% decrease in sulfhydryl group content in brain mitochondrial fractions following in vitro exposure to free radical-generating systems, but no change in sulfhydryl group content following post-decapitative ischemia.

DISCUSSION

Exposure of brain mitochondria to free radical-generating systems clearly demonstrated that these organelles are sensitive to radical injury. There was no evidence of lipid or protein oxidation in the post-ischemic mitochondrial fractions, indicating that free radical reactions are not accountable for the injury sustained by the mitochondria during complete cerebral ischemia. These findings are consistent with those of Rehncrona (11) and Jemaa et al. (19), who used whole brain homogenates and demonstrated that lipid peroxidation does not occur under anaerobic conditions (12). Majewska et al. (4), in contrast, reported a 2-fold increase in malondialdehyde following 5 minutes of post-decapitative ischemia. Procedural differences may account for this discrepancy. While in the present study brains were maintained in a dearated solution at $37^{0}C$ for the ischemic period, in the study of Majewska et al. (4) the brain was exposed to air throughout the ischemic period.

While free radical mechanisms do not appear to account for mitochondrial damage during complete ischemia, the possibility that mitochondrial injury during post-ischemic reperfusion is due to free radical-induced lipid peroxidation is currently under investigation.

ACKNOWLEDGEMENTS

This project was supported by the Ontario Heart Foundation Grant No. 3-26 and the Brain Research Fund Foundation.

REFERENCES

1. Rehncrona S, Mela L and Siesjo BK (1979) Stroke 10:437-446
2. MacKinnon JA (1984) MSc Thesis, University of Western Ontario, London, Canada
3. Viereck C and Del Maestro RF (1984) Can J Neuro Sci 11(abstr):336
4. Majewska MD, Strosznajder J and Lazarewicz J (1978) Brain Res 158:423-434
5. Demopoulos HB, Flamm ES, Pietronigro DD and Seligman ML (1980) Acta Scand 492(Suppl):91-120
6. Slater TF (1984) Biochem J 222:1-15
7. Nohl H and Hegner D (1978) Eur J Biochem 82:563-567
8. Floyd RA, Zaleska MM and Harmon HJ (1984) In: Armstrong D (ed) Free Radicals in Molecular Biology, Aging and Disease. Raven Press, New York, pp 143-161
9. Halliwell B and Gutteridge JMC (1984) Biochem J 219:1-14
10. Rehncrona S, Mela L and Siesjo BK (1979) Stroke 10:437-446
11. Rehncrona S, Smith DS, Akesson B, Westerberg E and Siesjo BK (1980) J Neurochem 24:1630-1638
12. Hillered L and Ernster L (1983) J Cereb Blood Flow Metab 3:207-214
13. Bleasdale JE, Davis CS and Aquanoff BW (1978) Biochim Biophys Acta 528:331-343
14. Recknagel RO and Glende EA (1984) In: Packer L (ed) Methods in Enzymology, Vol. 105. Academic Press, New York, pp 331-337
15. Dillard CJ and Tappel AL (1984) In: Packer L (ed) Methods in Enzymology, Vol. 105. Academic Press, New York, pp 337-341
16. Khare S, Trivedi A, Kesavan PC and Prasad R (1982) Int J Radiat Biol 42:369-383
17. Jemaa R, Bezard J and Bralet J (1984) In: Bes A, Braquet P and Paoletti R (eds) Cerebral Ischemia: Proceedings of the Third International Symposium on Cerebral Circulation. Elsevier Science, Amsterdam, pp 233-238

© 1986 Elsevier Science Publishers B.V. (Biomedical Division)
Superoxide and Superoxide Dismutase in Chemistry, Biology and Medicine; G. Rotilio, editor

FORMATION OF REDUCED SPECIES OF OXYGEN IN HYPOXIC RAT LIVER: A QUANTITATIVE APPROACH

GERHARD GERBER, WERNER SIEMS, MARIANNE MUELLER
Institut fuer Biochemie, Humboldt-Universitaet zu Berlin,
1040 Berlin, Hessische Strasse 3-4 (DDR)

INTRODUCTION

The importance of oxygen-derived free radicals as mediators of postischemic injury in various tissues is commonly accepted (1-7). Nevertheless all prerequisites for a burst of activated oxygen species and H_2O_2 are fulfilled during incomplete ischemia as well as during hypoxic perfusion.

The rat liver was choosen as an appropriate model to study the contribution of nucleotide degradation via cytosolic xanthine oxidase and peroxisomal urate oxidase to the generation of $O_2^{\cdot -}$ and H_2O_2 under oxygen deprivation.

MATERIAL AND METHODS

The liver was removed, initially perfused and prepared for the experiments as communicated earlier (8). The organ was perfused hemoglobin-free at 25°C in a recirculating mode with modified Krebs-Henseleit-bicarbonate buffer (pH 7.4; 2 mM lactate, 0.3 mM pyruvate). The perfusate was equilibrated with O_2/N_2-mixtures to get oxygen tensions of 80 kPa (normoxia) or 5.2 kPa (severe hypoxia). The flow rate was 2 ml x g liver^{-1} x min^{-1}.

GSSG, GSH, uric acid and allantoin were determined as described (8). The shares of type D (dehydrogenase) and of type O (oxidase) of xanthine oxidoreductase activity were estimated as reported (9).

In some experiments pooled washed erythrocyte suspensions were added to the perfusion medium (hematocrit 15%; 37°C). GSH and osmotic fragility of red blood cells and hemolysis were determined (10).

RESULTS AND DISCUSSION

The biliary GSSG efflux rate was choosen as an indicator of the oxidative loading of the liver (11). In the first hour of normoxic perfusion experiments a rate of 1.9 nmol GSSG x min^{-1} x g liver^{-1} was found (Table I). It increased with the extent of hypoxia and approached 2.7 nmol GSSG x min^{-1} x g liver^{-1} in severe hypoxia. Allopurinol as well as formate prevented the increase in GSSG efflux (8). GSH was released at a rate of 15.5 nmol x min^{-1} x g^{-1} which was not changed in hypoxia. The origin of excreted GSH and GSSG should be cytosolic

TABLE I
GSSG EFFLUX AND URIC ACID FORMATION BY XO-TYPE OF XANTHINE OXIDOREDUCTASE
Values are given as μmol x g w.w.$^{-1}$. Number of experiments in parentheses

	Normoxia	Severe hypoxia
GSSG efflux		
0 - 60 min	0.114 (7)	0.164 (7)
60 - 180 min	0.054 (7)	0.107 (7)
Uric acid formation		
0 - 60 min	0.093 (6)	0.563 (4)
60 - 180 min	0.130 (6)	0.441 (4)

because of the very low exchange between mitochondrial and cytosolic glutathione. The intracellular GSSG : GSH showed only minor deviations; therefore the GSSG accumulation in the medium is a more sensitive marker of oxidative stress in hypoxia.

The cytosolic generation of $O_2^{\cdot -}$ and H_2O_2 by xanthine oxidoreductase was estimated on the basis of (i) shares of O-type and D-type of the enzyme and (ii) the inhibition of D-type activity by free NADH calculated from the lactate : pyruvate ratio. The relative contribution of the O-type increased from 20% to 27% in one hour of severe hypoxia. The D-type was hardly inhibited by the increased free NADH concentration. However in absolute terms the production of uric acid by XO-type of xanthine oxidoreductase increased about sixfold (Table I). That is due to the increased hypoxanthine supply. Marubayashi et al. (12) observed the rapid decline of the adenine nucleotide pool in oxygen-deprived liver. In severe hypoxia more than 85% of uric acid were degraded further to allantoin by the peroximal urate oxidase (13). In Table II a synopsis of hydrogen peroxide formation via nucleotide degradation is shown. Both the cytosolic as well as the peroxisomal part are tremendously accelerated in severe hypoxia and account for half of the oxygen consumption.

TABLE II

H_2O_2 PRODUCTION IN RELATION TO OXYGEN CONSUMPTION

Values are given as μmol x g w.w.$^{-1}$ x h^{-1}.

	O_2-consumption	Formation of hydrogen peroxide total	cytosolic	peroxisomal
Normoxia	26.4	2.6-3.9^{+}	0.2	0.4
Severe hypoxia	5.5	2.9	1.1	1.8

+ From the estimate that 10-15% of total oxygen consumption are used for H_2O_2 formation (14)

TABLE III

OXIDATIVE DAMAGE OF RAT ERYTHROCYTES IN THE LIVER PERFUSION SYSTEM

	Without liver	Normoxic perfusion	Hypoxic perfusion	n
GSH (% of initial)				
60 min	93.8 ± 6.2	89.8 ± 0.8	91.3 ± 3.9	8
120 min	90.0 ± 2.9	85.3 ± 2.0	65.4 ± 7.8	8
50% hemolysis (g NaCl x dl^{-1})				
0 min	0.38 ± 0.01	0.38 ± 0.01	0.38 ± 0.01	2
60 min	0.38 ± 0.00	0.40 ± 0.01	0.41 ± 0.01	2
120 min	0.39 ± 0.01	0.41 ± 0.00	0.45 ± 0.00	2
Hemolysis (μM Hb)				
0 min	34	34	34	2
60 min	42	43	111	2
120 min	95	60	150	2

In further experiments was shown that reduced species of oxygen generated in the hypoxic liver initiate changes of the plasma membrane and in the glutathione system of red blood cells perfusing the organ (Table III/ also see ref. (10)). These deteriorations were diminished by allopurinol and formate, respectively. Thus these parameters may be appropriate criteria for oxidative damage of liver and other organs.

The data presented here clearly demonstrate that under hypoxic conditions the enhanced rate of nucleotide degradation is accompanied with elevated cytosolic and peroxisomal production of activated oxygen. This stresses forward the conclusion about the formation of oxygen radicals as an early and decisive step in progression of hypoxic tissue injury. This mechanism is obviously related to the development of damage in organs preserved for transplantation.

REFERENCES

1. Haglund U, Lundgren O (1978) Fed Proc Fed Am Soc Exp Biol 37:2729-2733
2. McCord JM, Roy RS (1982) Can J Physiol Pharmacol 60:1346-1352
3. Tappel AL (1973) Fed Proc Fed Am Soc Exp Biol 32:1870-1874
4. Hearse DJ (1977) J Mol Cell Cardiol 9:605-616
5. Siesjoe BK, Rehncrona S, Smith C (1980) Acta Physiol Scand Suppl 492:121-128
6. Paller MS, Hoidal JR, Ferris TF (1984) J Clin Invest 74:1156-1164
7. Granger DN, Rutili G, McCord J (1981) Gastroenterology 81:22-29
8. Raeder L, Siems W, Mueller M, Gerber G (1985) Cell Biochem Funct 3:in press
9. Della Corte E, Stirpe F (1972) Biochem J 126:739-745
10. Garbe S, Siems W, Mueller M, Gerber G (1984) Biomed Biochim Acta 43:S81-S82
11. Sies H, Summer K (1975) Eur J Biochem 57:503-512
12. Marubayashi S, Takenaka M, Dohi K, Ezaki H, Kawasaki T (1980) Transplantation 30:294-296
13. Siems W, Schmidt H, Mueller M, Henke W, Gerber G (1985) Free Radical Res Comm:accepted for publication
14. Oshino N, Jamieson D, Sugano T, Chance B (1975) Biochem J 146:67-77

Superoxide and Superoxide Dismutase in Chemistry, Biology and Medicine; G. Rotilio, editor

HYPEROXIA CAUSES LOSS OF THE EXTRACELLULAR CALCIUM DEPENDENCE OF THE RAT ALVEOLAR MACROPHAGE RESPIRATORY BURST

HENRY JAY FORMAN, JUNE NELSON AND GERALD HARRISON
Graduate Hosp. & Univ. of Pennsylvania, 415 S. 19 St. Philadelphia, PA 19146 USA

INTRODUCTION

Several alveolar macrophage (AM) functions are lost early in the course of hyperoxic exposure (1). Activation of the respiratory burst involves integration of many cell functions. Study of the inhibition of the activation of the respiratory burst may therefore reveal how O_2 toxicity alters cellular regulation and function. Ca^{2+} mobilization is apparently involved in initiation of the respiratory burst by some stimuli (2, 3). We will report here the effects of hyperoxia upon concanavalin A (con A) stimulated $O_2^{\cdot-}$ production, which is partially enhanced, but not dependent upon the presence of extracellular Ca^{2+}.

MATERIALS AND METHODS

Adult male specific pathogen free rats were exposed to 100% O_2 for 36 or 48 hours. AM were prepared by lavage from controls and O_2 exposed rats by a modification of Myrvik, et al. (4). Cells were then suspended in an isotonic Hepes-phosphate buffer without Ca^{2+}, pH 7.4 (HPB). Typically, yields were 12×10^6 cells/rat, 90-95% of which excluded a vital dye, with 98% being AM for both controls and O_2-exposed rats.

The respiratory burst was measured by a modification of the method of Cohen and Chovaniec (5). Ca^{2+} was present at 1.3 mM where indicated. $O_2^{\cdot-}$ production was stimulated by addition of 250 μg/ml con A.

Free intracellular $[Ca^{2+}]$ ($[Ca^{2+}]_i$) was measured in AM loaded with quin2 (6). Cells were suspended in HPB containing 1.3 mM Ca^{2+}. Fluorescence was measured at baseline followed by sequential addition of con A (to 250 μg/ml), Triton X-100 (to 0.1%), and finally, buffered EGTA (to 2.5 mM). $[Ca^{2+}]_i$ was calculated using the equation:

$$[Ca^{2+}]_i = K_d (F - F_{min})/(F_{max} - F)$$

where K_d is the dissociation constant for quin2-Ca^{2+} (115 nM (6)), F is fluorescence in arbitrary units, F_{max} is the fluorescence after Triton-X-100 addition, and F_{min} is the fluorescence after addition of EGTA.

RESULTS

The con Astimulated respiratory burst was measured in the presence or absence of extracellular Ca^{2+} or a saturating concentration of verapamil. O_2 exposure caused a decline in the respiratory burst regardless of the assay medium contents

(Fig. 1). The decline was significant for all three curves by analysis of variance or polynomial least square regression. Verapamil inhibited the con A-stimulated respiratory burst when added prior to stimulation and if Ca^{2+} was present. Determinations of the percent dependence upon extracellular Ca^{2+} or the percent inhibition by verapamil indicated that the enhancing effect of extracellular Ca^{2+} on $O_2^{\overline{\cdot}}$ production declined from 2.0 fold in controls to negligible enhancement after 48h of hyperoxic exposure.

Measurements of Ca^{2+} mobilization were made with quin2. Loading with 10 μM quin2/AM for 15 min produced an internal [quin2] that did not inhibit the respiratory burst. In quin2 loaded, control macrophages, con A causes a rapid rise in $[Ca^{2+}]_i$ followed by a rapid decline and a second rise and decline towards the baseline. In the absence of quin2 incubation, no changes occur in autofluorescence. Stirring alone did not produce a change in quin2 fluorescence. Fig. 2 shows the calculated average $[Ca^{2+}]_i$ for control, 48 h O_2-exposed, and verapamil treated AM. No differences in baselines were observed. Resting $[Ca^{2+}]_i$ was 124.9 ± 5.2 nM for air controls, 125.9 ± 9.5 nM for 48h O_2-exposed, and 118.8 ± 16.6 nM for verapamil treated cells.

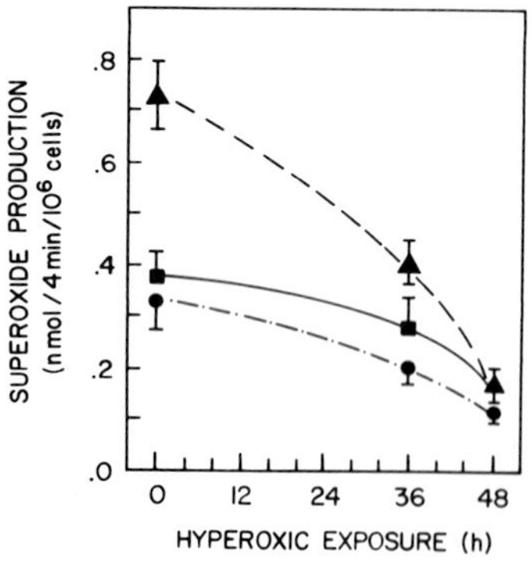

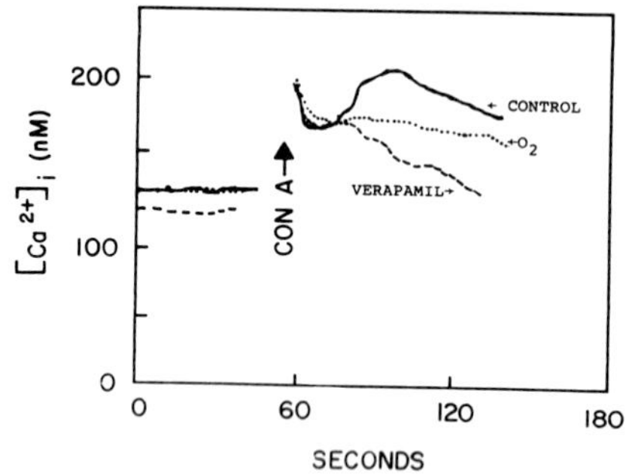

Fig. 1. Con A stimulated superoxide production. Assays contained zero extracellular Ca^{2+} (■), 1.3 mM Ca^{2+} (▲), or 1.3 mM Ca^{2+} (●) plus 20 μM verapamil. Mean and standard errors are indicated with 4-13 determinations for each point.
Fig. 2. Calculated $[Ca^{2+}]_i$.

The second "wave" was observed in every control (n=8), in only 3 of 8 O_2 exposed preparation and in only 1 of the 4 verapamil treated preparations. The area between 12 and 72 seconds after addition of con A was integrated above or below the value at 12 seconds. This area was 23.6 ± 4.7 for 8 controls, 2.9 ± 4.8 for 8 48h O_2-exposed preparations ($P < 0.005$ vs air) and -2.4 ± 4.0 for verapamil treated cells ($P < 0.005$ vs air, NS vs O_2 exposed cells).

DISCUSSION

The loss of respiratory burst capacity caused by hyperoxic exposure of rat AM involves at least two separate components: (1) decreased permeability of the cell to extracellular Ca^{2+} and (2) decreased activity independent of Ca^{2+} entry. At present, we do not know what accounts for the decrease in respiratory burst in the absence of extracellular Ca^{2+} or presence of verapamil. We can however, draw a conclusion about how O_2 exposure causes a decrease in the enhancement by extracellular Ca^{2+}.

Previous studies have shown that blocking of Ca^{2+} channels inhibits con A or f-met-leu-phe stimulation of the respiratory burst in AM (3, 7). It appears that O_2 exposure causes a progressive decrease in the ability to stimulate the respiratory burst (Fig. 1), which is partially due to a decrease in the ability of Ca^{2+} to enter the cell from the medium. This was more directly indicated by both the decreased effectiveness of verapamil following O_2 exposure.

In control cells, con A stimulated temporal changes in $[Ca^{2+}]_i$ (Fig. 2). Verapamil significantly reduced the second elevation in $[Ca^{2+}]_i$ suggesting that the second elevation represented influx of extracellular Ca^{2+}. Oxygen exposure did not alter the resting $[Ca^{2+}]_i$ but did suppress the second elevation. Thus, it can be concluded that O_2 exposure also suppressed Ca^{2+} entry from the extracellular medium. The immediate increase in $[Ca^{2+}]_i$, which is probably due to release of internal stores of Ca^{2+}, appeared to be similar in control, O_2-exposed, and verapamil-treated cells.

Our conclusion concerning the effect of O_2 toxicity on Ca^{2+} channels is thus supported by three different observations on the effect of O_2 exposure: decreased enhancement by extracellular Ca^{2+}, decreased inhibition by verapamil, and the loss of the ability of con A to stimulate a delayed transient increase in $[Ca^{2+}]_i$ as did verapamil.

ACKNOWLEDGEMENTS

Supported by Grant HL31831 from the National Institutes of Health.

REFERENCES

1. Rister, M, Baehner, RL (1977) Br J Haematol 36:241-248
2. Lew, PD, Stossel, TP (1981) J Clin Invest 67:1-9
3. Forman, HJ, Nelson, J (1983) J Appl Physiol 54:1249-1253.
4. Myrvik, QN, Leake, ES, Fariss, B (1961) J Immunol 86:128-132
5. Cohen, HJ, Chovaniec, M (1981) J Clin Invest 61:1081-1087
6. Tsien, RY, Pozzan, T, Rink, TJ (1982) J Cell Biol 94:325-334.
7. Sweeney, TD, Castranova, V, Bowman, L, Miles, PR (1981) Exper Lung Res 2:85-96.

Superoxide and Superoxide Dismutase in Chemistry,
Biology and Medicine; G. Rotilio, editor

SUPEROXIDE DEPENDENT ADRENOCHROME FORMATION IN BOVINE HEART SARCOLEMMAL MEMBRANES: EFFECT OF ARACHIDONIC ACID

CARLO GUARNIERI, CARLO VENTURA, CLAUDIO MUSCARI and CLAUDIO M. CALDARERA

Istituto di Chimica Biologica, Centro Studi e Ricerche sul Metabolismo Cardiaco, Università di Bologna, via Irnerio 48, 40126 Bologna (Italy)

INTRODUCTION

Cardiac microsomal and mitochondrial membranes are able to produce superoxide radicals ($O_2^{\cdot-}$) especially when incubated with polyunsatured fatty acids (1) or with doxorubicin (2). Our previous research has demonstrated that also sarcolemmal membranes which represent a component of the cardiac microsomes can produce $O_2^{\cdot-}$ when incubated with NADH or NADPH (3). This formation which was evaluated by monitoring the oxidation of adrenaline to adrenochrome was inhibited by superoxide dismutase and partially by catalase, while mannitol was ineffective. Moreover, the formation of adrenochrome was reduced incubating the sarcolemmal membranes with the β-receptor antagonist propranolol, while the α-antagonist phentolamine was without effect (4). The purpose of this study was to examine the role played by some polyunsaturated fatty acids on the formation of $O_2^{\cdot-}$ by heart sarcolemmal membranes.

MATERIAL AND METHODS

Bovine cardiac sarcolemmal was prepared and highly purified as described by Lamers and Stinis (5). The formation of adrenochrome was monitored at 480 nm at 25°C in a double beam Perkin Elmer spectrophotometer mod. 559 ($E480 = 4.02 \cdot 10^3 M^{-1} cm^{-1}$). The assay mixture (vol. 1 ml) consisted of 50 mM Tris/HCl pH 7.5, 1 mM EDTA 1 mM NADPH and 30-40 µg of sarcolemmal proteins. Adrenaline was added at 1 mM final concentration. Proteins were determined by the method of Bradford (6) using bovine albumine as standard.

RESULTS AND DISCUSSION

Table I shows the effect of some polyunsaturated fatty acids on the formation rate of adrenochrome by heart sarcolemmal membranes.

TABLE I

PRODUCTION RATE OF ADRENOCHROME BY CARDIAC SARCOLEMMAL MEMBRANES

The fatty acids were added before NADPH addition.

Numbers in parenthesis indicate the final concentrations. The values are the means $\pm$ S D of three samples from three hearts.

	nmol/min·mg prot
Control	132.4 $\pm$ 4.8
Linolenic acid (500 M)	129.6 $\pm$ 3.3
Arachidonic acid (500 M)	812.4 $\pm$ 4.8
Methylarachidonic acid (500 M)	136.6 $\pm$ 2.6
Arachidonic acid + superoxide dismutase (2 M)	29.8 $\pm$ 2.6
Arachidonic acid + catalase (2 M)	87.3 $\pm$ 5.3
Arachidonic acid + indomethacin (40 M)	101.2 $\pm$ 4.6
Arachidonic acid + Nordihydroguaiaretic acid (40 M)	698.0 $\pm$ 3.2
Phospholipase A_2 + Ca^{2+}	326.5 $\pm$ 4.5
Phospholipase A_2 + Ca^{2+} + indomethacin	82.9 $\pm$ 5.1

The presence of arachidonic acid produced a marked stimulatory effect, whilst linolenic acid or methylarachidonic acid were ineffective. The addition of superoxide dismutase and catalase strongly inhibited the rate of adrenaline oxidation, while indomethacin blocked the reaction at the control level. On the contrary, nordihydroguaiaretic acid was without effect. The addition of sarcolemmal membranes of phospholipase A_2 + Ca^{2+} produced a 3 fold increase in the rate of adrenochrome formation; such stimulatory effect was blocked by the simultaneous presence of indomethacin.

These results indicate that very likely arachidonic acid stimulates the formation of sarcolemmal adrenochrome partecipating in the reaction catalyzed by cyclooxygenase, which is known to involve the formation of radical intermediates (7). Such a conclusion mainly results from the inhibitory effect of indomethacin, while the absence of a marked inhibitory effect shown by the nordihydroguaiaretic acid seems to exclude the involvement of a lipooxygenase reaction. Moreover, since linolenic acid or methylarachidonic acid, which is not a substrate for cyclooxygenase reaction, does not activate the rate of adrenaline oxidation, it is likely that the arachidonic-stimulatory effect it is not caused by the autooxidation of the fatty acid.

An indirect confirmation of these conclusions can also be derived from the experiments with phospholipase A_2 and indomethacin which indicate an involvement of endogenous sarcolemmal arachidonic acid on the stimulation of adrenaline oxidation.

REFERENCES

1. Gudbjarnason S. and Doell B. (1978) In: De Duve C., Hayaishi O. (eds.) Tocopherol, Oxygen and Biomembranes. Elsevier, Amsterdam, pp. 297-310.
2. Doroshow J.H., Reeves J. (1981) Biochem. Pharmacol. 30, 259-262.
3. Guarnieri C., Ventura C. (1984) Biochem. Biophys. Res. Commun. 120, 21-27.
4. Guarnieri C., Ventura C., Georgountzos A., Muscari C., Budini R. (1985) Biochim. Biophys. Acta 838, 355-360.
5. Lamers J.M.J., Stinis J.T. (1983) Cell Calcium 4, 281-294.
6. Bradford M.M. (1976) Anal. Biochem. 72, 246-254.
7. Mason R.P., Kalyanaraman B., Tainer B.E., Eling T.E. (1980) J. Biol. Chem. 255, 5019-5022.

Superoxide and Superoxide Dismutase in Chemistry,
Biology and Medicine; G. Rotilio, editor

INCREASE OF GLUTATHIONE BIOSYNTHESIS AFTER OXIDATIVE STRESS INDUCED BY THIOLS

R. ISSELS[1,2], A. NAGELE[2], S. BOURIER[2], B. BÖNING[2], AND W. WILMANNS[1,2]
[1]Institut für Hämatologie, GSF and [2]Medizinische Klinik III, Klinikum Großhadern, University of Munich, West-Germany

INTRODUCTION

It has been previously shown that activated oxygen species play an important role in the cytotoxic effects of thiols like cysteamine (1). During autoxidation thiols consume oxygen and produce H_2O_2. The reaction rate of these processes is dependent upon environmental factors (e.g. pH, metal ions, medium) (2). We also found that the cytotoxic effects of thiols are synergistically enhanced by hyperthermia (3) and that intracellular glutathione (GSH) is increased after thiol treatment dependent on temperature (4). The purpose of the present study was to measure the ability of cysteamine to increase GSH biosynthesis by promotion of cyst(e)ine uptake from the medium into cells.

MATERIALS AND METHODS

Chinese hamster ovary (CHO) cells were plated 24hr prior to cysteamine exposure. For technical details of drug preparation and conditions of cell culture see (3). Experiments designed to measure total intracellular GSH were initiated by inoculating approx. 10^6 cells per flask (T_{25}flasks) containing 4.5ml fresh medium (Mc Coy's 5a Medium supplemented with 10%(v/v) newborn calf and 5%(v/v) fetal calf serum). After cysteamine exposure (0.4mM) at 37° for the indicated length of time, the medium was removed, cells were washed twice with PBS, and then trypsinized (0.25% trypsin not supplemented with EDTA for 2min, 37°). At the time of GSH measurement 98% of such treated cells were metabolically viable assessed by trypan blue exclusion. After dilution in cold PBS (5ml) the cells were counted and centrifuged at 4°. The pellets were resuspended in 0.3ml of 0.04M EDTA and 0.6ml 7.5%TCA, vortexed and centrifuged again at 4°. The cold supernatant was removed and assayed for total GSH by the method of Tietze (5). Protein determinations were made by the Lowry method (6). GSH content of control cells was (n=3) 17.6 +/- 5.1 (S.D.) nmol/mg protein. Determination of GSH content of CHO cells after pretreatment with 5µM buthionine sulfoximine (BSO) - an inhibitor of GSH biosynthesis - for 15 hr at 37° showed a reduction of GSH (4.70 +/- 1.1 nmol/mg protein).

Experiments designed to measure ^{35}S-cyst(e)ine uptake were initiated by plating approx. $5x10^5$ cells per 35mm diameter dish containing 2.0ml modified medium (Mc Coy's Medium 5a free of GSH and methionine but supplemented with ^{35}S-labeled cystine (0.1mM). Immediately before cysteamine exposure (0.4mM at 37°), the growth medium was replaced by 0.5ml of the modified medium and the drug was added in a small volume (5-20μl). At the indicated time intervals the uptake-medium was removed, cells were washed twice with PBS, trypsinized, and the supernatant (acid soluble) and the pellet (acid insoluble) of cell homogenates were obtained by centrifugation after treatment with 10% TCA. ^{35}S-cyst(e)ine uptake was determined in both fractions using a scintillation counter with an external standard for quench correction.

RESULTS AND DISCUSSION

Cysteamine exposure of CHO cells at 37° resulted in marked increase in ^{35}S-cyst(e)ine uptake from the medium into the cells. This increase parallels a pronounced elevation in intracellular GSH content (Table I).

TABLE I

UPTAKE OF CYST(E)INE AND INTRACELLULAR GSH CONTENT AFTER 1 HOUR CYSTEAMINE EXPOSURE OF CHO CELLS AT 37° AND 5°

Addition to the incubation medium	Uptake of ^{35}S-Cystine (nmol 1/2 cystine/mg protein)		Intracellular GSH (nmol/mg protein)
	Supernatant	Pellet	Supernatant
None at **37°**	12.1 (1.1)*	4.4 (3.0)*	17.6 (5.1)*
+ BSO**	2.0 (0.2)	4.2 (3.0)	4.7 (1.1)
+ 0.4mM cysteamine	36.2 (2.3)	4.6 (2.8)	35.1 (6.0)
+ catalase***	41.5 (3.5)	5.6 (2.7)	30.6 (5.1)
+ 0.4mM cysteamine + BSO	4.4 (0.2)	4.5 (3.2)	4.6 (2.5)
None at **5°**	1.2 (1.1)	2.4 (1.9)	20.8 (0.8)
+0.4mM cysteamine	10.4 (1.9)	3.3 (1.8)	16.2 (1.1)

*The results are averages (S.D) of two experiments with duplicate assays for each.

**Buthionine sulfoximine (5μM) was added 15hr prior to cysteamine.

***Catalase (SIGMA) 50μg/ml was added prior to cysteamine.

Both effects of cysteamine were completely blocked by incubation of cells at 5° during the time of drug treatment. After pretreatment of cells with BSO for 15hr, the cellular GSH content was reduced to approximately 20-25% of the control value. Neither an increase in ^{35}S-cyst(e)ine uptake nor in GSH biosynthesis could be observed when cysteamine was added to the medium in the presence of BSO. This inhibition of ^{35}S-cyst(e)ine uptake was observed to a similar extent when cells were not pretreated with BSO, but BSO was present during cysteamine exposure only (data not shown). Catalase which blocks completely the cytotoxic effects (1) of thiols as well as the mechanisms of thermosensitization (3), did not reduce the promotion of cystine uptake and the increase of GSH biosynthesis. Preliminary results of experiments using a cystine-free medium during thiol treatment showed a reduction of cysteamine toxicity. A ten-fold increase in the clonogenic survival of such treated cells with cysteamine compared to cells treated in complete medium has been observed (R. Issels et al. manuscript in preparation).

In conclusion the results suggest that thiols like cysteamine promote cyst(e)ine uptake in CHO cells followed by an increase in GSH biosynthesis. The increased generation of activated oxygen species during thiol autoxidation which further react with elevated GSH levels, could be an important step in the expression of the biological effects of thiol induced oxidative stress.

ACKNOWLEDGEMENTS

This work was supported by Grants Is 31/1-2 from the Deutsche Forschungsgemeinschaft.

REFERENCES

1. Issels RD, Biaglow JE, and Gerweck L (1984) In: Bors W, Saran M and Tait D (eds.) Oxygen Radicals in Chemistry and Biology. Walter de Gryter, New York, pp 665-671
2. Biaglow JE, Issels RD, Gerweck L, Varnes ME, Jacobson B, Mitchell JB, and Russo A (1984) Radiat. Res. 100: 298-312
3. Issels RD, Biaglow JE, Epstein L, and Gerweck L (1984) Cancer Res. 44: 3911-3915
4. Issels RD, Bourier S, Biaglow JE, Gerweck L, and Wilmanns W (1985) Manuscript submitted to Cancer Res.
5. Tietze F (1969) Anal. Biochem. 27: 502-522
6, Lowry OH, Rosebrough NJ, Farr AL, and Randall RJ (1951) J. Biol. Chem. 193: 265-275

Superoxide and Superoxide Dismutase in Chemistry,
Biology and Medicine; G. Rotilio, editor

STIMULATION OF HYDROXYL RADICAL PRODUCTION BY DOXORUBICIN-TREATED EHRLICH TUMOR CELLS

JAMES H. DOROSHOW

Department of Medical Oncology and Therapeutics Research, City of Hope National Medical Center, Duarte, CA 91010 (U.S.A.)

INTRODUCTION

Doxorubicin (DOX)+ is a quinone-containing antibiotic that is one of the most important agents currently used in cancer chemotherapy (1). Although the mechanism of tumor killing by this drug has previously been related to its effects on DNA structure or synthesis as a consequence of DNA intercalation (1), recent evidence suggests that cytotoxicity may also be due to free radical formation occurring at extranuclear sites (2). Thus, in this study we investigated whether treatment of intact Ehrlich tumor (ET) cells with doxorubicin resulted in the production of the hydroxyl radical (·OH), a potent biological oxidant.

MATERIALS AND METHODS

Doxorubicin hydrochloride was purchased from Adria Laboratories, Inc., Wilmington, Del. Other reagents were obtained from commercial sources as previously described (3). ET cells were maintained by weekly passage of one million cells i.p. in 20 gram female Swiss-Webster mice. Cells were harvested 5-6 days after implantation, and resuspended in phosphate buffered saline (PBS), pH 7.4, containing 10 mM glucose. Production of ·OH by drug-treated cells was measured by CH_4 release from dimethyl sulfoxide (DMSO) as previously reported (3). Treatments were performed in gas-tight 3 ml tubes for 3 hr at 37°C in a shaking water bath; 500 µl of the headspace gas was sampled for determination of CH_4 content by gas chromatography (3). Data were analyzed by the two-tailed Student's _t_ test for independent means.

+Abbreviations used in this study: Doxorubicin, DOX; hydroxyl radical, ·OH; dimethyl sulfoxide, DMSO; superoxide dismutase, SOD; Ehrlich tumor, ET.

RESULTS AND DISCUSSION

As shown in Table I, DOX enhanced CH_4 production by ET cells in the presence of reduced pyridine nucleotides; this process was superoxide- and hydrogen peroxide-dependent (inhibition by SOD and catalase) and was significantly enhanced by reduction of the oxygen content of the PBS. Furthermore, CH_4 production was stimulated by either $FeSO_4$ or NAD(P)H and was significantly inhibited by deferoxamine, suggesting that iron-dependent ·OH formation had occurred in the presence of DOX.

TABLE I

HYDROXYL RADICAL FORMATION BY DOXORUBICIN-TREATED TUMOR CELLS

Investigations were performed in a 2 ml volume of phosphate-buffered saline, pH 7.4, with 10 mM glucose, 1 mM NADPH, 100 mM DMSO, and 10^7 tumor cells. Data represent the mean ± S.E. of 3 to 12 experiments.

Experimental System	CH_4 Production ($pmol/3hr/10^7$ cells)
Control	12 ± 5
- NADPH + NADH (1 mM)	10 ± 6
- NADPH + $FeSO_4$ (50 µM)	25 ± 5
Doxorubicin (10 µM)	37 ± 6[a]
Doxorubicin (100 µM)	66 ± 18[a]
Doxorubicin (250 µM)	117 ± 8[a]
- cells	0[b]
+ catalase (3000 units/ml)	22 ± 6[b]
+ boiled catalase (3000 units/ml)	105 ± 13
+ SOD (20 µg/ml)	55 ± 5[b]
+ boiled SOD (20 µg/ml)	121 ± 25
+ deferoxamine (100 µM)	25 ± 8[b]
+ $FeSO_4$ (50 µM)	165 ± 10[a]
- NADPH + $FeSO_4$ (50 µM)	143 ± 5[a]
- NADPH + $FeSO_4$ (5 µM)	140 ± 2[a]
- NADPH + $FeSO_4$ (0.5 µM)	101 ± 8[a]
- NADPH + $FeSO_4$ (0.05 µM)	24 ± 24
- NADPH + NADH (1 mM)	73 ± 7[a]
- NADPH + NADP+ (1 mM)	17 ± 13[b]
- NADPH + NAD+ (1 mM)	28 ± 7[b]
Using PBS bubbled for 30 min with N_2	317 ± 41[b]

[a]Significantly different from control, $p < 0.05$.
[b]Significantly different from complete system containing doxorubicin, $p < 0.05$.

These results suggest that intact tumor cells are capable of activating DOX to its semiquinone free radical intermediate. In the presence of reducing equivalents and/or trace quantities of iron, reaction of the semiquinone (or an iron-doxorubicin conjugate) with molecular oxygen leads to the formation of .OH. Whether .OH production occurs via a Fenton-type reaction or, as suggested by Bates and Winterbourn (4), by reaction of the semiquinone with H_2O_2 has not been definitively addressed here. However, our results obtained under hypoxic conditions favor the latter hypothesis. In any event, these studies taken together with previous experiments demonstrating inhibition of doxorubicin-related tumor cell killing by ·OH scavengers (2) strongly suggest that production of oxyradicals in vivo contributes to the antineoplastic action of doxorubicin.

ACKNOWLEDGEMENTS

This study was supported by the Leukemia Society of America and grant CA 31788 from the National Cancer Institute.

REFERENCES

1. Young R, Ozols R, Myers C (1981) N Engl J Med 305: 139
2. Doroshow J (1985) Proc Amer Assoc Cancer Res 26: 225
3. Doroshow J (1983) Cancer Res 43: 4543
4. Bates D, Winterbourn C (1982) FEBS Lett 145: 137

SELECTIVE DAMAGE OF Se-GLUTATHIONE PEROXIDASE BY OXY RADICALS GENERATED BY DAUNOMYCIN.

IRENE MAVELLI°, MARIA ROSA CIRIOLO[x], LUCIANA DINI[x] AND GIUSEPPE ROTILIO[x]
°Institute of Applied Biochemistry University of Rome "La Sapienza",
[x]Department of Biology II University of Rome, Rome, Italy.

INTRODUCTION

Erythrocyte GSH-peroxidase is inactivated in a reaction involving H_2O_2 and hemoglobin, when red blood cells are exposed to an autoxidizable aminophenol, divicine, the redox cycle of which is kept working by excess reductants(1). In the present work we studied the effect of daunomycin, a quinone redox cycling antibiotic, on Se-dependent GSH-peroxidase activity of erythrocytes and hepatocytes.

MATERIALS AND METHODS

Daunomycin was a kind gift of Dr. Malatesta from Farmitalia, Milan, Italy. All other chemicals were reagent grade.

Washed human red blood cell(HRBC) suspensions (5% v/v) and hemolysates were incubated in phosphate-buffered saline at pH 7.4, in the presence of 5 mM glucose, with daunomycin for 3 hours at 37°C. Hemoglobin was determined as the cyano-met derivative (2). Production of H_2O_2 in hemolyzed HRBC's was assayed (3) measuring the inactivation of catalase in the presence of 0.05 M 3-amino-1,2,4 triazole (AT).

Hepatocytes were isolated from rat livers by the method of collagenase perfusion (4). Hepatocyte suspensions ($2x10^6$ cells/ml) were incubated in Krebs-Henseleit buffer at pH 7.4 supplemented with 12.6 mM Hepes at 37°C in rotating round bottom flasks in an atmosphere of 95% air and 5% CO_2.

Enzyme activities were assayed after removal of the drug by dialysis and Sephadex G-25 chromatography. Superoxide dismutase was determined by polarography (5), and expressed as µg/mg total protein, with reference to a standard sample of pure bovine enzyme, GSH-peroxidase (6), catalase (7), GSSG reductase and lactate dehydrogenase (8), cytochrome oxidase (9) and NADPH cytochrome P_{450} reductase (10) were assayed spectrophotometrically and expressed as Units (µmoles/min/mg total protein). Protein was determined according to Lowry et al. (11).

RESULTS AND DISCUSSION

Se-GSH peroxidase of HRBC's was significantly inactivated upon treatment with daunomycin of either intact or hemolyzed cells, while neither superoxide dismutase nor catalase activities were substantially modified under the same conditions. The inactivation was dose-dependent between 1 mM and 5 mM drug (Fig. 1) and it was not reversed by removal of the drug. Daunomycin induced a marked hemolysis also at the lowest concentrations used, suggesting that membrane damage was a primary event (Fig. 1, panel A). Hemoglobin oxidation always occurred, in particular in hemolysates (Fig. 1, panel B) with the same dose-dependent pattern of the GSH-peroxidase inactivation, suggesting that the inactivation of the enzyme was mediated by the interaction of the drug with hemoglobin. Generation of H_2O_2 in hemolysates treated with 3 mM daunomycin was followed by inactivation of catalase with AT (Table I). In the presence of AT GSH-peroxidase inactivation by daunomycin was enhanced, indicating that H_2O_2 was actually involved in inactivation of the enzyme.

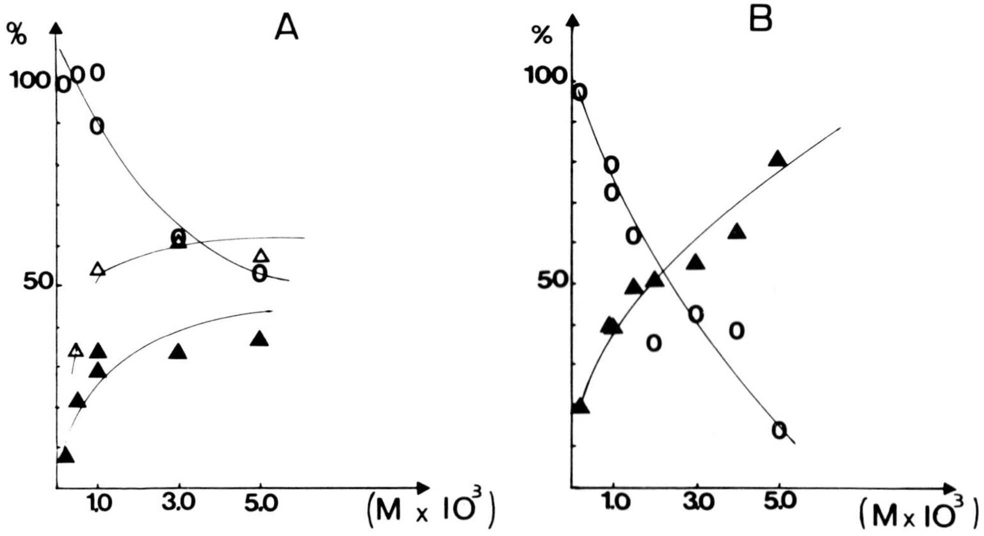

Fig. 1. Effect of daunomycin on the GSH-peroxidase activity and metHb content of intact (panel A) and hemolyzed (panel B) HRBC's.
(O–O) GSH-peroxidase activity (% of control);
(▲–▲) metHb (% of total Hb);
(Δ–Δ) hemolysis: Hb in 2,000 rpm supernatant expressed as % of total Hb.

TABLE I

DETECTION OF H_2O_2 GENERATED BY DAUNOMYCIN IN HEMOLYZED HRBC's.

ADDITIONS	CATALASE ACTIVITY %	GSH-PEROXIDASE ACTIVITY %
NONE	100	100
AT (50 mM)	96	100
DAUNOMYCIN (3 mM)	77	47
AT + DAUNOMYCIN	2	8

A similar inactivation of Se-GSH-peroxidase was observed upon 60 min incubation of isolated rat hepatocytes in the presence of 0.25-0.50 mM daunomycin. A decrease of cell viability also occurred (Table II).

TABLE II

ENZYME ACTIVITIES, RELATED TO THE OXIDATIVE METABOLISM, AND CELL DEATH UPON INCUBATION OF ISOLATED RAT HEPATOCYTES WITH DAUNOMYCIN.

	NONE	DAUNOMYCIN	
ENZYME ACTIVITIES		0.25 mM	0.50 mM
Se-GSH-peroxidase	0.23±0.22	0.18±0.03	0.14±0.01
GSSG reductase (10^2xUnits)	8.27±0.99	9.00±1.11	7.70+0.15
catalase	1.60±0.12	1.83±0.49	1.46±0.35
superoxide dismutase	4.01±0.17	3.71±0.20	3.89±0.25
lactate dehydrogenase	1.95±0.24	1.76±0.02	2.23±0.19
cytochrome oxidase	0.20±0.01	0.18±0.01	0.17±0.03
NADPH cytochrome P_{450}-reductase (10^2xUnits)	1.67±0.34	1.86±0.06	1.88±0.11
CELL DEATH*	10±5	23±10	45±12

*Cellular lactate dehydrogenase leakage into the medium, expressed as % of total cellular activity.
The results represent M ± SD of 3 separated experiments.

The effect is apparently quite specific with respect to GSH-peroxidase and depends on the presence of particulate enzyme systems. In fact, addition of daunomycin to the hepatocyte cytosol, containing the Se-GSH-peroxidase, in the presence of NADPH, resulted in no inactivation. No oxygen uptake by NADPH was observed with the cytosol even after addition of daunomycin. This result indicates that the enzyme inactivation is not due to a direct reaction between the drug and the enzyme and that redox activation of drug is required. The specific sensitivity of the active site of Se-GSH-peroxidase to daunomycin activated either by reaction with hemoglobin or redox cycling with electron transfer enzymes may be an important factor in the cell damage occurring upon incubation of erythrocytes and hepatocytes with daunomycin. In particular, it will make the normal cell more sensitive to oxy radicals as already shown with cells naturally deficient of GSH-peroxidase (12).

ACKNOWLEDGEMENTS

This work was supported by the Special Project "Oncologia" and "Genetic Engineering and Molecular Basis of Inherited Disease" of the Italian National Research Council (C.N.R.)

REFERENCES

1. Mavelli I., Ciriolo M.R. and Rotilio G. (1985) Biochem.Biophys. Acta in press.
2. Drabkin D.L. and Austin J.M. (1953) J. Biol. Chem. 112:52-65.
3. Cohen G. and Hochstein I. (1964) Biochemistry 3: 895-900.
4. Moldeus P., Hogberg J. and Orrenius S. (1978) Methods Enzymol. 52: 60-71.
5. Rigo A., Viglino P, and Rotilio G. (1975) Anal. Biochem. 68: 1-8.
6. Paglia D.E. and Valentine W.N. (1967) J. Lab. Clin. Med. 70: 158-169.
7. Lück H. (1963) in: Methods in Enzymatic Analysis, 2nd ed. Bergmeyer H.U. ed. Verlag Chemie, Acad. Press, New York, pp. 885-888.
8. Bergmeyer H.U. Gawehn K. and Grassl M. (1974) in: Methods Enzymatic Analysis vol. 1 Bergmeyer H.U. ed. Verlag Chemie Acad. Press New York pp. 425-522.
9. Cooperstein S.J. and Lazarow A. (1951) J. Biol. Chem. 189: 665-670.
10. Strobel H.W. and Dignam J.D. (1978) in: Methods in Enzymology vol. LII Fleischer S. and Packer L. eds. Acad. Press New York pp. 89-96.
11. Lowry O.H., Rosebrough N.J., Farr L., Randall R.J. (1951) J. Biol. Chem. 193: 265-275.
12. Bozzi A., Mavelli I., Mondovì B., Strom R. and Rotilio G. (1979) Cancer Biochem. Biophys. 3: 135-141.

Superoxide and Superoxide Dismutase in Chemistry,
Biology and Medicine; G. Rotilio, editor

ACTIVE OXYGEN PRODUCTIONS OF MITOCHONDRIA FROM TUMORS AND LIVER OF ADRENALECTOMIZED RODENTS

YOSHIHIKO ŌYANAGUI* AND BUNJI HAGIHARA

*Central Lab.Fujisawa Pharmaceut.Co.,Kashima,Yodogawa-ku,Osaka 532 and School of Medicine, Osaka Univ. Kita-ku, Osaka 533 (Japan)

INTRODUCTION

Decrease of tumor cytochrome aa_3 content is evident(1,2,3). Moreover, the other cytochromes in tumors seemed not to function as in normal tissues. Loschen et al(4) proved O_2^- production in SOD-free beef heart mitochondrial sonicated particles(MSP). The sites of O_2^- production on cytochrome chain were investigated in rodent tumors.

MATERIAL AND METHODS

Tumor cells were suspended in 0.25mM sucrose followed by digestion with 0.002% Nagase(5). Fitted-pestle Teflon homogenizer was used for mitochondrial preparation. Sonicated mitochondria were sedimented(78,800g, 90min.) to obtain MSP. Cytochromes were calculated from difference spectra at room-temperature obtained by high amplitude split-beam spectrophotometer(2). Extinction coefficients were; $E(aa_3)$ 605-630nm=16.5, E(b) 563-575nm=17.9 and $E(c+c_1)$ 551-540nm =19.0 $mM^{-1}.cm^{-1}$. Liquid nitrogen(low) temperature spectra gave the ratio of c and c_1. Oxygen consumption was measured by oxygen electrode. O_2^- production was generally assayed under 5mM succinate as Sigma bovine Cu,Zn-SOD(10ng/ml) inhibitable NADH oxidation coupled with lactate dehydrogenase(LDH)(6).

RESULTS AND DISCUSSIONS

O_2 consumption of 3'-DAB induced hepatoma was less sensitive than that of normal liver(Fig.1). Low-temperature spectra of anaerobe(solid line) showed higher peaks of cytochromes than 1mM KCN containing aerobe(dotted line) in tumor(Fig.2). This KCN-insensitive electron leakage was evident at cytochrome b and c_1 sites and may cause free radical formation. Regenerating rat liver leaked only little electrons. This means that the changes of tumor cytochrome reductions are not reflection of its rapid growth. As adrenalectomized rat liver showed tumor-like pattern, the changes of tumor cytochrome reduction is possible to be due to the hormonal disorder of tumor bearing host. Normal heart, kidney and thymus

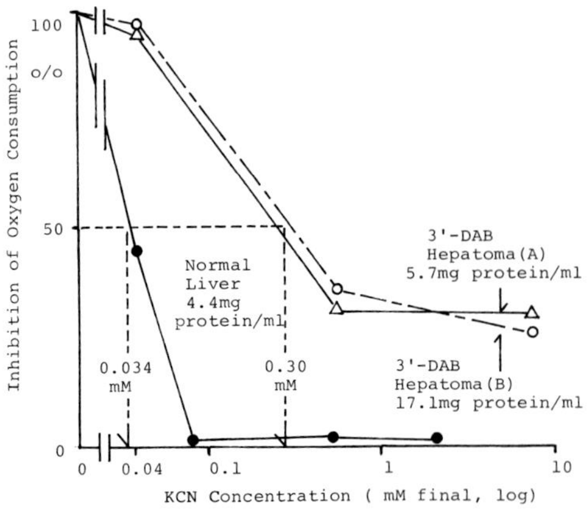

Fig.1. Sensitivities to KCN of mitochondrial O_2 consumption(3mM succinate, pH 7.4)

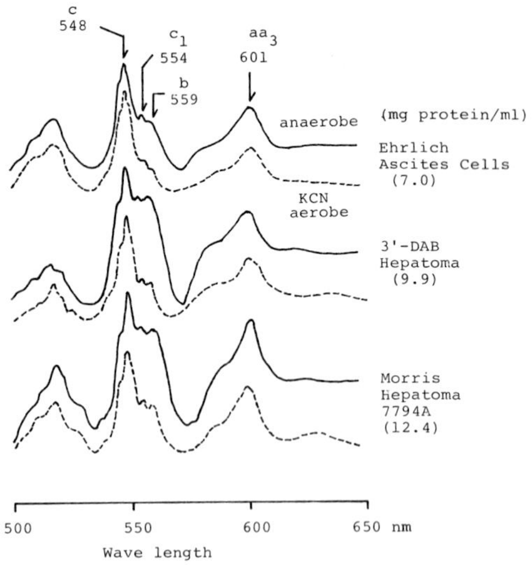

Fig.2. Low-temperature difference spectra of tumor mitochondria(5mM succinate at pH 7.4)

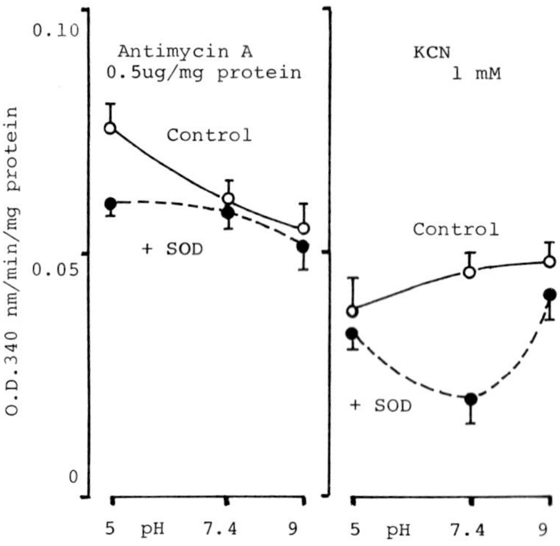

Fig.3. O_2^- production by Ehrlich tumor MSP (5mM succinate, 0.5uM catalase, 60mM Na-phosphate)

produced almost no free radical, but brain showed a little tumor-like mode of reduction. Half inhibitory concentration of KCN on O_2^- production was 0.6 mM for normal and regenerating livers and 1.8mM for Ehrlich tumor MSP which was observed only at pH 7.4(Fig.3). Tumor MSP O_2^- production was not remarked under antimycin A. Substitution of succinate by pyruvate resulted similiar O_2^- production. O_2^- formation at NADH site, is thus not important in tumor mitochondria as reported in normal tissues. Mitochondria contain much unsaturated fatty acids which are sensitive to oxidation. Increased O_2^- production may contribute to swell or deform the mitochondria. This autolytic tendency decreases mitochondrial number. O_2^- generated by tumor cells may also serve for its invasion acting on vascular permeability or lysosomal enzyme release.

Tissue	cyt.aa_3 pmol / mg protein	cyt. aa_3/c ratio	e-leakage percentage at each cytochrome sites (b, c_1, c, aa_3)	e-flow arrived to O_2
Normal	179	1.08	100 o/o → 5, 2, 2, 5 o/o	87 o/o
Regenerating (2 days)	20	1.27	8, 3, 2, -3 o/o	90 o/o
Adrenalectomized (4 days)	147	0.80	46, 48, 15, -2 o/o	28 o/o
Ehrlich Ascites Cells	13	0.20	100 o/o → 43, 52, 4, 8 o/o	25 o/o
3'-DAB Hepatoma	96	0.68	66, 70, 27, 3 o/o	7 o/o
Morris Hepatoma 7794A	112	0.69	57, 52, 30, 4 o/o	14 o/o

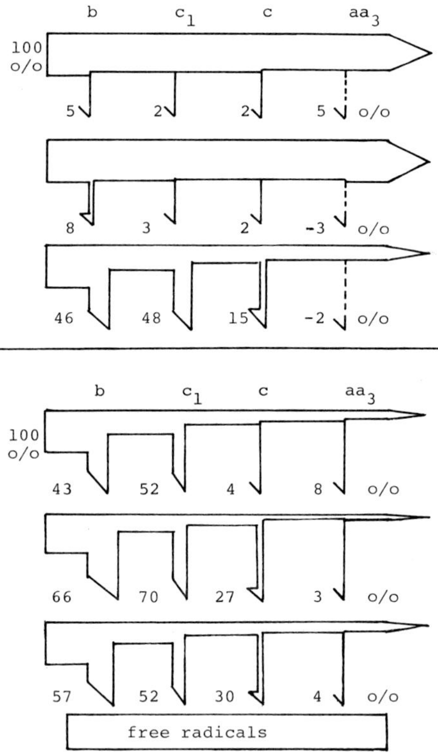

Distribution of electron flow in liver and tumor respiratory cytochrome chains. Electron flow entering to cytochrome b under 5mM succinate, was set as 100%.

REFERENCES

1. Chance B, Hess B (1959) J.Biol.Chem. 234:2404
2. Hagihara B, Sato N, Fukuhara T, Tsutsumi K, Ōyanagui Y (1973) Cancer Res. 33:2947
3. Ōyanagui Y, Sato N, Hagihara B (1974) Cancer Res. 34: 458
4. Loschen G, Azzi A, Richter C, Flohé L (1974) FEBS Lett. 42:68
5. Kobayashi S, Hagihara B, Masuzumi M, Okunuki K, (1966) Biochem. Biophys. Acta 113:421.
6. Chan PC, Bielski BHJ (1974) J. Biol.Chem. 249:1317

Superoxide and Superoxide Dismutase in Chemistry, Biology and Medicine; G. Rotilio, editor

FACTORS INFLUENCING O_2^- AND t-BuOOH DEPENDENT LIPID PEROXIDATION OF TUMOUR MICROSOMES

SILVIA BORRELLO, GIORGIO MINOTTI AND TOMMASO GALEOTTI
Institute of General Pathology, Catholic University, L.go F.Vito 1, 00168 Roma (Italy)

INTRODUCTION

High mitotic activity and degree of dedifferentiation in experimental tumours are accompanied by a loss of susceptibility of microsomal membranes to peroxidative agents (1). Such a feature, which is shared by other tumour membranes, e.g. mitochondria (1) and plasmalemmas (2), has been ascribed to phospholipid and polyunsaturated fatty acid (PUFA) deficiency (2, 3), rather than high levels of membrane antioxidants (e.g. vitamin E) (4). However, tumour microsomes with a normal lipid substrate content, such as those from the slow-growing, well differentiated, Morris hepatoma 9618A, still exhibit an impairment of lipid peroxidation, whose extent is dependent on the nature of the peroxidative agent employed (4, 5).

A membrane constituent involved in microsomal lipid peroxidation as a prooxidant is cytochrome P-450; its function is that of a hydroperoxide oxidase which activates ROOH or LOOH to initiating or propagating radicals, respectively (6, 7). Thus, lack of this hemoprotein in tumour microsomes might also represent a factor limiting lipid peroxidation.

In this report we correlate the rates of lipid peroxidation, induced by superoxide radicals and t-butyl hydroperoxide (t-BuOOH), with the PUFA and cytochrome P-450 content of microsomes isolated from two Morris hepatomas with different growth rate and degree of differentiation.

MATERIALS AND METHODS

Tumour and normal tissue preparations, as well as isolation of microsomes and assay methods have been reported in detail elsewhere (4).

RESULTS AND DISCUSSION

The extent and profile of microsomal lipid peroxidation is dependent on the experimental conditions employed, that is, the peroxidative agent (NADPH, O_2^-, hydroperoxide, ascorbate), the presence of iron and its chelators (ADP, EDTA), as well as other factors (protein concentration, pH, ionic strength, temperature). Microsomes from fast-growing, poorly differentiated, tumours always exhibit lower rates of lipid peroxidation than normal membranes, regardless of the peroxidative system used, provided that it is functioning under optimal conditions. This has been shown previously (1) and it is also reported in Table I. Hepatoma 3924A

TABLE I

RELATIONSHIP BETWEEN LIPID PEROXIDATION, PUFA AND CYTOCHROME P-450 CONTENT OF RAT LIVER AND HEPATOMA MICROSOMES

O_2^- system: 0.2 mg protein/ml, 0.05 mM $FeCl_3$, 1 m ADP, 0.33 mM xanthine, 50 μg/ml xanthine oxidase (Sigma, type I) in 0.15 M KCl, 50 mM Tris-HCl (pH 7.5), at 25°C under an oxygen atmosphere. t-BuOOH system: 0.2 mg protein/ml, 0.11 mM EDTA, 0.25 mM t-BuOOH in 0.15 M KCl, 50 mM Tris-KCl (pH 7.5), at 37°C under an air atmosphere. The rates of malondialdehyde are initial velocities calculated from the linear portion of the curve (first 5 min). For other experimental conditions see ref. 4. Malondialdehyde and cytochrome P-450 values are means ± S.E. of 4-10 and 10-11 experiments, respectively. The PUFA content is given as the percentage (w/w) present in each sample (mean of 3-4 preparations).

Microsomes	Malondialdehyde (nmol/min/mg protein)		PUFA (w/w)	Cytochrome P-450 (nmol/mg protein)
	O_2^-	t-BuOOH		
Rat liver	6.86 ± 0.30	0.64 ± 0.04	38.7	0.78 ± 0.02
Hepatoma 9618A	6.27 ± 0.45	0.30 ± 0.04	31.3	0.30 ± 0.02
Hepatoma 3924A	1.48 ± 0.12	0.08 ± 0.02	10.0	nil

(fast-growing) microsomes produce very low amounts of malondialdehyde (MDA) both in the presence of O_2^- and t-BuOOH. The values of MDA are initial velocities, but the resistance to peroxidation in these hepatoma microsomes is also observable during incubations carried out for long time. Conversely, hepatoma 9618A (slow-growing) microsomes show normal rates of MDA production in the presence of O_2^-; peroxidation is significantly reduced only in the t-BuOOH system. Hepatoma 3924A microsomes contain about 25% of the liver PUFA and are deprived of cytochrome P-450. Hence, both initiation and propagation of lipid peroxidation are expected to be remarkably inhibited. In hepatoma 9618A microsomes the nearly normal PUFA content allows O_2^--induced lipid peroxidation to proceed at a regular rate, at least during the early phase (initiation). On the other hand, when peroxidation is induced by t-BuOOH the lower availability of cytochrome P-450 (60% deficiency) limits the generation of the initiating radicals t-BuOO$^\cdot$ and t-BuO$^\cdot$, and the resistance to peroxidation is observed immediately. Time course experiments with O_2^- indicate a late inhibition of MDA and LOOH formation in hepatoma 9618A microsomes, suggesting that the cytochrome P-450 deficiency becomes the rate-liming factor also in the oxidative cleavage of LOOH to propagating lipidoxy radicals (4).

The present and previous results have allowed to identify two major factors influencing hepatoma microsomal lipid peroxidation: the lipid substrate and cytochrome P-450. The known resistance to peroxidation of tumour membranes may thus be ascribed to the deficiency either of both factors or of cytochrome P-450 alone.

ACKNOWLEDGEMENTS

The work was supported by a grant of the Italian National Research Council, Special Project "Oncology", contract number 84.00593.44.

REFERENCES

1. Bartoli GM and Galeotti T (1979) Biochim Biophys Acta 547: 537-541
2. Galeotti T, Borrello S, Palombini G, Masotti L, Ferrari MB, Cavatorta P, Arcioni A, Stremmenos C and Zannoni C (1984) FEBS Lett 169: 169-173
3. Bartoli GM, Bartoli S, Galeotti T and Bertoli E (1980) Biochim Biophys Acta 620: 205-211
4. Borrello S, Minotti G, Palombini G, Grattagliano A and Galeotti T (1985) Arch Biochem Biophys 238: 588-595
5. Bartoli GM, Galeotti T, Borrello S and Minotti G (1982) In: Galeotti T, Cittadini A, Neri G and Papa S (eds) Membranes in Tumour Growth. Elsevier Biomedical Press, Amsterdam, pp 461-470
6. Svingen BA, Buege JA, O'Neal FO and Aust SD (1979) J Biol Chem 254: 5892-5899
7. White R and Coon MJ (1980) Annu Rev Biochem 49: 315-356

Superoxide and Superoxide Dismutase in Chemistry,
Biology and Medicine; G. Rotilio, editor

SUPEROXIDE DISMUTASE DEPLETION INDUCES LIPID MODIFICATION IN RAT LIVER MICROSOMAL MEMBRANES

GIANNA MARIA BARTOLI, BARTOLO GIANNATTASIO, PAOLA PALOZZA and ACHILLE CITTADINI.
Institute of General Pathology, Catholic University, Rome, Italy.

INTRODUCTION

Many data indicate that lipid peroxidation can be initiated by the reactive species of oxygen, although it is not definitely established which single species shows damaging effect in living organisms. The oxygen toxicity may be caused either by an increase of oxygen metabolism or by a decrease of the defensive systems i.e. glutathione peroxidase (GSH-Px) and superoxide dismutase (SOD).

We have previously observed that tumour cells lack defensive enzymes against reactive species of oxygen (1,2). This loss may cause a situation of oxidative stress for the tumour cell which in turn may provoque an enhanced rate of membrane lipid fatty acid peroxidation. Actually the analysis of tumour membrane composition shows that the unsaturated fatty acids are strongly decreased and that the ratio saturated-unsaturated fatty acids is markedly increased (1). This apparent contraddiction observed in tumour membranes may be explained as a consequence of a compensatory response of the membranes to the endogenous oxidative stress.

The aim of the present work is to verify whether it is possible to correlate the modification of cell membrane fatty acid composition to the oxidative stress caused in vivo by a decrease of defensive enzymes (SOD, GSH). For this purpose we have used an experimental model which consists of SOD deficiency obtained in rat liver with a copper deficient diet, comparing it to the neonatal liver that physiologically exhibits a low level of SOD.

MATERIAL AND METHODS

Copper deficiency has been obtained with a previously described method (3). All experimental conditions have been performed as standard procedures and have been reported elsewhere (2).

RESULTS

Table 1 shows the content of cytosolic SOD in liver from copper deficient rats and from 1 day and 7 days newborn rats. The Cu-Zn SOD decreases by 85% in liver of rats on copper-deficient diet for eight weeks. The SOD in neonatal liver does not show any difference as far as the animal age is concerned; it remains stable at a level of about 30% of the adult SOD content. Upon copper deficiency the liver microsomal membranes show a different sensitivity to in vitro peroxidation.

TABLE 1. SUPEROXIDE DISMUTASE CONTENT OF LIVER FROM ADULT, COPPER DEFICIENT AND NEWBORN RATS.

	Adult	Cu-deficient	1 day	7 days
SOD (ug/g.w.w.)	186.9 ± 26.9 (3)	29.5 ± 6.6 (3)	51.7 (2)	58.5 (2)
%	100	15.8	27.7	31.3

Values are mean ± SEM (n° of experiments).

In fact the malondialdehyde production from Cu deficient membranes is reduced after 30 min of incubation by 30% with respect to the control (43.6 ± 3.5 vs 61.7 ± 6.7 nmoles/mg prot). The microsomal membrane conjugate diene content shows the same behaviour: after 30 min of incubation it decreases to 7.9 ± 1.1 compared to 11.3 ± 1.8 nmoles/mg prot. of the control. Measurements of microsomal membrane fatty acid composition substantiate the data reported about the in vitro lipope-

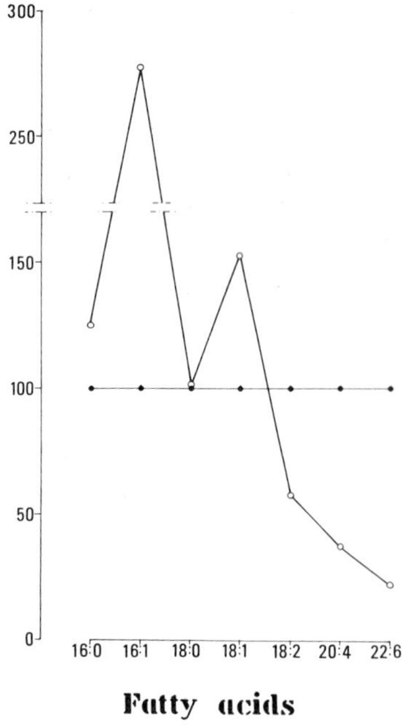

Fig. 1. Fatty acids composition of control (•—•) and Cu-deficient rats (○—○). The individual fatty acids are reported as % of the control. Mean of 3 experiments.

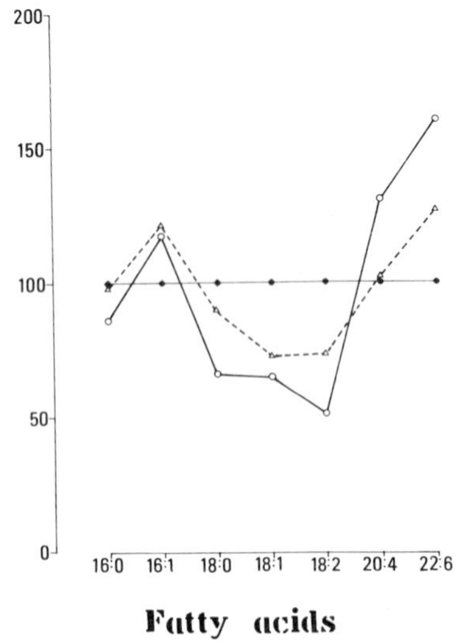

Fig. 2. Fatty acid composition of adult (•—•) 1 day (○—○) and 7 days (△- - -△) newborn rats. The individual fatty acids are reported as % of the adult. Mean of 2 experiments.

roxidation sensitivity of the Cu-deficient membranes. As shown in Fig. 1 copper deficiency determines strong decrease of polyunsaturated fatty acids and significant increase of mono and saturated fatty acids, thus reversing polyenes- monoenes ratio. That this situation is specifically determined by SOD deficiency is supported by the result of experiments performed in GSH-Px deficient rats. The induced diminution of Se-GSH-Px determines neither different peroxidation sensitivity in vitro nor fatty acid changed profile (not shown). As reported in Fig. 2 neonatal liver membranes show a different behaviour: polyunsaturated fatty acids seem to slightly increase, probably due to a more unsaturated fatty acid content of the milk diet, and the ratio between saturated-monounsaturated fatty acids is similar to that found in copper deficiency. It is noteworth to notice that neonatal SOD is not fully synthetized and its low level is not due to copper deficiency which can be induced in foetal liver and can determine a further SOD diminution (4).

CONCLUSIONS

From the data reported above we can conclude that SOD depletion causes deep modifications in endogenous cell membranes and that the alterations observed in copper deficiency are very similar to that observed in tumour cells (1). This resemblance suggests that the loss of defenses against oxygen toxicity may truly represent a significant step during carcinogenesis. In relation to the concept of carcinofoetalism we have examined the neonatal liver membranes, but inaspectately we did not find important parallelism with the experimental model. We are inclined to believe that the differences observed are due to a physiological homeostasis in the neonatal liver and to an imbalance of oxydative metabolism occurring in copper deficiency.

ACKNOWLEDGEMENTS

Work supported by Special Project "Oncology" of the CNR; grant N. 84.00593.44.

REFERENCES

1. Bartoli GM, Bartoli S, Galeotti T and Bertoli E (1980) Biochim Biophys Acta 620: 205-211

2. Bartoli GM, Galeotti T, Borrello S and Minotti G (1982) In: Galeotti T, Cittadini A, Neri G and Papa S (eds) Membranes in Tumour Growth. Elsevier, Amsterdam, pp 461-470

3. Paynter DI, Moir RJ and Underwood EJ (1979) J Nutr 109: 1570-1576

4. Keen CL, Cohen NL, Hurley LS and Lonnerdal B (1984) Biochem Biophys Res Commun 118: 697-703

Superoxide and Superoxide Dismutase in Chemistry, Biology and Medicine; G. Rotilio, editor

PHYSIOLOGICAL ASPECTS OF EXTRACELLULAR-SUPEROXIDE DISMUTASE

STEFAN L MARKLUND

Department of Clinical Chemistry, Umeå University Hospital, S-901 87 Umeå, Sweden.

INTRODUCTION

Extracellular-superoxide dismutase (EC-SOD) is a tetrameric, slightly hydrophobic glycoprotein with heparin-affinity (1). It is composed of four non-covalently bound subunits, has a molecular weight of about 135,000, and contains four copper atoms and possibly also four zinc atoms. EC-SOD is the major SOD in most extracellular fluids (2), but occurs also in tissues (3). The present paper reports on further investigations on physiological properties of the enzyme.

Acute phase reaction

Infections, trauma and various idiopathic inflammatory states induce a reaction called the acute phase response. It encompasses changes in many physiological parameters, including several plasma proteins. Since superoxide radical producing leukocytes accumulate at the site of lesions, the reaction of plasma EC-SOD was of interest to investigate. Surgical trauma induces a typical

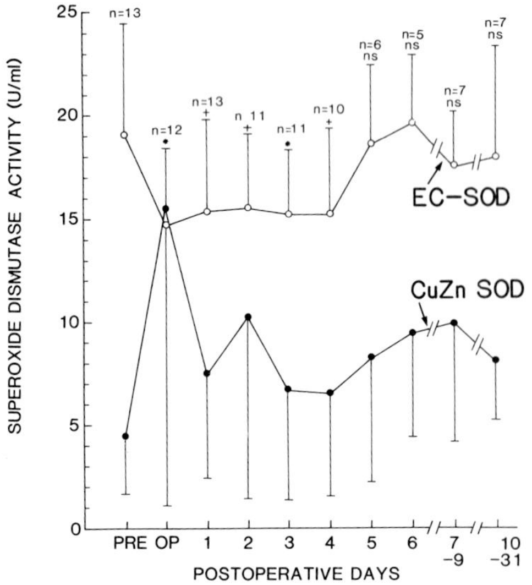

Fig. 1. Changes in plasma EC-SOD and CuZn SOD after surgical trauma
Op = operating day a few hours after surgery.

acute phase response with an accurately timed onset. Plasma was collected from 13 patients before and after surgery. EC-SOD, CuZn SOD and a number of other plasma proteins were determined. A typical acute phase response occurred as evidenced from increases in α_1-antitrypsin, haptoglobin, orosomucoid and C-reactive protein (data not shown). Plasma EC-SOD decreased immediately after surgery, remained low for 4-5 days and then returned to the preoperative level (Fig 1). Plasma IgG and albumin decreased almost as much (data not shown) probably due to hemodilution. Most or all of the decrease in EC-SOD probably has the same cause. It appears that plasma EC-SOD has a negligible or possibly weakly negative acute phase response. CuZn SOD increased postoperatively (Fig. 1) most likely due to leakage from traumatized tissues. In collaboration with M Öhman and PA Kling.

SOD isoenzymes in synovial fluid

There is evidence for a participation of toxic oxygen reduction products in inflammatory joint disease (e.g. refs 4-7). With that background it appeared worthwhile to assess the protection against superoxide radicals in normal synovial fluid (SF) as well as in SF from patients with various types of inflammatory joint disease. Table 1 collects the results.

TABLE I

SOD ISOENZYME ACTIVITIES IN SF AND PLASMA

The results are presented as means ± S.D. Statistical comparison between "controls" and the RA and ReA groups was performed with the Wilcoxon rank sum test.

	EC-SOD U/ml	CuZn SOD U/ml	Cyanide resistant SOD U/ml	Protein mg/ml
"Controls" (n=17)	67.3±39.4	11.1±7.7	4.3±1.2	25.6±3.7
Rheumatoid arthritis (n=21)	29.1± 9.0 $p<0.001$	21.3±8.9 $p<0.001$	3.5±0.92 $p<0.001$	43.3±9.5 $p<0.001$
Reactive arthritides (n=9)	63.8±55.0 n.s.	10.2±3.9 n.s.	3.2±1.0 n.s.	45.4±8.5 $p<0.001$
Plasma, healthy controls (n=16)	19.6± 3.0	2.8±0.9	5.4±0.5	

EC-SOD is the major isoenzyme in normal ("control") synovial fluid. The activity is more than three times higher than the activity in plasma (Table I). The CuZn SOD activity is much lower, but still higher than in plasma. SF from patients with reactive arthritides did not differ significantly from control SF. However, in SF from patients with rheumatoid arthritis (RA) the EC-SOD activity was less than half the normal activity. Whether this is a primary deficiency or is secondary cannot be judged at present. Whatever the reason for the lower activity, it will enhance the susceptibility of SF components and the synovia to superoxide radicals in RA patients. CuZn SOD was higher in RA SF, most likely due to tissue damage and leakage from the intracellular space. Collaboration with A. Bjelle and L.G. Elmqvist.

EC-SOD in disulfiram-treated alcoholics

Disulfiram is in vivo partially reduced to diethyldithiocarbamate (DDTC), which is an efficient inhibitor of CuZn SOD both in vitro and in vivo (8). EC-SOD was recently shown to be more than an order of magnitude more DDTC-sensitive than CuZn SOD (9). The finding raised the possibility that plasma EC-SOD would be inhibited in alcoholics on long-term aversion therapy with disulfiram.

TABLE II
PLASMA SOD ACTIVITY IN DISULFIRAM TREATED ALCOHOLICS AND CONTROLS

	Alcoholics, disulfiram treated (n=12)	Alcoholics, no disulfiram treatment (n=11)	Healty controls (n=19)
Plasma EC-SOD (U/ml)	15.1± 4.1	18.3±7.5	17.8±3.7
Plasma CuZn SOD (U/ml)	10.8±10.7	8.8±7.5	3.1±2.2
Plasma cyanide resistant SOD (U/ml)	6.7± 1.9	6.4±1.6	4.3±1.0

As seen in Table II EC-SOD was 15-20 % lower in the disulfiram-treated alcoholics as compared to control alcoholics and to healthy controls. The moderate reduction is probably due to the disulfiram treatment, but should not pose any health hazard. Larger reductions are likely in disulfiram overdosage and might then contribute to the toxic effects. In collaboration with M. Öhman.

Effect of heparin-injection and possible binding to blood vessel endothelium

Human lung EC-SOD was upon chromatography on Heparin-Sepharose divided into three roughly equally large fractions. A without affinity, B with intermediate affinity, and C which desorbed relatively late in a NaCl gradient (1). In subsequent, more rapidly and efficiently prepared batches of EC-SOD, the C fraction is more dominant. The findings pointed to the possibility that EC-SOD binds to heparin-analogues in vivo, like heparan sulphate on endothelial cells and other cell-types. To test for this possiblility, heparin was injected iv into two healthy males and plasma collected for SOD determination (Fig. 2).

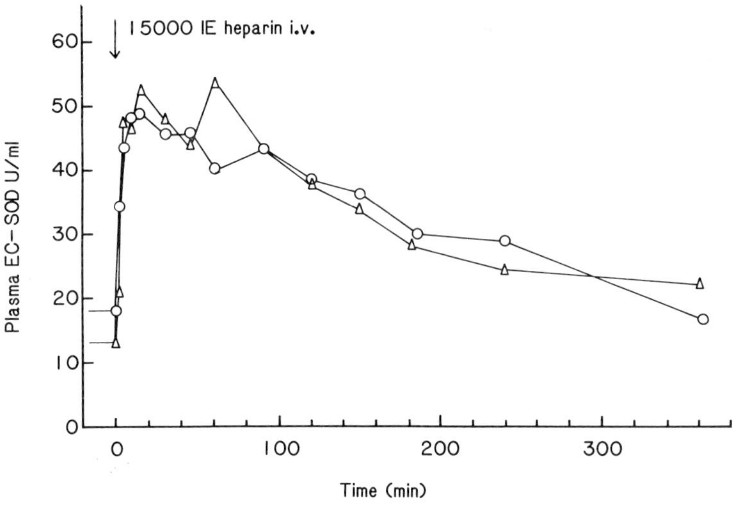

Fig. 2. Effect of heparin-injection on plasma EC-SOD activty.

As seen there is a rapid 2 to 3-fold rise in plasma EC-SOD activity. The rapidity of the rise indicates that the released EC-SOD must have been located on vessel endothelium. The finding points to the possibility that the EC-SOD in the vascular system forms an equilibrium between the plasma phase and receptors (heparan sulphate?) on endothelial cells. EC-SOD might be especially designed to protect the endothelium and other heparan sulphate-containing cells against superoxide radicals. In this context it is interesting to note the efficient protection against self-inactivation of activated PMN´s by a CuZn SOD modified to facilitate association with the plasma membrane (10). The protective ability of the membrane bound SOD in Nocardia asteroides against PMN also demonstrates the importance of membrane-association for cell-protection against superoxide (11). In collaboration with K Karlsson.

ACKNOWLEDGEMENTS

The studies were supported by the Swedish Medical Research Concil, Grant no 04761.

REFERENCES

1. Marklund SL (1982) Proc Natl Acad Sci, USA, 79:7634-7638
2. Marklund SL, Holme E, Hellner L (1982) Clin Chim Acta, 126: 41-51
3. Marklund SL (1984) J Clin Invest, 74:1398-1403
4. Blake DR, Hall ND, Bacon PA, Dieppe PA, Halliwell B, Gutteridge JMC (1981) Lancet, No 21:1142-1144
5. Blake DR, Hall ND, Bacon PA, Dieppe PA, Halliwell B, Gutteridge JMC (1983) Ann Rheum Dis 42:89-93
6. Lunec J, Blake DR, Stone R, Bacon PA (1984) Brit J Rheumatol 23: 145-146
7. Goebel KM, Stork U (1983) Amer J Med, 74:124-128
8. Heikkila RE, Cabbat FS, Cohen G (1976) J Biol Chem 251:2182-2185
9. Marklund SL (1984) Biochem J, 220:269-272
10. Salin ML, McCord JM (1977) In: Michelson AM, McCord JM, Fridovich I (eds) Superoxide and Superoxide Dismutases. Academic Press, London, New York, San Fransisco, pp 257-270
11. Beaman BL, Black CM, Doughty F, Beaman L (1985) Infect Immun 47:135-140

Superoxide and Superoxide Dismutase in Chemistry, Biology and Medicine; G. Rotilio, editor

THE ROLE OF INTRACELLULAR PROTEOLYTIC SYSTEMS IN ANTIOXIDANT DEFENSES

KELVIN J. A. DAVIES

Institute for Toxicology and Department of Biochemistry, The University of Southern California, 1985 Zonal Avenue, Los Angeles, California 90033, U.S.A.

INTRODUCTION

Oxygen radicals and other activated oxygen species appear to be produced in all aerobic cells and elegant defensive systems have evolved to protect against them (1-4). The superoxide anion radical ($O_2^{\bar{\cdot}}$), the hydrodioxyl radical ($HO_2\cdot$, sometimes called the perhydroxyl radical), hydrogen peroxide (H_2O_2), and photo-induced singlet molecular oxygen ($^1O_2$) are well documented, and the biological significance of the hydroxyl radical (•OH) has gained wide support in recent years (2). Similarly, the possible importance of oxygen radical-induced oxo-heme oxidants or other transition metal oxidants has received growing attention (2). Certain activated oxygen species can (directly or indirectly) induce lipid peroxidation in biological membranes, and cellular carbohydrates and nucleic acids have also been shown to be targets of oxygen radical damage. It is less widely recognized, however, that proteins also undergo extensive modification by reaction with oxygen radicals, or other activated oxygen species.

Primary antioxidant roles have been argued for enzymes such as superoxide dismutase (SOD), glutathione peroxidase, and catalase (1) and for compounds such as vitamin E (3) and uric acid (4). Recently, I and my colleagues have gained evidence that novel intracellular proteolytic systems may conduct the specific recognition and rapid degradation of oxidatively damaged proteins (5-10). Such proteolytic systems (proteinases, proteases, and peptidases) prevent the accumulation of inactive and potentially harmful oxidized proteins. It is, thus, proposed that previously unknown intracellular proteolytic systems may play an important role in overall intracellular antioxidant defenses. To support this hypothesis I present experiments, with intact cells and with cell-free extracts, which indicate that a wide variety of cells can recognize and degrade oxidatively damaged proteins.

MATERIALS AND METHODS

Human and rabbit erythrocytes (RBC) were collected in 0.9% heparinized saline. Mitochondria were prepared as previously described (3). *E. coli* (K12 wild-type) were grown to confluence on glucose minimal medium. Cell-free RBC extracts were prepared from cells washed three times by centrifugation in 0.9% saline. The washed cells were suspended to 2.5 times their volume in ice-cold water containing 0.5 mM dithiothreitol and stirred vigorously for 30 min (on ice) in order to induce cell lysis. The lysed cells were next centrifuged for 60 min at 20,000xg in order to remove the membrane fraction (as well as any unbroken cells). The supernatant (opproximately 100 ml) was then dialyzed for 20 hr (10,000 mw cut-off membrane) against a buffer (6.0 L) consisting of 50 mM HEPES (pH 7.8) + 10% glycerol + 5mM ditiothreitol, with three changes of buffer solution. RBC extracts were stored at -20 C for up to three months until needed. Cell-free *E. coli* extracts were prepared from washed cells suspended in 50 mM triethanolamine + 10% sucrose (pH 8.0). Cells were frozen in dry-ice and ethanol, thawed in ice-water, and then further lysed by treatment with lysozyme (0.1 mg/ml) for 30 min at 4 C. β-mercaptoethanol and KCl were added to 10 mM and 500 mM final concentrations, and the cell lysates were centrifuged for 40 min at 40,000 x g to remove membranes, cell debris, and unbroken cells. The supernatants were finally dialyzed as described above (for RBC) to produce cell-free *E. coli* extracts. Alanine and tyrosine release from proteins was measured by fluorescence techniques (5,10). Bovine serum albumin, superoxide dismutase, and catalase were ^{3}H-labeled by reductive methylation using ^{3}H-borohydride and ^{3}H-formaldehyde. The degradation of these ^{3}H-labeled proteins was measured by release of acid-soluble counts following treatment with trichloroacetic acid. All results are means ± SE of 3-5 independent experiments.

RESULTS

Exposure of human RBC, rabbit RBC, or rat skeletal muscle to oxygen radicals generated by xanthine + xanthine oxidase (X + XO) resulted in rapid rates of protein degradation as reflected by release of amino acids (table I). The complete radical generating system of X + XO was necessary to elicit this response and neither X alone nor XO alone induced protein degradation. Mature RBC do not have the capacity to synthesize alanine *de novo*, and skeletal muscle cannot synthesize tyrosine. The appearance of these free amino acids is, therefore, a true reflection of release from existing proteins *i.e.* protein degradation.

TABLE I

PROTEIN DEGRADATION IN ERYTHROCYTES AND SKELETAL MUSCLE

Human and rabbit RBC (5% v/v) and individual soleus muscles from rats were incubated for one hour at 37 C in a shaking water bath. Where used, xanthine and xanthine oxidase concentrations were 5 mM and 0.02 units/ml respectively.

Additions	Human RBC*	Rabbit RBC*	Rat Muscle**
None	0.7 ± 0.1	0.6 ± 0.1	0.2 ± 0.03
X + XO	19.7 ± 0.9	14.3 ± 0.4	0.9 ± 0.06
Increase	28 x	24 x	5 x

* Alanine release: $nmol \cdot ml^{-1} \cdot hr^{-1}$ ** Tyrosine release: $nmol \cdot min^{-1} \cdot mg^{-1}$

Rapid rates of protein degradation were also observed when mitochondria or *E. coli* were exposed to free radicals (table II). For these experiments mitochondria and *E. coli* were first incubated with ^3H-leucine, using pulse-chase procedures, in order to allow the synthesis of ^3H-labeled proteins. The labeled mitochondria or *E. coli* were then incubated with X + XO, or with the redox cycling quinones menadione or adriamycin. Protein degradation was greatly increased by all free radical generating agents in a concentration-dependent manner. Although not shown, it is important to note that neither X alone, nor XO alone had any effect on protein degradation, and that no agent had any effect when oxygen was excluded from the system and replaced with argon or nitrogen. These findings point to a clear role for oxygen radicals and, perhaps, other activated oxygen species in initiating intracellular protein degradation.

TABLE II

PROTEIN DEGRADATION IN MITOCHONDRIA AND *E. COLI*

Rat liver mitochondria and *E. coli* were incubated for one hour at 37 C. Where used, X and XO concentrations were 5 mM and 0.02 units/ml respectively. Menadione and adriamycin were added to a final concentration of 10 µM. Protein degradation was measured by release of acid-soluble counts, and is expressed as percent of total (acid-soluble + acid-precipitable) counts.

Additions	Mitochondria	*E. coli*
None	16.6 ± 2.1	0.1 ± 0.01
X + XO	83.2 ± 7.9	9.5 ± 0.84
Menadione	76.4 ± 9.1	10.7 ± 1.04
Adriamycin	85.3 ± 8.5	8.8 ± 0.65

One model which could explain the results of tables I and II would be protein damage by oxygen radicals, followed by recognition and degradation by intracellular proteolytic systems. To test this hypothesis, a simple purified protein (bovine serum albumin or BSA) was first exposed to damage by •OH. The hydroxyl radical was implicated (by other experiments) in the protein degradation reported in tables I and II, and the effects of •OH on BSA have been extensively studied in my laboratory (to be published elsewhere). Extracts of RBC and of *E. coli* were prepared from lysed cells and were extensively dialyzed as described in Materials and Methods. These extracts of RBC and *E. coli* contain the cells' soluble proteolytic systems (proteinases, proteases, and peptidases) which, when incubated with suitable protein substrates, will catalyze the complete degradation of proteins to amino acids. As shown in table III, incubation of •OH-treated BSA with RBC or *E. coli* extracts resulted in rapid rates of BSA degradation whereas untreated BSA was degraded at a very slow rate. By manipulation of experimental conditions it has been possible to increase BSA degradation to more than 60% (not shown) but such results are difficult to interpret since substrate limitations may become significant. It is important to recognize that only BSA was exposed to •OH for the experiments of table III, that all •OH production had ceased prior to incubation with RBC or *E. coli* extracts, and that the extracts were suspended in an "antioxidant buffer" consisting of HEPES, glycerol, and dithiothreitol which was previously shown to prevent protein damage and degradation by oxygen radicals (5). Neither ATP nor ubiquitin, which have previously been reported to stimulate the degradation of certain proteins (11,12), were present in the dialyzed extracts and the addition of ATP had no effect on rates of protein degradation. The results of table III demonstrate that soluble intracellular proteolytic systems can recognize and degrade proteins which have been modified or damaged by oxygen radicals. The results further indicate that no activation of these proteolytic systems is required, other than the provision of suitable substrates.

TABLE III

DEGRADATION OF BSA BY RBC AND *E. COLI* EXTRACTS

^{3}H-BSA was either exposed to •OH generated by ^{60}Co radiation (10 krad) under 100% N_2O, or was untreated. Percent ^{3}H-BSA degradation was measured by release of acid-soluble counts during a subsequent one hour (37 C) incubation with rabbit RBC or *E. coli* extracts, in the presence or absence of 5 mM ATP.

Cell Extract	Untreated BSA	•OH-Treated BSA	Increase
RBC	0.13 ± 0.02	5.5 ± 0.4	42 x
E. coli	0.21 ± 0.06	9.2 ± 0.7	46 x

In order to examine the broader biological significance of these findings it was necessary to test the effects of other activated oxygen species. Hydrogen peroxide is, perhaps, the most ubiquitous activated form of oxygen although it is also the least reactive. It has been known for some time that the enzyme superoxide dismutase (SOD) is inactivated by its product H_2O_2, and it has been reported that the inactivation involves oxidative modification (or damage) to an active-site histidine (13). In my laboratory we have also found that the enzyme catalase undergoes a more modest degree of inactivation when incubated with H_2O_2, which is its natural substrate. Following exposure to H_2O_2, both ^3H-labeled SOD (bovine Cu/Zn) and catalase displayed greatly increased proteolytic susceptibility when incubated with RBC or *E. coli* extracts (table IV). For these experiments, ^3H-labeled SOD and catalase were incubated at pH 7.4 with H_2O_2 for 30 min and then dialyzed extensively prior to incubation with RBC or *E. coli* extracts. As reported for BSA above, the addition of 5 mM ATP had no effect on the degradation of SOD or catalase pre-exposed to H_2O_2 (not shown). The results of table IV, therefore, represent true recognition and degradation of damaged SOD and catalase by intracellular proteolytic systems rather than a more generalized activation of proteases. It was also found that enzyme inactivation and proteolytic susceptibility were highly correlated, and that both increased with H_2O_2 concentration (to be published elsewhere). Cu/Zn SOD is not an *E. coli* enzyme yet *E. coli* extracts preferentially degraded the oxidatively modified enzyme. This may suggest that certain key features of oxidative protein damage are universal signals for intracellular proteolysis.

TABLE IV

DEGRADATION OF SOD AND CATALASE BY RBC AND *E. COLI* EXTRACTS

^3H-labeled SOD and catalase were exposed to 1.0 mM H_2O_2 for 30 min at pH 7.4 and then dialyzed. Percent ^3H-SOD and ^3H-catalase degradation was measured by release of acid-soluble counts during a subsequent 1 hour (37 C) incubation with RBC or *E. coli* extracts. S.E.'s were less than 5% in all cases below.

Cell Extract	Superoxide Dismutase Untreated	Superoxide Dismutase H_2O_2 Treated	Catalase Untreated	Catalase H_2O_2 Treated
RBC	0.03	1.66	0.02	0.78
E. coli	0.12	3.63	0.01	0.94

DISCUSSION

The results presented demonstrate that proteins which have been damaged by oxygen radicals and other activated oxygen species can be recognized and rapidly degraded by (soluble) intracellular proteolytic systems. Previous reports from this laboratory (5-10), as well as studies of glutamine synthetase damage and degradation published by Levine et al. (14) and Fucci et al. (15) provide further support for this conclusion. More recent and more extensive work suggests that these proteolytic systems consist of proteinases, proteases, and peptidases which are inhibited to varying degrees by transition metal chelators, by serine protease inhibitors, and by hemin (to be published elsewhere). The identity of each proteolytic enzyme, as well as the mechanistic relationship(s) between protein damage and proteolysis are now under detailed study in my laboratory.

Oxidatively damaged enzymes and other proteins typically loose their normal functions and, thus, play no useful role in maintaining cellular homeostasis. Furthermore, extensively oxidized proteins may become highly insoluble (partially due to cross-linking) and cause cellular damage. It is proposed that intracellular proteolytic systems, which recognize and degrade oxidatively damaged proteins, may play a major role in protecting cells against the harmful effects which such damaged proteins would produce. Intracellular proteolytic systems may, thus, be considered as an antioxidant "second line of defense" against oxygen radicals and other activated oxygen species. The detailed roles of proteolytic systems within overall antioxidant defenses are under continuing study.

ACKNOWLEDGEMENTS

I am indebted to Ms. Sharon Wu Lin for her expert technical assistance. This research was supported by grant #ES 03598 and by grant #ES 03785 from the National Institutes of Health. Further support was provided by a Faculty Research and Innovation Award from the University of Southern California.

REFERENCES

1. Fridovich I. (1975) Ann. Rev. Biochem. 44:147-159
2. Aust, S.D., Morehouse, L.A., and Thomas, C.E. (1985) J. Free Radicals Biol. Med. 1:3-25
3. Quintanilha, A.T., Packer, L., Davies, J.M.S., Raccanelli, T.L., and Davies, K.J.A. (1982) Ann. N.Y. Acad. Sci. 393:32-46
4. Sevanian, A., Davies, K.J.A., and Hochstein, P. (1985) J. Free Radicals Biol. Med. 1:117-124
5. Davies, K.J.A., and Goldberg, A.L. (1983) Fed. Proc. 42:1812
6. Davies, K.J.A., and Goldberg, A.L. (1984) Fed. Proc. 43:606
7. Davies, K.J.A., and Delsignore, M.E. (1984) Fed. Proc. 43:1858
8. Lin, S.W., and Davies, K.J.A. (1985) Fed. Proc. 44:1093
9. Davies, K.J.A. (1985) Fed. Proc. 44:1894
10. Davies, K.J.A. (1985) In: Eaton JW, Konzen DK, White JG (eds) Cellular and Molecular Aspects of Aging: The Red Cell as a Model. Alan R. Liss, New York, pp 15-24
11. Goldberg, A.L., and St. John, A.C. (1976) Annu. Rev. Biochem. 45:747-803
12. Hershko, A., and Ciechanover, A. (1982) Annu. Rev. Biochem. 51:335-364
13. Bray, R.C., Cockle, S.A., Fielden, E.M., Roberts, P.B., Rotilio, G., and Calabrese, L. (1974) Biochem. J. 139:43-48
14. Levine, R.L., Oliver, C.N., Fulks, R.M., and Stadtman, E.R. (1981) Proc. Natl. Acad. Sci. U.S.A. 78:2120-2124
15. Fucci, L., Oliver, C.N., Coon, M.J., and Stadtman, E.R. (1983) Proc. Natl. Acad. Sci. U.S.A. 80:1521-1525

Superoxide and Superoxide Dismutase in Chemistry,
Biology and Medicine; G. Rotilio, editor

THE PROOXIDANT STATE IN INITIATION AND PROMOTION BY CARCINOGENS AND MODALITIES USED IN CANCER THERAPY AND THE ANTICARCINOGENIC ACTION OF ANTIOXIDANTS.

CARMIA BOREK

Departments of Radiology and Pathology, College of Physicians and Surgeons,
Columbia University, New York, N.Y. 10032 USA

INTRODUCTION

Recent experimental findings and some current thoughts have underscored the important role of free radical reactions (1,2) in the process of carcinogenesis (3-6). Much of the compelling evidence nas emanated from the observations that antioxidants and enzymes which scavenge free radicals or which react with their products serve as powerful anticarcinogens and inhibit the neoplastic process at its various stages (3-8).

An initiating event in many free radical reactions may be the production of the superoxide radical (O_2^-) from dissolved molecular oxygen by the addition of a single electron (2). Superoxides are formed naturally in cellular metabolic processes and their production may be enhanced in living cells by the action of radiation and a variety of chemicals (1,8).

Superoxide dismutates in the presence of superoxide dimutase (SOD) to form hydrogen peroxide which can react with reduced metals (Fenton Chemistry) to produce the more toxic reactive hydroxyl radical (OH·). Hydroxyl radical reacts with biologically important molecules such as lipids and DNA to produce additional chain reactions with oxygen. For example, lipid radicals react with oxygen to produce a variety of products such as aldehydes which cross-link cellular macromolecules including DNA (9). The results are damage to the integrity of the cell and a plausible role in the carcinogenic process.

Catalase and glutathione peroxidase with its coordinating mineral, selenium, detoxify cellular peroxides and protect the cells from subsequent deleterious effects (1).

Other protection from free radical mediated events is afforded by the action of the fat soluble membrane antioxidants, Vitamin E, β carotene and Vitamin A (7).

Our studies using cultured cells have focussed on the role of cellular permissive and protective factors in the neoplastic process induced by radiation and chemicals; factors which control cellular metabolism and the prooxidant state of the cell (3,7,10,11).

Cell transformation in vitro

Cell cultures offer powerful tools in carcinogenesis research. In these *in vitro* systems cells are grown under defined conditions free from complex homeostatic mechanisms that prevail in vivo. These systems afford us the opportunity to assess at a cellular level the carcinogenic potential of physical and/or chemical agents (10-14). We expose single cells to the agent(s) and can then determine the ensuing short-term toxic actions on the cells such as cell killing, DNA and chromosome damage and cellular membrane damage, as well as their long-term effects in causing malignant transformation (10-13). Having these defined models we can identify modulating factors which enhance transformation rates (co-carcinogens, promotors) or inhibit and at best eliminate the induction and development of neoplastic transformation (7, 10-15).

Examples of such systems are the hamster embryo primary cultures and the C3H/10T½ mouse line (10,12-14).

Neoplastic transformation is at first recognized morphologically by the altered morphology of the transformed cells as compared to the normal (Fig. 1) and at later stages by the ability of the transformed cells to grow in agar and form tumors in appropriate hosts (10, 12, 13).

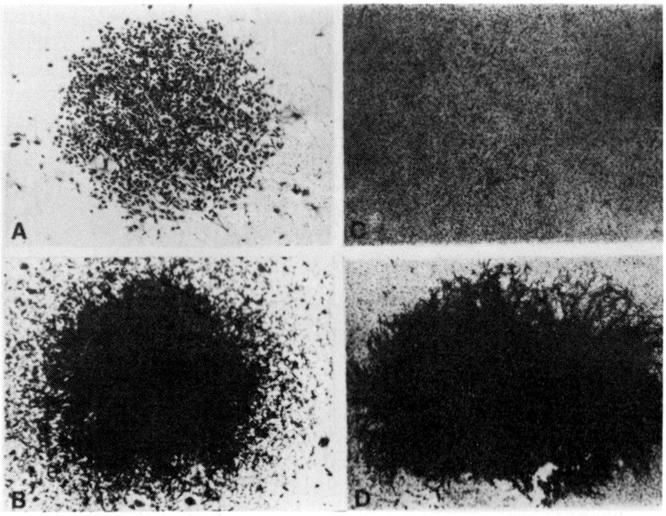

Fig. 1. Morphological criteria for transformation. (A and B) Clonal assay in hamster embryo cell cultures; (C and D) focus in mouse embryo 10T½ cells. (A) Normal hamster embryo cell clone, 14 days old. (B) A 14-day-old clone of hamster cellstransformed in vitro by 300 rad of X rays. (C) A monolayer of untransformed 10T½ cells 6 weeks in culture. (D) A type III focus of 10T½ cells transformed by 300 rad of X rays, 6 weeks in culture. Note the flatness of the normal cells and their regular orientation. In contrast, the transformed cells form multilayers and display a lack of orientation that is clearly distinguishable at the periphery. From Borek.

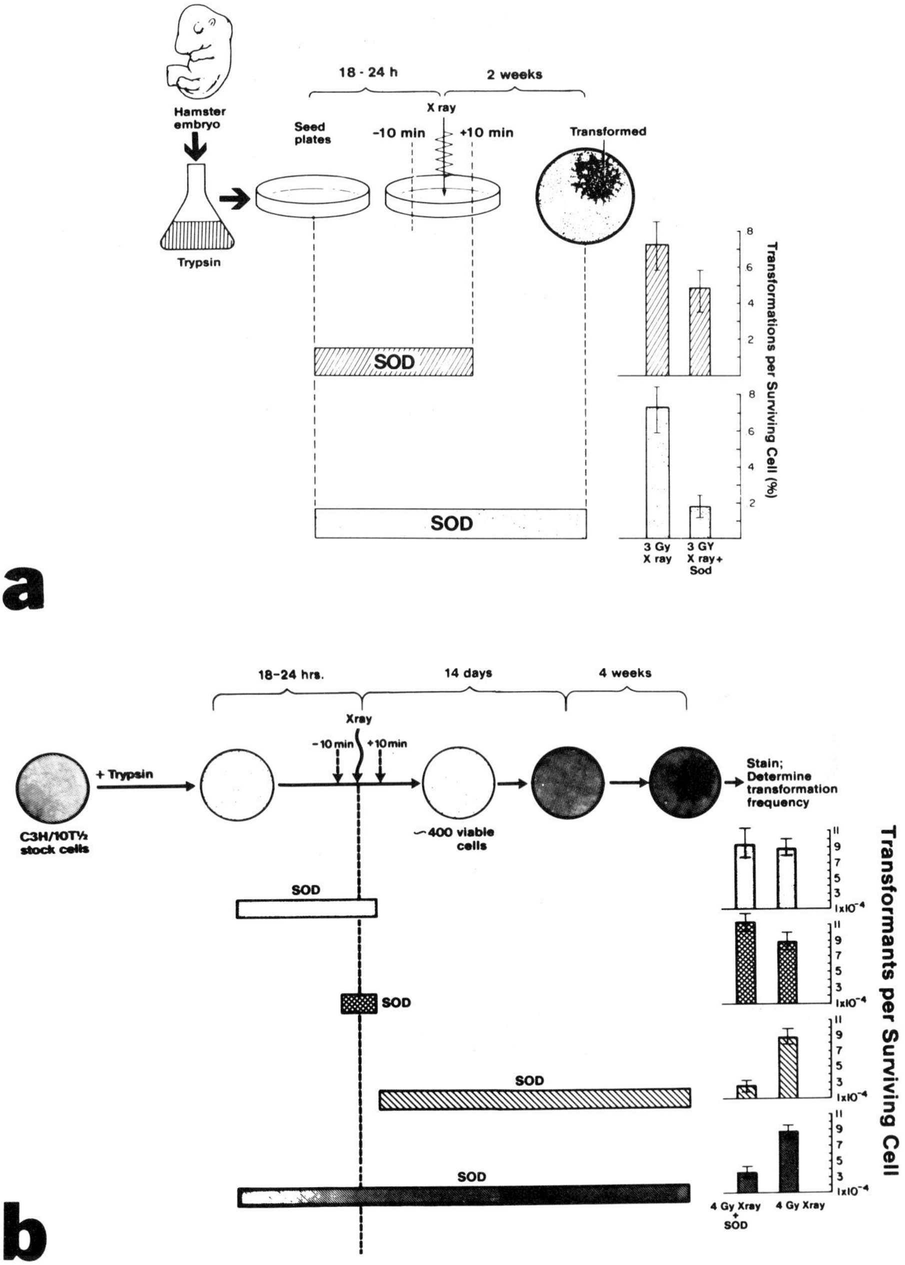

Fig. 2. The inhibition of radiogenic transformation by SOD in hamster embryo cells (a) and in C3H/10T½ mouse cells (b) (3, 12).

SOD, an inhibition of initiation and promotion.

Using the hamster embryo cell system and the C3H/10T½ mouse cells we have found that SOD is a potent inhibitor of cell transformation by radiation (Fig. 2) (3,16), bleomycin (3,16) and the hypoxic sensitizer Misonidazole (12, 16). The effectivenss of SOD was most dramatic upon prolonged contact with the cells (Fig. 2). While SOD inhibited initiation in the hamster cells (Fig. 1) (3) it predominantly acted on promotion by the oncogenic agents or by the tumor promotor TPA (Table I) (Fig. 2).

In the mouse C3H/10T½ cells SOD inhibited only late stages in transformation (Fig. 2). Catalase had little effect on inhibiting transformation in the hamster embryo cell system (Table I) (3). Since the cells are relatively rich in catalase (Table II) their content may serve as adequate protection against the oncogenic effects of radiation and bleomycin.

TABLE I

INHIBITION OF CELL TRANSFORMATION BY SOD AND CATALASE[a] FOLLOWING HAMSTER EMBRYO CELLS EXPOSED TO X RAYS AND TPA

Treatment	Surviving fraction[b] ± SE (plating efficiency[c] ± SE)	Percentage transformation ± Se
Control	(4.6 ± 0.12)	0
SOD	0.95 ± 0.04	0
Catalase	0.97 ± 0.03	0
TPA	0.92 ± 0.05	0
DMSO	0.95 ± 0.04	0
3 Gy	0.75 ± 0.12	0.74 ± 0.14
3 Gy + SOD	0.91 ± 0.09	0.17 ± 0.06
3 Gy + catalase	0.80 ± 0.11	0.66 ± 0.09
3 Gy + SOD + catalase	0.87 ± 0.12	0.13 ± 0.10
3 Gy + TPA	0.69 ± 0.14	1.62 ± 0.12
3 Gy + SOD + TPA	0.89 ± 0.10	0.55 ± 0.08
3 Gy + catalase + TPA	0.82 ± 0.13	1.32 ± 0.15
3 Gy + catalase + SOD + TPA	0.93 ± 0.12	0.41 ± 0.09
3 Gy + SOD[d]	0.80 ± 0.10	0.21 ± 0.07
3 Gy + TPA[d]	0.64 ± 0.12	1.56 ± 0.08
3 Gy + SOD[d]	0.82 ± 0.13	0.61 ± 0.11

[a]SOD and/or catalase where present on the cells throughout the experiment.

[b]Surviving fraction = number of colonies counted(number of cells plated x plating efficiency).

[c]Plating efficiency = (number of colonies counted/number of cells plated) x 1000.

[d]SOD and/or TPA where added 24 hr after irradiation, and left until the end of the experiment.

SOD can also modify the homeostatic regulation of radiogenic transformation by thyroid hormones (17) (Table III). Thyroid hormones have long been known to control cellular metabolism. When cells in vitro are depleted of thyroid hormones they become refractory to the transforming action of radiation, chemical carcinogens (15), viruses (18) and tumor promotors (17,19). By adding triodothyrodine (T_3) to the cells at 12 hours prior to their exposure to the carcinogen, cell transformation is restored at a frequency which is T_3 dose dependent over 10^{-10}- 10^{-12} μm (10, 15,18).

TABLE II

CELLULAR CONTENT OF CATALASE AND SOD* IN HAMSTER EMBRYO CELLS (HE) PER MG OF PROTEIN

HE Cultures	Catalase	SOD
Primary	6 units/96 μg ± 2%	37 ng/72 μg ± 2%
Secondary	7.5 units/130 μg ± 2%	37 ng/97.5 μg ± 2%

*No SOD or catalase were detected in fetal calf serum, within the sensitivity of the assay (100 ng of SOD; 1 unit of catalase).

Thyroid hormones thus serve as a permissive factor in transformation. While the hormone action appears to be mediated in part via its role in the synthesis of interval "transforming protein"(15) it is possible that reduced levels of thyroid hormones may be associated with a reduced generation of active oxygen species following oxidant stress induced by the oncogenic agents such as x rays. This possibility is underscored by the fact that the addition of SOD prevents the restoration of transformation by T_3 in hypothyroid cultures (Table III).

The role of Selenium in Transformants.

Cellular protection against carcinogenic action of chemicals and radiation can be imparted by preincubation of cells with Vitamin E or by selenium added as sodium selenite (9,10,11,20).

The inhibition of transformation by selenium is mediated via the induction of selenium dependent glutathione peroxidase, glutathione and catalase, resulting in enhanced levels of these scavenging systems in the cells and doubling peroxide breakdown. Glutathione-S-transferase - a selenium independent enzyme was not modified.

While Vitamin E acted in additive fashion to inhibit transformation it did not affect any of the scavenging systems tested nor did it enhance the inducing effect of selenium (20).

Thus, while selenium acts as an anticarcinogen by inducing cellular protective systems Vitamin E appears to act via alternate mechanisms (20).

TABLE III

THE INHIBITION OF RADIOGENIC TRANSFORMATION IN VITRO BY SOD[a] UNDER VARIOUS THYROID HORMONE CONDITIONS.[b]

C3H/10T½ Cells

	Treatment	Transformation Frequency/(10^{-4}) surviving cell
$-T_3$	Control	0
	SOD[a]	0
	400 rads	0.60
	400 rads+SOD[a]	0
$+T_3$	Control	0
	SOD[a]	0
	400 rads	7.78
	400 rads+SOD[a]	1.58
Eu	Control	0
	SOD[a]	0
	400 rads	8.55
	400 rads+SOD[a]	2.01

a)SOD conc.= 3 units/ml added at the time of irradiation. b) see text for designation of +T3 Eu.

CONCLUSION

Our studies show that the prooxidant state of the cell plays a role in cellular transformation by radiation, the promotor TPA by a variety of chemicals used in cancer therapy or are known as environmental carcinogens. These oncogenic agents are toxic substances which produce enhanced levels of reactive oxygen species upon interaction with cellular molecules. The resulting effect which include damage to DNA and cell transformation can be inhibited by SOD which catalytically scavenges these reactive species. Though SOD action may be mediated at the cell surface one cannot exclude the possibility of its internalization in part via endocytosis.

SOD can inhibit initiation and/or promotion depending on the cell system and possibily on the species and origin of tissue. The activity of the enzyme as an inhibitor is eliminated by heat inactivation.

While SOD acts directly, selenium exerts cellular protection against environmental carcinogens by enhancing internal levels of glutathione, peroxidase, catalase and GSH and doubling the breakdown of toxic peroxides. Vitamin E, an inherent membrane associated antioxidant complements the action of selenium as an anticarcinogen but apparently acts via different mechanisms than that of selenium. The protective factors can override permissive factors required for transformation as seen by the action of SOD in inhibiting thyroid hormone potentiation of radiogenic transformation.

REFERENCES

1. Pryor WA, ed. (1976) Free Radicals in Biology, NY Acad. Press, 1, pp. 1-49.
2. Fridovich I (1978) Science 201:875-880.
3. Borek C, Troll W (1983) Proc. Natl. Acad. Sci. USA 80:1304-1307.
4. Ames BN (1983) Science 221:1256-1264.
5. Zimmerman R, Cerutti PA (1984) Proc. Natl. Acad. Sci. USA 81:2085-2087.
6. Totter JR (1980) Proc. Natl. Acad. Sci. USA 77:1763-1767.
7. Borek C (1982) in Molecular Interrelation of Nutrition and Cancer, Arnot MS, Van Eyes J, Wang YM (eds), Raven Press, NY, pp. 337-350.
8. Nygaard A, Simic M (eds.) (1983) Radioprotectors and Anticarcinogens, Academic Press, N.Y.
9. Lesko SA, Lorentzen RJ, Ts o POP (1980) Biochemistry 19:3023-3028.
10. Borek C (1985) Pharmac. Ther. 27:99-142.
11. Borek C, Biaglow JE (1984) Proc. Am. Assoc. Cancer Res. 25:125.
12. Borek C (1984) in Methods in Enzymology, Packer L (ed) 105:464-479.
13. Borek C (1982) in Advances in Cancer Research, 37:159-232.
14. Borek C, Sachs L (1966) Nature 210:276-278.
15. Borek C, Guernsey DL, Ong A, Edelman IS (1983) Proc. Natl. Acad. Sci. USA 80:5749-5752.
16. Borek C, Hall EJ (1982) Annals of NY Acad. Sci. 193-210.
17. Borek C, Cleaver JE, Fujiki H (1984) in Cellular Interaction by Environmental Tumor Promotors and Relevance to Human Cancer. Fujiki H et al (eds) pp. 195-206, Japan. Soc. Press.
18. Borek C, Ong A, Rhim JS (1985) Cancer Res. 45:1702-1703.
19. Borek C, Cleaver JE (submitted 1985) Antagonistic action of tumor promotor and poly(adenosine diphosphoribose) synthesis inhibitor in radiation-induced transformation in vitro.
20. Borek C, Ong A, Donahue L, Biaglow JE, (submitted 1985) Selenium and Vitamin E Inhibit Radiogenic and Chemically Induced Transformation In Vitro via different mechanisms.

© 1986 Elsevier Science Publishers B.V. (Biomedical Division)
Superoxide and Superoxide Dismutase in Chemistry,
Biology and Medicine; G. Rotilio, editor

LACK OF CROSS-RESISTANCE TO X RAYS AND FREE RADICAL GENERATING DRUGS IN OXYGEN-RESISTANT CHINESE HAMSTER OVARY CELLS

H JOENJE, HG KEIZER, JJP GILLE, P VAN DER VALK and J VAN RIJN*
Institute of Human Genetics and *Department of Radiotherapy, Free University
P.O. Box 7161, 1007 MC Amsterdam (The Netherlands)

INTRODUCTION

According to the free radical theory of oxygen toxicity adverse effects of atmospheric oxygen are mediated by partially reduced or "activated" oxygen species. During normal aerobic metabolism these species are kept at nontoxic levels by the antioxygenic defense system of the cell, but under hyperoxic conditions their excessive production supposedly overwhelmes these defenses (1). A number of agents, such as X rays and certain antitumor drugs, are also thought to induce cell killing by the excessive formation of activated oxygen species. This common toxicity mechanism of O_2, X rays and free radical generating drugs would predict that cells with increased defenses against activated oxygen are likely to be less sensitive to the toxic effects of such agents. We tested this prediction by using an oxygen-resistant substrain of Chinese hamster ovary (CHO) cells obtained by longterm adaptation to stepwise increased levels of atmospheric oxygen; this substrain is capable of proliferating at 99% O_2/1% CO_2 , an oxygen level that is ordinarily lethal to cells in tissue culture (2). The increased oxygen tolerance appeared to be associated with increased cellular activities of CuZn superoxide dismutase (SOD), MnSOD, catalase and glutathione peroxidase (GSHPX). Using reproductive cell killing as an end point we tested the oxygen-resistant cells for possible cross-resistance to X rays and the free radical generating antitumor drugs mitomycin C, adriamycin, VP 16-213 and bleomycin.

RESULTS AND DISCUSSION

Antioxygenic enzymes. Figure 1 summarizes the specific activities of CuZnSOD, MnSOD, catalase and GSHPX in normal and oxygen-resistant CHO cells. All four enzyme activities were significantly elevated in the resistant cells, suggesting that these cells are better protected against intracellularly generated superoxide, hydrogen peroxide, lipoperoxides and OH radical (3).

X ray sensitivity. The oxygen-resistant CHO strain showed a definite lack of cross-resistance to X rays, as tested by clonogenic cell survival; if anything, they tended to be more radiosensitive than the normal controls (fig. 2).

Drug-sensitivity. As shown by the survival curves in figure 3 the oxygen-resistant cells failed to exhibit cross-resistance to the antitumor drugs mitomycin C, adriamycin, VP 16-213 and bleomycin. In the case of adriamycin dose-dependent drug uptake and cyanide-resistant respiration in whole cell susp-

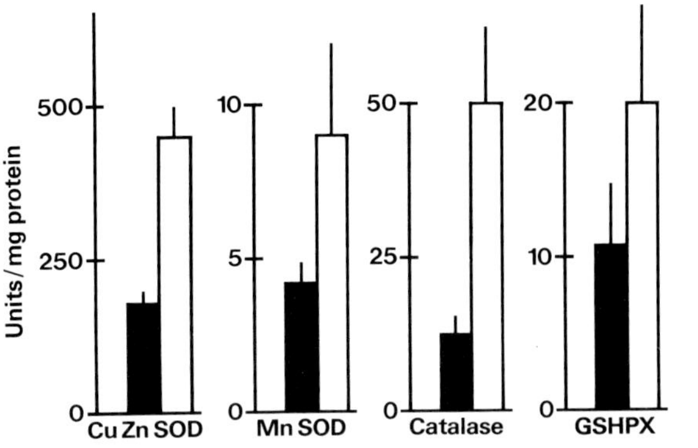

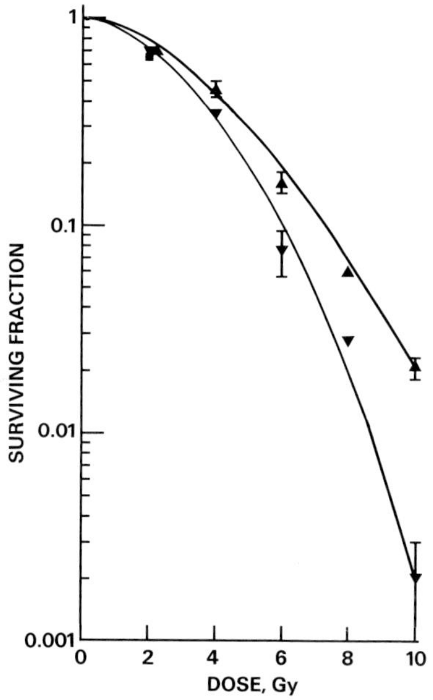

Fig. 1 (above). Antioxygenic enzymes in normal (black bars) and O_2-resistant (empty bars) CHO cells

Fig.2 (right). Clonogenic cell survival as a function of X ray dose in normal controls (▲) and O_2-resistant cells (▼)

Fig. 3 (below). Clonogenic cell survival of normal control (○) and O_2-resistant (●) CHO cells as a function of the indicated drug treatment. Drug exposure was for 1 h at the concentrations shown.

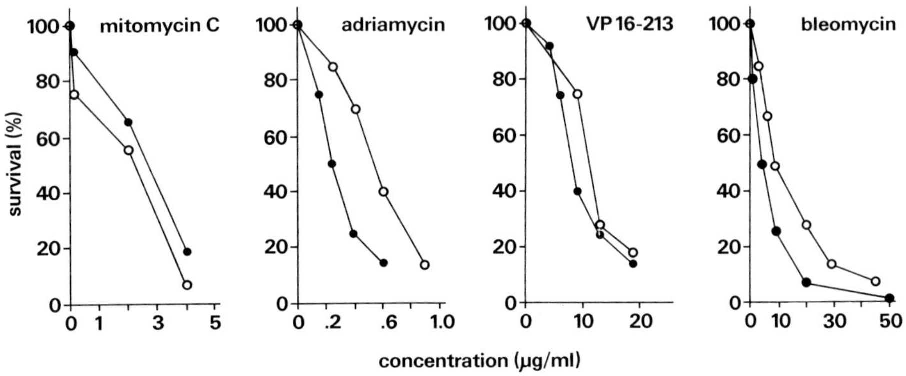

ensions, as well as adriamycin semiquinone radical formation in cell homogenates (as assessed by electron spin resonance) were essentially the same for both cell types (H.G. Keizer, unpublished data).

If activated oxygen species are causative in the toxicity of X rays and the drugs tested, it should be explained why elevated intracellular levels of antioxygenic enzymes fail to mitigate toxicity. We think that (i) site-specificity of free radical production, (ii) subcellular location of the critical target(s), and

(iii) compartmentalization of antioxygenic defenses, should be considered to explain this result. In the case of X rays free radical production is randomly distributed throughout the cell, but the critical target in reproductive cell killing is known to be nuclear DNA, the main damaging radical species being the OH radical. For the 4 drugs tested there is good evidence that they bind to DNA and produce various kinds of DNA lesions such as adducts, crosslinks and strand breaks, the latter type of lesion presumably resulting from site-specific generation of OH radicals. Thus the primary target involved in reproductive cell killing by the agents tested is most likely nuclear DNA. On the other hand, there is substantial evidence that the antioxygenic enzymes are localized outside the nucleus. MnSOD is mitochondrial (4), although in human cells a substantial part may be found in the cytoplasm (5). Catalase is mainly peroxisomal (2), while GSHPX is both mitochondrial and cytoplasmic (6). CuZnSOD has been localized in the cytosol and in lysosomes (7), and EC-SOD is extracellular (8). Patton et al. (9) have found CuZnSOD activity to be associated with nuclear preparations from hamster liver homogenates, but this is no conclusive evidence for intranuclear localization of the enzyme, especially since by immunocytochemical vizualization of the enzyme in human leukocytes we failed to detect intranuclear enzyme antigen, whereas the nuclear envelope was strongly positive (H. Joenje and R.R. Frants, unpublished).

Our data suggest that cellular concentrations of antioxygenic enzymes are relatively unimportant for cellular sensitivity to active-oxygen-generating agents that primarily act through the destruction of intranuclear targets.

AKNOWLEDGEMENTS-This work was supported by the Netherlands Cancer Foundation.

REFERENCES

1. DiGuiseppi, Fridovich I (1984) CRC Crit Rev Toxicol 12:315-342
2. Van der Valk P, Gille JJP, Oostra AB, Roubos EW, Sminia T, Joenje H (1985) Cell Tissue Res 239:61-68
3. Halliwell B (1982) Trends Biochem Sci 7:270-272
4. Weisiger RA, Fridovich I (1973) J Biol Chem 248:4793-4796
5. McCord JM, Boyle JA, Day Jr ED, Rizzolo LJ, Salin ML (1977) In: Michelson AM, McCord JM, Fridovich I (eds) Superoxide and Superoxide Dismutases. Academic Press, London, pp 129-138
6. Flohé L (1982) In: Pryor WA (ed) Free Radicals in Biology, Vol V. Academic Press, New York, pp 223-254
7. Geller BL, Winge DR (1984) Meth Enzymol 105:105-114
8. Marklund S (1984) J Clin Invest 74:1398-1403
9. Patton SE, Rosen GM, Rauckam EJ (1980) Mol Pharmacol 18:588-593

Superoxide and Superoxide Dismutase in Chemistry,
Biology and Medicine; G. Rotilio, editor

INVOLVEMENT OF PROOXIDANTS AND ANTIOXIDANTS IN CELLULAR DIFFERENTIATION AND AGING PROCESS: A HYPOTHESIS

R.S. SOHAL, K.J. FARMER AND R.G. ALLEN
Department of Biology, Southern Methodist University, Dallas, Texas 75275, USA

The following is a summary of the evidence, based on several lines of investigation, which, taken together, suggests that oxygen free radicals may play a causal role in the induction of cellular differentiation and senescence.

There is an inverse relationship between metabolic rate (rate of oxygen consumption) and life span in homeotherms and poikilotherms. This relationship is, however, most evident in poikilotherms, which do not maintain a stable basal metabolic rate. Studies in this laboratory on insects have shown that experimental regimes that decrease metabolic rate invariably extend life span of organisms (1-3). For example, the average and the maximum life span of houseflies is prolonged by: (i) removal of wings to prevent flying, (ii) confinement of individual flies in vials, where there is not enough space to fly, (iii) increase in the relative number of females in the population, which reduces the flying activity of males in pursuit of sexually receptive females, and (iv) decrease in ambient temperature, which also reduces the level of physical activity. Similarly, in hibernating mammals, such as Turkish hamsters, life span is prolonged by hibernation (4).

The rate of age-related physiological and biochemical changes is modulated by metabolic rate. Lipofuscin, Schiff-base like chloroform-soluble fluorescent material (SFM), thiobarbituric acid (TBA)-reactants and n-pentane production, whose levels have been shown in insects and mammals to increase with age, were used as markers of physiological age (1-3,5). A decrease in metabolic rate of houseflies retards the rate of accumulation of lipofuscin, SFM and TBA-reactants, and of n-pentane production (1-3). There is convincing evidence indicating that oxygen-derived free radicals are causally involved in the genesis of lipofuscin, SFM, TBA-reactants and n-pentane. Increase of physical activity in rats has been shown to elevate the levels of free radicals and products of lipid peroxidation (6). Physical exercise also increases the exhalation of n-pentane in humans (7).

Products of free radical reactions, e.g. H_2O_2 (8) and n-pentane (5,9), are produced at faster rates in old organisms than in young, which suggests either an increase in the generation of free radicals and/or a decline in antioxidant defenses in the old. An increase in cyanide-resistant respiration has been noted in heart mitochondria of old rats (10). A decline in SOD activity and glutathione levels has been reported in certain tissues of rodents (for references, see 3,11). A comprehensive examination of the antioxidant defenses in the housefly has indicated a significant decline in SOD activity, catalase activity, and levels of glutathione and lipid-soluble antioxidants in old organisms (8).

One mechanism by which metabolic rate may affect the aging process may involve the deleterious effects of free radicals on cellular constituents. Damage to cellular membranes and genome would have particularly serious consequences. Increased peroxidizability of membranes and cross-linking of chromatin proteins in senescent cells lends credence to this hypothesis (3,11).

Another mechanism may be that decline in the efficiency of antioxidant mechanisms and the increase in the rate of free radical generation, together, induce a shift in the redox state of the cell in favor of prooxidants. Evidence in support of this hypothesis is provided by the finding that ratios of GSSG/GSH, NAD/NADH and NADP/NADPH increase in tissues of old animals (12). Changes in redox state have been shown to alter ionic balance of cells, which, in turn, is known to alter chromatin configuration and chromosomal puffing patterns, indicative of altered gene expression (for references, see 3,11).

Oxygen free radicals also seem to play an inductive role in cellular differentiation. Phenotypic differentiation of cells is demonstrably affected by oxygen tension (3). Our studies on differentiation (spherulation) in slime mold, *Physarum polycephalum*, indicate that cyanide-resistant respiration, and concentration of inorganic and organic peroxides increases during differentiation (13). There is a dramatic increase (up to 45-fold) in Mn-SOD activity, and a sharp decrease (up to 80%) in glutathione concentration during differentiation (13). Experimental increase and decrease in glutathione levels retards or accelerates the rate of differentiation, respectively (3,11). Paraquat accelerates differentiation. A similar pattern of changes occurs during larval development and metamorphosis in the housefly (14). Glutathione content declines during larval development and is relatively low in pupae. Cyanide-resistant respiration is relatively high in pupae. We have also found that *in vitro* differentiation of mammalian myoblasts into myocytes is accompanied by a 2-fold increase in SOD activity and a 30% decrease in glutathione concentration. On the basis of this and other evidence in the literature, it is apparent that the cellular enviornment becomes less reducing during differentiation and senescence.

We have proposed a unified theory of aging and differentiation which postulates that during differentiation and senescence cellular environment becomes progressively more oxidizing, altering gene expression (3,11). Senescence is believed to be due to a loss in gene regulatory efficiency. The postulated mechanism involves either direct damage to the regulatory sites of the genome and/or changes in gene expression due to the shift in redox state of cells.

REFERENCES

1. Sohal RS (1981) In: Sohal RS (ed) Age Pigments. Elsevier/North Holland, Amsterdam, pp 303-316
2. Sohal RS (1985) In: Kerkut GA, Gilbert LI (eds) Comprehensive Insect Physiology, Biochemistry and Pharmacology. Pergamon Press, Oxford, vol 10, pp 596-631
3. Sohal RS, Allen RG (in press) In: Woodhead AD (ed) The Molecular Basis of Aging. Plenum Press, New York
4. Lyman CP, O'Brien RC, Greene GC, Papafrangos ED (1981) Science 212:668-670
5. Sohal RS, Muller A, Koletzko B, Sies H (1985) Mech Ageing Dev 29:317-326
6. Davies KJA, Quintanilha AT, Brooks GA, Packer L (1982) Biochem Biophys Res Comm 107:1198-1205
7. Dillard CJ, Litov RE, Savin WM, Dumelin EE, Tappel AL (1978) J Appl Physiol 45:927-932
8. Sohal RS, Farmer KJ, Allen RG, Cohen NR (1984) Mech Ageing Dev 24: 185-195
9. Sagai M, Ishinose T (1980) Life Sci 27:731-738
10. Hegner D (1980) Mech Ageing Dev 14:101-118
11. Sohal RS, Allen RG (in press) In: Pryor WA (ed) Advances in Free Radicals in Biology and Medicine. Pergamon Press, Oxford
12. Noy N, Schwartz H, Gafni A (1985) Mech Ageing Dev 29:63-69
13. Allen RG, Newton RK, Sohal RS, Shipley GL, Nations C (in press) J Cell Physiol
14. Allen RG, Sohal RS (in press) In: Collatz KG, Sohal RS (eds) Aging in Insects: Strategies and Mechanisms. Springer-Verlag, Heidelberg

© 1986 Elsevier Science Publishers B.V. (Biomedical Divison)
Superoxide and Superoxide Dismutase in Chemistry,
Biology and Medicine; G. Rotilio, editor

DIFFERENT EFFECTS OF AGE AND PHYSICAL EXERCISE ON THE SUPEROXIDE DISMUTASE ACTIVITY IN SKELETAL AND HEART MUSCLE.

ELISABETH STEINHAGEN-THIESSEN, ABRAHAM REZNICK*, JAN-DIRK RINGE, DAVID GERSHON*
Medizinische Klinik, Universität Hamburg (FRG) and *Technion, Haifa (Israel)

Free radical damage involving oxygen may contribute to the aging process in cells and tissues. Superoxide dismutase as an defense enzyme against oxygen toxicity has become an important enzyme in cancer-, injury- and inflamatory research and also in research of aging processes (1,2). Therefore, senescence may be accompanied with a decline in one of the enzymes which provide anti-oxidant protection. Because SOD is the major intracellular enzyme, we analyzed the amount of SOD in skeletal and heart muscle tissue according to aging and, additionally, according to short- and longterm physical exercise in mice. All measurements were done in two genetically different strains of mice to show that our data are of generally significance and not an observation of one unique strain of mice.

MATERIALS AND METHODS

Female CWI and C57 Bl mice were kept in groups of 10 in polypropylene cages 15 x 42 x 26 cm and exposed to ambient daylight at a room temperature of 20 $\pm$ 3°C and fed with Altromin (Altromin,Lage Westfalen) and water ad libitum. The animals were weighed weekly. The mice were forced to run in electrically driven running wheels starting 5 min per day with increasing training time until a maximum of 30 min/day. The mice were running 4,5 m min^{-1}. At the age of 9,14,18,21-22 and 27 months 8 to 10 mice/group were sacrificed and the skeletal and heart muscle were taken immediately, weighed and frozen in liquid nitrogen. Groups of mice of the same ages were served as controls. In the shortterm training program the mice were trained for 5 weeks. During the longterm training program the mice were trained for 2-3,8,15,21 months and throughout life. The muscle material was immediately homogenized and prepared as described previously (3). Cytosolic SOD was estimated according to the method of Marklund and Marklund (4). The protein was determined according the Lowry method (5) and the DNA content according to the Burton method (6).

RESULTS

The specific activity of the cytosolic SOD in the skeletal muscle of the CWI mice showed age dependent differences. An adaptive increase in the specific activity of SOD in the young (6 mo.) and middle age (22 mo.) animals during the 5 weeks training period (Fig. 1) was estimated. But a significant decrease of SOD activity was observed for the 27 mo. olf mice after this shortterm training period. We describe this as a"negative adaptation to training and aging". The enzyme

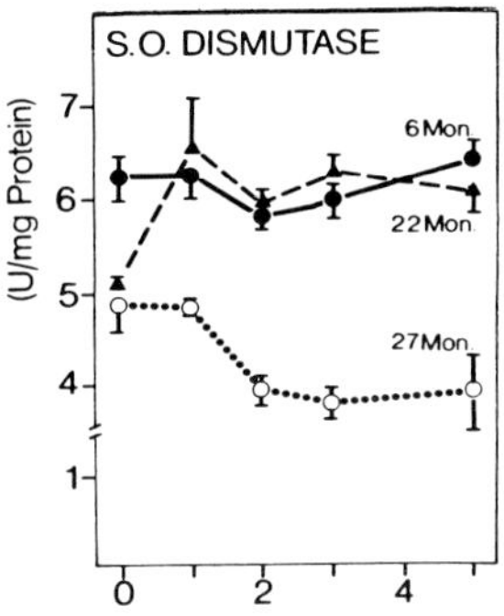

Fig. 1. Weeks of training.

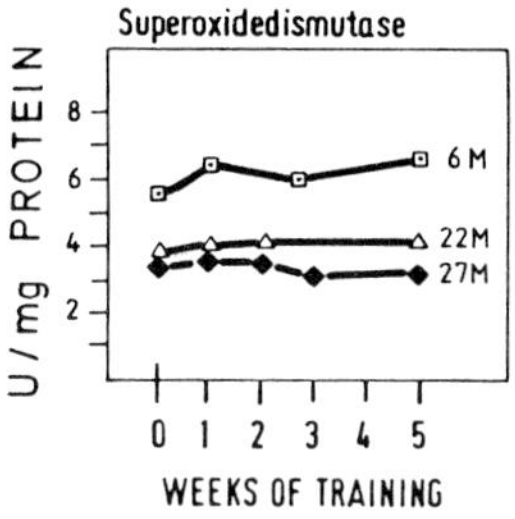

Fig. 2.

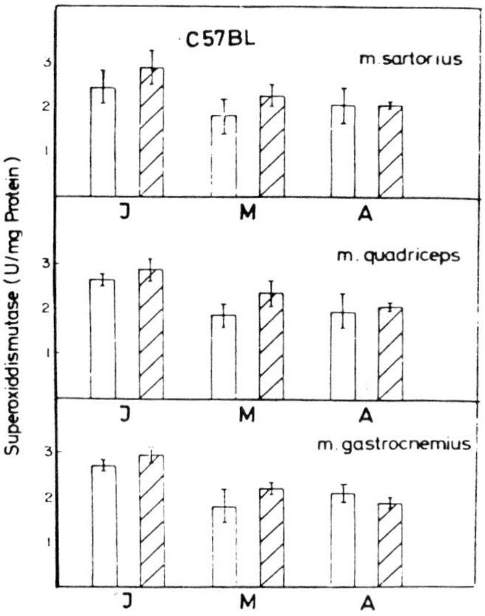

Fig. 3. J = 6 mo., M = 22 mo., A = 27 mo.

pattern of the heart muscle material in these animals is shown in Fig. 2. Again a partial adaptation to the training was measured in the young animals while there was a lower activity in the middle age and senile animals. There were no significant changes in these groups after the shortterm training period.

In addition the C57/Bl mice showed similar results of the SOD activity during the shortterm training period in the separated muscles (m.sartorius, m.gastrocnemius, m.quadriceps) (Fig. 3).

In contrast to these protocol animals of both strains were trained for a lifelong period (30 min/day). The SOD activity in the skeletal muscle showed no signi-

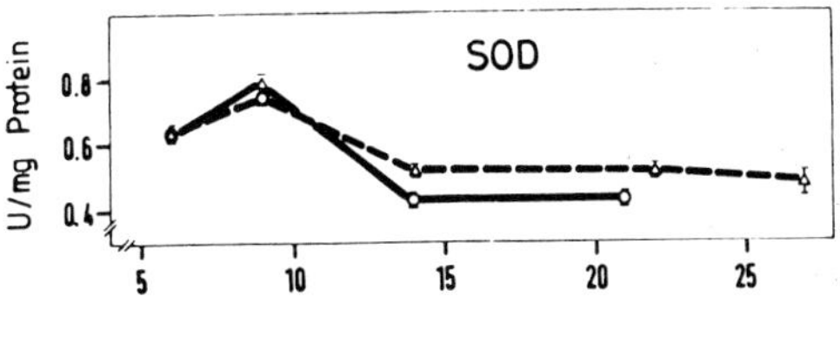

months of age and training

Fig. 4. ——— trained
- - - untrained

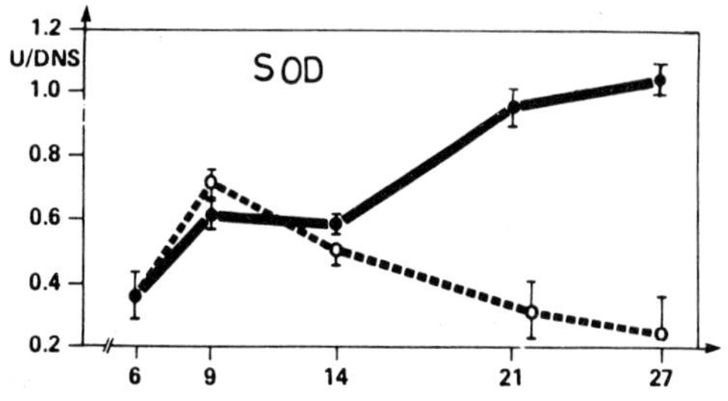

weeks of age and training

Fig. 5. ——— trained
- - - untrained

ficant differences of trained and untrained animals. In both groups was a significant age dependent decline of these specific activity between the 9 and 15 months of age (Fig. 4). But in the heart muscle tissue a significant difference of the specific activity of SOD between lifelong trained and untrained animals was observed (Fig. 5). Here a positive adaptation of aging and training in heart muscle tissue was estimated.

DISCUSSION

Controverse data exist concerning the beneficial or determental effects of physical exercise during aging. In our animal model we could show that in the young animals there was a positive adaptation to the shortterm training, but a negative adaptation in the very old animals. The old muscle tissue had no ability to regulate the SOD activity during this stress period in the senile animals. Similar enzyme pattern were observed for creatine kinase and aldolase in this tissue (3,7,8). Also no adaptation to training and aging was seen in the skeletal muscle tissue of the longterm trained animals.

But in contrast to this observation a significant increase in the heart muscle SOD activity during aging and the longterm exercise regimen was seen.

We explained this as an ability of the heart muscle tissue to regulate the SOD activity during daily work load. Also these findings are representative for the metabolic difference in these two muscle tissues. The heart muscle tissue in the longterm training conditions showed compensatory changes with aging and training which was not seen in the skeletal muscle tissue of these animals. This might be due to the fact that the two different muscle tissues respond metabolically different to physical exercise.

REFERENCES

1. Fridovich I (1983) Ann Rev Pharmacol Toxicol 23:239-257
2. Harman D (1981) Proc Nat Acad Sci USA 78:7124-7128
3. Steinhagen-Thiessen E, Reznick A, Hilz H (1980) Mechanism of Aging and Development 12:231-236
4. Marklund S, Marklund G (1974) Eur J Biochem 47:469
5. Lowry OH, Rosebrough NJ, Farr AL, Randall RJ (1951) J Biol Chem 193:265
6. Burton K (1968) In: Grossman L, Moldave K (eds) Methods in enzymology, Academic Press, New York, pp 163-166
7. Steinhagen-Thiessen E, Reznick A, Hilz H (1981) Mechanism of Aging and Development 16:363-369
8. Reznick A, Steinhagen-Thiessen E, Gellersen B, Gershon D (1983) Mechamism of Aging and Development 23:253-258

© 1986 Elsevier Science Publishers B.V. (Biomedical Division)
Superoxide and Superoxide Dismutase in Chemistry,
Biology and Medicine; G. Rotilio, editor

CIGARETTE SMOKE, FREE RADICALS, AND BIOLOGICAL DAMAGE*

Edward T. Borish,[+] John P. Cosgrove,[+] Daniel F. Church,[+] Walter A. Deutsch,[#] and William A. Pryor[+#*]

Departments of [+]Chemistry and [#]Biochemistry, Louisiana State University, Baton Rouge, Louisiana 70803

INTRODUCTION

Cigarette smoke is a complex, dynamic system consisting of thousands of compounds (1,2), many of which are known carcinogens, cocarcinogens, or promoters. Experimental (9-11) and epidemiological data (12) indicate that cigarette tar is primarily responsible for the carcinogenicity of cigarette smoking. Smoke also contains high concentrations of organic free radicals, both in the tar and the gas phase smoke (3,4). A large body of evidence supports the involvement of these radicals in smoking-related diseases, particularly cancer (5-8). In this paper we will discuss the production of active oxygen species by tar and the role of these active oxygen species in causing DNA damage.

DISCUSSION

The free radical in cigarette tar was originally suggested to be an odd electron delocalized over a PAH molecule (13). However, we have shown that the predominate organic radical present in tar is actually a semiquinone ($QH^{\bullet}$) held in a polymeric matrix (4). Since synthetic quinone/hydroquinone (Q/QH_2) polymers are potent redox catalysts (14,15), it is reasonable to assume that the Q/QH_2 radical in cigarette tar could similarly be redox active. Many Q/QH_2 play an important role in controlling the levels of oxyradicals in biological systems (16). For example, quinone anti-cancer drugs have been shown to be capable of reducing dioxygen to form superoxide and related active oxygen species (17). The quinone nature of the radical in cigarette tar suggests that tar also might be capable of reducing dioxygen to superoxide. Indeed, we have recently found that aqueous extracts of cigarette tar consume oxygen, and spin trapping experiments using DMPO indicate that these aqueous extracts produce hydroxyl

* This work was supported in part by a grant from the National Institutes of Health (HL-25820) and a contract from the National Foundation for Cancer Research, both to William A. Pryor. Address reprint requests and correspondence to Dr. Pryor.

ABBREVIATIONS: DMPO, 5,5-dimethylpyrroline-N-oxide; EDTA, ethylenediamine tetraacetic acid; DTPA, diethylenetriamine pentaacetic acid; ESR, electron spin resonance; PAH, polynuclear aromatic hydrocarbon; SOD, superoxide dismutase.

radicals, as shown in the Table below (18). The involvement of hydroxyl radicals was confirmed by the addition of the hydroxyl radical scavenger ethanol, causing a decrease in the intensity of the ESR signal due to the hydroxyl spin adduct and the concommitant formation of the 1-hydroxyethyl radical adduct. These aqueous extracts of tar also cause single strand breaks in covalently closed circular DNA (19). The results of adding protective enzymes, metal chelators and hydroxyl radical scavengers indicate that the hydroxyl radicals are responsible both for the DMPO spin adduct ESR signal and for the DNA damage (Table).

TABLE: Effects of various additives on the degree of DNA nicking and on the HO-DMPO• spin adduct intensity

	DNA Nicks		ESR Signal	
Additives	fmole	% Change	Intensity*	% Change
Tar	8.4	---	11	---
Tar + DTPA	1.7	-80	0	-100
Tar + EDTA	16.3	+94	181	+1500
Tar + EDTA + catalase	2.4	-85	2	-99
Tar + EDTA + SOD	8.6	-47	88	-51
Tar + EDTA + benzoate	1.8	-89	---	---
Tar + EDTA + ethanol	---	---	56	-69

* Due to the hydroxyl radical adduct of DMPO, in arbitrary units

The results in the Table can best be explained by the following scheme, where Q, QH_2, and $QH^{\bullet}$ are quinone, hydroquinone and semiquinone groups, respectively, in the tar.

$$Q + QH_2 \rightleftharpoons 2QH^{\bullet} \quad [1]$$

$$QH^{\bullet} + O_2 \longrightarrow Q + O_2^{\bar{\bullet}} + H^+ \quad [2]$$

$$O_2^{\bar{\bullet}} + Fe^{III} \longrightarrow O_2 + Fe^{II} \quad [3]$$

$$QH^{\bullet} + Fe^{III} \longrightarrow Q + Fe^{II} + H^+ \quad [4]$$

$$2O_2^{\bar{\bullet}} + 2H^+ \longrightarrow H_2O_2 + O_2 \quad [5]$$

$$H_2O_2 + Fe^{II} \longrightarrow HO^{\bullet} + HO^- + Fe^{III} \quad [6]$$

$$\text{DMPO} + HO^{\bullet} \longrightarrow \text{HO-DMPO}^{\bullet} \quad [7]$$

$$\text{DNA} + HO^{\bullet} \longrightarrow \text{DNA Damage} \quad [8]$$

Metals are certainly implicated in these reactions, but we have no evidence that iron is the specific metal involved. Also, it is not likely that free metal ions are participating; it is more probable that the metals are complexed, either with an added chelator such as EDTA or with components of the cigarette tar such as the Q/QH_2 functionalities.

The ability of iron complexes to catalyze hydroxyl radical production (eq 6) is well-established (20,21). In the absence of added chelator, iron is presumably complexed with some component of the tar, and this tar-iron complex can decompose hydrogen peroxide. Our results suggest that the tar-iron complex is less effective at catalyzing eq 6 than is the iron-EDTA complex, since the addition of EDTA leads to a much higher yield of hydroxyl spin adduct. Of course, an EDTA-metal complex could also accelerate any of the other reactions that involve metal ions, leading to more hydrogen peroxide.

Interestingly, addition of EDTA increases hydroxyl-mediated damage to DNA much less than it increases the concentration of $HO\text{-}DMPO^{\bullet}$ spin adduct (see the third line in the Table). We have previously shown that an ESR signal becomes associated with DNA when the DNA is incubated with tar in vitro (22), suggesting the possibility of a complex between ESR-active species in cigarette tar and DNA similar to that formed by anti-cancer drugs with DNA (23,24). While addition of EDTA greatly enhances the concentration of $HO^{\bullet}$ in the bulk solution, it may disrupt a DNA-metal-tar complex that leads to hydroxyl radical production in the immediate vicinity of the DNA and is unusually effective in producing DNA damage (25). Thus, less $HO^{\bullet}$ would be formed in the vicinity of the DNA, resulting in a smaller increase in DNA damage (eq 8) than in the yield of spin trappable hydroxyl radicals (eq 7).

By scavenging $O_2^{\overline{\bullet}}$, SOD keeps $O_2^{\overline{\bullet}}$ from reducing Fe(III) to Fe(II), preventing the cycling of iron (eq 3) and reducing the yield of $HO^{\bullet}$ from H_2O_2 (eq 6). However, SOD inhibits DNA damage by only about 50% in our system, even at high SOD concentrations. Therefore, our results suggest the presence of another species in tar capable of reducing Fe(III). Butler et al. (26) have shown that mitomycin C and adriamycin semiquinone radicals can reduce Fe(III) as efficiently as can the superoxide ion. Thus it seems likely that the tar semiquinone radical is capable of reducing Fe(III) to Fe(II) in the absence of $O_2^{\overline{\bullet}}$, as shown in eq 4. If tar were the only reducing agent for Fe(III), the addition of SOD would have no effect on the amount of $HO^{\bullet}$ formed. Since about 50% inhibition is observed with SOD, eq 3 is operative in the absence of SOD, and the rates of eqs 3 and 4 must be similar.

Inhibition by SOD, catalase and hydroxyl radical scavengers clearly indicates that $O_2^{\overline{\bullet}}$, H_2O_2 and $HO^{\bullet}$ are involved in these reactions. It is generally accepted that superoxide leads to hydrogen peroxide and that H_2O_2 is the precursor of the hydroxyl radical, as shown in eqs 5 and 6. Therefore, the fact that both catalase and benzoate cause inhibition (Table) implies that superoxide is not directly responsible for the observed DNA damage and that the hydroxyl radical is the immediate damage-producing agent (Table).

Gas-phase cigarette smoke and tar have very different properties; gas-phase smoke contains strong oxidants whereas tar is a reducing system (3). It is particularly interesting to note that the damage to DNA produced by tar results from oxidation of the DNA. This damage occurs because tar reduces dioxygen to the superoxide radical, which ultimately leads to hydroxyl radicals that cause the damage to DNA that is observed. Hydroxyl radicals also have been shown to be responsible for cellular damage produced by the quinone anti-tumor antibiotics (23). However, in contrast to the quinone anti-cancer drugs or to PAH, which require activation by either reducing or oxidizing enzyme systems (17,27), cigarette tar by itself can lead to DNA damage. That is, cigarette tar contains the reducing equivalents necessary to produce superoxide, as well as metals and metal-chelating agents that can convert the resulting hydrogen peroxide to hydroxyl radical. It is possible that endogenous reductases *in vivo* could provide even more reducing equivalents and thus increase the yield of active oxygen species from tar.

After our demonstration that the tar free radical becomes associated with DNA when tar and DNA are incubated *in vitro* (22), we initiated a program to identify the specific component in tar that binds to DNA and the base to which binding occurs. To date, our results have been inconclusive, suggesting that there may not be a single discrete species in tar that covalently binds to a specific DNA base. Rather, it seems more likely that the radicals in tar might be involved in tar-induced tumors in the following way. Water-soluble tar components containing the Q/QH_2 functionality could be carried to the locus of cellular DNA or RNA and form a complex. This complex might consist of a metal "sandwich," with the metal holding the DNA and tar quinone-hydroquinone radical together. If the metal in this sandwich were iron (or a similar transition metal capable of redox cycling), conditions would exist for site-specific DNA damage to occur. The Q/QH_2 functionality in tar could reduce dioxygen to superoxide, which would then dismutate to form hydrogen peroxide. The transition metal would decompose the hydrogen peroxide to form hydroxyl radicals in the immediate vicinity of the DNA, causing oxidative DNA damage. This damage could be due either to base modifications typical of free radical reactions (such as thymine glycol or hydroxymethyl uracil) (28) or to generalized damage that initiates non-error-free DNA repair. In either case, circumstances may arise in which the initial damage is translated into a point mutation, resulting in the initiation of tumor growth.

REFERENCES

1. U.S. Public Health Service. in: Smoking and Health: The Report of the Surgeon General, ch. 14, p. 35, U.S. Department of Health, Education and Welfare, Washington, D.C. (1979).

2. M.R. Guerin. in: Banbury Report. A Safe Cigarette (G.B. Gori and F.G. Bock, eds), p. 192, Cold Spring Harbor Laboratory, New York (1980).

3. D.F. Church and W.A. Pryor. Environ. Health Perspect. (in press).

4. W.A. Pryor, B.J. Hales, P.I. Premovic and D.F. Church. Science 220: 425-427 (1983).

5. R.A. Floyd, ed, Free Radicals and Cancer Marcel Dekker, New York (1982).

6. W.A. Pryor. Annals New York Acad. Sci. 393: 1-30 (1982).

7. H. Thaler-Dao, A.C. de Paulet and R. Paoletti, eds, Icosanoids and Cancer Raven Press, New York (1984).

8. P.O.P. Ts'o, W.J. Caspery and R.J. Lorentzen. in: Free Radicals in Biology, Vol. III (W.A. Pryor, ed) pp. 251-300, Academic Press, New York (1971).

9. D. Hoffmann, E.L. Wynder, A. Rivenson and S.S. Hecht. Prog. exp. Tumor Res. 26: 43-67, (1985).

10. D. Hoffman, and E.L. Wynder. in: The Chemistry of Tobacco and Tobacco Smoke (I. Schmeltz, ed), p. 123, Plenum Press, New York (1972).

11. U.S. Public Health Service. The Health Consequences of Smoking, CANCER: The Report of the Surgeon General p. 193, U.S. Department of Health and Human Services, Washington, D.C. (1982).

12. U.S. Public Health Service. The Health Consequences of Smoking, THE CHANGING CIGARETTE: The Report of the Surgeon General, p. 79, U.S. Department of Health and Human Services, Washington, D.C. (1981)

13. M.J. Lyons, J.F. Gibson and D.J.E. Ingram. Nature 181: 1003-1004 (1958).

14. Y. Iwasawa and S. Ogasawara. Chem. Lett. 845-848 (1974).

15. Y. Iwasawa and S. Ogasawara. J. Catal. 46: 132-142 (1977).

16. D.C. Borg and K.M. Schaich. Isr. J. Chem. 24: 38-53 (1984).

17. K. Hamana, K. Kawada, K. Sugioka, M. Nakano, S. Tero-Kubota and Y. Ikegami. Biochemistry International 10: 301-309 (1985).

18. J.P. Cosgrove, E.T. Borish, D.F. Church and W.A. Pryor. Biochem. Biophys. Res. Comm. (in press).

19. E.T. Borish, J.P. Cosgrove, D.F. Church, W.A. Deutsch and W.A. Pryor. Biochem. Biophys. Res. Comm. (in press).

20. B. Halliwell. FEBS Lett. 92: 321-326 (1978).

21. J.M. McCord and E.D. Day, Jr. FEBS Lett. 86: 139-142 (1978).

22. W.A. Pryor, K. Uehara and D.F. Church. Oxygen Radicals in Chemistry and Biology (W. Bors, M. Saron and D. Tait, eds), pp 193-201, Walter de Gwyter and Co., Berlin (1984).

23. K. Sugioka, H. Nakano, J. Tsuchija, M. Nakano, Y. Sogioka, S. Tero-Kubota and Y. Ikeyame. Biochemistry International 9: 237-242 (1984).

24. G.M. Ehrenfeld, L.O. Rodriguez, S.M. Hecht, C. Chang, V.J. Basus and J.J. Oppenheimer. Biochemistry 24: 81-92 (1985)

25. P.J. Angiolio, L.A. Donoso, R. Folberg, J.J. Augsberger and M. Wax. Cancer Biochem. Biophys. 8: 61-65 (1985).

26. J. Butler, B.M. Hoey and A.J. Swallow. FEBS Lett. 182: 95-98 (1985).

27. O. Perin-Roussel, A. Croisy, B. Ekert and F. Zajdela. Cancer Lett. 22: 289-298 (1984).

28. R. Cathcart, E. Schwiers, R.L. Saul and B.N. Ames. Proc. Natl. Acad. Sci. USA 81: 5633-5637 (1984).

Superoxide and Superoxide Dismutase in Chemistry,
Biology and Medicine; G. Rotilio, editor

PHOSPHOLIPASE A_2 ACTIVATION FOLLOWING MEMBRANE LIPID PEROXIDATION

ALEX SEVANIAN and EUNJOO KIM. University of Southern California, Institute for Toxicology, Los Angeles, CA 90033

INTRODUCTION

The peroxidation of membrane phospholipids is commonly encountered following free radical induced oxidation in cells. During the past decade several investigators have reported that membrane lipid peroxidation is accompanied by increased phospholipase A_2 activity (1-3). This activity has been correlated with the extent of free radical induced membrane lipid peroxidation and exerts a measurable effect on the fate of lipid peroxidation products (3,4). Relevant to a major theme within this series of monographs is the observation that ischemia-reperfusion injury, and other forms of tissue shock injury of a purported free radical basis, is accompanied by elevations in phospholipase A_2 activity (5-8).

We have examined the nature of this apparent phospholipase activation using artificial membranes in the form of unilamellar vesicles (liposomes), to determine if the effects of peroxidation and phospholipase action resemble those encountered in in vivo systems. Aspects of phospholipase A_2 selectivity towards oxidized versus intact fatty acyl components of phospholipids were examined and these effects were related to alterations in membrane structure following lipid peroxidation.

MATERIAL AND METHODS

Liposomes were prepared as described previously (3) using a 4:1 molar ratio of phosphatidylcholine (PC) and phosphatidylethanolamine (PE). The liposomal suspensions were subjected to peroxidation by incubation in the presence of Fe^{2+}-ADP (20 and 120 uM, respectively) and 200 uM ascorbate. Incubation intervals up to 30 min. at 37°C were used, and the extent of peroxidation was determined via measurement of conjugated dienes and thiobarbituric acid reaction products (TBAR) as described (9).

Following peroxidation samples were removed, mixed with 1/10 volume of 1 mM EDTA and incubated for 2 hrs in the presence of a 5 uM diphenylhexatriene (DPH). Fluorescence polarization measurements were then made as described by Shinitzky and Barenholtz (10). The remainder of each sample was incubated for intervals up to 15 min. at 37°C in the presence of 1 mM Ca^{+2} or 1 mM Ca^{+2} plus 0.4 U/ml phospholipase A_2 (from c. ademanteus). Samples were extracted with chloroform/methanol and processed for measurement of free fatty acids and phospholipids as previously described (3). Characterization and quantitation of fatty acids and fatty acid oxidation products was conducted by means of gas and high pressure liquid chromatography, respectively, after hydrolysis and derivatization (11). All calculations were based on the use of NIH fatty acid standards, freshly prepared fatty acid conjugated dienyl alcohols from linoleic, linolenic and arachidonic acids, and were normalized to the original phospholipid content determined by phosphorus analysis (12).

RESULTS

The liposomal phospholipids consisted of 20% polyunsaturated fatty acids by weight with arachidonic acid as the most common component (41% of PUFA). The extent of lipid peroxidation in the liposomes and a comparison to control incubations is shown in Table I. Under these conditions 20-40% of the PUFA were lost with the largest decrease seen for 22:6. No loss of saturated fatty acids was noted. Subsequent incubation of liposomes in the presence of phospholipase A_2 resulted in the release of up to 50% of the peroxides which became associated with the free fatty acid pool. Incubation of oxidized liposomes without phospholipase A_2 revealed that 90% of the peroxides remained associated with the phospholipid pool. These results are detailed in Table II.

TABLE I

Conditions	O.D. 234 nm	O.D. 532 nm
A) control	0.015 $\pm$ 0.002	0.050 $\pm$ 0.008
B) Fe^{+2} + ascorbate	0.626 $\pm$ 0.094	0.383 $\pm$ 0.025
A) + control (15 min)	0.022 $\pm$ 0.010	0.055 $\pm$ 0.008
A) + PLase A_2 (15 min)	0.040 $\pm$ 0.012	0.088 $\pm$ 0.009
B) + control (15 min)	0.591 $\pm$ 0.101	0.341 $\pm$ 0.031
B) + PLase A_2 (15 min)	0.570 $\pm$ 0.090	0.371 $\pm$ 0.055

TABLE II

Treatment	Fraction (O.D. 234 nm) FFA	PL	% of total O.D. 234 as free fatty acids
control	0.002	0.023	8.0 $\pm$ 2.9
+ PLase A_2	0.012	0.019	38.7 $\pm$ 5.7
Fe^{+2} -ascorbate	0.008	0.125	6.0 $\pm$ 3.1
+ PLase A_2	0.070	0.068	50.7 $\pm$ 7.4

Significant hydrolysis of intact fatty acids was also noted in these experiments and a comparison of phospholipase A_2 -mediated release of various fatty acids for control vs. oxidized liposomes is shown in Figure 1. No effect on the hydrolysis of saturated fatty acids, and a minimal augmentation of mono- and dienoic fatty acids was found in the oxidized samples as compared to controls. The effects of oxidation on the release of polyunsaturated fatty acids, on the other hand, was quite pronounced. For example, as much as 13% of the remaining arachidonic acid was release by phospholipase A_2 in peroxidized liposomes as compared to 6% hydrolysis for control liposomes. Thus, polyunsaturated fatty acids appear to be particularly susceptible to phospholipase A_2 attack when membranous phospholipids are subjected to peroxidation.

Associated with the findings described above was a marked increase in the fluorescence polarization of DPH incorporated into the liposomes. Polarization values increased from control levels of 0.202 $\pm$ 0.018 to 0.289 $\pm$ 0.024 for peroxidized samples. Extraction of the lipids and probe into hexane revealed that equal amounts of DPH probe were incorporated into the control and peroxidized samples.

Fig. 1

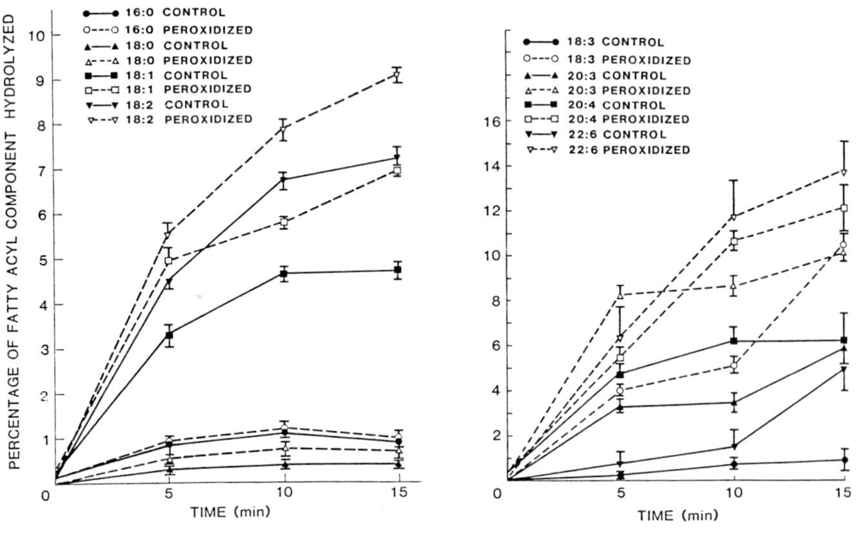

DISCUSSION

On the basis of studies using model membranes it appears that a significant augmentation of phospholipase A_2 activity is brought about through the peroxidation of the component phospholipids. This phospholipase A_2 activity is found to release oxidized fatty acids with a high order of selectivity such that a minimum quantity of peroxides remain in the membrane phospholipids. Associated with this release of oxidized fatty acids is the hydrolysis of significant quantities of intact fatty acids, where the order of hydrolysis correlates with the degree of fatty acid unsaturation. It is noteworthy that arachidonic acid is a major fatty acid affected, and suggests that relatively low levels of peroxidation in membranes may bring about a facile release of arachidonic acid. As similar phenomenon have been reported in vivo following oxidative injury (5,6,8,13), a connection between lipid peroxidation and eicosanoid synthesis is a tempting speculation and the regulatory effect on phospholipase A_2 during this process seems plausible.

A correlation between lipid peroxidation, phospholipase A_2 activity and membrane structure is also evident. Increases in the fluorescence polarization of DPH in peroxidized membranes suggests that peroxidation of membranes causes the order of membrane phospholipids to increase. The creation of membrane environments with order parameters different from the bulk of the membrane suggests that the structure of the membrane experiences widespread transitions and this may be the basis for increased phospholipase susceptibility. The preferred incorporation of arachidonic acid bearing phospholipids into these transitional regions, created by peroxidized lipids, may thereby predispose these phospholipids to phospholipase A_2 attack.

REFERENCES

1. Kagan, V.Y., Shvedova, A.A. and Novikov, K.N. Participation of phospholipases in the repair of photoreceptor membranes exposed to peroxide oxidation. Biofizika 2:279-283 (1978).

2. Yasuda, M. and Fujita, T. Effect of lipid peroxidation on phospholipase A_2 activity of rat liver mitochondria. Japan. J. Pharmacol. 27:429-435 (1977).

3. Sevanian, A., Muakkassah-Kelly, S.F. and Montestruque, S. The influence of phospholipase A_2 and glutathione peroxidase on the elimination of membrane lipid peroxides. Arch. Biochem. Biophys. 223:441-452 (1983).

4. Tan, K.H., Meyer, D.J., Belin, J. and Ketterer, B. Inhibition of microsomal lipid peroxidation by glutathione and glutathione transferase B and AA: Role of endogenous phospholipase A_2. Biochem. J. 220:243-252 (1984).

5. Weglicki, W.B., Dickens, B.F. and Mak, T. Enhanced lysosomal phospholipid degradation and lysophospholipid production due to free radicals. Biochem. Biophys. Res. Commun. 124:229-235 (1984).

6. Chan, P.H., Yurko, M. and Fishman, R.A. Phospholipid degradation and cellular edema induced by free radicals in brain cortical slices. J. Neurochem. 38:525-531 (1982).

7. McCord, J.M. Oxygen-derived free radicals in post-ischemic tissue injury. New Engl. J. Med. 312:159-163 (1985).

8. Sobel, B.E., Corr, P.B., Robison, A.K., Goldstein, R.A., Witkowski, F.X. and Klein, M.S. Accumulation of lysophosphoglycerides with arrythmogenic properties in ischemic myocardium. J. Clin. Invest. 62:546-553 (1978).

9. Beuge, J.A. Aust, S.D. Microsomal lipid peroxidation. In: Methods in Enzymology vol. 52 (S. Fleisher and L. Packer, eds.) pp. 302-310, Academic Press, New York (1978).

10. Shinitzky, M. and Barenholtz, Y.J. Biol. Chem. 249:2652-2660 (1974).

11. Wu, G.S., Stein, R.A. and Mead, J.F. Autoxidation of phosphatidylcholine liposomes. Lipids 17:403-413 (1982).

12. Bartlett, G.R. Phosphorus assay in column chromatography. J. Biol. Chem. 234:466-468 (1959).

13. Chien, K.P., Pfau, R.G. and Farber, J.L. Ischemic myocardial cell injury. Amer. J. Pathol. 97:505-530.

Superoxide and Superoxide Dismutase in Chemistry,
Biology and Medicine; G. Rotilio, editor

OXYGEN FREE RADICALS STIMULATE THE SYNTHESIS OF LIPID DERIVED MEDIATORS: A CRITICAL ROLE OF THIOL GROUP REDOX STATUS.

Isabelle Maridonneau-Parini, A. Fradin, L. Touqui and F. Russo-Marie
Unité des Venins, Unité associée Pasteur/INSERM U285, Institut PASTEUR, 75015 Paris, France.

INTRODUCTION

Interactions between oxygen free radicals and arachidonic acid metabolism can occur in various physiopathological conditions such as tissue ischemia, or inflammation in which oxygen free radicals are generated (1). We proposed to study the effect of intracellular oxygen free radical generated by phenazine methosulfate (PMS) (2) on the synthesis of lipid-derived mediators in a cell line (rat renomedullary interstitial cells, RMIC) in which the arachidonic acid metabolism has been well characterized (3).

We report in this paper that PMS induces the production of oxygen free radicals and increases the synthesis of pharmacologically active lipids. H_2O_2 rather than $O_2^{\bullet -}$ is involved in the prostaglandin E_2 (PGE_2) stimulating process which results, for one part, from the activation of phospholipase A_2 (PLA_2) activity and, for another part, from the cellular depletion of thiol groups which enhances cyclooxygenase activity.

METHODS

Monolayer tissue cultures of rat renomedullary interstitial cells were obtained as described previously (3) and incubated, for experimental procedures, in a PBS medium (pH 7.4 at 37°C). These cells synthesize mainly PGE_2 after PLA_2 stimulation (4). The superoxide anion production was measured spectrophotometrically by the SOD-inhibitable nitroblue tetrazolium (NBT) reduction in formazan using the wavelength 560 nm. In some experiments, the intracellular Cu-Zn SOD was inhibited by pre-incubating cells for 30 min with the copper-chelator diethyldithiocarbamate (DDC). PGE_2 was measured by radioimmunoassay and a lipoxygenase derived product, 15-HETE was detected by reverse phase-high performance liquid chromatography (5). The spectrophotometric measurement of intracellular soluble thiol groups was performed at 535 nm according to the method described by Saville (6). Lyso-platelet activating factor (lyso-PAF) was measured once converted by acetylation to PAF-acether which was assessed by its ability to aggregate rabbit platelets (7). Cellular protein measurement was performed using the technique described by Lowry (8).

RESULTS

In renomedullary interstitial cells, PMS (50 to 300 µM) induced the formation of superoxide anions ($O_2^{\bullet -}$) as demonstrated by the reduction of NBT which was inhibited by exogenous SOD (50 µg/ml, from bovine erythrocytes) and increased in cells pre-treated by DDC (0.5 or 1 mM). The arachidonic acid metabolism was not involved in the PMS-induced NBT reduction since indomethacin (10 μM); NDGA (10 μM); mepacrine (1 μM) and dexamethasone (1 μM) inhibitors of cyclooxygenase, lipoxygenase and phospholipase A_2 respectively, were unable to affect the formation of formazan. PGE_2 synthesis was increased in PMS-treated cells. A kinetic study showed that the PGE_2 formation reached the maximal value 30 min after the addition of 75 μM PMS (untreated cells: 2.4 $\pm$ 0.4 ng/mg of cellular protein; PMS-treated cells: 21.2 $\pm$ 1.3 ng/mg of cellular protein). In dose-response experiments (ranging from 1 to 1000 μM) the maximal stimulation was obtained for 100 μM PMS. The role of 75 μM PMS on the formation of lipoxygenase derivatives was also investigated. The synthesis of 15-HETE was about 2 fold enhanced and inhibited by 10 μM NDGA. Addition of SOD (20, 50, 100 μg/ml) in the incubation medium enhanced the stimulatory effect of PMS on the PGE_2 production whereas catalase (5 to 100 μg/ml) decreased it. The amount of soluble thiol groups was decreased in PMS-treated cells. Stimulation of PGE_2 synthesis was observed when cells were treated by 0.2 mM diamide, a thiol oxidative agent or by 75 μM PMS, an oxygen free radical generator. The activity of PLA_2 was determined in PMS-treated cells by measuring the formation of lyso-PAF. Its formation was 3 fold increased by 75 μM PMS and unaffected by 0.2 mM diamide. The formation of PGE_2 in PMS-treated cells was investigated in presence of added arachidonic acid in order to bypass its release from membrane lipids by PMS-stimulated PLA_2, thus reflecting the activity of cyclooxygenase. At low arachidonic acid concentrations (from 0.5 to 2 μg/ml), the cyclooxygenase activity was enhanced in presence of PMS, whereas for higher concentrations (10 μg/ml), values of PGE_2 in control and PMS experiments were not significantly different. Such a stimulating effect of PMS for low arachidonic acid concentrations was not reproduced when 75 μM PMS was substituted by the calcium ionophore A23187 (1 μM) which we had shown to activate PLA_2 to a similar extent than 75 μM PMS.

DISCUSSION

In this paper, we report that in RMIC in culture, an endogenous production of oxygen free radicals by PMS leads to an increase of PGE_2, a cyclooxygenase derived product, of 15-HETE, a lipoxygenase derived product, and of lyso-PAF, a PLA_2-derived product, precursor of PAF-acether. All these pharmacologically active lipids, result from phospholipase A_2 activation. The PMS stimulatory effect on prostaglandin synthesis is enhanced by SOD and decreased by catalase, suggesting the involvement of H_2O_2, as it has already been reported by other groups (9,10). Once formed, $O_2{\cdot}^-$, and H_2O_2 are involved in the generation of hydroxyl radical •OH which oxidizes thiol groups and unsaturated lipids (1). Therefore, cellular depletion in thiol groups observed in PMS treated cells may be the result of •OH formation plus an increase of GSH-peroxidase activity. In physiological conditions, the cyclooxygenase activity is controlled by GSH-peroxidase, which removes both H_2O_2 and lipid peroxides (11). A decrease in the cellular GSH content reflects a modified GSH-peroxidase activity which leads to the accumulation of peroxides with a subsequent stimulation of cyclooxygenase (11). Indeed diamide, a thiol oxidative agent mimicked the effect of PMS on PG synthesis whereas this same oxidative agent had no effect on lyso-PAF release, indicating that PLA_2 was not activated by the same mechanism. Nevertheless, the stimulation of lyso-PAF in PMS-treated cells indicates that oxygen free radicals also stimulate PLA_2. The mechanism for PLA_2 activation is not well understood, it could be the consequence of membrane lipid peroxidation. Indeed there is circumstantial evidence to support the association of lipid peroxidation and phospholipase activation, such as the production of membrane phase discontinuities at the sites of peroxidation, the increased influx of calcium and the creation of a negative charge surface (12). The biochemical event linked to the stimulation of lipoxygenase derived products was not investigated in the present study. It might be the consequence of the activation of PLA_2 and/or a direct effect on lipoxygenase activity as suggested by the previously reported similarity of this enzyme with cyclooxygenase.

In conclusion, the intracellular generation of oxygen free radicals enhances the synthesis of lipid derived mediators through different mechanisms of enzyme activations. The increase in PG synthesis results in part from PLA_2 activation and in other part from cyclooxygenase activation.

REFERENCES

1. Freeman BA, Crapo JD (1984) In : Rubin E, Damjanov I (eds) Advances in the Biology of Diseases. Williams and Wilkins, Baltimore, London, pp 26-40
2. Maridonneau I, Braquet P, Garay RP (1983) J Biol Chem 258:3107-3113
3. Russo-Marie F, Paing M, Duval D (1979) J Biol Chem 254:8498-8505
4. Dunn MJ, Staley RS, Harrison M (1976) Prostaglandins 12:37-42
5. Maridonneau-Parini I, Fradin A, Touqui L, Russo-Marie F (1985) Biochem Pharmacol (in press)
6. Saville B (1958) Analyst 83:670-674
7. Touqui L, Jacquemin C, Dumarey C, Vargaftig BB (1985) Biochim Biophys Acta 833:111-118
8. Lowry OH, Rosebrough NJ, Farr AL, Randall RJ (1951) J Biol Chem 193:265-271
9. Baud L, Nivez MP, Chansel D, Ardaillou R (1981) Kidney Int 20:332-339
10. Taylor L, Menconi MJ, Polgar P (1983) J Biol Chem 258:6855-6859
11. Lands WEM, Kulmacz RJ, Marshall PJ (1984) In : Pryor WA (ed) Free radicals in Biology , Academic Press, New york, 6:pp39-61
12. Scott JA (1984) J theor Biol 111:659-665

Superoxide and Superoxide Dismutase in Chemistry, Biology and Medicine; G. Rotilio, editor

MONOSACCHARIDE AUTOXIDATION: A PROSPECTIVE ROLE IN DISEASE PROCESSES

PAUL J THORNALLEY

Department of Pharmaceutical Sciences, Aston University, Aston Triangle, Birmingham B4 7ET, United Kingdom.

INTRODUCTION

Following the discovery of superoxide dismutase, much research effort has been devoted to the study of the biological distribution and role of the enzyme, and to the potential sources of superoxide formation in biological systems. After two decades of endeavour in superoxide/superoxide dismutase-related research, it now appears that superoxide dismutase is found widespread in aerobic life, and the dismutase activity is necessary to prevent the toxic effects of superoxide formation in biological metabolism. Superoxide is formed by many biological processes: by phagocytosing polymorphonuclear leukocytes, by subcellular organelles (e.g. chloroplasts and mitochondria), by isolated enzymatic reactions (e.g. xanthine/xanthine oxidase), and by the spontaneous autoxidation of reducing metabolites (e.g. leukoflavins, hydroquinones and thiols) (1). It has recently been disclosed that superoxide is also formed by the slow non-enzymatic reaction of monosaccharides with oxygen (2,3) - herein termed monosaccharide autoxidation.

Monosaccharides undergo a slow reaction with oxygen via their ene-diol tautomer. The major products isolated are hydrogen peroxide (formed from intermediary superoxide) and α-oxoaldehydes.

$$RCH(OH)CHO \longrightarrow RC(OH)=C(OH)H \xrightarrow{O_2} RCOCHO + (O_2^{-}),\ H_2O_2$$

Superoxide dismutase rapidly detoxifies the superoxide formed in this process, yet there remains the potentially noxious α-oxoaldehyde product - this may be detoxified by the glyoxalase system.

In contrast to the historical development of superoxide dismutase, the glyoxalase system has been studied since 1913 (4), yet its biological function and the source(s) of its substrates - α-oxoaldehydes - remains unclear. It comprises two enzymes, glyoxalase I and II, and a catalytic amount of reduced glutathione (5). Glyoxalase I (EC4.4.1.5) catalyses the formation of α-hydroxy-S-glutathione esters from reduced glutathione and α-oxoaldehyde: $RCOCHO + GSH \longrightarrow RCH(OH)CO\text{-}SG$. Glyoxalase II (EC3.1.2.6) is a thiolesterase which catalyses the hydrolysis of product from the glyoxalase I reaction to α-hydroxy acid with the regeneration of reduced glutathione:

$RCH(OH)CO\text{-}SG + H_2O \longrightarrow RCH(OH)CO_2H + GSH$

In this report, I describe the involvement of superoxide in monosaccharide autoxidation, the importance of the glyoxalase system in the detoxication of α-oxoaldehyde autoxidation product, and the prospective role of monosaccharide autoxidation in disease processes.

MECHANISM OF MONOSACCHARIDE AUTOXIDATION

A simple monosaccharide, RCH(OH)CHO, may enolise to an ene-diol, RC(OH)=C(OH)H. The ene-diol reduces oxygen to superoxide and is itself oxidised to a semidione radical, RC(O•)=C(OH)H. Further electron transfer may occur to form α-oxoaldehyde and hydrogen peroxide (6).

$RC(OH){=}C(OH)H + O_2 \rightarrow RC(O^{\bullet}){=}C(OH)H + O_2^- + H^+ \rightarrow RCOCHO + H_2O_2$

The hydrogen peroxide formed may be reduced by the ene-diol to hydroxyl radicals, yet evidence for this reaction is only detected above pH 8. At physiological pH, the parent monosaccharide scavenges the hydroxyl radicals and only the 1-hydroxyalkyl radical, formed by the scavenging reaction, is detected (7). These processes are catalysed by trace metal ions (ferric, cupric) but only up to a maximum limit controlled by the rate of enolisation of the parent monosaccharide. The most reactive physiological monosaccharides are the trioses, glyceraldehyde-3-phosphate and dihydroxyacetone phosphate, which are both oxidised to hydroxypyruvaldehyde phosphate $Na^{+-}OPO_3CH_2COCHO$ (6).

THE BIOLOGICAL EFFECTS OF MONOSACCHARIDE AUTOXIDATION

Monosaccharide autoxidation produces a low flux of superoxide and hydrogen peroxide: in cells, this is expected to produce a small stimulation in the hexose monophosphate shunt - this has been observed in glyceraldehyde-treated human red blood cells (8). Autoxidising monosaccharides may also: nonenzymatically oxidise NADPH, oxidise oxyhaemoglobin, and provide α-oxoaldehyde substrates for the glyoxalase system (8,9,10). α-Oxoaldehydes may otherwise cross-link protein, DNA and RNA - when undergoing replication (11), and may also be involved in the formation of browning pigments (12).

The autoxidation of monosaccharides is an extremely slow reaction yet the chronic exposure of biological tissue to this process may suggest a role in diseases associated with the ageing process - particularly where the concentrations of glycolytic intermediates are above normal, e.g. diabetic pathogenesis (13). Simple monosaccharides also exhibit oxygen-dependent mutagenic and carcinostatic activities (11,14) - this is thought to be due to the effects of monosaccharide autoxidation (15).

It is now clear that a novel role for the glyoxalase system is to protect cells from the oxoaldehydes formed in monosaccharide autoxidation - hydroxypyruvaldehyde phosphate has been detected in cell systems (16). It is therefore important that glycolytic enzymes operating through ene-diol intermediates have evolved to exclude trace metal ions and oxygen from the active site to prevent α-oxoaldehyde formation and inactivation of the enzyme by monosaccharide autoxidation (17).

ACKNOWLEDGEMENTS

PJ Thornalley thanks the Medical Research Council, UK for financial support for his research.

REFERENCES

1. Fridovich I (1975) Ann Revs Biochem 44:147

2. Kasimura N, Morita J and Komano T (1979) Carbohydr Res 70:C3-C7

3. Wolf SP, Crabbe MJC and Thornalley PJ (1984) Experientia 40:244

4. Dakin HD and Dudley HW (1913) J Biol Chem 155:243

5. Racker E (1951) J Biol Chem 190:685

6. Thornalley PJ, Wolf SP, Crabbe MJC and Stern A (1984) Biochim Biophys Acta 797:276

7. Thornalley PJ and Stern A (1985) Carbohydr Res 134:191

8. Thornalley PJ and Stern A (1984) Biochim Biophys Acta 804:308

9. Thornalley PJ, Wolf SP, Crabbe MJC and Stern A (1984) Biochem J 217:615

10. Wolf SP and Crabbe MJC (1985) Biochem J 226:625

11. Schauenstein E, Esterbauer H and Zollner H (1977) In: Aldehydes in Biological Systems, Pion Ltd, London, pp 101-111

12. Monnier VM and Cerami A (1983) Biochim Biophys Acta 760:97

13. Warren S (1962) The Pathology of Diabetes Mellitus, Academic Press, New York.

14. Yamaguchi T and Nakagawa K (1983) Agric Biol Chem 47:2461

15. Thornalley PJ (1985) Environmental Health Perspectives (in press)

16. Cogoli-Greuter M and Christen P (1981) J Biol Chem 256:5708

17. Christen P, Anderson TK and Healy MJ (1974) Experientia 30:603

Superoxide and Superoxide Dismutase in Chemistry, Biology and Medicine; G. Rotilio, editor

IS ACTIVATION OF THE ALPHA AND BETA ADRENERGIC RECEPTOR A REDOX PHENOMENON?

D.A. PETERSON AND J.M. GERRARD

Veterans Administration Medical Center, Minneapolis, MN 55417 U.S.A.; and Manitoba Institute of Cell Biology, Winnipeg, Manitoba RA3EOV9 Canada

INTRODUCTION

The reducing properties of catacholamines have been known for many years. We have investigated the possible relevance of these reactions to their biologic activities in several systems. Our initial studies were with model systems for prostaglandin synthetase, a heme containing enzyme, which we felt required ferrous heme for activity. Our results were consistent with epinephrine activating the enzyme by reducing ferric heme to ferrous heme.[1]

The alpha two adrenergic receptor of the platelet was next investigated. A series of agonists were reducing agents, as measured by heme reduction, while antagonists were inactive. Several heme chelating agents inhibited alpha two initiated platelet aggregation. Thus, it was suggested that alpha two adrenergic activation is initiated by a reduction of a heme group in the receptor while agonists inhibit by occupying the receptor but not effecting the necessary reduction.[2]

The beta adrenergic receptor adenyl cyclase complex consists of the receptor, a regulatory protein, and adenyl cyclase.[3] Following activation of the receptor-regulatory protein complex by a beta agonists, there is exposure of a sulfhydryl group which is necessary for adenyl cyclase activation. The latter protein complex also contains a sulfhydryl group which is essential for activity.[4,5] We used a model system to evaluate whether beta adrenergic agonists could activate the receptor-adenyl cyclase complex by reducing a disulfide bond directly or indirectly.

MATERIALS AND METHODS

Organic chemicals were obtained from Sigma Chemical Co. DTNB reduction was followed spectrophotometrically at 412 nm. Cytochrome c reduction was followed spectrophotometrically at 550 nm. Reductions were carried out in 50 mM TRIS buffer, pH 7.4 at room temperature.

RESULTS AND DISCUSSION

Potent beta agonists were able to reduce a relatively small amount of DTNB under the above experimental conditions. This reduction was markedly enhanced in the presence of ferric chloride (Table I).

TABLE I

REDUCTION OF DTNB BY BETA AGONISTS AND ANTAGONISTS (300 μM) AT 10 MINUTES

Compound	Reduction x 10^{-7} M
Isoproterenol	12.2 $\pm$ 2.4
Epinephrine	10.5 $\pm$ 1.3
Norepinephrine	10.5 $\pm$ 1.2
Metaproterenol	2.43 $\pm$ 0.2
Synephrine	4.78 $\pm$ 1.0
Propranolol	0
Beta-phenylethylamine	0
Dichloroisoproterenol	0

While the amount of reduction of DTNB is small in this system, one could expect this could be increased if the disulfide groups, metal, and agonist, were tightly coupled, a situation which could occur as a result of tight binding of the agonist to the receptor. In support of this concept, beta agonists were potent reducing agents of cytochrome c, while beta antagonists were inactive (Table II). Ellman has previously observed that heme is quantitatively oxidized while reducing DTNB.[6] The above observations define how each step in the reductive sequence might be effected in high yield under optimal conditions.

TABLE II

REDUCTION OF CYTOCHROME C BY BETA-ADRENERGIC AGONISTS

Agonist (30×10^{-6} M)	Cytochrome c Reduced ($\times 10^{-6}$ M)
Isoproterenol	11.5 $\pm$ 1.5
Epinephrine	10.5 $\pm$ 2.2
Norepinephrine	11.8 $\pm$ 4.5

In further support of this mechanism of receptor activation is the observation that cytochrome c reduction by isoproterenol is inhibited by propranolol (Table III). The molar range of this inhibition is consistent with what is seen clinically with intravenous doses of these medications.

TABLE III
CYTOCHROME C REDUCTION (μM)

Isoproterenol (6.1 μM)	6.8 ± 0.8
Isoproterenol (6.1 μM) with propranolol (6 x 103 μM)	5.5 ± 0.4

In summary:

1) At low concentrations of isoproterenol approximately 1 mole of cytochrome c is reduced per mole of agonist (this is also true of other potent agonists).

2) Beta-adrenergic agonists but not antagonists can reduce the disulfide bond of DTNB.

3) Less than 0.01 mole of DTNB is reduced per mole of isoproterenol.

4) In view of the established importance of a disulfide bond to protein configuration, we speculate that reduction of a disulfide bond in the beta-adrenergic receptor complex by beta-adrenergic agonists is a critical step in transmission of the agonist signal.

5) Additional studies are required to establish how agonist activity is closely coupled to reduction of disulfide bonds (i.e., possibly via an intermediate heme reduction step).

6) Combined with our previous work on the alpha-adrenergic receptor, these results suggest a general molecular mechanism for adrenergic receptor activation.

REFERENCES

1. Peterson DA, Gerrard JM, Rao GHR, White JG (1980) Prostaglandins and Medicine 5:357-364
2. Peterson DA, Gerrard JM, Glover SM, Rao GHR, White JG (1982) Science 215:71-73
3. Lefkowitz RJ, Caron MG, Stiles G (1984) New Eng J Med 310:1570-1579
4. Stadel JM, Lefkowitz RJ (1979) Mol Pharmacol 16:709-718
5. Vanquelin G, Goltari S, Strosberg AD (1980) Mol Pharmacol 17:163-171
6. Ellman GL (1959) Arch Biochem Biophys 82:70-77

Superoxide and Superoxide Dismutase in Chemistry,
Biology and Medicine; G. Rotilio, editor

THE POSSIBLE PARTICIPATION OF BETA-CAROTENE IN THE DEFENCE MECHANISM AGAINST SUPEROXIDE IN PLANT CELLS.

HD RABINOWITCH AND D SKLAN, Faculty of Agriculture, Hebrew University, Israel.

INTRODUCTION

Beta-carotene, together with many other carotenoids, can quench singlet oxygen to its relatively harmless ground state level. In addition, carotenoids may prevent certain free-radical-induced chain reactions, thus protecting the green cells against photo-oxidative damage (9).

Superoxide, is slowly generated in the chloroplasts during normal photosynthetic activity (1). There, it is efficiently inactivated by the enzyme superoxide dismutase (SOD), which is concentrated mainly in the chloroplasts. When the photosynthetic electron flow is impaired, superoxide production is accelerated, then the amount of SOD present in the cells may not be sufficient to scavenge the sudden increase in superoxide concentration. Consequently, severe damage to cell components may occur, followed by cell death.

We hereby provide evidence that beta-carotene reacts with the superoxide *in vitro*, and in addition, we show that the pigment may provide the green cells with some protection against the damage caused by the superoxide.

MATERIALS AND METHODS

In vitro procedures Pure beta-carotene was dispersed in water using Tween 80 (3), and thereafter was added to either of the two non-related systems which were used to produce the superoxide. (4, 10). The change in the absorption spectra was recorded, and the reaction mixtures were extracted with methanol-petroleum ether (PE) and separated by either TLC (11) or by HPLC on a reversed phase C-18 column, eluting with 5% H_2O in methanol (12).

Plant material The two youngest, fully expanded leaves of two tomato isolines (6th backcross) differing in their high-pigment (*hp*) locus (2), were used throughout. Ten mm diameter discs from freshly detached leaves, were dipped in 1% aqueous solution of diethyldithiocarbamate (DDTC) (13). After 2 h dark and 12 h light, the leaf discs were washed thrice with distilled water and transferred to 5×10^{-4} M methyl-viologen (MV) solution until the first changes in pigmentation intensity became visible. Control samples from opposite leaflets were dipped in distilled water. In another experiment, similar freshly cut leaf discs were dipped in 5×10^{-3} M MV.

Assays Chlorophyll (Chl) and beta carotene, were extracted by the cold acetone procedure (6) and with methanol-PE mixture, (11). The amounts of these pigments were calculated from spectrophotometric readings (6, 7). Protein and superoxide dismutase were measured as described (4, 5).

RESULTS AND DISCUSSION

In vitro tests It is apparent from Table 1 that the presence of carotene in the reaction mixtures, reduced the rates of both NBT and cytochrome c reduction. A similar if greater effect was produced by SOD, and addition of both SOD and beta-carotene resulted in a further decrease in the reaction rates, indicating a possible reaction between these two compounds. Tocopherol, mannitol, dimethylfuran, catalase and H_2O_2 had no effect on the reaction rates, with or without carotene addition (Table 1). None of these additives showed interaction with SOD (not shown). Absorbance curves of beta carotene in a xanthine-xanthine oxidase system without the cytochrome c, showed both a decrease in intensity, and transient changes in the short UV spectrum (not shown). Following the extraction of the reaction products, their separation on TLC with 20% acetone in PE, revealed compounds which remained at the origin of the plates. HPLC, with a reversed phase column, revealed the presence of substances eluting before beta carotene (not shown). It appears that none of the following: singlet oxygen, hydroxy radicals and H_2O_2 are generated in these systems. It is therefore proposed that the changes in both the absorption spectrum and the structure of beta carotene, result from a direct reaction with O_2^-.

TABLE 1. RATE OF REDUCTION OF NITROBLUE TETRAZOLIUM (NBT) AND CYTOCHROME C BY SUPEROXIDE AS AFFECTED BY VARIOUS ADDITIVES.
Results are means of 5 replications ± SD. Numbers within column not followed by the same letter differ significantly ($P<0.05$).

Additive	NBT	Cytochrome c
	-------- Changes in OD	--------
None	40.8 ± 2.2a	17.5 ± 1.8a
SOD 10ug/ml	8.4 ± 0.3c	5.4 ± 0.4c
Carotene 80ug/ml	18.3 ± 1.5b	13.1 ± 0.6b
Carotene 80ug/ml + SOD 10ug/ml	5.2 ± 0.2d	3.2 ± 0.2d
Tocopherol 100ug/ml	40.9 ± 2.3a	17.4 ± 1.1a
Carotene 80ug/ml + Tocopherol 500ug/ml	17.8 ± 0.8b	13.4 ± 1.0b
Mannitol 0.1M	40.3 ± 2.1a	17.8 ± 1.2a
Carotene 80ug/ml + Mannitol 0.1M	17.5 ± 1.1b	13.4 ± 1.1b
Catalase 8.25 ug/ml	40.2 ± 3.0a	17.7 ± 1.6a
Carotene 80ug/ml + Catalase 8.25ug/ml	17.6 ± 1.2b	13.3 ± 0.9b
Carotene 80ug/ml + H_2O_2	-	13.5 ± 0.9b
Dimethylfuran 1.0mM	-	17.8 ± 1.4a
Carotene 80ug/ml + Dimethylfuran 1.0mM	-	13.5 ± 0.8b

Plant experiments The above reactions were performed in an aqueous dispersion of beta carotene, using detergent for producing micellar suspension, and this precludes accurate determination of Km. It may be speculated that similar effects exist *in vivo* and may play a significant biological role. Leaves of the two plant genotypes, did not differ significantly in their SOD specific activity. The '*hp*' leaves, however, contained 25-30% higher levels of the two major plastid pigments than those of the normal type. When treated with MV at $5x10^{-3}$ M, the leaf discs of the latter were always the first to show reduction in pigmentation intensity. The amount of Chl a in the treated leaf discs was 51±4.6 and 73±6.3% of the non-treated controls for the normal and the '*hp*' genotypes, respectively. DDTC-treated leaf discs with approximately 60% SOD activity of the respective controls, were treated with $5x10^{-4}$ M MV. Leaf discs of the normal type were the first to show signs of decreased pigmentation intensity. The amounts of Chl a, were 47±5.1 and 67±5.4% of the non-treated controls for the normal and '*hp*' genotypes, respectively. At that time, no significant reduction in the Chl content was found in those leaf discs not previously exposed to DDTC.

MV is readily reduced within the green cells and then reacts with O_2 to generate superoxide, and SOD provides protection (8). Leaf discs, with high levels of beta carotene, were more resistant to photo-oxidative damage caused by MV then those of the normal type in both with normal or with reduced SOD activity. It is therefore proposed that beta carotene may provide some protection to the green cells against injuries caused by O_2^-.

REFERENCES

1. Asada K, Kanematsu S, Takahashi M, Kono Y (1976) In: Yasunobu KT, Mower MF, Hayaishi O (eds) Iron and Copper Proteins. Plenum, NY, pp.551-565
2. Baker LR, Tomes ML, (1964) Proc. Amer. Soc. Hort. Sci. 85:507-513.
3. Ben-Aziz A, Grossman S, Ascarelli I, Budowski P (1971) Phytochem. 10:1445-1452.
4. Beauchamp C, Fridovich I (1971) Anal. Biochem. 44:276-287.
5. Bradford MM (1976) Anal. Biochem. 72:248-254.
6. Bruinsma J (1967) Photochem. Photobiol. 2:241-249.
7. Foppen FH (1971) Chromtogr. Rev. 14:133-298.
8. Harper DB, Harvey BMR (1978) Plant Cell Environ. 1:211-215.
9. Krinsky NI,(1979) Pure & Applied Chem. 51:649-660.
10. McCord JM, Fridovich I (1969) J. Biol. Chem. 244:6049-6055.
11. Rabinowitch, H. D., Budowski, P. and Kedar, N. 1975. Planta 122:91-97.
12. Sklan D, Halevy O (1984) British J. Nutr. 52:107-114.
13. Tanaka K, Sugahara K, (1980) Plant & Cell Physiol. 21:601-611.

PHARMACOLOGICAL AND CLINICAL ASPECTS OF SUPEROXIDE AND SUPEROXIDE DISMUTASE

Superoxide and Superoxide Dismutase in Chemistry,
Biology and Medicine; G. Rotilio, editor

ORGOTEIN, THE DRUG VERSION OF BOVINE COPPER, ZINC SUPEROXIDE DISMUTASE: III. PROPERTIES OF INJECTABLE SUPEROXIDE DISMUTASE.

M.G.P. SAIFER, L.D. WILLIAMS, P.H.-C. DANG, R. SOMACK, C. SLOAN AND Q.R. MIRANDA

DDI Pharmaceuticals, Inc., Mountain View, CA 94043 U.S.A.

INTRODUCTION

Superoxide dismutase (SOD) for parenteral administration is marketed in five European countries, where it is prescribed principally for treatment of musculoskeletal inflammation, especially osteoarthritis. Safety and efficacy have been demonstrated in animals (1,2) and in clinical studies. Appropriately prepared SOD is essentially nontoxic at clinically useful doses, with rare hypersensitivity reactions being the only noteworthy drug-related side-effect reported.

Quality specifications for orgotein have been published previously (1); however, those specifications are inadequately stringent to assure the quality of SOD for safe and effective large-scale parenteral use in humans. Recognition of the potential importance of pharmacologically active contaminants has led to more stringent specifications, as well as acceptance limits based on animal tests for irritancy and sensitization potential (3).

The molecular weight of superoxide dismutase can influence its bioavailability after injection (4,5,6,7) and might alter its effectiveness and safety. The distribution of molecular weights of SOD in preparations of purified bovine copper, zinc superoxide dismutase and some pharmacologic implications of size heterogeneity are discussed below.

MATERIALS AND METHODS

Superoxide Dismutase

Superoxide dismutase was prepared from USDA inspected, bovine liver by previously described methods (8) which avoid the use of organic solvents. SOD was over 99.5% pure by SDS-polyacrylamide gradient gel electrophoresis and by HPLC on Waters' I-125 columns. SOD was stored as a frozen solution except as otherwise indicated.

HPLC Size-exclusion Chromatography

I-125 columns from Waters Associates (0.78 x 30 cm) were equilibrated with 0.025 M phosphate buffer (pH 6.8) and run at 0.5 ml/min at 23°C. Columns were monitored at 214 nm. Integration was performed by a Spectra-Physics model 4270 integrator.

Bioavailability and Serum Clearance

SOD was assayed in serum of mice at intervals after tail vein injection. Total bovine SOD was measured by immunoassay. In order to differentiate the various forms of bovine SOD in serum, the method of Beauchamp and Fridovich (9) as adapted to thin film agarose gels (1) was used. Quantitation was by visual comparison to a series of standards.

Immunoassay

A solid-phase enzyme-linked immunoassay used rabbit anti-SOD bound to polystyrene tubes. SOD in samples competed with SOD covalently linked to horse-radish peroxidase (HRP) for binding sites on coated tubes. Bound HRP was measured using o-phenylenediamine as indicator. Above 100 ng/ml, results were similar to total SOD measured by the Beauchamp-Fridovich method.

RESULTS

Aggregation of Lyophilized SOD

When SOD was subjected to lyophilization in the absence of sucrose, aggregates formed during storage. The principal new species had the apparent molecular volume of dimerized SOD (63,000 Daltons). The amount of dimer formed increased with the time of storage after lyophilization and the storage temperature (Figure 1 and Table I). The extent of dimerization depended on the pH of the solution which was lyophilized (Table II).

TABLE I
DIMERIZATION OF LYOPHILIZED SOD DURING STORAGE AT ROOM TEMPERATURE

STORAGE	PERCENT DIMER	
DAYS	W/SUCROSE	NO SUCROSE
0	0.23	0.35
2	0.25	0.48
5	ND	0.90
7	ND	1.20
11	ND	1.70
25	0.30	2.20
34	0.30	3.50

TABLE II
EFFECT OF pH ON DIMERIZATION OF LYOPHILIZED SOD STORED 15 DAYS AT ROOM TEMP. AND 40°C.

	PERCENT DIMER	
pH	RT	40°C.
4.5	1.1	2.1
5.0	1.2	2.5
6.0	1.2	2.7
7.0	1.3	2.8
8.0	1.3	3.1

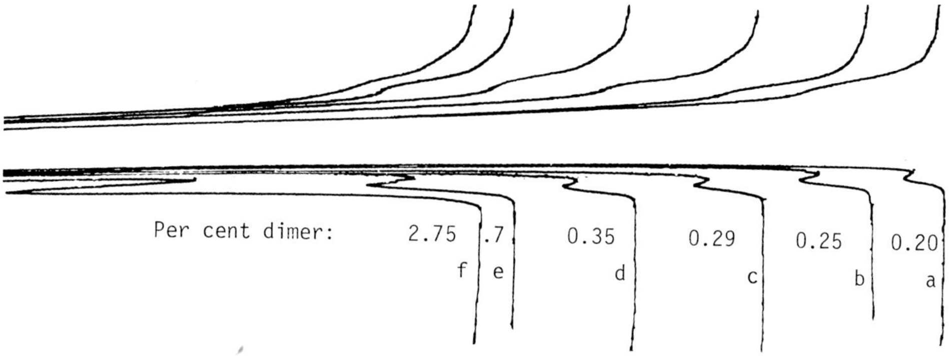

Fig. 1. Typical HPLC chromatograms of SOD stored as a frozen liquid (a) or lyophilized (b-f). Freshly thawed (a); lyophilized with sucrose, then 48 hrs. at 40°C (b) or 60°C (c); lyophilized without sucrose, then 0 hrs. (d) and 48 hrs. at 40°C (e) or 60°C (f).

When 2 mg Sucrose USP for each 1 mg protein was included in the solutions to be lyophilized, the extent of dimerization upon lyophilization was reduced and the tendency to form dimers during storage was nearly eliminated (Figure 1 and Table I).

Clearance of Aggregated SOD from Serum

Native SOD is cleared from the blood by the kidneys with a half-life which in rats is approximately 6 minutes (1). SOD which has been dimerized by lyophilization shares with chemically cross-linked SOD (4) an extended persistence in the serum (Table III). In mice, SOD with a 63,000 molecular weight displayed a half-life of one hour or more, i.e. over ten times that of the native enzyme.

TABLE III

CLEARANCE RATES FOR NATIVE AND DIMERIZED SOD IN MICE

Form of SOD	Interval	Half Life
Native	0-60 min.	4 min.
Dimer	5-180 min.	60-100 min.
	180-2900 min.	400-600 min.

DISCUSSION

While intravenous administration of orgotein is not commonly employed clinically, many published pharmacology studies have used intravenous administration of SOD in laboratory animals. Since most commercially available SOD preparations are lyophilized in the absence of sucrose or other effective stabilizers, their dimer contents are expected to be high and variable. The difference in serum clearance rates between the native and dimerized forms will result in dimer eventually becoming the predominant circulating species. In fact, with the difference in clearance rates reported here, dimer will represent more than 50% of the circulating SOD within an hour for any preparation initially containing 0.4% dimer or more. The higher the initial dimer content, the sooner dimer will become predominant.

In addition to its demonstrated effect on the rate of clearance of SOD activity from the circulation, it would be expected that the state of aggregation of SOD might influence the rate of SOD absorption from injection sites. Aggregation might also influence the penetration of SOD into its site of therapeutic action. Aggregation of proteins is known to enhance their immunogenicity (10). These possibilities deserve further study.

REFERENCES

1. Huber W, Saifer MGP (1977) In: Michelson AM, McCord JM, Fridovich I (eds) Superoxide and Superoxide Dismutases. Academic Press, London, pp 517-536
2. Huber W, Menander-Huber KB, Saifer MGP, Dang PH-C (1977) In: Willoughby DA, Giroud JP, Velo GP (eds) Perspectives in Inflammation: Future Trends and Developments. University Park Press, Baltimore, pp 527-540
3. Carson S, Vogin EE, Huber W, Schulte TL (1973) Toxicol Appl Pharmacol 26:184
4. Huber W, Saifer MG, Williams LD (1977) Cross-linked Orgotein, US Patent 4,009,267
5. McCord JM, Stokes SH, Wong K (1979) In: Weissmann G, et al (eds) Adv in Infl Res 1:273
6. Wong K, Cleland LG, Poznansky MJ (1980) Agents and Actions 10:231
7. Pyatak PS, Abuchowski A, Davis FF (1980) Res Comm Chem Path Pharm 29:113
8. Huber W, Chow SH, Saifer MG (1974) Enzymatic Treatment of Protein Mixtures Containing Orgotein, US Patent 3,806,411
9. Beauchamp C, Fridovich I (1971) Anal Biochem 44:276
10. Bellanti JA (1971) Immunology, Saunders, Philadelphia, pp 167-168

Superoxide and Superoxide Dismutase in Chemistry,
Biology and Medicine; G. Rotilio, editor

QUALITY CRITERIA OF Cu/Zn-SUPEROXIDE DISMUTASE (SOD) FOR CLINICAL USE.

DEREK SAUNDERS, WOLFGANG BLOCK, SUNG-MAN KIM, FRITZ ÖTTING, KLAUS WELLMANN and LEOPOLD FLOHE.

Center of Research, Grünenthal GmbH, Zieglerstraße 6, D-5100 Aachen (FRG)

The use of any protein for the treatment of disease requires stringent requirements of purity to be met. This is particularly true of Cu/Zn superoxide dismutase used to reduce deleterious effects of superoxide radicals in inflammations, as these conditions would be aggravated by application of impure preparations. Quality control of the enzyme isolated from beef tissue needs not only to confirm the integrity of the enzyme, but also to define the levels of impurities present: obviously these levels have to be extremely low.

The enzyme is analysed by various means. Specific dismutase activity is measured by the standard xanthine oxidase/cytochrome c assay (1); only preparations with > 3000 U/mg are acceptable. Copper and zinc metal contents are measured, as are spectral characteristics. The latter are determined in part by the metal ion ligands, which are in turn responsible for the activity. Quantitative SDS-PAGE (Fig. 1a) and exclusion HPLC using a Waters I-125 column (Fig. 1b) examine the molecular homogeneity of the enzyme, and ensure

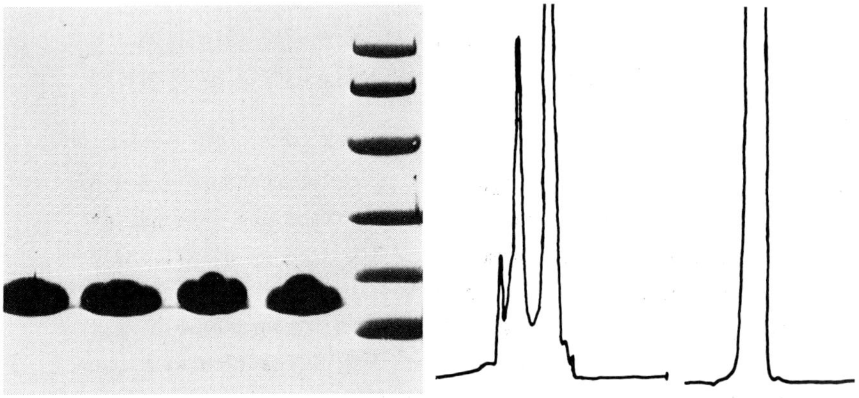

Fig. 1. (a) SDS-PAGE gradient (4-20% PAA) gel electrophoresis of reduced beef SOD
(b) Exclusion HPLC of beef SOD using I-125 column (Waters), crude and pharmacological quality.

that aggregates, which are readily formed on storage or mistreatment, are absent, as are any other visible contaminants.

Agarose gel electrophoresis (Fig. 2a) and thin-layer isoelectric focussing (Fig. 2b) define the electrical homogeneity of preparations. It is well known that Cu/Zn superoxide dismutase can exist in a variety of electrophoretically distinct forms of the same molecule. Nevertheless, methods which isolate only one form offer lower risks of contamination with other substances.

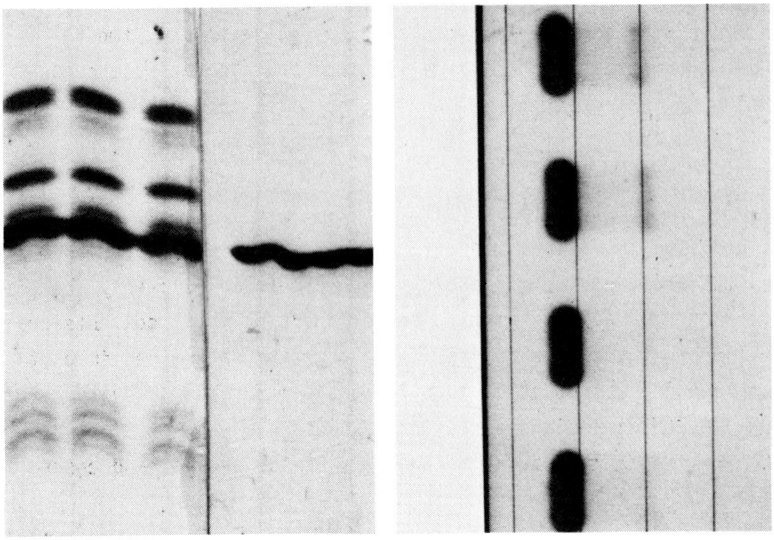

Fig. 2. (a) Agarose gel electrophoresis at pH 8.5, and (b) thin layer isoelectric-focussing (pH 4 - 6) of partially and fully purified beef SOD.

Specific contaminants are also investigated. Immunoelectrophoresis against polyclonal antibodies to whole blood extracts and to fractions obtained at very early stages of the purification demonstrate the absence of all antigens other than the required enzyme (Fig. 3). Proteases are never detected in our preparations, although other methods give materials with considerable residual proteolytic activity. Carbohydrates are detected by reaction with phenol and sulphuric acid at early stages of purification but not with the final material; such molecules can be surprizingly intransigent to removal. The sterility and absence of pyrogens is assessed by accepted bacteriological and pharmacological assays.

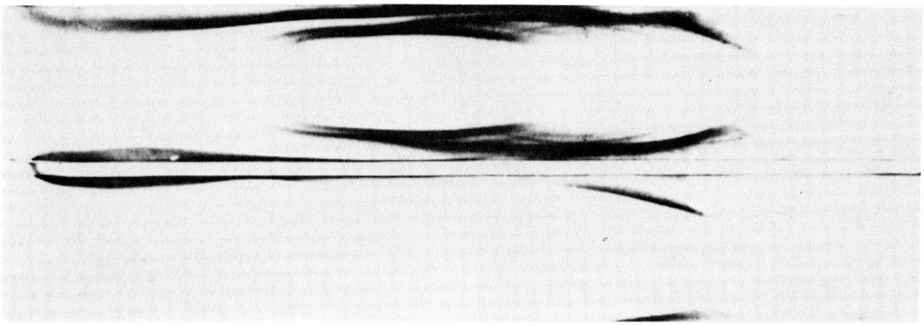

Fig. 3. Immunoelectrophoresis of (a)crude and (b) pharmacological quality beef SOD.Antibody raised in rabbits to crude beef SOD.

To ensure that preparations are stable, various relevant quality parameters are measured as a function of time at normal and elevated temperatures. Traces of partially denatured enzyme, which would not be readily observed by the methods described above, act as denaturing catalysts, and substantially reduce the shelf-life.

We have assured ourselves of the reproducibility and high performance of our purification method and the quality and stability of our preparation over several years. The incidence of adverse reactions to our Peroxinorm^{R} is so low that it cannot be correlated with any quality parameter. It is known however that less stringently controlled preparations lead to increased episodal risks, which justifies the very high quality of our beef Cu/Zn superoxide dismutase preparation.

REFERENCES

McCord J, and Fridovich I. (1969) J Biol Chem 244: 6049 - 6055.

Superoxide and Superoxide Dismutase in Chemistry, Biology and Medicine; G. Rotilio, editor

TEN YEARS OF CLINICAL EXPERIENCE WITH SOD TREATMENT OF INFLAMMATORY DISORDERS

KLAUS M. WILSMANN
Grünenthal GmbH, P.O.Box 129, D-5190 Stolberg, West Germany

For several years a group of scientists (HUBER et al. 1968) investigated the possibility of using a hyaluronidase-derived protein as an anti-inflammatory agent. Independent of these investigators an other group (FRIDOVICH, McCORD, 1969) discovered a copper-zinc protein they called SOD because of its ability to convert O_2^--radicals into O_2 + H_2O_2. It finally became obvious that both proteins were identical, and in the following years the role of O_2^--radicals in inflammatory processes was clarified more and more. This opened a new insight into inflammatory processes leading to a new antiphlogistic principle.

First clinical experiences in different inflammatory conditions, e. g. osteoarthritis, rheumatoid arthritis, cystitis, and Peyronie's disease, were reported in 1974 (LUND-OLESEN, MENANDER; MARBERGER et al.). These reports with SOD encouraged further evaluation of this new antiphlogistic principle which led to an increasing number of investigations on the effectiveness of SOD.

A review of the past 10 years shows that the clinical use of SOD is centred on three fields of application: Inflammatory diseases of the locomotor system, of the connective tissues, and of the urinary bladder.

West Germany was the first country where highly purified bovine SOD was launched as a registered drug for human use (1981).

The first double blind controlled study in osteoarthritis of the knee with SOD versus placebo (PUHL et al., 1981) and a further trial versus corticosteroid (GAMMER, BROBÄCK, 1984) showed the following results: The superiority of SOD (4 mg intra-articularly/ week for 8 weeks; n = 45) to placebo (n = 40) was statistically significant and clinically relevant in all the assessed parameters; in a comparison of methylprednisolone acetate (40 mg intra-articularly every two weeks for 6 weeks, n = 14) and 8 mg SOD (n = 12, administered at the same regime) SOD showed a comparable effect to that of the corticoid, whereas 16 mg SOD (n = 10) was superior to the corticoid with a longer lasting improvement. An overall evaluation of a total of 2817 patients with inflammatory osteoarthritis in 11 clinical trials shows a responder rate of about 86 % (total remission 20 %; marked improvement 50 %; moderate improvement 16 %).

In rheumatoid arthritis SOD was first given i. m., and in later investigations intra-articularly. In a double blind controlled trial versus placebo SOD i. m. (4 times 8 mg/week for 4 weeks followed by 3 times 8 mg/week for 8 weeks) led to a slight additional clinical improvement in patients under a basic therapy regimen with ASA or ASA + corticoid (MENANDER-HUBER, HUBER, 1977). GOEBEL et al. (1981, 1983) observed in two double blind studies (SOD versus ASA and corticoid, resp.) that intra-articular SOD (4 mg/week for 6 weeks) was superior to intra-articular ASA (4 mg/week for 6 weeks) and comparable to intra-articular methylprednisolone acetate (4 mg/week for 6 weeks). These results suggest that SOD administered intra-articularly might be preferable to intramuscularly.

Local infiltration of SOD in extra-articular inflammatory processes (e. g. tendinitis, tendovaginitis, bursitis, epicondylitis, periarthritis) has proved to be effective. In these conditions acute cases obviously respond better than chronic ones: In a field trial (n = 237, unselected cases) with humeroscapular periarthritic patients 39 % experienced marked improvement, and 29 % became pain-free. In patients with acute humeroscapular periarthritis or epicondylitis, however, 64 % and 70 % resp. became pain-free (ZYLKA et al., 1982; MÜLLER, MOLL, 1983).

In a number of clinical studies, starting with MARBERGER (1974), very impressive results were obtained with SOD in radiation and interstitial cystitis. The reason why these results are so convincing is that most of the patients had already been treated unsuccessfully for long periods with other therapeutical means. The rate of responders averaged from six open trials with 165 patients was higher than 90 % in the parameters pain, ulceration, micturition frequency, and bladder capacity. 66 % of the patients became pain- free, in 75 % the ulcerations healed. The most frequently used SOD dose was 8 - 16 mg/session for up to 9 sessions, the usual route of administration was by infiltration into the bladder wall (SCHILLING et al., 1980; FRICK et al., 1981; KADRNKA, 1981; MARBERGER et al., 1981; HAUBENSAK, KONRAD, 1982; REUSS, CARL, 1983).

In 1974 MARBERGER also first reported on the basis of 8 patients that Peyronie's disease could be successfully treated with intrafocal SOD. A survey of the clinical studies carried out so far (10 open studies with a total 228 patients) confirms the favourable effect on the clinical symptoms. 50 - 70 % of the patients became pain-free during erection. In 80 % a clear cut improvement in the induration was seen with complete remission of the induration in 20 - 25 %. Improvement of the deviation was observed in 60 % with disappearance of this symptom in about 20 %. The usual dosage was 4 - 8 mg intrafocally/session for up to 15 sessions (MARBERGER et al., 1974; BARTSCH et al., 1981; GUSTAFSON et al., 1981; SCHILLING et al., 1981, 1985; RUGENDORFF, SCHNEIDER, 1983; HAUTUMM, STEFFENS, 1985; HASELBERGER, LUDWIG, 1985; MARR, 1985; SCARPA, 1985; PISANI, 1985; SCHNEIDER et al., 1985).

First communications on the use of SOD in other inflammatory diseases of the connective tissues e. g. Dupuytren's contracture, keloids and hypertrophic scars strongly suggest that SOD might be a useful tool in these conditions as well (BRÜNING, SCHRAMM, 1984; GRÖSSER et al., 1984).

The incidence of adverse reactions with highly purified bovine SOD compiled from 57 clinical trials with a total of 6885 patients can be considered as being low: local reactions occurred in 5.5 % of cases, allergic reactions in 0.8 % including 3 cases (= 0.04 %) of anaphylactic shock.

CONCLUSION

In inflammatory osteoarthritis and extra-articular inflammatory processes SOD proved to be an effective antiphlogistic agent and can therefore be considered a valuable alternative to the corticosteroids and NSAIDs.

In rheumatoid arthritis intra-articular SOD should be considered as adjuvant treatment in cases of mono-arthritis and in cases of polyarthritis where basic treatment cannot control the process in all the joints. Further trials with high intramuscular doses and with short intervals between injections (e. g. daily) might bring further progress in this indication.

In radiation and interstitial cystitis as well as in Peyronie's disease SOD appears to be becoming the drug of choice. After some pilot studies in Dupuytren's contracture, hypertrophic scars and keloids, SOD is also gaining a high and valuable position in the treatment of these conditions.

SOD treatment may cause allergic/anaphylactic reactions. However, with the highly purified bovine SOD the total allergy rate of 0.8 % or 0.04 % shock incidence, resp., is relatively low. Further progress in this field may be expected from (recombinant) human SOD.

REFERENCES

Bartsch, G., Menander-Huber, K.B., Huber, W., Marberger, H.: Orgotein, a new drug for the treatment of peyronie's disease. Europ. J. Rheumatol. Inflamm. 4, 250 - 259 (1981)

Brüning, W., Schramm, G.: Therapieversuch der Dupuytren'schen Kontraktur mit Superoxid-Dismutase, 2 Fallbeispiele. Rheuma, Schmerz & Entzündung 4, 30 - 33 (1984)

Frick, J., Danner, Ch., Kunit, G.: Klinische Erfahrungen mit Orgotein bei urologischen Erkrankungen. Europ. J. Rheumatol. Inflamm. 4, 260 - 263 (1981)

Gammer, W., Brobäck, L.-G.: Clinical comparison of orgotein and methylprednisolone acetate in the treatment of osteoarthrosis of the knee joint. Scand. J. Rheumatol. 13, 108 - 112 (1984)

Goebel, K.M., Storck, U., Neurath, F.: Intrasynovial orgotein therapy in rheumatoid arthritis. Lancet I, 1015 - 1017 (1981)

Goebel, K.M., Storck, U.: Effect of intra-articular orgotein versus a corticosteroid on rheumatoid arthritis of the knees. Am. J. Med. 74, 124 - 128 (1983)

Grösser, A., Stark, F., Landthaler, M.: Behandlung von hypertrophen Narben und Keloiden mit Orgotein. Hautarzt 35, 377 - 378 (1984)

Gustafson, H., Johansson, B., Edsmyr, F.: Peyronie's disease (induratio penis plastica) experience of local treatment with orgotein. Europ. J. Urol. 7, 346 - 348 (1981)

Haselberger, J., Ludwig, G.: Klinische Erfahrungen mit p-Aminobenzoat und Orgotein bei der Induratio penis plastica. In: Berichtsband vom IPP-Symposium, Bonn 1984 (Weißbach, L., Boedefeld, E.A., Widmann, T., eds.). Brimberg, Aachen, 74 - 81 (1985)

Haubensak, K., Konrad, G.: Die Behandlung der Schrumpfharnblasen mit Orgotein. Urologe B. 22, 134 - 137 (1982)

Hautumm, B., Steffens, L.: Durchführung der Orgotein-Injektion bei der Induratio penis plastica und Ergebnisse bei 20 Patienten. In: Berichtsband vom IPP-Symposium, Bonn 1984 (Weißbach, L., Boedefeld, E.A., Widmann, T., eds.). Brimberg, Aachen, 93 - 98 (1985)

Huber, W., Saifer, T.L., Carson, S., Goldhammer R.E., Vogin, E.E.: Some chemical and pharmacological properties of a novel antiinflammatory protein. Toxicol. Appl. Pharmacol., 12: 308 (1968)

Kadrnka, F.: Ergebnisse einer multizentrischen Orgotein-Prüfung bei Strahlen- und interstitiellen Cystitiden. Europ. J. Rheumatol. Inflamm. 4, 237 - 243 (1981)

Lund-Olesen, K., Menander, K.B.: Orgotein: a new anti-inflammatory metalloprotein drug: preliminary evaluation of clinical efficacy and safety in degenerative joint disease. Curr. ther. Res. 16, 706 - 716 (1974)

Marberger, H., Huber, W., Bartsch, G., Schulte, T., Swoboda, P.: Orgotein: a new antiinflammatory metalloprotein drug evaluation of clinical efficacy and safety in inflammatory conditions of the urinary tract. Int. Urol. Nephrol 6, 61 - 74 (1974)

Marberger, H., Huber, W., Menander-Huber, K.B., Bartsch, G.: Orgotein: a new drug for the treatment of radiation cystitis. Europ. J. Rheumatol. Inflamm. 4, 244 - 249 (1981)

Marr, K.-U.: Modalitäten der Orgotein-Behandlung in der Facharztpraxis. In: Berichtsband vom IPP-Symposium, Bonn 1984 (Weißbach, L., Boedefeld, E.A., Widmann, T., eds.). Brimberg, Aachen, 99 - 104 (1985)

McCord, J.M., Fridovich, J.: Superoxide dismutase: an enzymatic function for erythrocuprein (hemocuprein). J. Biol. Chem., 244: 6049 - 6055 (1969)

Menander-Huber, K.B., Huber, W.: Orgotein, the drug version of bovine Cu-Zn superoxide dismutase: II. A summary account of clinical trials in man and animals. In: Superoxide and Superoxide Dismutase (Michelson, A.M., McCord, J., Fridovich, I., eds.) Academic Press, London, 537 - 549 (1977)

Müller, U., Moll, G.: Über die Behandlung der Epicondylitis mit lokal injiziertem Orgotein (Doppelblindstudie). Rheumatologie 42, 21 - 24 (1983)

Pisani, E.: Orgoteina per infiltrazione ed ionoforesi associata al laser nella terapia dell'induratio penis plastica In: Simposio Internazionale Sull'Induratio Penis Plastica, Milano 1984. (Pisani, E., ed.) Masson Italia, Milano, 47 - 54

Puhl, W., Biehl, G., Kölbel, R., Hofer, H.: Ergebnis einer multizentrischen Orgotein-Prüfung bei Gonarthrose. Europ. J. Rheumatol. Inflamm. 4, 264 - 270 (1981)

Reuß, K., Carl, P.: Behandlung der Ulcus simplex vesicae und der ulcerierenden radiogenen Cystitis mit Superoxiddismutase. Urologe A 22, 290 - 293 (1983)

Ruggendorff, E.W., Schneider, H.-J.: Die Behandlung der Induratio penis plastica mit Superoxiddismutase. Urologe B 23, 118 - 120 (1983)

Scarpa, R.: Il trattamento ionoforetico con orgoteina della malattia di la peyronie: Nostra esperienza clinica e valutazione con ecografia e autofotografia peniena. In: Simposio Internazionale Sull'Induratio Penis Plastica, Milano 1984. (Pisani, E., ed.) Masson Italia, Milano, 63 - 71 (1985)

Schilling, A., Chaussy, Ch., Schüller, J., Staehler, G., Walther, V.: Superoxiddismutase (SOD): Erste Erfahrungen bei der Anwendung eines neuen anti-inflammatorischen Prinzips in der Behandlung der ulzerierenden Zystitis und der Induratio penis plastica (IPP). Verhandlungsband der Deutschen Gesellschaft für Urologie. XXXII. Tagung, Berlin, 404 - 405 (1980)

Schilling, A.: Erfahrungen bei der Bahandlung der Induratio penis plastica mit Orgotein und ihre Sicherheit. In: Berichtsband vom IPP-Symposium, Bonn 1984 (Weißbach L., Boedefeld, E.A., Widmann, R., eds). Brimberg, Aachen, 88 - 92 (1985)

Schneider, H.-J., Rugendorff, E.W., Röhrborn, C.: Die Behandlung der Induratio penis plastica mit Orgotein. In: Berichtsband vom IPP-Symposium, Bonn 1984 (Weißbach, L., Boedefeld, E.A., Widmann, T., eds.). Brimberg, Aachen, (1985)

Zylka, N., Zylka, V.: Orgotein (Peroxinorm^R) bei der Periarthritis humero-scapularis. Rheumamedizin 4, 175 - 185 (1982)

Superoxide and Superoxide Dismutase in Chemistry, Biology and Medicine; G. Rotilio, editor

RADIATION CYSTITIS : TREATMENT OF LATE LESIONS BY SUPEROXIDE DISMUTASE ADMINISTERED LOCALLY AND SYSTEMICALLY

A. SCHILLING

Abteilung für Urologie, Städt. Krankenhaus München- Bogenhausen,Englschalkingerstraße 77, 8ooo München 81 (West Germany)

INTRODUCTION

Radiogenic late lesions are defined to be such lesions as occur at the earliest 2-3 months, but also up to 2-2o years after radiation therapy (4,5,8). The metalloprotein orgotein reduces oxygen radicals toxic to tissue, which are released in inflammatory processes, such as radiation cystitis. For this reason, the efficacy of this substance in the treatment of radiation cystitis was to be examined.

MATERIAL AND METHODS

Depending on the lesion existing, the radiation cystitis was classified in cases of diffuse radiation cystitis and in cases of radiation ulcers.

Diffuse radiation cystitis: A total group of 56 patients, consisting of 14 males and 42 females of an average age of 61,5 or 59,9 years, respectively, was treated. Main complaints stated in this group included urge symptoms, followed by urge incontinence, vesical spasms, painful micturition and recurrent haematuria.

In all patients of this group, urinary infection and infravesical obstructions were excluded or had been fully treated. Thus a total of 41 patients had to undergo primary treatment. Primary treatment included 14 patients having undergone antibiotic treatment for an urinary tract infection. 15 patients had received a type of routine therapy of the bladder. In addition 6 had to be curetted and 6 patients had to undergo urethrotomy.

Orgotein treatment of diffuse radiation cystitis was carried through with 8 mg orgotein which was administered 3 times i.m. per week, for 7 weeks.

Radiation ulcers: Until today, a total of 27 female patients was treated locally with orgotein. The average age of this group was 56,7 years. The main complaints stated were pains and cramps of the bladder followed by urge symptoms. As above described in these cases, too, urinary tract infection as well as infravesical obstructions were excluded or fully treated.

8 mg orgotein was dissolved in 1o cc n. saline and injected submucosally around the ulcer through an endoscope using a special needle (Storz Comp.West-Germany). The injections were repeated every other week up to a total of 8 injections.

RESULTS

Diffuse radiation cystitis: In order to judge the efficacy, subjective criteria such as complaints in the bladder region or micturition difficults , were used as well as objective, measurable criteria, such as the flow rate, residual urine and vesical capacity. As agreed, such patients who did not subjectively improve after 3 weeks therapy, as well as those patients who showed a side - effect, were taken out of the study. Because of side-effects 4 patients were dropped from the study, 6 patients for non-response and 1 patient because of a new urinary tract infection.

The outcome of the subjective complaints are summarized in table 1 and 2.Whereas primary treatment did not produce any statistically significant effect to the complaints in the bladder region, it was significantly outdone by orgotein. The X^2 test was used for statistical evaluation (table 1).

TABLE 1:

Complaints in the bladder region; no improvement after primary treatment significant improvement after orgotein i.m.
(statistically significant in 5 % level using the X^2- test)

complaints	primary treatment before	primary treatment after	orgotein treatment before	orgotein treatment after
none	13	13	17	27
none-minor	0	0	0	2
minor	8	11	15	12
minor-evident	2	3	3	3
evident	16	12	16	7

Both treatments produced significant improvement of micturition difficulties. Orgotein treatment produced, as far as micturition difficulties were concerned,a significantly better effect than primary treatment (table 2).

In micturition frequency, too, both treatments were effective. Orgotein treatment also produced additional improvement here (table 3).

In the objectively measurable criteria, such as flow rate and vesical capacity, the same result was found statistically: Both forms of treatment were effective, but orgotein produced an additional effect. On the contrary, formation of residual urine in the bladder was not positively influenced by any form of therapy (table 4).

TABLE 2:

Dysuric complaints:
Both treatments were effective, orgotein was superior to the primary therapy (significant at 5 % level, X^2- test)

complaints	primary treatment before	after	orgotein treatment before	after
none	0	0	0	8
none-minor	0	0	0	3
minor	2	9	7	28
minor-evident	1	7	7	3
evident	35	22	37	9

TABLE 3 :

Micturition frequency:
Both treatments are significantly effective (X^2- test p< o,o5).
Orgotein is significantly superior to the primary treatment (p< o,o5).

frequency	n	primary treatment before	after	n	orgotein treatment before	after
day	36	11,9± o,5	1o,7± o,5	5o	1o,7± o,4	8,1± o,4
night	35	5,1± o,4	4,6± o,3	47	4,9± o,3	3,o± o,2

TABLE 4:

Flow rate, vesical capacity, residual urine:

	n	primary treatment before	after	n	orgotein treatment before	after
flow rate (ml/min)	39	16,7± o,8	18,6± o,7	49	18,7± o,6	21,2± o,8*
residual urine (ml)	39	36,3± 7,6	37,1± 9,8	5o	28,5± 7,5	23,6± 6,7
capacity (ml)	39	164,8± 9,8	179,2± 11,5	51	177,8± 1o,7	216,9± 1o,5*

* Both treatments are significantly effective (X^2- test p < o,5 orgotein is significantly superior to the primary treatment (p<o,o5).

Radiation ulcers: Until today, a total of 27 patients with radiogenic ulcerations was treated with orgotein. A decrease of the ulcer size and thus of the accompanying complaints was reached in all patients. In 18 cases, the ulcerations were fully cured. The average time of observation of this group of patients is 2,3 years (table 5).

TABLE 5 :

Local application of orgotein in radiation ulcers of the bladder: Observation time 4 month - 5 years, mean 2,3 years

decrease in size	27/27
complete healing	18/27

DISCUSSION

In inflammatory processes oxygen radicals toxic to tissue are released (1). The metalloprotein orgotein reduces oxygen radicals (3). Radiation lesions of the bladder are inflammatory conditions. Thus the use of orgotein in this affection seems justified. Preliminary promising reports were already published (6). In a first small series we were able to realise the same striking results (1o). Up to now our results in radiation ulcers continue to be very satisfying. These data are confirmed by others (9). In an other case with radiation urethritis even after one local application a dramatic improvement of the patients condition could be achieved (7). In one case with Peyronie`s disease we observed a significant improvement of a concomitant palmar contraction(Dupuytren) after application of orgotein to the shaft of the penis. Out of this experience we started with systemic orgotein application although being aware that only few of the substance injected will reach the point of interest because of the high renal clearance of this drug (2). The systemic application was limited to those conditions with diffuse lesions of the bladder where the injection could not be performed around a circumscript lesion. Even using the systemic application our results demonstrate that orgotein provokes some improvements in diffuse radiation cystitis concerning the dysuric symptoms, regional bladder complaints as well as the frequency, bladder capacity and flow rate. Relying upon our own results and the experience described in the literature we feel that orgotein is the drug choice in treatment of radiation lesions of the bladder.

LITERATURE

1) Badwey, J.A. et al: Ann. Rev. Biochem. 49 (198o), 695

2) Flohe,L. et al: Eur. J. Rheumtol. Inflamm.

3) Hassan, H.M. et al: In: Enzymatic Basis of Detoxication,Vol.1, p 311,Jakoby W.B. (ed.), Academic Press, New York 198o

4) Kümper, H.J. et al: Urologe B 1o (197o) ,277

5) Lenz, P. et al: Z. Urolo. 63 (197o), 411

6) Marberger, H. et al: Intern. Urol. Nephrol. 6 (1974) 61

7) Mayer, P. et al: Urologe A 23 (1984) 65

8) Remê, H.: Internist (Berl.) 4 (1963), 168

9) Reuß, K. et al: Urologe A 22 (1983) 29o

1o) Schilling, A. et al: Verhandlungsber. 87, bes. Urol. 32. Tagung (198o), 4o4.

Superoxide and Superoxide Dismutase in Chemistry, Biology and Medicine; G. Rotilio, editor

TREATMENT WITH LIPOSOMAL SUPEROXIDE DISMUTASE OF PATIENTS WITH KAWASAKI DISEASE.

KYOICHI SOMIYA[1,5], YUKIE NIWA[2], KOSUKE SHIMODA[3], SHIGEKO FUKAMI[4], KRYSTYNA PUGET[5] and A. MICHAEL MICHELSON[5]
[1]Dept. of Pediatrics, Hamamatsu Medical Center ; [2]Niwa Institute for Immunology ; [3]Dept. of Pediatrics, Chigasaki Tokushukai Hospital ; [4]Dept. of Pediatrics, Hamamatsu Red-Cross Hospital ; [5]Institut de Biologie Physico-Chimique, 13, rue Pierre et Marie Curie, 75005 Paris, France.

INTRODUCTION

Kawasaki disease (KD mucocutaneous lymph node syndrome) an acute febrile vasculatis of early childhood[1] first described in 1967 occurs often in Japan and sporadic cases have also been reported in other countries. Coronary occlusion resulting from enhanced platelet aggregation and thromboarteritis may be a main cause of sudden death in the patients. Approximately 15% of the patients develop coronary arteritis leading to aneurysm or stenosis. Although the presence of immune complex has been confirmed in these patients, the cause and pathogenesis remain unclear. Niwa and collaborators have recently reported a marked increase in oxygen radical generation by polymorphonuclear neutrophils from patients within five days after onset of the disease[2]. Liposomal-encapsulated superoxide dismutase (LIPSOD) was used to treat KD patients and its clinical efficiency investigated.

PATIENTS AND METHODS

LIPSOD (100-300 μg/kg, twice or three times a week, intramuscular or intravenous injection) was administered to seventeen patients with KD until remission. As a control study, 20 age- and sex-matched KD cases receiving aspirin (50-90 mg/kg per day) were studied, in comparison with the LIPSOD group. Among the patients receiving LIPSOD, 5 cases had been given a single injection of γ-globulin (200 mg/kg) before LIPSOD administration. All patients fulfilled the major criteria of Kawasaki et al., having at least an increase in body temperature (above 38° C) for more than five days, bilateral hyperemia of the bulbar conjunctiva, non purulent swelling of cervical lymphnodes, characteristic skin changes (erythema of the palms and soles, indurative edema of hand and feet, redness and dryness of lips and strawberry tongue with subsequent membranous desquamation from fingertips and lips at the convalescent stage).

Examination of the coronary artery was performed by two-dimensional echocardiography (done three times a week for 30 days after onset of disease) and diagnosed according to the criteria of the Kawasaki Disease Research Committee in which patients with a 1.5 times or more increase in coronary lumen diameter in comparison with the adjacent vessel diameter or a diameter of coronary arteries 3 mm or more are diagnosed as having coronary artery damage.

RESULTS

In most of the patients treated with LIPSOD, fever disappeared within three days after administration. Eruption, conjunctive hyperemia and lymphnode swelling faded away within a few days. There were significant differences in duration of elevated complement-reactive protein (CRP) levels and most cardinal signs between LIPSOD and aspirin groups (Table 1). Coronary aneurysm was demonstrated in 6 (30%) patients of the aspirin group but in only 1 (8%) of the L-SOD group at 30 days after onset of the disease. The clinical course of a representative patient under therapy with LIPSOD and γ-globulin is shown in Fig. 1. Fever and elevated CRP levels were decreased more rapidly in this patient, compared with those in the aspirin or LIPSOD groups.

TABLE I

COMPARISON OF THE DURATION IN DAYS OF CLINICAL SIGNS AND ELEVATED CRP LEVELS BETWEEN L-SOD AND ASPIRIN GROUPS

Clinical signs	Aspirin group	LIPSOD group
Fever	6.7 ± 4.6 (days)	2.5 ± 1.5 (days)
Rash	4.9 ± 3.4	2.4 ± 1.5
Conjunctival hyperemia	6.9 ± 4.1	3.3 ± 1.2
Oropharyngeal change	8.3 ± 3.4	5.1 ± 1.3
Cervical lymphadenopathy	6.8 ± 4.5	4.0 ± 1.2
Indurative edema	5.5 ± 3.5	3.3 ± 1.8
Elevated CRP levels (above 0.5 mg/dl)	17.7 ± 7.9	10.8 ± 3.9

Aspirin-treated group : 13 males and 7 females mean age 23.0 ± 4.2 months
L-SOD-treated group : 7 males and 5 females mean age 29.3 ± 19.1 months.
The durations of cardinal clinical symptoms and elevated CRP level in each group were expressed as mean ± SD. Except for indurative edema (non-significant), statistical differences between groups were $p < 0.01$.

DISCUSSION

As conventional treatment of KD, aspirin has been recommended by the Japanese committee and has generally been used with the aim of inhibiting thrombotic complication. However, Koren et al.[3] reported the ineffectiveness of aspirin in preventing platelet coagulation probably because plasma in children cannot be saturated with administered acetyl salicylate. This report seems to be supported by that of the Kawasaki Disease Research Committee in which coronary aneurysm was found in 22% of aspirin-treated cases on day 30 of the disease. Recently, high dose γ-globulin therapy has been applied as an immunomodulator and clinical efficacy of this therapy in reducing the coronary abnormalities has been observed[4]. However, there are still many problems in treatment with high doses of γ-globulin because of the possible inability of the infant's own immune system to produce endogenous γ-globulin due to a feedback mechanism by large amounts of exogenous γ-globulin. In addition, this therapy is not very effective for infants younger than 12 months, and a risk of viral infection is important. LIPSOD is effective in various inflammatory disorders in which the pathogenesis is related to an increase in oxygen radical generation[5]. LIPSOD treatment of Kawasaki disease is also effective and shows a significant improvement compared with aspirin treatment. Clinical use of LIPSOD can be recommended not only because of its efficiency but also the lack of toxicity.

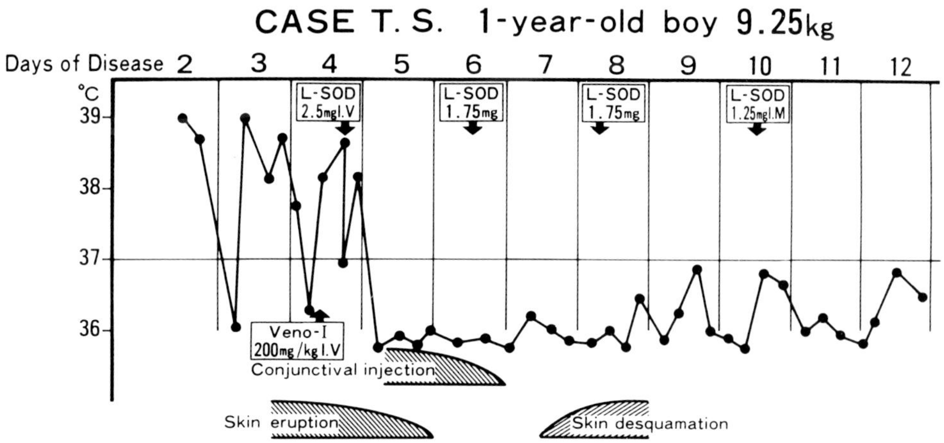

CRP	(mg/dℓ)	9.06	3.58	0.92	0.26	0.01↓
ESR	(mm/h)	24	32		24	12
WBC		16700	9300	9700	11800	10600
IgG	(mg/dℓ)	565	742		884	
Albumin	(g/dℓ)	4.82	3.69	4.31	4.83	5.36

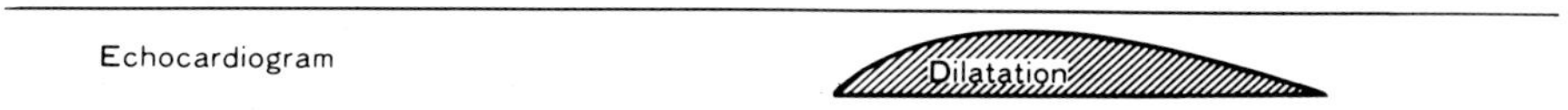

REFERENCES

1. Kawasaki T, Kosaki F, Okawa S, Shigematsu I, Yanagawa H (1974) Pediatrics 54 : 273-276
2. Niwa Y, Somiya K (1984) J Pediatr 104 : 56-60
3. Koren G, MacLeod SM (1984) J Pediatr 105 : 991-995
4. Furusho K, Nakano H, Shinomiya K, Tamura T, Manabe Y, Kawarano M, Baba K, Kamiya T, Kiyosawa N, Hayashidera T, Hirose O, Yokiyama T, Baba K, Mori C (1984) Lancet ii : 1055-1058
5. Niwa Y, Somiya K, Michelson AM, Puget K (1985) Free Radical Research Communs (in press)

Superoxide and Superoxide Dismutase in Chemistry, Biology and Medicine; G. Rotilio, editor

NEUTROPHILS ADHERENT TO A NON-PHAGOCYTOSABLE SURFACE PRODUCE OXIDANTS ONLY AT THE SITE OF ATTACHMENT

MARGRET C.M. VISSERS, W. ANTONY DAY AND CHRISTINE C. WINTERBOURN
Pathology Department, Christchurch Clinical School of Medicine, Christchurch (New Zealand)

INTRODUCTION

The site of oxygen radical production by the neutrophil is an important consideration for its directed action against a target cell or tissue. In this paper, we examine the respiratory burst and the site of oxidant production by neutrophils stimulated by IgG aggregates trapped within glomerular basement membrane matrix. We have previously shown that neutrophils adhere to this surface and that degranulation results in proteolytic degradation of the membrane matrix (1).

METHODS

Neutrophils were isolated from the peripheral blood of normal human donors (2). Oxygen consumption was measured continuously using a Clark-type oxygen electrode, superoxide production was determined by measuring superoxide dismutase-inhibitible cytochrome c reduction, and H_2O_2 production was measured after incubation in the presence of horseradish peroxidase and scopoletin (3).

Localisation of H_2O_2 production was carried out after incubation in the presence of Ce^{3+} according to Briggs et al. (4). Electron dense cerium perhydroxide precipitates were viewed by electron microscopy.

RESULTS

The amounts of oxygen consumed by neutrophils stimulated with basement membrane containing IgG, with IgG aggregates in suspension or with PMA, are shown in Table I. There was a marked increase in the oxygen uptake by neutrophils on addition of basement membrane containing IgG aggregates, but the amount of O_2^- detected in the medium with cytochrome c represented only 5% of the oxygen consumed. The amount of H_2O_2 recovered was 25% of the oxygen consumed and was always higher than the O_2^-. By comparison, when the cells were stimulated by IgG aggregates or by PMA, 50-90% of the O_2 was recovered as O_2^- and 20-25% as H_2O_2. With these stimuli the amount of O_2^- always exceeded the H_2O_2 produced (Table I). Basement membrane alone did not stimulate the cells (Table I); nor did it interfere with cytochrome c reduction by PMA-stimulated neutrophils (not shown).

TABLE I

O_2 CONSUMPTION, O_2^- PRODUCTION AND H_2O_2 PRODUCTION BY NEUTROPHILS

Addition to neutrophils	O_2 Consumed	O_2^- Produced	H_2O_2 Produced
	nmoles/10^7 cells in 5 min		
None	5.5 ± 3.0 (16)	0.9 ± 0.4 (7)	0.4 ± 0.2 (4)
Basement membrane containing IgG aggregates	43 ± 4 (5)	2.8 ± 2.8 (8)	5.5 ± 1.4 (5)
Basement membrane	6.3 ± 0 (2)	0.5 ± 0.3 (3)	N.D.
IgG aggregates	166 ± 15 (6)	67 ± 16 (6)	21 ± 2 (5)
PMA	213 ± 55 (5)	185 ± 15 (6)	21 ± 7 (5)

10^7 Neutrophils/mL were stimulated at 37°C with 1.5 mg/mL basement membrane, 1 mg/mL IgG aggregates or 0.1 μg/mL PMA. O_2 consumed, O_2^- and H_2O_2 produced were measured in separate experiments carried out concurrently. Results shown are the mean ± SD of (n) experiments. N.D. = not done. Taken from Ref. (5).

When nitroblue tetrazolium was used as a probe for O_2^- production by neutrophils on basement membrane containing IgG aggregates, it was readily reduced to the blue formazon, which was detectable by eye.

When neutrophils are incubated in the presence of cerium, an electron dense hydroxide precipitate forms at the site of H_2O_2 production (4). Electron microscopic examination of neutrophils stimulated on basement membrane in the presence of Ce^{3+} showed electron dense deposits only in the contact area between the cell and the basement membrane, and not on the rest of the cell surface (Fig. 1). When a soluble stimulus (PMA) was used, cerium precipitate was seen over the entire plasma membrane and inside vacuoles (5).

DISCUSSION

While there is marked O_2 consumption by neutrophils stimulated on basement membrane containing IgG aggregates, O_2^- is virtually undetectable in the medium by cytochrome c reduction; nitroblue tetrazolium reduction, however, does occur. The results show that O_2^- is produced at a site accessible to nitroblue tetrazolium, a small diffusible molecule, but not to cytochrome c, and, together with the electron microscopic studies, imply that

O_2^- and H_2O_2 are formed only where the plasma membrane is in contact with the basement membrane surface. Almost all the O_2^- reacts at this site, whereas some H_2O_2 is able to diffuse into the medium. Adherent neutrophils are thus able to direct their oxidant attack at components of the target surface.

The inability to measure cytochrome c reduction also suggests that proteins in the medium cannot gain access to the area of cell adherence and O_2^- production. This may be important in vivo, since it would prevent control of oxidative reactions by scavenging enzymes such as superoxide dismutase or catalase.

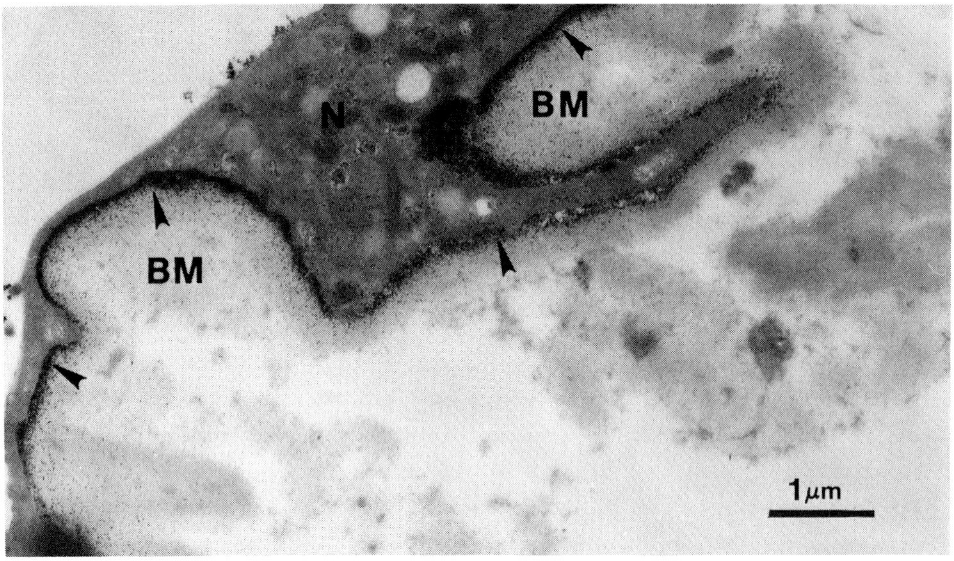

Fig. 1. A neutrophil (N) adherent to basement membrane (BM) containing IgG aggregates. Electron dense cerium perhydroxide deposits (arrows) are seen in the area of contact.

REFERENCES

1. Vissers MCM, Winterbourn CC, Hunt JS (1984) Biochim Biophys Acta 804:154-160

2. Böyum A (1968) Scand J clin Lab Invest 21 (Suppl 97):77-89

3. Root RK, Metcalf J, Oshino N, Chance B (1975) J Clin Invest 55:945-955

4. Briggs RT, Drath DB, Karnovsky ML, Karnovsky MJ (1975) J Cell Biol 67:566-586

5. Vissers MCM, Day WA, Winterbourn CC (1985) Blood (In Press)

© 1986 Elsevier Science Publishers B.V. (Biomedical Division)
Superoxide and Superoxide Dismutase in Chemistry, Biology and Medicine; G. Rotilio, editor

EXAMINATION OF EXHALED PENTANE AS A MEASURE OF LIPID PEROXIDATION

M. J. THOMAS, T. R. HINSON, N. E. ADAIR, L. HINER, and B. D. McLEES

Departments of Biochemistry and Medicine, The Bowman Gray School of Medicine, Wake Forest University, Winston-Salem, North Carolina 27103 (U.S.A.)

INTRODUCTION

Free radicals may be involved in causing the lung injury that results from a condition known as adult respiratory distress syndrome (ARDS) and after exposure to 100% oxygen [1]. Peroxidation of polyunsaturated fatty acids (PUFA) by a free radical chain mechanism may be a major contributor to the injury. However, lipid peroxidation has not been conclusively demonstrated *in vivo*.

We are attempting to determine whether lipid peroxidation plays a major role in lung injury induced by hyperoxia and by phorbol myristate acetate (PMA) [a model for ARDS]. To do this we are examining the evolution of pentane from intact rabbits and isolated perfused lungs and attempting to correlate pentane evolution to changes in lung lipid composition. Current information suggests that pentane evolution from biologic systems results from the decomposition of ω-6 conjugated PUFA hydroperoxides. Decomposition involves β-scission of an alkoxy radical.

MATERIALS AND METHODS

Gas collection. A 4 L teflon bag equipped with a barbed on/off GC valve and septum was repeatedly flushed with "zero grade" air and evacuated until GC analysis showed the gas to be free of pentane. A timed collection of gas was obtained from the ventilator after a 10 min washout period.

Gas chromatography. The procedure was modified from Lawrence *et al.* [2]. A 50 cc sample was trapped on a column packed with Tenax and activated charcoal. The sample was then backflushed onto a 6 ft x 1/8 in column of n-octyl Porasil C.

Animal preparation for collection of gases. After treatment the rabbits were anesthetized with xylazine and ketamine and tracheostomy performed. The animals were ventilated with "zero grade" air. Samples were taken for analysis at 30, 60, and 120 min after tracheostomy. The right lung was used to establish the extent of damage and the left lung used for lipid analysis.

Preparation of isolated perfused lungs. After tracheostomy, a median sternotomy was performed. The left ventricle and main pulmonary artery were cannulated, the animal exsanguinated, and the lungs perfused with modified Krebs-Henseleit buffer. The heart and lungs were dissected free and placed in a humidified 37°C hood. After 20 min equilibration normal saline or saline with various additives was injected into the pulmonary artery cannula. Exhaled gas was collected at baseline and at 30, 60, and 120 min post-injection.

Preparation of lipid samples. After grinding the lung in liquid nitrogen the residue was extracted with chloroform-methanol. Analysis for peroxidized lipids was carried out by high pressure liquid chromatography using a 15 cm x 4.6 mm column containing 3 μm ODS packing. The eluant was 77/22 v/v methanol/water plus 0.01% acetic acid programmed up to 88/12 v/v methanol/water.

RESULTS

Analysis of exhaled gases after injection of PMA (40 μg/kg) showed an increase in pentane production (6.52 ± 3.03 pmol/kg/min) compared to baseline levels found in control animals (1.68 ± 0.53 pmol/kg/min) at 120 min. Both normal saline and *E. coli* endotoxin caused no increase over background levels. Isolated perfused lungs treated with PMA did not display an increased pentane evolution compared to controls. After acute exposure to PMA no PUFA hydroperoxides were detected in the extracts and the fatty acid composition was not statistically different from that found in normal animals. After chronic exposure to PMA no statistically significant changes in fatty acid composition were found. Based on our limits of detection, we estimate that less than 1% of the PUFA was converted to isolable conjugated diene hydroperoxide.

Exposure to 100% oxygen resulted in a 1.5-fold increase in pentane production at 48 h as compared to 24 h and 72 h time points. However, control animals that breathed air showed 1.2-fold greater pentane levels at 24 h as compared to 48 h and 72 h. The pentane levels for the hyperoxic animals at 24 h and 72 h were identical to those of the air-breathing controls at 48 h and 72 h, Table I. No PUFA hydroperoxides were detected by HPLC. There was no change in fatty acid composition of hyperoxic animals as compared to their air-breathing controls, Table II. The level of peroxidation is estimated to be <1%.

DISCUSSION

Our studies with PMA indicate that the pentane measured after PMA injection into intact rabbits does not originate from the lung, but instead is derived from some other organ. This implies free radical damage is occurring at some other site and that the pentane is transported to the lung via the circulatory system. Hyperoxia, too, does not cause the release of pentane from the lung.

Our results raise the question of whether free radical peroxidation of lung lipids takes place under the conditions we have used. It may be true that lipid peroxidation is confined only to certain membrane subfractions and attempts to examine the total membrane may mask subtle changes in certain subfractions. Since there is presumptive evidence for the participation of strong oxidants in hyperoxic induced lung injury [3-5], a more selective study of lung membranes should yield insight into the participation of free radicals.

TABLE I

PENTANE EVOLUTION FROM RABBITS EXPOSED TO AIR OR TO 100% OXYGEN

Time (h)	Pentane (pmol/kg/min)[a]	
	O_2	Air
24 [4]	1.46 ± 0.47	3.08 ± 0.93
48 [8]	3.45 ± 0.36	1.26 ± 0.55
72 [4]	1.25 ± 0.48	1.51 ± 0.52

[a]Mean ± SEM. The number of animals at each time point is shown in brackets.

TABLE II

PERCENT COMPOSITION OF THE PREDOMINANT PUFA IN ANIMALS EXPOSED TO AIR OR TO 100% OXYGEN

Time (h)	18:1	18:2	18:3	20:4
		OXYGEN[a]		
24 [4]	21.4 ± 0.7	21.3 ± 2.4	1.80 ± 0.38	7.1 ± 1.7
48 [8]	21.4 ± 1.3	22.1 ± 2.8	1.54 ± 0.28	6.4 ± 1.5
72 [4]	19.8 ± 0.6	22.6 ± 1.7	1.82 ± 1.83	7.6 ± 1.1
		AIR		
24 [4]	20.1 ± 2.6	22.7 ± 1.8	1.53 ± 0.64	6.7 ± 2.1
48 [8]	20.0 ± 1.2	19.7 ± 2.2	1.19 ± 0.36	7.8 ± 0.9
72 [4]	21.1 ± 1.3	24.2 ± 2.1	2.12 ± 0.29	5.4 ± 1.0

[a]Mean ± SEM. The number of animals is given in brackets.

ACKNOWLEDGEMENTS

This work was supported by NIH grant HL33532.

REFERENCES

1. McCord JM, Fridovich I (1978) Ann Intern Med 89:122-127
2. Lawrence GD, Cohen G (1982) Anal Biochem 122:283-290
3. Steinberg H, Mock SA, Greenwald RA, Das DK (1983) In: Greenwald RA, Cohen G (eds) Oxy Radicals and Their Scavenger Systems, Vol II. Elsevier, New York, pp 159-165
4. Forman HJ, Rotman EI, Fisher AB (1983) Lab Invest 49:148-153
5. Dumelin EE, Dillard CJ, Tappel AL (1978) Arch Environ Health 33:129-134

Superoxide and Superoxide Dismutase in Chemistry,
Biology and Medicine; G. Rotilio, editor

POLYETHYLENE GLYCOL CONJUGATED SUPEROXIDE DISMUTASE AND CATALASE PREVENTS ACCUMULATION OF LUNG OXIDIZED GLUTATHIONE, LUNG INFLAMMATION, LUNG INJURY AND MORTALITY IN RATS EXPOSED TO HYPEROXIA

CARL W. WHITE AND JOHN E. REPINE

Departments of Pediatrics and Medicine and the Webb-Waring Lung Institute, University of Colorado Health Sciences Center, Box C322, 4200 East Ninth Avenue, Denver, Colorado, 80262 (USA)

INTRODUCTION

A method for measuring O_2 metabolites in biological systems is needed to help determine their importance in health and disease. Unfortunately, nearly all of the available approaches for measuring O_2 metabolites in biological systems are limited by their lack of sensitivity and/or specificity under conditions of physiologic oxidant stress. Measurement of O_2 metabolites is difficult because of the very short half lifes of the highly reactive O_2 metabolites and their relative inaccessibility within tissues. We addressed this problem by postulating that increases in tissue oxidized glutathione (GSSG) might reflect tissue oxidant stress. Studies using recently developed methods for measuring oxidized (GSSG) and reduced glutathione (GSH) indicate that virtually all of the glutathione in tissues and body fluids exists in the reduced form under normal conditions (1-3). Accordingly, an elevation in the amount of oxidized glutathione would most likely represent a perturbation in systems which generate or transfer reducing equivalents to glutathione peroxidase, and/or an increased burden of unscavenged oxidants.

To test our premise, we hypothesized that exposure to hyperoxia promotes the generation of excessive oxygen metabolites and, secondarily, lipid peroxidation products, which cause the accumulation of oxidized glutathione

(GSSG) in lung tissue. This model of normobaric oxygen toxicity in rats has been widely employed as one suitable for studies of O_2 metabolite induced disease.

MATERIALS AND METHODS

Male Sprague-Dawley rats (350-400g) were injected with polyethylene glycol-attached superoxide dismutase plus catalase (PEG-SOD plus PEG-CAT) or saline and then exposed to hyperoxia (>99% O_2, 1 atm) or normoxia (21% O_2) for 54 hours. Subsequently, rats underwent anesthesia, tracheotomy, thoracotomy and cannulation of their pulmonary arteries and left atria. Lungs from rats were then isolated and perfused blood-free and freeze-clamped at liquid nitrogen temperature. Lung tissue was prepared and analyzed for GSH and GSSG (3). Lung lavage albumin concentrations, pleural effusions, lung lavage neutrophils and the durations of survival of rats exposed to hyperoxia were also assessed in separate experiments.

RESULTS

Rats which were exposed to hyperoxia for 54 hours had lung GSSG contents which were increased (33 ± 2 nmoles/2 lungs, $p<0.05$) compared to GSSG contents of lungs from rats exposed to normoxia (5 ± 0.4 nmoles/2 lungs). In contrast, GSH contents of lungs from rats exposed to hyperoxia (3.4 nmoles/2 lungs) were increased only slightly compared to GSH contents of normoxia (2.6 nmoles/2 lungs) exposed lungs. In parallel experiments, lungs of rats that had been treated with PEG-SOD plus PEG-CAT before exposure to hyperoxia contained less GSSG (19 ± 3 nmoles/2 lungs) than lungs from saline-injected, hyperoxia-exposed control rats (33 ± 2 nmoles/2 lungs). Pleural effusion volumes, lung lavage albumin concentrations, absolute lavage neutrophil counts and mortality were also decreased in PEG-SOD plus PEG-CAT pretreated compared to saline pretreated rats which were then exposed to hyperoxia for 54 hours.

DISCUSSION

We found that very little oxidized glutathione (GSSG) is normally present in the lung. However, exposure to hyperoxia leads to striking accumulations of oxidized glutathione (GSSG) in lung tissues while the contents of total glutathione (GSH+2GSSG) change relatively little. In addition, accumulations of lung oxidized glutathione are, at least in part, related to excesses of O_2 metabolites generated during exposure to hyperoxia. This impression is based primarily on the observation that lung GSSG accumulations decreased when rats were pretreated with PEG-SOD plus PEG-catalase before exposure to hyperoxia. Moreover, the parallel reductions in pleural effusions, lavage albumin concentrations and mortality noted in PEG-SOD plus PEG-CAT treated rats further suggests that both lung injury and oxidation of lung tissue sulfhydryls are related to O_2 metabolites in the setting of normobaric hyperoxia.

Reactions which can lead to GSSG formation have been well reviewed (4,5). All involve direct or indirect sulfhydryl oxidation. GSSG generation is fairly specific for conditions of oxidative stress. GSSG can then form mixed disulfides with cell proteins which, in turn, can lead to dysfunction of important systems (energy production, membrane transport, etc). Since GSSG can be toxic it is expelled by most cells. This threshold for discharge from cells varies in different tissues, but during oxidant stress, intracellular GSSG levels in liver closely parallel GSSG levels in bile and blood (6).

In summary, our findings suggest then that levels of lung tissue oxidized glutathione (GSSG) may provide a useful marker of physiologic oxidant stress during pulmonary oxygen toxicity and that long-lived polyethylene glycol-attached antioxidant enzymes may be useful probes for investigating the contribution of O_2 metabolites in lung and other tissue injuries.

ACKNOWLEDGEMENTS

The authors acknowledge A. Abuchowski, G.M. Kazo, J.H. Jackson, R.F.

Mimmack, B.A. Freeman and J.M. McCord who participated in the original full study. Dr. White is an American Heart Association Clinician-Scientist awardee and this work was supported in part by a grant from the American Lung Association.

REFERENCES

1. Smith AC, Boyd MR (1984) J Pharmacol Exp Ther 229:658-663
2. Sies H, Brigelius R, Akerboom TPM (1983) In: Larsson A, Holmgren A, Orrenius S, Mannervik B (eds) Functions of Glutathione: Biochemical, Physiological, Toxicological and Clinical Aspects. Raven Press, New York, pp 51-64
3. Adams JD, Jr, Lauterburg BH, Mitchell JR (1984) J Pharm Exp Ther 237:749-754
4. Kosower NS, Kosower EM (1978) Int Rev Cytol 54:109-160
5. Smith CV, Hughes H, Lauterburg BH, Mitchell JR (1983) In: Larsson A, Holmgren A, Orrenius S, Mannervik B (eds) Functions of Glutathione: Biochemical, Physiological, Toxicological and Clinical Aspects. Raven Press, New York, pp. 125-137
6. Akerboom TPM, Bilzer M, Sies H (1982) J Biol Chem 237:4248-4252, 1982.

Superoxide and Superoxide Dismutase in Chemistry, Biology and Medicine; G. Rotilio, editor

MODIFICATION OF PULMONARY OXYGEN TOXICITY BY BLEOMYCIN TREATMENT: ROLE OF LUNG ANTIOXIDANT ENZYMES

SADIS MATALON, RENEE BUSH AND WAYNE V. HARPER
Department of Physiology, State University of New York at Buffalo, Buffalo NY. 14214, USA.

INTRODUCTION

High concentrations of oxygen are often administered to patients with various forms of lung injury to maintain adequate arterial oxygenation. Numerous studies on animals with healthy lungs have shown that prolonged exposure to 100% O2 at one atmosphere causes progressive injury to the components of the blood gas barrier leading to death from respiratory failure (1,2). The purpose of this study was to investigate whether the existence of fibrotic lung disease would affect the appearance of sublethal and lethal hyperoxic injury in rabbits and, if so, to identify a possible mechanism that may account for this effect.

MATERIALS AND METHODS

New Zealand White rabbits, initially weighing 1.9-2.2 kg and showing normal patterns of eating, drinking, and activity, were used for these experiments. Under anesthesia, a catheter was inserted in the trachea and bleomycin (5 units/kg of body weight), dissolved in 5 ml of sterile saline, was instilled in the lungs of 30 rabbits. Fifteen animals received an equal volume of sterile saline. All animals breathed air for the next 35 days whereupon they were either sacrificed or exposed to 100% O2. All procedures have been described in detail previously (3). In brief, in six experimental and five control animals we measured Pa_{O2}, Pa_{CO2} and pH weekly after the injection and static lung compliance (C_L), total lung capacity (TLC) and hydroxyproline content (OH-Pro) at sacrifice. Five experimentals and three controls were exposed to 100% O2 in individual environmental chambers until death. Pa_{O2}, Pa_{CO2}, pH and hematocrit were measured throughout the exposure and OH-Pro shortly after death. In the remaining 19 rabbits (12 experimental and seven control) we measured the levels of catalase, superoxide dismutase and DNA either after sacrifice or after breathing 100%O2 for 64 hr.

RESULTS

Bleomycin injected rabbits showed a 30% decrease in TLC and C_L and a 100% increase in the value of OH-Pro . Histologically, there was an increased cellularity and thickening of the alveolar septa. Seven days post-injection their mean Pa_{O2} was 59 ±3 but returned to its preinjection value 35 days later. Their survival time in 100% O2 was 120 ± 4 hr (mean and SE) vs 77 ± 5 for the saline group ($p<0.01$). Mean values for the Pa_{O2} and Pa_{CO2} are shown in Fig. 1. Intratracheal injection of bleomycin resulted in a doubling of catalase activity, but no significant changes in CuZn or MnSOD levels. Exposure of bleomycin injected animals to 100% O2 for 64 hr produced three and four fold increases in MnSOD and catalase levels from saline injected controls respectively (Table 1). When the units of these enzymes were expressed per ug of DNA however, only the catalase level was significantly different from control (Fig. 2).

Table 1

Antioxidant enzyme and DNA levels in rabbits injected with bleomycin or saline (control). Animals were either sacrificed or exposed to 100% O2 35 days later. Means ± 1 SEM ; n = number of animals

# units	Control	Bleomycin	Bleomycin & 64 h in 100% O2
CuZn SOD	849 ± 104	1002 ± 123	1057 ± 160
Mn SOD	979 ± 193	1469 ± 135	2817 ± 354 *
Catalase	238 ± 98	538 ± 170*	903 ± 21 *
DNA (ug)	1426 ± 158	1952 ± 286	2177 ± 378
	(n=5)	(n=8)	(n=4)

* = significantly different from control ($p<0.05$)

DISCUSSION

Our data indicate that a single intratracheal injection of bleomycin resulted in a moderate degree of lung fibrosis and increased tolerance to the toxic effects of 100% O2. The onset

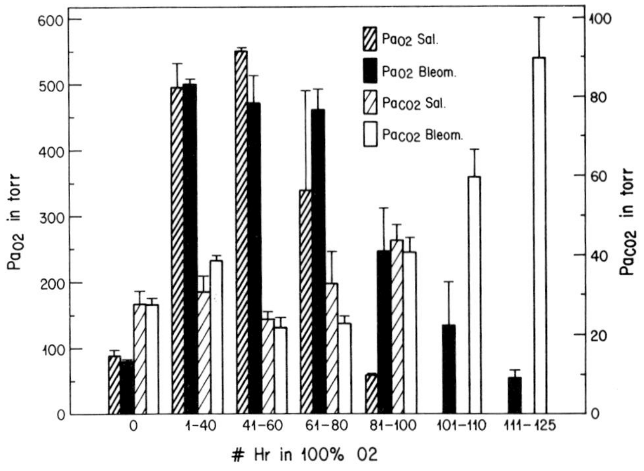

Fig. 1. Serial measurements of arterial oxygen (Pa_{O2}) and carbon dioxide (Pa_{CO2}) partial pressures in rabbits that received either bleomycin or saline and exposed to 100% O2 35 days later.

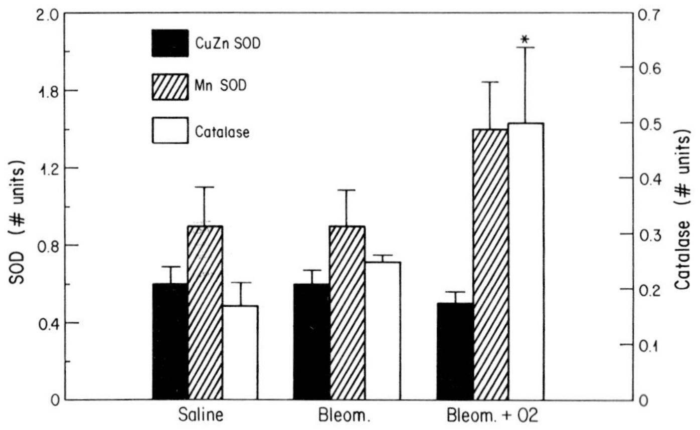

Fig. 2. Superoxide dismutase and catalase levels per µg of DNA in lung homogenates of rabbits that received either bleomycin or saline and were either sacrificed (saline, bleomycin) or exposed to 100% O2 for 64 hr. 35 days later.

of both the lethal and sublethal damage, as indicated by mean survival time and decrease in Pa_{O2}, were delayed in this group as compared to control. This may be due to the higher levels of MnSOD and catalase in the lungs of these rabbits. Furthermore, although the increase in MnSOD could be due to a higher number of inflammatory cells, since its value normalized for the lung DNA content was not different from control, the equivalent value of catalase doubled indicating a higher level of this enzyme in the pulmonary cells.

The oxygen induced damage is thought to be mediated by the increased production of superoxide anions, hydrogen peroxide, singlet oxygen and other toxic oxygen metabolites that cause lipid peroxidation, protein denaturation and the consequent destruction of cell membranes (4). Jackson and Frank (5) and Crapo et al.(6) have shown that exposure of rats to either sublethal O2 or ozone gas offers some protection against a subsequent exposure to hyperoxia by increasing the amount of antioxidant enzymes in the lungs. In this respect the question as to whether the existence of lung disease may protect an animal from the toxic effects of oxygen breathing may depend on the ability of the organism to increase the levels of lung antioxidant enzymes in response to the initial injury.

ACKNOWLEDGMENTS

This project was supported by a National Heart, Lung, and Blood Institute grant (HL-31197) and a grant from the Cummings Foundation, Buffalo, NY.

REFERENCES

1. Matalon S, Egan EA (1981) Effects of 100% O2 breathing on permeability of alveolar epithelium to solute. J Appl Physiol 50:859-863
2. Matalon S, Cesar MC (1985) Effects of 100% oxygen breathing on the capillary filtration coefficient in rabbit lungs. Microvasc Res 29:70-80
3. Matalon S, Harper WV, Goldinger JM, Nickerson PA, Olszowka J (1985) Modification of pulmonary oxygen toxicity by bleomycin treatment. J Appl Physiol 58: 1802-1809
4. Freeman, BA, Crapo JD (1982) Biology of disease. Free radicals and tissue injury. Lab Invest 47:412-426
5. Jackson RM, Frank L (1984) Ozone-induced tolerance to hyperoxia in rats. Am Rev Respir Dis 129:425-429
6. Crapo JD, Barry BE, Foscue HA, Shelbourne J (1980) Structural and biochemical changes in rat lungs occuring during exposure to lethal and adaptive doses of oxygen. Am Rev Resp Dis 122:123-143

Superoxide and Superoxide Dismutase in Chemistry, Biology and Medicine; G. Rotilio, editor

ANTI-INFLAMMATORY PROPERTIES OF SUPEROXIDE DISMUTASE DERIVATIVES (SOD-PEG-3 AND 18) IN CARRAGEENAN PLEURISY

CONFORTI A., MARRELLA M., MILANINO R., MORETTI U., VELO G.P., LARGAJOLLI R., VERONESE F.M., BOCCU', E., SCHIAVON O.
Institute of Pharmacology, University of Verona, Italy and Department of Pharmaceutical Sciences, University of Padova, Italy

INTRODUCTION

The anti-inflammatory activity of superoxide dismutase has been demonstrated in many inflammatory conditions both in man and in animals (1). However its clinical use is limited by the very short plasma half-life (≃ 5 minutes). Recently it was found that conjugation with different macromolecules significantly lengths the plasma half-life of SOD, unmasking its anti-inflammatory properties (2, 3). Elsewhere we reported the anti-inflammatory activity on carrageenan oedema of two derivatives of SOD with monomethoxypolyethyleneglycol (PEG) having different molecular weights and pharmacokinetic characteristics (4). We here report on the effect of the same derivatives on carrageenan pleurisy in rats.

MATERIAL AND METHODS

Preparation of PEG-superoxide dismutase derivatives

The PEG-bound superoxide dismutase was prepared following a method recently devised by us: the polimer is activated by trichlorophenylchloroformate to give a phenylcarbonate which is reactive towards the protein amino groups (5).

Evaluation of SOD activity

SOD activity was evaluated using the method described by McCord and Fridovich (6), in plasma and exudate collected from rats. Samples of fresh solution of native SOD were simultaneously assayed to ascertain the reliability of assay. The endogenous superoxide dismutase activity in untreated rats was also evaluated using the same procedure and then subtracted from that observed in treated animals.

Carrageenan pleurisy

Sprague-Dawley female rats weighing 200-210 g were used. Pleurisy was induced by an intrapleural injection of 0.15 ml of 1% carrageenan suspended in sterile saline. SOD and its derivatives were dissolved in sterile saline and injected into the tail vein at the same time as carrageenan (0.4 ml/100 g). Indomethacin was adminstered orally one hour before the irritant (0.5 ml/100 g). Two experiments with different drug doses were performed as indicated in table I and II.

RESULTS

The anti-inflammatory activity of SOD derivatives is showed in table I.

TABLE I

Treatment	Exudate volume ml (S.D.) 3 h	6 h	24 h
Control	0.75 (0.35)	1.24 (0.27)	1.37 (0.39)
Indomethacin 5 mg/kg	0.12 (0.12)** I = 84	0.60 (0.30)** I = 52	1.68 (0.55) I = 0
Native SOD 3,500 U/kg	0.58 (0.18) I = 23	1.21 (0.277) I = 2	
SOD-PEG 3 3,500 U/kg	0.61 (0.22) I = 19	1.04 (0.36) I = 16	1.87 (1.28) I = 0
SOD-PEG 18 3,500 U/kg	0.28 (0.23)** I = 63	0.18 (0.14)** I = 85	0.79 (0.59)* I = 42

**P < 0.01
*P < 0.05 Student's t test; I = % of inhibition
5-8 animals were used each time

In table II is showed the superoxide dismutase activity in plasma and exudate of rats injected with 35,000 U/kg of SOD-PEG 18 measured 3, 6 and 24 hours after carrageenan administration. The pharmacological effect was the same of the 3,500 U/kg dose.

TABLE II

	UNITY OF SOD (S.D.)					
	3 h (n = 10)		6 h (n = 10)		24 h (n = 9)	
	U/ml	U/Tot	U/ml	U/Tot	U/ml	U/Tot
Plasma	406.8 (110.9)	3420 (926.6)	279.3 (84.9)	2334.4 (773.1)	114.6 (18.2)	966.8 (138.9)
% of Dose	47	47	34	34	14	14
Exudate	67.4 (36.9)	83.17	184.9	152.7	121.7	91.2

Pleural cavity was injected with 1 ml saline
n = number of rats

To evidence the diffusion of SOD-PEG 18 into the pleural cavity, the ratio of exudate to plasma concentration is plotted against time. Its values increase from a minimum of 0.18 at the 3rd hour to a maximum of 1.07 at the 24th hour, with a correlation coefficient (R) of 0.61 ($P < 0.001$).

Conclusion

Our findings clearly demonstrate that the SOD-PEG derivatives have a greater anti-inflammatory activity than the native enzyme. Moreover for the longer-lasting compound, we found an effective presence in plasma and exudate as late as 24 hours after carrageenan injection.

REFERENCES

1. Huber W (1980) In: Willoughby DA and Giroud JP (eds) Inflammation: Mechanisms and Treatment. MTP Press Limited, pp 27-42
2. Wong K, Cleland LG, and Poznawsky MJ (1980) Agents and Actions 10 (3):231-239
3. Pyatak PS, Abuchowski A, Davis FF (1980) Res Comm Chem Path Pharmac 29(1):113-127
4. Veronese FM, Boccù E, Schiavon O, Velo GP,Conforti A, Franco L, Milanino R (1983) J Pharm Pharmacol 35:757-758
5. Veronese F.M., Largajolli R, Boccù E, Benossi CA, Schiavon O (1985) J Appl Biochem Biotechnol 10, in press
6. McCord J and Fridovich I (1969) J Biol Chem 25:6049-6055

Superoxide and Superoxide Dismutase in Chemistry,
Biology and Medicine; G. Rotilio, editor

SUPEROXIDE DISMUTASE OF *GLUCONOBACTER CERINUS*

KOJI TSUKUDA, TOSHIKO KIDO*, SUEYOSHI UEDA, YOSHIYUKI SHIMASUE, AND KENJI SODA*
Pias Co. LTD., 3-21-3 Toyosaki, Oyodoku, Osaka 531(Japan)
*Laboratory of Microbial Biochemistry, Institute for Chemical Research, Kyoto University, Uji, Kyoto-Fu 611(Japan)

INTRODUCTION

The acetic acid bacteria, *Acetobacter* and *Gluconobacter* actively oxidize carbohydrates and alcohols to organic acids such as gluconic acid and acetic acid. Some have been used to produce vinegar and L-sorbitol in food and pharmaceutical industry. They require a high concentration of dissolved oxygen in a medium as strict aerobes, and thus probably have effective oxygen scavenging systems. In this study, we investigated the distribution of superoxide dismutase in various strains of *Acetobacter* and *Gluconobacter* to demonstrate a thermostable superoxide dismutase in *G. cerinus*, and characterize it enzymologically. We modified the enzyme chemically with polyethylene glycol, and compared its *in vivo* stability with that of the native enzyme.

EXPERIMENTAL PROCEDURES

Acetobacter and *Gluconobacter* were grown in the medium containing 0.5% yeast extract; 1.0% glycerol; and 0.1% K_2HPO_4 at 30 °C for 24 hrs under vigorous reciprocal shaking. The cells harvested were washed with 0.85% NaCl. Superoxide dismutase activity was determined at 25 °C and pH 7.8 with a system of xanthine, xanthine oxidase and ferricytochrome c.[1] One unit of activity is defined as the amount of enzyme that causes a 50% inhibition in the rate of cytochrome *c* reduction. Specific activity is expressed as units per mg of protein. Primary amino groups were determined according to the method of Satake *et al.*[2]

Superoxide dismutase(5.33 mg) was modified with monomethoxypolyethylene glycol 5000(10 mg) activated in the presence of cyanuric chloride according to the procedure of Abuchowski *et al.*[3] About 23.8% of amino groups of the native enzyme was modified with polyethylene glycol 5000.

RESULTS

Purification of superoxide dismutase from **Gluconobacter cerinus***(IFO 3268).* We examined the distribution of superoxide dismutase(EC 1. 15. 1. 1) in various strains of *Acetobacter* and *Gluconobacter* at 70 °C. The extracts of *G. cerinus* showed a high activity of superoxide dismutase, and the enzyme was found to be relatively thermostable(Table I). The enzyme was purified from the extracts by ammonium sulfate fractionation, DEAE-Toyopearl column chromatography and Sephadex G-150 column chromatography. The purified enzyme was found to be homogeneous upon disc gel electrophoresis and ultracentrifugation.

TABLE I

THERMOSTABLE SUPEROXIDE DISMUTASE ACTIVITY OF CRUDE EXTRACTS FROM *GLUCONOBACTER* AND *ACETOBACTER*

Enzyme Source	Relative Activity*	
	Heat Treatment at 70°C	
	for 5 min	for 20 min
G. cerinus (IFO 3264)	85.3	
G. cerinus (IFO 3270)	42.0	
G. cerinus (IFO 3269)	81.0	
G. cerinus (IFO 3268)	140.1	45.4
G. cerinus (IFO 3267)	80.1	
G. cerinus (IFO 3265)	51.6	
G. cerinus (IFO 3263)	46.9	
G. dioxyacetonicus (IFO 3272)	51.4	
G. dioxyacetonicus (IFO 3274)	53.5	
G. suboxydans (IFO 3291)	35.6	
G. suboxydans (IFO 3258)	46.2	
G. suboxydans (IFO 3254)	62.8	
G. albidus (IFO 3253)	111.8	23.4
G. oxydans (IFO 3189)	38.7	
G. gluconicus (IFO 3171)	93.8	17.9
G. gluconicus (IFO 3285)	124.3	26.6
G. gluconicus (IFO 3286)	39.1	
G. industrius (IFO 3260)	135.3	30.3
G. nonoxygluconicus (IFO 3275)	42.2	
G. melanogennes (IFO 12258)	62.6	
A. ascendens (IFO 3188)	18.6	
G. lquefaciens (IFO 12388)	93.8	35.3

* The native enzyme activity was 100.

General properties.

The enzyme has a molecular weight of approximately 47,000, and consists of two subunits identical in molecular weight. The enzyme exhibits absorption maxima at 280 and 470 nm. Its general properties are similar to those of Mn-superoxide dismutase of *Bacillus subtilis*.[4] The study by atomic absorption spectrophotometry indicates the presence of 1.22 gatoms of Mn and 0.19 gatom of Cu per mol of the enzyme. No significant amount of Fe is found in the enzyme. The enzyme was affected by neither 5 mM cyanide nor 10 mM hydrogen peroxide. Those show that the superoxide dismutase of *G. cerinus* contains Mn as a catalytically essential metal. The enzyme is stable when heated at 70 °C for 5 min (pH 7.2), and also between pH 6.0 - 11.0 (at 37 °C for 30 min).

Effect of modification on circulating life of superoxide dismutase.

The native superoxide dismutase and polyethylene glycol-superoxide dismutase were injected intravenously into feet of female guinea pigs, and their stability was examined. As shown in Fig. 1, the modified enzyme showed a half-life of 330 min,

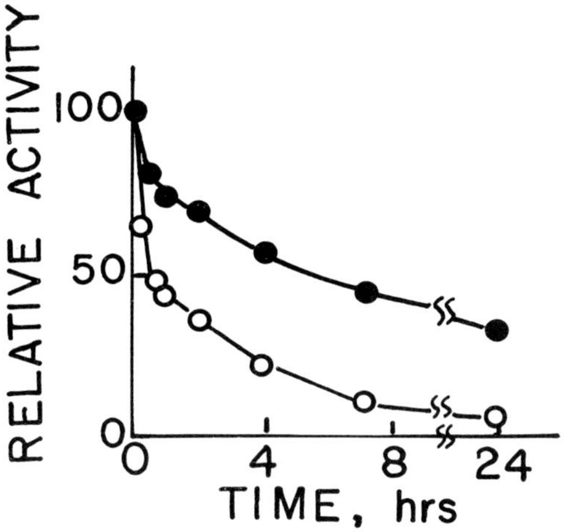

Fig. 1. Blood circulating life of native(o)- and polyethylene glycol 5000(●) superoxide dismutase in guinea pigs. Five animals were used in each group.

which is higher 7.5-times that that of the native one (45 min). In addition, the native and modified enzymes retained 6% and 30% of the original activity 24 hrs after the injection of enzyme. Thus, the modification of enzyme enhances the circulating life effectively.

DISCUSSION

Enzymes from thermophilic microorganisms generally are more resistant to heat treatment than mesophilic ones. However, we have found that the strictly aerobic and mesophilic bacterium, *G. cerinus* produces the thermostable superoxide dismutase abundantly. The enzyme effectively scavenges $O_2^{\overline{\cdot}}$ and shows the anti-inflammatory activity. It is therapeutically important to maintain the concentration of superoxide dismutase above the functionally efficient level. The Cu, Zn-superoxide dismutase from bovine erythrocyte is labile, and its half life time is 6 min in rat blood.[5] The half life of Mn-superoxide dismutase from *G. cerinus* was 45 min in guinea pig. McCord and Wong[5] have reported that the polyethylene glycol -bovine superoxide dismutase increases a plasma half life time effectively. The modification of enzyme from *G. cerinus* with polyethylene glycol markedly increases the half life. The polyethylene glycol bound-superoxide dismutase probably is useful as a potent and stable anti-inflammatory agent. *G. cerinus* is widely used in the vinegar production. Thus, its superoxide dismutase is utilized as a safe and effective antioxidant in food industry.

ACKNOWLEDGEMENTS

We thank Dr. K. Asada for his valuable advice throughout the work.

REFERENCES

1. McCord J M, Fridovich I (1969) J Biol Chem 244:6049-6055
2. Satake K, Okuyama T, Ohashi M, Shinoda T (1960) J Biochem 47:654-660
3. Abuchowski A, Nichols T E, Palczuk C, Davis F F (1977) J Biol Chem 252:3578-81
4. Tsukuda K, Kido T, Shimasue Y, Soda K (1983) Agric Biol Chem 47:2865-2870
5. McCord J M, Wong K (1979) In: Oxygen Free Radicals and Tissue Damage. Excerpta Medica, Amsterdam, pp 343-360

Superoxide and Superoxide Dismutase in Chemistry,
Biology and Medicine; G. Rotilio, editor

COPPER CHELATES OF SLOW ACTING ANTIRHEUMATIC DRUGS AND NON-STEROIDAL ANTI-INFLAMMATORY DRUGS : THEIR SOD-LIKE ACTIVITY AND STABILITY

N A ROBERTS & P A ROBINSON

Roche Products Ltd., Biology Department, Welwyn Garden City, Herts, England
AL7 3AY

INTRODUCTION

The antirheumatic activity of D-penicillamine may be partly attributable to copper chelation and the ability of its chelate(s) to catalyse the dismutation of superoxide radicals $O_2^{\dot{-}}$ (1). However, similar superoxide dismutase (SOD)-like activity is also seen _in vitro_ with copper chelates of other ligands, e.g. amino acids (2), NSAIDs, which do not possess disease modifying activity in RA. The clinical activity of a ligand may therefore reflect its ability to form copper chelates with SOD-like activity _in vivo_ in competition with endogenous ligands such as albumin. We have therefore designed a simple _in vitro_ test system which simultaneously detects SOD-like activity of copper chelates and differentiates between strong and weak ligands.

TEST DESIGN

Superoxide radicals are generated by hypoxanthine/xanthine oxidase (XOD) and detected by the reduction of nitroblue tetrazolium (NBT). Introduction of copper (as $Cu(Gly)_2$) catalyses superoxide dismutation and hence decreases the rate of NBT reduction. The copper complex of EDTA, a stronger ligand than glycine, has no SOD-like activity. Hence if excess EDTA is present all the copper will bind to EDTA and superoxide dismutation will not be catalysed. If however, copper is introduced into the assay bound to a stronger ligand than EDTA, then that complex will remain intact in the presence of excess EDTA, and any SOD-like activity of the complex will be observed. An I_{50} for that inhibition with respect to copper concentration can be determined. The SOD-like activity of all weak chelates will be seen only when the copper concentration is increased to saturate the EDTA.

METHOD

Solutions of copper (II) bisglycinate, and test compound are prepared at concentrations of 10^{-2}M. The copper and test compound in the molar ratio of 1:4 are mixed and diluted in 100 mM Hepes buffer, pH 7.8, to a final copper concentration of 10^{-3}M. This solution is further diluted tenfold in buffer, mixed in varying known proportions with EDTA and the mixture left to equilibrate at room temperature for 1 hour. Aliquots calculated to give a final EDTA concentration of 20 μM are taken and tested for inhibition of the reduction of nitroblue tetrazolium by the superoxide radical.

The assay solution contains 2 x 10^{-5} hypoxanthine, 5 x 10^{-4}M nitroblue tetrazolium, 1 mg/ml gelatin and 100 mM oxygenated Hepes buffer, pH 7.8, to a total assay volume of 3.0 ml.

The reaction is initiated by the addition of 30 μl of 0.1 units/ml xanthine oxidase and allowed to go to completion at room temperature. Reduced nitroblue tetrazolium is measured photometrically at 550 nm.

Active compounds are checked for a null effect on the activity of XOD.

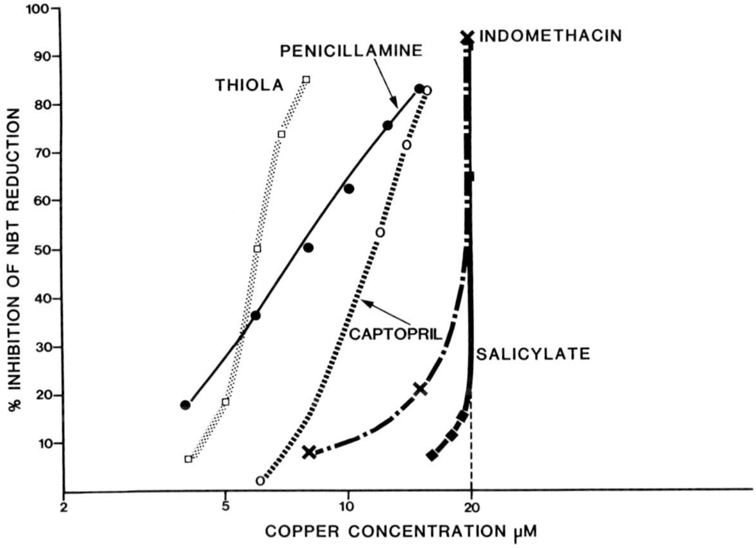

Fig. 1. Comparative activity profiles of SAARDs and NSAIDs

RESULTS AND DISCUSSION

In the absence of EDTA most copper chelates give an I_{50} of about 0.5 μM in this assay. In the presence of 20 μM EDTA some (stable) chelates have an I_{50} significantly below 20 μM and hence the ligand competes effectively against EDTA for copper binding (ACTIVE). Weak chelates lose all their copper to EDTA until this is saturated (i.e. with 20 μM Cu). Their SOD-like activity is then apparent and they give an I_{50} value very close to 20 μM (INACTIVE).

Some ligands form copper chelates without SOD-like activity (like EDTA). These do not inhibit NBT reduction even at chelate (copper) concentrations far in excess of 20 μM (EDTA-like).

The tabulated results show that D-penicillamine, captopril and thiola are all ACTIVE in this test (as defined above) while indomethacin, salicylate and other NSAIDs are INACTIVE. This distinction correlates well with clinical anti-rheumatic activity.

TABLE 1

COMPARISON OF STABLE SOD-LIKE ACTIVITY IN VITRO WITH CLINICAL ANTIRHEUMATIC ACTIVITY

Ligand	I_{50} μM	Test Activity Classification	Clinical Antirheumatic Activity
D-Penicillamine	8	Active	+
Thiola	6	Active	+
Captopril	11.5	Active	+
Aspirin	20	Inact	−
Indomethacin	19	Inact	−
Carprofen	19	Inact	−
Trien	$\gg$20	EDTA-like	−
L-Cysteine methyl ester	$>$20	EDTA-like	−
Azathioprine	$\gg$20	EDTA-like	+
Chloroquine	19.5	Inact	+
Sulphasalazine	20	Inact	+

Trien is a good copper ligand in vivo and is used clinically in the treatment of Wilson's disease where it increases cupriuresis. However, its copper chelate is without SOD-like activity. It is also clinically inactive as an antirheumatic agent. The same is true of the mucolytic agent and thiol L-cystine methyl ester. Thus the assay does not indiscriminately detect all good copper chelators or all thiols. Antirheumatic agents which may have a different mode of action to D-penicillamine are not detected.

SUMMARY

An assay is described which simultaneously tests for the SOD-like activity of copper chelates and their stability in the presence of excess EDTA. The EDTA is introduced to simulate the competitive endogenous ligands a copper chelating drug will encounter in vivo. Under the conditions of the assay D-penicillamine and like drugs are clearly distinguished from NSAIDs. The test systems may therefore be useful in the identification of novel antirheumatic agents of the D-penicillamine-type.

REFERENCES

1. Lengfelder E & Elstner E F (1978) Hoppe-Seyler's Z Physiol Chem 359 : S.751
2. Joester K-E, Jung G, Weber U & Weser U (1972) Febs Letters 25 : 25

© 1986 Elsevier Science Publishers B.V. (Biomedical Division)
Superoxide and Superoxide Dismutase in Chemistry,
Biology and Medicine; G. Rotilio, editor

IN VIVO INHIBITION OF SUPEROXIDE PRODUCTION OF GRANULOCYTES BY THE ANTI-RHEUMATIC DRUG PIROXICAM. AN INTERFERENCE WITH THE ACTIVATION OF NADPH-OXIDASE

P. BIEMOND[1,2], A.J.G. SWAAK[3], J.M.A. PENDERS[4], C.M. BEINDORFF[1] AND J.F. KOSTER[1]
[1]Dept. of Biochemistry I, [2]Dept. of Internal Medicine III, Erasmus University Rotterdam, [3]Dept. of Rheumatology, Daniël den Hoed Clinic and [4]Medical Service, Pfizer, Rotterdam, The Netherlands.

INTRODUCTION

In rheumatoid arthritis (RA), as well as in other inflammation diseases, polymorphonuclear cells (PMN) and macrophages are stimulated. This results in the secretion of inflammatory mediators, including large amounts of superoxide (O_2^-) and hydrogen peroxide (H_2O_2) (1). These activated oxygen intermediates and the secondary formed radicals,like hydroxyl radicals ($OH^\cdot$), are able to destroy membrane lipids, proteins, DNA, hyaluronic acid and cartilage (2,3). For these reasons the importance of oxygen-free radicals and related activated oxygen intermediates in the pathogenesis of various diseases is increasingly recognized. The PMN cells produce O_2^- by the activity of NADPH oxidase, an enzyme complex located in the plasma membrane and phagosome. In the non-stimulated cell this oxidase is inactive, however, upon stimulation an activation mechanism for NADPH oxidase is triggered. Little is known about the molecular basis of this activation mechanism. It has been shown for different anti-rheumatic drugs that they are able to inhibit the production of free oxygen radicals (4). Therefore, it is not unlikely that the beneficial effect of the non-steroidal anti-inflammatory drugs (NSAID) is not only based on the inhibition of cyclooxygenase (5), but also by preventing the formation of oxygen-free radicals.

Encouraged by the reports (6,7) we focussed our study on the *in vivo* inhibition of O_2^- production from PMN by piroxicam. The O_2^- production of PMN is tested before and different times after the administration of piroxicam in RA, osteoarthritis and controls. Further a start has been given to investigate the molecular level at which piroxicam exerts its action.

MATERIALS AND METHODS

PMN were harvested from defribinated blood by sodium metrizoate-ficoll centrifugation as described in ref. 8.

To measure the O_2^- production of PMN the cytochrome *c* reduction method was used.

Membrane fragments were isolated from PMN according to Tauber et al. (9). Isolated PMN were preincubated with or without 100 µM piroxicam for 5 min followed by stimulation of the cells by incubation with 100 mg/l PMA for 5 min at 37°C.

The *in vivo* effect of piroxicam was tested in patients with classical or definite RA, patients with osteoarthritis and in healthy controls. Piroxicam was given orally, 40 mg at time zero and another 20 mg 24 h later. Blood was taken for PMN isolation and to obtain a serum sample at the indicated times. Serum piroxicam levels were measured with high performance thin layer chromatography (10).

RESULTS

O_2^- production of PMN isolated from blood from patients suffering from RA and OA and healthy controls, before treatment with piroxicam was measured. The individual values are the mean value of two preparations obtained with a time interval of 24 h. Fig. 1 shows the O_2^- production after stimulation with either PMA or serum treated zymosan. Stimulation with PMA reveals no significant difference in O_2^- production between RA and control, but OA produced 129% more O_2^- than controls ($p<0.05$). Using serum treated zymosan O_2^- production was significantly higher in RA and OA, 212% ($p<0.02$) and 154% ($p<0.02$), respectively.

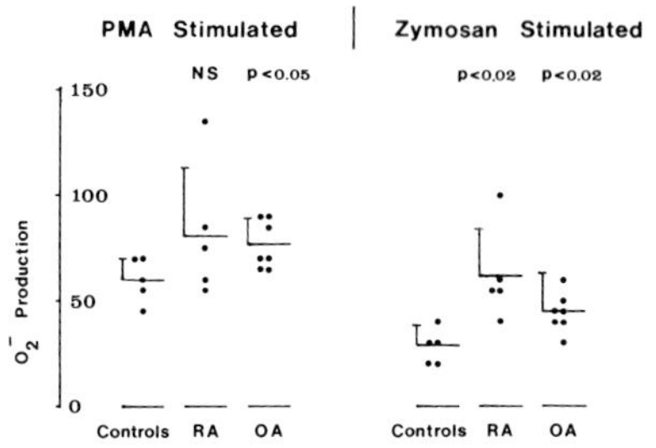

Fig. 1. O_2^- production of PMN.

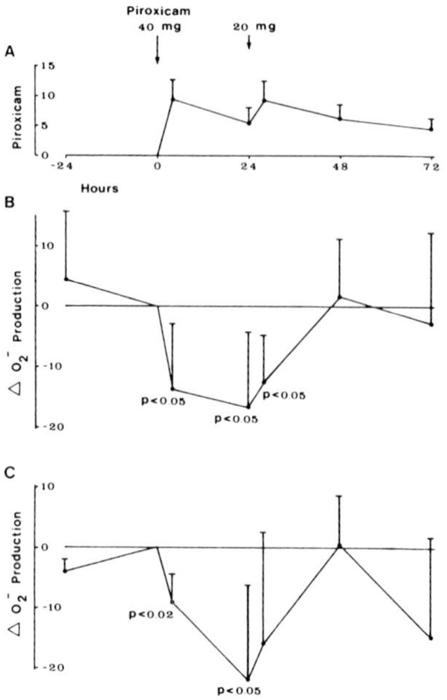

Fig. 2. Effect of piroxicam on O_2^- production (controls).

The effect of administrating piroxicam orally on the O_2^- production of isolated PMN from RA, OA and controls was investigated. Piroxicam was administered as described in the Methods section. The O_2^- production was measured in isolate PMN obtained at the indicated times. PMA and serum treated zymosan was used as stimulus (Fig. 2). PMN preparations were washed five times to eliminate serum piroxicam. For each individual the difference in O_2^- production at various times is expressed in comparison with th O_2^- production at time zero, immediately before the first dose of piroxicam. Without stimulation with PMA or serum treated zymosan the PMN produced little O_2^- and this was not changed by treatment with piroxicam. Stimulation of the cells in the control group (n=5, Fig. 2) there was a significant decrease in O_2^- production 4 h after taken 40 mg piroxicam independently of PMA or serum treated zymosan as stimulus. At 28 h, 4 h after the second dose of piroxicam (20 mg), the O_2^- production was only significantly reduced when serum treated zymosan was used as stimulus. At 48 h and 72 h the O_2^- reached the initial level. The t=-24 h indicate the variation of O_2^- production without treatment. For RA patients (n=5) the same pattern was obtained. After piroxicam treatment at 4 and 28 h a significant decrease in O_2^- production was seen, followed by a return to the initial O_2^- production. Also OA followed the same pattern (n=5), but only at 28 h a significant decrease in O_2^- production could be measured. Without piroxicam treatment control showed no significant change in O_2^- production over a period of 4 days and also no diurnal variation was found.

The effect of piroxicam was essentially the same in all three groups and lead in general to an average decrease of O_2^- production of 25%.

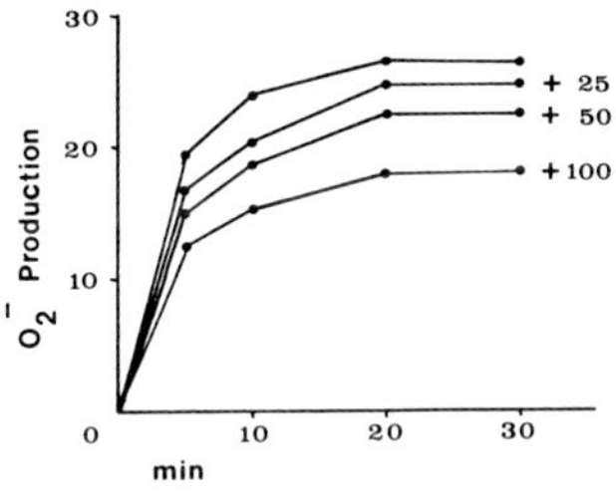

Fig. 3. "In vitro" inhibition by piroxicam of O_2^- production.

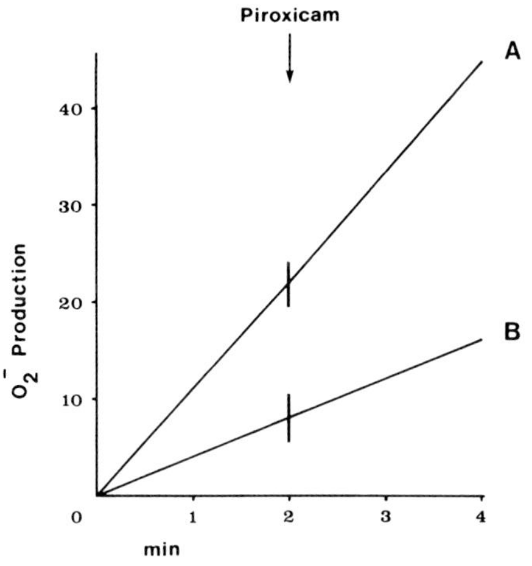

Fig. 4. O_2^- production of membrane fragments. A control, B piroxicam.

Piroxicam was able to inhibit the O_2^- production of intact PMN in a concentration dependent matter (Fig. 3). To identify the mode of action on the O_2^- producing NADPH oxidase complex, PMN were homogenized after stimulation of the NADPH oxidase with PMA. The capacity of the membrane fragments to produce O_2^- was not changed by the addition of 100 µM piroxicam during the O_2^- production (Fig. 4). Membrane fragments obtained from PMN, stimulated with PMA in the presence of piroxicam produced smaller amounts of O_2^- in comparison with membrane fragments obtained from PMN stimulated in the absence of piroxicam. Addition of piroxicam to these membrane fragments did not result in an enhanced inhibition of O_2^- production. These results indicate that piroxicam has no effect on the NADPH oxidase complex itself if it is already stimulated, but that piroxicam interferes in the mechanism by which the NADPH oxidase complex is activated.

DISCUSSION

Previous investigations have revealed that NSAID like indomethacin, ibuprofen, phenylbutazone and piroxicam (4,11) are able *in vitro* to decrease the release of O_2^- from PMN. Abramson et al. (12) showed that piroxicam did also inhibit O_2^- production of PMN *in vivo* in healthy volunteers. In four RA patients no significant effect was found (12). In these studies only at two times, before and during piroxicam treatment, the O_2^- production was measured. In our study two piroxicam doses were given and the O_2^- production was measured at seven different times. In contrast to Abramson et al. (12) we found for all three groups significant decreases in O_2^- production. Probably *in vivo* PMN will be inhibited to a larger extent, because the isolation procedure lasts some time and involved frequent washing of the cells. The conclusion can be drawn that *in vivo* piroxicam is able to decrease the O_2^- production of PMN to a considerable extent.

From the experiments with membrane fragments the conclusion has to be drawn that piroxicam has no inhibitory effect on the activated NADPH oxidase, but that piroxicam is interfering in the activation mechanism. Little is known about the mechanism by which NADPH oxidase is activated. Therefore, piroxicam can potentially be very useful for investigations designed to elucidate this mechanism.

REFERENCES

1. Babior BM, Kipnes RS, Curnutte JT (1973) J Clin Invest 52:741-744
2. Halliwell B, Gutteridge JMC (1984) Biochem J 219:1-14
3. Freeman BA, Crapo JD (1982) Lab Invest 47:412-426
4. Gay JC, Lukens JN, English DK (1984) Inflamm 8:209-222
5. Vane JR (1971) Nature new biol. 231:232-235
6. Abramson S, Edelson H, Kaplan H, Given W, Weissmann G (1984) Inflamm 8:S103-S108
7. Kaplan HB, Edelson H, Korchak HM, Given W, Abramson S, Weissmann G (1984) Biochem Pharm 33:371-378
8. Biemond P, Van Eijk HG, Swaak AJG, Koster, JF (1984) J Clin Invest 73:1576-1579
9. Tauber AI, Goetzl EJ (1979) Biochemistry 18:5576-5584
10. Riedel KD, Laufen H (1983) J Chromatogr 276:243-248
11. Edelson HS, Kaplan HB, Korchak HM, Smolen JE (1982) Biochem Biophys Res Commun 104:247-253
12. Abramson S, Edelson HS, Kaplan HB, Ludewig R, Weissmann G (1984) Am J Med 77(4B):3-6

Superoxide and Superoxide Dismutase in Chemistry,
Biology and Medicine; G. Rotilio, editor

ROLE OF OXYGEN RADICALS IN RAT ADJUVANT ARTHRITIS

T. YOSHIKAWA, H. TANAKA, N. YOSHIDA, T. TANIGAWA, M. TANIGAWA, H. MIYAGAWA, S. SUGINO AND M. KONDO

First Department of Medicine, Kyoto Prefectural University of Medicine, Kamikyo-ku, Kyoto 602, Japan.

INTRODUCTION

Our previous study described the increase in lipid peroxide level in the synovial fluids of patients with rheumatoid arthritis (RA) (1) and in the synovia of rats with adjuvant arthritis (2). In addition, the aggravation of adjuvant arthritis could be reduced by vitamin E (2). These observations show the close relationship between free radical lipid peroxidation and arthritis. McCord speculated that superoxide radicals are an important factor in RA (3), and Fridovich suggested that superoxide dismutase (SOD), injected into an inflamed area, minimizes the damage caused by superoxide anion secreted by phagocytes (4).

In this study, we investigated the effects of SOD and indomethacin on lipid peroxidation in rat adjuvant arthritis.

MATERIALS AND METHODS

Wistar strain female rats 6 weeks of age were used. Experimental adjuvant arthritis was induced by the injection of heat-killed *Mycobacterium tuberculosis.* Each animal received 0.1 ml of a 0.6% (6 mg/ml) adjuvant suspension in heavy mineral oil injected subcutaneously into the tail.

On the day of adjuvant administration, the animals were divided into four groups; injected subcutaneously once daily with 0.5 ml of physiological saline, 5.0 mg/kg of SOD or 10.0 mg/kg of SOD, or administered orally once daily with 1.5 mg/kg of indomethacin.

The rat knee joint synovia were isolated, mixed with cold saline solution in a 1:100 ratio (w/v), and homogenized in a Teflon Potter-Elvehjem homogenizer.

The serum level of acid phosphatase was determined by the method of Shibko and Tappel (5), and the serum β-glucuronidase, by the method of Kato *et al.* (6) the serum and tissue levels of thiobarbituric acid (TBA)

reactive substances were determined by the methods of Yagi (7), and Ohkawa et al. (8), respectively.

The degree of swelling in the extremities was evaluated by immersing them to a specified depth in a solution containing 0.04% aerosol and measuring the degree of swelling as the percentage increase in solution level after the injection of adjuvant.

Results

Swelling of legs. The swelling reached a peak level at 3 and 6 weeks after the adjuvant injection in the saline-treated control rats, but was slight at 3 weeks in SOD (5.0 mg/kg)-injected rats, 3, 4 and 5 weeks in SOD (10.0 mg/kg)-injected rats, and 2, 3, 4, 5, 6 weeks in indomethacin-injected rats.

Albumin/globulin ratio. The albumin/globulin (A/G) ratio is diminished in severe arthritis (9). It was found that the A/G ratio was most depressed at 2 weeks after the adjuvant injection in the control group. The depression was inhibited significantly by the injection of 10.0 mg/kg of SOD and 1.5 mg/kg of indomethacin.

Serum levels of acid phosphatase and β-glucuronidose. The serum levels of acid phosphatase rose at 1 week after the adjuvant injection. The increase was reduced significantly at 1 or 2 weeks after the adjuvant administration in rats injected with 10.0 mg/kg of SOD and at 2 weeks in rats treated with 1.5 mg/kg of indomethacin. A similar pattern was observed in the changes of β-glucoronidase level.

Serum levels of TBA reactive substances. The serum levels of TBA reactants increased significantly after the adjuvant injection. The level in the control rats rose steeply at 2 weeks after the adjuvant injection and showed a gradual decline thereafter. In the SOD and indomethacin administered rats, the increase was inhibited significantly.

Knee joint synovial levels of TBA reactive substances. The rat knee joint synovial membrane was isolated, and the synovial levels of TBA reactants were determined. The bilateral synovial membranes were isolated simultaneously and homogenized together for use in the determination of synovial levels of TBA reactants. The synovial levels of TBA reactants in the control rats rose sharply at 2 weeks after the adjuvant injection and showed a decline thereafter. The increase in TBA reactants was reduced significantly by the administration of SOD and indomethacin.

DISCUSSION

In the present investigation, inflammatory reaction indicated by leg swelling and a depressed A/G ratio were inhibited by the SOD and indomethacin administration. The serum levels of lysosomal enzymes and TBA reactive substances, and the synovial levels of TBA reactants rose with the development of adjuvant arthritis. The elevation of the levels of lysosomal enzymes were reduced to some extent, and the increase of TBA reactants was inhibited significantly by the SOD and indomethacin administration. These findings suggest that lipid peroxidation might promote lysosomal membrane destruction and arthritis, and that SOD and indomethacin may, in part, exert their anti-inflammatory effect either directly or indirectly by preventing lipid peroxidation damage of the synovial membrane.

REFERENCES

1 Yoshikawa T, Yokoe N, Takemura S, Kato H, Hosokawa K, and Kondo M (1979) Japan. J. Med. 18: 199-204

2 Yoshikawa T, Tanaka H, and Kondo M, (1983) Biochem. Med. 29: 227-234

3 McCord JM, (1974) Science (Washington, D.C.) 185: 529-531

4 Fridovich I, (1975) Annu. Rep. Med. Chem. 10: 257

5 Shibko S, and Tappel AL (1963) In: Biochim. Biophys. Acta 73: 76-86

6 Kato K, Yoshida K, Tsukamoto H, Nobunaga M, Masuya T, and Sawada T, (1969) Chem. Pharm. Bull. 8: 239-242

7 Yagi K, (1976) Biochem. Med. 15: 212-216

8 Ohkawa H, Ohishi N, and Yagi K, (1979) Anal. Biochem. 95: 351-358

9 Piliero SJ, Graeme M L, Sigg EG, Chines G, and Colombo C, Life Sci. 5: 1057-1069

Superoxide and Superoxide Dismutase in Chemistry, Biology and Medicine; G. Rotilio, editor

ORGOTEIN SUPPOSITORIES IN ACUTE IDIOPATHIC PROCTITIS

T. RANZI, L. PALMISANO*, L. PIODI, P. VELIO, P. BIANCHI

Cattedra di Patologia Medica III, Clinica Medica I, Università di Milano
*Istituto di Ricerca C. Serono, Roma

INTRODUCTION

Idiopathic proctocolitis and, in general, inflammatory bowel disease (IBD), is a chronic inflammatory condition with sole recto-colic involvement, unknown etiology and relapsing-remitting course. Pathological aspects are marked by inflammatory changes, generally considered to be immunologically mediated.

There are at least two observations which suggest that the lesions, in IBD, could be, at least in part, oxygen-mediated. First, the presence of leukocyte infiltration may account for the production of an excess of oxygen radicals; indeed, lymphocytes and macrophages (largely represented even in healthy gastrointestinal tract) are potential sources of oxygen free radicals (1). Secondly, an increased chromosomal breakage has been documented in lymphocytes from patients with IBD; the protective effect of in vitro added Superoxide Dismutase (SOD) seems to support a role of active oxygen in the mechanisms of the disease. In Crohn's disease, particularly, such chromosomal aberrations can be transferred to lymphocytes in healthy subjects both by intact lymphocytes and serum ultraphiltrates from people affected; the incubation of cells with SOD is able to inhibit chromosomal breakage (2-3).

On the basis of these observations, we evaluated the in vivo effect of Orgotein, a Cu-Zn SOD (Orgoten Serono), in patients with idiopathic proctitis or proctosigmoiditis. These conditions were chosen because they are easy to treat with topical therapy and to monitor by endoscopy.

MATERIALS AND METHODS

Six patients, aged between 24-34 years, entered the study. In 3 cases (pts N° 1, 5,6) sigmoid colon was involved, too, whereas in the remaining three only rectum was affected. All of them fulfilled diagnostic criteria (clinical, endoscopic, bioptical) for IBD (4), other inflammatory diseases of known etiology having been ruled out.

TABLE I

ORGOTEIN SUPPOSITORIES IN ACUTE IDIOPATHIC PROCTITIS

Patients' features at entry and results

Patient	Age (yrs)	Present involvement (*)	Duration of the disease (mo)	Days of therapy	Results Clinics (C) Endoscopy (E) Biopsy (B)	Simultaneous administration of oral Salazopyrin (gr/die)
1 C.F.	25	PS	30	15	Unchanged (C,E,B)	3
2 T.T.	32	P	36	15	Resolution (C,E,B)	3
3 D.F.F.	28	P	42	15	Unchanged (C,E,B)	3
4 P.G.	34	P	/	15	Resolution (C,E,B)	/
5 D.A.G.	44	PS	64	20	Resolution (C,E)	3
6 V.L.	24	PS	24	20	Marked improuvement (C,E)	2.5

* P = Proctitis.
PS = Proctosigmoiditis.

Only one patient (N° 4) was in his first episode of disease, whilst the others had already suffered from at least 2 acute attacks, with a duration of the disease ranging from 24 to 64 months. The latter subjects were all receiving oral maintainance therapy with salazopyrin (2.5-3g/die) which was not discontinued during the trial. No patient showed abnormalities, when routine laboratory tests (blood cell count, ESR, blood nitrogen and glucose, transaminases and urine analysis) were performed.

Treatment schedule with Orgotein was as follows: patients N° 1,2,3,4: 24 mg suppositories. 1 supp./die for 15 days. Patients N° 5,6: 12 mg suppositories. 2 supp./die for 20 days.

Endoscopy was performed at time 0 and after 7 and 15 days or 7, 15 and 20 days in the first 4 patients and in the remaining two, respectively. In addition, biopsy was performed in the first 4 patients at the end of the treatment.

These data are summarized in Table I.

RESULTS AND DISCUSSION

In 2 patients (N° 1 and 3) the disease persisted activly at the end of the treatment; in one patient (N° 6) a marked improuvement was observed; finally, in 3 patients (N° 2,4,5) the complete resolution of the acute episode was noted (Table I). No local or systemic side effects of Orgotein therapy were observed.

In the year which followed this pilot study, all the patients were still receiving salazopyrin and underwent one or more recurrences.

In conclusion, in 4 out of 6 patients topic treatment with Orgotein resulted in a favourable action, free from side effects. Obviously, these preliminary results need to be confirmed; for this purpose a double blind, randomized clinical trial in 40 patients is presently underway.

REFERENCES

1. Kagnoft MF (1981) In: Johnson LR (ed) Physiology of the gastrointestinal tract. Raven Press, New York
2. Emerit I (1983) Lymphokines 8:413
3. Emerit I, Michelson AM (1981) In: Pena A.S. (ed) Recent advances in Crohn's disease. Martinus Nijhoff Pub., Hague, pp 225
4. (1983) In: Allan RN, Keghley MRB, Alexander Williams J., Hawkins C (eds) Inflammatory bowel disease. Edimburg, Churchill Livingstone

Superoxide and Superoxide Dismutase in Chemistry,
Biology and Medicine; G. Rotilio, editor

HEME - ACTIVATION BY TETRACHLORODECAOXIDE (TCDO)

E.F.ELSTNER, F,GOETZ, K.HOLLFELDER, H.HEINISCH, K.H.SCHLEIFER and F.W.KÜHNE*
Techn. Universität München, Inst. für Botanik und Mikrobiologie
* OXO- Chemie, Heidelberg, (W- Germany)

INTRODUCTION

Oxoferin[R] as an aqueous 3 mM solution of TCDO ($Cl_4O_{10} \cdot H_2O$)$^{n-}$ has been shown to accelerate secundary wound healing (1). This process is dependent on a) collagen synthesis and angiogenesis and b) activation of blood cells such as macrophages, granulocytes and neutrophils (2,3). Since the processes a) and b) are directly or indirectly dependent on the availability of oxygen, wound healing is strongly impaired by hypoxic conditions and can be stimulated by leucocyte activation, reoxygenation and treatment with bactericidal compounds. TCDO stimulates phagocytosis and exhibits bactericidal activities (4). Here we report on some biochemical and microbiological properties of TCDO.

RESULTS AND DISCUSSION

I) Heme - TCDO as an activated oxygen species

As recently reported (5,6)TCDO activates certain heme compounds such as peroxidase, hemoglobin, myoglobin or cytochrome c which are converted into oxidants similar to " compound I ". This activation can be monitored as ethylene release from S-methyl-α-ketobutyrate (KMB) or 1-aminocyclopropane-1-carboxylate (ACC). In contrast to $OH^{\cdot}$ or $^1O_2$ - generators, the TCDO - heme reaction does not produce ethylene from methionine (6). The detection limit for TCDO with ACC as monitor molecule is ca. 200 pM.

II) Heme - dependent oxygen release from TCDO

Molecular oxygen is released from TCDO (determined polarographically) with different rates, dependent on the individual heme-compound (table 1). Besides oxygen, the only other breakdown product observed is chloride (Cl^-). During the course of oxygen release, the heme compound is oxidized as shown for cytochrome c in figure I. 1µl of concentrated TCDO (= 158 mM $Cl_4O_{10}{}^{n-}$), corresponding to 0,6 µmol, complete the oxidation of 2.5 x 10^{-5} M cytochrome c (horse heart, Sigma) within 30 min.Superoxide dismutase has no influence on this reaction.

TABLE I

HEME DEPENDENT OXYGEN RELEASE FROM TCDO (10 µmol/ ml)

	Heme activators * (10^{-5}M)							
	none	Hb	Mb	Myo	POD	Hem	Cyt.c	ProP IX
nmol O_2 release/ ml in 5 min.	0	150	140	25	25	16	5-10	5-10

* Hb= hemoglobin (Sigma), Mb= methemoglobin (Sigma);Myo= myoglobin (Sigma); POD= peroxidase (horse radish, Boehringer); Hem= hemin (Serva); Cyt.c= cytocrome c (horse heart, Boehringer); ProP IX= Protoporphyrin IX (Sigma).

FIGURE I

OXIDATION OF REDUCED CYTOCHROME C BY TCDO

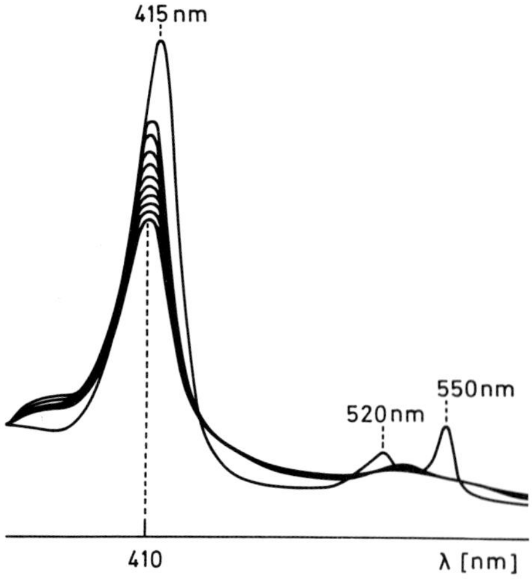

Method: In 3 ml cuvette:
0.1 M phosphate buffer pH 7.8
2.5 x 10^{-5} M cytochrome c
addition of 0.6 µmol TCDO; mix
scanning time: 30 min.

Analogous observations are made with the other " active " heme compounds.

III) Reaction of TCDO - peroxidase with linolenic acid

TCDO enhances the peroxidase- (POD, Boehringer, 1 U/ ml) driven diene conjugation of linolenic acid, observed as the increase of absorbance at 234 nm. Catalase, superoxide dismutase (100 U) or mannitol (10 mM) have no influence on this reaction (figure 2).

FIGURE II

OXIDATION OF LINOLENIC ACID BY TCDO- POD

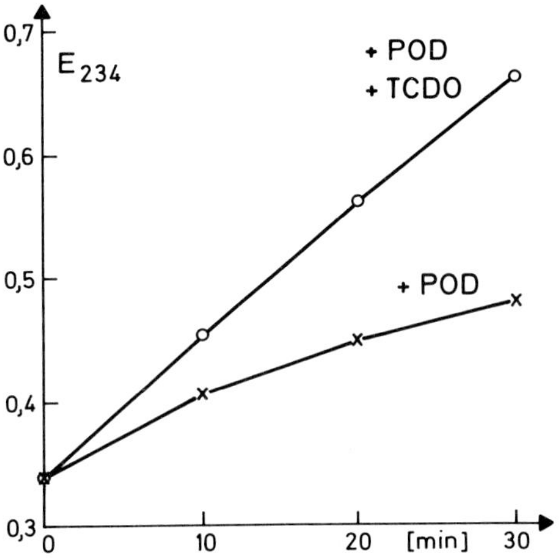

Method: In 3 ml cuvette:
linolenic acid: 7.5 x 10^{-6}M
POD: 3 U (Boehringer)
TCDO: 6 µmol
in: 0.1 M phosphate buffer pH 7.8

Analogous observations are made with myoglobin instead of POD.

IV) Spectrum of bactericidal activities of TCDO-activated hemin

In table II, maximal survival times of several bacterial cultures in TCDO (10 mM) or TCDO + 10 µ M hemin are listed:

TABLE II

BACTERICIDAL ACTIVITIES OF TCDO - HEMIN (SURVIVAL TIMES: min)

addition	culture				
	E. coli	S.aureus	Mc.luteus	Ps.fluorescens	C.pasteurianum
TCDO	60	60	60	60	15
TCDO + hemin	5	1	2.5	2.5	0.5

TCDO in the presence of hemin exhibits highly significant bactericidal activity, especially against anaerobes such as clostridia. Hemoglobin or myoglobin can substitute for hemin.

V) The active oxidant

In approximation to the ClO_2^- - deuteroporphyrin IX - Fe^{III+}-complex reported by Kelly et al (7), the TCDO- heme complex might decay as follows:

$$4\ Fe^{III+}\text{-(heme)} + (\ Cl_2O_{10} \cdot H_2O\ )^- \longrightarrow 4\ Fe^{V+}O + 4\ OCl^- + O_2 + H_2O$$

$$4\ Fe^{III+}\text{-(heme)} + 4\ OCl^- \longrightarrow 4\ Fe^{V+}O + 4\ Cl^-.$$

$Fe^{V+}O$ might be seen as the active species (oxidant), responsible for the observed biochemical and biological effects.

REFERENCES

1. Hinz,J.,Hautzinger,H.,Helling,J.,Schirren,G., Sell,G., Stahl, K.W. and Kühne, F.W.(1984) Fortschritte der Medizin 18:523-528

2. Hunt,T.K., Knighton,D.R., Thakral,K.K., Goodson, W.H. and Andrews, W.S. (1984) Surgery 96: 48-54

3. Kühne,H.H., Ullman,U. and Kühne,F.W. (1985) Infection 13: 52-56

4. Ullmann,U. and Kühne, F.W. (1984) Infection 12: 225-229

5. Elstner,E.F. and Youngman,R.J. (1984) In: Rotilio, G., Bannister,J.V. (eds) Oxidative damage and related enzymes. Life Chem. Reports suppl. 2, Harwood Acad. Publ. Chur-London-Paris-New York. pp 48-54

6. Youngman,R.J., Wagner,G.R., Kühne,F.W. and Elstner,E.F. (1985) Z. Naturforsch. 40c: 409-414

7. Kelly,H.C., Parigi,K.J., Wilson,I., Davies,D.M., Jones,P. and Roettger,L.J. (1981) Am. Chem. Soc. 20: 1086-1090.

© 1986 Elsevier Science Publishers B.V. (Biomedical Division)
Superoxide and Superoxide Dismutase in Chemistry,
Biology and Medicine; G. Rotilio, editor

GLY-L-HIS-L-LYS:COPPER(II) - A HUMAN PLASMA GROWTH FACTOR WITH SUPEROXIDE DISMUTASE-LIKE AND WOUND-HEALING PROPERTIES

LOREN PICKART[1], D. DOWNEY[2], STEVE LOVEJOY[3] AND BORIS WEINSTEIN[3]
[1] Iama, 15232 S.E. 48th, Bellevue, WA 98006 (USA), [2] Virginia Mason Hospital, Seattle, WA 98101 (USA), [3] Department of Chemistry, University of Washington, Seattle, WA 98195 (USA)

INTRODUCTION

GHL-Cu (Gly-L-His-L-Lys:Copper(II)) is a growth factor isolated from human plasma (1,2). We have found GHL-Cu to accelerate the healing of superficial wounds in rats, mice, and pigs. The biochemical basis of this action appears to reside in GHL-Cu's possession of tissue-protective superoxide dismutase-like activities and its facilitation of angiogenesis by chemoattractive effects on capillary endothelial and mast cells.

GHL has an affinity for copper(II) equivalent to that of albumin for copper(II) and GHL-Cu should be a stable complex *in vivo* (1). The addition of GHL alone to biological systems is likely to result in a significant portion being converted to GHL-Cu. An analog of GHL, glycyl-L-histidine (GH), forms a similar copper(II) complex (3) and possesses significant superoxide dismutase-like activity (4). Preliminary measurements on GHL-Cu using pulse radiolysis techniques have given a rate constant of about 5×10^8 at pH 6.8 for the dismutation of superoxide anion (5), a rate similar to that found for copper(II) complexes of GH and glycyl-L-histidyl-L-leucine (4). This observation, coupled with reports of angiogenic activity in GHL-Cu (6), prompted testing of the complex on the healing rate of superficial wounds.

MATERIALS AND METHODS

GHL-Cu was prepared by published methods (2). In rats, a circular (2 cm diameter) patch of skin was excised from the back. The outer circumference of the wound was injected with 50 micrograms GHL-Cu in 0.05 ml phosphate buffered saline (or saline only for controls, 10 rats each group) three times - after wounding and 24 and 48 hours later. The wounds were inspected on days 5, 10, 15, 20 and 25 by debridement of the wound to a sharp wound margin, but preservation of the proliferating granulation tissue. Wounds were photographed along with a scale bar. The wound circumference was used as the measure of healing and quantified by computerized digitization of the perimeter of the wound in the photograph, then corrected with the scale bar. Time points and group information were coded to eliminate subjective bias.

In an additional blinded study of these effects, the photographs of wounds at 25 days were judged by four surgeons as fully-healed or not fully-healed, then the judgment statistically analyzed by the Fisher Exact Test.

The chemoattractant effect of GHL-Cu on capillary endothelial cells (BACE line) and human fibroblasts was assayed by standard methods in Boyden chamber with growth media and 10% serum (6).

RESULTS

A statistically significant enhancement of healing could be demonstrated 15 days after injury which increased at days 20 and 25 (Fig. 1). At 25 days after surgery, 6 of 10 of the GHL-Cu treated wounds were judged to be fully-healed while no controls were fully-healed ($p<0.03$). Qualitatively similar results were obtained on wounds in mice and pigs.

GHL-Cu induced a doubling of migration of capillary endothelial cells in the Boyden chamber assay, having a half-maximal effect at about 5×10^{-11} M, but was five orders of magnitude less active on fibroblasts (Fig. 1).

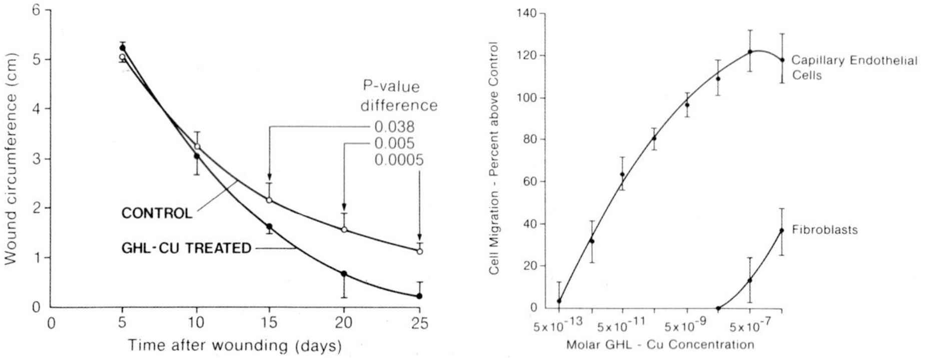

Fig. 1. Left: Effect of GHL-Cu on wound healing in rats. Statistical significance (T-statistic) between control and treated wounds is given at 15, 20, and 25 days. Right: Effect of GHL-Cu on migration of capillary endothelial and fibroblastic cells.

DISCUSSION

These results plus other evidence suggest a role for GHL-Cu in healing processes. Raju et al (6) found GHL-Cu induced the synthesis of a family of six proteins of molecular weights ranging from 35,000 to 66,000 in rabbit corneas which was immediately followed by non-inflammatory new capillary formation directed toward the implanted GHL-Cu. Protein extracts of the GHL-Cu treated corneal tissue induced a seven-fold increase in the migration of bovine capillary cells in the Boyden chamber assay. GHL is also a potent chemoattractant for mast cells, the assemblage of which around the growing tips of capillary sprouts is felt to be obligatory for proper blood vessel ingrowth (7).

Neurogensis is critical to proper healing and GHL induces axonal and dendrite outgrowth in cultured central and peripheral neurons (1). Kidney deterioration and loss of glomerular function after severe injury is common; GHL serves as the sole growth factor necessary for the growth of glomerular cells (8). GHL administered to mice increases the circulating level of erythropoietin, the hormone that stimulates red blood cell production (9). GHL may be produced by tissues to facilitate their vascularization since rapidly growing normal and neoplastic cultured cells produce either GHL or a GHL-like peptide (7).

We postulate GHL-Cu's activities to arise both from the molecules' tissue-protective superoxide dismutase-like activities that minimize tissue damage after wounding and its chemoattractant actions on cells critical for neovascularization. GHL-Cu did not significantly enhance the migration of fibroblastic cells which secrete the structural protein collagen into the wound area during early phases of healing. Thus, GHL-Cu may function in a secondary phase of healing during which the initial inflammation is suppressed and the wounded area is reintegrated with the vascular and neuronal systems. Superoxide dismutase activity may be biologically linked to the induction of angiogenesis and neurogenesis.

REFERENCES

1. Pickart L (1983) Lymphokines 8:425-446
2. Perkins CM, Stenkamp R, Rose NJ, Weinstein B, Pickart L (1984) Inorg Chim Acta 67:93-99
3. Kwa EY, Lin BS, Rose NJ, Weinstein B, Pickart L (1983) Peptides: Structure and Function 8:805-808
4. Weser U, Schubotz LM (1980) In: KD Rainsford (ed) Trace Elements in the Pathogenesis and Treatment of Inflammation. Birkhauser, Basel, pp 103-120
5. Beilski BHJ, Rose NJ, Pickart,L , work in progress.
6. Raju K, Alessandri G, Gullino P (1984) Cancer Res 44:1579-1584
7. Poole TJ, Zetter BR (1983) Cancer Res 43:5857-5861
8. Oberley TD, Murphy PJ, Steinert BW, Albrecht RM (1982) Virchows Arch Cell Pathol 41:145-170
9. Naughton B, Naughton G, Lui P, Zuckerman GB, Gordon AS (1982) J Surg Oncol 21:97-103

Superoxide and Superoxide Dismutase in Chemistry,
Biology and Medicine; G. Rotilio, editor

IRREVERSIBLE INHIBITORS OF IRON-SUPEROXIDE DISMUTASE: POTENTIAL ANTI-TRYPANOSOMAL AGENTS.

STEVEN MESHNICK*, MICHELE SIMON+, ABIGAIL FALK+, ROBERT GRADY+ AND NGUYEN LE TRANG#. *City University of New York Medical School, 137th St. and Convent Ave., New York, NY 10031, +Cornell University Medical College, New York, NY 10021 and #Rockefeller University, New York, NY 10021.

INTRODUCTION

Trypanosomes are protozoan parasites which are responsible for widespread disease in Latin America and Africa. We have been looking for unusual biochemical features in these parasites in order to design chemotherapeutic agents (1). We found that trypanosomes were unusual for eukaryotic animals in that they had iron-containing superoxide dismutases (Fe-SODs). We have subsequently purified an Fe-SOD from a related organism (Crithidia fasciculata) and have attempted to use it to search for potential chemotherapeutic agents (2,3). Recently, we found a biscatechol, N^1,N^6-bis(2,3-dihydroxybenzoyl)1,6-diaminohexane (bis-C6-DHB), that inhibits purified C. fasciculata Fe-SOD better than a mammalian Cu,Zn-SOD (4). In this paper, we report on our studies of the inhibitory and antitrypanosmal activities of a series of related quinones.

METHODS

SOD was prepared and assayed as previously described (4). All experiments were run in triplicate. Analogs of bis-C6-DHB were prepared as described (5): $N^{1,3}$-bis(2,3-dihydroxybenzoyl)-1,3-diaminopropane (bis-C3-DHB), $N^{1,4}$-bis(2,3-dihydroxybenzoyl)-1,4-diaminobutane (bis-C4-DHB), $N^{1,9}$-bis(2,3-dihydroxybenzoyl)-1,9-diaminononane (bis-C9-DHB). All other compounds were obtained from Aldrich Chemical Co. (Milwaukee, WI) or Sigma (St. Louis, MO). Antitrypanosomal activity was assessed in vivo against Trypanosoma brucei-infected mice as in earlier studies (6). Compounds were solubilized by sonication and then administered intraperitoneally 24 hours after infection. Mice were given either 1/4 or 1/2 of the calculated LD_{50}. Survival data was analyzed using the Wilcoxon rank sum test (7).

RESULTS

Four biscatechols were evaluated as inhibitors of crithidial Fe-SOD and mammalian Cu,Zn-SOD *in vitro* (Table 1). Activity against the Fe-SOD was dependent on the length of the diaminoalkane linker, with optimal inhibition achieved by bis-C6-DHB (with a six-carbon linker). None of the four compounds had substantial activity against the bovine erythrocyte SOD.

TABLE 1. INHIBITION OF CRITHIDIAL FE-SOD AND MAMMALIAN CU,ZN-SOD BY BISCATECHOLS

COMPOUND*	PERCENT INHIBITION (± S.D.)	
	FE-SOD	CU,ZN-SOD
bis-C3-DHB	22±19	10±5
bis-C4-DHB	33±8	4±5
bis-C6-DHB	60±16	7±9
bis-C9-DHB	33±4	8±8

*Enzyme preparations were incubated with 50 uM inhibitor for 24 hours.

Eight naphthoquinone-derivatives were also tested at concentrations of 50 uM. Five of these compounds (gossypol, lapachol, lawsone, menadione and menoctone) had little or no inhibitory activity (<20%) against crithidial Fe-SOD. On the other hand, three of the naphthoquinones tested completely inhibited the crithidial Fe-SOD (>98%): 1,2-naphthoquinone, 1,4-naphthoquinone and 1-hydroxy-4-amino-naphthalene (HAN). These three compounds were then tested for activity at 10 uM (Table 2). All three proved to be highly effective against the Fe-SOD (70 - 95% inhibition) and to have little effect at this concentration against the Cu,Zn-SOD (0 - 22% inhibition). These naphthoquinone-derivatives were then tested for antitrypanosomal activity *in vivo* (Table 3).

TABLE 2. INHIBITION OF CRITHIDIAL FE-SOD AND MAMALIAN CU,ZN-SOD BY NAPHTHOQUINONE-DERIVATIVES

COMPOUND*	PERCENT INHIBITION (± S.D.)	
	FE-SOD	CU,ZN-SOD
1,2-Naphthoquinone	95±11	22±16
1,4-Naphthoquinone	79±12	6±7
HAN	70±5	0±0

*Enzyme preparations were incubated with 10 uM inhibitor for 24 hours.

TABLE 3. EFFECTS OF NAPHTHOQUINONE-DERIVATIVES ON SURVIVAL OF *T. BRUCEI*-INFECTED MICE

TREATMENT	SURVIVALS	MEAN	SIGNIFICANCE
Control	3,4,4,4,4,4,4, 5,5,5	4.2	
1,2-Naphthoquinone			
2.5 mg/kg	4,5,5,5,5	4.8	>.05
5.0 mg/kg	5,5,5,5,5	5	<.05
1,4-Naphthoquinone			
1.25 mg/kg	5,5,5,6,6	5.4	<.01
2.5 mg/kg	3,4,5,5,5	4.4	>.05
HAN			
2.25 mg/kg	4,4,5,5,5	4.6	>.05
4.5 mg/kg	5,5,5,5,6	5.2	<.05

DISCUSSION

In an earlier publication, we have shown that the biscatechol, Bis-C6-DHB, inhibits crithidial Fe-SOD by alkylation. It is first oxidized to a quinone and then appears to covalently attach to a nucleophilic moiety in the vicinity of the active site (4). In this paper, we have studied the effect of related biscatechols as well as eight naphthoquinone-derivatives on both crithidial Fe-SOD and mamalian Cu,Zn-SOD. Three of the naphthoquinones tested, in fact, proved to be more effective that Bis-C6-DHB as inhibitors of Fe-SOD, while remaining weak inhibitors of the Cu,Zn-SOD. These compounds were then tested in vivo and found to have slight, but statistically significant, antitrypanosomal activity. We hope that further structure-activity studies on these compounds might lead to the development of an effective antitrypanosomal agent.

ACKNOWLEDGEMENTS

This work was supported by grants from the NIH (AI 20197), the World Bank/UNDP/WHO Special Programme for Research and Training in Tropical Diseases and the Irma T. Hirschl foundation. We thank J. Eaton, J. Haldane and A. Swiecicki for their assistance.

REFERENCES

1. Meshnick SR (1984) Pharmacol Ther 25:239-254.
2. Trang NL, Meshnick SR, Kitchener K, Eaton JW, Cerami A (1983) J Biol Chem 258:125-130.
3. Meshnick SR, Trang NL, Kitchener K, Cerami A, Eaton JW. (1983) In: Cohen G, Greenwald R (eds.) Oxy Radicals and Their Scavenger Systems: Molecular Aspects. Elsevier, New York, pp. 348-352.
4. Meshnick SR, Kitchener KR, Trang NL (1985) Biochem Pharmacol (in press).
5. Bhargava KK, Grady RW, Cerami A (1980) J Pharm Sci 69:986-989.
6. Meshnick SR, Grady RW, Blobstein SH, Cerami A (1978) J Pharmacol Exp Ther 207:1041-1050.
7. Colton T (1974) Statistics in Medicine. Little, Brown & Co., Boston.

Superoxide and Superoxide Dismutase in Chemistry, Biology and Medicine; G. Rotilio, editor

THE ANTIMALARIAL ACTIVITY OF DDC-TREATED SUPEROXIDE DISMUTASE

*STEVEN MESHNICK, +MARIA FEBBRAIO, #SANDRA EDWARDS, +AVIVA ABOSCH, +ALEXANDRA FAIRFIELD AND #JOHN EATON. *City University of New York Medical School, 137th St. and Convent Ave., New York, NY 10031, +Cornell University Medical College, New York, NY 10021 and the University of Minnesota Medical School, Minneapolis, MN 55455.

INTRODUCTION

Malaria, caused by protozoan parasites of the genus *Plasmodium*, represents one of the most serious public health problems in the world today (1). These parasites, which are intraerythrocytic for much of their lifecycle, appear to be particularly sensitive to agents which are generators of activated oxygen (2,3). This sensitivity may, in part, be due to a deficiency of endogenous parasite superoxide dismutase (SOD) (4,5). Indeed, intraerythrocytic malaria appear to import and utilize host SOD (4-6). We have been interested in using inhibitors of host SOD to enhance the antimalarial activity of oxidant drugs. In the course of these investigations, we made the unusual finding that bovine erythrocyte SOD, which has been pretreated with the inhibitor diethyldithiocarbamate (DDC), appears to have antimalarial activity.

METHODS

Plasmodium falciparum (strain FCR 3) was cultured in candle jars (7). Bovine erythrocyte SOD, DDC and bovine serum albumin (BSA) were obtained from Sigma Chemical Co, St. Louis, MO. DDC-treatment of SOD and BSA were performed by dissolving both proteins to 4 mg/ml in Hank's buffer containing 0.1 M DDC and incubating for 2 hours at 37°. Treated protein was then separated from free DDC by gel filtration on PD-10 columns (Pharmacia, Piscataway, NJ). SOD was assayed by the method of McCord and Fridovich (8). Antimalarial activity was assessed by adding treated and untreated protein to triplicate wells of microtiter plates and then counting parasitemias after 48 hours.

RESULTS

DDC-treated SOD had no detectable enzyme activity (< 10 units/mg protein) whereas sham-treated enzyme contained 4200 units/mg. When added to cultures of

P. falciparum in human red blood cells, the DDC-treated enzyme (and not the untreated SOD) cleared cultures of parasites (Table 1). Concentrations of DDC-treated SOD as low as 100 ug/ml (3 uM) had significant antimalarial activity (Table 2).

TABLE 1. EFFECTS OF DDC-TREATED SOD ON P. FALCIPARUM IN VITRO

TREATMENT	CONCENTRATION	PARASITEMIA (PER 500 RBCs)
Control		35.3 $\pm$ 7.2
SOD	1 mg/ml	18.7 $\pm$ 3.7
DDC-SOD	1 mg/ml	0 $\pm$ 0

While DDC-treated SOD exhibited antimalarial activity, DDC-treated BSA did not affect P. falciparum growth in culture. Parasitemias were 15 per 500 RBCs in the presence of DDC-treated BSA (0.1 mg/ml) when they were 16.8 $\pm$ 1.8 per 500 RBCs in control cultures. Thus the observed antimalarial activity of DDC-treated SOD was not a result of a nonspecific effect of DDC-treatment and was not due to incomplete removal of DDC by gel filtration.

We initially suspected that DDC-treated SOD might inhibit malaria growth by chelating and sequestering free copper or zinc in the culture medium. However, the antimalarial activity of DDC-treated SOD could not be reversed by the addition of excess copper or zinc to the culture medium (Table 2).

DISCUSSION

DDC-treated SOD appears to inhibit the growth of P. falciparum in vitro. We have, as yet, not been able to elucidate the mechanism by which this effect occurs nor to ascertain whether this effect has physiological significance.

We have recently been evaluating the effects of primaquine, an oxidant drug, on P. falciparum grown in DDC-treated erythrocytes (6). Since intraerythrocytic plasmodia depend to some extent on host SOD, DDC-induced depletion of host SOD makes plasmodia more vulnerable to the effects of primaquine. Thus, primaquine and DDC are synergistic both in vitro and in vivo (6). We now report that DDC-treated SOD, by itself, has antimalarial activity. Accordingly, the direct antimalarial activity of DDC-treated SOD may, in part, be responsible for the synergistic effect observed.

TABLE 2. EFFECTS OF METALS ON THE ANTIMALARIAL ACTIVITY OF DDC-TREATED SOD

TREATMENT	CONCENTRATION	PARASITEMIA
Experiment 1		
Control		13 ± 1.7
Copper acetate	1 uM	8.3 ± 1.5
DDC-SOD	100 ug/ml (3 uM)	5.3 ± 2.5
+copper acetate	1 uM	3.0 ± 1.0
Experiment 2		
Control		19.3 ± 3.2
Zinc acetate	1 uM	14.7 ± 2.5
DDC-SOD	100 ug/ml	3.7 ± 2.9
DDC-SOD +	100 ug/ml (3 uM)	
zinc acetate	1 uM	3.7 ± 2.1

ACKNOWLEDGEMENTS

This work was supported by grants from the NIH (AI 20197), the World Bank/UNDP/WHO Special Programme for Research and Development in Tropical Diseases and the Irma T. Hirschl Foundation. We thank J. Haldane for his helpful discussions.

REFERENCES

1. Walsh JA (1984) In: Warren K, Mahmoud AA (eds.) Tropical and Geographical Medicine. McGraw-Hill, New York, pp. 1073-1085.
2. Friedman MJ (1979) Nature 280:245-247.
3. Allison AC, Eugui EM (1983) Ann Rev Immunol 1:361-392.
4. Fairfield AS, Meshnick SR, Eaton JW (1983) Science 221:764-766.
5. Fairfield AS, Meshnick SR, Eaton JW (1984) In: Brewer G, Eaton JW (eds.) Malaria and the Red Cell. Alan Liss, New York, pp. 13-23.
6. Febbraio M, Fairfield AS, Meshnick SR (submitted for publication).
7. Trager W, Jensen, J (1976) Science 193:673-675.

© 1986 Elsevier Science Publishers B.V. (Biomedical Division)
Superoxide and Superoxide Dismutase in Chemistry,
Biology and Medicine; G. Rotilio, editor

ROLE OF OXYGEN-DERIVED FREE RADICALS AS MEDIATORS OF POST-ISCHEMIC INJURY: A CLINICALLY ORIENTED OVERVIEW

GREGORY B. BULKLEY, M.D., F.A.C.S. and JON B. MORRIS, M.D.
Department of Surgery, The Johns Hopkins Medical Institutions, Blalock 685, 600 N. Wolfe Street, Baltimore, Maryland 21205, USA (301) 955-8500

In the past five years there has been a virtual explosion of information available about the mechanisms and control of free radical-mediated reperfusion injury. This progress has led us to the threshold of what will probably be broad clinical applications in the next decade. Because many excellent reviews of the area have been published,[1,2,3,4,5,6] and further information is available elsewhere in this volume, we will not attempt to provide an in depth analysis of this field here. Instead, it is perhaps appropriate to step back and view what has been done, and what is about to be done from the perspective of its potential for clinical application.

The Clinical Problem of Ischemia

Ischemic tissue injury is a major cause of human disease, particularly in industrialized societies. Atherosclerosis remains the primary cause of death in technologically advanced countries, manifest as ischemic heart disease and stroke.[7] Moreover, ischemic injury to other organs, including the kidney, intestine, and extremities as the result of atherosclerosis, comprises a major contribution to mortality and morbidity caused by vascular disease. Furthermore, the primary cause of failure of cardiopulmonary resuscitation is irreversible ischemic injury to the brain. Finally, post-ischemic damage to organs preserved for transplantation is a major factor limiting not only the success of organ grafting, but the availability of organs for transplantation.

Small Intestine

Most of the recent interest in free radical-mediated reperfusion injury has been generated by the studies of Parks, Granger, McCord, and their colleagues in the Department of Physiology at the University of South Alabama.[2,8,9,10,11,12,13] Using a feline model of partial intestinal ischemia, they have provided a large body of indirect evidence, mostly through the use of specific and somewhat specific inhibitors of free radical generation (allopurinol, soybean trypsin inhibitor) or activity (superoxide dismutase, catalase, mannitol, dimethyl sulfoxide), that much of the early mucosal injury seen in the cat small intestine following relatively brief periods of ischemia is mediated by free radicals generated from oxygen by activated xanthine oxidase at reperfusion. Unfortunately, attempts to modify more severe degrees of injury with this approach have been a failure.[14] It appears that much of the advanced tissue injury that leads to lethal transmural infarction in clinical **intestinal ischemia** is caused by ischemia itself, rather than by reperfusion injury. Consequently, the potential for clinical application in this area is probably limited.

There may perhaps be a related condition of intestinal ischemia that will prove amenable to this approach, however, Neonates with **necrotizing enterocolitis** begin their disease with an early mucosal lesion like that described by Parks and his colleagues. Perhaps due to the immunologic niavetè of the host, this early loss of the intestinal mucosal barrier to lumenal bacteria can result in a progression of this early lesion to a lethal transmural infarction.[15] If this early lesion is caused by a free radical injury,[16] and if its prevention can block the subsequent progression of the disease, there may be a potential for its successful modification by free radical-directed therapy.

Stomach

Experimental studies suggest that either superoxide dismutase or allopurinol can prevent the formation of superficial mucosal ulcerations in the rat gastric mucosa in response to severe physiologic stress[14] a model of clinical **stress gastric ulceration**[17]. This disease is probably initiated by ischemia due to splanchnic vasospasm,[18] but the actual tissue injury may well be mediated by free radicals at reperfusion.

Liver

The liver is similarly susceptible to **ischemic injury**, either directly, or via a similar splanchnic vasospastic mechanism.[19] Recent evidence, also from the group at South Alabama, suggests that much of the injury sustained by a rat liver as the result of ischemia can be prevented with treatment directed at free radical generation during reperfusion.[20]

Pancreas

The pancreas is susceptible to a form of ischemic injury that appears similar to that described for the stomach and liver. In an isolated canine model of the early stages of acute ischemic pancreatitis, we have found that treatment with SOD or allopurinol markedly attenuates the resulting pancreatic injury.[21,22] Even more interesting is the finding that similar treatment blocks the manifestations of pancreatitis in models of gallstone and alcohol-induced hyperlipidemic pancreatitis. This suggests that xanthine oxidase activation by proteolytic enzymes could be the final common pathway to the development of **acute pancreatitis** in man.[22] Indeed, clinical trials testing this hypothesis in man are currently being designed.

Heart

A large body of evidence suggests that myocardial damage following both global and regional ischemia can be strikingly ameliorated by the blockade of free radical-mediated reperfusion injury.[4] Following global ischemia such as that sustained following **cardiac arrest** or elective hypothermic **cardioplegia** during cardiopulmonary bypass, treatment with either allopurinol or free radical scavengers markedly improves contractility,[23,24,25] and decreases the incidence and severity of post-ischemic arrhythmias. In models of regional ischemia, mimicking **myocardial infarction**, infarct size can be reduced as much as two thirds by this approach, and it appears this is true even when there is no opportunity for gross reperfusion.[26] Anticipated clinical applications therefore include the treatment of acute myocardial infarction, not only as an adjuvant to reperfusion therapy with balloon dilatation, streptokinase, and tissue plasminogen activator, but as primary treatment alone.

Brain

Perhaps the most exciting potential area of application of this approach is the treatment of cerebral ischemia. Currently the human brain appears to be limited to a 4-5 minute period of total ischemia before irreversible injury is sustained. Thereafter, cardiopulmonary resuscitation is thought not to be worthwhile. Indeed our current legal and medical definition of death is related to this concept of "brain death." If much of this injury, which we have thought to be irreversible at the time of our first opportunity for resuscitation, is generated subsequently at reperfusion, then the therapeutic and even medicolegal implications should not be underestimated.

Skin

Studies in the rat abdominal skin have demonstrated improved viability of **free flaps**[27] transferred from one site to another, of **skin grafts** stored overnight and then revascularized, and the total surface area of skin surviving elevation in a **pedicle flap** model.[28] Although the clinical heterogeneity of these problems makes them difficult to approach from the standpoint of a randomized, controlled clinical trial, it is reasonable to anticipate early application in this area on an episodic basis.

Kidney

Studies from several different laboratories following 40-60 minutes of warm ischemia suggest substantial amelioration of **acute ischemic renal injury** with modification of free radical injury at reperfusion.[29,30] Although these patients are also heterogenous, free radical therapy may find application in people in shock states as prophylaxis for acute renal failure, as well as for ischemic injuries to other organs.

Organ Preservation For Transplantation

The amelioration of reperfusion injury following 24 or more hours of cold ischemia, as in renal preservation for transplantation, has even more dramatic effects in thisporcine model of human cadaveric renal transplantation.[31] It appears that here the component of post-ischemic injury due to reperfusion is proportionately greater than the component due to ischemia itself. In the rabbit lung, a similar approach has dramatically extended the period of successful organ preservation from 4.5 to 12 hours.[32] Clinical trials of this approach are currently underway or planned in several centers, and it is not unlikely that the first demonstration of the successful application of this approach to man will be in this area. This is partly because the prospective nature and somewhat uniform presentation of this clinical problem lends itself particularly well to a controlled clinical trial.

Summary

A great deal is still unknown regarding the fundamental pathophysiology of free radical-mediated reperfusion injury. For example, the exact role of leukocytes in the lesions reviewed above is far from fully understood. On the other hand, these animal studies published in the past five years have provided a solid groundwork for the initiation of clinical applications. The next decade will undoubtedly see these two areas progress simultaneously, as clinicians faced with common and lethal diseases seek to apply these principles as quickly as possible. On the other hand, it will be primarily the studies of the fundamental aspects of this process that will provide the real basis for advance. It is unusual to see such rapid applications of basic scientific findings to the clinical arena. This is perhaps a reflection of the fundamental importance of free radical-mediated tissue injury in human disease.

Acknowledgments

Supported by NIH grant #AM31764. Dr. Morris is a Dudley P. Allen Fellow from the Department of Surgery, Case Western Reserve University, Cleveland, Ohio.

References

1. Bulkley, G.B. (1983): The role of oxygen free radicals in human disease processes. Surgery 94, 407-411.

2. Parks, D.A., Bulkley, G.B., and Granger, D.N. (1983a): Role of oxygen-derived free radicals in digestive tract diseases. Surgery 94, 415-422.

3. Parks, D.A., Bulkley, G.B., and Granger, D.N. (1983b): Role of oxygen free radicals in shock, ischemia, and organ preservation. Surgery 94, 428-432.

4. Gardner, T.J., Stewart, J.R., Casale, A.S., Downey, J.M., and Chambers, D.E. (1983): Reduction of myocardial ischemic injury with oxygen-derived free radical scavengers. Surgery 94, 423-427.

5. McCord, J.M. (1983): The superoxide free radical: its biochemistry and pathophysiology. Surgery 94, 412-414.

6. McCord, J.M. (1985): Oxygen-derived free radicals in postischemic tissue injury. New Eng. J. Med. 312, 159-163.

7. Havlic, R.J. and Feinleib, M., ed. (1978): Proceedings of hte conference on the decline in coronary heart disease mortality. Bethesda, Maryland: National Institutes of Health.

8. Granger, D.N., Rutili, G., and McCord, J.M. (1981): Superoxide radicals in feline intestinal ischemia. Gastroenterology 81, 22-29.

9. Parks, D.A., Bulkley, G.B., Granger, D.N., Hamilton, S.R., and McCord, J.M. (1982a): Ischemic injury in the cat small intestine: role of superoxide radicals. Gastroenterology 82, 9-15.

10. Grogaard, B., Parks, D.A., Granger, D.N., McCord, J.M., and Forsberg, J.O. (1982): Effects of ischemia and oxygen radicals on mucosal albumin clearance in intestine. Am. J. Physiol. 242, G448-G454.

11. Parks, D.A. and Granger, D.N. (1983): Ischemia-induced vascular changes: role of xanthene oxidase and hydroxyl radicals. Am. J. Physiol. 245, G285-G289.

12. Parks, D.A., Granger, D.N., Bulkley, G.B., and Shah, A.K (1985): Soybean trypsin inhibitor attenuates ischemic injury to the feline small intestine. Gastroenterology 89, 6-12.

13. Parks, D.A., Shah, A.K., and Granger, D.N. (1984): Oxygen radicals: effects on intestinal vascular permeability. Am. J. Physiol. 247, G167-G170.

14. Parks, D.A., Granger, D.N., and Bulkley, G.B. (1982b): Superoxide radicals and mucosal lesions in the ischemic small intetine. Fed. Proc. 41, 1742.

15. Bailey R.W. and Bulkley, G.B.: Role of the circulation in neonatal necrotizing enterocolitis. In Pathophysiology of the Splanchnic Circulation, ed P.R. Kvietys, J.A. Barrowman, and D.N. Granger, Boca Raton, FL: CRC Press, Inc., in press.

16. Oshima, A., Bulkley, G.B., and Hamilton, S.R. (1984): Role of oxygen-derived free radicals in the development and progression of reperfusion necrosis in the rat small intestine, a model of neonatal necrotizing nterocolitis. Gastroenterology 86, 1202.

17. Itoh, M., and Guth, P.H. (1985): Role of oxygen-derived free radicals in hemorrhagic shock-induced gastric lesions in the rat. Gastroenterology 88, 1162-1167.

18. Bulkley, G.B., Oshima, A., Bailey, R.W., and Horn, S.D.. (1985): Control of gastric vascular resistance in cardiogenic shock. Surgery 98, 213-223.

19. Bulkley, G.B., Oshima, A., and Bailey, R.W. (1986): Pathophysiology of hepatic ischemia in cardiogenic shock. Am J. Surg., in press.

20. Adkinson, D., Hollworth, M.E., Benoit, J.N., Parks, D.A., McCord, J.M., and Granger, D.N.: Role of free radicals in ischemia-reperfusion injury to the liver. Acta. Physiol. Scand., in press.

26. Chambers, D.E., Parks, D.A., Patterson, G., Yoshida, S., Burton, K., Parmley, L.F., McCord, J.M., and Downey, J.M. (1983): Role of oxygen derived free radicals in myocardial ischemia. Fed. Proc. 47, 1093.

21. Sanfey, H., Bulkley, G.B., and Cameron, J.L. (1984): The role of oxygen-derived free radicals in the pathogenesis of acute pancreatitis. Ann. Surg. 200, 405-413.

22. Sanfey, H., Bulkley, G.B., and Cameron, J.L. (1985): The pathogenesis of acute pancreatitis: the source and role of oxygen-derived free radicals in three different experimental models. Ann. Surg. 201, 633-639.

23. Casale, A.S., Bulkley, G.B., Bulkley, B.H., Flaherty, J.T., Gott, V.L., and Gardner, T.J. (1983): Oxygen free-radical scavengers protect the arrested, globally ischemic heart upon reperfusion. Surg. Forum 34, 313-316.

24. Shlafer, M., Kane, P.F., and Kirsh, M.M. (1982): Superoxide dismutase plus catalase enhances the efficacy of hypothermic cardioplegia to protect the globally ischemia, reperfused heart. J. Thorac. Cardiovasc. Surg. 83, 830-839.

25. Shlafer, M., Kane, P.F., Wiggins, V.Y., and Kirsh, M.M. (1982): Possible role for cytotoxic oxygen metabolites in the pathogenesis of cardiac ischemic injury. Circulation 66, I85-I92.

27. Manson, P.N., Narayan, K.K., Im, M.J., Bulkley, G.B., and Hoopes, J.E.: Improved survival in free skin flap transfers in rats. Surgery, in press.

28. Manson, P.N., Anthenelli, R.M., Im, M.J., Bulkley, G.B., and Hoopes, J.E. (1983): The role of oxygen-free radicals in ischemic tissue injury in island skin flaps. Ann. Surg. 198, 87-90.

29. Hannson, R., Gustafsson, B., Jonnson, O., Lundstum, S., Petterson, S., Schersten, T., and Waldenstrom, J. (1982): Effect of xanthene oxidase inhibition on renal circulation after ischemia. Transplant. Proc. 14, 51-58.

30. Paller, M.S., Hoidal, J.R., and Ferris, T.F. (1984): Oxygen free radicals in ischemic acute renal f ailure in the rat. J. Clin. Invest. 74, 1156-1164.

31. Koyama, I., Bulkley, G.B., Williams, G.M., and Im, M.J.: The role of oxygen free radicals in mediating the reperfusion injury of cold preserved ischemic kidneys. Transplantation, in press.

32. Stuart, R.S., Baumgartner, W.A., Borkon, A.M., Bulkley, G.B., Brawn, J.D., DeLaMonte, S.M., Hutchins, G.M., and Reitz, B.A. (1985): Five-hour hypothermic lung preservation with oxygen free-radical scavengers. Transpl. Proc. 17, 1454-1456.

© 1986 Elsevier Science Publishers B.V. (Biomedical Division)
Superoxide and Superoxide Dismutase in Chemistry,
Biology and Medicine; G. Rotilio, editor

REOXYGENATION INJURY IN THE ISOLATED PERFUSED WORKING RAT HEART: ROLES OF XANTHINE OXIDASE AND TRANSFERRIN

MATTHEW B. GRISHAM, WILLIAM J. RUSSELL, RANJAN S. ROY AND JOE M. McCORD
Department of Biochemistry, University of South Alabama, Mobile, AL 36688, USA

INTRODUCTION

Much evidence suggests that superoxide and related active oxygen species play important roles in the tissue injury associated with ischemia and subsequent reperfusion of a variety of organs (1). Furthermore, the enzyme xanthine oxidase, produced in energy depleted tissues by proteolytic conversion of the native xanthine dehydrogenase, appears to be the major source of the radicals, via the scheme depicted in Figure 1 (1-4).

Most of the work designed to test this hypothesis has been performed with models of ischemia rather than hypoxia. The reduction or cessation of perfusion *per se* almost surely causes insult and injury beyond that caused by the so-called "oxygen paradox" (e.g., as a result of the buildup of tissue lactic acid and the concomitant drop in pH). In an effort to examine the contribution of the xanthine oxidase-dependent component of reoxygenation injury in relative isolation from other aspects of ischemic injury, we have studied the buffer perfused isolated working rat heart, subjected to 40 minutes of anoxia and substrate deprivation followed by reoxygenation. We examined the abilities of superoxide dismutase and catalase, as well as two inhibitors of xanthine oxidase, to protect the reoxygenated heart. We also examined the ability of iron-loaded transferrin to exacerbate the radical-mediated injury, presumably by its ability to catalyze the formation of OH· from O_2^-· and H_2O_2.

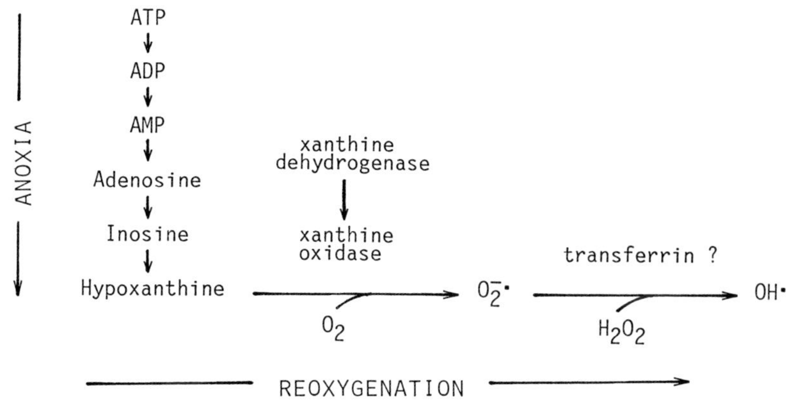

Fig. 1. Proposed mechanism for hypoxia/reoxygenation-induced free radical injury.

MATERIALS AND METHODS

Male Wistar rats (220-260 g) were decapitated and hearts were cannulated on a standard working heart apparatus within 1 min of excision. All hearts were perfused from a reservoir placed 100 cm above the aortic cannula, while the left atrium received fluid from an atrial filling vessel maintained at a filling pressure of 13 cm H_2O. When the left atrial cannula was open, there was a net ejection of fluid against the aortic pressure head. Coronary flow under these conditions was approximately 15 ml/min. The standard perfusate was 37°C Krebs-Henseleit buffer gassed with either 95% O_2 + 5% CO_2 or 95% N_2 + 5% CO_2 to yield a final pH of 7.4. As indicated, the buffer contained 11 mM glucose and/or other enzymes or xanthine oxidase inhibitors. All hearts were paced at 300 beats/min. Control hearts were perfused for a 15 min stabilization period with oxygenated, glucose-containing buffer. They were then switched to anoxic, glucose-free buffer for 40 min, after which time they were returned to the original oxygenated buffer. Coronary perfusate was collected in 1 min fractions during hypoxia and reoxygenation, and myocardial damage was assessed by assaying perfusate for creatine kinase according to the method of Hughes (5).

To examine the effects of allopurinol (4-hydroxypyrazolo-[3,4-d] pyrimidine, Sigma) on cellular damage resulting from an anoxic insult, animals were pretreated with allopurinol at 30 mg/kg/day, IP, for 4 days. Hearts were then treated as described above, except that all buffers contained 50 uM allopurinol. The effects of pterinaldehyde (2-amino-4-hydroxy-6-pteridine carboxaldehyde, a gift of C.M. Baugh), superoxide dismutase (Pharmacia), catalase (Sigma), and transferrin (Sigma), were assessed without pretreatment of the animals, by simple addition of the substances to the perfusion buffers as indicated. Human apotransferrin was iron-loaded according to the method of Zapolski (6).

RESULTS

Figure 2 illustrates the time course of the damage incurred by the myocardium following reoxygenation of the hypoxic tissue as measured by the release of creatine kinase. There was no significant release of creatine kinase until oxygen was reintroduced following 40 min of glucose-free anoxia. Addition of superoxide dismutase (20 ug/ml) to the perfusion buffers resulted in a dramatic reduction (77%) of creatine kinase release, indicating that $O_2^{-}\cdot$ is involved in the cytotoxic reaction (Fig. 2 and Table I). Likewise, catalase (20 ug/ml) provided substantial protection, inhibiting creatine kinase release by 63%. The combination of superoxide dismutase and catalase provided even greater protection (87%) than either enzyme alone.

Hearts from animals pretreated with the xanthine oxidase inhibitor allopurinol and perfused with buffers containing this inhibitor also released markedly less

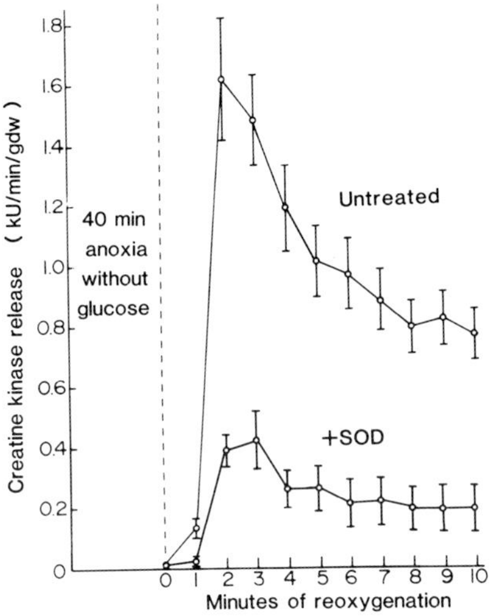

Fig. 2. Prevention of creatine kinase release from the reoxygenated rat myocardium by superoxide dismutase. Where indicated, all buffers contained superoxide dismutase at 20 ug/ml. Upon reoxygenation, perfusate was collected in 1 min fractions and assayed for creatine kinase. Points represent means ± S.E.M. For the control group, n = 9; for the treated group, n = 5.

TABLE I

CREATINE KINASE RELEASED DURING TEN MINUTES OF REOXYGENATION

Treatment*	CPK Units ± SEM	% of Control	Significance
None (n=9)	10,900 ± 940	100%	
+ SOD (n=5)	2,500 ± 690	23%	$p < 0.001$
+ Catalase (n=4)	3,900 ± 420	36%	$p < 0.001$
+ SOD + Catalase (n=4)	1,450 ± 470	13%	$p < 0.001$
+ Allopurinol (n=5)	1,380 ± 70	13%	$p < 0.0001$
+ Pterinaldehyde (n=4)	4,950 ± 1100	45%	$p < 0.05$
+ Transferrin (n=9)	22,500 ± 2160	206%	$p < 0.05$

* Superoxide dismutase (SOD) and catalase were present in the perfusion buffers at 20 ug/ml. Allopurinol, pterinaldehyde, and transferrin were present at 50 uM, 1 uM, and 15 uM, respectively.

creatine kinase when reoxygenated following the anoxic insult, implicating xanthine oxidase as the major source of the radical under these conditions (Table I). The protective effect of 1 uM pterinaldehyde in the perfusion buffers (without pretreatment of animals) is also shown in Table I.

The inclusion of the plasma iron-binding protein transferrin in the perfusion buffers produced a two-fold exacerbation of creatine kinase release in response to the reoxygenation insult. This result is also shown in Table I.

DISCUSSION

The mechanism depicted in Fig. 1 was first proposed to explain the protective effect of superoxide dismutase on capillary permeability in a model of ischemia/reperfusion injury in the feline intestine in vivo (7). Many variations have since been reported, utilizing isolated organ systems as well as in vivo models, and examining as "end-points" a number of physiological functional parameters in addition to biochemical markers of cellular damage (1). The present study was designed to eliminate any non-radical dependent mechanisms of injury which might result from the buildup of toxic metabolites or drop in tissue pH, resulting simply from the lack of mechanical perfusion. These factors aside, we have attempted to determine what portion of the injury to the heart resulting from anoxia/reoxygenation is attributable to the production of $O_2^{-}\cdot$ by xanthine oxidase. Table I shows that the combination of superoxide dismutase and catalase prevented the release of creatine kinase by 87% relative to the untreated control, and that the same degree of inhibition was seen with allopurinol treatment, suggesting that xanthine oxidase is responsible for nearly all the injury sustained in this model. The less-than-maximal inhibition seen with either superoxide dismutase or catalase alone suggests that three mechanisms of injury are operative: 1) superoxide-specific injury; 2) H_2O_2-specific injury; and 3) $OH\cdot$-dependent injury, which requires the simultaneous presence of both $O_2^{-}\cdot$ and H_2O_2, and presumably a transition metal catalyst (such as transferrin).

Allopurinol is a highly specific drug, having been designed as a xanthine oxidase inhibitor for use in treating victims of gout. Nevertheless, it might possess other pharmacological activity which could account for its protective effect to the heart. Because the evidence identifying xanthine oxidase as the source of the radical in all previous studies rests on the ability of allopurinol to block the reperfusion or reoxygenation injury, we deemed it necessary to examine another potent inhibitor of xanthine oxidase, choosing for this purpose pterinaldehyde. We have recently shown it to be a potent competitive inhibitor of feline intestinal xanthine oxidase with a K_i of 1 nanomolar (8). The presence of pterinaldehyde at 1 uM inhibited creatine kinase release by 55%, supporting the role ascribed to xanthine oxidase. Possible reasons for less inhibition than was obtained with allopurinol include the fact that pterinaldehyde is a competitive inhibitor, whereas oxypurinol, the more active metabolite of allopurinol, is a non-competitive inhibitor.

The apparent interaction of $O_2^{-}\cdot$ and H_2O_2 in the presence of a suitable transition metal chelate to produce a more potent oxidizing species (9) has been observed by many. The plasma iron-transporting protein transferrin has been shown capable of supporting this reaction in an isolated perfused rabbit interventricular septal preparation exposed to an exogenous $O_2^{-}\cdot$ producing system (10). We now report that 15 uM transferrin in the perfusion buffers caused a two-fold exacerbation of injury in the isolated perfused rat heart, where the source of radical generation is endogenous.

ACKNOWLEDGEMENT

This work was supported by grant #83 997 from the American Heart Association.

REFERENCES

1. McCord JM (1985) New Engl J Med 312:159-163

2. Parks DA, Bulkley GN, Granger DN, Hamilton SR, McCord JM (1982) Gastroenterology 82:9-15

3. Roy RS, McCord JM (1983), In: Greenwald R, Cohen G (eds) Oxy Radicals and Their Scavenger Systems: Vol II Cellular and Molecular Aspects. Elsevier Science, New York, pp 145-153

4. Chambers DE, Parks DA, Patterson G, Roy RS, McCord JM, Yoshida S, Parmley L, Downey JM (1985) J Mol Cell Cardiol 17:145-152

5. Hughes BP (1962) Clin Chim Acta 7:597

6. Zapolski EJ, Princiotto, JV (1980) Biochemistry 19:3599-3603

7. Granger DN, Rutili G, McCord JM (1981) Gastroenterology 81:22-29

8. Granger DN, McCord JM, Parks DA, Hollwarth ME (1985) Gastroenterology 85 (in press)

9. McCord JM, Day ED (1978) FEBS Lett 86:139-142

10. Burton KP, McCord JM, Ghai G (1984) Am J Physiol 246 (Heart Circ Physiol 15):H776-H783

Superoxide and Superoxide Dismutase in Chemistry, Biology and Medicine; G. Rotilio, editor

ALLOPURINOL, XANTHINE OXIDASE AND MYOCARDIAL INJURY

DAVID J. HEARSE, ALLAN S. MANNING, JAMES M. DOWNEY*, DEREK M. YELLON.

The Heart Research Unit, The Rayne Institute, St Thomas' Hospital, London SE1, UK. *The Dept of Physiology, University of Alabama, Mobile, Alabama, U.S.A.

INTRODUCTION

Myocardial ischaemia, initiates a complex sequence of progressively more severe cellular reactions which, if not halted, may lead to cell death. In the early stages of myocardial ischaemia cellular injury is reversible, such that if adequate coronary flow is restored any metabolic imbalance is rapidly normalized and the tissue will return to normal contractile function. As the duration of ischaemia increases and the damage becomes more severe reperfusion may not result in an immediate return of function and recovery may be delayed for some time. For example, after 15 minutes of ischaemia adenine nucleotide content (1) and contractile activity (2) may be depressed for almost a week after the onset of reperfusion. However, as long as cellular injury is in the reversible phase a full recovery of function will ultimately occur. Eventually however ischaemic injury will become so severe, that irreversible damage will occur and from this point on the tissue is destined to cell death and reperfusion has no beneficial effect. The exact mechanism for the critical transition from reversible to irreversible injury remains unknown but a number of factors have been suggested as important determinants of cell death or survival; these include myocardial high energy phosphate depletion, intracellular calcium overload, loss of ionic homeostasis, labilization of lysosomes and phospholipase-mediated membrane injury. Recently, oxygen-derived free radicals such as superoxide (O_2^-) and the hydroxyl radical (OH^-), have been suggested (3,4,5) as important mediators of ischaemic injury.

Reperfusion of ischaemic tissue, either by surgical intervention or the natural growth of collaterals, is a necessary step if one wishes to halt the advance of ischaemic injury and promote tissue recovery. However, despite its obvious benefits reperfusion is not without hazard (6). In moderately ischaemic tissue in which injury is in its reversible phase reperfusion can trigger potentially lethal arrhythmias such as ventricular fibrillation (7) and in severely injured tissue, reperfusion can alter and accelerate the processes of cell death (8) and may lead to haemorrhage and other forms of severe tissue injury. As with ischaemia-induced injury, controversy exists over the mechanisms underlying various unfavourable reperfusion-induced changes. Calcium overload and cell swelling have received considerable attention and more recently free radicals have been proposed (7,9,10) as important contributors.

In discussing therapeutic interventions for the prevention of injury during ischaemia and reperfusion two main objectives can be defined: firstly, to minimize the number of cells which become irreversibly injured and die (ie, in the case of regional ischaemia reducing the size of an evolving infarct) and secondly, to prevent the occurence of arrhythmias which may impair contractile performance or more importantly may be lethal. During the last decade there have been many studies of myocardial protection in which a wide variety of drugs and biochemical interventions have been claimed (11,12,13,14) to reduce infarct size and/or combat arrhythmias. To this list we can now add "anti free radical" agents such as allopurinol (which inhibits xanthine oxidase and hence limits free radical production during hypoxanthine degradation) and the enzyme superoxide dismutase (which along with a number of organic agents such as mannitol and ascorbate can scavenge free radicals once they are formed). In 1971 DeWall et al (15) reported that allopurinol treatment reduced arrhythmias in dog hearts with regional ischaemia. The anti arrhythmic effects of allopurinol were originally (15) thought to reflect purine conservation in these hearts, but we would now suggest that allopurinol may act by suppressing superoxide production via xanthine oxidase. We, and others (9,10) have now also shown that allopurinol and a variety of scavengers can effectively combat reperfusion-induced arrhythmias. In addition to the promising anti arrhythmic properties of anti free radical interventions their potential clinical importance has recently been reinforced by the observation that, in a number of experimental models, infarct size can be considerably reduced by interventions such as allopurinol (16), superoxide dismutase and catalase (17) and a variety of agents and procedures which limit leucocyte infiltration (an additional potential source of free radicals) (18) into ischaemic tissue.

The objective of the present article is to review some recent studies of the anti infarct and anti arrhythmic properties of anti free radical interventions.

INFARCT SIZE AND FREE RADICALS

Since there is a direct relationship between mortality and the size of an evolving myocardial infarct considerable efforts have been made to develop interventions which prevent cell death and so reduce the size of an evolving infarct (19).

Allopurinol and Infarct Size

In recent studies (16) we have used a closed chest dog coronary artery embolization procedure (20) to assess to what extent allopurinol can influence the size of an evolving myocardial infact. Dogs (n=10 in each group) received 400 mg of allopurinol in their food 24 hours before surgery. They also received a bolus (25mg/kg) of the drug intravenously at the time of surgery and every 8 hours thereafter. Control dogs received saline. All dogs received 141-cerium labeled microspheres (10 micron) injected into the left ventricle immediately after coronary occlusion. These microspheres are distributed in the heart in proportion to coronary flow and lodge in the capillaries where they remain for the duration of the experiment. After 24 hours of coronary artery embolization hearts were removed and sliced into 5mm sections which were then incubated in triphenyltetrazolium to define the area of necrosis. The volume of the infarcted tissue was then calculated and expressed as a percent of the zone at risk (the zone of underperfusion as defined by autoradiography of the radioactive microspheres).

The results in figure 1 show that infarct size in the drug-treated group was approximately 40% smaller than in the control group. The amount of the risk zone that infarcted in the drug group was 20.0$\pm$2.3% whereas for the control group the value was 58.0±2.7%.

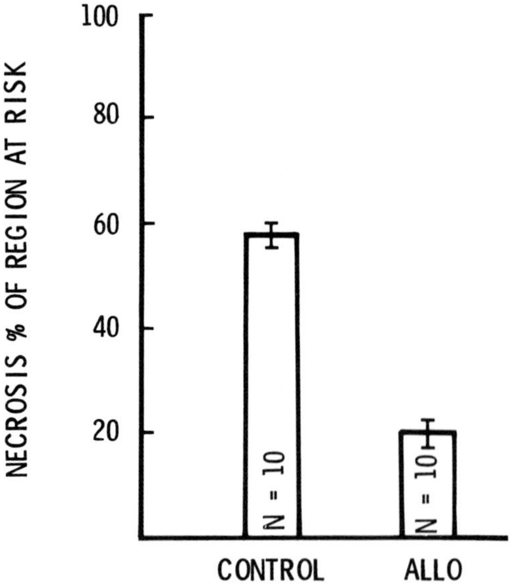

Fig. 1. Allopurinol and infarct size. Effect of allopurinol on the extent of tissue necrosis in the dog heart with 24 hours of coronary artery embolization. Infarct size is expressed as a percent of the region at risk. Bars represent the SEM.

Possible mechanisms for allopurinol protection

Ischaemia results in the net degradation of high energy phosphaes, a process which terminates in the xanthine oxidase-mediated conversion of hypoxanthine to uric acid. Conversion to uric acid results in the irreversible loss of purine base from the nucleotide pool. It has been postulated that protection of the nucleotide pool could explain the beneficial effect of allopurinol during periods of oxygen depletion (15) and recent reports do indicate a beneficial effect when the purine base is replenished exogenously (21, 22). For our preparation there are however two arguments against the nucleotide pool theory. Firstly, the unphosphorylated purines (adenosine, hypoxanthine and xanthine) are all lipid soluble and can be washed out of the tissue. Since collateral flow in the ischaemic tissue in our preparation was in the order of 0.1 to 0.2 ml/min/g of tissue, this would result in a half life for the unphosphorylated purines of less than 7 minutes. A blockade of xanthine oxidase alone would not be expected to keep the purines from being lost from the ischaemic zone. Secondly, there is little evidence that the loss of unphosphorylated purines represents a lethal event during ischaemia. In the absence of any other proposed mechanisms we would therefore suggest that the ability of allopurinol to limit infarct size in both the permanently ischaemic and the reperfused myocardium, is most likely to be due directly to its ability to inhibit xanthine oxidase and thereby prevent the production of superoxide.

ARRHYTHMIAS AND FREE RADICALS

As discussed at the beginning of this article, the two main consequences of ischaemia and reperfusion are cellular injury leading to cell death and disorders of cardiac rhythm. In the latter instance arrhythmias may lead to cell death in what is otherwise potentially viable tissue. The mechanisms underlying the genesis and control of ischaemia- and reperfusion-induced arrhythmias have been reviewed extensively elsewhere (7,23,24). Similarly detailed information is available to describe the various types of arrhythmias which may arise (fibrillation, tachycardia, ectopics etc), the different times at which they arise (early and late arrhythmias), and the differences in origin and pharmacological control between reperfusion- and ischaemia- induced arrhythmias (7,24). Consideration of the above cited literature would indicate that a number of factors might be responsible for the genesis of various arrhythmias, these include stimulation of alpha and/or beta adrenergic receptors (25,26); elevation of intracellular cyclic AMP content (27); the formation and release of lysophosphatides (28); disturbances of ionic homeostasis (particularly transsarcolemmal shifts of potassium and calcium) (29,30); and the metabolism of free fatty acids (31,32). More recently it has been suggested that free radical production may play an important role (9,10) and that inhibition of free radicals, paticularly those derived via the xanthine oxidase pathway, might pave the way to a new and important therapeutic principle.

Allopurinol and Ischaemia- and Reperfusion-induced arrhythmias

We have recently carried out studies (33) in the anaesthetized rat to assess the extent to which allopurinol can prevent arrhythmias during both ischaemia and reperfusion. Using a modification of the technique of Selye et al (34) coronary arteries were ligated in vivo to generate an area of regional ischaemia. Twenty four hours prior to study rats were orally treated with allopurinol (20 mg/kg body wt). A second, similar, dose was given, this time intravenously, after the animal had been anaesthetized, 15 minutes prior to coronary occlusion. For studies of ischaemia-induced arrhythmias the occlusion was maintained for 30 minutes, for studies of reperfusion-induced arrhythmias the occlusion was maintained for a 5 minute period and this was followed by a 10 minute period of reperfusion.

Allopurinol and Ischaemia-induced Arrhythmias

In control animals (n=18) severe ventricular arrhythmias began after 4-5 minutes of coronary artery occlusion, these peaked after 10 minutes and then normally subsided by 18-20 minutes. The mean time to onset of ventricular fibrillation was 303$\pm$101 seconds and the mean time to onset of VT was 344$\pm$32 seconds. Fifty percent of animals exhibited ventricular fibrillation and almost 90% exhibited ventricular tachycardia.

Allopurinol treatment had no significant effect on the incidence of ventricular fibrillation, the duration of ventricular fibrillation, the duration of ventricular tachycardia, and mortality. However, the incidence of ventricular tachycardia was reduced from 89% to 50% ($p<0.05$) and the total number of premature ventricular complexes (in those animals surviving the 30 minute period of coronary artery occlusion) was reduced from 471$\pm$128 to 116$\pm$46 ($p<0.02$).

Allopurinol and Reperfusion-induced Arrhythmias

Figure 2 shows that in placebo-treated control animals, the severity of reperfusion-induced arrhythmias tended to be greater than that observed during

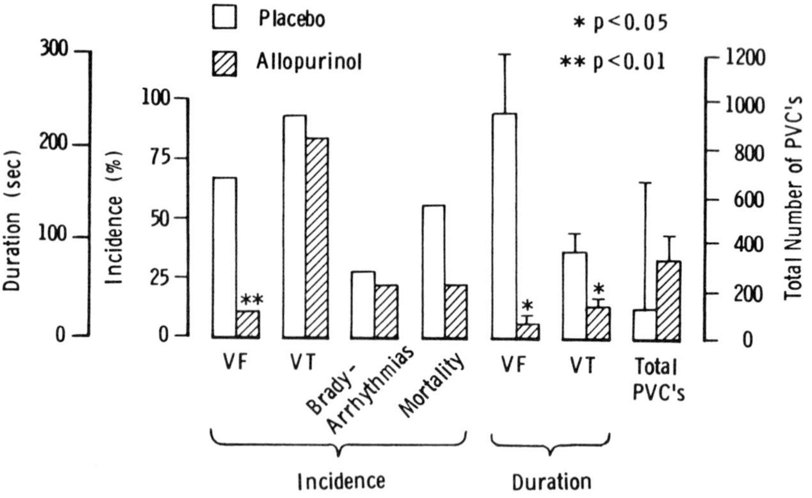

Fig. 2. Effect of allopurinol on reperfusion-induced arrhythmias in the in vivo anaesthetized rat.

Rats were subjected to 5 minutes of regional ischaemia followed by reperfusion during which time cardiac rhythm was monitored. VF = ventricular fibrillation, VT = ventricular tachycardia, PVC's = premature ventricular complexes. Values of duration of VF and VT are shown as the mean ± SEM, n = 18 animals in each group. Values of duration of VF and VT are shown as the mean ± SEM, n = 18 animals in each group.

ischaemia. Thus, almost 70% of animals exhibited ventricular fibrillation during reperfusion, and this lasted much longer than during ischaemia. As a consequence over 50% of rats died, mainly from ventricular fibrillation. Allopurinol treatment resulted in a major reduction in the incidence of ventricular fibrillation (67% to 11%; $p<0.01$) and the mean duration of ventricular fibrillation (230±70 to 14±1 seconds). Allopurinol also reduced the duration of ventricular tachycardia to approximately one-third of that seen in the control group (93±26 to 34±8 seconds; $p<0.05$) and halved mortality from 56% to 22%.

We have thus shown that the administration of allopurinol prior to coronary artery occlusion and reperfusion in an anaesthetized in vivo rat preparation greatly reduced the incidence and severity of reperfusion-induced arrhythmias and also has some effect upon ischaemia-induced arrhythmias. Association of this observation directly with free radical formation must of course be undertaken with great caution since we have not been able to measure these short-lived molecular intermediates and in this study we have not measured or demonstrated any changes in the activity or ratio of xanthine dehydrogenase to xanthine oxidase. Despite this limitation, we would argue that certain other factors support our association between an anti arrhythmic effect and a reduction of free radical formation. Firstly, allopurinol is a highly specific inhibitor of xanthine oxidase (3). Secondly, free radical formation will occur under conditions of ischaemia despite the reduced availability of oxygen (35). However we would expect there to be far greater production of superoxide during reperfusion because of the return of large quantities of oxygen, this being especially so in very severely ischaemic tissue. This gains further support from our observation that arrhythmias are more severe during reperfusion and that they are more amenable to reduction by allopurinol. Thirdly, in a recent study, Woodward et al (10) reported that the free radical scavenger, superoxide dismutase, can effectively reduce reperfusion arrhythmias in the rat.

MECHANISMS FOR FREE RADICAL INJURY AND PROTECTION

The mechanisms underlying free radical-induced damage in the heart are poorly understood. Meerson (36) proposed that superoxide could induce damage by lipid peroxidiation, in oxidizing the lipid in membranes a lipid peroxide radical is formed and this would act to break down the structure of the membrane. In addition, the lipid peroxide itself would be a strong oxidant which could initiate further damage. Lipid peroxidation is difficult to detect and at present the only available method has been to assay for malonyl dialedehyde, a bi-product of lipid peroxidation. We, as well as others (37,38), have found that in isolated rat hearts a modest rise in malonyl dialdehyde levels accompanies a 40 minute period of global ischaemia or hypoxia.

Recent studies by Rowe et al (39) have suggested that the hydroxyl radical is the real culprit in ischaemic damage. Superoxide can combine with hydrogen peroxide (another product of the xanthine oxidase reaction) and via the Haber-Weiss reaction it can form a hydroxyl radical (40). It was originally thought that this reaction was too slow to be of pathological significance but it is now known that trace metals in tissue can catalyse the reaction at a much higher rate (41). Hydroxyl radical scavengers such as, mannitol, or catalase, which remove the hydrogen peroxide substrate required for the reaction, have been shown to be able to prevent free radical damage to the sarcoplasmic reticulum (39).

Myocardial tissue contains endogenous scavengers such as superoxide dismutase, glutathione and catalase, and under normal conditions these provide valuable protection against free radical damage. However during ischaemia and reperfusion where large quantities of free radicals may be produced these endogenous scavenging systems may be swamped and thus be unable to prevent substantial tissue injury. This problem may be exacerbated by the possibility that in ischaemia these endogenous scavenging systems are themselves inhibited (39).

If free radical formation is able to initiate a sequence of events leading to the development of an arrhythmia, then it is necessary to ask whether this is some direct electrical effect of these charged intermediates or whether they act via some intermediate process? At the present time this question cannot be definitively answered, but we would advance our view that some perturbation of membrane integrity and function may be a key link in the induction of injury and the development of an electrophysiological abnormality. Since it is well known (3) that free radicals can cause major membrane damage the most likely effect of free radicals is to alter membrane structure and the membrane's ability to control ionic shifts.

In conclusion, evidence is presented that superoxide radical production from xanthine oxidase may play an important role in the genesis of reperfusion-induced arrhythmias and a lesser role in the genesis of ischaemia-induced arrhythmias and that effective pharmacological control of reperfusion arrhythmias, particularly the potentially lethal rhythm disturbance ventricular fibrillation, might be accomplished either by inhibitors of free radical formation or by the use of appropriate scavengers.

CONCLUDING COMMENTS

A variety of oxygen-derived free radicals are produced as a consequence of normal cellular metabolism. In healthy tissue these potentially toxic metabolic bi-products are removed by endogenous scavengers, thus avoiding any serious cellular injury. Under conditions of ischaemia and reperfusion a number of factors may combine to increase the extent of free radical production, initiate new sources of production (eg xanthine oxidase) and limit the activity of natural scavenging mechanisms. The resulting increase in free radical activity may well represent an important component in the complex process of tissue injury. It would therefore seem reasonable to include anti free radical agents in any intervention designed to afford tissue protection.

Evidence has been presented that superoxide and other oxygen-derived free radicals may contribute to the development of myocardial infarction and to the occurence of arrhythmias. We have also shown that interventions capable of scavenging, or preventing the formulation of, free radicals can reduce infarct size and limit the severity of arrhythmias. At this stage the association between free radicals and these events is circumstantial, a problem which arises from the difficulty of measuring these short lived, highly reactive intermediates. Further studies are required to strengthen this association, to define mechanisms of action and to identify the nature and sources of the free radicals. At the present time however we believe that there is significant evidence to warrant further investigations of xanthine oxidase and its ability to produce superoxide.

REFERENCES

1. Reimer KA, Hill ML, Jennings RB (1981). Prolonged depletion of ATP and of the adenine nucleotide pool due to delayed resynthesis of adenine nucleotides following reversible myocardial ischemic injury in dogs. J Mol Cell Cardiol 13: 229-239.
2. Bush LR, Buja LM, Samowitz W, Rude RE, Walthan M, Tilton GP, Willerson JT (1983). Recovery of left ventricular segmental function after long-term reperfusion following temporary coronary occlusion in conscious dogs. Circ Res 53: 248-263.
3. McCord JM (1984). Are free radicals a major culprit? In: Therapeutic approaches to myocardial infarct size limitation. Eds DJ Hearse and DM Yellon. Raven Press New York pp 209-218.

4. Chambers DE, Parks DA, Patterson G, Yoshida S, Burton K, Parmley LF, McCord JM, Downey JM (1983). Role of oxygen derived radicals in myocardial ischemia. Fed Proc 42: 1093.
5. Parks DA, Bulkley GB, Granger DN, Hamilton SR, McCord JM (1977). Ischemic injury in the cat small intestine. Role of superoxide radicals. Gastroenterology 82: 9-15.
6. Hearse DJ (1977). Reperfusion of the ischemic myocardium. J Mol Cell Cardiol 9: 605-615.
7. Manning AS, Hearse DJ (1984). Reperfusion-induced arrhythmias: mechanisms and prevention. J Mol Cell Cardiol 16: 497-518.
8. Hearse DJ (1984). Reperfusion of the ischemic myocardium. Clinical Research Reviews 4: 58-61.
9. Manning AS, Coltart DJ, Hearse DJ (1984). Ischemia and reperfusion-induced arrhythmias in the rat. Effects of xanthine oxidase inhibition with allopurinol. Circ Res 55: 545-550.
10. Woodward B, Zakaria M (1984). The effects of some free radical scavengers on reperfusion-induced arrhythmias in the isolated rat heart. J Mol Cell Cardiol.
11. Braunwald E, Maroko PR (1974). The reduction of myocardial infarct size. N Engl J Med 291: 525-526.
12. Kloner RA, Braunwald E (1980). Observations on experimental myocadial ischemia. Cardiovasc Res 14: 371-395.
13. Opie LH (1980). Myocardial infarct size II. Comparison of anti-infarct effects of beta blockade, glucose-insulin-potassium, nitrates and hyaluronidase. Am Heart J 100: 531-552.
14. Bache RJ (1984). Can drugs really limit infarct size? In: Therapeutic approaches to myocardial infarct size limitation. Eds DJ Hearse and DM Yellon. Raven Press New York pp 185-208.
15. DeWall RA, Vasko KA, Stanley EL, Kezdi P (1971). Responses of the myocardium to allopurinol. Am Heart J 82: 362-370.
16. Akizuki S, Yoshida S, Chambers DE, Eddy L, Parmley L, Yellon DM, Downey JM (1985). Infarct size limitation by the xanthine oxidase inhibitor allopurinol in closed chest dogs with small infarcts. Cardiovasc Res In Press.
17. Jolly SR, Kane WJ, Baillie MB, Abrams GD, Lucchesi BR (1984). Canine myocardial reperfusion injury; its reduction by the combined administration of superoxide plus superoxide dismutase. Circ Res 54: 277-286.
18. Romson J, Hook B, Kunkel S, Abrams G, Schork A, Lucchesi B. Reduction in the extent of myocardial ischemic injury by neutrophil depletion in the dog. Circulation 67: 1016-1023.
19. Hearse DJ, Yellon DM (1984). Therapeutic approaches to myocardial infarct size limitation. Raven Press New York.
20. Chambers DE, Yellon DM, Hearse DJ, Downey JM (1983). Flurbiprofen limits infarct size 6 hours but not 24 hours following coronary occlusion in the dog. Am J Cardiol 51: 884-890.
21. Zimmer HG (1983). Effects of inosine on cardiac adenine nucleotide metabolism in rats treated with isoproterenol. J Mol Cell Cardiol 15: Suppl 1-296.
22. Juhasz-Nagy A, Papp L, Syabo Z (1983). Experimental evidence for the cardioprotective action of inosine. J Mol Cell Cardiol 15: Suppl 1-297.
23. Opie LH, Nathan D, Lubbe WF (1979). Biochemical aspects of arrhythmogenesis and ventricular fibrillation. Am J Cardiol 43: 131-148.
24. Corr PB, Witowski FX (1983). Potential electrophysiologic mechanisms responsible for dysrhythmias associated with reperfusion of ischemic myocardium. Circulation 68: suppl 1 16-24.
25. Corr PB, Shayman JA, Kramer JB, Kipnis RJ (1981). Increased alpha-adrenergic receptors in ischemic cat myocardium: a potential mediator of electrophysiological derangements. J Clin Invest 67: 1232-1226.

26. Fitzgerald JD (1982). The effects of beta-adrenoceptor blocking drugs on early arrhythmias in experimental and clinical myocardial ischemia. In: Early Arrhythmias Resulting from Myocardial Ischemia. Ed. by J R Parratt (publ. Macmillan Press, London) 295-315.
27. Podzuweit T, Dalby AJ, Cherry GW, and Opie LH (1978). Cyclic AMP levels in ischemic and non-ischemic myocardium following coronary artery ligation: relation to ventricular fibrillation. J Mol Cell Cardiol 10: 81-94.
28. Corr PB, Gross RW, Sobel BE (1982). Arrhythmogenic amphiphilic lipids and the myocardial cell membrane. J Mol Cell Cardiol 14: 619-626.
29. Hirche H, Friedrich R, Kebell U, McDonald F, Zylka V (1982). Early arrhythmias, myocardial extracellular potassium and pH. In: Early Arrhythmias Resulting from Myocardial Ischemia. Ed. by J R Parratt (publ. Macmillan Press Ltd, London) 139-154.
30. Clusin WT, Buchbinder M, Harrison DC (1983). Calcium overload, 'injury current', and early ischemic cardiac arrhythmias - a direct connection. Lancet 272-273.
31. Kurien VA, Yates PA, Oliver MF (1971). The role of free fatty acids in the production of ventricular arrhythmias after acute coronary artery occlusion. Eur J Clin Invest 1: 225-241.
32. Most AS, Capone RJ, Mastrofrancesco PA (197). Free fatty acids and arrhythmias following coronary artery occlusion in pigs. Cardiovasc Res 11: 198-205.
33. Manning AS, Coltart DJ, Hearse DJ (1984). Ischaemia- and reperfusion-induced arrhythmias in the rat: effects of xanthine oxidase inhibition with allopurinol. Circ Res 55: 545-550.
34. Selye H, Bayusz E, Graslo S, Mendell P (1960). Simple technique for the surgical occlusion of coronary vessels in the rat. Angiology 11 395-409.
35. Rao PS, Cohen MV, Mueller HS (1983). Production of free radicals and lipid peroxides in early experimental myocardial ischemia. J Mol Cell Cardiol 15: 713-716.
36. Meerson FZ, Kagan VE, Kozlor YP, Belkina LM, Arkhipinko YP (1982). The role of lipid peroxidation in pathogenesis of ischemic damage and the anti-oxidant protection of the heart. Basic Res Cadiol 77: 465-485.
37. Gaudel Y, Duvelleroy MA (1984). Role of oxygen radicals in cardiac injury due to reoxygenation. J Mol Cell Cardiol 16: 459-470.
38. Guarnieri C, Flamigni F, Caldarera CM (1980). Role of oxygen in the cellular damage induced by reoxygenation of hypoxic heart. J Mol Cell Cardiol 12: 797-808.
39. Rowe TG, Manson NH, Caplan M, Hess ML (1983). Hydrogen peroxide and hydroxyl radical mediation of activated leucocyte depression of cardiac sarcoplasmic reticulum. Circ Res 53: 584-591.
40. Haber F, Weiss J (1934). The calalytic decomposition of hydrogen peroxide by iron salts. Proc Royal Soc London A147: 332-351.
41. McCord JM, Day DE (1978). The superoxide dependent production of hydroxyl radical catalyzed by iron-EDTA complex. FEBS Letters 86: 139-142.

© 1986 Elsevier Science Publishers B.V. (Biomedical Division)
Superoxide and Superoxide Dismutase in Chemistry,
Biology and Medicine; G. Rotilio, editor

FREE RADICALS AND ISCHEMIC INJURY: ALLOPURINOL PRETREATMENT OR FREE RADICAL SCAVENGERS IMPROVE MYOCARDIAL PROTECTION DURING CARDIOPLEGIC ARREST AND REPERFUSION.

D.J. CHAMBERS, M.V. BRAIMBRIDGE, D.J. HEARSE.
Heart Research Unit, The Rayne Institute, St. Thomas' Hospital, London, UK.

INTRODUCTION

Oxygen-derived free radicals such as superoxide (O_2^-), arising from a number of different pathways (including the postischemic degradation of hypoxanthine by the ischemia-induced activation of xanthine oxidase), have been implicated in ischemia and reperfusion-induced myocardial injury (1).

In this study we have used an isolated perfused working rat heart model of cardiopulmonary bypass, involving cardioplegic arrest and global ischemia, to investigate whether anti-free radical interventions (allopurinol pretreatment and/or the addition of free radical scavengers to the cardioplegic and reperfusion solutions) could improve postischemic myocardial recovery.

MATERIAL AND METHODS

Animals and drugs

Hearts were obtained from male Wistar rats (250-300 gm body wt). Allopurinol (10mg/ml) was administered orally as a suspension in distilled water at a dose of 20mg/kg body wt daily for 2 days. On day 3, a solution (prepared at pH 10.8 to facilitate solubility) was injected i.v. 1 hour prior to removal of the heart. Superoxide dismutase and catalase (20 and 100 IU/ml respectively) were added to the cardioplegic and/or the aerobic reperfusion solution given immediately after the ischemic period.

Experimental preparation and time-course

An isolated perfused working rat heart preparation (2) was used. Hearts were subjected to 5 min of aerobic perfusion in the Langendorff mode (3), and then converted to aerobic preischemic control perfusion in the working mode for 20 min. During this time control values of aortic and coronary flow, cardiac output, heart rate, aortic pressure, stroke volume and stroke work were measured. The hearts were arrested by a 3 min infusion (at 37°C or 21°C) of St. Thomas' Hospital cardioplegic solution (4), followed by global ischemia for 30 minutes at 37°C or 300 minutes at 21°C (reinfusions of cardioplegic solution every 30 minutes). The hearts were then aerobically reperfused in the Langendorff mode for 15 min and converted to aerobic postischemic perfusion in the working mode (when postischemic functional indices were again measured and expressed as a percentage of their preischemic control values). At least 6 hearts were used for each condition studied, and the results were expressed as means $\pm$ SEM. Statistical significance was established using the Student's t-test.

Experimental groups

1. Control

Pretreated rats

2. Allopurinol pretreatment
3. Allopurinol pretreatment - hypothermic ischemia
4. Allopurinol pretreatment with superoxide dismutase (SOD) and catalase (CAT) added to the cardioplegic solution (CS) and reperfusion solution (RS)

Non-pretreated rats

5. SOD plus CAT added to CS and RS
6. SOD added to CS and RS
7. CAT added to CS and RS
8. SOD plus CAT added to CS only
9. SOD plus CAT added to RS only

RESULTS

Postischemic functional recovery in hearts from allopurinol pretreated (group 2) and control rats subjected to 30 min of normothermic global ischemia are shown in Fig 1 and Table 1. As can be seen, allopurinol pretreatment significantly improved myocardial function above that of control. However, allopurinol pretreatment was not effective in improving postischemic function following hypothermic ischemia (Table 1).

Hearts from allopurinol pretreated rats with SOD and CAT added to cardioplegic and reperfusion solutions (group 4) also had a significantly improved postischemic recovery of function above control (Table 1). However, this improvement was the same as that of allopurinol pretreatment alone (group 2), and no additional protection was observed.

Significantly improved postischemic functional recovery in hearts from non-pretreated rats was seen when SOD plus CAT were present during the initial aerobic Langendorff reperfusion period (groups 5 and 9). The extent of the recovery in these hearts was comparable to that in allopurinol pretreated hearts following normothermic ischemia (group 2).

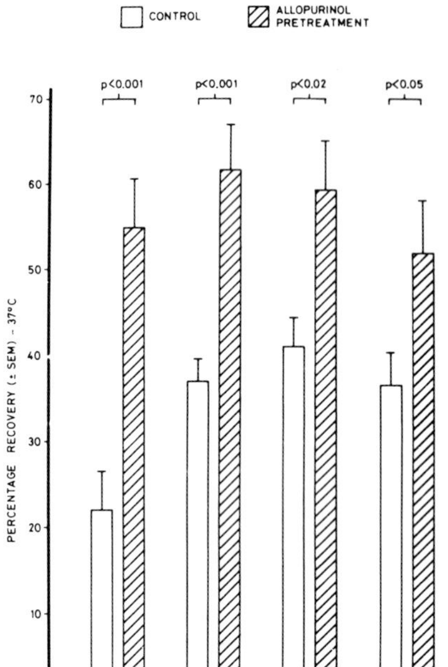

Fig 1. Postischemic functional recovery in hearts from control and allopurinol pretreated rats (group 2) following normothermic ischemia.

DISCUSSION

Ischemia and reperfusion-induced tissue injury in a variety of organs has been shown (5,6) to be reduced by pretreatment with the xanthine oxidase inhibitor, allopurinol. The mechanism by which allopurinol was protecting the tissue was thought to be by a conservation of purines. A more recent suggestion, however, has been that protection of ischemic tissue may result from an inhibition of oxygen-derived free radicals produced as a byproduct of or hypoxanthine degradation of the ischemia-induced activation of xanthine oxidase (1).

In seeking to define which of these mechanisms would best explain the significantly improved myocardial protection seen in these studies, the comparable functional recoveries obtained in hearts from allopurinol pretreated rats and in hearts treated with a combination of free radical scavengers would suggest the involvement of oxygen-derived free radicals, rather than a conservation of purines. Also, the absence of an additive effect when xanthine oxidase inhibition was combined with scavenging of free radicals (which can act on free radicals generated by sources other than xanthine oxidase) suggests that xanthine oxidase is the major source of free radicals, for the conditions used in this study. In addition, the results suggest that free radical scavengers should be present during the aerobic reperfusion period immediately following ischemia, which is the most likely period for oxygen-derived free radical generation.

TABLE 1.
Postischemic functional recovery in hearts from experiment groups 1-9.

	Recovery %			
Group	Aortic Flow	Cardiac Output	Stroke Volume	Stroke Work
1	25±3	38±2	41±4	35±4
2	48±6****	57±6***	60±6***	56±6***
3	25±8	32±9	36±8	30±9
4	47±4****	60±7***	63±8***	57±7***
5	48±6**	55±5**	58±5**	52±5**
6	28±3	38±2	41±2	37±2
7	16±6	27±6	28±6*	23±6*
8	39±9	45±8	50±8	44±8
9	51±4****	56±3***	59±4***	53±5***

*$p < 0.05$, **$p < 0.02$, ***$p < 0.01$, ****$p < 0.001$

Thus, in this model of cardiopulmonary bypass, improved myocardial protection was achieved either by preventing the generation of oxygen-derived free radicals (by inhibition of ischemia-induced activation of xanthine oxidase), or by removing these radicals after their production (by effective use of free radical scavengers). Therefore, these results suggest that anti-free radical interventions may have a role in improving myocardial protection afforded by cardioplegia.

REFERENCES

1. McCord JM (1985) New Eng.J.Med. 312:159-163.

2. Hearse DJ, Braimbridge MV, Jynge P. In: Protection of the Ischemic Myocardium: Cardioplegia' Raven Press, Amsterdam, p59.

3. Langendorff O. (1958) Pflugers Arch. 61:291.

4. Jynge P, Hearse DJ, Feuvray D, Mahalu W, Cankovic-Darracott S, O'Brien K, Braimbridge MV (1981) Scand J Thorac Cardiovasc Surg 83:830-839.

5. DeWall RA, Vasko KA, Stanley EL, Kezdi P (1971) Am Heart J 82:362-370.

6. Cunningham SK, Keaveny TV, Fitzgerald P (1974) Br J Surg 61:562-565.

© 1986 Elsevier Science Publishers B.V. (Biomedical Divison)
Superoxide and Superoxide Dismutase in Chemistry,
Biology and Medicine; G. Rotilio, editor

REPERFUSION-INDUCED ARRHYTHMIAS AND OXYGEN-DERIVED FREE RADICALS : STUDIES WITH "ANTI-FREE RADICAL" INTERVENTIONS IN THE ISOLATED PERFUSED RAT HEART.

A.S.MANNING, M.BERNIER, D.J.HEARSE

The Heart Research Unit, The Rayne Institute, St Thomas' Hospital, London SE1 7EH, U.K.

INTRODUCTION

We have suggested recently (1) that xanthine oxidase-derived free radicals, in particular the superoxide radical ($O_2{\cdot}^-$), may play an important role in the genesis of reperfusion-induced arrhythmias. We have further assessed this possibility by studying the effect of a variety of agents known to possess free radical scavenger properties (superoxide dismutase, catalase, methionine, glutathione, and mannitol).

These studies have been carried out in the isolated, perfused rat heart preparation with transient coronary artery occlusion.

METHODS

Animals, Perfusion Technique and Perfusion Media.

Male rats of the Wistar strain (220-280 g body weight) were used for all studies. Isolated hearts were perfused in the Langendorff mode with Krebs-Henseleit bicarbonate buffer, pH 7.4, (modified to contain 11.1 mM glucose, 4.3 mM K^+ and 2.5 mM Ca^{2+}) which was gassed with 95% oxygen plus 5% carbon dioxide. Prior to use, the perfusion fluid was filtered through a cellulose acetate membrane (5.0 µm pore size).

Experimental Time Course:

Hearts were perfused for an initial 10 minute period. At the end of this time, a ligature was placed around the left anterior descending coronary artery close to its origin. Both ends of the ligature were passed through a small plastic cylinder which was then pressed against the artery. The resulting arterial occlusion was maintained for 10 minutes by clamping the plastic cylinder and ligature. At the end of the 10 minute period of coronary occlusion, the artery was reperfused by removing the clamp and the tube. Reperfusion rhythm disturbances were then monitored for a 3 minute period.

Addition of Pharmacological Interventions:

The following compounds were included separately in the perfusion medium throughout the experimental time course: L-methionine (10 mM), superoxide dismutase, SOD (EC 1.2.3.2, 0.1 Unit/liter, Pharmacia, Sweden), catalase (EC 1.11.1.6, 1 Unit/liter, Sigma, U.K.), reduced glutathione (10 μM), and mannitol (50 mM).

Indices Measured and Evaluation of Rhythm Disturbances:

An epicardial ECG recording was obtained via two silver electrodes attached to the heart and the ECG trace was permanently recorded on a Gould Physiograph. Heart rate (calculated from the ECG trace) was measured continuously and coronary flow were recorded at 5 minute intervals.

High-speed electrocardiograms were analyzed for: (i), the incidence of ventricular fibrillation (and also whether fibrillation was spontaneously reversible or whether hearts remained in irreversible fibrillation); and (ii), the incidence of ventricular tachycardia (defined as five or more consecutive PVC's). In addition, we have quantified the total duration during which each heart remained in normal sinus rhythm during the reperfusion phase.

RESULTS

Shown in Table 1 is the effect of the addition of free radical scavengers to the perfusion fluid on reperfusion-induced arrhythmias. The incidence of reperfusion-induced ventricular fibrillation was significantly reduced by the addition of all agents ($p < 0.01$). Similarly reperfusion-induced ventricular tachycardia was reduced (to between 40% and 20%) and associated with this reduction in arrhythmias was an increase in the total duration of normal sinus rhythm during the 180 second reperfusion phase from 58 ± 22 seconds to between 165 to 178 seconds ($p < 0.001$). Heart rate was generally unaltered by the addition of free radical scavengers.

	% INCIDENCE			
	Ventricular Fibrillation	Irreversible Ventricular Fibrillation	Ventricular Tachycardia	Time in Normal Sinus Rhythm (sec)
CONTROL	80	60	80	59 ± 22
L-METHIONINE 10 mM	7***	0**	20**	177 ± 2***
SOD 0.1 Unit/liter	0***	0**	40*	175 ± 2***
CATALASE 1 Unit/liter	7***	0**	27**	178 ± 1***
GLUTATHIONE 10 uM	20**	7**	27**	165 ± 12***
MANNITOL 50 mM	7***	7***	40*	168 ± 11***

TABLE 1 : The effect of free radical scavengers on reperfusion-induced arrhythmias in the isolated, perfused rat heart with temporary artery occlusion (n=15 in each group). $*p < 0.05$, $**p < 0.01$, $***p < 0.001$.

DISCUSSION:

In the present study, we have endeavoured to obtain more evidence for an involvement of free radicals in the genesis of reperfusion-induced arrhythmias. Using the isolated, perfused rat heart with regional ischemia, we have been able to show that a range of free radical scavengers all reduce the incidence of reperfusion arrhythmias.

Since intermediates such as the superoxide and the hydroxyl free radical may have very short half lives and can travel only short distances from their site of formation, it must be questioned how events occuring in the myocyte could be influenced by scavengers in the vascular space. A possible explanation may be derived from a study by Jarasch et al (2) who demonstrated that the majority of myocardial xanthine oxidase is located in the vascular endothelium and not in the myocyte. Thus, superoxide produced via xanthine oxidase in endothelial cells during reperfusion, and hydroxyl radicals derived from it via the Haber-Weiss reaction, might well cause injury to the endothelium or adjacent contractile or conducting cells, but this could be prevented by scavengers or inhibitors of free radical production which are present in the vascular and extravascular space.

In conclusion, although we have presented further evidence that free radicals may play a role in the genesis of reperfusion arrhythmias, the evidence is circumstantial and must remain so until techniques such as electron spin resonance can be exploited to identify and measure free radical production in the beating heart. However, we believe that our results do justify further consideration of the role of free radicals in myocardial injury and protection and also the possibility that free radical scavengers, or inhibitiors may represent a new approach in the treatment of ventricular arrhythmias induced by myocardial ischemia and reperfusion.

ACKNOWLEDGEMENTS

This work was supported in part by grants from the British Heart Foundation and St Thomas' Hospital Research Endowment Fund. The assistance of Mrs C. Erlebach is gratefully acknowledged.

REFERENCES

1. Manning AS, Coltart DJ, Hearse DJ (1984) Ischemia and reperfusion-induced arrhythmias in the rat. Effects of xanthine oxidase inhibition with allopurinol. Circ Res 55: 545-548.

2. Jarasch ED, Grund C, Bruder G, Heid HW, Keenan TW, Franke WW (1981) Localization of xanthine oxidase in mammary gland epithelium and capillary endothelium. Cell 25: 67-82.

Superoxide and Superoxide Dismutase in Chemistry,
Biology and Medicine; G. Rotilio, editor

SUPEROXIDE DISMUTASE ENHANCES RECOVERY FOLLOWING MYOCARDIAL ISCHEMIA/REPERFUSION *IN VIVO* AND *IN VITRO*

MONICA HUHN-EDHOLM, ELISABETH ÖSTLING AND ERIK FELLENIUS
Department of Pharmacology, Pharmacia AB, S-751 82 Uppsala (Sweden)

The purpose of this study was to evaluate the ability of superoxide dismutase (SOD) to protect the heart from ischemia and reperfusion induced injury *in vivo* and *in vitro* both in the presence and absence of leucocytes.

MATERIAL AND METHODS

For the *in vitro* studies, hearts of pentobarbital-anaesthetized male rats of the Sprague-Dawley strain, weighing 230-300 g, were perfused according to Langendorff with a modified Krebs-Henseleit bicarbonate buffer, 1.25 mM Ca^{2+} and 11 mM glucose in a non-recirculating system (1). The pressure on the heart was 80 cm H_2O. The pressure development was monitored and set to 100% in each experiment.

Leucocytes, containing polymorphonuclear (PMN) cells, were isolated according to Axelsson et al. (2). Activation of the leucocytes by PMA (phorbol 12-myristate 13-acetate) was checked by measuring chemiluminescence (3).

In the *in vivo* studies male Sprague-Dawley rats, weighing 200 g, were pretreated with pethedine, 5 mg/kg, and ether anaesthetized. The hearts were exteriorized through the intracostal space (4). The left coronary artery was either ligated permanently, or with a loop which was pulled away after varying times. After ligation, the heart was positioned back into the chest. Mortality was followed for one week from the time of SOD administration.

Student t-tests were used to estimate statistical significance of differences, with the exception of the experiment with mortality, where life tables were used. Statistical significance ($p<0.05$) denotes difference between groups with and without SOD. Mean ± S.D. with the number (n) of observations are given. Bovine SOD was obtained from Grünenthal GmbH (Aachen-Eilendorf, West Germany) (~7000 U/mg).

RESULTS

<u>SOD enhances work performance following ischemia in *in vitro* perfused heart</u>. The work performance recovered faster in hearts perfused with SOD, added before the ischemia or added before and after the ischemia, as compared with control hearts without SOD (Fig. 1). Postischemic addition of SOD had no effect. Thirty min after the ischemic period no difference on heart work was observed between any of the different treatments.

Activated leucocytes impaired work performance, while non-activated cells were without effect in perfusion experiment with well-oxygenated hearts (Fig. 2). When

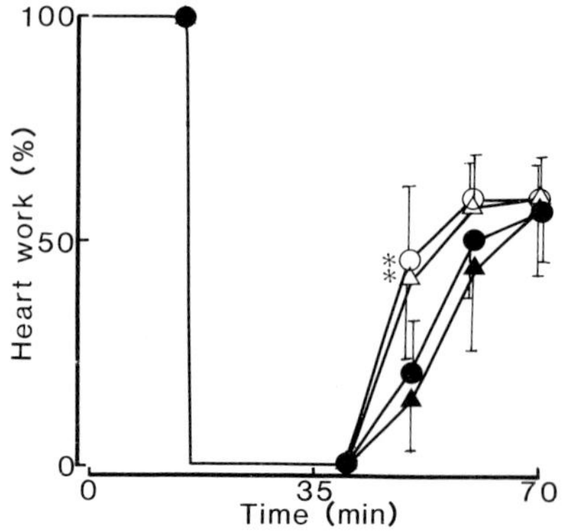

Fig. 1. The effect of SOD on work performance after ischemia in isolated perfused hearts. Control hearts were perfused without SOD (● n=9). Sod, 300 000 U/L, was added 5 min preischemically (△ n=5), 5 min preischemically and 10 min postischemically (o n=5) and 10 min postischemically (▲ n=6).

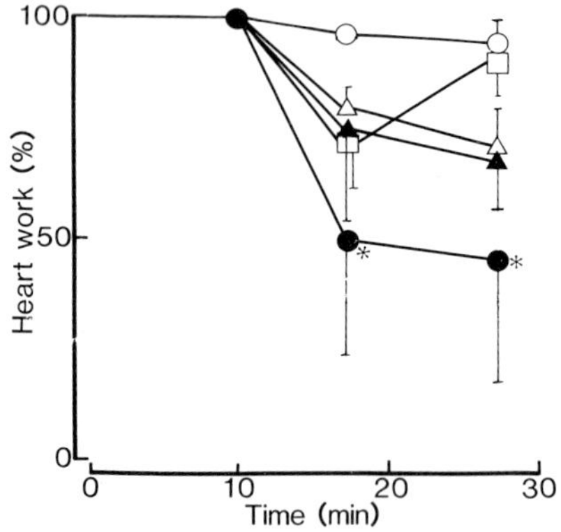

Fig. 2. The effect of SOD on work performance of isolated perfused hearts in presence of activated leucocytes. The hearts were perfused for 10 min with buffer. Thereafter cells were infused for 7 min with or without PMA followed by 10 min perfusion with buffer. Buffer (o n=4); PMA 0.65 µM (△ n=5); leucocytes ($20 \cdot 10^6$ PMN/ml) (□ n=3); leucocytes+PMA (o n=8); leucocytes+PMA+SOD, 300 000 U/L (△ n=8) were infused at a rate of 1/10 of the coronary flow.

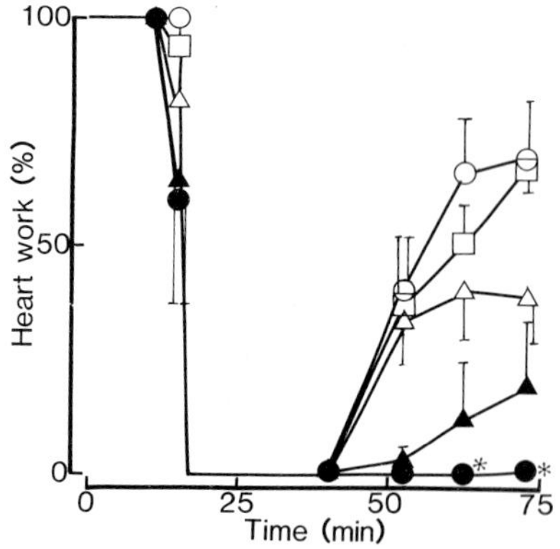

Fig. 3. The effect of SOD on work performance of isolated perfused hearts after ischemia in presence of activated leucocytes. The cells were infused 4 min before ischemia and 3 min after ischemia. Buffer (o n=5); PMA 0.65 µM (△ n=6); leucocytes ($20 \cdot 10^6$ PMN/ml) (□ n=3); leucocytes+PMA (● n=7); leucocytes+PMA+SOD, 300 000 U/L (▲ n=7).

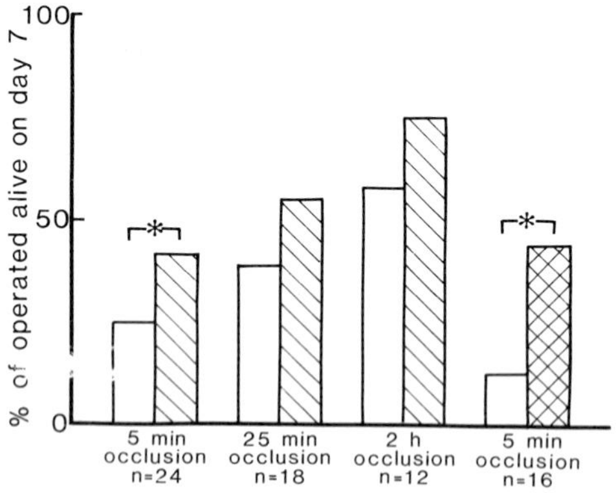

Fig. 4. Effect of SOD and lidocain on mortality after reperfusion-induced injury of the rat heart *in vivo*. Control animals (□) were given 0.15 M NaCl 2 ml/kg i.v. and treated animals (▧) were given SOD 10 mg/kg (70 000 U/kg) dissolved in 0.15 M NaCl i.v. 30 s before reperfusion or lidocain 8.5 µM (▩) i.v. 2 min before reperfusion.

activated cells were infused before, during and after ischemia none of the hearts were able to recover (Fig. 3). SOD improved function in all these experiments.

SOD decreases mortality following infarction and reperfusion *in vivo*. In the *in vivo* studies with the left coronary artery permanently ligated, SOD was injected i.v. immediately before the operation or 30 min after the operation. The mortality frequence from the time of SOD administration till the 7th day was 74% in the control and 68% in the animal receiving SOD before the ligation. The mortality was 50% in the control and 50% in animal receiving SOD 30 min after ligation. No statistically significant differences were seen between SOD treatment and control groups in the above studies.

In the studies where the coronary artery was ligated with a loop, which was pulled away after 5 min occlusion, 25 min occlusion and 2 h occlusion, there was a statistically significant difference between lidocain or SOD-treated animals as compared with control in the experiment with 5 min occlusion but not during longer ischemia (Fig. 4). In these experiments the mortality was again measured from the time of drug administration till the 7th day.

DISCUSSION

The majority of the animals died within 30 min after the infarction or infarction plus reperfusion. The decrease in mortality frequency by both SOD and lidocain was also found within this time period (data not shown), indicating the prevention by the drugs of mortal arrhythmias (5). In addition, better work performance may also explain the beneficial effect by SOD treatment. Thus, there was a more rapid improvement of function by SOD within the perfused heart (Fig. 1). However, there are several mechanisms, such as activation of leucocytes (6) that may contribute to the outcome in the *in vivo* situation which are absent in the used *in vitro* perfused system. Activated leucocytes were also shown to have a dramatic influence on work-performance even in the well-oxygenated perfused heart with (Fig. 2) or without (Fig. 3) ischemia. Whether the improvement seen by SOD-treatment in the perfused system also explain the beneficial effect of SOD on mortality *in vivo* remains to be elucidated.

REFERENCES

1. Neely JR, Liebermeister H, Battersby EJ, Morgan HE (1967) Am J Physiol 212:804-814
2. Axelsson LG, Nilsson G, Björkstén B (1981) J Immunol Methods 46:251-258
3. Hodgson EK, Fridovich I (1973) Photochemistry and Photobiology 18:451-455
4. Johns TNP, Olson BJ (1954) Ann Surg 140:675-682
5. Woodward B, Zakaria MNM (1985) J Mol Cell Cardiol 17:485-493
6. Romson JL, Hook BG, Kunkel SL, Abrams GD, Schork MA, Lucchesi BR (1983) Circulation 67:1016-1023

Superoxide and Superoxide Dismutase in Chemistry, Biology and Medicine; G. Rotilio, editor

CORRELATION BETWEEN PROSTAGLANDINS RELEASE AND OXIDATIVE STRESS IN THE ISOLATED PERFUSED RAT HEART

RENATA BARSACCHI*, NICLA NANNI, PAOLO CAMICI, GUALTIERO PELOSI, ANTONIO BENASSI, DANIELA GIANNESSI and FULVIO URSINI**

C.N.R. Institute of Clinical Physiology, University of Pisa

* Institute of Biological Chemistry, University of Pisa

** Institute of Biological Chemistry, University of Padova, Italy.

INTRODUCTION

Chemiluminescence (CL) detection from exposed or perfused organs, originally described by Chance et al., is utilized in our laboratory to monitor the oxidative stress in the heart (1). The isolated perfused rat heart emits a spontaneous CL. This CL emission is oxygen dependent and is stimulated by an oxidative stress induced either by the perfusion with micromolar doses of organic hydroperoxides or by gluthatione depleting agents (2, 3). Recently, the activatory effect of hydroperoxides on prostaglandin (PG) synthase has been demonstrated and these compounds seem to play a role in cell function and pathology (4). In the present paper the relationship between an oxidative stress, induced by cumene hydroperoxide (CHP) administration, and PG release was studied in isolated perfused rat hearts.

MATERIALS AND METHODS

Chemiluminescence emitted from the isolated heart was monitored using a single photon counting apparatus (2). The initial emission rate of CL during CHP perfusion was measured by linear regression analysis of the emission values integrated at two second intervals.

Prostaglandin (PG) release from the heart was evaluated as 6-KetoPGF_1 by commercially available radioimmunoassay (NEN).

Vitamin E deficient rats were fed on a Selenium/Vitamin E deficient diet (Laboratori Piccioni, Milano) for six weeks. All chemicals were of analytical grade.

RESULTS AND DISCUSSION

The kinetics of the initial rate of CL emission vs CHP concentration is reported in Fig. 1. The dramatic effect of Vitamin E deficiency on the susceptibility to the oxidative stress is also shown for comparison.

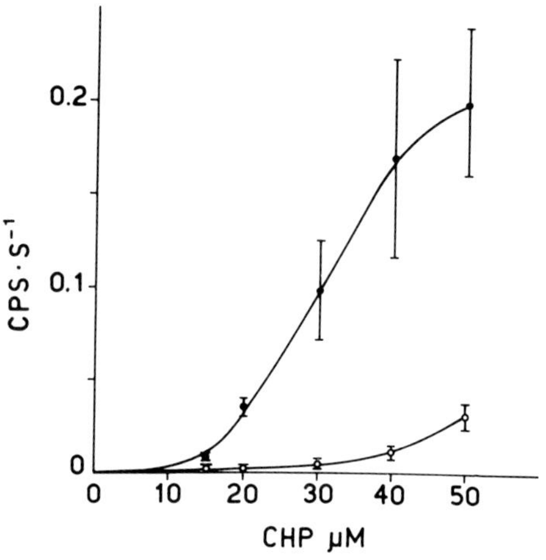

Fig. 1. Kinetics of chemiluminescence emission induced by CHP in controls experiments (open circles) and in Vitamin E deficient rat heart (closed circles).

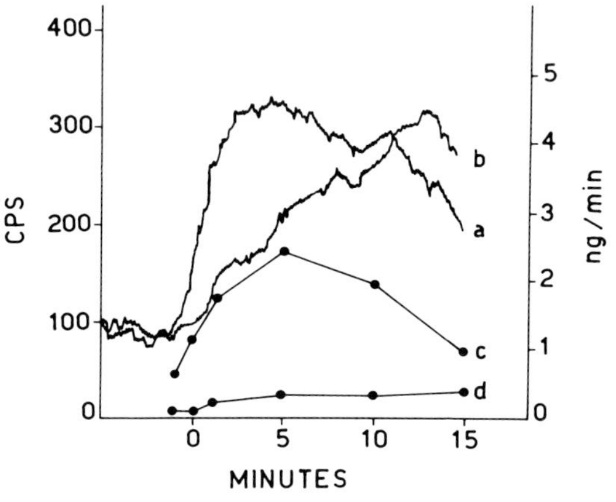

Fig. 2. Effect of aspirin on CL emission and 6-KetoPGF$_{1}\alpha$ release from the rat heart during perfusion with 70 μM CHP: a) CL emission in the control experiment; b) CL emission in the presence of 2.7 mM aspirin; c) 6-KetoPGF$_{1}\alpha$ release in the control experiment; d) 6-KetoPGF$_{1}\alpha$ release in the presence of aspirin. CHP perfusion started from time 0.

Cumene hydroperoxide perfusion in the isolated rat heart induced a dose dependent increase of PG release that paralleled CL emission. The presence of 2.7 mM aspirin in the perfusate completely inhibited the release. Aspirin, on the other hand, stimulated CL emission (Fig. 2).

These data support that: a) PG release is stimulated by the oxidative stress, b) aspirin dependent cyclooxigenase inhibition increases the steady state concentration of hydroperoxidic radicals generated by lipid peroxidation or, possibly, by activation of lipooxigenase.

The latter hypothesis is also supported by the evidence reported in Fig. 3. The antioxidant nordihydroguaiaretic acid (NDGA), which is known to inhibit lipooxigenase in vitro, completely abolished, at 2.5 μM concentration, the CL induced by 100 μM CHP.

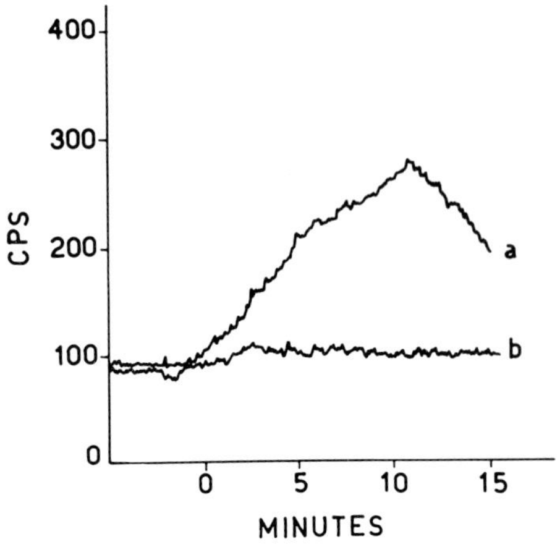

Fig. 3. Effect of 2.5 μM NDGA on the CL emission in the heart following 70 μM CHP perfusion: a) CL emission in control experiment, b) CL emission in the presence of 2.5 μM NDGA. CHP perfusion started from time 0.

REFERENCES

1. Boveris A, Cadenas E, Reiter R, Filipkowski M, Nakase Y, Chance B (1980) Proc Natl Acad Sci USA, 77: 343-351
2. Barsacchi R, Camici P, Bottigli U, Salvadori PA, Pelosi G, Maiorino M, Ursini F (1983) Biochim Biophis Acta 762: 241-247
3. Barsacchi R, Pelosi G, Camici P, Bonaldo L, Maiorino M, Ursini F (1984) Biochim Biophis Acta 804: 356-360
4. Lands VEM, Kulmacz RJ, Marshall PJ (1984) In: Pryor A (ed) Free radicals in Biology, vol VI, pp 39-57, Academic Press New York

Superoxide and Superoxide Dismutase in Chemistry,
Biology and Medicine; G. Rotilio, editor

PRESERVATION OF SUPEROXIDE-DISMUTASE (SOD) - SPECIFIC ACTIVITY IN THE REPERFUSED ISCHEMIC MYOCARDIUM BY A PROSTACYCLIN - MIMETIC (ILOPROST OR ZK 36374)

CHRISTOPH THIEMERMANN, ELISABETH STEINHAGEN-THIESSEN*, KARSTEN SCHRÖR
Pharmakologisches Institut der Universität Köln, D-5000 Köln 41 (FRG) and *Medizinische Klink der Universität Hamburg, 2000 Hamburg 20 (FRG)

INTRODUCTION

The stable prostacyclin-mimetic iloprost was previously reported to exert cardioprotective actions in an in-vivo model of acute myocardial ischemia plus reperfusion [1-3]. Reperfusion of ischemic myocardium initially leads to aggravation of tissue damage [4] that is possibly related to a loss of antioxidative enzymes, such as superoxide dismutase (SOD)[5]. Recent investigations of our group have demonstrated a membrane-stabilizing action of iloprost in the reperfused, globally ischemic rabbit heart in vitro [6]. This would tend to reduce the leakage of large molecules, such as enzyme proteins. The present study was designed to investigate whether this mechanism operates in the reperfused ischemic myocardium in vivo.

MATERIALS AND METHODS

Animal preparations and measurements

Adult cats were thoracotomized and subjected to ligation of the left anterior decending coronary artery for 3 h, followed by 2 h of reperfusion by removal of the ligature. Arterial blood pressure, heart rate and a limb ECG were monitored continuously. At 5 h, the hearts were removed, divided into severely ischemic and non-ischemic myocardium and subjected to determination of creatine phosphokinas (CK) and cytosolic SOD-specific activities [1-3].

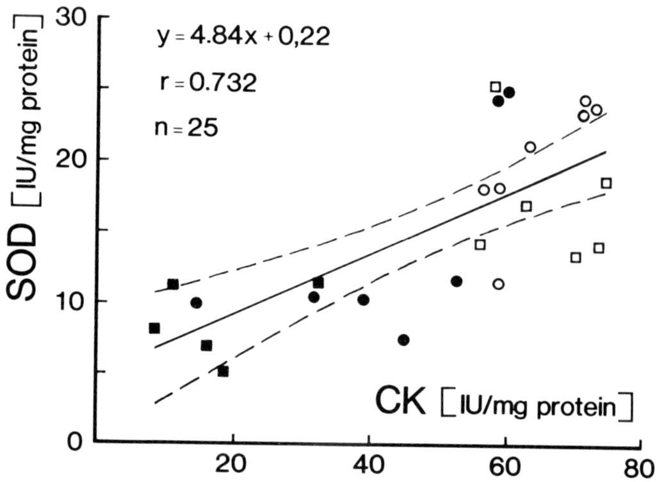

Fig.1. Linear correlation between CK- and SOD-specific activities in cat hearts subjected to ischemia + reperfusion (filled symbols) or sham-operation (open symbols) and treated with iloprost (0.6 µg/kg x min) (circles) or vehicle (squares).

Treatment

Iloprost (0.6 µg/kg x min) or vehicle were infused iv., starting at 0.5 h after LAD occlusion. Infusions were maintained throughout the experiment. Sham-operated cats, subjected to all of the same procedures except ligation of the LAD, served as controls.

RESULTS

Figure 1 depicts CK- and SOD-specific activities of left ventricular center ischemic myocardium (filled symbols) and non-ischemic myocardium of sham-operated animals (open symbols). It is evident that ischemia + reperfusion results in a considerable reduction of both CK and SOD in vehicle-treated (squares) cats, by 50-70% of control, which is significantly improved by iloprost ($P < 0.01$)(circles). There is no change in these parameters in non-ischemic hearts but a linear-correlation between SOD and CK in all groups of animals studied ($P < 0.001$).

There was a considerable rise in the ST-segment during ischemia and appearance of a Q-wave (suggestive for myocardial necrosis) during reperfusion in vehicle-treated animals that were largely antagonized by iloprost ($P < 0.01$). However, iloprost did not modify general hemodynamics or heart rate at the dose used.

DISCUSSION

The present study confirms our earlier findings on improved preservation of reperfused ischemic cat myocardium by iloprost [1-3]. The linear correlation between cytosolic SOD and CK suggests that the beneficial effects of iloprost on these parameters and the ECG probably result from a membrane stabilizing action, thus extending data from an in vitro model of myocardial ischemia [6] to the intact animal. This action on the plasma membrane prevents the leakage of large molecules during reperfusion, is selective for injured myocardial tissue and can clearly be separated from hemodynamic effects of the compound.

REFERENCES

1. Schrör K, Gallenkämper W (1983) Fed Proc 42:633 (1983)
2. Thiemermann C, Schrör K (1984) Biomed Biochim Acta 43: S 151 - S 154
3. Thiemermann C, Steinhagen-Thiessen E, Schrör K (1984) J Cardiovasc Pharmacol 6: 365-366
4. Lucchesi BR (1985) In: Schrör K (ed) Prostaglandins and Other Eicosanoids in the Cardiovascular System. Karger, Basel, pp 160-171
5. Meerson FZ, Kagan VE, Kozlov Yu P, Belkina LM, Arkhipenko Yu V (1982) Basic Res Cardiol 77:465-485
6. Smith EF III, Kloster G, Stöcklin G, Schrör K (1984) Biomed Biochim Acta 43: S 155 - S 158

Superoxide and Superoxide Dismutase in Chemistry,
Biology and Medicine; G. Rotilio, editor

SUPEROXIDE DISMUTASE IN CARDIAC TRANSPLANTS

K. MAILER[1], M. KLOC[2], AND S. STEPKOWSKI[3]
[1.] Department of Chemistry, St. Mary's University, Halifax, Canada; [2.] Department of Biochemistry, and [3.] Department of Surgery, Dalhousie University, Halifax, Canada.

INTRODUCTION

Genetic dissimilarity between donor and host determines the fate of a foreign organ transplanted into a human or animal host. When the two are very dissimilar, normal tissue architecture in the allograft is destroyed and the graft is rejected (1,2). The molecular mechanism of graft rejection is unknown, but large inflammatory changes accompany, e.g., cardiac tissue rejection. The coupling mechanism between inflammatory leukocyte infiltration and graft destruction is very poorly understood. There is, however, much evidence that activated PMNLs and monocytes can generate superoxide in a respiratory burst, and that superoxide and the hydroxyl radical (formed from $O_2^{\bar{\cdot}} + H_2O_2$) are toxic to biomolecules (3). Since the transplanted organ itself is provided with an efficient anti-oxidative enzyme defence system, we reasoned that inflammatory damage to transplanted tissue and the eventual destruction of this tissue may involve inactivation of one or more of the anti-oxidative defence enzymes in the transplanted tissue.

MATERIAL AND METHODS

Male rats (250-300g) from two inbred strains, Lewis (RT-1^{ℓ}) and BN (RT-1^{n}) were used to perform syngeneic (Lewis → Lewis) and allogeneic (BN → Lewis) heart transplantations. Heterotopic cardiac allografts and isografts were transplanted to the great abdominal vessels of recipient animals (4). Cardiac isografts survive indefinitely while the allografts are totally rejected within 7 days. Since lymphocytic infiltration of the allograft peaks at 5 ± 2 days, organs were removed from transplanted and control animals on the 5th day post-transplantation: cardiac allograft and isograft, the graft recipient's own heart, the hearts of non-transplanted BN and Lewis control rats, and the spleens and livers of all graft recipients and Lewis control rats. Enzyme activities were assayed in the supernatants of the Polytron treated tissues (12,000xg supernatant used for GSH.Px and SOD; 800xg supernatant for CAT) using standard methods (5,6,7). To measure protein synthesis, graft recipients and controls were injected, i.v. 5 days post transplantation, with ℓ-[^{35}S]-methionine (800μCi/rat). After two hours the animals were sacrificed and the organs excised, homogenized and centrifuged at 800xg for 10 min at 4°C. Supernatant aliquots, after incubation with RNAse, were loaded on gradient polyacrylamide gels. After electrophoresis, the gels were processed for autoradiography.

RESULTS

Organ	SOD Activity 5 Days Post-Transplantation	
	U/mg protein	Total Activity
Lewis control heart	13.1 ± 1.7[a] (6)[b]	237 ± 12
BN control heart	14.0 ± 5.5 (4)	* 278 ± 11
Isograft recipient's own heart	13.5 ± 1.9 (5)	** 264 ± 43 ↑
Cardiac isograft	* 10.7 ± 2.4 (5) ↓	* 207 ± 18 ↓
Allograft recipient's own heart	12.4 ± 1.5 (6)	* 264 ± 20 ↑
Cardiac allograft	* 8.0 ± 2.6 (6) ↓	* 185 ± 38 ↓
Lewis control liver	11.4 ± 2.5 (5)	2439 ± 312
Isograft recipient's liver	10.3 ± 1.6 (5)	2648 ± 277
Allograft recipient's liver	11.1 ± 3.0 (6)	2514 ± 223
Lewis control spleen	11.8 ± 4.0 (5)	189 ± 29
Isograft recipient's spleen	** 8.8 ± 2.6 (5) ↓	173 ± 40
Allograft recipient's spleen	* 5.6 ± 2.1 (6) ↓	157 ± 44

a Mean ± S.D.

b Number of rats tested

Values significantly different at P = 0.05 (*) or P = 0.1 (**)

↓ Decrease, ↑ Increase in activity relative to Lewis control

SOD is decreased by 40% in the cardiac allografts at the height of the host rejection response. SOD is decreased also by 53% in the spleens of allograft recipients. These decreases are not a general result of the trauma of surgery; they are seen only in the transplanted hearts of the genetically dissimilar transplantations and not in the recipient's own heart nor in its liver. There is a much smaller decrease in SOD in the isograft. Glutathione peroxidase (GSH.Px) and catalase (CAT), two other anti-oxidative enzymes, were unchanged or very little affected in the transplanted hearts. Patterns of protein synthesis were comparable in the allografts, isografts, and control hearts. Finally, preliminary experiments show that decreasing SOD in the allografts is accompanied by membrane destruction as seen by increasing levels of malonaldehyde.

DISCUSSION

Intra abdominal cardiac allografts between appropriate strains of inbred rats, as used here, are an ideal *in vivo* system for controlled investigation of the pathogenesis of graft rejection. The rejection phenomenon involving the transplanted heart is isolated as far as possible from whole body deterioration since the animal's circulation (and general health) is maintained by its own heart.

Both specific and total SOD activity are greatly reduced in the transplanted hearts and spleens of the (BN → Lewis) transplantations while such activities are reduced only slightly in (Lewis → Lewis) transplantations. The animals' own hearts and livers, organs not intimately involved in rejection of the allograft, have comparable specific SOD activities to isograft and non-operated control animals. There appears to be an increase in total SOD activity in the transplant recipients' own hearts as if in compensation for the general inflammatory response to transplantation surgery.

Host response to an allograft involves mobilization of inflammatory cells at the graft site and a subsequent respiratory burst; the $O_2^{\bar{}}$ thus generated can ultimately cause a cascade of oxidized products both in membranes and in the cytosol. We postulate that SOD in the allograft may be inactivated by continuous toxic-oxygen bombardment, and that radical and peroxide damage can then proceed unchecked by the enzyme defence system. This is supported both by increased MDA levels accompanying SOD decrease and previously reported inactivation of SOD by high levels of H_2O_2 *in vitro* (8).

We have demonstrated decreased activity of the primary anti-oxidative defence enzyme SOD in rat cardiac allografts, and also approximately normal levels of GSH.Px and CAT. There is no indication, 5 days post-transplantation, that abnormally high or low protein synthesis is associated with the decreased enzyme activity in the allograft, but rather that the enzyme is inactivated *in situ*. Such inactivation is accompanied by increased membrane breakdown and it may be caused by prolonged exposure to H_2O_2 or other products of a prolonged inflammatory response. These results demonstrate a direct link between the inflammatory response of the host animal and destruction of the transplanted organ; they indicate that deregulation of an anti-oxidative defence enzyme, superoxide dismutase, is an early molecular event in cardiac allograft destruction.

REFERENCES

1. Forbes RDC, Guttmann RD (1984) Immunol Rev 77:5-29
2. Häyry P, von Willebrand E, Parthenais E, Nemlander A, Soots A, Lautenschlager I, Alfoldy P, Renkonen R (1984) Immunol Rev 77:85-142
3. Halliwell B (1982) Cell Biol Int'l Rpts 6:529-542
4. Ono K, Lindsey ES (1969) J Thorac Cardiovasc Surg 57:225-229
5. Cohen C, Dembiec C, Marcus J (1970) Anal Biochem 34:30-38
6. McCord J, Fridovich I (1969) Superoxide dismutase. J Biol Chem 244:6049-6055
7. Paglia DE, Valentine WN (1967) J Lab Clin Med 70:158-169
8. Hodgson EK and Fridovich I (1975) Biochemistry 14:5294-5299

Superoxide and Superoxide Dismutase in Chemistry,
Biology and Medicine; G. Rotilio, editor

INVOLVEMENT OF SUPEROXIDE AND XANTHINE OXIDASE WITH DEATH DUE TO CEREBRAL ISCHEMIA-INDUCED SEIZURES IN GERBILS

JOSEPH S. BECKMAN, GERALD A. CAMPBELL, CHARLES J. HANNAN, JR., CYNTHIA S. KARFIAS, AND BRUCE A. FREEMAN.

Departments of Anesthesiology and Biochemistry, University of Alabama in Birmingham, Birmingham, AL, Department of Pathology, Duke University Medical Center, Durham, NC; Madigan Army Medical Center, Tacoma, WA

INTRODUCTION

The partial reduction of oxygen to more reactive intermediates may contribute to cerebral ischemic injury. Endogenous antioxidants such as α-tocopherol, ascorbate, and glutathione decrease in ischemic tissue during reperfusion, while lipid breakdown products accumulate, which are indicative of oxygen radical production[1,2]. If oxidative damage is a determinant of ischemia, then antioxidants and scavengers of oxygen free radicals should ameliorate ischemic brain injury. A growing body of evidence suggests that the degradation of xanthine dehydrogenase (XDH) to xanthine oxidase (XO) by Ca^{++}-activated proteases may be a significant source of superoxide in ischemic tissue[3]. Protection has been obtained in an intestinal ischemia model by both allopurinol, which inhibits xanthine oxidase, and superoxide dismutase (SOD), which scavenges superoxide at diffusion-limited rates[4,5].

Because mongolian gerbils (Meriones unguiculatus) lack a complete Circle of Willis, temporary bilateral occlusion of the common carotid arteries produces ischemia in the entire cerebrum and yields a high mortality rate[6]. After recovery from anesthesia, the gerbils exhibited cerebral damage as evidenced by circling behavior and paresis of the limbs[6]. The gerbils suffered increasingly severe clonic-tonic seizures, which ended in respiratory arrest between 4-10 hr[7]. Jarrot and Domer[6] have shown that survival at 24 hrs is indicative of long-term recovery and may be used as a criteria for successful pharmacological intervention. We have found that both polyethylene glycol (PEG) conjugated-SOD and allopurinol will protect gerbils from death due to ischemia-induced seizures, and have measured the appearance of xanthine oxidase activity in reperfused ischemic brain. Our results suggest that superoxide generated by XO may be a significant factor in cerebral ischemic injury.

MATERIALS AND METHODS

Bilateral occlusion of the common carotid arteries was performed as described by Jarrot and Domer[6] on 50-60 g gerbils obtained from Tumblebrook Farms, Westbrook, MA. Body temperature was maintained during the operation by a lamp, because we found that pentobarbital anesthesia dropped the body temperature to about 30°C, which significantly increased control survival rates. Gerbils surviving to 3 days or those observed to die were perfused with formalin. Four cresyl violet-stained sections were examined to determine the extent and nature of infarction.

Monomethoxy PEG (MW 5,000) was obtained from Aldrich, while all other reagents and enzymes were obtained from Sigma. PEG was conjugated by the cyanuric chloride method to bovine blood Cu,Zn SOD (Sigma or Diagnostic Data Inc.) and to catalase (Worthington)[8,9]. PEG-SOD was inactivated with H_2O_2[10]. Pharmacokinetic studies showed that IP injections of PEG-SOD reached maximal activity in 3-4 hr in blood and that the circulating half-life was 36 hr for PEG-SOD compared to 6 minutes for native SOD[11]. The activity of XO and XDH was measured by a fluorescent pterin assay[12]. Total XDH + XO activity was estimated by adding 25 μM methylene blue. Both XO and XDH rates were inhibited with 5 μM allopurinol.

RESULTS

PEG-SOD Treament. Occlusion times from 20 to 60 minutes produced mortality rates greater than 85% among controls, whereas PEG-SOD, injected intraperitoneally (IP) 3 hr prior to the occlusion, protected 39% of the animals after 40 minutes of occlusion (Fig. 1). The protection by PEG-SOD was not statistically significant after 20 or 60 minutes of occlusion. Protection was

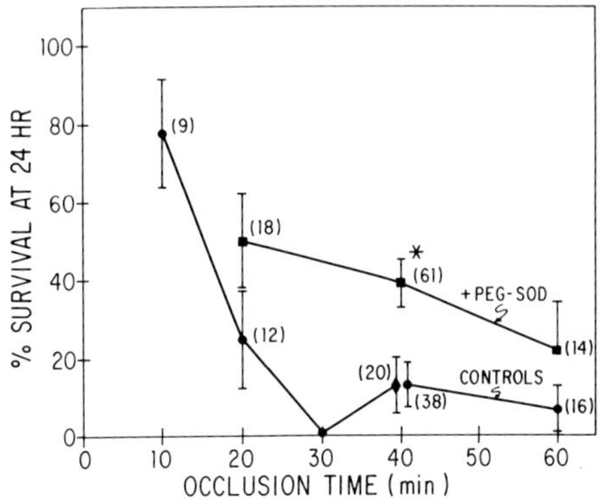

Fig. 1. The effect of time of bilateral occulsion on survival. (■) PEG-SOD (20 U/g BW) IP 3 hr prior to occlusion, (●) no injection, and (◆) H_2O_2-inactivated PEG-SOD. Bars indicate standard deviations and sample sizes are shown in parentheses. Stars indicate statistical significance at p<.05.

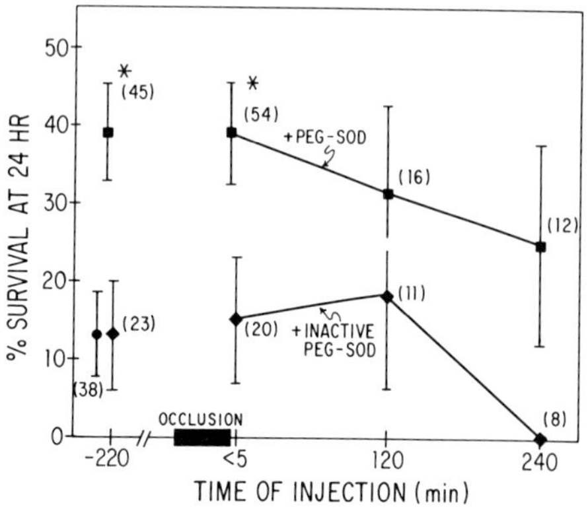

Fig. 2. The effect of time of injection (IV) following a 40 minute bilateral occlusion. Results of IP injection at -220 minute are shown for comparison. Symbols are as in Fig. 1. Gerbils injected at 2 or 4 hr received an additional 50 mg/kg pentobarbitol.

possible if PEG-SOD was injected immediately after the release of the occlusion into the femoral vein, and a positive though not statistically significant effect was observed even several hrs later (Fig. 2). Native SOD was less effective than PEG-SOD (Table I). While this could be simply due to the rapid clearance of SOD by kidneys, we have shown that, unlike native SOD, PEG-SOD may become internalized by cultured endothelial cells, and can protect these cells from oxygen radical stress[13]. PEG-catalase did not have a protective effect alone, and decreased the protection afforded by PEG-SOD, though the change was not significant by Fisher's exact test ($p<.4$). No protection was obtained from injecting either free PEG IV (Table I) or H_2O_2-inactivated PEG-SOD (Fig. 2). SOD inactivated by H_2O_2 differs from the native enzyme by only a single broken histidyl bond adjoining the catalytically active copper[10]. Thus, the protective effect of PEG-SOD is due to superoxide scavenging rather than an osmotic or other nonspecific effect of the PEG-SOD.

Allopurinol Treatment. Pretreatment of gerbils with oral allopurinol, a potent inhibitor of XO and XDH[14], yielded a 42±14% survival rate (Table I). Allopurinol inhibited more than 99% of liver XO and XDH activity and eliminated brain activity. IP injections of the alkaline allopurinol solutions were also protective, but caused abdominal cramps in the gerbils,

TABLE I.

Survival of gerbils at 24 hr following a 40 minute bilateral occlusion. IP injections were made 3 hr prior to the operation, while IV injections were made immediately post-occlusion. Allopurinol (50 µg/g BW) was injected 3x IP daily starting 2 days before the operation for the first group of animals and was given by gastric lavage in the second group.

TREATMENT	% SURVIVAL	SAMPLE SIZE
No Treatment	15± 4	32
PEG (0.1 ml 1% w/v; IV)	10± 9	10
PEG-SOD (20 U/g BW; IP)	39± 7	61
SOD (unmodified, 100 U/g BW)		
0 min prior IP	10± 9	10
60 min prior IP	0± 0	10
180 min prior IP	25±15	10
5 min post-occlusion	25±15	12
PEG-catalase (1000 U/g BW; IP)	6± 6	16
PEG-catalase + PEG-SOD (IP)	25±15	12
Allopurinol (IP)	37± 9	30
Allopurinol + PEG-SOD	17± 9	18
Allopurinol (Oral)	42±14	12
Allopurinol + PEG-SOD	50±16	10

which may have decreased the survival of gerbils also treated with PEG-SOD. No negative effect on survival was observed in PEG-SOD treated gerbils given oral allopurinol.

Xanthine Oxidase. Assay on brain homogenates that had been chromatographed to remove hypoxanthine[15] showed only XDH activity in control gerbils or in gerbils that had suffered a 40 min occlusion without reperfusion (Table II). After 2 hr of reperfusion, 36% of total activity was XO. No significant change occurred in the total XO + XDH activity, consistent with the conversion of XDH to XO.

TABLE II.

Brain XDH and XO activity (±S.D.) before and following a 40 minute bilateral occlusion. One unit is defined as 1 μmol/min of isoxanthopterin formed. * denotes p<.05 by the *a posteriori* Bonferroni test after ANOVA.

	μU/g brain			SAMPLE
TREATMENT	XO	XO + XDH	%XO	SIZE
control	0.00±0.00	1.23±0.11	0	3
0 hr reperfusion	0.11±0.22	1.46±0.56	8	5
2 hr reperfusion	0.56±0.22*	1.57±0.33	36	3

Neuropathology. When long-term survivors were sacrificed after 3-7 days, the majority possessed bland, often unilateral infarcts in watershed areas of the cerebrum accompanied by a laminar necrosis in the adjacent cerebral cortex and a characteristic necrosis of pyramidal cells in the hippocampus. These infarcts extended into the deep nuclei, including the neostriatum, thalamus, zona incerta, subthalamic nucleus, and the substantia nigra, and tended to have an anterior distribution. The infarcts showed no evidence of neutrophilic infiltration, although foamy macrophages and capillary proliferation were observed in infarcts over a week old. In 10 untreated animals sacrificed at 5 hr post-occlusion, the histological changes in the cerebral cortex consisted of pyknosis of neuronal nuclei and vacuolation of the neuropil in layer III and occasionally in layers V and VI. Similar changes were observed in localized areas of the deep gray structures of the cerebrum. These changes occurred bilaterally and were more extensive than the infarcts seen in long-term survivors. Although the extent of vacuolation in deep structures after 5 hr was highly variable from animal to animal, the same structures were affected in all animals with no qualitative differences between PEG-SOD treated and untreated animals.

DISCUSSION

Cerebral ischemia affects many processes, but is often considered in terms of its impact upon neurons and supporting glial structures. While the classical pattern of cerebral infarction occurs in the gerbil, the index measured in the bilateral occlusion model of stroke is death due to seizures. Seizure activity in gerbils subjected to a bilateral occlusion is related to a breakdown of the blood-brain barrier (BBB) which develops over several hours during reperfusion[16,17]. In contrast, cytotoxic edema, due to the intracellular swelling of glia and neurons, was more evident after 1 hr of reperfusion and was resolving at the time when the gerbils were dying.

Previous studies of intestinal ischemia have found the endothelium to be intimately involved in superoxide-dependent injury[3,4]. Our data are remarkably consistent with these studies, even though the total XO + XDH activity in gerbil brain is about 1000 times less than in liver or intestine, and 100 times less than in heart on a wet weight basis. If brain XO produces superoxide at the same rate as bovine milk XO, then only 10-20 pmol/min of superoxide would be produced per g brain. However, XO appears to be located in brain capillary endothelial cells, which could make this rate of superoxide formation particularly significant. The specific activity of XO is enriched about 5-fold in rat brain capillaries[18], supporting the immunocytochemical localization of XO in capillary endothelial cells[19]. Furthermore, the direct injection of xanthine oxidase into brain is known to produce edema and vascular permeability changes[20].

The conversion of XDH to XO during the reperfusion phase offers an explaination for why PEG-SOD is protective when injected immediately post-occlusion. The endothelium is the one tissue in brain which is in direct contact with PEG-SOD injected into the circulation. SOD protects endothelium from such diverse stresses as oxygen toxicity[21,22], acute inflammation[23,24], intestinal ischemia[4,5], and concussive brain injury[25]. Recently, SOD has been shown to protect beagle puppies from intraventricular hemorrhage, which is an ischemia-induced hemorrhaging from immature capillaries into brain ventricles[26]. Our data indicates that PEG-SOD may be beneficial even after an ischemic event has occurred, suggesting that PEG-SOD and allopurinol may have therapeutic value in treating such diverse cerebrovascular diseases as stroke, neonatal intraventricular hemorrhage, and decompression sickness[27], where the etiology involves acute injury to the brain microvasculature.

ACKNOWLEDGEMENTS

We thank Drs. I. Fridovich and K. V. Rajagapolan for advice, Gustavo Arcia for help with statistics, Diagnostic Data Inc, Mountainview, CA for providing SOD, the Army Reserve for a reserve tour (JSB) to perform initial experiments, and the support of NIH grant R01-NS-21782 (BAF).

REFERENCES

1. Watson BW, Busto R, Goldberg WJ, Santiso M, Yoshida S, Ginsberg MD (1984) J Neurochem 42:269
2. Flamm, ES, Demopoulos HB, Seligman ML, Poser RG, Ransohoff J (1977) Stroke 9:445
3. McCord JM (1985) New Eng J Med 312:159
4. Granger DN, Rutili G, McCord JM (1981) Gastroenterology 81:22
5. Parks DA, Bulkley GB, Granger DN, Hamilton ST, McCord JM (1982) Gasteroenterology 82:9
6. Jarrot DM, Domer FR (1980) Stroke 11:203
7. Loskota WJ, Lomax P (1982) Electroenceph Clin Neurophysiol 38:597
8. Pyatak PS, Abuchowski A, Davis FF (1980) Res Commun Chem Pathol Pharmacol 15:641
9. Abuchowski A, McRoy JR, Palczuk NC, Es TV, Davis FF (1977) J Biol Chem 252: 3582
10. Hodgson EK, Fridovich I (1975) Biochemistry 14:5299
11. Petkau A, Chelack WS, Kelly K, Barefoot C, Monasterski L (1976) Res Commun Chem Pathol Pharmacol 15:641
12. Markley HG, Faillace LA, Mezey E (1973) Biochim Biophys Acta 309:23
13. Minor RL, Beckman JS, White C, Rosen G, Freeman BA. (1985) In prep.
14. Hille R, Massey V (1981) Pharmacol Ther (1981) 14:249
16. Picozzi P, Todd NV, Crockard HA, (1985) J Cereb Blood Flow Metab 5:10
17. Avery S, Crockard HA, Russell RR (1984) J Neurol Neurosurg Psych 47:604
18. Jarasch ED, et al. (1981) Cell 25:67
19. Betz AL, J Neurochem (1985) 44:574
20. Chan PH, Schmidley JW, Fishman RA, Longar SM (1984) Neurology 34:315
21. Freeman BA, Young SL, Crapo JD (1983) J Biol Chem 258:12534
22. Turrens JF, Crapo JD, Freeman BA (1984) J Clin Invest 73:87
23. Del Maestro RF (1982) Can J Physiol Pharmacol 60:1406
24. Ley K, Arfors KF (1982) Microvas Res 24:25
25. Kontos HA, Wei EP, Povlishock JT, Dietrich WD, Magiera CJ, Ellis EF (1980) Science 209:1242
26. Ment LA, Steward WB, Duncan CC (1985) J Neurochem 62:563 (1985).
27. Bove AA (1982) Undersea Med Soc 9:91

Superoxide and Superoxide Dismutase in Chemistry, Biology and Medicine; G. Rotilio, editor

INVOLVEMENT OF NEUTROPHILS IN THE CORTICAL BLOOD FLOW IMPAIRMENT AFTER CEREBRAL ISCHEMIA IN THE RAT; EFFECTS OF ANTINEUTROPHIL SERUM AND SUPEROXIDE DISMUTASE

BJARNE GRÖGAARD, LUDWIG SCHÜRER, BENGT GERDIN & KARL-E. ARFORS

Department of Experimental Medicine, Pharmacia AB, Uppsala and Department of Surgery, University Hospital, Uppsala (Sweden)

INTRODUCTION

In several organs, for example the brain, kidney, heart and skeletal muscle, a period of ischemia is followed by impairment of blood flow, a finding which has been believed to represent a common pathophysiologic postischemic response. In the brain, two different types of flow depression have been described, the no-reflow phenomenon and delayed hypoperfusion (DHP; for survey see 1). In the former type there is no recovery of flow and a permanent occlusion of the microvascular bed has been shown (2). In DHP, an initial recovery is followed by a subsequent decrease in flow. The mechanism of DHP is not clear. In the present investigation we have explored the possibility that this phenomenon is mediated by an intravascular aggregation of polymorphonuclear leukocytes (PMNLs) and/or by microvascular damage caused by generation of free radicals. For this purpose we have used a rat model for reversible incomplete forebrain ischemia, which reproducibly results in a DHP with local cerebral blood flow (lCBF) values that are about 40 % of the normal.

MATERIALS AND METHODS

A modification of a method described by Smith _et al_. (3) was employed. Fasted male rats of the Wistar strain, 300-400 g, were anesthetized, intubated and maintained on N_2O/O_2 anesthesia under muscle relaxation and controlled ventilation. Catheters were inserted into the tail artery and vein for blood pressure recording, blood sampling and infusions, into the right atrium for exsanguination and into the right brachial artery for blood sampling during ^{14}C-iodoantipyrine infusion. Both carotid arteries were dissected free and prepared for clamping. The blood glucose concentration, hematocrit and acid-base status were followed during the experiment. EEG and body temperature were monitored continuously.

Preparation of antineutrophil serum (ANS)

This was done as previously described by Lundberg _et al_. (4). The immunization, however, took place in sheep. PMNLs (50×10^6, 0.5 ml) were mixed with an equal volume of Freund's complete adjuvant and injected intramuscularly every second week. Blood was collected and the antiserum was then processed as described previously (4). Normal sheep serum (NSS) for control experiments was collected prior to immunization and treated in the same way as the ANS.

Experimental protocol

Rats were injected intraperitoneally with 1 ml of ANS or NSS at -24 and -12 h. Alternatively, 0.5 ml of ANS or NSS was infused i.v. at a rate of 0.1 ml/min, commencing 2 min after recirculation. One group of animals was given 4 mg/kg body weight (b.w.) (about 15 000 U) of yeast superoxide dismutase (SOD) (Pharmacia AB) or bovine SOD (Grünenthal) i.v. immediately before recirculation and another group of animals was given 50 mg/kg b.w. (about 185 000 U) of bovine SOD one minute prior to induction of ischemia. In the ANS/NSS experiments orbital blood samples were taken for PMNL counts before treatment and before and after experimentation.

Cerebral ischemia was produced by bilateral clamping of the carotid artery and simultaneous induction of hemorrhagic hypotension to 50 mm Hg. After 15 min of ischemia the clamps were released and the shed blood was retransfused. Sixty minutes thereafter lCBF was measured with the ^{14}C-iodoantipyrine technique (5,6). Forty µCi of ^{14}C-iodoantipyrine in 1 ml of saline was infused i.v. for 45 s and blood samples were drawn every 5 s during the isotope infusion. At the end of the infusion period the animals were decapitated and the brain was immediately frozen in isopentane chilled to -50 to -60°C. Cryostat sectioning and quantitative autoradiography were performed as described elsewhere (6).

RESULTS

During ischemia there was virtual cessation of flow to the rostral cortical part of the brain (Fig. 1). The blood flow to anterior cortical structures was well below 1 % and the flow to the hippocampus was about 3 % of the preischemic flow. Five minutes after restitution of flow there was hyperperfusion, with lCBF more than 200 % of the normal, and 60 min thereafter this flow was 30-50 % of the normal; thus the ischemic insult in this model was followed by DHP and not by a no--reflow phenomenon as described by Ames (2).

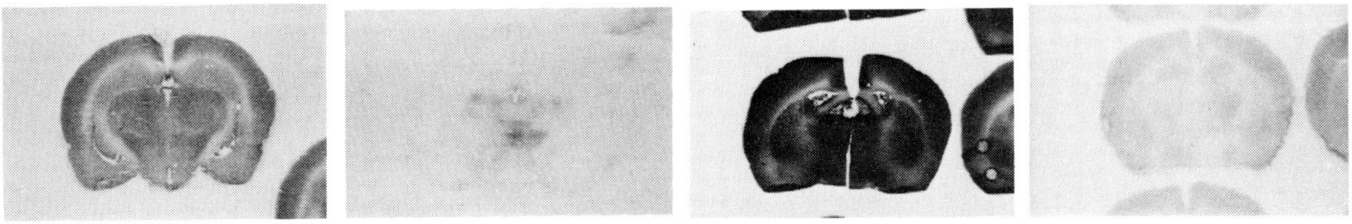

Fig. 1. Autoradiographs showing, from left to right, normal cerebral blood flow (CBF), CBF during ischemia, during postischemic hyperperfusion at 5 min of recirculation, and during delayed postischemic hypoperfusion.

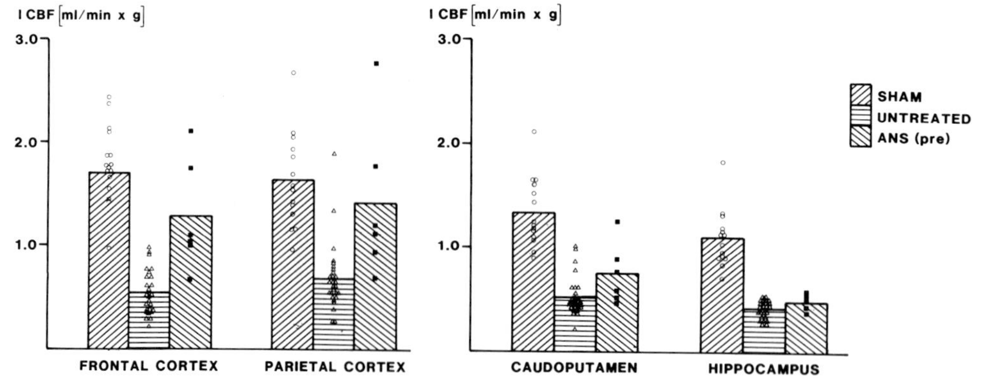

Fig. 2. Local cerebral blood flows in various structures after 15 min of ischemia and at 60 min of recirculation in animals pretreated with antineutrophil serum and in controls.

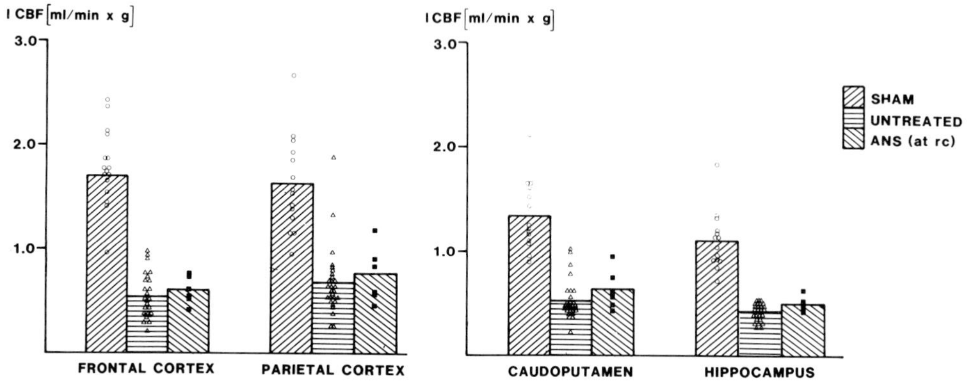

Fig. 3. Local cerebral blood flows in various structures after 15 min of ischemia and at 60 min of recirculation in animals given antineutrophil serum 2 min after recirculation and in controls.

Pretreatment with ANS resulted in a depletion of circulating PMNLs to less than 5 % of the normal value before the induction of ischemia. The ANS-pretreated animals, which during and after ischemia had very low peripheral PMNL counts, exhibited a distinct improvement in blood flow 60 min after recirculation compared with those not treated with ANS. This tendency to normalization of lCBF was most pronounced in the frontal and sensorimotor cortex. It was slightly less marked in the parietal cortex and was not seen in the visual cortex, caudoputamen or hippocampus (Fig. 2). Animals which had a normal PMNL count during ischemia and were given ANS and thus were depleted of their circulating PMNLs after recirculation, had a peripheral PMNL count of less than 5 % of that of the non-ANS-treated animals between 15 and 60 min after recirculation. In contrast to the animals given ANS prior to ischemia, however, they did not exhibit any sign of improvement in their lCBF, either in cortical or in subcortical structures (Fig. 3). Furthermore, animals given 4 mg/kg b.w. of bovine or human SOD at recirculation or 50 mg/kg b.w. of bovine SOD immediately before ischemia did not show any improvement in cerebral circulation at the 60-min measurement time (Fig. 4).

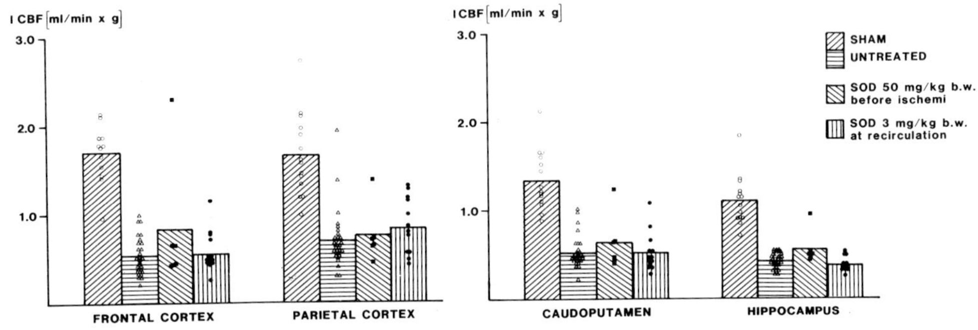

Fig. 4. Local cerebral blood flows in various structures after 15 min of ischemia and at 60 min of recirculation in SOD-treated animals and in controls.

DISCUSSION

This study has shown that the DHP seen after 15 min of cerebral ischemia in the present model is counteracted in animals depleted of their PMNLs before induction of the ischemia.

As it has recently been claimed that postischemic perfusion disturbances are due to organ trapping of PMNLs in the recirculation phase, it seemed relevant from an explanatory standpoint to determine whether the damage could be counteracted by inducing depletion of the circulating PMNLs immediately after recirculation. In the present study this was done by giving ANS i.v. starting 2 min after recirculation. Fifteen minutes later the peripheral PMNL count was very low and the flux of PMNLs to the brain was obviously greatly hampered. Nevertheless, there was no sign of improvement in the cerebral blood flow in these animals. The most reasonable interpretation for this finding is that the damage caused by PMNLs had already occurred during ischemia or immediately on reperfusion. The requirement of PMNLs at or very close to the time of ischemia in the present model, however, gives reason to suspect that the pathophysiologic mechanisms involved in DHP might be similar to those seen in the no-reflow phenomenon. To judge from our blood flow data, however, the phenomenon of "no reflow", in sensu strictu, is not seen. In "no reflow", a microvascular obstruction has been observed and among suggested causes capillary occlusion due to sticking of PMNLs to the endothelium has been mentioned. This has been demonstrated in the heart (7). Whether such microvascular occlusion occurs during ischemia or extremely early during recirculation, has not been reported. If the pathophysiologic mechanism is similar in DHP, the following sequence of events would seem reasonable: during ischemia or within the first minutes of recirculation, an intracapillary occlusion occurs due to an altered PMNL-endothelial interaction. Secondary to this there is an upstream stasis of red cells and a progressive increase in the microvascular hematocrit in the capillary in question. If this occlusion cannot be overcome on recirculation by an adequate perfusion pressure, a no-reflow phenomenon in that capillary will develop. If, however, the occlusion can be relieved, a period of normal or high flow will supervene. With time, however, the endothelial damage caused by the PMNLs will result in an endothelial cell swelling, which will gradually occlude the capillaries and cause DHP. Such endothelial swelling has been observed in DHP (7).

The inability of high doses of SOD to normalize the cerebral blood flow is contrary to findings in other organs. In other microvascular beds it has been shown that the noxious effect of PMNLs is largery attributable to their activation and generation of free radicals and the damage caused has been abolished by SOD. This was not the case in our experimental system. In the present study a very high dose of bovine SOD, 50 mg/kg b.w., was given immediately before the ischemia to ensure a high blood level even during the ischemic period. As the cerebral ischemia was accompanied by arterial hypotension to 50 mm Hg, SOD would not be excreted by the kidneys and we therefore anticipated that a sufficient concentration of SOD would be present in the circulation both during the ischemic period and at least during the first 10-15 min of recirculation. As the noxious damage is obviously caused within that time period, to judge from the ANS results, the inability of SOD to improve lCBF could not reasonably be explained by renal elimination of the drug.

The positive effects of SOD in certain organs has been taken as evidence of free-radical production in the extracellular environment during the recirculation phase. The origin of the radicals produced has not been convincingly demonstrated, however. The PMNL is a possible source. Apart from that it has been suggested that in the intestine and kidney free radicals are formed as a result of simultaneous conversion of the enzyme xanthine dehydrogenase to xanthine oxidase and accumulation of a suitable substrate, hypoxanthine; the latter being formed by pronounced degradation of energy-rich purines. The distribution of the radical-producing enzyme xanthine oxidase in the body is not completely known. The existence of xanthine oxidase in the brain is disputed, in contrast to that in the kidneys and intestines.

The importance of DHP for the ultimate neuronal damage seen after cerebral ischemia is not known. Certain structural areas in the brain are much more vulnerable to ischemic damge than can be explained by the degree of DHP alone. Studies have shown that transection of inhibitory tracts to the hippocampus results in an increased cell death but not in an altered degree of DHP after ischemia (8). This indicates that the degree of stimulation and the metabolic activity of the neurons are of importance and in a flow situation with a high need and low supply it is reasonable to assume that the degree of perfusion is a limiting factor. In the present context, however, the beneficial effect of pretreatment with ANS on DHP has to be coupled to an increased neuronal survival in order to be of functional interest. Studies to elucidate this question are in progress.

REFERENCES

1. Siesjö BK (1984) J Neurosurg 60:883-908
2. Ames III A (1968) Am J Pathol 52:437-453
3. Smith ML, Bendek G, Dahlgren N, Rosen I, Wieloch T, Siesjö BK (1984) Acta Neurol Scand 69:385-401
4. Lundberg C, Lebel L, Gerdin B (1984) Lab Invest 50:726-732
5. Sakurada O, Kennedy C, Jehle J, Brown JD, Carbin GL, Sokoloff L (1978) Am J Physiol 234:H59-H66
6. Abdul-Rahman A, Dahlgren N, Ingvar M, Rehncrona S, Siesjö BK (1978) Acta Phys Scand 106:53-60
7. Engler RL, Schmid-Schönbein GW, Pavelec RS (1983) Am J Pathol 111:98-111
8. Paljärvi L, Rehncrona S, Söderfeldt B, Olsson Y, Kalimo H (1983) Acta Neuropathol (Berl) 60:232-240
9. Wieloch T, Lindvall O, Blomqvist P, Gage FH (1985) Neurol Res. In press.

© 1986 Elsevier Science Publishers B.V. (Biomedical Division)
Superoxide and Superoxide Dismutase in Chemistry, Biology and Medicine; G. Rotilio, editor

ROLE OF OXYGEN RADICALS IN GASTROINTESTINAL ISCHEMIA

DALE A. PARKS and D. NEIL GRANGER

Department of Anesthesiology, University of Alabama at Birmingham, Birmingham, AL 35294 and Department of Physiology, University of South Alabama, Mobile, AL 36688

Ischemia-induced tissue injury is a well established factor in the pathogenesis of serious, often life threatening, diseases. There is a rapidly growing body of evidence that implicates oxygen free radicals, cytotoxic reduction products of molecular oxygen, as the mediators of the microvascular and parenchymal damage that occurs upon reperfusion of ischemic tissues. We have used both microvascular permeability and morphologic evaluations of the mucosa as indexes of tissue injury to examine the role that oxygen free radicals play in the pathogenesis of ischemia-reperfusion injury in the small intestine. This discussion will be limited to only the vascular permeability studies.

A major consequence in the pathogenesis of ischemic bowel disorders is enhanced transcapillary filtration, interstitial edema, and ultimately, net fluid movement into the lumen of the bowel. A comparison of vascular permeability estimated in control intestinal preparations with values obtained in preparations subjected to one hour of ischemia and reperfusion clearly indicated that ischemia-reperfusion significantly increased vascular permeability (1). Several endogenous substances have been implicated in the pathogenesis of intestinal ischemia, including histamine, prostaglandins, lysosomal enzymes and bacterial endotoxins. However, none of these substances appear to be responsible for the increased vascular permeability produced by ischemia in the small intestine (2). We next considered the possibility that oxygen radicals were responsible for the ischemia-induced increase in intestinal vascular permeability. To test this possibility, we pretreated cats with 15,000 U/kg superoxide dismutase (SOD) approximately 15 minutes prior to reperfusion of the ischemic small bowel (2). Since the circulating half life of SOD is quite short, about 15 minutes in the cat, a second group of animals were studied in which both renal pedicles were ligated. SOD pretreatment significantly attenuated the ischemia-induced increase in vascular permeability (Fig. 1). The protective effect of SOD was greatly improved by renal ligation.

Based on the SOD studies and circumstantial evidence already in the literature, a mechanism was proposed to account for oxygen radical production in post-ischemia small intestine (Fig. 2). During the ischemic period, ATP is catabolized to hypoxanthine, which accumulates in the tissues. As a result of the low energy charge, there is an influx of Ca^{++} into the cells. The intracellular Ca^{++} then triggers the conversion of normally occurring xanthine dehydrogenase (D) to the oxygen radical-producing xanthine oxidase (O) via a protease. When the intestine is reperfused, molecular oxygen is reintroduced into the tissues and reacts with hypoxanthine and xanthine oxidase to produce a burst of superoxide anion (O_2-), hydrogen peroxide (H_2O_2) and hydroxyl radicals (OH·). These highly reactive and cytotoxic free radicals can then result in peroxidation of the lipid components of cellular, mitochondrial, and lysosomal membranes (3). It has also been suggested that oxygen radicals may degrade hyaluronic acid and collagen, alter phospholipase activity, modulate nucleotide cyclase activities, and damage nucleic acids and DNA (3,4).

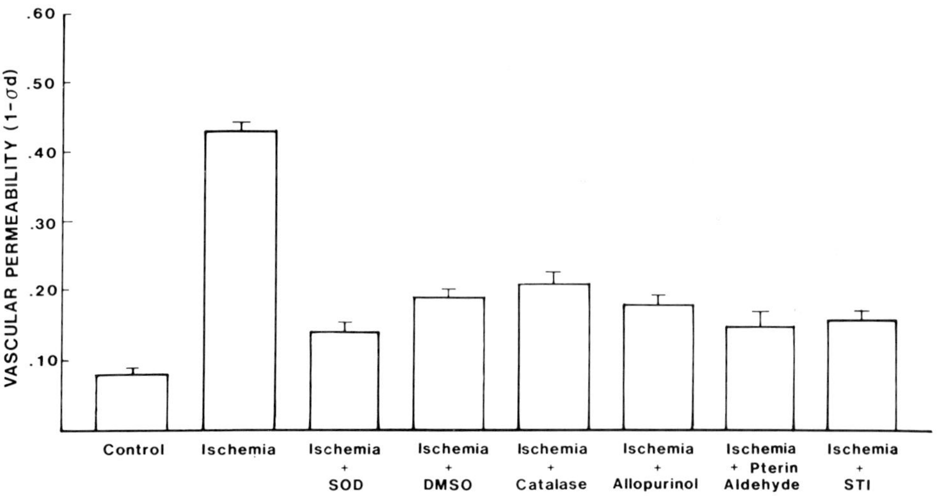

Figure 1. Effect of superoxide dismutase (SOD), dimethylsulfoxide (DMSO), catalase, allopurinol, pterin aldehyde and soybean trypsin inhibitor (STI) on the increase in vascular permeability induced by one hour of ischemia and reperfusion in the small intestine.

There are several important features of intestinal xanthine oxidase which lend support to its role as the major source of oxygen radicals with ischemia-reperfusion. First, xanthine oxidase is a well documented source of oxygen radicals (5). Furthermore, the activity of xanthine oxidase in the intestine is very high (6). Another important feature of intestinal xanthine oxidase is the rapidity with which the dehydrogenase form is converted to the oxidase form with ischemia; only about one minute of ischemia is required for complete conversion (7). The effects of xanthine oxidase inhibitors, allopurinol and pterin aldehyde, on the increased vascular permeability induced by ischemia-reperfusion have also been examined (8,9). Both xanthine oxidase inhibitors were very effective in attenuating the ischemia-induced increase in vascular permeability (Fig. 1), which supports a role for xanthine oxidase in the pathogenesis of intestinal ischemia.

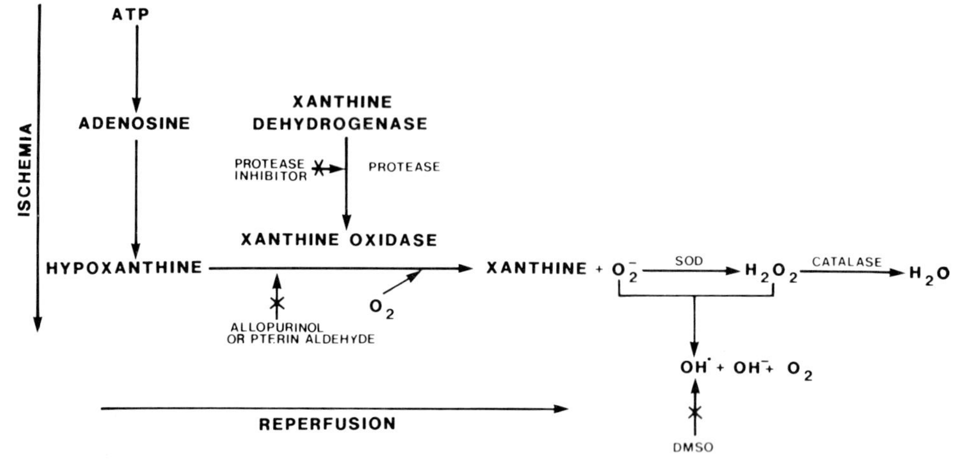

Figure 2. Mechanism of oxygen radical production during reperfusion of ischemic small bowel.

It has recently been demonstrated that a tungsten diet decreased xanthine oxidase activity to approximately 10% of control in rat small intestine (10). In preliminary studies, we have been able to decrease xanthine oxidase activity by 75% in cat small intestine with a tungsten diet. When the intestine of tungsten-treated cats was subjected to ischemia, it was found that the increase in vascular permeability associated with ischemia was greatly attenuated. These results provide further support for a role of xanthine oxidase in the pathogenesis of ischemic injury to the small intestine. Since pretreatment with soybean trypsin inhibitor (STI) had been shown to prevent the conversion of xanthine dehydrogenase to xanthine oxidase (7) in the ischemic rat intestine, we examined the effect of STI on the ischemia-induced increase in vascular permeability (11). Pretreatment with STI significantly attenuated the increased vascular permeability associated with one hour of ischemia and reperfusion. These results (Fig. 1) are consistent with the finding that a protease inhibitor completely prevents D-to-O conversion in the small intestine and thereby prevents oxy radical formation via xanthine oxidase.

Although the SOD studies provided strong support for the involvement of oxygen radicals in ischemia-induced injury, it remained unclear whether superoxide anion per se or an oxygen radical derived from O_2- (e.g., hydrogen peroxide or hydroxyl radical) mediated the ischemia-induced increase in vascular permeability. Therefore, we performed additional studies to determine if dimethylsulfoxide (DMSO), a hydroxyl radical scavenger, or catalase protected against the increased vascular permeability associated with ischemia (8,12). The results from these studies (Fig. 1) indicate that both DMSO and catalase provide significant protection against ischemic injury. These results are consistent with the highly cytotoxic hydroxyl radical being the primary mediator of the microvascular permeability change.

This review presented evidence to indicate that oxygen-derived free radicals play a major role in the microvascular injury associated with reperfusion of ischemic small intestine. There is considerable evidence presented that xanthine oxidase is the major source of the oxygen radicals produced with ischemia-reperfusion. Taken as a whole, these results suggest that oxygen free radicals are generated during reperfusion of ischemic small intestine and that the enzymatic source is xanthine oxidase.

REFERENCES

1. Granger DN, Sennett M, McElearney P, Taylor AE (1980) Gastroenterology 79:474-480
2. Granger DN, Rutili G, McCord JM (1981) Gastroenterology 81:22-29
3. Fridovich I (1978) Science 201:875-880
4. Slater TF (1984) Biochem J 222:1-15
5. McCord JM, Fridovich I (1968) J Biol Chem 243:5753-5760
6. Batelli MG, Della Corte E, Stirpe E (1972) Biochem J 126:747-749
7. Roy RS, McCord JM (1982) Can J Physiol Pharm 60:1346-1352
8. Parks DA, Granger DN (1983) Am J Physiol 245:G285-289
9. Granger DN, McCord JM, Parks DA, Hollworth ME (1986) Gastroenterology 91:000-000 (in press)
10. Topham RW, Walker MC, Calisch MP (1982) Biochem Biophys Res Comm 109:1240-1246
11. Parks DA, Granger DN, Bulkley GB, Shah AK (1985) Gastroenterology 89:6-12
12. Parks DA, Granger DN (1984) Gastroenterology 84:1207

© 1986 Elsevier Science Publishers B.V. (Biomedical Division)
Superoxide and Superoxide Dismutase in Chemistry, Biology and Medicine; G. Rotilio, editor

ONE-ELECTRON REACTIONS IN CONNECTION TO THE ISCHEMIA/REPERFUSION SYNDROME - - DESCRIPTION OF A MODEL

ULF NILSSON[1,3], HJÖRDIS THOR[2], PETER MOLDEUS[2] and ANN-CHRISTIN BYLUND-FELLENIUS[3]

1) Dept. of Biochemistry/Biophysics, Chalmers University of Technology and the University of Gothenburg, S-412 96 Gothenburg

2) Dept of Toxicology, Karolinska Institutet, S-104 01 Stockholm

3) Dept. of Research and Development, Pharmacia AB, S-751 82 Uppsala

INTRODUCTION

In this study we present the first steps in the development of a model for studying one-electron processes on the cellular level in connection to the ischemia/reperfusion syndrome. As indication of these reactions is taken the oxidation of the spin probe OXANOH , reported by Rosen et al. a few years ago (1), to its corresponding stable nitroxide radical OXANO. The cell membrane is shown to be highly permeable to both forms of the probe, making OXANOH well suited to monitor intracellular one-electron processes.

MATERIALS AND METHODS

OXANO was synthesized by the method of Keana as modified by Rauckman et al. (2) and was a precious gift from Dr Lars-Inge Olsson at the synthetic division of Pharmacia AB. OXANOH was prepared by bubbling 10 mM OXANO with hydrogen in the presence of a noble-metal catalyst (3). Hepatocytes (4) and cells from the intestinal mucosa of the rat (5) were prepared as previously described. Glutathione and glutathione disulfide were analyzed according to Reed et al. (6).

Incubation and EPR-analyses of intestinal cells

$1\text{-}4\times10^7$ cells were spun briefly to form a loose pellet, the supernatant was sucked off and discarded, and the cells were flushed with either nitrogen or oxygen, and incubated in sealed tubes for 25 min at 37^0 C. The tubes were then opened and the pellets resuspended in 5 mM OXANOH pH 7.4 to a final volume of 1 ml. The samples were equilibrated with air by gentle bubbling through a Pasteur pipette, and analyzed approximately 90 s after reopening the tubes by measuring the height of the leftmost peak in the EPR spectrum of OXANO. The increase in peak height was then monitored for two minutes. The measured concentrations and rates were normalized to 4×10^7 cells and 90 s after reopening the tubes and were corrected for autoxidation of OXANOH. The EPR instrument (Varian E-9) was equipped

---

Abbreviations used:
OXANOH: 2-ethyl-1-hydroxy-2,5,5-trimethyl-3-oxazolidine
OXANO: 2-ethyl-2,5,5-trimethyl-3-oxazolidinoxyl

with a TM_{110} cylindrical cavity and a capillary cell. Instrumental settings were: microwave power, 10 mW; microwave frequency, 9433 MHz; modulation amplitude, 0.2G; scan rate, 12.5 G/min; time constant, 0.3 s.

RESULTS

Characterization of OXANO/H

The stock solution of OXANOH is stable for a few hours, provided that the appropriate chelators are used (1) and that it is kept on ice. In isolated hepatocytes OXANO/H caused a slow depletion (50% within 45 min) of intracellular glutathione without any concomitant increase of the disulfide. The depletion could be completely preventedby selectively blocking cytochrome P450 with metyrapone. A 2 h incubation of hepatocytes with OXANOH (0-5 mM) decreased viability in a dose-dependent fashion from 85% to 65%. OXANO/H penetrate the hepatocyte membrane and associate with cytochrome P450 as judged by characteristic changes in the 390-420 nm region of the absorbance spectrum of the cells (7). The effect was dose dependent and completely inhibitable by metyrapone. In a previous study by Ross et al. (8) it was concluded that OXANOH did not react with sulphur-centered radicals and may be selective towards certain types of radicals. Further characterization have been made in a series of studies by Rosen et al. (1,9-11).

Intestinal cells

Cells incubated under nitrogen produced approximately 100% more OXANO during the first 90 s after reoxygenation than cells incubated in the same manner under oxygen. This increase was approximately 50% inhibited by addition of SOD (200ug) at the time of reoxygenation (TABLE I). These results are statistically significant at or above the 95% confidence level as determined by Studentš t-test. The rate of formation of OXANO during the period 90-150 s after reoxygenation varied in the same manner (data not shown).

CONCLUSION

Results are presented, indicating increased formation of species, capable of one-electron transfer during reperfusion following ischemia. The fact that SOD inhibits part of this increase strongly suggests that superoxide is at least one of these species.

Admittedly the model bears, as best, a crude resemblance to the in vivo phenomenon of ischemia and is perhaps more correctly described in terms of hypoxia and hyperoxia. We believe, however, that with further development it has potential for providing data on the time course and location of events leading to increased one-electron activity in connection to ischemia/reperfusion as well as some information concerning the quantitative aspects.

TABLE I

RESULTS FROM INCUBATIONS OF INTESTINAL CELLS

Data are given as mean values ± S.E.M.

Experimental condition	Number of samples	[OXANO] (μM)
Normoxia	10	4.85 ± 0.70
Ischemia	10	9.56 ± 0.94
Normoxia(SOD)	9	5.56 ± 0.49
Ischemia(SOD)	9	7.88 ± 0.82
Ischemia-Normoxia	10	5.02 ± 0.99
Ischemia(SOD)-Normoxia(SOD)	9	2.32 ± 0.47
Ischemia/Normoxia	10	2.31 ± 0.31
Ischemia(SOD)/Normoxia(SOD)	9	1.41 ± 0.07

REFERENCES

1. Rosen GM, Finkelstein E and Rauckman EJ (1982) Arch Biochem Biophys 215: 367-378
2. Rauckman EJ, Rosen GM and Abou-Donia MB (1975) Syn Commun 5:409-413
3. Rozantsev EG (1979) Free Nitroxyl Radicals p.93-94, Plenum, New York
4. Moldeus P, Högberg J and Orrenius S (1978) Meth Enzymol 51:60-71
5. Dawson JR and Bridges JW (1979) Biochem Pharmacol 28:3299-3305
6. Reed PJ, Babson JR, Beatty PW, Brodie AE, Ellis WW and Pollis DW (1980) Anal Biochem 106:55-62
7. Schenkmann JB, Remmer H and Estabrook RW (1967) Mol Pharmacol 3:113
8. Ross D, Albano E, Nilsson U and Moldeus P (1984) Biochem Biophys Res Commun 125:109-115
9. Rosen GM, Rauckman EJ and Hanck KW /1977) Toxicol Lett 1:71-74
10. Rauckman EJ, Rosen GM and Kitchell BB (1979) Mol Pharmacol 15:131-137
11. Finkelstein E, Rosen GM and Rauckman EJ (1984) Biochem Biophys Acta 802: 90-98

Superoxide and Superoxide Dismutase in Chemistry,
Biology and Medicine; G. Rotilio, editor

THE ROLE OF OXYGEN FREE RADICALS FOR POSTISCHEMIC KIDNEY DAMAGE

TORE SCHERSTÉN, ROLAND HANSSON AND SILAS PETTERSSON
Department of Surgery I, University of Göteborg, S-413 45 Göteborg, Sweden

The ultrastructural lesions of the ischemic kidney cells have been thoroughly described (1,2,3). Recent studies have indicated that the ischemic lesions become worse at recirculation of the kidney (1,2,3,4). The most commonly given explanation for this phenomenon is cell swelling in certain parts of the kidney leading to impaired re-establishment of blood flow and a prolonged warm ischemia time - the "no-reflow" phenomenon (5,6). Another possible explanation is the formation and release of toxic amounts of free radicals at recirculation, a theory that has been suggested as an explanation for the aggravation of the ischemic lesions in brain, liver and intestine (7,8)

In recent studies from our laboratory, evidence has been presented supporting the view that oxygen free radicals contribute to the functional and morphological lesions observed in the rabbit kidney at recirculation after ischemia (9,10,11).

Oxygen free radicals are formed by reduction of molecular oxygen in the univalent pathway in the mitochondria. In this pathway the one electron reduction results in the formation of the superoxide anion radical, the two electron reduction product is the hydrogen peroxide and the three electron reduction product is the highly reactive hydroxyl radical. These intermediates in the univalent pathway are extremely toxic and therefore protective and controlling mechanisms are operating in the cells. These mechanisms consist of superoxide dismutase which catalyses the dismutation of the superoxide to hydrogen peroxide and by catalase and peroxidases that can reduce hydrogen peroxide to water and thereby avoiding the formation of the toxic hydroxyl radical.

Another pathway for univalent reduction of molecular oxygen to superoxide is the oxidation of hypoxanthine to xanthine by the enzyme xanthine oxidase.

The reasons for the formation of oxygen free radicals in amounts exceeding the capacity of the cellular defence systems during recirculation are not known. It might be caused by impaired mitochondrial function or by a decreased concentration of scavenger enzymes during ischemic conditions as shown for instance for heart muscles. As mentioned above another mechanism might be an accumulation of hypoxanthine because of an increased degradation of nucleotides during the ischemic period.

In our research group in Gothenburg we have tried to achieve evidence that oxygen free radicals are produced at a rate exceeding the capacity of the cellular defence system during the reperfusion phase after renal ischemia. In most of our studies we have used rabbits exposed to renal ischemia for one hour. The effects of xanthine oxidase inhibition, or pretreatment with scavenger enzymes or mannitol on hypoxanthine metabolism, postischemic renal blood flow, erythrocyte distribution, renal morphology and function have been analysed in a series of experiments.

MATERIAL AND METHODS

In all studies rabbits weighing 2-5 kg were used. They were deprived of food for 12 hrs before the operation but had free access to water. The central ear artery was catheterized for continous monitoring of the arterial blood pressure. The kidneys were exposed under sterile conditions via a long midline incision. Ischemia was achieved by clamping the renal artery and ureter with artery foreceps.

Renal blood flow was measured using an electromagnetic flowmeter after stabilisation of blood flow and pressure.

Determination of hypoxanthine and xanthine in tissue and blood was performed by isotachophoresis using an LKB Tachophor with UV detector as desribed by Gustavsson et al (12).

Erythrocyte distribution was determined after injection of labelled erythrocytes obtained from the rabbit. The erythrocytes were labelled with ^{51}Cr-sodium chromate. Specimens from different zones of the kidney (cortex, outer stripe, inner stripe and the inner zone) were weighed and their radioactivity, as well as the radioactivity of blood samples was determined in a gamma counter.

The renal function of the rabbits was estimated using serum creatinine values, the urine osmolality, the urine excretion of potassium and sodium and the ^{51}Cr-EDTA-clearance.

Light and electron microscopy was performed as described in reference (11).

RESULTS AND COMMENTS

Different series of experiments were performed with pretreatment of the rabbit with allopurinol, superoxide dismutase, catalase or mannnitol as described by Hansson (11).

The concentration of hypoxanthine and xanthine in arterial blood and in renal venous blood increased during the ischemic period and reached maximal value 3 min after the start of recirculation in the rabbits not pretreated with allopurinol. In rabbits pretreated with allopurinol (50 mg/kg body weight) the hypoxanthine concentration was high 3 min after recirculation and decreased then to normal level within 30 min. The xanthine concentration in the pretreated animals was zero both before and after the ischemic period indicating a total block of hypoxanthine oxidation. Consequently the production of oxygen free radicals in this pathway was diminished.

In order to examine the effect of inhibition of the xanthine oxidase activity by allopurinol and of scavenger enzymes (SOD and catalase) on renal blood flow during recirculation a series of experiments was performed. In kidneys from non-pretreated animals a transient increase in total renal blood flow was observed within the first minute after recirculation. Thereafter the blood flow return to preischemic values. After pretreatment with allopurinol the same pattern was seen but generally the total flow was higher in these animals. In rabbits pretreated with superoxide dismutase (SOD) the blood flow increased significantly during the first minutes of recirculation and remained then at this high level during 30 min. The same was found for catalase pretreatment. In none of these kidneys the water content changed after ischemia or after pretreatment with these scavenger substances.

In these experiments it was found that the vascular resistance decreased initially during recirculation in both pretreated and non-pretreated kidneys. In the pretreated kidneys the vascular resistance remained low during the 30 min recording in contrast to the non-pretreated kidneys. The experiments thus showed that pretreatment with allopurinol and scavenger enzymes caused an increase of the renal blood flow at recirculation after ischemia as compared with non-pretreated animals. This increase of blood flow might be ascribed to lower vascular resistance in the kidneys of these pretreated animals.

In order to analyse the mechanism for the altered circulation during reperfusion after ischemia the erythrocyte distribution was studied by the aid of ^{51}Cr-labelled erythrocytes given before the ischemi period. In the postischemic recirculated kidney (30 min of recirculation) there was a statistically significant trapping of the erythrocytes both in the outer and the inner stripe of the non-pretreated kidneys, whereas the erythrocyte content in the other parts of the kidney was unchanged. Pretreatment with allopurinol, SOD or catalase prevented this trapping completely.

The trapping of erythrocytes in the outer stripe of the medulla may have several explanations. One is endothelial cell swelling and damage secondary to ischemia as such. Another is release of oxygen free radicals during recirculation. Our results support the hypothesis of endothelium cell or blood cell damage due to formation of oxygen free radicals. The observation that both SOD and catalase have the potential to reduce a trapping of erythrocytes and the damage of the kidney during recirculation suggests that both superoxide and hydroxyle radicals are formed during recirculation.

In five groups of rabbits the kidney morphology was studied after 60 min of warm ischemia and after up to 30 min of recirculation. A marked aggrevation of the ischemic lesions at recirculation was noticed. Accumulation and extravasation of erythrocytes at the cortico-medullary junction endothelial cell swelling and swelling of the cells in the collecting ducts were observed in the non-pretreated kidneys. These changes were to a large extent prevented by pretreatment with allopurinol or scavenger enzymes. The morphological examination showed the most pronounced effect on vessels and collected ducts and also to some extent in the proximal tubuli.

Allopurinol seemed to give the best protection against the morphological lesions.

The renal function after 60 min ischemia with and without pretreatment of the rabbits with allopurinol, scavenger enzymes or mannitol was studied in an experimental series containing five groups of rabbits with 8 animals in each group. At the time of induced ischemia contralateral nephrectomy was performed. Two rabbits were subjected to a unilateral nephrectomy without an ischemic damage superimposed on the remaining kidney. This rabbits showed maximal serum creatinine values of 140 and 170 µmol/l during the observation period. In the non-pretreated control group two rabbits died, 6 and 12 days after the operation. The serum creatinine concentration had increased steadily during the observation period up to 800 µmol/l when they died. The surviving 6 rabbits had a mean maximum serum creatinine value of 370 µmol/l. After two weeks the mean serum creatinine value was 140.

All the animals pretreated with superoxide dismutase, mannitol and catalase survived the observation period. During the first three days after operation the serum creatinine value was significantly lower than in the control group. After two weeks the Cr-EDTA-clearance was identical in the treated and non-treated rabbits. It was obvious from our results that the protective effect of mannitol was equal to that of catalase and superoxide dismutase. The most likely explanation to this finding is the scavenger effect of mannitol but may in part be an effect of prevention of endothelial cell swelling.

Our experimental results indicate that the formation and release of oxygen free radicals are important for the post-ischemic renal lesions and functional impairment. The results have also indicated that pretreatment with scavenger enzymes or substances may be of importance in the clinical work when we induce renal ischemia, for instance in transplantation of kidneys as well as in extracorporeal renal surgery and renal vascular reconstructions. The experimental results may also have relevance for other organs subjected to ischemia and reperfusion.

REFERENCES

1 Glaumann B, Trump BF (1975) Virchows Arch B Cell Path 19:303-323

2 Glaumann B, Glaumann H, Berezesky JK, Trump BF (1977) Virchows Arch B Cell Path 24:1-18

3 Venkatachalam MA, Ernard DB, Donohoe JF, Levinsky NG (1978) Kidney Int 14:31-49

4 Harvig B (1979) Acta Univ Upsal 324

5 Karlberg L (1981) Acta Univ Upsal 381

6 Koletsky S, Gustafsson GE (1947) J Clin Invest 26:1072-1078

7 Demopoulos HB, Flamm ES, Pietronigro DO, Seligman ML (1980) Acta Physiol Scand Suppl 492:91-119

8 Granger DN, Rutili G, McCord J (1981) Gastroenterol 81:22-29

9 Hansson R, Gustafsson B, Jonsson O, et al (1982) Transplant Proc 14:51-58

10 Hansson R, Jonsson O, Lundstam S, Pettersson S, Scherstén T, Waldenström J (1985) Clin Scien in press

11 Hansson R (1983) Thesis, University of Göteborg

12 Gustavsson B, Olsson R-M, Waldenström J (1982) Anal Biochem 122:1-5

Superoxide and Superoxide Dismutase in Chemistry,
Biology and Medicine; G. Rotilio, editor

PREVENTION OF ACUTE RENAL FAILURE BY THE USE OF SCAVENGERS TOWARDS OXYGEN-FREE RADICALS.

MATS WOLGAST, ALFRED BAYATI, ÖRJAN KÄLLSKOG and BO ODLIND
Department of Physiology & Medical Biophysics, Biomedical Center, Box 572 S-751 23
Uppsala Sweden

INTRODUCTION

Kidneys exposed to 45 min of warm or 12 hr of cold ischemia will not show any major cellular dysfunction. After recirculation, however, both red and white blood cells will become trapped into the medullary vascular bed with subsequent sustained ischemia in this region of the kidney. This additional ischemia causes detoriation of the medullary segments of the nephron, i.e. the loop of Henle and the medullary collecting ducts with subsequent cell swelling and obliteration and later (due to trapping of rejected cell debris) obstruction of the tubular lumen i.e. a blockade to the passage of urine. Since glomerular ultrafiltration proceeds the early proximal tubular segment will dilate and the pressure rise until the filtration eventually ceases. The clearance function will be reduced towards 5% of that under normal conditions, i.e. an "acute renal failure" develops. Here, it would seem not unlikely that the trapping is due to the action of oxygen-free radicals generated during the early recirculation phase. In order to test this possibility, scavengers were administered to animals, where the kidneys were exposed to ischemia with the aim to prevent the present blood cell aggregation and hence the development of acute renal failure.

METHODS AND RESULTS

The red cell volume in the cortex and the medulla was determined using 51-Cr labelled red cells and expressed percentage red cell volume. Fig. 1 shows the red cell volume under 1) control conditions , 2) 20 min after recirculation on kidneys exposed to 45 min of warm ischemia and 3) kidneys exposed to the same ischemia but pre-treated with allopurinol. The figure demonstrates that ischemia leads to a massive trapping of red cells primarily in the renal medulla and that allopurinol is able to substantially prevent this aggregation. Superoxidedismutase will give the same positive result.

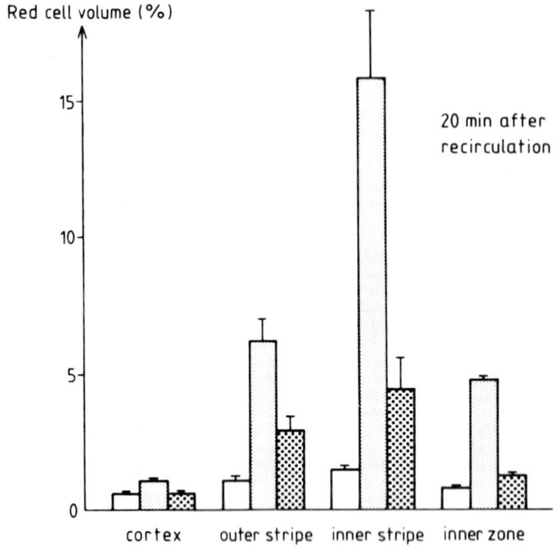

Fig 1 Red cell volume in the cortex, the outer and inner stripe of the outer medulla and the inner medulla during 1) control conditions (bars to the left), 2) 20 min after onset of circulation from 45 min of warm ischemia (bars in the middle) and 3) kidneys exposed to the same ischemia but treated with allopurinol (bars to the right).

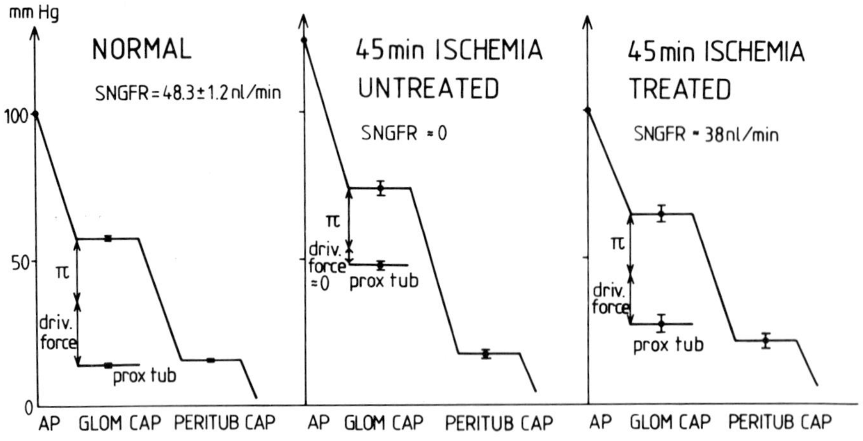

Fig 2 Micropuncture studies of nephrons of 1) normal nephrons , 2) kidneys exposed to 45 min of warm ischemia and 3) kidneys exposed to the same ischemia ; the rats were pre-treated with allopurinol orally in a dose of 5 mg/day for 5-7 days .

The net driving force for glomerular ultrafiltration (calculated by subtracting the proximal tubular pressure and the colloid osmotic pressure of systemic plasma from the glomerular capillary pressure) is about 23 mm Hg under normal conditions (panel to the left) and the resulting single nephron filtration rate (SNGFR) 48.3 ± 1.2 nl/min. In kidneys subjected to 45 min of warm ischemia both the net driving force and the SNGFR were essentially zero. Treatment with allopurinol was able to improve the net driving force to 18 mm Hg and the corresponding SNGFR next to normal . The total filtration rate (GFR) did, at the same time, improve towards 0.4 ml/min after 1 hour, which constitutes a 50 % of that under normal conditions, i.e. allopurinol was able to prevent the development of acute renal failure (Fig 3).

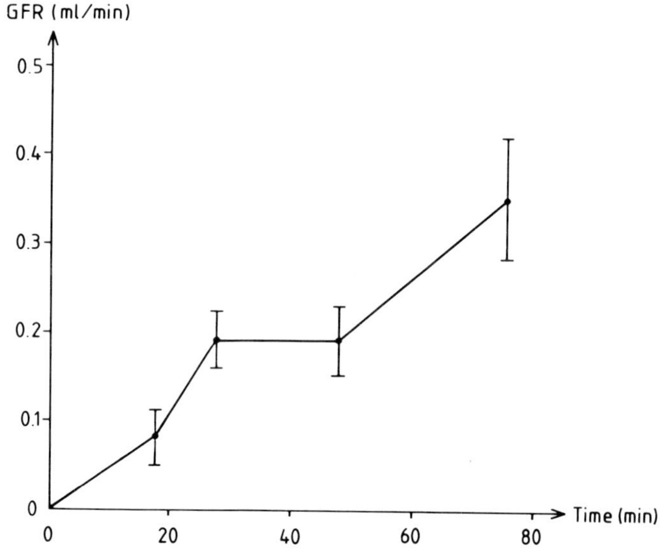

Fig 3. Restoration of the glomerular filtration rate (GFR) up to 80 min after recirculation.

DISCUSSION

Although the function of the kidney epithelial cells will remain largely unaffected by 45 min of warm ischemia, the trauma causes profound alterations to the capillary endothelial membranes resulting in the trapping of red bloodcells in the early recirculation phase in primarily the renal medulla. The study also shows that scavengers are able to prevent red cell trapping and thereby also prevent the development of acute renal failure.

Superoxide and Superoxide Dismutase in Chemistry,
Biology and Medicine; G. Rotilio, editor

A SPIN TRAPPER PROTECTS RATS AGAINST LETHAL SHOCK

GIAN PAOLO NOVELLI, LORENZO BORDI, GUGLIELMO CONSALES

Institute of Anesthesiology and Intensive Care, University of Florence
Policlinico Careggi, 50134 Florence (Italy)

INTRODUCTION

There is evidence of a massive involvement of oxygen radicals in the pathogenesis of circulatory shock [1-4].

The proofs, however, are mostly indirect and derived from experiments directed to observe the consequences of increasing or decreasing scavenging activities.

In the hypothesis of an "in vivo" action of nitrones similar to that exerted on oxygen radicals "in vitro", the spin trapper phenyl-t-butyl-nitrone (PBN) was administered to rats searching for prevention or reversion of circulatory shock.

MATERIALS AND METHODS

All experiments were performed on male Wistar strain rats (200 g) fasted 12 hours before experiments and forbidden to drink water between 2 hours before and 8 hours after shock inducing maneuvers.

Shock was produced by trauma (rotating drum, 1200 revolutions), by endotoxin (Escherichia Coli, Difco 0111:B_4, 20 mg/kg i.p.) or by superior mesenteric artery occlusion (50 min). Severity of shock was assessed by the degree of metabolic acidosis and of increased hematocrit (Hct), by typical pathology and by survival (after 3, 6, 12 and 24 hours). Animals surviving after this time were considered permanent survivors.

To prevent shock, the spin-trapper PBN was injected intraperitoneally in doses of 150 or 50 mg/kg, each single dose being dissolved in olive oil, so to prevent the hydrolisis (out of contact with air) that might be consequent to acqueous solution [5].

To reverse shock, experiments were performed on traumatic shock alone, injecting the PBN (150 mg/kg i.p.) 30 or 60 minutes after the moment of appraisal of well demonstrated shock.

The blood samples for measurement of acid-base status and of Hct were obtained in all experimental groups by cardiac puncture of animals different from those used for control of survival.

Control experiments were performed injecting olive oil alone.

Statistical evaluation of survival was performed with the "chi square" test.

RESULTS

Here reported tables are concerning survival only being clear that other data parallel it.

It evident that administration of the spin trapper PBN is highly effective as prevention and as reversion of experimental shock. Administration of PBN later than one hour seems to have a non significant activity.

No one adverse effect of PBN was observed.

TABLE I

Survival of rats pretreated with PBN (150 or 50 mg/kg i.p.) in olive oil 10 minutes before lethal endotoxin, trauma or SMAO.

Statistical significance against controls: * $p < 0.001$; ** $p < 0.0001$

		survivors/examined at			
		3 hrs	6 hrs	12 hrs	24 hrs
Olive oil	endotoxin	10/15	4/15	1/15	1/15
	trauma	8/13	5/13	1/13	0/13
	SMAO	10/15	8/15	0/15	0/15
PBN 150 mg/kg	endotoxin	15/15	15/15**	15/15**	15/15**
	trauma	14/14	14/14**	14/14**	14/15**
	SMAO	15/15	15/15**	14/15**	15/15**
PBN 50 mg/kg	endotoxin	11/15	11/15	10/15*	10/15*
	trauma	10/12	10/12	9/12*	9/12**
	SMAO	13/15	12/15	9/15*	7/15*

DISCUSSION

Here reported data are confirmative of a favourable action exerted by the spin trapper PBN on three models of circulatory shock in rats, although there is no element to attribute oxygen radicals trapping. However, no other action of PBN has been reported or observed, so that the scavening effect of the single dose seems to be the most probable mechanism involved. Therefore, it might be hypothesized that scavenging of oxygen radicals breakes off the chain of lipoperoxidations and consequent damages. If these are too much advanced and severe, the PBN is deprived of effect.

TABLE II

Survival of rats receiving PBN (150 mg/kg i.p.) 30 or 60 minutes after severe traumatic shock, as confirmed by metabolic acidosis and high hematocrit. Statistical significance against controls: * $p < 0.001$; ** $p < 0.0001$

	survivors / examined at			
	3 hrs	6 hrs	12 hrs	24 hrs
Olive oil (n = 16)	8/16	3/16	0/16	0/16
PBN after 30' (n = 14)	14/14*	14/14**	14/14**	14/14**
PBN after 60' (n = 16)	16/16*	16/16**	16/16**	16/16**

REFERENCES

1. Del Maestro RF (1980) Acta Physiol Scand 492:153
2. Lefer AM, Araki H, Okamatsu S (1981) Circ Shock 8:273
3. Novelli GP, De Gaudio AR (1983) In: Lewis D, Haglund H (eds) Shock research. Elsevier, Amsterdam, pp 31-42
4. Saez JC, Ward PH, Gunther B, Vivaldi E (1984) Circ Shock 12:229
5. Rosen GM, Rauckman EJ (1984) Methods in Enzymology 105:198

Superoxide and Superoxide Dismutase in Chemistry,
Biology and Medicine; G. Rotilio, editor

ROLE OF OXYGEN RADICALS IN ENDOTOXIN-INDUCED SHOCK AND DISSEMINATED INTRAVASCULAR COAGULATION IN RATS.

T. YOSHIKAWA, Y. KAKIMI, O. SETO, M. MURAKAMI, Y. FURUKAWA, S. SUGINO, AND M. KONDO.

First Department of Medicine, Kyoto Prefectural University of Medicine, Kamikyo-ku, Kyoto 602, Japan

It was reported that an exposure to endotoxin can prime neutrophils for increased release of superoxide (1), and that an endotoxin challenge results in superoxide generation in mice at levels that may ultimately results in death (2). The generation of oxygen free radicals seems to be implicated in the pathogenesis of shock and disseminated intravascular coagulation (DIC) caused by endotoxin. To elucidate the hypothesis, the effects of SOD and catalase on endotoxin-induced shock and DIC were examined in rats.

MATERIALS AND METHODS

Female rats of Wistar strain with weights ranging from 200 to 220 g were used. Experimental DIC was induced by sustained infusion of 100 mg/kg of endotoxin, into the femoral vein for 4 hr using a syringe. Experimental shock was induced by a single injection of 100 mg/kg of endotoxin.

The effects of SOD and catalase on the experimental DIC was examined in rats injected with SOD from bovine liver (3500 U/mg) or catalase from bovine liver (35,000 U/mg protein), subcutaneously before induction of DIC. The effect of SOD (3,050 U/mg protein) and catalase (40,000 U/mg protein) on experimental shock, was examined in the rats injected subcutaneously with the agents 12 and 1 hr before induction of shock.

RESULTS

1.Endotoxin-induced DIC

The fibrin (ogen) degradation products (FDP) level increased in rats given endotoxin, however, pretreatment with 0.5, 5.0 or 50.0 mg/kg of SOD, or, 0.01, 0.1 or 1.0 mg/kg of catalase reduced significantly the FDP level. The levels of fibrinogen decreased after the infusion of endotoxin. The decrease in fibrinogen level was inhibited significantly in rats administered 50.0 mg/kg of SOD or 1.0 mg/kg of catalase. Prothrombin time (PT) was prolonged by infusion of endotoxin. The prolongation of PT was inhibited significantly in rats pretreated with 50.0 mg/kg of SOD, or 0.1 or 1.0 mg/kg of catalase. Partial thromboplastin time (PTT) was prolonged after infusion of endotoxin. Pretreatment with SOD at 0.5, 5.0 or 50.0 mg/kg, or catalase at 0.1 or 1.0 mg/kg inhibited the prolongation significantly.

The platelet counts were decreased in the rats infused with endotoxin. When rats were pretreated with 0.5, 5.0 or 50.0 mg/kg of SOD, or 0.01, 0.1 or 1.0 mg/kg of catalase, the reduction of platelet count was inhibited significantly in these animals compared to those pretreated with saline.

In the normal rats infused with physiological saline solution for 4 hr, the glomerular fibrin deposits (%GFD) was 0%; that is, no fibrin thrombi were formed in any of the glomeruli. When the rats were infused with endotoxin, %GFD was 87.6 ± 12/4% (mean ± SD). However, in rats pretreated with 0.5, 5.0 or 50.0 mg/kg of SOD or 0.01, 0.1 or 1.0 mg/kg of catalase, the formation of fibrin thrombi was reduced significantly.

2. Endotoxin-induced shock.

Systolic blood pressure of rats injected with 1.0 ml of physiological saline alone was 121 ± 1 mmHg (mean ± SD). The blood pressure was reduced significantly 45 min after the injection of endotoxin. The reduction was significantly inhibited by 10 or 50 mg/kg of SOD, or 1.0 mg/kg of catalase.

Serum acid phosphatase activity was increased 45 after the injection of endotoxin. The increase in the activity was inhibited significantly by the administration of SOD (10 or 50 mg/kg) or catalase (0.1 or 1.0 mg/kg).

Serum β-glucuronidase activity was increased significantly after the injection of endotoxin. The increase in the activity was inhibited by the pretreatment of 50 mg/kg of SOD, but could not by catalase.

Serum levels of cathepsin B was increased significantly after the injection of endotoxin. The increase in the level was inhibited by the administration of 50 mg/kg of SOD or 1.0 mg/kg of catalase.

Serum TBA reactants were increased significantly 45 after the injection of endotoxin. The increase was inhibited by the administration of SOD but not by catalase.

DISCUSSION

The present study demonstrated the protective effects of SOD and catalase against the aggravation of DIC and shock in rats induced by endotoxin. These findings suggest that superoxide and hydrogen peroxide can affect the experimental DIC and shock states.

Hydrogen peroxide and superoxide radicals have been reported to cause aggregation of blood platelets (3, 4). Lipid peroxidation mediated by active oxygens is believed to be one of the important causes of cell membrane destruction and cell damage (5). The injury to the endothelial cells, platelets or any kind of tissues may activate the coagulation system (6).

SOD and catalase can prevent the aggravation of endotoxin shock. Gray reported that this enzyme SOD enhances survival in endotoxin-challenged mice (2). These observations support the hypothesis that oxygen radicals are produced in endotoxin shock. We reported that the levels of TBA reactant, important and damaging products of free radical lipid peroxidation, in abdominal aortic wall and ileum mucosa were increased in rats infused with endotoxin (7). In shock states, these peroxidations may occur in several organs. Since the effectiveness of SOD and catalase against DIC and shock, these enzymes may be useful for the treatment of these pathological conditions.

REFERENCES

1 Johnston RB JR, Guthrie LA, McPhai LC, (1983) In: Greenwald RA, Cohen G (eds) Oxy Radicals and Their Scavenger Systems Vol II. Cellular and Medical Aspects, Elsevier Biomedical, New York, pp 69-73

2 Gray B, (1981) Toxicol Appl Pharmacol. 60: 479-484

3 Handin RI, Carabin R, Boxer GJ, (1977) J Clin Invest. 59: 959-965

4 Higashi O, Kikuchi Y, (1974) Tohoku J Exp Med. 112: 271-278

5 Tappel AL (1973) Fed Proc. 32: 1870-1874

6 Minna JD, Robboy SI, Colman, RW, (1983) In: Disseminated Intravascular Coagulation in Man, Charles C Thomas, Springfield, pp. 3-18

7 Yoshikawa T, Murakami M, Furukawa Y, Kato H, Takemura S, Kondo M. (1983) Thromb Haemostas. 49: 214-216

Superoxide and Superoxide Dismutase in Chemistry,
Biology and Medicine; G. Rotilio, editor

REACTIVE OXYGEN SPECIES INCLUDING SUPEROXIDE ION GENERATED FROM ACTIVATED ANTHRACYCLINES AND THE RELATIONSHIP TO CARDIOTOXICITY

J. WILLIAM LOWN*[1] JAMES A PLAMBECK[1], SHAM M. SONDHI[1] AND JOHN H. PETERS
[1]Department of Chemistry, University of Alberta, Edmonton, AB,
T6G 2G2 Canada; [2]Life Sciences Division, SRI International, Menlo Park, CA
94025 USA

INTRODUCTION

The biochemical basis for the cardiotoxicity which severely restricts the use of antitumor anthracyclines (1) is at present unknown. However there is growing evidence that it is related to the redox activity of the quinone moiety leading to the production of oxygen based radicals which can damage nucleic acids and cause lipid peroxidation (2,3). The latter effect induces cardiotoxicity in rats (4) and rat cardiac tissue accumulate anthracyclines more than other tissues (5). In addition cardiac tissue seems to be particularly susceptible to oxidative lesions owing to the relatively low levels of antioxidants or enzymes that normally protect against such attack (2,3).

We report the redox activity, as measured by activation of several anthracyclines by rat liver microsomes, by chemical reduction and by electrochemistry. Assays are applied to determine the relative production of $O_2^{\cdot -}$ and of $OH^{\cdot}$ from the activated drugs. Correlations are observed among the different parameters which may prove useful in the design of less cardiotoxic agents.

MATERIALS AND METHODS

Materials

The anthracyclines were obtained from NCI, Bethesda, MD; Farmitalia Carlo Erba, Milan; and Pfizer Co. Tokyo, Japan. Nitro-blue tetrazolium and calf thymus DNA were from Sigma PM2-CCC-DNA was prepared as described previously. The rat liver microsomes were prepared in a modification of the procedure of Omura and Takesue (6). Microsomal oxygen consumption data were obtained by the procedure described by Peters et al (7). The nitro-blue tetrazolium assay for $O_2^{\cdot -}$ generation was performed as described by Beauchamp and Fridovich (8). The assay for $OH^{\cdot}$ generation employing nicking of circular PM2 DNA by reduced anthracyclines in the presence of O_2 has been described (9). Electrochemical methods used have been described previously (10).

RESULTS

Several derivatives of doxorubicin, daunorubicin and other anthracyclines have been examined for relative abilities to augment oxygen consumption by rat liver microsomes. Measurement of parent drug disappearance and of metabolite appearance for anthracyclines with a broad range of activities indicates that a single reaction, conversion to the 7-deoxyaglycone, occurs. Multiple tests of selected derivatives showed that the system exhibits saturation kinetics and calculated values of V_{max}/K_m were comparable to results of the survey tests (7). Separate 13-dihydrodiastereoisomers of one of the anthracyclines yielded nearly identical results suggesting that the reaction is not stereoselective at that position. Associated measurements of the ability of a

TABLE

Microsomal O_2 Consumption and Production of $O_2^{\overline{\cdot}}$ and $OH^{\bullet}$ from Reduced Anthracyclines

Anthracycline	Microsomal O_2 uptake (% of DXR)	Superoxide Production (% of DXR)	DNA Nicking by ethidium assay (% DXR)
DXR	100	100	100
4-DeO-11-deDDXR	138	128	-
11-DeODXR	53	120	-
DAU	109	89	102
4DeOMe-11deODAU	120	220	-
11-DeDAU	44	120	-
N,NDiMeDXR	145	-	102
N,NDiBzDXR-ol	16	13	-
N,NDiMeDAU	108	-	102
NBzDAU	12	-	83
N,NDiBzDAU	2	-	11
3'-DeNH$_2$-3'-pipDAU	73	90	99
3'-DeNH$_2$-3'(4Mepip)DAU	73	8	-
3'-DeNH$_2$-3'(4MeOpip)DAU-ol	64	85	-
3'-DeNH$_2$-3'(4Morph)DAU	26	62	62
3'-DeNH$_2$-3'-(thiomorph)DAU	17	51	-
N(THP)DAU	68	90	-
AD-32	4	28	-
5-Imino-DXR	8	21	10
5-IminoDAU	8	-	62
5-IminoDAU-ol	11	10	49

representative number of the compounds after chemical reduction to convert oxygen to the reactive species, superoxide anion, measured colorimetrically, or to the hydroxyl radical, measured by the ability to nick circular-closed DNA, were also made (Table). The reactivities of the anthracyclines in these latter two assays were found to be positively correlated with the results obtained with the rat liver microsomes. Superoxide anion generation from chemically reduced anthracyclines was inhibited by the addition of calf thymus

DNA and the extent of inhibition was positively correlated with the measured DNA association constants of the anthracyclines (7). However, these latter two parameters were unrelated to superoxide anion generation in the absence of DNA or to the augmentation of oxygen consumption in liver microsomes. Half-wave potentials were negatively correlated with both the results of the microsomal oxygen consumption test and the product of superoxide anion in the chemical test system (7). No relationships were discerned among the DNA association constants, half-wave potentials or reoxidizibilities of the anthracyclines tested. Comparisons of the relatively low activities of certain of the anthracyclines in the microsome test with their known high activitites against murine tumors in vivo, but low cardiotoxicities in animal model systems, suggest that the separation of the cytotoxic antitumor and cardiotoxic actions of these derivatives may have been achieved.

In conclusion, the augmentation of oxygen consumption by rat liver microsomes and related chemical assays can be useful techniques for an initial assessment of one important aspect of the activity of anthracyclines. However, the results must be combined with pharmacological data, especially on drug disposition and metabolism, to provide pertinent basic information for the development of useful chemotherapeutic agents.

ACKNOWLEDGEMENTS

This work was supported in part by grant CA-32215 (to J.H.P.) from the National Cancer Institute, USPHS; and by grants from the National Cancer Institute of Canada (to J.W.L.) and from the Natural Sciences and Engineering Research Council of Canada (to J.W.L. and J.A.P.).

REFERENCES

1. Arcamone, F. (1984) Med. Res. Rev. 4:153-181.
2. Bachur, N. R., Gordon, S. L., Gee, M. V. (1978) Cancer Research 38:1745-1752.
3. Myers, C. E. (1982) In: Muggia, F. M., Young, C. W., Carter, S. K. (eds) Anthracycline Antibiotics in Cancer Therapy, M. Nijhoff, Boston, p. 297-306.
4. Mizukami, M., Aono, J., Sakai, K. Hata, S. Nakano, M. (1984) Arzneim-Forsch 34(1):569-577.
5. Peters, J. H., Gordon, G. R., Kashiwase, D. Acton, E. M. (1981) Cancer Chemother. Pharmacol. 7:65-69.
6. Omura, T., Takesue, S. (1970) J. Biochem. 67:249-257.
7. Peters, J. H., Gordon G. R., Kashiwase, D., Lown, J. W., Yen, S.-Y., Plambeck, J. A. (1985) Biochem. Pharmacol. in press.
8. Beauchamp, C., Fridovich, I. (1971) Anal. Biochem. 44:276-285.
9. Lown, J. W. (1984) Methods in Enzymol. 105:532-539.
10. Plambeck, J. A., Lown, J. W. (1984) J. Electrochem. Soc. 131:2556-2563.

© 1986 Elsevier Science Publishers B.V. (Biomedical Division)
Superoxide and Superoxide Dismutase in Chemistry,
Biology and Medicine; G. Rotilio, editor

PREVENTION OF DOXORUBICIN CARDIAC TOXICITY BY DIMETHYL SULFOXIDE

JAMES H. DOROSHOW* AND JOEL SCHECHTER**

*Dept. of Medical Oncology, City of Hope National Medical Center, Duarte, CA 91010 (U.S.A.), and **Dept. of Anatomy, University of Southern California, Los Angeles, CA 90033 (U.S.A.)

INTRODUCTION

Doxorubicin (DOX)+ is a quinonoid xenobiotic that plays a critical role in current cancer chemotherapy because of its therapeutic activity against a wide variety of human malignancies (1). Unfortunately, treatment with DOX can produce a characteristic form of cardiac injury that involves disruption of mitochondria and sarcoplasmic reticulum (1,2). Although the mechanism of DOX cardiac toxicity remains incompletely understood, several recent studies have demonstrated that reduction of the quinone moiety of DOX by cardiac organelles (3) results in a one-electron transfer to molecular oxygen forming superoxide anion and the hydroxyl radical (·OH) (4). Because of the oxidizing potential of ·OH, we examined in these studies the effect of the ·OH scavenger dimethyl sulfoxide (DMSO) on DOX cardiac toxicity.

MATERIALS AND METHODS

Materials. For these experiments, 20 gram male CDF_1 mice were obtained from Simonsen Laboratories, Gilroy, CA. Doxorubicin hydrochloride was obtained from Adria Laboratories, Columbus, OH and was reconstituted in 0.9% sterile NaCl on the day of administration. DMSO was purchased from Sigma Chemical Co., St. Louis, MO and was used as a 25% (v/v) solution in sterile, 0.9% NaCl.

Animal studies. DOX or DMSO were administered by i.p. injection; where indicated, control animals were treated simultaneously with identical volumes of 0.9% saline. All drug treatments occurred between 8-9 A.M. Pacific Standard Time. Morphologic studies were performed as previously described

---

+Abbreviations used in this study: Doxorubicin, DOX; hydroxyl radical, ·OH; dimethyl sulfoxide, DMSO; intraperitoneal, i.p.

(2). Survival data were compared by the log rank test; other data were analyzed by the two-tailed t-test.

RESULTS AND DISCUSSION

When CDF_1 male mice were pretreated with DMSO (10 gr/kg) 1 hr prior to a 20 mg/kg dose of DOX, 52% of the animals survived long-term compared to 2% of mice receiving saline pretreatment (Table I). DMSO alone in this dose produced no mortality. As shown in Table I, the efficacy of DMSO in ameliorating DOX toxicity was dose- and time-related; significant protective activity was not demonstrated when the ·OH scavenger was administered 1 hr after DOX. We also found that subcutaneous, rather than i.p., administration of DMSO in the same dose and schedule produced a statistically similar degree of survival enhancement (data not shown).

TABLE 1

EFFECT OF DMSO ON SURVIVAL AFTER DOXORUBICIN TREATMENT

Doxorubicin 20 mg/kg was given i.p.; DMSO was administered i.p. in the dose and at the times indicated. Data are expressed as % survival 56 days after treatment; control animals received equivolume dosages of 0.9% saline rather than DMSO.

DMSO DOSE	Timing of DMSO with Respect to Doxorubicin			
(gr/kg)	2 hr before	1 hr before	Simultaneous	1 hr after
0	0% (n=10)	2% (n=63)	0% (n=10)	0% (n=10)
1.0		0% (n=21)		
2.5		10% (n=29)		
5.0		13% (n=67)*		
10.0	35% (n=20)*	52% (n=50)*	36% (n=25)*	4% (n=25)

* Significantly different from respective saline pretreatment control, $p < 0.05$.

To assess the mechanism of protection by DMSO, cardiac tissue was examined by electron microscopy. Evaluation of the effect of DMSO pretreatment on DOX cardiac toxicity was performed using a quantitative scale (from 0 to 3) of heart damage for multiple electron micrographic sections as previously described (2). Cardiac toxicity scores for saline controls (mean ± S.E.) (0.0 ± 0.0, n=4), and DMSO controls (0.14 ± 0.02, n=4) were significantly different ($p < 0.005$) from those for animals treated with DOX (1.96 ± 0.18, n=4). Furthermore, the cardiac damage found in animals pretreated with DMSO (10 gr/kg) before an identical dose of DOX (20 mg/kg) (0.37 ± 0.07, n=4) was significantly different from animals pretreated with saline ($p < 0.001$). Thus, DMSO had a profound effect on the morphologic expression of DOX cardiac toxicity. Light microscopic studies, on the other hand, failed to reveal any amelioration of the gastrointestinal toxicity of DOX by DMSO (data not shown).

In these studies we have demonstrated that the ·OH scavenger DMSO specifically reduced the cardiac toxicity of DOX without inhibiting the toxic effect of this antibiotic on the gastrointestinal tract. Although this study does not, by itself, establish that oxyradical production is the major mechanism underlying the cardiac toxicity of DOX, taken together with previous investigations (3,4), it strongly indicates that free radical formation may play a critical role in this important clinical syndrome.

ACKNOWLEDGEMENTS

This study was supported by grant CA 31788 from the National Cancer Institute and by the Leukemia Society of America.

REFERENCES

1. Carter SK (1975) J Natl Cancer Inst 55:1265
2. Doroshow J, Tallent C, Schechter J (1985) Amer J Path 118:288
3. Doroshow J (1983) Cancer Res 43:460
4. Davies KJA, Doroshow J, Hochstein P (1983) FEBS Lett 153:227

Superoxide and Superoxide Dismutase in Chemistry,
Biology and Medicine; G. Rotilio, editor

METHYLENE BLUE AMELIORATES DOXORUBICIN TOXICITY

WILLIAM J. M. HRUSHESKY, RANDAL OLSHEFSKI, PATRICIA WOOD, STEVEN MESHNICK, JOHN R. BABSON, WILLIAM WOODS, JOHN W. EATON

INTRODUCTION

The clinical usefulness of doxorubicin is limited by the development of a dose-dependent cardiomyopathy. This toxicity has been blamed upon the production of free radical species, most notably superoxide and hydroxyl radical.[1-6] These radicals are thought to overwhelm cellular defenses and to lead to cytotoxicity. The process described may damage many intracellular targets including membrane lipids, nucleic acids and important constitutive proteins.[7-15]

In order to react with oxygen, doxorubicin must be enzymatically reduced to its semiquinone form. Methylene blue, a redox dye with little toxicity, is also enzymatically reduced by cytosolic, mitochondrial, and sarcosomal fractions of mouse heart homogenate.[16] Coadministration of methylene blue increases the survival of mice treated with otherwise lethal doses of doxorubicin and does not interfere with the doxorubicin anticancer activity. The dye may do this by lowering the concentration of intracellular NAD(P)H reducing equivalents, limiting the rate of enzymatic doxorubicin reduction, and therefore indirectly slowing subsequent oxygen-drug interactions.[16]

Methylene blue has been used as an electron shuttle for modifying biological reactions.[17] For example, methylene blue acts as an electron acceptor for the oxidation of pyridine nucleotides in glycolysis.[18] It can act as an electron acceptor and subsequent donor in the conversion of methemoglobin to hemoglobin, a reaction mediated in part by a NADPH-dependent enzyme, methemoglobin reductase.[19]

It is believed that the major part of doxorubicin anticancer activity resides in its ability to intercalate into the DNA molecule[20], form DNA-protein crosslinks[21] and to cause topoisomerase-dependent DNA single-strand breaks.[22,23] Abundant evidence also exists suggesting that the production of superoxide and hydroxyl radical, from the doxorubicin semiquinone, may play a role in directly damaging nuclear and cytosolic structures.[11-15] Some oxygen-dependent cytotoxicity is independent of nucleic acid interaction and resides in doxorubicin's ability to damage cell membranes.[7-10] The exact contribution of each of these mechanisms to cytotoxicity is less than clear. We have investigated how methylene blue modulates doxorubicin-induced toxicity by employing a number of methods.

This brief paper and the accompanying table summarizes the results of these investigations.

MATERIAL AND METHODS

In Vivo Animal Studies

In all studies, the concentration of doxorubicin and methylene blue was 18 mg/kg and 25 mg/kg, respectively.

Lethality Study 1: One-year-old female Balb/c mice received doxorubicin intraperitoneally followed within 5 minutes by methylene blue.

Lethality Study 2: Three-month-old Balb/c mice received doxorubicin intravenously or intraperitoneally, followed by methylene blue given by the same route.

Lethality Study 3: Three-month-old female CD_2F_1 received methylene blue intravenously immediately before and subcutaneously 1, 3, and 6 hours after intravenous doxorubicin.

Determination of cardiac damage: A staff pathologist examined one micron glycol methacrylate sections of hearts of mice that received either doxorubicin and methylene blue or doxorubicin and placebo without prior knowledge of treatment group. Hearts were scored by the method of Billingham.[24]

Anti-Neoplastic Study: Female Balb/c mice received 7.0×10^4 viable L1210 leukemia cells, followed three days later by a single treatment of doxorubicin. Methylene blue was administered .5, 1.5, 4, 24 and 48 hours after doxorubicin.

Cellular Uptake/Retention: CD_2F_1 female mice received 1.0 uCi of radiolabeled doxorubicin with a final dose of doxorubicin of 18 mg/kg. Methylene blue was given immediately after doxorubicin. Heart, spleen, and liver samples were prepared for scintillation counting by a modification of the method of Harrison and Wagner.[25]

GSH analysis: Thirty-week-old CD_2F_1 female mice received methylene blue or saline intravenously. Animals were killed four hours later and reduced glutatione content of trichloroacetic acid extracts of crude heart homogenate was measured.[26]

Malondialdehyde assay: Groups of four CD_2F_1 mice were given intravenous doxorubicin, methylene blue, or both. The animals were sacrificed two hours later, the hearts were removed and homogenized in a 0.02 M potassium phosphate buffer containing butylated hydroxytoluene to prevent further oxidation of lipids. Homogenized samples were then assayed for MDA by adding 50% thiobarbituric acid, heating for 90 minutes, and measuring the absorption of the adduct at 533 nm as described by Bernheim.[27] Protein determinations were performed as described by Lowry[28], and results were expressed as OD units/mg protein.

In Vitro Studies

L1210 cytotoxicity: L1210 cells grown in RPMI media containing 10% horse serum were exposed for one hour to doxorubicin (1.0 ug/ml) and methylene blue (5.0 ug/ml) or doxorubicin alone. Percent survival was measured by the method of back extrapolation.[29]

Strand breaks and crosslink assays: The method of alkaline elution described by Kohn[30] was used to measure doxorubicin-induced single-strand breaks and DNA-protein crosslinks in cultured L1210 cells. Concentrations of doxorubicin and methylene blue were 1.0 ug/ml and 5.0 ug/ml, respectively.

Oxygen consumption: The rate of oxygen consumption in a mouse heart homogenate system was determined at 27°C with a Rank Brothers oxygen electrode (Bottisham, England). Reducing equivalents (NADH and NADPH), doxorubicin, and methylene blue were added to the system in an order-dependent manner. The rate of oxygen consumption was measured following each addition.

Radiation Study

Female Sprague-Dawley rats were injected intraperitoneally with either 25 mg/kg methylene blue or saline 30 minutes before receiving 1100 rads of abdominal x-irradiation. Stool blood content was measured 48 hours prior to and up to 96 hours following treatment by the HemoQuant assay.[31] The animals were then sacrificed and intestinal denudation was measured by quantifying mucosal height.[32]

RESULTS

Methylene blue effectively protects healthy mice from doxorubicin-induced lethality and heart damage but does not interfere with the ability of doxorubicin to prolong the survival of animals innoculated with L1210 leukemia. Methylene blue also does not interfere with the killing of cultured L1210 by doxorubicin. An NADPH-dependent methylene blue diaphorase is present in murine heart. Methylene blue does not change the cellular uptake or retention of doxorubicin by liver spleen or heart cells. Methylene blue and doxorubicin treatment each lower intracellular reduced glutathione concentrations. Methylene blue lowers whole heart malondialdehyde concentration. Doxorubicin minimally and methylene blue markedly stimulate the rate and amount of oxygen consumption of heart homogenate in the presence of NADPH. An intermediate amount of oxygen is consumed at an intermediate rate when both agents are added. In vivo, under euoxic conditions doxorubicin potentiates while methylene blue protects cells against radiation-induced damage. Finally, the interaction of nucleic acids and doxorubicin is modified by methylene blue. Methylene blue alone causes a few DNA protein crosslinks and DNA single strand breaks. Doxorubicin alone causes moderate numbers of crosslinks and extensive DNA single strand breaks. The combination of methylene blue with doxorubicin markedly decreases the frequency of doxorubicin-induced strand breaks and results in a moderate increase in the number of DNA protein crosslinks.

TABLE 1.
TABULAR SUMMARY OF RESULTS

	MB	DOX	MB+DOX
Animal lethality	0	++++	+
In vivo cardiac damage	0	++++	++
In vivo anti-L1210 effect	0	++++	++++
In vivo cellular uptake/retention	++++	++++	no interaction
In vivo GSH decrease	+++	+	not done
In vivo cardiac membrane damage: MDA conc.	0[a]	+++	0[a]
In vitro L1210 cytotox	0	++++	++++[b]
In vitro L1210 strand breaks	+	++++	++
In vitro L1210 DNA X links	+	+++	++++
Heart homogenate O_2 consumption	++++	+	++[c]
In vivo intestinal radiation damage	protects	worsens	not done

[a]significantly below control value, $p < .05$
[b]low concentrations of methylene blue
[c]dependent on the order of reagent addition to the system

CONCLUSION

The basis of the protection by methylene blue against doxorubicin and radiation toxicity is not yet clear. Methylene blue may: 1) compete for reducing equivalents necessary for the activation of doxorubicin; 2) cause the oxygen concentration in specific doxorubicin-sensitive and irradiation-sensitive subcellular compartments to decrease to the point where less oxygen-dependent damage results; 3) scavenge active oxygen intermediates. In aggregate, our results emphasize but do not explain the variagate pharmacologic effects of methylene blue.

REFERENCES

1. Bachur NR, Gordon SL, Gee MV (1978) Cancer Res 38:1745-1750
2. Handa K, Santo S (1975) Gann 66:43-47
3. Doroshow J (1983) Cancer Res 43:460-472
4. Doroshow J (1982) Proc Am Assoc Cancer Res 23:172(abstr)
5. Doroshow J (1983) Cancer Res 43:4543-4551
6. Thayer WS (1977) Chem Biol Interact 19:265-278
7. Myers CE, McGuire WP, Liss RH et al (1977) Science 197:165-167
8. Goodman J, Hochstein P (1977) Biochem Biophys Res Commun 77:797-803
9. Minnaugh EG, Thrush MA, Bhatnager M, Gram TE (1984) Biochem Pharmaco 34:847-856
10. Minnaugh EG, Kennedy KA, Trush MA, Sinha BK (1985) Cancer Res 45:3296-3304
11. Lown SW, Simm SS, Majundar KC, Chang RY (1977) Biochem Biophys Res Commun 76:705-710
12. Levin M, Silber R, Israel M, Goldfeder A, Khetarpal VK, Potmesil M (1981) Cancer Res 41:1006-1010
13. Begleiter A, Blair GW (1984) Cancer Res 44:78-82
14. Berlin U, Haseltine WA (1981) J Biol Chem 256:4747-4756
15. Eliot H, Gianni L, Myers C (1984) Biochemistry 23:928-936
16. Hrushesky WJM, Olshefski R, Wood P, Meshnick W, Eaton JW (1985) Lancet i:565-567
17. Ehrlich P, Leppmann A (1890) Deutsche Med Wochenschr 16:493-494
18. Brin M, Yonemoto RH (1958) J Biol Chem 230:307-317
19. Sass MD, Caruso CJ, Axelrod DR (1969) Clin Chim Acta 24:77-85
20. Gabbay EJ, Grier D, Fingerle RE, Reimer R, Levy R, Pearce SW, Wilson WD (1976) Biochem 15:2062-2070
21. Ross WE, Glaubiger D, Kohn KW (1978) Biochem et Biophysica Acta 519:23-30
22. Ross WE, Glaubiger D, Kohn KW (1979) Biochem et Biophysica Acta 562:41-50
23. Tewey KM, Rowe TC, Yang L, Halligan BD, Liu LF (1984) Science 226:466-468
24. Billingham ME, Mason JW, Briston MR, Daniels SR (1978) Cancer Treat Rep 62:865-872
25. Harrison K, Wagner HN (1978) J Nucl Med 19:84-86
26. Prins HK, Loos JA (1969) In: Yunis J (ed) Biochemical methods in red cell genetics. New York, pp 115-135
27. Bernheim F, Berheim ML, Wilber KM (1948) J Biol Chem 174:257-264
28. Lowry O, Rosenbrough NJ, Farr AL, et al (1951) J Biol Chem 193:265-275
29. Alexander P, Mikulski ZB (1961) Biochem Pharmacol 5:275-282
30. Kohn KW (1979) In: Methods in Cancer Research, vol XVI. Academic Press, Inc. New York pp 291-345
31. Schwartz S, Dahl J, Ellefson M, Ahlquist D (1983) Clin Chem 29:2061-2067
32. Mulholland MW, Levitt SH, Song CW, Potish RA, Delaney JP (1984) Cancer 54:2396-2402

Superoxide and Superoxide Dismutase in Chemistry,
Biology and Medicine; G. Rotilio, editor

LOSS OF ENZYMATIC ACTIVITY OF LENS SOD WITH AGE

ROSALIE K. CROUCH AND MARY T. SIMPSON

Department of Ophthalmology, Medical University of South Carolina, 171 Ashley Avenue, Charleston, SC 29425

INTRODUCTION

The mechanism of aging is poorly understood. However, accumulating evidence indicates that "the sum of the deleterious free radical reactions going on continuously throughout the cells and tissues constitutes the aging process or is a major contributor to it" (1). The major free radicals in mammalian systems are oxygen radicals. If tissue becomes more susceptible to damage with age, a potential mechanism would be a decrease in the enzymes whose function is to scavenge for these radicals. The SOD enzyme is thus a particularly interesting model for studies on the process of aging.

The ocular lens is a unique organ for an investigation of aging of proteins as the nucleus of the lens contains protein molecules which are as old as the animal (2). The characterization of an inactivated SOD might well provide evidence as to the mechanism of the inactivation process. In the lens, removal of this vital protective enzyme has important implications for cataract formation. Indeed a role for SOD inactivation in cataract formation is reasonable since SOD specific activity, based on protein content, is reported to be decreased in the cortex and nucleus of human lenses having increased nuclear color (3).

SOD enzymatic activity has been reported to be reduced with age in a number of tissues based on immunoprecipitation studies: e.g. heart (4), liver (4), and whole lens (5). However, earlier reports detected no decrease based on activity related to soluble protein content (6). We report here results on the enzymatic activity of human and rat lens SOD based on immunoreactivity and soluble protein and on our studies demonstrating that the SOD synthesized in both the young and old lens in organ culture is enzymatically active.

MATERIALS AND METHODS

Lenses were obtained from hooded Long Evans rats or from human eyes donated to the South Carolina Eye Bank for research. Eyes were dissected with the precautions previously observed (7). Rat liver SOD was isolated as described (7) and human SOD was obtained commercially (Sigma Chemicals, St. Louis). SOD enzymatic activity was measured by the NBT method (8) after extensive dialysis of samples. Soluble protein was obtained by the Bradford method (9). Immunoreactivity was determined by radioimmune assay (7).

Biosynthesis of lens SOD was studied by incubating the whole lens in minimum essential media (MEM, Eagle's media with Earle's salts, 0.27% $NaHCO_3$, 10% fetal calf serum and 2.0 mM glutamine) supplemented with 25 µCi ^3H lysine/ml (0.5 Ci/mmole ^3H, 5-^3H lysine, Amersham) for 48-72 hours. The lenses were washed in cold media and frozen in 1 ml DTE buffer (2mM Tris-HCl, 80mM NaCl, 1mM EDTA and 10mM DTE, pH 7.3). Samples are freeze-thawed twice, centrifuged at 78,000 x g for 16 hours and assayed. For immunoprecipitation, samples are incubated with SOD antibody (1:1000 dilution) for 24 hours at 4°C, treated with Pansorbin (Calbiochem) beads (10) and spun through a step sucrose gradient (1.0 M to 2.5 M). The beads are collected at the interface. In some experiments, samples were eluted on a DEAE column by standard procedures.

RESULTS AND DISCUSSION

We have reported (7) that the SOD of the rat cortex and nucleus loses enzymatic activity with age whereas the capsular SOD remains active. We have found similar data in the human for males of the ages 5 and 68 years respectively (Figure 1). Therefore, this phenomenum of accumulation of inactivated SOD in the cortex and nucleus and loss of the protective action of SOD in those tissues with age is not species specific.

To determine if the SOD synthesized in the old lens has full enzymatic activity, we have studied the biosynthesis of SOD in the old and young rat lens. We have used the whole lens in organ culture incubating in media containing tritiated lysine. Labeled SOD is recovered from incubated lenses by immunoprecipitation. The enzyme was assayed for tritium content, immuno-activity and enzymatic activity.

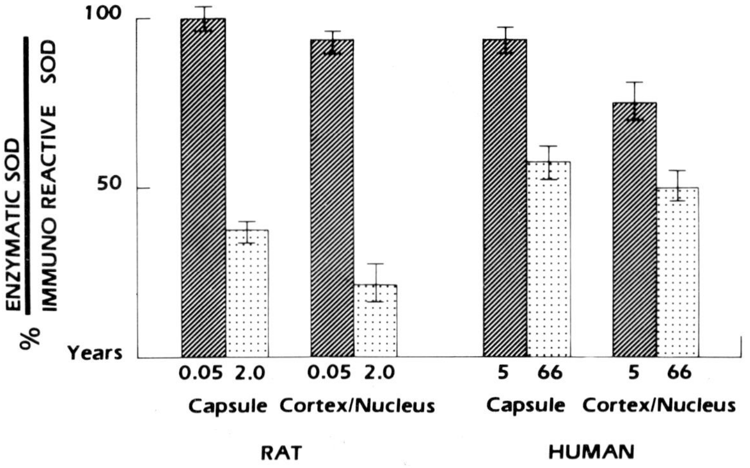

Figure 1: Enzymatic activity versus immunological activity of lenses from rats and humans. The human data represents triplicate assays on a single lens. Lens were free from cataracts or excessive coloration.

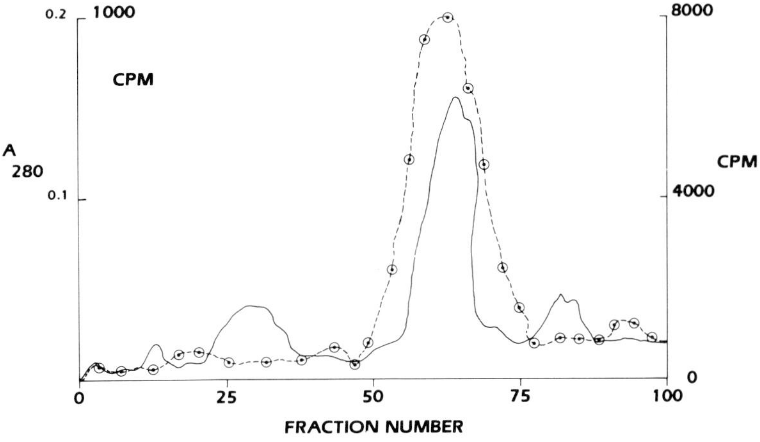

Figure 2: Elution pattern of DEAE column of rat lens SOD immunoprecipitated after organ culture in ^3H lysine. Buffer was linear gradient from 0.005 M to 0.18M Na_3PO_4. Lenses were from 6 month animals.

Tritiated SOD is recovered from the capsule of both old and young lenses. When assayed for enzymatic and immuno activity, there was no evidence of defective enzyme being synthesized. For confirmation, the isolated enzyme was run on a DEAE column. One major peak was obtained (Figure 2) and only a single peak contained tritium. When pure rat liver ^{125}I-SOD (7) was coeluted, the elution volume was found to be the same as for the tritiated sample. The tritiated enzyme had identical values (within experimental error) for immunoreactivity and enzymatic activity.

From these results we therefore conclude that the inactivation of the lens SOD of old animals has occurred as a post translational process.

ACKNOWLEDGEMENTS

This work was supported by NIH grant EY-03923.

REFERENCES

1. Harmon D (1981) Proc Nat Acad Sci USA 78:7124
2. Hockwin O, Hahn U, Swanson AA (1976) Ophthalmic Res 8:321
3. Fecondo JV, Augusteyn RC (1983) Exp Eye Res 36:15
4. Reiss U, Gershon D (1976) Biochem Biophys Res Comm 73:255
5. Dovrat A, Gershon D (1976) Exp Eye Res 33:651
6. Kellogg EW, Fridovich I (1976) J Gerontol 31:405
7. Crouch R, Patrick J, Goosey J, Coles W (1984) Cur Eye Res 3:119.
8. Beauchamp C, Fridovich I (1971) Anal Biochem 44:276
9. Bradford M (1976) Anal Biochem 72:248
10. Kessler SW (1975) J Immunol 115:1617

Superoxide and Superoxide Dismutase in Chemistry,
Biology and Medicine; G. Rotilio, editor

OXY RADICAL-INDUCED POLYMERIZATION OF AMINOPHOSPHOLIPIDS IN CATARACT

KAILASH C. BHUYAN, RAVIDATT W.P. MASTER AND DURGA K. BHUYAN
The Mount Sinai School of Medicine of the City University of New York, One Gustave L. Levy Place, N.Y. 10029, USA

INTRODUCTION

The role of reactive species of oxygen in peroxidation of plasma membrane lipids leading to malondialdehyde (MDA) accumulation has been demonstrated in human senile cataract(1) and in other experimental cataracts(2). MDA has been shown to form Schiff-base conjugates with aminophospholipids, such as phosphatidyl ethanolamine(PE) and phosphatidyl serine(PS) in tissues of vitamin E deficient animals(3) and in sickled erythrocytes(4). We here demonstrate the preponderance of PE•MDA•PS adduct in lipid extracts of human senile cataract by thin layer chromatography(TLC) and its formation in rabbit lens homogenate incubated in presence of MDA or lipid peroxidation promoters, H_2O_2 or tertiary-butyl hydroperoxide(TBHP).

MATERIALS AND METHODS

Cataractous lenses of the human were obtained after surgery from The Manhattan Eye, Ear and Throat Hospital and Lenox Hill Hospital, New York, and normal lenses were obtained from donor eyes received at the Eye Bank for Sight Restoration, Inc. N.Y. The lenses received in 0.15 M NaCl containing 22.6 µM 2,6-di*tert*-butyl-4-methyl-phenol(BHT), were weighed and extracted for lipids with $CHCl_3$:CH_3OH(2:1, V/V) (5) and $HClO_4$(1%, W/V). The extracts were centrifuged at 755xg for 15 min at 0-4°C and the residue was reextracted with $CHCl_3$:CH_3OH:H_2O::2:1:1. The $CHCl_3$ phase was dried under N_2 at 37°C, the residue was washed twice with the same extractant, the final $CHCl_3$ phase was dried under N_2 at 37°C, dissolved in a known volume of $CHCl_3$ and a 10 µl aliquot of each was used for TLC analysis. The solvents for TLC, $CHCl_3$, CH_3OH were of HPLC grade from Aldrich Chemical Co. and other chemicals were from Fischer Scientific Co. The silica gel HL TLC plates, 20x20 cm, 250µ thick were from Analtech. The solvent system for unidimensional TLC was $CHCl_3$:CH_3OH:CH_3COOH:H_2O::50:25:7:3(V/V).

After development and drying, the TLC plate was exposed to 6N HCl vapor at 40°C for 1h. Then the second dimension was run in a solvent system containing $CHCl_3$:CH_3OH:NH_3(7M)::65:25:4(6). For quantitative TLC, the spots were visualized by a brief exposure to iodine vapor; the boundary of each spot was marked, the gel was scraped into borosilicate glass tube, digested with 0.4 ml 70% $HClO_4$ and assayed for inorganic phosphorous(7). The aminophospholipids were visualized by heating the plate at 80°C for 2h or longer after spraying with 0.2% ninhydrin in ethanol. Phosphate containing spots were visualized by the molybdenum spray reagent (1.37% molybdenum oxide in 4M H_2SO_4). For a permanent record TLC plates were sprayed with 40% H_2SO_4 and heated at 110°C for 3h or longer.

Lenses used for in vitro experiments were from 4-5 wk-old Dutch-belted rabbits. Lens homogenate 20% (W/V) in 0.15 M NaCl was prepared. Homogenate, 1.75 ml was incubated with 1 mM H_2O_2, TBHP or MDA in a final volume of 2 ml, pH 7.2, at 25°C with shaking for 4h and 6h. Incubation mixture was centrifuged at 41,000xg for 30 min to obtain a crude membrane pellet, which was used for extration of lipids for TLC analysis as described above. Fluorescence of chloroform lipid extracts was measured at 470 nm when excited at 360 nm. For measurement of diene conjugates chloroform lipid extracts were dried under N_2, dissolved in a known volume of CH_3OH and absorption was recorded at 233 nm. (ε 233 nm conjugated dienes=$2.8 \times 10^4 \times M^{-1} \times cm^{-1}$)

RESULTS

The typical TLC profiles of phospholipids from normal and cataractous lenses of the human are shown in Fig. 1. The composition of phospholipids of normal and cataractous lenses was found to be similar. Both normal as well as cataractous lenses had the major phospholipid components, PE, PS, phosphatidyl choline(PC) and sphingomyelin(SP). An additional phospholipid with an R_f 0.66, intermediate between the R_f values 0.75 of PE and 0.57 of PS was present in both types of lenses. It was iodine and molybdate positive, ninhydrin negative and could be resolved into spots corresponding to PE and PS by two-dimensional TLC after hydrolysis with HCl.

Quantitatively this spot which is PE·MDA·PS was 97±18 μg/μmole PE (Mean±S.D.,n=4) in normal lenses while in cataract it was 1269±656 (n=4).

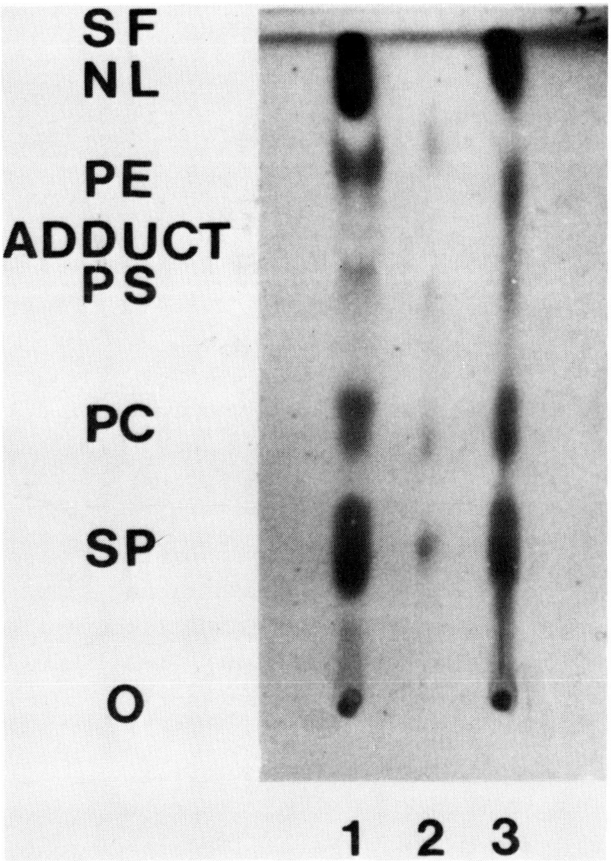

Fig. 1. TLC separation of phospholipids of normal and cataractous lenses of the human, visualized by spraying with H_2SO_4. SF, solvent front; NL, Neutral Lipids; O, origin. Lane 1, normal lens; Lane 2, standards and Lane 3, cataractous lens.

Fig. 2 shows the effects of 1 mM H_2O_2, TBHP or MDA on polymerization of aminophospholipids of rabbit lens homogenate after incubation for 4h at 25°C.

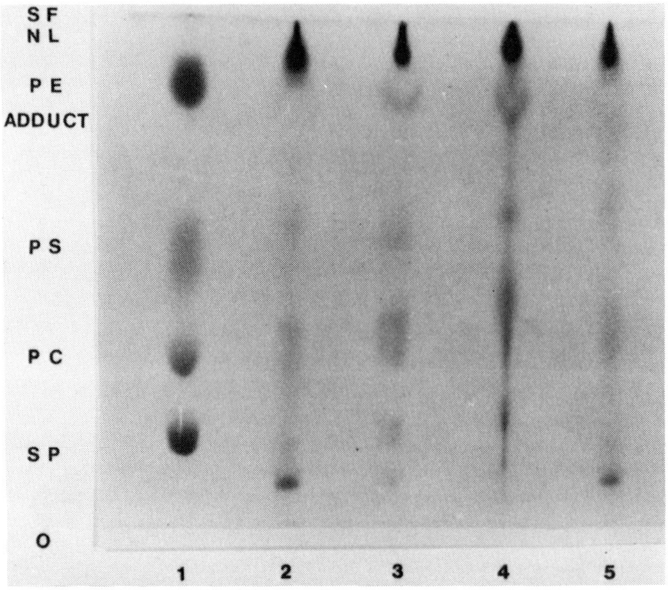

Fig. 2. TLC profiles of lipid extracts of crude membrane pellets obtained from rabbit lens homogenate incubated with 1mM H_2O_2, TBHP or MDA for 4h at 25°C. Lane 1, Standard phospholipids 10 µl mixture (2 mg of each/ml), Lens homogenate: Lane 2, control; Lane 3, with 1mM H_2O_2; Lane 4, with 1mM TBHP and Lane 5, with 1mM MDA.

Lane 1 is the standards of major phospholipid components showing absence of adduct. Lanes 2, 3, 4 and 5 show an additional spot between PE and PS which is more pronounced with TBHP (lane 4) and trace amount in lane 2. Lane 3 (H_2O_2) and lane 5 (MDA) have adduct spots of intermediate intensities. On quantitation, based on the estimation of inorganic phosphorus, lanes 2, 3, 4, 5 had 278, 392, 2156 and 602 µg adduct/µmole PE, respectively.

Results given in Table 1 show that PE·MDA·PS adduct in lens homogenate at 0h was 2.65 nmole/g. It increased by 6.7-fold after 6h without added oxidant, and further increased 8-9 fold with 1mM H_2O_2, TBHP or MDA. As compared to values at 0h, relative fluorescence(470nm emission/360nm excitation) of the lipid extract increased 6-fold with H_2O_2, 3.5-fold with TBHP and

1.5-fold with MDA after 6h incubation. Diene conjugates increased 1.8-fold after incubation of lens homogenate for 6h without oxidant or MDA, and increased 3.6-fold with H_2O_2, 2.7-fold with TBHP and 1.8-fold with MDA.

TABLE I. Polymerization of aminophospholipids of Rabbit lens induced by oxidative stress or MDA *in vitro*[a]

Compound added	Time of Incubation (hour)	PE•MDA•PS Adduct (nmole/g lens)	Relative fluorescence[b] (units/g wet wt)	Diene conjugates (nmole/mg lipid)
none	0	2.65	117	35
none	6	17.8	-	63
H_2O_2, 1mM	6	23.61	721	126
TBHP, 1mM	6	22.22	413	94
MDA, 1mM	6	21.81	179	63

[a]Details given in Materials and Methods. The data represent mean values of triplicate experiments. [b] Fluorescence of standard quinine sulfate, 1µg/ml in 1N H_2SO_4 = 658 units at excitation 360 nm/emission 470 nm.

DISCUSSION

Evidence presented shows an additional phospholipid component among the major phospholipids isolated from human senile cataract. From its R_f the unknown spot seems to be a Schiff-base conjugate formed as a result of cross-linking between PE and PS. The adduct showed fluorescence maximum at 470nm when excited at 360nm, which is a characteristic of Schiff-base conjugate(3). Trace amount of such product has also been observed in normal lenses of the human. It is possible that polymerization of aminophospholipids by cross-linking could be the result of aging, or a post-mortem effect. In order to clarify this, studies which were done using fresh lenses, showed no adduct spot in the phospholipid profile. However, its formation in lens homogenate increased when subjected to oxidative stress of H_2O_2 or TBHP. Incubation with MDA also formed a similar adduct. Thus, it is clear that the adduct seen in human senile cataract is not an artifact. PE•MDA•PS has been shown previously in tissues of animals deficient in vitamin E (3) and in sickled erythrocytes (4).

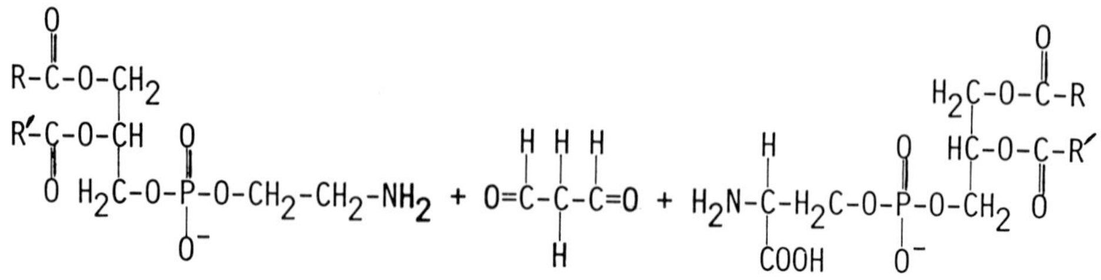

Fig. 3. A cross-linking reaction between carbonyl groups of MDA and amino groups of PE and PS forming Schiff-base conjugate PE•MDA•PS.

Thus, it is possible that polymerization of aminophospholipids of lens by a cross-linking reaction with the carbonyl groups of MDA (Fig. 3) might be one of the mechanisms involved in the pathogenesis of human senile cataract.

ACKNOWLEDGEMENTS

Supported by NEI Grant EY03012, in part by EY01867, and the Pack Foundation for Medical Research, New York.

REFERENCES

1. Bhuyan KC, Bhuyan DK, Podos SM (1981) IRCS Med. Sci. 9, 126-127
2. Bhuyan KC, Bhuyan DK (1983) In: Greenwald RA, Cohen G (eds) Oxy radicals and their scavenger systems, cellular and medical aspects. Elsevier Science Publishing Co., Inc., New York, Vol 2, pp. 349-356
3. Tappel AL (1978) In: Trump BF, Arstila AU (eds) Pathobiology of cell membranes. Academic Press, Inc., New York, Vol 1 pp 145-170
4. Jain SK, Shohet SB (1984) Blood 63:362-367
5. Folch J, Lees M, Sloane-Stanley GH (1957) J Biol Chem 226: 497-509
6. Skipski VP, Peterson RF, Barclay M (1964) Biochem J, 90: 374-377
7. Chen PS Jr, Toribara TY, Warner H (1956) Anal Chem 28: 1756-1758

© 1986 Elsevier Science Publishers B.V. (Biomedical Division)
Superoxide and Superoxide Dismutase in Chemistry,
Biology and Medicine; G. Rotilio, editor

ANTIOXIDANTS IN PREVENTION OF OXIDATIVE DAMAGE TO THE LENS AND CATARACT IN VIVO

DURGA K BHUYAN, STEVEN M PODOS AND KAILASH C BHUYAN
The Mount Sinai School of Medicine of the City University of New York, One Gustave L. Levy Place, N.Y. 10029, USA

INTRODUCTION

Cataract is the third major cause of blindness among people of advanced age. Although its exact molecular mechanism is not yet clear the current belief is that oxidative damage to the lens is the initiating event in the pathogenesis of cataract. Protein oxidation forming high molecular weight aggregates due to disulfide and nondisulfide cross-linking in human senile cataract and in many experimental cataracts has been observed(1,2). Recently, we have demonstrated enhanced lipid peroxidation in cataract of the human and in several experimental cataracts(3). It has also been observed that polymerization of aminophospholipids(4) and lens crystallins(5) occurs due to cross-linking reaction with the carbonyl groups of malondialdehyde(MDA). With this understanding it was expected that a group of compounds capable of scavenging oxidizing species, such as O_2 free radicals, H_2O_2 and lipid free radicals, might be effective in retarding the progression of cataract. Therapeutic trials of antioxidants were initiated in cataracts induced in animals, and the results are summarized in this communication.

MATERIALS AND METHODS

Cataracts were induced in Dutch-belted rabbits by 0.4% 3-amino-1H-1,2,4-triazole(3-aminotriazole) or 40% D(+)-galactose in diet, and in Sprague-Dawley rats by single subcutaneous injection of Na_2SeO_3(20 µmole/kg body wt). Therapeutic efficacies of antioxidants in cataracts were evaluated by periodic observation and photographic recording of the morphological changes of lenses by slit lamp examination. Among the biochemical parameters, H_2O_2 and ascorbic acid of aqueous humor and vitreous humor, and MDA and galactitol of lens were estimated at the end of the therapy. The various procedures used to measure these constituents were described previously(6,7).

RESULTS AND DISCUSSION

Antioxidant potencies of compounds were assessed by measuring their ability to decrease lipid peroxidative damage induced in lens by reactive species of O_2 in vitro(3). In lens cultured in Krebs-Ringer bicarbonate(KRB) medium(3) supplemented with xanthine-xanthine oxidase(XOD)-Fe^{3+}·ADP system(8,9), significantly higher MDA, 3.76 nmole/g wet wt was formed as compared to 1.37 nmole in control lenses cultured in KRB medium alone(footnote to Table I).

TABLE I. EFFECTS OF SOME ANTIOXIDANTS ON OXIDATIVE STRESS-INDUCED LIPID PEROXIDATION OF RABBIT LENS IN VITRO[a]

Antioxidants	Oxidizing Species	MDA of Lens[b] Inhibition(%)	(P)
Vitamin E, 2.32 mM	$O_2 \cdot^-$, $^1\Delta gO_2$, PUFA·	44	<0.001
Vitamin E & Vitamin C	$O_2 \cdot^-$, OH·, PUFA·	59	<0.001
Vitamin C, 2.5 mM	$O_2 \cdot^-$, OH·	13	<0.02
Trolox, 2 mM	$O_2 \cdot^-$, $^1\Delta gO_2$, PUFA·	31	<0.01
Centrophenoxine[c], 5 mM	OH·, PUFA·	56	<0.001
Sorbinil, 0.5 mM	$O_2 \cdot^-$	20	<0.01
Quercetin, 5 mM	PUFAOO·	35	<0.01
GSH, 5 mM	$O_2 \cdot^-$, PUFA·	37	<0.01
Orgotein, 11 µg/ml	$O_2 \cdot^-$	22	<0.01
Penicillamine, 5 mM	unknown	5	0.8
Propyl gallate, 0.2 mM	PUFAO·	39	<0.01
BHT, 0.1 mM	OH·, PUFA·	35	<0.001

[a]Normal lenses of healthy rabbits weighing 1-1.5 kg were used. The culture system consisted of modified KRB medium supplemented with 2.5 mM xanthine, 0.05 IU XOD per ml medium, 0.3 mM ADP and 0.12 mM $FeCl_3$. Other additions were as indicated in the Table. Each lens was placed in a vial containing the culture system except XOD which was then added to start the reaction. The final volume was 8 ml, pH 7.4, and gas phase 95% O_2 and 5% CO_2. Incubation was at 37°C for 3h in a water bath shaking at 72 oscillations/min. At the end of incubation, the lens was removed, rinsed with 0.15M NaCl containing 22.6µM 2,6-ditert-butyl-4-methyl-phenol(BHT) and assayed for MDA(3). [b]MDA in control lenses cultured in KRB medium alone was 1.37±0.35(n=14), with xanthine-XOD was 1.93±0.28(n=4) with Fe^{3+}·ADP it was 1.89±0.11(n=4) with xanthine-XOD-Fe^{3+}·ADP was 3.76±0.24(n=42). [c]dimethylaminoethyl p-chlorophenoxyacetate(CPA, Sigma Chemical Co.). PUFA, polyunsaturated fatty acid. Part of the data from Ref.(3,6).

This maximum (100%) lipid peroxidative damage to the lens was decreased 56-59% by centrophenoxine or vitamins E and C, 30-44% by BHT, propyl gallate, quercetin, vitamin E, trolox or reduced glutathione (GSH), 20% by orgotein or sorbinil. Penicillamine did not have any effect. Some of these compounds were used for therapeutic trials in experimental cataracts, and the results are given in Table II.

TABLE II. THERAPEUTIC EFFECTS OF VARIOUS ANTIOXIDANTS ON LIPID PEROXIDATION OF LENS IN EXPERIMENTAL CATARACTS

Experimemtal cataracts	Compounds administered	MDA of lens	
		(nmole/g wet wt)	% placebo
3-Aminotriazole	Vitamin E[a] (I.M.)	1.76±0.25 (8)[b]	67
-induced	Diluent "	2.63±0.24 (8)	100
(Rabbits)	Vitamin E[c],topical	1.38±0.15 (3)	62
	Peanut oil, "	2.21±0.40 (3)	100
	Sorbinil[d],topical	1.32±0.34 (4)	77
	Diluent, "	1.72±0.34 (4)	100
	Orgotein[e] (I.V.)	2.78±0.29 (6)	99
	Sucrose "	2.80±0.30 (6)	100
Galactose-	Special diet GE40[f]	1.25±0.19 (6)	55
induced	Special diet GE4[f]	2.26±0.10 (10)	100
(Rabbits)	CPA[g] (I.P.)	1.72±0.44 (10)	45
	NaCl,0.15M (I.P.)	3.81±0.63 (8)	100
Selenium-	Vitamin E(I.P.)	1.71±0.55 (3)	49
induced	Diluent "	3.52±0.41 (3)	100
(Rats)[h]	Vitamin C(I.P.)	2.72±0.51 (3)	83
	Diluent "	3.26±0.37 (3)	100
	Penicillamine	3.00±0.31 (5)	90
	Regular diet	3.32±0.33 (4)	100

[a]Injectable form (Hoffmann-La Roche, Inc., N.J.), 50 mg/kg body wt administered daily intramuscularly (I.M.) for 5-16 wks. [b]Mean ± S.D., n=lenses of rabbits or groups of pooled lenses of 4-7 rats. [c]680 IU (Sigma Chemical Co.)/ml peanut oil instilled in right eyes of rabbits (placebo in left eyes) with simultaneous feeding of 0.4% 3-aminotriazole in diet. [d]1%Ophthalmic suspension (Pfizer, Inc.) instilled in right eyes(placebo in left eyes) with simultaneous feeding of 3-aminotriazole in diet. [e]Orgotein (3300 units bovine liver SOD/mg protein, Diagnostic Data, Inc.) 5 mg/kg body wt, injected daily intravenously(I.V.) for 2 wks. [f]GE40 contains 40% galactose and 40 mg vitamin E/kg and GE4, 40% galactose and 4 mg vitamin E/kg. [g]CPA, 25 mg/kg body wt daily I.P., pretreatment for 1 wk followed by 40% galactose in diet and CPA therapy for 11-18 days. [h]Rats were injected (I.P.) with 2.5g vitamin E or vitamin C(Eli Lilly and Co.) /kg body wt or fed 2.5g penicillamine/kg diet for 3-6 wks.

Lenticular lipid peroxidation was decreased 33-50% after vitamin E therapy of cataracts induced in rabbits by 3-amino-triazole or galactose and in rats by selenium. Similar protection to the lens was afforded by centrophenoxine in galactose-induced cataract in rabbit. The compounds which did not have any effect on peroxidative damage to the lens lipids were orgotein in 3-aminotriazole-induced cataract, and vitamin C or penicillamine in selenium-induced cataracts.

Morphologically, vitamin E therapy arrested the progression of 3-aminotriazole-induced cataract in about 50% of rabbits. Cataract was also prevented by topical vitamin E. However, the effect was observed in both the eyes although one eye was treated with placebo. This could be due to the systemic effect of vitamin E. The altered biochemical parameters such as markedly increased H_2O_2 and decreased ascorbic acid of ocular humors and increased MDA of lens in cataract, were restored close to normal in rabbits on therapy (6,7). Vitamin E in galactose-induced cataract delayed its onset, retarded its progression and reversed it in some rabbits. Accumulation of galactitol in lenses of rabbits on GE40 was 43% less than (11.22±0.34 μmole/g wet wt, n=10) in those fed on GE4. Vitamin E was also found to be effective in retarding cataract in Emory mouse(10). These observations confirm the original findings of Creighton and Trevithick(11), who observed retardation of galactose-induced cataract by vitamin E in rat lens in vitro. Later from the same laboratory it was also shown that vitamin E prevented cataracts in diabetic rats(12). Centrophenoxine, which has been claimed to decrease lipofuscin accumulation in tissues(13), was therapeutically effective in preventing the onset or progression of galactose-induced cataracts in rabbits. H_2O_2 of aqueous humor and vitreous humor was 50% less as compared to 0.20±0.07mM(mean±S.D.,n=10), and 0.12±0.02mM(n=5) of placebo treated rabbits, respectively. Ascorbic acid (0.93±0.23mM,mean±S.D.,n=11) in aqueous humor was restored to normal. Galactitol in lenses of rabbits on therapy was 50% less than 15.13±1.83μmole/g wet wt(n=6) in placebo treated rabbits. Regarding the therapeutic efficacy of some other compounds in cataract, orgotein administered intravenously, intracamerally or subconjunctivally, was unable to prevent or retard the progression of 3-aminotriazole-induced cataract in rabbit. The results of topical sorbinil in this cataract were inconclusive. Vitamin E, vitamin C or penicillamine was therapeutically ineffective in selenium-induced cataracts in rats although some biochemical prevention by vitamin E was observed.

In conclusion, antioxidants afford protection to the lens from oxidative stress-induced damage. Vitamin E was effective in preventing 3-aminotriazole-induced and galactose-induced cataracts in rabbits and centrophenoxine in arresting the progression of galactose-induced cataracts. Vitamin E and centrophenoxine appear to be suitable compounds for clinical trials for cataract in the human.

ACKNOWLEDGEMENTS

Supported by NEI Grant EY03012, in part by EY01867, and the Pack Foundation for Medical Research, New York.

REFERENCES

1. Augusteyn RC (1981) In: Duncan G (ed) Mechanisms of cataract formation in the human lens. Academic Press, Inc., New York, pp 71-115
2. Spector A (1984) Invest Ophthalmol Vis Sci 25:130-146
3. Bhuyan KC, Bhuyan DK (1984) Curr Eye Res 3:67-81
4. Bhuyan KC, Master RWP, Bhuyan DK (1984) Invest Ophthalmol Vis Sci (Suppl) 25: Abstr 13 p 154
5. Azari P, Clarkson DP (1984) Invest Ophthalmol Vis Sci (Suppl) 25: Abstr 10 p 207
6. Bhuyan KC, Bhuyan DK (1985) Curr Eye Res: under publication
7. Bhuyan KC, Bhuyan DK (1982) Ann NY Acad Sci 393:169-171
8. Fong KL, McCay PB, Poyer JL, Keel BB, Misra H (1973) J Biol Chem 248:7792-7797
9. Kellogg III EW, Fridovich I (1975) J Biol Chem 250:8812-8817
10. Varma SD, Chand D, Sharma YR, Kuck JFR Jr, Richards RD (1984) Curr Eye Res 3: 35-57
11. Creighton MO, Trevithick JR (1979) Exp Eye Res 29:689-693
12. Ross WM, Creighton MO, Stewart-DeHaan PJ, Sanwal M, Hirst M, Trevithick JR (1982) Can J Ophthalmol 17:61-66
13. Bourne GH (1965) Progr Brain Res 20:187-201

Superoxide and Superoxide Dismutase in Chemistry, Biology and Medicine; G. Rotilio, editor

THE EFFECTS OF DESFERRIOXAMINE AND ASCORBATE ON HAEMOGLOBIN FROM THALASSAEMIC ERYTHROCYTES

CATHERINE RICE-EVANS (1), E BAYSAL (1), D FLYNN (2), G KONTOGHIORGHES (3).
Departments of Biochemistry (1), Paediatrics (2), Haematology (3), Royal Free Hospital and School of Medicine, Rowland Hill Street, London. NW3 2PF

INTRODUCTION

The iron-catalysed production of oxygen derived free radicals is considered to play a major role in various pathological conditions. In thalassaemia the sources of excess storage iron are transfused red cells and gastro-intestinal iron absoption. In severe iron overload the plasma transferrin becomes saturated and the unbound iron may be an important factor in iron toxicity (1). Desferrioxamine is a complex hydroxylamine capable of chelating iron in the ferric state and currently used therapeutically in the treatment of thalassaemia major.

The damaging effects of iron arise from its ability to catalyse the formation of hydroxyl radicals $OH^{\bullet}$ from hydrogen peroxide, which itself can be derived from the dismutation of the superoxide radical $O_2^{\bullet -}$ (2).

$$Fe^{3+} + O_2^{\bullet -} \longrightarrow Fe^{2+} + O_2$$

$$Fe^{2+} + H_2O_2 \longrightarrow Fe^{3+} + {}^{\bullet}OH + OH^-$$

$$O_2^{\bullet -} + H_2O_2 \xrightarrow[\text{catalyst}]{\text{Fe salt}} O_2 + {}^{\bullet}OH + OH^-$$

Our previous study (3) showed that the increased methaemoglobin formation and decreased intracellular glutathione levels after iron-induced oxidative stress could be substantially inhibited by desferrioxamine in both control and thalassaemic erythrocytes. However, desferrioxamine was less effective in inhibiting methaemoglobin formation in red cell haemolysates treated under identical oxidative conditions. Furthermore, haemoglobin in intact erythrocytes was more susceptible to oxidation via iron-induced free radical effects than red cell haemolysates at the same concentration, suggesting that the membrane has an important role in enhancing oxidative damage in these iron-catalysed systems.

The purpose of this work is to study the effects of desferrioxamine on haemoglobin in thalassaemic haemolysates, since haemolysis is part of the pathology of erythrocyte disorders of iron overload. Ascorbate is included since it aids iron uptake by desferrioxamine (and by ferritin) in thalassaemic patients.

MATERIALS AND METHODS

Fresh human erythrocytes were obtained from normal, healthy donors and from splenectomised patients with β-thalassaemia major. After centrifugation, the plasma and buffy coat were removed and the red cells washed three times with isotonic phosphate buffer, pH 7.4. In the experiments involving the haemolysates the original red cells were lysed in 20 volumes of ice-cold distilled water, the pH was adjusted to 7.4 and incubations in the various experimental systems were carried out at 37°C for various time intervals up to 24 hours. The experimental systems consisted of iron (II) sulphate (100μM), L-ascorbic acid (1mM), hydrogen peroxide (200μM) with and without desferrioxamine of concentrations ranging up to 600μM.

Desferrioxamine-iron complexes were prepared by mixing equimolar concentrations of desferrioxamine with iron II or iron III.

Methaemoglobin formation was measured by observing the decrease in absorbance at 620nm after the addition of cyanide (4). The degree of haemolysis was estimated spectrophotometrically at 575nm using the method in (5).

RESULTS AND DISCUSSION

The results (Table 1) show that exposing thalassaemic erythrocyte haemolysates to oxidative stress by incubation with the iron II/ascorbate/hydrogen peroxide - system results in methaemoglobin formation after a 24-hour incubation at 37°C to the extent of 55% (n=6) compared with 15% (n=6) in the control thalassaemic haemolysates. The presence of 400μM desferrioxamine in the iron II/ascorbate/hydrogen peroxide incubation medium suppresses haemoglobin oxidation but does not totally inhibit methaemoglobin formation.

Desferrioxamine alone exerts a damaging effect on haemoglobin in thalassaemic red cell haemolysates incubated at 37°C compared with its action on thalassaemic erythrocytes (3). Incubation with desferrioxamine in the absence of the iron-catalysed oxygen radical system induced haemoglobin oxidation which increases as a function of time to the extent of 53% (n=6) methaemoglobin after a 24-hour incubation (Table 1) and also increases with the concentration of desferrioxamine (Fig. 1). The combination of 1mM ascorbate with 400μM desferrioxamine considerably suppresses the toxic effects of the iron chelator down to 20% methaemoglobin (n=6) after a 24-hour incubation.

TABLE 1

THE EFFECTS OF 400μM DESFERRIOXAMINE (DFO) ON METHAEMOGLOBIN PRODUCTION IN THALASSAEMIC HAEMOLYSATES

			% Methaemoglobin			
Time	Control	+DFO	Ascorbate +DFO	Iron (II)/ Ascorbate/ H_2O_2	Iron (II)/ Ascorbate/ H_2O_2 +DFO	Ascorbate
5hr	3.2±0.5 n=6	15.6±2 n=6	5.7±0.8 n=3	4.0±2 n=6	6.0±0.7 n=3	2.4±1 n=3
8hr	5.0±0.4 n=6	23.1±3 n=6	8.2±0.6 n=3	11.0±5 n=6	8.3±0.7 n=6	3.3±0.3 n=2
10hr	6.0±0.4 n=6	27.3±3 n=6	9.7±1 n=3	11.0±3 n=6	9.7±0.9 n=6	3.8±1 n=3
24hr	15.0±1 n=6	53.0±7 n=6	20.3±2 n=6	55.3±3 n=6	24.0±3 n=6	11.0±2 n=3

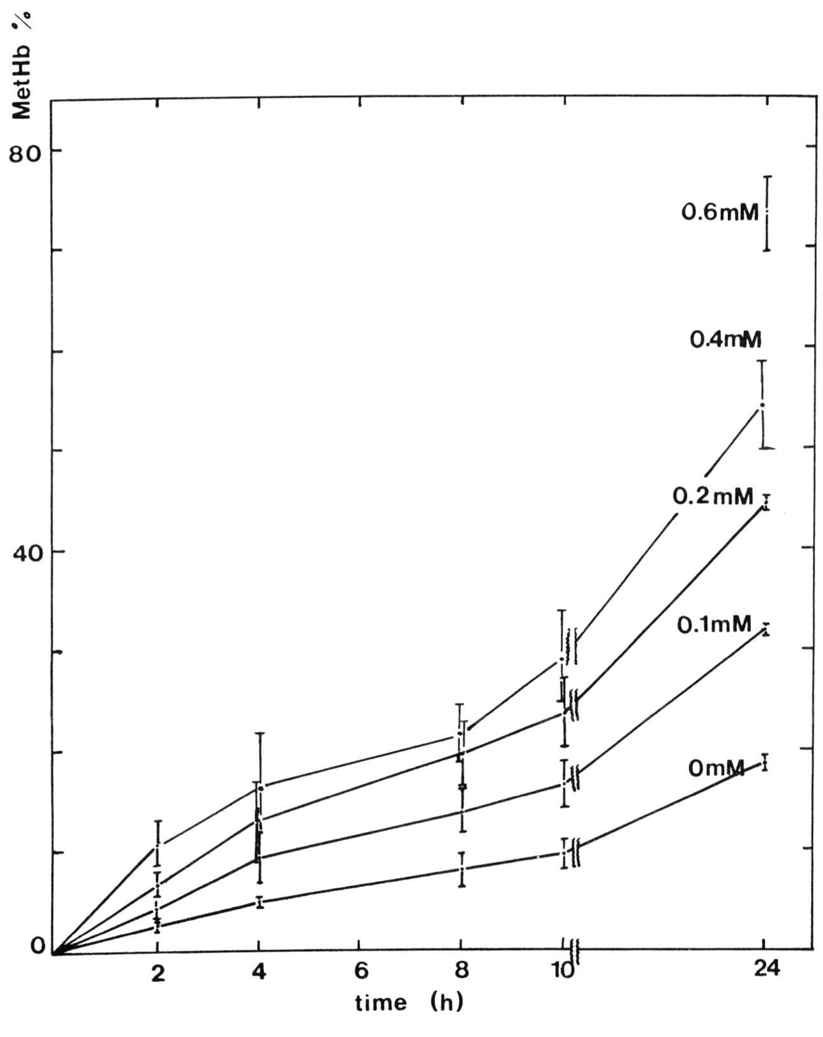

Fig. 1

The effects of desferrioxamine concentration on methaemoglobin production in erythrocyte haemolysates as a function of time of incubation at 37°C.

Haemoglobin from thalassaemic erythrocytes is more susceptible to interaction with desferrioxamine alone than by the iron-catalysed oxygen radicals generated from the iron II/ascorbate/hydrogen peroxide system over the range of incubation times studied up to 10h. This implies that desferrioxamine itself can act as an oxidising agent on the haemoglobin from the thalassaemic erythrocytes. Similar results have been obsererved in normal human erythrocytes under identical conditions (6).

That the desferrioxamine- iron III complex is not an oxidising agent is shown by the inhibition of haemoglobin oxidation by the presence of desferrioxamine in the iron II/ascorbate/hydrogen peroxide system. This suggests that the binding site for iron on desferrioxamine plays an important role in its oxidising capacity. These conclusions are supported by the results on the iron-chelating drug in the presence of the free radical scavenger and reducing agent, ascorbate, which suppresses the oxidative potential of desferrioxamine by its own reducing properties.

ACKNOWLEDGEMENTS

We gratefully acknowledge financial assistance from the Turkish-Cypriot authorities, U.K. Thalassaemia Society and the Peter Samuel-Royal Free Fund.

REFERENCES

1. Rachmilewitz EA, Lubin BH, Shohet SB (1976) Blood 47:495-505.
2. Walling C (1982) In: King TE, Mason HS, Morrison M (eds) Oxidases and related systems, Pergamon Press, Oxford, pp 85-97.
3. Rice-Evans C, Baysal E, Kontoghiorghes G, Flynn DM, Hoffbrand AV (1985) Free Radical Research Communications. In press.
4. Harley JO, Mauer AM (1960) Blood 16:1722-1735.
5. Brownlee NR, Huttner JJ, Panganamala RV, Cornwell DG (1977) Journal of Lipid Research 18:635-644.
6. Rice-Evans C, Baysal E, Flynn D, Kontoghiorghes G (1985) Biochem. Soc. Trans. In press.

© 1986 Elsevier Science Publishers B.V. (Biomedical Divison)
Superoxide and Superoxide Dismutase in Chemistry,
Biology and Medicine; G. Rotilio, editor

ERYTHROCYTE ANTIOXIDANT SYSTEM IN α-THALASSAEMIA.

GERLI G.C.*, MONGIAT R.*, GUALANDRI V.°, ORSINI G.B.°, MOSCHINI G.^`, BUSO G.P.`, AGOSTONI A.*

*Clinica Medica V Ospedale S.Paolo, Università di Milano
°Cattedra di Genetica Umana, Università di Milano
^Dipartimento di Fisica dell'Università di Padova
`Laboratori Nazionali dell'INFN di Legnaro - Padova

INTRODUCTION

α-thalassaemia can be considered a condition of augmented oxidant stress, more than ß-thalassaemia, because of the presence in the erythrocytes of free ß-globin chains which autoxidise in a more slightly rapid rate (1) generating reactive oxygen free radicals (2).

We have already demonstrated an increase of erythrocyte antioxidant enzymes (superoxide dismutase (SOD), catalase and glutathione peroxidase (GSH-Px)) in subjects with ß-thalassaemia minor (3).

In the present investigation we have studied the behaviour of antioxidant enzymes, glucose-6-phosphate dehydrogenase (G-6-PD) and reduced glutathione (GSH) in subjects with α-thalassaemia trait. The serum levels of the trace metals related to antioxidant enzymes have been also assayed.

MATERIAL AND METHODS

Erythrocyte SOD, catalase, GSH-Px, G-6-PD and GSH were assayed as previously described (3, 4) in 19 females with α-thalassaemia trait. Enzyme activities were expressed per ml of red blood cells (this type of expression, according to Sagone and Balcerzak is the least ambiguous) (5). The diagnosis of α-thalassaemia was based on haematological and genetic evidence and was confirmed by globin chain synthesis. 35 normal females were studied as controls. Serum iron was determined by colorimetric assay; serum copper and zinc were assayed by atomic spectrophotometric procedure, while a proton induced X-ray fluorescence technique (6) was used to determine the selenium concentration in serum and in 6 (3 normal and 3 thalassaemic) erythrocyte samples.

RESULTS

Thalassaemic women showed the mild haematological abnormalities related to α-thalassaemia trait (Hb = 12.8 ± 1.8 g/dl; MCH = 23.8 ± 3.2 pg; MCV = 72 ± 8 fl); reticulocytes were 9 ± 4 ‰ ; erythrocyte GSH was higher in α-thalassaemic women (2305 ± 678 μM/l RBC) than in normal controls (2007 ± 313 μM/l RBC, $p<0.05$), while G-6-PD was in the normal range.

Erythrocyte antioxidant enzyme activities were significantly increased in α-thalassaemic females. Among the trace metals, copper, zinc and iron serum levels were similar in both groups whereas the selenium was significantly reduced in α-thalassaemic females (Table I).

TABLE I

ERYTHROCYTE ANTIOXIDANT ENZYME ACTIVITIES AND TRACE METAL SERUM LEVELS IN α-THALASSAEMIC AND NORMAL FEMALES.

	SOD μg/mlRBC	Catalase K/mlRBC	GSH-Px IU/mlRBC	Copper ng/ml	Zinc ng/ml	Iron ng/ml	Selenium ng/ml
α-thal. ♀ (n=19)							
Mean	421	173	13.47	960	650	990	53
SD	110	35	3.44	290	70	220	21
Normal ♀ (n=35)							
Mean	294	129	7.26	1150*	680*	970	73*
SD	62	22	1.71	400	60	240	21
p	<0.001	<0.001	<0.001	ns	ns	ns	<0.025

* mean of 10 cases

Red cell selenium concentration was significantly higher in α-thalassaemic subjects (187 ± 16 ng/g RBC) than in controls (144 ± 6 ng/g RBC, $p<0.02$).

DISCUSSION

The increase of antioxidant enzymes and GSH that we have observed in erythrocytes of α-thalassaemic subjects can be considered a defence mechanism against oxidant threat which is very stressing in this condition as a consequence of the high flux of oxygen radicals due to autoxidation of free β globin chains (1, 2).

Very high levels of SOD have been previously reported by Yenchitsomanus and Wasi (1983) (7) in haemoglobin H disease, while an increase of GSH-Px was already referred by Beutler et al. (1977) (8) both in subjects with haemoglobin H disease and α-thalassaemia trait.

There are no published data on selenium status in α-thalassaemic subjects. Because selenium is the essential metal of GSH-Px and this enzyme correlates with selenium serum concentration (9, 10), we studied selenium status in these subjects and found a significant reduction of serum selenium, while red cell selenium concentration was increased in parallel with augmented levels of the erythrocyte selenium dependent GSH-Px.

REFERENCES

1. Winterbourn CC, Mc Grath BM, Carrel RW (1976) Biochem J 155:493-502
2. Brunori M, Falcioni G, Fioretti E, Giardina B, Rotilio G (1975) Eur J Biochem 53:99-104
3. Gerli GC, Beretta L, Bianchi M, Pellegatta A, Agostoni A (1980) Scand J Haematol 25:87-92
4. Gerli GC, Beretta L, Bianchi M, Agostoni A, Gualandri V, Orsini GB (1982) Scand J Haematol 29:135-140
5. Sagone AL Jr, Balcerzak SP (1970) Am J Med Sci 259:350-357
6. Berti M, Buso G, Colautti P, Moschini G, Stievano BM, Tregnaghi C, (1977) Anal Chem 49:1313-1315
7. Yenchitsomanus P, Wasi P (1983) J Clin Pathol 36:329-333
8. Beutler E, Matsumoto F, Powars D, Warner J (1977) Blood 50:647-655
9. Perona G, Guidi GC, Piga A, Cellerino R, Milani G, Colautti P, Moschini G, Stievano BM (1979) Brit J Haematol 42:567-574
10. Pleban PA, Munyani A, Beachum J (1982) Clin Chem 28:311-316

© 1986 Elsevier Science Publishers B.V. (Biomedical Division)
Superoxide and Superoxide Dismutase in Chemistry,
Biology and Medicine; G. Rotilio, editor

ERYTHROCYTE ANTIOXIDANT ENZYMES IN ALCOHOLIC LIVER DISEASE.

MONGIAT R.*, GERLI G.C.*, SANDRI M.T.*, ALMINI D.*, BARETTA A.*, MOSCHINI G.°, BUSO G.P.°, BRUNO S.§, AGOSTONI A.*

* Clinica Medica V, Ospedale S.Paolo, Università di Milano
^ Dipartimento di Fisica dell'Università di Padova
° Laboratori Nazionali dell'INFN di Legnaro - Padova
§ Divisione di Medicina I, Ospedale S.Paolo, Milano

INTRODUCTION

It is well known that the erythrocytes of subjects with alcoholic liver disease present structural and metabolic abnormalities. In fact altered membrane lipid composition have been described (1); besides an impairment of methylene blue--stimulated hexose monophospate shunt associated with increased Heinz body formation and low levels of reduced glutathione (GSH) have been reported (2). It has also been suggested that above mentioned erythrocyte abnormalities in subjects with alcoholic liver cirrhosis may induce an increase sensitivity to the oxidant stress (2).

More recently it has been emphasized that oxygen free radicals can play a significant rolein mediating the permanent features of alcoholic liver cirrhosis (3). The aim of the present paper has been to study antioxidant enzyme activities (superoxide dismutase(SOD), catalase and glutathione peroxidase (GSH-Px)) in erythrocytes of subjects with alcoholic liver disease of different severity, in order to evaluate if these cells are protected against oxidant stress. Besides we have assayed erythrocyte GSH content and serum trace metals related to antioxidant enzymes.

MATERIAL AND METHODS

10 males with alcoholic liver disease without cirrhosis and 40 patients with alcoholic cirrhosis of different severity were studied. The diagnosis of alcoholic liver disease and cirrhosis was based on clinical and biochemical data and, in some cases, was confirmed also by a histological examination. Cirrhotic patients were grouped according to the criteria for assessment of hepatic reserve proposed by Child and Turcotte (4) and so classified: <u>group A</u>: 10 males with minimal cirrhosis; <u>group B</u>: 19 subjects (11 males and 8 females) with moderate cirrhosis; <u>group C</u>: 11 subjects (6 males and 5 females) with advanced cirrhosis. 36 normal

controls (19 males and 17 females) were studied in the same period.

Erythrocyte antioxidant enzymes were assayed as previously described (5); GSH was determined according to Beutler et al. (1963) (6). Serum iron was determined by colorimetric assay and serum copper by an atomic spectro-photometric procedure while a proton induced X-ray fluorescence technique (7) was used for selenium concentration in serum.

RESULTS

All patients had macrocytosis and an increase of reticulocytes, group C presented the lower levels of Hb (10.6 ± 1.9 g/dl). Total bilirubin was augmented in group B (2.6 ± 1.4 mg/dl) and C (3.8 ± 1.4 mg/dl) because of the higher indirect bilirubin levels (group B: 1.7 ± 0.9 mg/dl; group C: 2.8 ± 0.8 mg/dl). Erythrocyte antioxidant enzymes activity, GSH content and serum metals are shown in Table I.

TABLE I

ERYTHROCYTE ANTIOXIDANT ENZYME ACTIVITIES, GSH CONTENT AND SERUM TRACE METALS IN CONTROLS AND IN PATIENTS WITH ALCOHOLIC LIVER DISEASE WITHOUT CIRRHOSIS AND WITH ALCOHOLIC CIRRHOSIS.

	SOD µg/gHb	Catalase K/gHb	GSH-Px IU/gHb	GSH µM/lRBC	Cu§ ng/ml	Fe ng/ml	Se§ ng/ml
alcoholic liver disease without cirrhosis (n=10)							
Mean	770	384	24.90	2007	1180	920	59
SD	248	62	6.24	335	250	450	7
cirrhosis group A (n=10)							
Mean	701	385	20.52	1663*	1400°	1090	76
SD	73	73	8.35	300	120	720	18
cirrhosis group B (n=19)							
Mean	674	337*	18.88^	1673*	1220^	1230	67
SD	161	59	4.40	374	230	590	10
cirrhosis group C (n=11)							
Mean	652	325*	13.84*	1489*	1110	1110	52^
SD	142	43	3.44	382	380	600	11
controls (n=36)							
Mean	739	405	22.55	2216	1020	1010	72
SD	144	43	5.36	349	320	250	20

§ Mean of 20 cases

^ $p<0.05$ - ° $p<0.01$ - * $p<0.001$

DISCUSSION

Our results indicate that the erythrocytes of subjects with moderate and advanced alcoholic liver cirrhosis seem less protected against oxidant stress as a consequence of the presence of unbalanced antioxidant system. In fact, normal SOD levels could dismutate superoxides yet resulting in hydrogen peroxides which cannot be sufficiently scavenged for low levels of catalase and GSH-Px. On the other hand hydrogen peroxides can react with superoxides so generating hydroxyl radicals whose toxic effect could be emphasized because of the decrease of its scavenger catalase.

The unbalanced erythrocyte antioxidant system could predispose these cells to haemolisis. In fact, the decreased catalase and GSH-Px cannot protect erythrocyte membrane and haemoglobin from oxidant damage, so helping haemoglobin denaturation with increased Heinz body formation and augmented susceptibility to destruction in the spleen. Moreover the low levels of GSH could support the oxidant damage.

The low levels of catalase and GSH could be the result of inhibitory action of the free radicals (8) while the reduction of GSH-Px (a selenium dependent enzyme) could be due to the low serum selenium levels (9). Normal SOD activity could confirm the highest stability of this enzyme as already observed in other contitions of augmented oxidant stress as hypobaric hypoxia (10) and in cord blood (11).

REFERENCES

1. Cooper RA (1970) Semin Hematol 7:296-322
2. Smith JR, Kay NE, Gottlieb AJ, et al. (1979) Am J Hematol 6:313-321
3. Lewis KO, Paton A (1982) Lancet 2:188-189
4. Child CG, Turcotte JG (1964) In: Child CG (ed.) The liver and portal hypertension. W B Saunder Co, Philadelphia, pp 49-51
5. Gerli GC, Beretta L, Bianchi M, et al. (1980) Scand J Haematol 25:87-92
6. Beutler E, Duron O, Kelly BM (1963) J Lab Clin Med 61:882-888
7. Berti M, Buso G, Colautti P, et al. (1977) Anal Chem 49:1313-1315
8. Freeman BA, Crapo JD (1982) Lab Invest 47:412-426
9. Perona G, Guidi GC, Piga A, et al. (1979) Brit J Haematol 42:567-574
10. Agostoni A, Gerli GC, Beretta L, et al. (1983) Clin Chim Acta 133:153-157
11. Agostoni A, Gerli GC, Beretta L, et al. (1980) J Clin Chem Clin Biochem 18:771-773

© 1986 Elsevier Science Publishers B.V. (Biomedical Division)
Superoxide and Superoxide Dismutase in Chemistry,
Biology and Medicine; G. Rotilio, editor

LIPID PEROXIDATION AND CELLULAR DAMAGE IN EXTRAHEPATIC TISSUES OF BROMOBENZENE--INTOXICATED MICE

ALESSANDRO F. CASINI, MARCO FERRALI, ALFONSO POMPELLA, EMILIA MAELLARO and MARIO COMPORTI

Istituto di Patologia Generale dell'Università di Siena, Via del Laterino 8, 53100 Siena (Italy)

INTRODUCTION

Recent studies from our laboratory (1) have shown that following bromobenzene (BrB) or iodobenzene administration to mice liver necrosis develops only when the hepatic glutathione (GSH) depletion reaches a threshold value. Lipid peroxidation, measured by different methods, was found to occur at a very extensive level when the hepatic GSH depletion reached the same critical values as those observed for liver cell death.

It has been observed that BrB, in addition to causing hepatic necrosis, can also exert toxic effects on extrahepatic tissues. These include lung, where a necrosis of the bronchiolar epithelium has been reported (2); and kidney, where BrB produces necrosis of the proximal convoluted tubules (3). Therefore it was investigated whether BrB induces in extrahepatic tissues alterations similar to those seen in the liver, i.e., whether GSH depletion and lipid peroxidation are general phenomena occurring, besides in the liver, in many other tissues after BrB intoxication.

MATERIALS AND METHODS

Male NMRI albino mice (in which the hepatic GSH content had been decresed by nearly 50% by means of a pure glucose diet for two days) were used. BrB was administered intragastrically at the dose of 15 mmol/Kg body wt. Lipid peroxidation of cell membranes was assessed by measuring the tissue content of malonic dialdehyde (MDA), one of the end products of lipid peroxidation. Liver damage was assessed by measuring the serum glutamate-pyruvate transaminase (SGPT) activity. Kidney damage was evaluated by measuring the blood urea nitrogen. Lung damage was assessed by measuring the lactate dehydrogenase (LDH) activity in the lung lavage fluid.

RESULTS AND DISCUSSION

A time-course study of hepatic GSH depletion, liver necrosis and lipid peroxidation after BrB poisoning showed that both liver necrosis and lipid peroxidation appeared after 9 hr, even though at 9 hr the GSH content was already decreased by 84% as compared with that of glucose-fed non-intoxicated animals. The plot of the individual values obtained at 15-20 hr for transaminase or MDA levels against the corresponding hepatic GSH contents, indicated that severe lipid peroxidation and necrosis occurred only when GSH levels reached a threshold limit (2.2-0.9 nmol/mg protein).

A time-course study of the alterations produced by BrB in the kidney showed that the GSH content decreased linearly with the intoxication time up to 15 hr. The blood urea nitrogen levels (kidney damage) increased after 9 hr. Similarly lipid peroxidation, absent through 9 hr, was found to occur since the 15th hr. The mean MDA value was however much lower than that found in the liver. As in the case of liver damage, when the individual blood urea nitrogen values obtained at 15-20 hr were plotted against the corresponding GSH contents (Fig. 1A), a severe cellular damage was evident only when the GSH level reached a threshold value (2.2-0.8 nmol/mg protein). Such a threshold was also evident for the onset of lipid peroxidation (Fig. 1B). These results are therefore similar to those

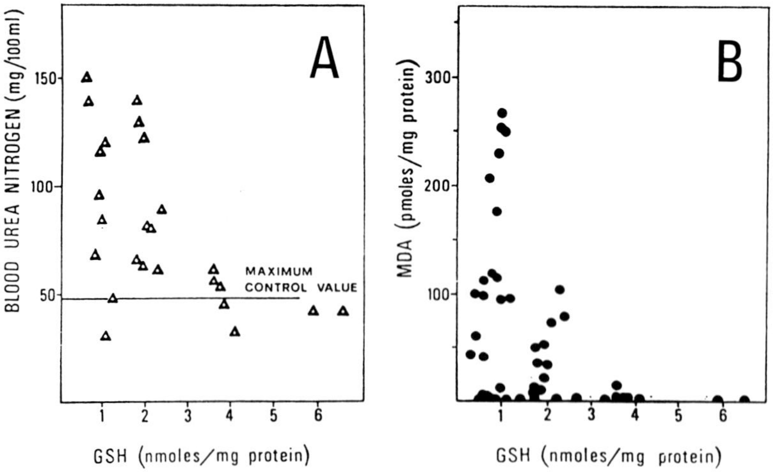

Fig. 1. Plot of blood urea nitrogen levels (A) and of lipid peroxidation (B) against the corresponding GSH contents of the kidney in mice intoxicated with BrB.

obtained for the liver.

A time-course study of the alterations produced by BrB in the lung showed that the GSH content of the lung decreased markedly during the first 6 hr after the intoxication; subsequently the decrease was more gradual. Lipid peroxidation appeared since the 6th hr. The LDH activity in the lung lavage fluid (cellular damage) was increased at 18-20 hr. When the individual values obtained at 15-20 hr for both LDH (Fig. 2A) and MDA (Fig. 2B) were plotted against the corresponding GSH contents, the values appeared more dispersed, so that critical GSH threshold could not be defined. However, high levels of lipid peroxidation and cell damage are accompanied by GSH levels. Additional studies showed that BrB induces GSH depletion and lipid peroxidation in other extrahepatic tissues (heart, skeletal muscle and brain). Thus lipid peroxidation may be the cause of the alterations seen in liver and extrahepatic tissues after BrB intoxication.

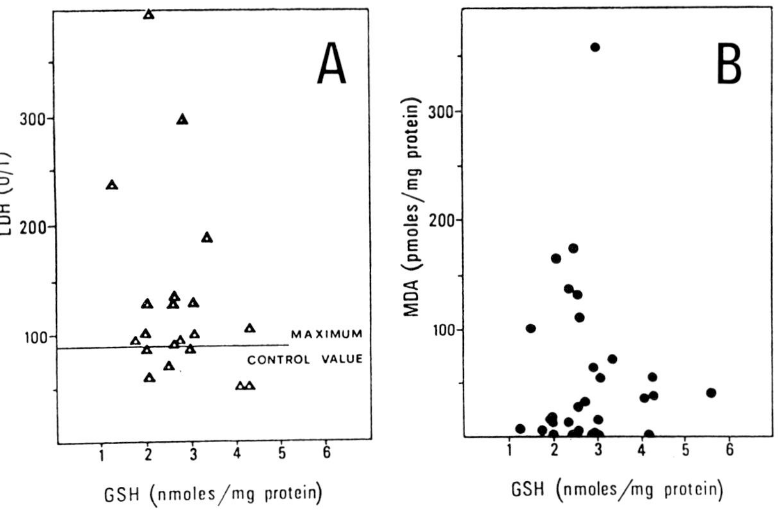

Fig. 2. Plot of lactate dehydrogenase (LDH) activities in the lung lavage fluid (A) and of lipid peroxidation values (B) against the corresponding GSH contents of the lung in mice intoxicated with BrB.

REFERENCES

1. Casini AF, Pompella A, Comporti M (1985) Am J Pathol 118:225-237
2. Reid WD, Ilett KF, Glick JM, Krishna G (1973) Am Rev Respir Dis 107:539-551
3. Reid WD (1973) Exp Mol Pathol 19:197-214

Superoxide and Superoxide Dismutase in Chemistry, Biology and Medicine; G. Rotilio, editor

LIPID PEROXIDATION IN CEREBELLUM OF RATS ACUTELY TREATED WITH ETHANOL

HELENE ROUACH, MARIE THERESE ORFANELLI, BRIGITTE JANVIER AND ROGER NORDMANN
Service de Biochimie de la Faculté de Médecine de Paris-Ouest et INSERM U 72; 45 rue des Sts Pères, 75006 Paris (France)

INTRODUCTION

Chronic or acute ethanol intoxication has been reported to increase hepatic lipid peroxidation. These effects have been claimed to be specific for the liver tissue, since no peroxidation was detected in brain and lung following acute ethanol administration (1,2). However the brain is vulnerable to lipoperoxidative damage due to its high content of unsaturated fatty acids and to a relatively low activity of catalase and glutathione peroxidase which protect tissues from lipid peroxidation. The high ascorbic acid concentration in the brain may therefore have an important role in the prevention of peroxidative damage in this tissue. Free oxygen radicals could be formed in the brain through different pathways such as, among others, the mitochondrial respiratory chain or monoamine oxidase. Some of the effects of ethanol on the central nervous system are similar to those produced by an increased lipid peroxidation. It has been shown, for instance, that acute ethanol administration to mice reduces the relative proportion of unsaturated fatty acids in the synaptosomal phospholipids (3). We therefore reconsidered in the present study the effects of acute ethanol administration on lipoperoxidation and ascorbate content in the cerebellum. Cerebellum was chosen because lipid peroxidation is higher in that brain area than in others (4). Cerebellum has also been suggested to be more susceptible than other brain areas to a deficiency of vitamin E, a natural antioxidant (5).

MATERIAL AND METHODS

Ethanol (2.3g/kg b.wt) as a 20 p.cent (v/v) aqueous solution was administered (i.p.) to starved (16 hours) male Sprague-Dawley rats (200-220g). Cerebellum, cleaned off adhering blood, was homogenized (1:40) in phosphate buffer (20mM, pH7.4). Aliquots of the homogenates were removed immediately for zero time malondialdehyde (MDA). The homogenate was then centrifuged during 10 min. at 1000g. Samples of the supernatant were incubated without added substrates during 30 min. under air at 37°C in a shaking water bath. Aliquots of the incubated mixture were taken each 5 min. to determine the lipid peroxidation rate. Lipid peroxides were quantified by the thiobarbituric acid test according to (6) and ascorbic acid according to (7).

RESULTS

The concentrations of lipid peroxides in freshly homogenized cerebellum after ethanol intoxication are shown in Table 1. The cerebellum obtained 2 or 4 hours after the treatment showed a significant higher lipid peroxide level than controls. The lipid peroxide levels were no more significantely different at 6 hours. Such an increase in the cerebellum lipid peroxide level is however slighter than the one observed in liver (Table 1).

The rate of cerebellar lipid peroxydation was also studied during incubation of tissue homogenates (without addition of any inducer). The rates of lipid peroxide formation were increased 2 and 4 hours after ethanol and returned to the original rate after 6 hours (Table 2) whereas the in vitro addition of ethanol (100mM) before starting the incubation resulted in a slight inhibition [about 10 p.cent] of the peroxidation rate.

The ascorbic acid content of the cerebellum was significantly decreased by ethanol administration (30.6 ± 0.9 vs 27.8 ± 2 nmol/mg protein in controls and 4 hours after ethanol, respectively; $0.01 < p < 0.02$; 6 animals in each group).

TABLE I

EFFECTS OF ETHANOL ON LIPID PEROXIDE LEVELS IN RAT CEREBELLUM AND LIVER.

Values expressed as mean S.E. ***$p < 0.01$, **$0.01 < p < 0.02$, *$0.02 < p < 0.05$ † $p < 0.05$ vs. control animals. Number of animals in parentheses.

	nmol Malondialdehyde per g tissue			
	Liver		Cerebellum	
	Control	Ethanol	Control	Ethanol
2 hours	106 ± 25(8)	191 ± 22(5)***	99 ± 17(7)	130 ± 12(10)**
4 hours	145 ± 23(4)	348 ± 42(5)***	90 ± 9(11)	118 ± 12(7)***
6 hours	107 ± 13(5)	174 ± 42(4)*	86 ± 14(5)	109 ± 18(6)†

TABLE II

PEROXIDATION RATE IN RAT CEREBELLUM FOLLOWING ETHANOL TREATMENT

The homogenates were incubated at 37°C under an air atmosphere.

	Rate of lipid peroxide formation (nmol MDA/min/g protein)	
	Control	Ethanol
2 hours	78 ± 6 (12)	99 ± 10 (10)***
4 hours	72 ± 8 (7)	94 ± 9 (7)***
6 hours	55 ± 8 (6)	55 ± 8 (8)

DISCUSSION

The results of the present study show that an acute ethanol administration induces an increased lipoperoxidation in the rat cerebellum. This increase is, however, less marked than in the liver, where ethanol is actively oxidized.

The increase in cerebellar peroxidation may be due to an enhanced production of oxygen radicals and/or to a decrease in endogenous protectants against active oxygen species.In this study we have shown that ascorbic acid, a free radical scavenger and an antioxydant, is decreased. This may indicate a consumption of ascorbic acid in an attempt to quench pathological free radical reactions. The previously reported decrease of Cu-Zn superoxide dismutase following acute ethanol (8) may also play a role in the increase in lipid peroxidation. It appears that lipid peroxidation may be one of the contributing mechanisms in the toxic effects of ethanol on the central nervous system.

ACKNOWLEDGEMENTS

This work was supported by INSERM , IREB (grant 83-05) and EPHE.

REFERENCES

1. DiLuzio NR, Hartman AD (1967) Fed Proc 26:1436-1442
2. Corongiu FP, Lai M, Milia A (1983) Biochem J 212:625-631
3. Littleton JM, John G (1977) J Pharm Pharmac 29:579-580
4. Floyd RA, Zaleska MM, Harmon HJ (1984) In: Armstrong D, Sohal RS, Cutler RG, Slater TF (eds) Free Radicals in Molecular Biology, Aging, and Disease. Raven Press, New York, pp143-161
5. Vatassery T, Angerhofer CK, Knox CA, Deshmukh (1984) Biochim Biophys Acta 792: 118-122
6. Mihara M, Uchiyama M, Fukuzawa K (1980) Biochem Med 23: 302-311
7. Cooper AJL, Pulsinelli WA, Duffy TE (1980) J Neurochem 35: 1242-1245
8. Ledig M, M'Paria JR, Mandel P (1981) Neurochem Res 6: 385-391

© 1986 Elsevier Science Publishers B.V. (Biomedical Division)
Superoxide and Superoxide Dismutase in Chemistry,
Biology and Medicine; G. Rotilio, editor

SOD ACTIVITY OF HUMAN FETAL BLOOD BY A CHEMILUMINESCENT MICROMETHOD.

TRUNG PHAM HUU*, YVES DUMEZ**, CLAUDE MARQUETTY*, NORMA AMIT* and JACQUES HAKIM*.

*Lab. Immuno-hématologie, Université Paris 7, Faculté X. Bichat
16, rue Huchard 75018 Paris (France)
** Maternité Port-Royal, Hôpital Cochin, 27 rue du Faubourg St-Jacques
75014 Paris (France).

INTRODUCTION

For prenatal diagnostic of some hereditary diseases i.e. C.G.D., S.C.I.D., small amount of fetal blood is still necessary for biological investigations. It is obtained by venous umbilical puncture under fetoscopy (1). On the other hand, an available micromethod for SOD determination using luminol-enhanced chemiluminescence (CL) is fitted for assay on very small material (2). It has

- a high sensitivity using only 5 µl of SOD blood extract for each determination
- a high specificity according to the parallel-line analysis which detects external interference with the enzymatic determination (3).

Six fetal blood of 20-24 gestational weeks were evaluated by this technique for their SOD activity.

MATERIAL AND METHODS

Fetal blood was obtained by venous umbilical puncture under fetoscopy for prenatal diagnosis of SCID CGD. Maternal blood contamination was excluded by the red cell volume distribution curve obtained with a Coulter-counter Channelanalyzer C 1000 and by a Kleikauer test.

20 µl of blood was hemolyzed with 980 µl of distilled water containing 1 mM DTPA

SOD was extracted from hemolysate by a ethanol-chloroform procedure described by Mc Cord and Fridovich (4) at 4°C.

SOD determination : Luminol assay was monitored in a luminometer Picolite (Packard comp.) model 6500 at 37°C with continuous magnetic stirring and set on program 2, counting time 30 sec, sequence number 5, background 0 ; the final conditions for each assay were :

- 95 µl of 0.1 M glycine-NaOH buffer at pH 9.4 1 mM DTPA
- 5 µl of blood extract or diluted human erythrocyte SOD (Sigma) extract.
- 100 µl of 6 mM xanthine in distilled water. The reaction was started by injection of:
- 50 µl of a solution of xanthine-oxidase (0.416 µg/ml) and 50 µl of luminol 6×10^{-6}M both in glycine buffer.

- Standard was determined by human erythrocyte SOD (Sigma) dissolved in 0.15 M NaCl. 45 μg/ml and submitted to ethanol-chloroform extraction. Four different dilutions of standard SOD and blood homogenate were measured. Data were analyzed by converting the percentage response to probit response.

RESULTS

Luminol in presence of $O_2^{\bar{\cdot}}$ produced by a xanthine-xanthine oxidase reaction gives a large quantity of luminescence (CL); then, the inhibition of luminol-enhanced CL has been used to detect SOD activity. Fig. 1 represents various CL curve according to the indicated conditions. Conditions of curve "b" , a steady-state of $O_2^{\bar{\cdot}}$ production for a short time (2,5 min),were chosen. A SOD standard scale is obtained with four concentrations of human erythrocyte SOD (0,31-2,5 μg/ml) and by plotting the probit of CL inhibition percentage versus log SOD concentration (Fig. 2 a).

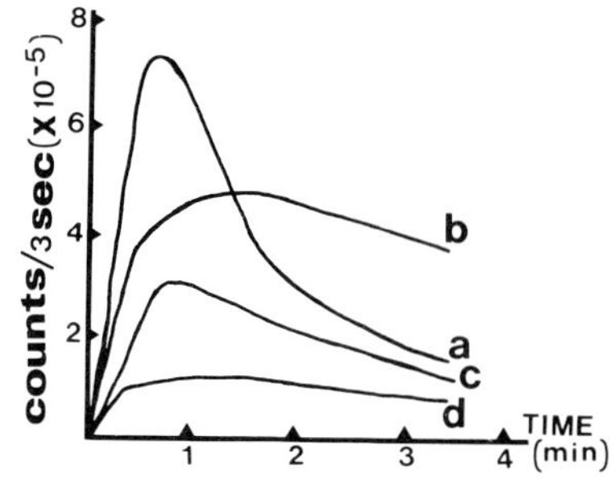

Fig. 1. Representative kinetic CL during xanthine-xanthine oxidase reaction. Reaction medium was : 95 μl of 0.1 M glycine-NaOH buffer (pH 9.4), 1mM DTPA, 100 μl of 6mM xanthine, 5 μl of blank extract. The reaction was started by injection : 50 μl of 6 x 10^{-6}M luminol in glycine buffer, 50 μl of xanthine-oxidase in glycine buffer at concentration 4.16 μg/ml (a), 0.416 (b), 0.083 (c), 0.042 (d).

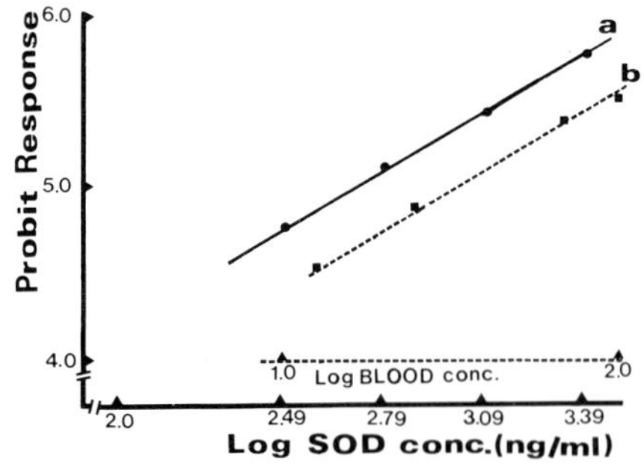

Fig. 2. "a" line was obtained from probit of percentage CL inhibition of 4 dilutions of standard SOD (linear with slope = 1,11 ; r = 0,99) "b" line obtained with blood.

Table 1 : Standard scale obtained with four dilutions of human erythrocyte SOD (Sigma)

Table 2 : SOD activity of six fetal bloods.

TABLE 1

Scale with four dilutions of human erythrocytes SOD (Sigma Chem. Co)

SOD dilution	Log of SOD (ng/ml)	Percentage ihnibition	Probit
SOD 1/44	2.49	40.5	4.76
SOD 1/72	2.79	55.9	5.14
SOD 1/36	3.09	66.6	5.44
SOD 1/18	3.39	78.0	5.77

TABLE 2

SOD activity of fetal blood

nb	Gestational age (week)	SOD activity (μg/gr Hb)
1	20	736
2	22	899
3	24	686
4	21	911
5	20	938
6	22	918
		m = 848 ± 108

CONCLUSION

It may be concluded that at 20th gestational week human fetal blood has acquired a SOD activity and its level, (m = 848 ± 108 μg/gr of Hb) is similar to adult blood (m = 836 ± 112/ gr of Hb). Its measurement may be useful in some pathological states (Down syndrome, Sickle cell, intra-uterin growth retardation, spontaneous abortion).

ACKNOWLEDGEMENTS

We thank Miss B. Boitte and Mr P. Sourbier for their excellent assistance.

REFERENCES

1. Dumez Y (1982) Ann Radiol 8 : 567-571
2. Pham Huu T, Marquetty C, Pasquier C, Hakim J (1984) Anal Biochem 142 : 467-472
3. Mc Cord JM, Fridovich IJ (1969) J Biol Chem 244 : 6049-6053
4. Eldred GE, Hoffert JR (1981) Anal Biochem 110 : 137-143

Author index